Springer Monographs in Mathematics

For further volumes:
http://www.springer.com/series/3733

Haruzo Hida

Elliptic Curves and Arithmetic Invariants

Springer

Haruzo Hida
Department of Mathematics
University of California
Los Angeles, CA, USA

ISSN 1439-7382
ISBN 978-1-4614-6656-7 ISBN 978-1-4614-6657-4 (eBook)
DOI 10.1007/978-1-4614-6657-4
Springer New York Heidelberg Dordrecht London

Library of Congress Control Number: 2013937713

Mathematics Subject Classification: 11F11, 11F25, 11G18, 11G15, 11F80, 11R23, 11F67, 11G05, 11F80

Printed on acid-free paper

Springer is part of Springer Science+Business Media (www.springer.com)

Preface

Number theorists study arithmetic objects by attaching invariants to them to make them clearly visible. Each important research object in number theory has its L-functions. As they are functions of complex or p-adic variables, we can evaluate them at integers, getting concrete numbers from the starting object.

The Archimedean L-function (in this book) has a functional equation. If the value of its Γ-factor and its counterpart under the functional equation neither vanish nor have a pole at an integer m, assuming that the L-function at m is finite, we call such an m critical with respect to the L-function. If m is critical for the L-function, the value is expected to give an algebraic number up to a canonical transcendental factor (called a period, which often behaves compatibly under algebraic operations, for example, direct product of the objects, tensor product, and so on). Such phenomena were first found by Euler in the mid-18th century for the Riemann zeta function and were rediscovered in the 1970s by Shimura for many automorphic/geometric L-functions. Today we benefit from a good analysis both of the automorphic period of automorphic L-values through the work of Shimura and others as well as that of the geometric period of algebraic varieties and motives from the work of Shimura, Deligne, and others.

Here arises a typical number-theoretic question comparing the automorphic/geometric definitions of a period: Once an automorphic form is attached to a geometric object, such as an abelian variety or, more generally, a motive, do the automorphic period and the geometric period coincide? Assuming holomorphy of the L-function and the Γ-factor at an evaluating integer s_0, if s_0 is not critical, the Γ-factor has a pole at the reflex point s'_0 of s_0, and the L-value at s'_0 would simply be 0. Thus, its derivative at s'_0 is related by the functional equation to the value at the original point s_0 up to a correction factor in addition to the period. The correction factor is an interesting new invariant, the *regulator*, which depends crucially both on the arithmetic object and on the point of evaluation.

Once the algebraic part of critical L-values is well defined, we might want to relate it to a new arithmetic object produced from the starting object. For example, if we are given an imaginary quadratic field, from its reciprocity law we can produce a Dirichlet L-function whose value at the critical point 0 is essentially the order of the class group of the imaginary quadratic field (Dirichlet's celebrated class number formula). Dirichlet's formula is the identity between two arithmetic invariants: One is given by the value at 0 of the L-function associated with the imaginary quadratic field, and the other is the order of a finite abelian group specifically produced out of the integer ring of the field.

Going further, we might want to search a p-adic analytic L-function having the same value (up to a simple *p-adic modification factor* similar to an Euler factor) at critical evaluation points. Again, we therefore get a p-adic period and a p-adic regulator and an arithmetic object (now somehow p-adically made) whose size is (often conjecturally) given by the value or the derivative (possibly up to the p-adic regulator).

The p-adic modification factor rarely (but sometimes) vanishes at some integer points (producing an exceptional zero of the p-adic L-function), making the p-adic critical value close to an Archimedean noncritical value. The p-adic derivative of the p-adic L-function is equal up to a factor $\mathcal{L}$, similar to the regulator, to the (algebraic part of) Archimedean L-value, getting another invariant, called the $\mathcal{L}$-invariant. Mazur–Tate–Teitelbaum systematically studied the $\mathcal{L}$ although the value of the $\mathcal{L}$ was studied and even determined for some specific L-functions earlier than their work.

One may look into the growth of L-values over points densely populated in the domain of the L-function. Suppose that a p-adic L-function L_p is defined on a unit disk of p-adic numbers. If its p-adic absolute values $|L_p(s)|_p$ are bounded by the supremum $p^{-\mu}$ for $0 \leq \mu \in \mathbb{Q}$, this μ is the μ-invariant first studied by Iwasawa in the 1960s and 1970s. Iwasawa then defined his λ-invariant to be the number of zeros of the p-adic L-function (counted with multiplicity) in order to measure the growth of a p-adic Dirichlet L-function and the growth of the p-part of the class numbers of the cyclotomic fields of p-power roots of unity.

If we are given a compatible system of l-adic Galois representations, its trace a_p of the p-Frobenius element at unramified primes p outside l is algebraic and independent of l. This a_p itself is an invariant describing the structure of the Galois group (and also the local Galois group at p) of the splitting Galois extension of the representation, and we may uniquely form an Archimedean L-function $\sum_{n=1}^{\infty} c_n n^{-s}$ with Euler product converging for $\mathrm{Re}(s) \gg 0$ having $c_p = a_p$ for primes p from it. Or we can study the field generated by a_p over the rational numbers $\mathbb{Q}$, which is called the Hecke field of the system if the system comes from an automorphic/modular form (as we expect that every system comes from an appropriately chosen automorphic form). Once a Hecke field is given, its degree over the rational numbers is a most primitive example of the invariants attached to the system.

Often the definition of the invariant itself does not give us a straight way of understanding it. So number theorists seek a more direct or theoretical way to compute the invariant. This is done in many cases through heuristic understandings of the invariant. For example, the critical/noncritical L-value is supposed to compute the order of an arithmetically defined abelian group (an arithmetic cohomology group, called a Selmer group) similarly to the Dirichlet class number formula. If we could prove this heuristic, it would be a generalization of Dirichlet's class number formula. Some mathematicians have tried to specify what kind of arithmetic cohomology group should appear this way (i.e., the Bloch–Kato conjecture for general motives; more specifically, the Birch–Swinnerton-Dyer conjecture if the L-function is associated with a rational elliptic curve and the evaluation point of interest is $s = 1$). This is an identification problem of two different invariants, one analytically defined and the other arithmetically defined. Similarly, Shimura and Deligne have studied the equality of the automorphic period and the geometric period. All of these problems (which are central to present-day number theory) can be summarized into a set of problems targeting an identity of two or more different definitions/incarnations of the invariant.

Since miraculous (and possibly profound) formulas are scattered throughout number theory, mathematicians working in this area tend to seek a new one directly or indirectly relating different arithmetic invariants. Most of the celebrated conjectures in this research area fall into this category out of the desire to convince ourselves of the harmony of our universe. I am not at all against the idea. However, before setting out to see the God-given harmony of numbers, we would do better to worry about the nontriviality or triviality of the invariant. For example, Ferrero–Washington proved the vanishing of the (original) Iwasawa μ-invariant long ago in an elementary way. The vanishing is equivalent to the nontriviality of the p-adic L-function modulo p. From their result, if Iwasawa's λ-invariant vanishes (i.e., is trivial) very often, we get the infinity of regular primes (does the name indicate that regular primes must be "regularly" found?). This type of banal outcome, though possibly deep, is often more intriguing in its proof than in the proof of a nice formula connecting far-reaching arithmetic objects. Furthermore, as the outcome is banal, one may not be able to develop a conjectural formula from the problem (which satisfies mathematicians' aesthetic desires).

The critical L-values must be generically nonzero (with interesting and rare exceptions) unless the vanishing is forced by some trivial reason, and its first derivatives must be nonzero most of the time if the vanishing is forced. This is basically telling us that normally we are working in an easy environment (without the need to worry about the triviality of the invariant). We should be working in "regular" cases (of nonvanishing L-values) to produce the formulas in most instances, rather than working against hostility, requiring heroic efforts to prove an identity of higher derivatives. The regulators must not vanish in most of the cases (nonvanishing is often assumed by the lack of its proof), but to my knowledge, nobody seems to have made a systematic algebraic/arithmetic theory aimed at showing the nontriviality/triviality of the invariant.

In summary, there appear to be two fundamental problems concerning the invariants in number theory:

A. Find (even conjecturally) a relation (or an identity) of two or more arithmetic invariants of a different nature.
B. Prove the nontriviality/triviality of important arithmetic invariants.

In the early 1990s, I somehow realized the importance of problem B (and of making an algebraic theory dealing with it) and started working on it, although I still continue to work out things belonging to problem A. A principle (useful in attacking problem B) I came up with is rather ordinary:

If we want to show algebraically the nonvanishing of some value, in practice, the only way is to spread or interpolate the values as a good function over a geometric object (say, an algebraic variety that parameterizes the different values we study). If we could show that the function is nonconstant, there are not many zeros; hence, its specialization to the value we want is often nonzero. If we are able to go farther, we can even show that the zero set of the function somehow avoids all specific points of our interest (thus, getting the nontriviality of "each" desired value).

This book is the first report on my effort for the past 20 years, and I hope to present my progress report covering more new topics in the near future (as I believe that the method should be amenable to generalizations and has much room for improvement, because of the simplicity of the idea). We limit ourselves to topics directly related to elliptic modular forms (to make this book accessible to graduate students), though many results can be generalized (at least) to Hilbert modular forms (such generalizations are often given in my research articles quoted in the text).

Here is an outline of the book. In Chap. 1, after giving definitions of Iwasawa's invariants and the $\mathcal{L}$-invariants, as an example of the above principle, we reproduce Sinnott's proof of the nonvanishing of Dirichlet L-values modulo p (originally proven by L. Washington in 1978) under twists by most characters modulo l-power for a prime $l \neq p$. Then we go on to describe an outline of the proof of the rationality of Hecke L-values by Shimura and its p-adic version by Katz; hence, we get p-adic Hecke L-functions (of imaginary quadratic fields). For rationality, we need the invariant differential operators acting either on C^∞ modular forms or on p-adic modular forms. Shimura prominently resurrected the use of such operators (going back to Maass) to prove the rationality results of modular/automorphic L-values up to "periods," and Katz made its p-adic "avatar." We will give perhaps the simplest construction (which I found about 20 years ago) of the p-adic operator, which connects directly with Archimedean operators and which requires only the deformation theory of ordinary elliptic curves as a tool (and should be easily generalized to the ordinary locus of Shimura varieties of PEL type).

In Chap. 2, without becoming too technical, we recall the geometric definition of elliptic modular forms (and moduli problems of elliptic curves over fields) just for our use in Chap. 3 to state the main results.

Almost all the main themes of the book appear in Chap. 3, possibly without a detailed proof. We prove the fast growth of the degree of Hecke fields for non-CM analytic families of slope 0, the nonvanishing (in most cases) of the $\mathcal{L}$-invariant of the modular adjoint L-function at $s = 1$ (restricting ourselves to simplest cases), and a sketch of a proof of the vanishing of the μ-invariant of Katz p-adic L-functions. At the end, we state a principle of distinguishing mod-p Shimura subvarieties of a Shimura variety as a subvariety stable under Hecke correspondences (Chai–Oort's rigidity principle, an example of problem A), which is heavily used in the proof of the vanishing of the μ-invariant. This includes a brief outline of the theory of modular curves as a Shimura variety (and the Igusa tower over it).

Chapter 4 is an introduction to functorial scheme theory (and tools we need for fully proving some results illustrated in Chap. 3). Proofs of some theorems are given in a different way from other books for algebraic geometers (which is often unilluminating for number theorists). Some ring-theoretic results and more geometric results are quoted from different sources. Chapter 5 is a continuation of Chap. 4 but gathers more classical geometric results for varieties over a field. Readers who are familiar with functorial algebraic geometry or who just want to know the results for problem B without going into technicalities may skip these two chapters.

Chapter 6 is a detailed description of modular curves of individual levels over general rings. In particular, it contains a detailed description of deformation coordinates of the modular curve around an ordinary elliptic curve with complex multiplication. The study of the move of the deformation coordinate under the natural action of the stabilizer of the origin of the deformation space (carrying the CM elliptic curve) is a main tool to prove the rigidity principle.

In Chap. 7, we interpret the theory given in Chap. 6 in Shimura's way as the simplest of Shimura varieties. In addition, we complete and legitimize the sketchy construction of p-adic invariant differential operators first given in Chap. 1 (in Sect. 7.2.6). We also prove the irreducibility of the Igusa tower in the way given in [PAF] in more down-to-earth terms (which is the basis of the q-expansion principle essential in computing μ-invariants).

In Chap. 8, we generalize Sinnott's method (reproduced in Chap. 1) of proving the nonvanishing modulo p of Dirichlet L-values to Hecke L-values via the theory of Shimura varieties.

Chapter 9 is devoted to the proof of the vanishing of the μ-invariant of the p-adic Hecke L-functions (of each imaginary quadratic field) via a detailed analysis of the q-expansion of Eisenstein series. The rigidity principle and the

q-expansion principle (based on the irreducibility of the Igusa tower) play a prominent role.

In Chap. 10, we give a proof of Chai's local rigidity lemmas that is essential to prove the rigidity principle. In the last chapter, Chap. 11, we prove the rigidity principle for the elliptic Shimura curve and products of the copies of the curve (making this book logically complete).

This book's target audience encompasses graduate students (who already know the basics of elliptic modular forms, for example, as described in Chaps. 1–5 of [IAT]), postdoctoral scholars, and senior researchers working in the field of number theory.

In 2011, my graduate students Bin Zhao and Ashay Burungale (respectively in their third and second year of graduate school at UCLA) read most of the manuscript and suggested many improvements. I personally thank them for their careful reading.

In the hope of giving a good introductory account of my earlier (but more difficult) book [PAF], I started writing the manuscript while I was visiting l'Institut Henri Poincaré (Paris, France) in January–March of 2010 and continued writing while I visited Kyoto University in Japan from September to December 2010, partially supported by Clay Mathematics Institute as a senior scholar. I gave many lectures on the topics of this book at these institutions. I thank the audiences of my lectures for their enthusiasm/criticism and also thank these institutions for their hospitality and support. I hope that the original purpose of an introductory account of my book [PAF] has been accomplished to a good extent.

While preparing the manuscript, I was partially supported by the NSF grants DMS 0753991 and DMS 0854949.

Los Angeles, CA, USA Haruzo Hida

Suggestions to the Reader

In the text, articles are quoted by abbreviating the author's name; for example, articles by Hida–Tilouine are quoted as [HT1] and [HT2]. There is one exception: Articles written by the author are quoted, for example, as [H04a] and [H98], indicating also the year published (or the year written in the case of preprints). For these examples, [H04a] and [H98] are published in 2004 and in 1998, respectively. The articles by the author in a preprint form are quoted, for example, as [H13c] (though their publication year may differ from 2013). Books are quoted by abbreviating their title. For example, one of my earlier books with the title *Geometric Modular Forms and Elliptic Curves* is quoted as [GME]. Our style of reference is unconventional but has been used in my earlier books [MFG], [GME], and [PAF], and the abbreviation is (basically) common in all three books.

As for the notation and the terminology, the symbol $\mathbb{Z}_p$ denotes the p-adic integer ring inside the field $\mathbb{Q}_p$ of p-adic numbers, and the symbol $\mathbb{Z}_{(p)}$ is used to indicate the valuation ring $\mathbb{Z}_p \cap \mathbb{Q}$. Throughout the book, we indicate an algebraic closure $\overline{\mathbb{Q}}$ of $\mathbb{Q}$. A subfield E of $\overline{\mathbb{Q}}$ is called a number field (often assuming $[E:\mathbb{Q}] := \dim_{\mathbb{Q}} E < \infty$ tacitly). For a number field E, O_E denotes the integer ring of E, $O_{E,p} = O_E \otimes_{\mathbb{Z}} \mathbb{Z}_p \subset E_p = E \otimes_{\mathbb{Q}} \mathbb{Q}_p$, and $O_{E,(p)} = O_E \otimes_{\mathbb{Z}} \mathbb{Z}_{(p)} \subset E$. A quadratic extension M/F is called a CM field if F is totally real and M is totally imaginary. For the integer ring of an imaginary quadratic field M (the simplest case of CM fields), we often use $\mathfrak{O}$ for its integer ring (in place of O_M).

The symbol $\mathcal{W}$ is exclusively used to indicate a discrete valuation ring inside $\overline{\mathbb{Q}}$ with residual characteristic p. The ring $\mathcal{W}$ could be of infinite rank over $\mathbb{Z}_{(p)}$ but with finite ramification index over $\mathbb{Z}_{(p)}$; hence, it is still discrete. The p-adic completion $\varprojlim_n \mathcal{W}/p^n\mathcal{W}$ is denoted by W, and we write $W_m = W/p^mW = \mathcal{W}/p^m\mathcal{W}$. We denote the algebraic closure $\overline{\mathbb{Q}}_p$ with p-adic absolute value $|\cdot|_p$ with $|p|_p = p^{-1}$ and write $\mathbb{C}_p$ for the completion of $\overline{\mathbb{Q}}_p$ under this absolute value. We fix two embeddings $i_\infty : \overline{\mathbb{Q}} \hookrightarrow \mathbb{C}$ and $i_p : \overline{\mathbb{Q}} \hookrightarrow \overline{\mathbb{Q}}_p \subset \mathbb{C}_p$ throughout the book. We assume that $\mathcal{W} \hookrightarrow \overline{\mathbb{Q}} \xrightarrow{i_p} \mathbb{C}_p$ is continuous under the topology of the valuation ring $\mathcal{W}$; thus, we regard W as a closed subring of $\mathbb{C}_p$. For a local ring B, $\mathfrak{m}_B$ denotes its maximal ideal. We often write $\mathbb{F}$ for the residue field $\mathcal{W}/\mathfrak{m}_{\mathcal{W}} = W/\mathfrak{m}_W$.

The symbol $\mathbb{A}$ denotes the adele ring of $\mathbb{Q}$. For a subset Σ of rational primes, we set $\mathbb{A}^{(\Sigma\infty)} = \{x \in \mathbb{A} | x_\infty = x_p = 0 \text{ for } p \in \Sigma\}$. If Σ is empty, $\mathbb{A}^{(\infty)}$ denotes the ring of finite adeles. We put $\mathbb{Z}_\Sigma = \prod_{p\in\Sigma} \mathbb{Z}_p$ and define $\mathbb{Z}_{(\Sigma)} = \mathbb{Z}_\Sigma \cap \mathbb{Q}$. If $\Sigma = \{p\}$ for a prime p, we write $\mathbb{A}^{(p\infty)}$ for $\mathbb{A}^{(\Sigma\infty)}$. For a vector space of a number field E, we write $V_{\mathbb{A}} = V(\mathbb{A})$ and $V_{\mathbb{A}^{(\Sigma\infty)}} = V(\mathbb{A}^{(\Sigma\infty)})$ for $V \otimes_{\mathbb{Q}} \mathbb{A}$ and $V \otimes_{\mathbb{Q}} \mathbb{A}^{(\Sigma\infty)}$, respectively. We identify $\mathbb{A}^{(\Sigma\infty)\times}$ with the subgroup of ideles $x \in \mathbb{A}^\times$ with $x_v = 1$ for $v \in \Sigma \sqcup \{\infty\}$ in an obvious way. The maximal compact subring of $\mathbb{A}^{(\infty)}$ is denoted by $\widehat{\mathbb{Z}}$, which is identified with the profinite ring $\prod_p \mathbb{Z}_p = \varprojlim_N \mathbb{Z}/N\mathbb{Z}$. We put $\widehat{\mathbb{Z}}^{(\Sigma)} = \widehat{\mathbb{Z}} \cap \mathbb{A}^{(\Sigma\infty)}$ and $\widehat{\mathbb{Z}}^{(p)} = \widehat{\mathbb{Z}} \cap \mathbb{A}^{(p\infty)}$. For a module L of finite type, we write $\widehat{L} = L \otimes_{\mathbb{Z}} \widehat{\mathbb{Z}} = \varprojlim_N L/NL$, $\widehat{L}^{(\Sigma)} = L \otimes_{\mathbb{Z}} \widehat{\mathbb{Z}}^{(\Sigma)}$, and $\widehat{L}^{(p)} = L \otimes_{\mathbb{Z}} \widehat{\mathbb{Z}}^{(p)}$.

An algebraic group T (defined over a subring A of $\overline{\mathbb{Q}}$) is called a torus if its scalar extension $T_{/\overline{\mathbb{Q}}} = T \otimes_R \overline{\mathbb{Q}}$ is isomorphic to a product $\mathbb{G}_m^r$ of copies of the multiplicative group $\mathbb{G}_m$. The character group $X^*(T) = \mathrm{Hom}_{\text{alg-gp}}(T_{/\overline{\mathbb{Q}}}, \mathbb{G}_{m/\overline{\mathbb{Q}}})$ is sometimes simply denoted by $X(T)$, and elements of $X(T)$ are often called weights of T. We write $X_*(T)$ for the cocharacter group $\mathrm{Hom}_{\text{alg-gp}}(\mathbb{G}_{m/\overline{\mathbb{Q}}}, T_{/\overline{\mathbb{Q}}})$. Similarly, for a split formal torus $\widehat{T}$ (see Sect. 4.4.3 for formal groups), we write $X^*(T) = \mathrm{Hom}_{\text{formal-gp}}(\widehat{T}_{/B}, \widehat{\mathbb{G}}_{m/B})$ (formal character group) and $X_*(T) = \mathrm{Hom}_{\text{formal-gp}}(\widehat{\mathbb{G}}_{m/B}, \widehat{T}_{/B})$ (formal cocharacter group). Here B is either an algebraic closure of the finite field $\mathbb{F}_p$ with p elements or the discrete valuation ring W as above, and $\widehat{\mathbb{G}}_m$ is the formal multiplicative group over B we describe later in this book.

As we will describe in Chap. 4, a scheme X can be considered either a local ringed space or a functor from a category of algebras or schemes into sets. If we want to indicate that we are thinking of X as a functor, we often write the functor as $\underline{X}$ (although we remove the underline if it is clear from the context). Similarly, for a morphism $f : \mathrm{Spec}(A) \to \mathrm{Spec}(B)$ of affine schemes, the corresponding algebra homomorphism $B \to A$ is sometimes written as $\underline{f}$ or f^* or $f^\#$, depending on the setting.

Contents

1

Nontriviality of Arithmetic Invariants

This chapter is an introductory and historical discussion of problems concerning arithmetic invariants and L-values. Often a number-theoretic object has an associated L-function. A number field has its Dedekind zeta function, an algebraic variety has its Hasse–Weil zeta function, an automorphic form/representation has its Langlands L-functions, and we have L-functions associated with Galois representations, compatible systems of Galois representations and motives. Number theorists all agree that L-functions and L-values are useful invariants for studying the object with number-theoretic goals in mind. From L-functions, number theorists have created more invariants, for example, λ- and μ-invariant from p-adic L-functions and the $\mathcal{L}$-invariant from exceptional zeros of p-adic L-functions.

An equally important invariant is the value of a modular form. An elliptic modular form can be defined as a "function" of elliptic curves with additional structure; so, it gives an invariant, giving a label to each elliptic curve. For example, we have the J-invariant telling us the isomorphism class of each elliptic curve over an algebraically closed field and Weierstrass's g_2- and g_3-functions, giving the coefficients of the canonical equation defining the curve (and a differential on the curve). Often, special values of a modular form are related to an L-value of the elliptic curve.

By the solution of Serre's modulo p modularity conjecture by Khare and Wintenberger (with some help from Kisin), we now know that each regular rank-2 compatible system of (odd) Galois representations with coefficient field T appears as the system attached to an elliptic Hecke eigenform with coefficient field T (a generalization of the Shimura–Taniyama conjecture; see [KhW] II and [Kh2] Theorem 7.1). The original conjecture was proven earlier by Wiles/Taylor et al. (e.g., [GME] Chap. 5). The coefficient field T, called the Hecke field, is generated by the eigenvalues of Hecke operators. The eigenvalue itself is an arithmetic invariant. In the Hecke field, the eigenvalue a_p of the Hecke operator $T(p)$ [or $U(p)$] is the trace of the p-Frobenius substitution of the Galois representation of the motive. Thus, a_p is again an important arithmetic invariant, including the simple extension $\mathbb{Q}(a_p) \subset T$.

H. Hida, *Elliptic Curves and Arithmetic Invariants*, Springer Monographs in Mathematics, DOI 10.1007/978-1-4614-6657-4_1,

In this book, we study many arithmetic invariants, particularly their nontriviality [i.e., nonvanishing of invariant values and growth of the absolute degree of the field $\mathbb{Q}(a_p)$]. We restate the *geometric principle* in the Preface:

If one wants to show algebraically the nonvanishing of some value, in practice, the only way is to spread the value as a function (often an algebraic/polynomial function or a power series) over a geometric object (say, an algebraic variety that parameterizes the different values we study). If we could show the function is nonconstant, there are not many zeros; thus, its specialization to the value we want is often nonzero.

This is the main principle we follow in this book. This principle can also be applied to the field $\mathbb{Q}(a_p)$, as we may regard a_p as a function on the spectrum $\mathrm{Spec}(\mathbf{h})$ of the "big" Hecke algebra $\mathbf{h}$, and we study the growth of $\mathbb{Q}(a_p)$ over $\mathrm{Spec}(\mathbf{h})$ (see Chap. 3).

The solution to nontriviality problems often finds striking applications; for example, a result of Vatsal–Cornut [V3] is a key to the proof of the p-adic Birch–Swinnerton-Dyer conjecture in [SU]. We are certain that our results will also find important applications. Indeed, the anticyclotomic main conjecture has been proven in almost all cases in [H06a] and [H09b] based on the nonvanishing modulo p of Hecke L-values in [H04b] and [H07a], exposed in Chap. 8 in this book, and the proof of cases of the full main conjecture of CM fields appears imminent via results in Chaps. 8 and 9 and their generalization (see [Hs1] and [Hs2]) and from the proof of the anticyclotomic cases [H06a] (see [Hs3]).

In this introductory chapter, newcomers to number theory may encounter terminology and symbols new to them, for example, the notion of "functor" and the symbol $\mathrm{Spec}(A)(R)$. We suggest they just take the new notion as explained briefly there and go through the chapter to learn the overall structure of the theory. These technical terminologies will be defined/explained in full in later chapters, as indicated at each occurrence. Often in this book, our argument is based on a fixed choice of a prime p for which we assume $p > 2$ in the statements (but often we give proofs assuming that $p \geq 5$). In this book, we assume a basic knowledge of algebraic number theory (including class field theory in [BNT] and Iwasawa theory in [ICF]), Riemann/Dirichlet L-functions as in [EDM] (or [LFE]) and classical theory of modular forms as in [IAT] and [LFE]. If the reader is familiar with slightly more advanced algebraic geometry (for example, as exposed in Chap. 1 of [GME]), it would be easier to read through the heart of this book after Chap. 7, though we review a substantial amount of algebro-geometric tools in Chap. 4.

We write $\mathbb{Z}$ for the ring of integers and $\mathbb{Z}_l = \varprojlim_n \mathbb{Z}/l^n\mathbb{Z}$ for the ring of l-adic integers with its l-adic topology (for a prime l). The letters A and R often denote commutative rings with identity (hence, a $\mathbb{Z}$-algebra), and A and R could be "variables" running over the category of B-algebras (for a fixed base ring B). See Sect. 4.1 for a formal definition of categories (though for the first reading, it might be good not to care much about formal definitions).

1.1 Arithmetic Invariants in Iwasawa Theory

Here we give prototypical examples of our object of study: arithmetic invariants. We consider the group scheme $\mathbb{G}_m$ (see Chap. 4 for a survey of the theory of schemes and group schemes). For each commutative ring A, we put $\mathbb{G}_m(A) = \mathrm{Hom}_{\mathrm{alg}}(\mathbb{Z}[t,t^{-1}],A)$ (the set of all ring homomorphisms from $\mathbb{Z}[t,t^{-1}]$ into A), which is canonically isomorphic to $A^\times$ by sending an algebra homomorphism $\phi : \mathbb{Z}[t,t^{-1}] \to A$ to its value $\phi(t) \in A^\times$. Here $\mathbb{Z}[t]$ is the polynomial ring of one variable t with integer coefficients, and $\mathbb{Z}[t,t^{-1}] = \bigcup_n t^{-n}\mathbb{Z}[t]$ (the ring of fractions t inverted). Thus, we may regard $\mathbb{G}_m$ as a functor from the category of commutative algebras to the category of groups sending each algebra A to its multiplicative group $A^\times$. See Sect. 4.1.2 for categories and Sect. 4.1.3 for a precise definition of functors.

If the reader is not familiar with group schemes, for the moment, just take $\mathbb{G}_m$ to be an association $A \mapsto \mathbb{G}_m(A) = A^\times$ for any commutative ring A. Each algebra homomorphism $\phi : A \to A'$ automatically induces the homomorphism of groups $A^\times = \mathbb{G}_m(A) \ni a \mapsto \phi(a) \in \mathbb{G}_m(A') = A'^\times$ without changing the direction of the arrow. We call a *covariant* functor such an association keeping the direction of the arrows. This type of functor is called a "group" functor (as it has values in the category of groups), and if such a group functor is defined in purely an algebraic fashion, such as $\mathbb{G}_m$, it has a lot of structure useful for answering number-theoretic questions.

Since the ring $\mathbb{Z}[t,t^{-1}]$ defines $\mathbb{G}_m$, we write $\mathbb{G}_m = \mathrm{Spec}(\mathbb{Z}[t,t^{-1}])$ and $\mathrm{Spec}(\mathbb{Z}[t,t^{-1})(A) := \mathrm{Hom}_{\mathrm{alg}}(\mathbb{Z}[t,t^{-1}],A) = \mathbb{G}_m(A)$ (see Sect. 4.1.4 for the spectrum of rings). We also denote $\mathcal{O}_{\mathbb{G}_m} = \mathbb{Z}[t,t^{-1}]$, which is called the *affine ring* of $\mathbb{G}_m$. Similarly, we could consider $\mathbb{G}_a(A) = \mathrm{Hom}_{\mathrm{alg}}(\mathbb{Z}[T],A) \cong A$ by $(\phi : \mathbb{Z}[T] \to A) \mapsto \phi(T) \in A$, which is again a covariant group functor (sending a ring A to its additive group). Each element $\phi(t) \in \mathcal{O}_{\mathbb{G}_m}$ induces a function $\mathbb{G}_m(A) \ni a \mapsto \phi(a) = \phi(t)|_{t=a} \in A = \mathbb{G}_a(A)$; hence, we may regard $\phi : \mathbb{G}_m \to \mathbb{G}_a$ as a morphism of functors (and hence a morphism of schemes; see Lemma 4.17).

Here is another example of group functors: Fix a base ring B (for $\mathbb{G}_m$, we have taken B to be the ring of integers $\mathbb{Z}$). For a positive integer $N > 1$, let μ_N be the group of Nth power roots of unity; so, for any B-algebra A, we can think of $\mu_N(A) = \{\zeta \in A | \zeta^N = 1\}$, which is again canonically a group (a multiplicative group). This association of a group $\mu_N(A)$ to a p-adic B-algebra A is a functor; thus, μ_N is a group functor from the category of B-algebras to groups. For a prime p, we write $\mu_{p^\infty}(A) = \bigcup_n \mu_{p^n}(A)$, which is again a group functor from the category of B-algebras to groups. Note that μ_N is the kernel of $x \mapsto x^N$ in $\mathbb{G}_{m/B}$ (in aggregate indexed by B-algebras A). Clearly, $\mu_N(A) \cong \mathrm{Hom}_{\mathrm{alg}}(\mathbb{Z}[t]/(t^N - 1), A)$ by sending t to the image of t by an algebra homomorphism $\mathbb{Z}[t]/(t^N - 1) \to A$. Here $\mathbb{Z}[t]/(t^N - 1)$ is the residue ring $\mathbb{Z}[t]$ modulo the principal ideal $(t^N - 1)$. In this sense, $\mu_N = \mathrm{Spec}(\mathbb{Z}[t]/(t^N - 1))$, and $\mathcal{O}_{\mu_N} = \mathbb{Z}[t]/(t^N - 1)$. Abusing the language, we often write $\mathbb{Q}(\mu_N)$ for the subfield of $\mathbb{C}$ generated by all Nth roots of unity.

1.1.1 Iwasawa Invariant

Let p and l be primes. Studying the p-primary part p^{e_n} of the "minus" part h_n^- of the class number h_n of the cyclotomic field $\mathbb{Q}(\mu_{l^n})$ is a classic problem. When $p = l$, there is a formula by Iwasawa (cf. [ICF] Theorem 7.14):

$$e_n = \lambda n + \mu p^n + \nu \text{ for three integers } \lambda,\ \mu, \text{ and } \nu.$$

In 1979, Ferrero and Washington proved a conjecture of Iwasawa, asserting $\mu = 0$ ([FW] and [ICF] Chap. 7).

The Kubota–Leopoldt p-adic L-function $L_p(s,\chi)$ for each nontrivial character $\chi : (\mathbb{Z}/p\mathbb{Z})^\times/\{\pm 1\} \to \overline{\mathbb{Q}}_p^\times$ was originally defined as a continuous function of $s \in \mathbb{Z}_p$ interpolating Dirichlet L-values in the following sense:

$$L_p(1-k,\chi\omega^k) = (1-\chi\omega^k(p)p^{k-1})L(1-k,\chi\omega^k)$$

for all positive integers k. Iwasawa showed this function to be analytic of the form $L_p(s,\chi) = \Phi_\chi(\gamma^s - 1)$ for a power series $\Phi_\chi(T) \in \mathbb{Z}_p[[T]]$, where $\gamma = 1+p$ $(p > 2)$ and $(1+p)^s$ is given by the p-adic limit $\sum_{n=0}^{\infty}\binom{s}{n}p^n$. The product $\Phi(T) = \prod_{\chi\neq 1;\chi(-1)=1}\Phi_\chi(T)$ can be decomposed into $p^m\phi(T)$ in $\mathbb{Z}_p[[T]]$ with $p \nmid \phi(T)$. By Weierstrass's preparation theorem, $\phi(T)$ has finitely many zeros α_j in $\overline{\mathbb{Q}}_p$ with multiplicity m_j. By the class number formula, we conclude $\lambda = \sum_j m_j$ and $\mu = m$. See [EDM] Sect. 5 and [ICF] for more details.

Once a p-adic L-function $s \mapsto L_p(\gamma^s - 1)$ for a power series $L_p(T)$ interpolating good L-values is given, we may ask what the *μ-invariant* and *λ-invariant* of the power series $L_p(T) \in W[[T]]$ are. Here we allow a more general coefficient ring W, which is a complete discrete valuation ring over $\mathbb{Z}_p$. In this book, we study the vanishing of the μ-invariant for some p-adic L-functions [other than $L_p(s,\chi)$]. The vanishing of Iwasawa's μ-invariant is nonvanishing modulo $p\mathbb{Z}_p[[T]]$ of $\Phi(T)$; thus, this is a nontriviality problem for $\Phi(T)$ modulo p.

When $p \neq l$, in 1978, Washington ([Ws]) proved that e_n is bounded independently of n. The proof in the two cases $p = l$ and $p \neq l$ are philosophically close, although technically there is a difference. Washington achieved this by showing that Dirichlet L-values $L(0,\chi)$ do not vanish modulo p for almost all odd Dirichlet characters χ of the l-power conductor. A new proof of these facts later given by Sinnott [Sn1] and [Sn2] in the mid-1980s was found more amenable to generalization. In Sect. 1.2, we describe an "automorphic" version of the proof of Sinnott in the case of $p \neq l$.

In Chaps. 2 and 6, we recall the moduli theory of elliptic curves and the theory of modular curves. Then in Sect. 1.3 in this chapter and Sect. 3.5 in Chap. 3, and in Chaps. 8 and 9, we will discuss how to generalize the argument applied to Dirichlet L-values to Hecke L-values of imaginary quadratic fields and to more general L-values via the moduli theory. We fix an odd prime p and denote by l another odd prime different from p.

1.1.2 $\mathcal{L}$-Invariant

We describe the $\mathcal{L}$-invariant. We start with a brief definition of *compatible systems* of $\mathfrak{l}$-adic Galois representations with coefficients in a number field $T \subset \overline{\mathbb{Q}}$ (see [ALR] I.2.3–4 and [GME] Sect. 4.1.7 for some more details), where $\overline{\mathbb{Q}} \subset \mathbb{C}$ is the field of all algebraic numbers inside the complex field $\mathbb{C}$. Write $O = O_T$ for the integer ring of T (the integral closure of $\mathbb{Z}$ in T). For a prime ideal $\mathfrak{l}$ of O, we define $O_{\mathfrak{l}} = \varprojlim_n O/\mathfrak{l}^n$ by the $\mathfrak{l}$-adic completion of O. Its field of fractions (i.e., the $\mathfrak{l}$-adic field) is denoted by $T_{\mathfrak{l}}$.

A system $\rho := \{\rho_{\mathfrak{l}}\}_{\mathfrak{l}}$ of Galois representations is a collection of representations $\rho_{\mathfrak{l}} : \mathrm{Gal}(\overline{\mathbb{Q}}/\mathbb{Q}) \to GL_d(T_{\mathfrak{l}})$ acting on $V = T_{\mathfrak{l}}^d$ indexed by primes $\mathfrak{l}$ of T. The field T is called the coefficient field of ρ. Write $l(\mathfrak{l})$ for the residual characteristic of $\mathfrak{l}$ and $Frob_q$ for the Frobenius element at a prime q. Let $E_q(X) = \det(1 - \rho_{\mathfrak{l}}(Frob_q)|_{V_{I_q}} X)$ $(l(\mathfrak{l}) \neq q)$, where V_{I_q} is the space of coinvariants defined by $V_{I_q} = V/(\sum_{\sigma \in I_q}(\rho_{\mathfrak{l}}(\sigma) - 1)V)$ (the maximal quotient on which the inertia subgroup I_q acts trivially). Here we write $\rho_{\mathfrak{l}}(Frob_q)|_{V_{I_q}}$ for the action of the Frobenius element induced on the space of coinvariants. By a *compatible system* ρ, we mean:

1. Each $\rho_{\mathfrak{l}} : \mathrm{Gal}(\overline{\mathbb{Q}}/\mathbb{Q}) \to GL_d(T_{\mathfrak{l}})$ is continuous semisimple and unramified outside $S \cup \{l(\mathfrak{l}), \infty\}$ for a fixed finite set S of primes, where a representation of $\mathrm{Gal}(\overline{\mathbb{Q}}/\mathbb{Q})$ is *unramified* at a prime p (resp., at ∞) if an inertia subgroup I_p at p (resp., complex conjugation) is in the kernel of the representation [see Sect. 4.2.1 for the topology of $\mathrm{Gal}(\overline{\mathbb{Q}}/\mathbb{Q})$].
2. The Frobenius characteristic polynomial $E_q(X)$ has coefficients in T and is independent of $\mathfrak{l} \nmid q$.

This definition extends to a system of representations of $\mathrm{Gal}(\overline{\mathbb{Q}}/F)$ for a number field F in an obvious manner.

We suppose that ρ is geometric (here, in an imprecise sense, just meaning that it comes from algebro-geometric objects somehow). For example, taking the group of all l-power roots of unity μ_{l^∞} inside $\overline{\mathbb{Q}}$, we have the l-adic cyclotomic character $\mathcal{N}_l : \mathrm{Gal}(\overline{\mathbb{Q}}/\mathbb{Q}) \to \mathbb{Q}_l^\times$ given by $\zeta^\sigma = \zeta^{\mathcal{N}_l(\sigma)}$ for all $\zeta \in \mu_{l^\infty}(\overline{\mathbb{Q}})$. Since $\mathbb{Q}(\mu_{l^\infty}) = \mathbb{Q}(\zeta | \zeta \in \mu_{l^\infty}(\overline{\mathbb{Q}}))$ is unramified at any prime different from l, $\mathcal{N}_l$ is unramified at $p \neq l$; hence, $S = \emptyset$. By taking a prime $\mathfrak{P}$ above (p) of the integer ring $\mathbb{Z}[\mu_{l^\infty}]$ of $\mathbb{Q}(\mu_{l^\infty})$, reduction modulo $\mathfrak{P}$: $\zeta \mapsto (\zeta \bmod \mathfrak{P})$ embeds μ_{l^∞} into $\mathbb{F}^\times$ for the residue field $\mathbb{F}$ of $\mathfrak{P}$, and the (arithmetic) Frobenius element $Frob_p \in \mathrm{Gal}(\mathbb{F}/\mathbb{F}_p)$ acts on $\mathbb{F}$ by $x \mapsto x^p$; hence, $\mathcal{N}_l(Frob_p) = p$. Thus, $\mathcal{N} := \{\mathcal{N}_l\}_l$ is a compatible system of Galois characters (with coefficients in $\mathbb{Q}$). By class field theory, it is not too difficult to show

Exercise 1.1 *Any compatible system of characters of* $\mathrm{Gal}(\overline{\mathbb{Q}}/\mathbb{Q})$ *is of the form* $\chi\mathcal{N}^w := \{\chi \cdot \mathcal{N}_l^w\}_l$ *for a finite order character* $\chi : \mathrm{Gal}(\overline{\mathbb{Q}}/\mathbb{Q}) \to \mathbb{Q}[\chi]^\times$ *and an integer* w*, where* $\mathbb{Q}[\chi]$ *is the field generated over* $\mathbb{Q}$ *by the values of* χ.

Since $\det(\rho)$ is a compatible system of Galois characters, it is equal to $\mathcal{N}^w$ for an integer w up to finite order characters. The integer w is called the *weight* of ρ. By Eichler–Shimura, Deligne, and Serre, any Hecke eigen cusp form f of weight k has an associated compatible system ρ_f of absolutely irreducible two-dimensional Galois representations whose Frobenius polynomial at p is given by the Hecke polynomial of f, that is, $X^2 - a_p X + \psi(p)p^{k-1}$ for the $T(p)$ eigenvalue a_p and the Neben character ψ of f (see [GME] Chap. 4); thus, the weight of ρ_f is $k-1$.

We fix a prime $p > 2$ and write $|\cdot|_p$ for the p-adic absolute value on $\overline{\mathbb{Q}}$ normalized so that $|p|_p = \frac{1}{p}$. Then $\mathbb{C}_p$ is the completion of $\overline{\mathbb{Q}}$ under $|\cdot|_p$. The field $\mathbb{C}_p$ is algebraically closed (e.g., [NAA] Sect. 3.4.1). We write $\overline{\mathbb{Q}}_p$ for the field of all numbers in $\mathbb{C}_p$ algebraic over $\mathbb{Q}_p$. We write $\mathfrak{p} = \{\xi \in O_T : |\xi|_p < 1\}$ and often write $W := O_{T,\mathfrak{p}}$ and $K = T_{\mathfrak{p}}$. If confusion is unlikely, we simply write ρ for $\rho_{\mathfrak{p}}$, which acts on $V = T_{\mathfrak{p}}^d$. If we admit a conjectural principle that any geometric p-adic Galois representation fits into a compatible system (a version of the conjecture of Fontaine–Mazur), we could have started with a geometric $\mathfrak{p}$-adic representation without introducing compatible systems (though the use of a compatible system is more elementary at this stage).

In this section, we assume that $\rho_{\mathfrak{p}}$ is irreducible and ordinary in the following sense: ρ restricted to $\mathrm{Gal}(\overline{\mathbb{Q}}_p/\mathbb{Q}_p)$ is upper triangular with diagonal characters $\mathcal{N}^{a_j}$ on the inertia I_p for the p-adic cyclotomic character $\mathcal{N}$ ordered from top to bottom as $a_1 \geq a_2 \geq \cdots \geq 0 \geq \cdots \geq a_d$. Thus,

$$\rho|_{I_p} = \begin{pmatrix} \mathcal{N}^{a_1} & * & \cdots & * \\ 0 & \mathcal{N}^{a_2} & \cdots & * \\ \vdots & \vdots & \ddots & \vdots \\ 0 & 0 & \cdots & \mathcal{N}^{a_d} \end{pmatrix}.$$

In other words, we have a decreasing filtration $\mathcal{F}^{i+1}V \subset \mathcal{F}^i V$ stable under $\mathrm{Gal}(\overline{\mathbb{Q}}_p/\mathbb{Q}_p)$ such that the Tate twists $gr^i(\rho)(-i) := (\mathcal{F}^i V/\mathcal{F}^{i+1}V)(-i)$ are unramified. Here for a p-adic Galois module $M_{/\mathbb{Q}_p}$, the jth *Tate twist* $M(j)$ has the same underlying space, but the new twisted action is given by $\mathcal{N}^j(\sigma)\sigma v$ for the original action $v \mapsto \sigma v$. Define, following Greenberg in [Gr],

$$H_p(X) = \prod_i \det(1 - Frob_p|_{gr^i(\rho)(-i)} p^i X) = \prod_{j=1}^d (1 - \alpha_j X).$$

It is believed to have the identity $E_p(X) = H_p(X)$ if $p \notin S$ and $E_p(X)|H_p(X)$ otherwise. In any case, $\mathrm{ord}_p(\alpha_j) \in \mathbb{Z}$. Let us define

$$\beta_j = \alpha_j \text{ if } \mathrm{ord}_p(\alpha_j) \geq 1, \text{ and } \beta_j = p\alpha_j^{-1} \text{ if } \mathrm{ord}_p(\alpha_j) \leq 0,$$

and set $e = |\{j|\beta_j = p\}|$. Then we define *modification factors* (which is conjectured to appear in front of the complex L-values divided by a period in the evaluation formula of the conjectural p-adic L-function of ρ) by

$$\mathcal{E}(\rho) = \prod_{j=1}^d (1 - \beta_j p^{-1}) \text{ and } \mathcal{E}^+(\rho) = \prod_{j=1, \beta_j \neq p}^d (1 - \beta_j p^{-1}).$$

The number e is the number of the vanishing factors $(1-\beta_j p^{-1})$, and $e>0$ if ρ has eigenvalue 1 for the Frobenius element $Frob_p$ at p. The complex L-function of ρ is defined by $L(s,\rho)=\prod_l E_l(l^{-s})^{-1}$.

The L-function $L(s,\rho)$ is conjectured to have a meromorphic continuation to the whole complex plane, and if $\rho_{\mathfrak{l}}$ does not have a subquotient isomorphic to a power $\mathcal{N}_l^k$ ($(l)=\mathfrak{l}\cap\mathbb{Z}$) for an integer k, the continuation must be analytic everywhere. Moreover, we have a well-defined Γ-factor $\Gamma_\rho(s)$ (a product of specific translates of Γ-functions; e.g., [HMI] Sect. 1.2.1) and a constant ε with complex absolute value 1; they are supposed to satisfy a functional equation of the form $\Gamma_\rho(w+1-s)L(s,\tilde{\rho})=\varepsilon N^{-s}\Gamma_\rho(s)L(s,\rho)$, where N is a positive integer called the conductor of ρ and $\tilde{\rho}={}^t\rho^{-1}$.

We assume that ρ (that is, each $\rho_{\mathfrak{l}}$) does not contain $\mathcal{N}^k$ as a subquotient. An integer m is called *critical* for ρ if $\Gamma_\rho(w+1-s)$ and $\Gamma_\rho(s)$ are both finite at $s=m$. We assume that the value at 1 is critical for $L(s,\rho)$ (this condition should follow from the condition $a_1>0\geq a_d$ conjecturally). Assume that the compatible system comes in an algebraic way as a factor of the étale cohomology of an algebraic variety (after twisting by the cyclotomic character). As conjectured by Deligne [D6] after the earlier discovery of the rationality/algebraicity of many automorphic L-values by Shimura in [Sh6], [Sh7], [Sh8] and [Sh9]; see Sect. 8.1 in the text and [HMI] Sect. 1.2.1), we expect to have $\frac{L(1,\rho\otimes\varepsilon)}{c^+(\rho(1))}\in\overline{\mathbb{Q}}$ for a well-defined period $c^+(\rho(1))\in\mathbb{C}^\times$ for all finite order characters $\varepsilon:\mathbb{Z}_p^\times\to\mu_{p^\infty}(\overline{\mathbb{Q}})$. Let $\gamma=1+p\in\mathbb{Z}_p^\times$ so that $1+p\mathbb{Z}_p=\gamma^{\mathbb{Z}_p}$. Then we should have the following (cf. [MzTT] and [Gr]):

Conjecture 1.2 *Suppose that $s=1$ is critical for ρ. Then there exist a power series $\Phi^{an}(X)\in W[[X]]$ and a p-adic L-function $L_p^{an}(s,\rho)=\Phi_\rho^{an}(\gamma^{1-s}-1)$ interpolating $L(1,\rho\otimes\varepsilon)$ for p-power order characters ε such that*

$$\Phi_\rho^{an}(\varepsilon(\gamma)-1)\sim\mathcal{E}(\rho\otimes\varepsilon)\frac{L(1,\rho\otimes\varepsilon)}{c^+(\rho(1))}$$

with the modifying p-factor $\mathcal{E}(\rho)$ as above [putting $\mathcal{E}(\rho\otimes\varepsilon)=1$ if $\varepsilon\neq 1$]. The L-function $L_p^{an}(s,\rho)$ has zero of order $e+\mathrm{ord}_{s=1}L(s,\rho)$ for a nonzero constant $\mathcal{L}^{an}(\rho)\in\mathbb{C}_p^\times$ (called the analytic $\mathcal{L}$-invariant); we have

$$\lim_{s\to 1}\frac{L_p^{an}(s,\rho)}{(s-1)^e}=\mathcal{L}^{an}(\rho)\mathcal{E}^+(\rho)\frac{L(1,\rho)}{c^+(\rho(1))},$$

where "$\lim_{s\to 1}$" is the p-adic limit, $c^+(\rho(1))$ is the transcendental factor of the critical complex L-value $L(1,\rho)$, and $\mathcal{E}^+(\rho)$ is the product of nonvanishing modifying p-factors. If $e=0$, we have $\mathcal{E}^+(\rho)=\mathcal{E}(\rho)$ and $\mathcal{L}^{an}(\rho)=1$.

There is some ambiguity in the conjecture. We need to divide $L(1,\rho\otimes\varepsilon)$ by $c^+(\rho(1)\otimes\varepsilon)$ to get equality in place of $\sim$. Though the definition of $c^+(\rho(1))$ is precise only up to nonzero numbers in T, the ratio $c^+(\rho(1)\otimes\varepsilon)/c^+(\rho(1))$ can be explicitly determined as a power of the Gauss sum of ε^{-1} (see [HMI]

Sect. 1.2.1, [D6] and [H94]). To avoid technical details of identifying the ratio, we just left it ambiguous, inserting "$\sim$" in place of "=." Thus, once $c^+(\rho(1))$ is fixed, the power series Φ^{an}_ρ is uniquely determined.

When Katz constructed his p-adic Hecke L-functions in [K2], as we will see later, he astutely avoided this ambiguity of choosing $c^+(\rho(1))$ when ρ is an induced representation of a Hecke character of a CM field, and conjecturally his trick should also apply to general ρ (see [SGL] for some idea of how to make the p-adic L-function unique). In any case, Φ^{an}_ρ is unique up to nonzero constant in T; thus, the above conjecture determines $\mathcal{L}^{an}(\rho)$ uniquely.

When $e > 0$, by our definition of $\mathcal{E}$, we find $\mathcal{E}(\rho) = 0$; so, by the above interpolation formula, we have $L^{an}_p(1,\rho) = 0$; hence, in such a case, we say that $L^{an}_p(s,\rho)$ has an *exceptional zero* at $s = 1$. The exceptional zero appears if $\rho(Frob_p)$ has eigenvalue 1, and the number e is the multiplicity of the eigenvalue 1. Since e linear factors of $\mathcal{E}(\rho)$ vanish, we "naively" guessed that the order of zero at $s = 1$ is actually e, though each of these factors cannot be regarded as a special value at $s = 1$ of any analytic function dividing L_p (as the power function $n \mapsto p^n$ for integers n cannot be interpolated p-adically).

Example 1.3 Start with a Dirichlet character $\chi : (\mathbb{Z}/N\mathbb{Z}) \to \overline{\mathbb{Q}}^\times$ with $\chi(-1) = -1$. Then $c(\rho^+(1)) = 2\pi i$ for $i = \sqrt{-1}$ (as known from Euler's time). If we suppose $\chi = \left(\frac{-D}{\cdot}\right)$ for a square-free positive integer D, the modifying factor vanishes at $s = 1$ if the Legendre symbol $\left(\frac{-D}{p}\right)$ equals 1 ($\Leftrightarrow (p) = \mathfrak{p}\overline{\mathfrak{p}}$ in $O_{\mathbb{Q}[\sqrt{-D}]}$ with $\mathfrak{p} = \{x \in O_{\mathbb{Q}[\sqrt{-D}]} : |x|_p < 1\}$). By a work of Kubota–Leopoldt and Iwasawa (e.g., [LFE] Chaps. 3 and 4), we have a p-adic analytic L-function $L^{an}_p(s,\chi) = \Phi^{an}(\gamma^{1-s}-1)$ for a power series $\Phi^{an}(X) \in \Lambda = W[[X]]$ and $\gamma = 1+p$ such that for $\mathcal{E}(\chi\mathcal{N}^m) = (1-\chi(p)p^{m-1})$

$$L^{an}_p(m,\chi) = \Phi^{an}(\gamma^{1-m}-1) = \mathcal{E}(\chi\mathcal{N}^m)L(1-m,\chi) \sim \mathcal{E}(\chi\mathcal{N}^m)\frac{L(m,\chi)}{(2\pi i)^m}$$

for all positive integers m as long as $|n^m - n|_p < 1$ for all n prime to p.

If we have an exceptional zero at 1, it appears that we lose the exact connection of the p-adic L-value and the corresponding complex L-value. However, the conjecture says we can recover the complex L-value via an appropriate derivative of the p-adic L-function up to the factor $\mathcal{L}^{an}(\chi)$. We may regard χ as a Galois character $\mathrm{Gal}(\mathbb{Q}[\mu_N]/\mathbb{Q}) = (\mathbb{Z}/N\mathbb{Z})^\times \xrightarrow{\chi} \{\pm 1\}$, and we have the exceptional zero if $\chi(Frob_p) = 1$.

For our later use, for the class number h of $\mathbb{Q}[\sqrt{-D}]$, we write the generator of $\overline{\mathfrak{p}}^h$ as ϖ; hence, $\overline{\mathfrak{p}}^h = (\varpi)$ for $\varpi \in \mathbb{Q}[\sqrt{-D}]$. We then define $\log_p(\overline{\mathfrak{p}}) = \frac{1}{h}\log_p(\varpi)$ and $\log_p(\mathfrak{p}) = \frac{1}{h}\log_p(\overline{\varpi})$. Here $\log_p$ is the *Iwasawa logarithm*, which is an analytic function defined by the power series $t \mapsto \sum_{n=1}^\infty (-1)^{n+1}\frac{(t-1)^n}{n}$ on the open disk of radius 1 centered at 1 extended to the entire $\mathbb{C}_p^\times$ by $\log_p(p^\alpha) = 0$ for all $\alpha \in \mathbb{Q}$. Thus, we have $0 = \log_p(\varpi\overline{\varpi}) = \log_p(\varpi) + \log_p(\overline{\varpi})$, and hence $\log_p(\mathfrak{p}) = -\log_p(\overline{\mathfrak{p}})$. This number shows up as a factor of the $\mathcal{L}$-invariant for $L_p(s,\chi)$.

Assuming $p > 2$, we put $\langle x \rangle = \sqrt[p]{\exp_p(p\log_p(x))}$ for $x \in \mathbb{C}_p^\times$ (which is the *p-torsion-free projection* of x). Here $\exp_p$ is the p-adic exponential map, which is convergent at $p\log_p(x)$, and $\sqrt[p]{t} = \sum_{n=0}^{\infty}\binom{1/p}{n}(t-1)^n$, which is again convergent at $t = \exp_p(p\log_p(x))$ (see [LFE] Sect. 1.3). We have another example studied in [MzTT]:

Example 1.4 If $\rho_\mathfrak{p}$ is ordinary semistable, we can formulate a conjecture similarly. Here is such an example. Start with a rational elliptic curve $E_{/\mathbb{Q}}$, which yields a compatible system $\rho_E := \{T_l E\}_l$ given by the l-adic *Tate module* $T_l E = \varprojlim_n E[l^n](\overline{\mathbb{Q}})$ for $E[l^n] = \{x \in E(\overline{\mathbb{Q}}) | l^n x = \mathbf{0}_E\}$ (see Sect. 6.2.2). Suppose that E mod p removed singular points is isomorphic to $\mathbb{G}_{m/\mathbb{F}_p}$ (so, $p \in S$). Then

$$H_p(X) = (1-X)(1-pX),\ E_p(X) = (1-X),\ \mathcal{E}(\rho_E) = 0 \ \text{ and } \ \mathcal{E}^+(\rho) = 1.$$

By the solution of the Shimura–Taniyama conjecture by Wiles (see [GME] Sect. 5.2.4), there exists an elliptic cusp form $f = \sum_{n=1}^{\infty} a(n,f)q^n$ of weight 2 with $L(s,f) = \sum_{n=1}^{\infty} a(n,f)n^{-s} = L(s,E)$. This L-function has a p-adic analog $L_p^{an}(s,E)$ constructed by Mazur (see [LFE] Sect. 6.5). There exists $\Phi_E^{an}(X) \in \Lambda$ such that $\Phi_E^{an}(\varepsilon(\gamma)-1) = \mathcal{E}(\rho_E \otimes \varepsilon)\frac{G(\varepsilon^{-1})L(1,E,\varepsilon)}{\Omega_E}$ for all p-power order characters $\varepsilon : \mathbb{Z}_p^\times \to \mu_{p^\infty}(\overline{\mathbb{Q}}_p)$; in other words, $L_p^{an}(s,E) = \Phi_E^{an}(\gamma^{1-s}-1)$. Here Ω_E is a period of the nowhere vanishing differential of E (see Sect. 2.1.2 for differentials over a field and Corollary 2.33 in Sect. 2.2.1 for nowhere vanishing differentials; see also Sect. 6.1.4). Thus, if $Frob_p$ has eigenvalue 1 on $T_l E$, the exceptional zero appears at $s = 1$, as in the case of Dirichlet character. The $Frob_p$ has eigenvalue 1 if and only if E has split multiplicative reduction mod p.

The problem of $\mathcal{L}$-invariant is to compute explicitly the $\mathcal{L}$-invariant $\mathcal{L}^{an}(\rho)$ without recourse to the p-adic L-function. The $\mathcal{L}$-invariant in the above examples was computed in the 1970s through 1990s, i.e., the cases where $\rho = \chi = \left(\frac{-D}{\cdot}\right)$ and $\rho = \rho_E$ for E with split multiplicative reduction. The results are as follows:

Theorem 1.5 *Let the notation and the assumption be as given above.*

1. *We have* $\mathcal{L}^{an}(\chi) = \frac{\log_p(q)}{\mathrm{ord}_p(q)} = -\frac{\log_p(q)}{h}$ *for* $q \in \mathbb{C}_p$ *given by* $q = \varpi/\overline{\varpi}$ $(\overline{\mathfrak{p}}^h = (\varpi), \mathfrak{p}^h = (\overline{\varpi}))$ *and the class number* h *of* $\mathbb{Q}[\sqrt{-D}]$.
2. *For* E *split multiplicative at* p, *writing* $E(\mathbb{C}_p) = \mathbb{C}_p^\times/q^{\mathbb{Z}}$ *for the Tate period* $q \in \mathbb{Q}_p^\times$, *we have* $\mathcal{L}^{an}(\rho_E) = \frac{\log_p(q)}{\mathrm{ord}_p(q)}$.

See Sect. 2.4.2 for a description of Tate curves and the Tate period.

The first assertion is due to Gross–Koblitz [GsK] and Ferrero–Greenberg [FG]; the second was conjectured by Mazur–Tate–Teitelbaum [MzTT] and later proven first by Greenberg–Stevens [GS] and additionally by others later on. In Sect. 3.4.2, we give a brief description of Greenberg–Stevens' argument. In the theorem's first formula theorem, the sign "–" of $-\frac{\log_p(q)}{h}$ is correct, as explained in [Gr] (10) (and if we evaluate the value at $s = 0$, the minus sign disappears, as in [FG]).

Example 1.6 Starting with an ordinary p-adic Galois representation ρ : $\mathrm{Gal}(\overline{\mathbb{Q}}/\mathbb{Q}) \to GL_d(W)$, there is a systematic way to create many Galois representations whose eigenvalues of $Frob_p$ contain 1. Indeed, let ρ act on the $d \times d$ matrices $M_d(W)$ by conjugation. We have

$$M_d(W) = \mathfrak{sl}_d(W) \oplus \{\text{scalar matrices}\}$$

for the trace 0 space $\mathfrak{sl}_d(W)$, which is stable under the conjugation. We write $\mathrm{Ad}(\rho)$ for this Galois representation $\mathfrak{sl}_d(W) \ni x \mapsto \rho(\sigma)x\rho(\sigma)^{-1} \in \mathfrak{sl}_d(W)$. Then the action of $\mathrm{Ad}(\rho)(Frob_p)$ on $\mathfrak{sl}_d(W)$ has eigenvalue 1 with multiplicity $\geq d-1$. However, it is easy to check that the system $\mathrm{Ad}(\rho)$ is often not critical if $d > 2$ even if we start with geometric ρ. Thus, in this book, we study the case where $d = 2$. Take a representation $\rho_F : \mathrm{Gal}(\overline{\mathbb{Q}}/F) \to GL_2(W)$ associated with a p-ordinary Hilbert–Hecke eigenform of "weight ≥ 2" (i.e., belonging to a discrete series at ∞) over a totally real field F. We make $\mathrm{Ad}(\rho_F)$ and consider the induced representation $\rho_{\mathbb{Q}} = \mathrm{Ind}_F^{\mathbb{Q}} \mathrm{Ad}(\rho_F) : \mathrm{Gal}(\overline{\mathbb{Q}}/\mathbb{Q}) \to GL_n(W)$ for $n = 2[F : \mathbb{Q}]$. Then $\rho_{\mathbb{Q}}(Frob_p)$ has eigenvalue 1 with multiplicity $e > 0$. Here the number e is the number of prime factors of p in F. The system $\rho_{\mathbb{Q}}$ is critical at $s = 1$. The $\mathcal{L}$-invariant of $\rho_{\mathbb{Q}}$ is studied in [HMI] Sect. 3.4.

Returning to $\mathbb{Q}$ with $d = 2$, there could be two ways to find geometric objects parameterizing $\mathrm{Ad}(\rho)$. For example, each elliptic curve E defined over number fields F gives rise to a two-dimensional p-adic Galois representation $\rho_E : \mathrm{Gal}(\overline{\mathbb{Q}}/F) \to GL_2(\mathbb{Z}_p)$; so $E \mapsto \mathcal{L}(\mathrm{Ad}(\rho_E))$ could be considered as a function defined over the moduli space of elliptic curves (the modular curve). What is this function, a modular object? This point is interesting but has not yet been exploited to produce concrete results. In any case, the study of the moduli space is important, as we know by experience (as we see later in this book). Another way is to consider the universal Galois deformation space S of a given two-dimensional representation $\rho : \mathrm{Gal}(\overline{\mathbb{Q}}/\mathbb{Q}) \to GL_2(\mathbb{Q}_p)$; hence, for each point $P \in S$, we have a two-dimensional representation ρ_P : $\mathrm{Gal}(\overline{\mathbb{Q}}/\mathbb{Q}) \to GL_2(\kappa(P))$ for the coefficient field $\kappa(P)$. Then $P \mapsto \mathcal{L}(\mathrm{Ad}(\rho_P))$ is again a function on S. We study this path in Sect. 3.4.

1.2 Dirichlet L-Values

As another introductory example, we describe a way of showing the nonvanishing of Dirichlet L-values modulo p for most character twists. As was known from the days of Euler (see [EDM] Chap. II and [LFE] Preface), Dirichlet L-values can be obtained as an explicit sum of special values of a rational function $\Phi(t) = \frac{P(t)}{Q(t)}$ with $P(t), Q(t) \in \mathbb{Z}[t]$ at a certain set Ξ of roots of unity.

Regarding $\Phi(t)$ as a function defined on the multiplicative group $\mathbb{G}_m$, the idea (of Sinnott) is to regard the function $\Phi(t)$ as the spread-out object of the L-values over $\mathbb{G}_m$ (and a product of copies of $\mathbb{G}_m$). By the explicit formula, $\Phi(t)$ has a Laurent expansion of t nontrivial modulo p. On the other hand, we want to prove that the vanishing of most L-values modulo p implies the vanishing of Φ all over Ξ. If Ξ is a big set (for example, a dense set under a good topology), we may conclude that Φ modulo p must be zero, which is a contradiction against the nontriviality of the Laurent expansion.

1.2.1 A Nonvanishing Result for Dirichlet L-Values

Recall the group scheme $\mathbb{G}_m = \mathrm{Spec}(\mathbb{Z}[t, t^{-1}])$ as a covariant functor from the category of commutative rings (i.e., $\mathbb{Z}$-algebras) into the category of groups. We take it as a functor from commutative $\mathbb{Z}$-algebras into the category of abelian groups sending an algebra A to its multiplicative group. We have another functor $\mathbb{G}_a$ sending A to the additive group A. We may identify $\mathbb{G}_a(A)$ with the set of algebra homomorphisms from a polynomial ring $\mathbb{Z}[T]$ into A by sending $\varphi \in \mathrm{Hom}_{\mathrm{alg}}(\mathbb{Z}[T], A)$ to its value $\varphi(T)$. In this sense, we can write $\mathbb{G}_a = \mathrm{Spec}(\mathbb{Z}[T])$ and $\mathcal{O}_{\mathbb{G}_a} = \mathbb{Z}[T]$.

Each Laurent polynomial $\phi(t) \in \mathcal{O}_{\mathbb{G}_m} := \mathbb{Z}[t, t^{-1}]$ (the affine ring of the scheme $\mathbb{G}_m$) can be considered to be a function $\mathbb{G}_m(A) = A^\times \ni a \mapsto \phi(a) \in A = \mathbb{G}_a(A)$. Thus, we may regard $\phi : \mathbb{G}_m \to \mathbb{G}_a$ as a morphism of functors. Abusing the language, for any fraction $\phi(t) = \frac{P(t)}{Q(t)}$ with $P(t), Q(t) \in \mathbb{Z}[t]$, we regard $\phi : \mathbb{G}_m \to \mathbb{G}_a$, although the map is not well defined at a zero z of the denominator $Q(t)$ [i.e., if $Q(z) = 0$]. If we can write $\phi(t) = \frac{P(t)}{Q(t)}$ with $Q(z) = 0$ and $P(z) \neq 0$, we just simply put $\phi(z) = \infty$ (and call z a *pole* of ϕ). Such a function $\phi(t)$ is called a *rational function* of $\mathbb{G}_m$. Let $\mathbb{Q}(t)$ for the field of all rational functions on $\mathbb{G}_m$, and define the *stalk* at $1 \in \mathbb{G}_m(\mathbb{F}_p)$ by

$$\mathcal{O}_{\mathbb{G}_m,1} = \left\{ \phi(t) = \frac{P(t)}{Q(t)} \in \mathbb{Q}(t) \middle| P(t), Q(t) \in \mathbb{Z}[t],\ Q(1) \not\equiv 0 \mod p\mathbb{Z}_{(p)} \right\},$$

where $\mathbb{Z}_{(p)} = \left\{ \frac{a}{b} \middle| a, b \in \mathbb{Z},\ p \nmid b \right\}$. The ring $\mathbb{Z}_{(p)}$ is a local ring (or more specifically, a discrete valuation ring) with maximal ideal $p\mathbb{Z}_{(p)}$. Note that $\mathcal{O}_{\mathbb{G}_m,1}$ is the localization of the ring $\mathcal{O}_{\mathbb{G}_m}$ at maximal ideal $(p, t-1)$; so $\mathcal{O}_{\mathbb{G}_m,1} = \mathbb{Z}[t, t^{-1}]_{(p,t-1)}$ (see [CRT] Sect. 4 for localization of rings).

Set $1 < N \in \mathbb{Z}$ prime to p. We fix a Dirichlet character λ of $(\mathbb{Z}/N\mathbb{Z})^\times$ with $\lambda(-1) = -1$ and two embeddings $\mathbb{C} \overset{i_\infty}{\hookleftarrow} \overline{\mathbb{Q}} \overset{i_p}{\hookrightarrow} \overline{\mathbb{Q}}_p$. We regard λ as having values in any of the three fields. Consider a rational function:

$$\Phi(t) = \Phi_\lambda(t) = \sum_{n=1}^{\infty} \lambda(n) t^n = \frac{\sum_{a=1}^{N} \lambda(a) t^a}{1 - t^N} \in \mathbb{Z}[t, t^{-1}]_{(p,t-1)} = \mathcal{O}_{\mathbb{G}_m,1},$$

where $\mathcal{O}_{\mathbb{G}_m,1}$ is the stalk at the closed point $(p, t-1)$. Since the numerator $\sum_{a=1}^{N} \lambda(a)t^a$ is divisible by $(t-1)$, the rational function Φ is finite at $t=1$. Euler discovered in 1735 (a special case of) the following formula:

$$\Phi(1) = L(0, \lambda). \tag{1.2.1}$$

This is well known (see [EDM] Chap. II or [LFE] Sect. 2.1):

Exercise 1.7 *Using* (1.2.1), *prove that* $L(0,\lambda)$ *is a* p*-adic integer in the ring* $\mathbb{Z}_p[\lambda]$ *generated by the values of* λ *(as long as* $N > 1$ *is prime to* p*).*

Writing $\mathfrak{P}$ for the maximal ideal of the p-adic valuation ring of $\overline{\mathbb{Q}}_p$, the theorem of Washington [Ws] mentioned earlier can be stated as follows:

Theorem 1.8 *Let* l *be a prime different from* p*. We have* $L(0, \lambda\chi) \not\equiv 0 \mod \mathfrak{P}$ *for almost all characters* $\chi : \mathbb{Z}_l^\times \to \mu_{l^\infty}$*.*

Here "almost all" implies "except for finitely many." By Kummer's class number formula (e.g., [EDM] Sect. 5 or [ICF] Theorem 4.17), writing h_n^+ for the class number of the maximal real field of $\mathbb{Q}(\mu_{l^n})$ and $h_n^- h_n^+$ for the class number of $\mathbb{Q}(\mu_{l^n})$, we have

$$h_n^- = 2l^n \prod_{\chi \mod l^n : \chi(-1)=-1} \left(-\frac{1}{2}L(0,\chi)\right).$$

Then the boundedness of the p-primary part of h_n^- is clear from the theorem, and that of $h_n = h_n^+ h_n^-$ follows from this; see [ICF] Theorem 5.34 (an analytic proof) or [ICF] Theorem 10.9 (an algebraic proof via Kummer theory).

Write $\mathcal{C}(\mathbb{Z}_l, A)$ for the space of all continuous functions on the ring of l-adic integers $\mathbb{Z}_l$ with values in A. Continuity is with respect to the l-adic profinite topology on $\mathbb{Z}_l$. Often we equip A with the discrete topology; hence, $\phi \in \mathcal{C}(\mathbb{Z}_l, A)$ is a function factoring through $\mathbb{Z}/l^n\mathbb{Z}$ for sufficiently large n.

Our automorphic proof of this theorem has the following steps.

(Step1) *Hecke operators*: We introduce a Hecke operator $U(l)$ acting on rational functions on $\mathbb{G}_{m/\overline{\mathbb{F}}_p}$ and functions on μ_{l^∞}. For any given algebraic group, we can define Hecke operators as was first done by Mordell and Hecke for $GL(2)_{/\mathbb{Q}}$. These operators are very useful in characterizing algebraic/rational functions on the group and often have strong integrality properties. Because of the use of such operators, we call our method of the proof "automorphic."

(Step2) *Measure associated with Hecke eigenforms*: Choose a sequence of generators ζ_N in μ_N so that $\zeta_N^d = \zeta_{N/d}$ for $0 < d|N$ (to make things simple, we choose $\zeta_N = \exp\left(\frac{2\pi i}{N}\right)$). Fix an isomorphism $\mathbb{Z}_l \cong \mathbb{Z}_l(1) = \varprojlim_n \mu_{l^n}$ given by $\mu_{l^n} \ni \zeta_{l^n}^a \leftrightarrow a \in \mathbb{Z}/l^n\mathbb{Z}$. For a Hecke eigenform $f|U(l) = af$ with unit eigenvalue $a \in \overline{\mathbb{F}}_p^\times$, step 2 is the construction of a measure df on $\mathbb{Z}_l^\times$ with $\int_{a+l^n\mathbb{Z}_l} df \doteqdot f(\zeta_{l^n}^a)$

for the image $\zeta_{l^n}^a$ of ζ^a in μ_{l^n}. Here we note that Φ is a Hecke eigenform and that we will take $f = \Phi$. Such a construction was given for elliptic Hecke eigenforms in [Mz1] when $l = p$ (see [LFE] Sect. 4.4 for a version for p-adic Dirichlet L-functions), but we do this here for rational functions on $\mathbb{G}_m$ and $l \neq p$. Here a measure μ on $\mathbb{Z}_l$ with values in $\overline{\mathbb{F}}_p$ is an $\overline{\mathbb{F}}_p$-linear functional defined on the space $\mathcal{C}(\mathbb{Z}_l; \overline{\mathbb{F}}_p)$ of continuous functions: $\mathbb{Z}_l \to \overline{\mathbb{F}}_p$. Thus, $\mu : \mathcal{C}(\mathbb{Z}_l; \overline{\mathbb{F}}_p) \to \overline{\mathbb{F}}_p$ is an $\overline{\mathbb{F}}_p$-linear map. We often write $\mu(\phi) = \int_{\mathbb{Z}_l} \phi d\mu$.

(Step3) *Evaluation formula*: For a character $\chi : \Gamma = \mathbb{Z}_l^\times/\mu_{l-1} \to \overline{\mathbb{F}}_p^\times$, we prove

$$L(0, \lambda\chi^{-1}) \doteqdot \int_{\mathbb{Z}_l^\times} \chi d\Phi = \int_\Gamma \chi d\Psi \;\text{ for }\; \Psi(\zeta) = \sum_{\varepsilon \in \mu_{l-1}} \Phi(\zeta^\varepsilon).$$

(Step4) *Zariski density*: Consider Ψ as induced from the rational function $\tilde{\Psi}(t_\varepsilon) = \sum_{\varepsilon \in \mu_{l-1}/\{\pm 1\}} \left(\Phi(t_\varepsilon) + \Phi(t_\varepsilon^{-1})\right)$ on $G = \mathbb{G}_m^{\mu_{l-1}/\{\pm 1\}}$ (the product of copies of $\mathbb{G}_m$ indexed by $\varepsilon \in \mu_{l-1}/\{\pm 1\}$) by pullback under the embedding $i : \mu_{l^\infty} \hookrightarrow G$ given by $\zeta \mapsto (\zeta^\varepsilon)_\varepsilon \in G$. The Zariski density of $i(\mu_{l^\infty})$ in a big subvariety in G—if $\int_{\mathbb{Z}_l^\times} \chi d\mu_\Phi = 0$ for infinitely many characters—implies the constancy of $\tilde{\Psi}$ and hence that of Φ (a contradiction). For example, if $l = 7$, we have $\{1, \omega, \omega^2\} = \mu_3 \cong \mu_6/\{\pm 1\}$, and $1 + \omega + \omega^2 = 0$ gives a relation $t_{\omega^2} t_\omega t_1 = 1$; so the big subvariety is the one defined by $t_{\omega^2} t_\omega t_1 = 1$. The Zariski density is an essential input from Sinnott [Sn2].

We are going to describe each step separately in the following subsections.

1.2.2 Hecke Operators for $\mathbb{G}_m$

Slightly more generally, we start with $\mathbb{G}_m$ over a p-adic integral local domain B whose quotient field is K. This means that we regard the functor $\mathbb{G}_m$ as defined over the category of B-algebras (changing the base ring appears to be a trivial maneuver but has a strong impact on the structure of the corresponding scheme; see Sect. 4.1).

Take an algebraic closure $\overline{K}$ of K. We suppose that a prime l is invertible in B and $\mu_{l^\infty}(\overline{K}) \subset B$ (that is, all l-power roots of unity in $\overline{K}$ are contained in B). Fix a prime l prime to the characteristic of K. Let $K(\mathbb{G}_m) = K(t)$ be the rational function field of $\mathbb{G}_m = \mathrm{Spec}(B[t, t^{-1}])$ [i.e., $K(t)$ is made up of $\frac{P(t)}{Q(t)}$ for all polynomials $0 \neq Q(t), P(t) \in B[t]$]. Define a linear operator $U(l)$ acting on rational functions $\phi \in K(t)$ on $\mathbb{G}_m$ by

$$\phi|U(l)(t) = \frac{1}{l} \sum_{\zeta \in \mu_l} \phi(\zeta t^{1/l}). \tag{U}$$

We call this operator the *Hecke operator* of the prime l.

The Hecke operator has the following geometric interpretation: We have the l-power group homomorphism $[l] : \mathbb{G}_m \to \mathbb{G}_m$ given by $x \mapsto x^l$. The map $[l]$ is actually an endomorphism of the functor $\mathbb{G}_m$. In this sense, it is essentially surjective ($\mathbb{G}_m$ is a divisible group under fppf topology over B in the geometric terminology in Sect. 4.1.10), as $s \in \mathbb{G}_m(A)$ is in the image of $[l]$ from $\mathbb{G}_m(A[\sqrt[l]{a}])$. Its kernel is the group μ_l of lth roots of unity whose affine ring is a smaller ring $B[t]/(t^l - 1)$ free of finite rank l over B (we call the rank l the order of μ_l; so μ_l has finite "order" l).

For the moment, take B to be a field K. Then the affine ring $K[t, t^{-1}]$ of $\mathbb{G}_{m/K}$ has infinite dimension as a K-vector space and has Krull dimension 1 as a ring, hence far bigger than the affine ring $K[t]/(t^l - 1)$ of $\mu_{N/K}$ which has dimension l as K-vector space and Krull dimension 0 as a ring. In this sense, μ_N is "small" (see [CRT] Sect. 5 for Krull dimension).

Return to a general base ring B. Since $[l]$ is essentially surjective with small "finite" kernel, we call $[l]$ an *isogeny*. We will see later that $[l]$ is an étale isogeny if l is invertible in B (see Sect. 4.1.4 for the exact notion of étale map, but here "étale" basically means that for any maximal ideal $\mathfrak{m}$ of B, $B/\mathfrak{m}[\mu_l]$ is a separable extension of $B/\mathfrak{m}$; see [CRT] Sect. 26 for separability). Then taking t to be the variable on the target $\mathbb{G}_m$, we have $t^{1/l}$ in the source $\mathbb{G}_m$. Then $\phi \circ [l]^{-1} = \phi(t^{1/l})$ is an element of the function field of the source $\mathbb{G}_m$, which is a degree-l Kummer extension of the function field of the target $\mathbb{G}_m$.

Exercise 1.9 *Assume that B is a field K of characteristic different from l. Prove that $\{\zeta t^{1/l}\}_{\zeta \in \mu_l}$ is the full set of all conjugates of $t^{1/l}$ in the degree-l extension $K[t, t^{-1}][t^{1/l}]$ of $K[t, t^{-1}]$ inside an algebraically closed field containing $K[t, t^{-1}]$.*

The sum $\sum_{\zeta \in \mu_l} \phi(\zeta t^{1/l})$ is an element in the function field $K(t)$ of the target $\mathbb{G}_m$. Thus, the above operator $U(l)$ acts on the function fields of $\mathbb{G}_m$. Since $[l] \circ [l] = [l^2] : \mathbb{G}_m \to \mathbb{G}_m$ gives $x \mapsto x^{l^2}$, we can define $U(l^2)$ using $[l^2]$ in place of $[l]$. In other words,

$$\phi|U(l^2)(t) = \frac{1}{l^2} \sum_{\zeta \in \mu_{l^2}} \phi(\zeta t^{1/l^2}).$$

Then we have $U(l^2) = U(l)^2$. Similarly, we may define $U(l^n)$, which is equal to $U(l)^n$.

Expand ϕ into a Laurent series $\sum_{n \gg -\infty} a(n, \phi) t^n$. Then we have

$$\phi|U(l)(t) = \sum_n \frac{1}{l} \sum_{\zeta \in \mu_l} \zeta^n t^{n/l} = \sum_n a(nl, \phi) t^n,$$

because

$$\sum_{\zeta \in \mu_l} \zeta^n = \begin{cases} 0 & \text{if } l \nmid n, \\ l & \text{if } l | n. \end{cases}$$

This shows $a(n, \phi|U(l)) = a(nl, \phi)$; in particular,

$$a(n, \Phi|U(l)) = a(nl, \Phi) = \lambda(ln) = \lambda(l)\lambda(n) = \lambda(l)a(n, \Phi). \qquad \text{(UL)}$$

From this, we conclude $\Phi|U(l) = \lambda(l)\Phi$. Assume $\frac{1}{p} \notin B$. As long as K contains all Nth roots of unity (for example, $K = B = \overline{\mathbb{F}}_p$), regarding $\Phi \in K(\mathbb{G}_m)$, Φ is a Hecke eigenfunction in the stalk of $\mathcal{O}_{\mathbb{G}_m}$ at $1 \in \mathbb{G}_m$ given by

$$\mathcal{O}_{\mathbb{G}_m,1/B} = \{\phi = \frac{P(t)}{Q(t)} \in \overline{K}(\mathbb{G}_m)|P(t), Q(t) \in B[t] \text{ and } Q(1) \not\equiv 0 \mod p\}.$$

1.2.3 Measure Associated with a $U(l)$-Eigenfunction

We start with a brief description of measure theory on a profinite group G. Let B be a p-adic integral domain; thus, $B = \varprojlim_n B/p^nB$. For any profinite group $G = \varprojlim_n \overline{G}_n$ with finite groups $\overline{G}_n$, we write $\mathcal{C}(G, B)$ for the space of continuous functions: $G \to B$. Here the group G is a topological group whose system of open neighborhoods of $x \in G$ is given by $\{xG_n\}_n$ for normal subgroups G_n with $G/G_n = \overline{G}_n$.

Define a norm $|x|_p = p^{-n}$ on B if $x \in p^nB$ but $x \notin p^{n+1}B$. Then B is complete under this norm. If $pB = 0$, $|x|_p = 1$ if $x \neq 0$, and $|x|_p = 0$ if $x = 0$. Since G is compact, any continuous function $\phi : G \to B$ is uniformly continuous. In other words, for any given small positive real number δ, there exists a positive integer N such that $|\phi(x) - \phi(y)|_p < \delta$ if $xy^{-1} \in G_N$. If $pB = 0$, this just means that $\phi \in \mathcal{C}(G, B)$ is a locally constant function. A measure $\mu : \mathcal{C}(G, B) \to B$ is a B-linear map with $|\mu(\phi)|_p \leq |\phi|_p = \mathrm{Sup}_{x \in G} |\phi(x)|_p$. In other words, equipping $\mathcal{C}(G, B)$ with the uniform topology induced by the norm $|\cdot|_p$, a measure is a bounded B-linear functional. If $pB = 0$, this just means that μ is just a B-linear map. We often write $\mu(\phi) = \int_G \phi d\mu$.

Choosing a coset decomposition $G = \bigsqcup_x xG_n$ for every n and writing $\overline{x} = xG_n \in \overline{G}_n$, we define a locally constant function ϕ_n whose value at $y \in \overline{x}$ is given by $\phi(x)$. Then, by the uniform continuity of ϕ, $\phi = \lim_n \phi_n$ in $\mathcal{C}(G, B)$, and $\int_G \phi d\mu = \lim_n \int_G \phi_n d\mu$ by the continuity of μ. Writing $\chi_n^{(x)}$ for the characteristic function of $\overline{x}$, we define the volume of $\overline{x}$ to be $\int_{\overline{x}} d\mu = \int_G \chi_n^{(x)} d\mu$. Then we can decompose the constant function $\mathbf{1}$ with value 1 into the sum $\mathbf{1} = \sum_{\overline{x}} \chi_n^{(x)}$. Thus, $\phi_n = \mathbf{1}\phi_n = \sum_{\overline{x}} \phi(x)\chi_n^{(x)}$. This shows that $\int_G \phi_n d\mu = \sum_{\overline{x}} \phi(x) \int_{\overline{x}} d\mu$. Thus, the measure μ is determined by the values $\{\int_{\overline{x}} d\mu\}_{n,\overline{x}}$. Since $\overline{y} = \bigsqcup_{\overline{x}} \overline{x}$ for $\overline{x} = xG_n \subset \overline{y}$ if $n > m$, to have a well-defined measure μ, these values must satisfy the following distribution relation:

$$\sum_{\overline{x} \subset \overline{y}} \int_{\overline{x}} d\mu = \int_{\overline{y}} d\mu \quad \text{for all } n > m, \qquad \text{(Dist1)}$$

where $\overline{y} = yG_m$. The boundedness of μ follows from the fact that $\mu(\overline{x}) \in B$ for all n and j; so we can forget about it. The condition (Dist1) is equivalent to the following condition for all ϕ factoring through $\overline{G}_n$:

$$\sum_{\overline{x}} \phi(x) \int_{\overline{x}} d\mu = \int_G \phi d\mu = \sum_{\overline{y}} \phi(y) \int_{\overline{y}} d\mu \text{ for all } n > m, \tag{Dist2}$$

where $\overline{x}$ runs over $\overline{G}_n$ and $\overline{y}$ runs over $\overline{G}_m$.

Example 1.10 Take $G = \mathbb{Z}_l$ (additive group). Then we have $G_n = l^n\mathbb{Z}_l$ and $\overline{G}_n = \mathbb{Z}/l^n\mathbb{Z}$. The relation (Dist1) then becomes

$$\sum_{x \equiv y \mod l^m, x \in \mathbb{Z}/l^n\mathbb{Z}} \int_{x+l^n\mathbb{Z}_l} d\mu = \int_{y+l^m\mathbb{Z}_l} d\mu \quad \text{for all } n > m, \tag{1.2.2}$$

where x runs over a complete representative set in $\mathbb{Z}_l$ for $\mathbb{Z}/l^n\mathbb{Z}$.

Example 1.11 Take $G = \mathbb{Z}_l^\times$ (multiplicative group) for an odd prime l. We may regard a measure μ on $\mathbb{Z}_l^\times$ as that on the additive group $\mathbb{Z}_l$ by $\int_{\mathbb{Z}_l} \phi d\mu = \int_{\mathbb{Z}_l^\times} \phi d\mu$. In other words, for x divisible by l, we impose $\int_{x+l^n\mathbb{Z}_l} \phi d\mu = 0$. Under this vanishing condition, an assignment of values $\{\int_{x+l^n\mathbb{Z}_l} d\mu \in B\}_{n,x\in\mathbb{Z}/l^n\mathbb{Z}^\times}$ satisfying (Dist1) gives rise to a measure on $\mathbb{Z}_l{}^\times$. Splitting $\mathbb{Z}_l{}^\times$ into a product $\Gamma \times \mu_{l-1}$ for $\Gamma = 1 + l\mathbb{Z}_l$, a bounded measure on Γ is a system of values $\{\int_{x+l^n\mathbb{Z}_l} d\mu \in B\}_{n,x\in\Gamma/\Gamma^{l^n}}$ satisfying (Dist1).

We now give the definition of the measure associated with each eigenfunction $\phi \in \mathcal{O}_{\mathbb{G}_m,1/B}$ of $U(l)$. Write $D_\infty(\phi)$ for the polar divisor of ϕ (see Sect. 2.1.5 for divisor and polar divisor); thus, its support $|D_\infty(\phi)|$ is the finite set given by $\{\alpha \in \overline{K} | \phi(\alpha) = \infty\}$ for an algebraic closure $\overline{K}$ of K.

Lemma 1.12 *Let $0 \neq \phi \in \mathcal{O}_{\mathbb{G}_m,1/B}$ with $\phi|U(l) = a\phi$ for $a \in B^\times$. If $l \neq p$, then we have $|D_\infty(\phi)| = |D_\infty(\phi)^l|$, and $\phi(\zeta)$ is in B for all $\zeta \in \mu_{l^\infty}(\overline{K})$.*

Proof. We first show that $\phi(\zeta) \in K$. We have $|D_\infty(\phi)| = |D_\infty(\phi|U(l))|$ (as $\phi|U(l) = a\phi$). Note $\tau \in |D_\infty(\phi|U(l))| \Rightarrow \zeta\tau^{1/l} \in |D_\infty(\phi)|$ for some $\zeta \in \mu_l$. Thus, $|D_\infty(\phi|U(l))| = |D_\infty(\phi)| \subset |D_\infty(\phi)^l|$. This shows $|D_\infty(\phi)| = |D_\infty(\phi)^l|$. In particular, if $\zeta \in \mu_{l^n}(\overline{K}) \cap |D_\infty(\phi)|$, $1 = \zeta^{l^n} \in |D_\infty(\phi)|$, which is absurd. Therefore, ϕ is finite at all $\zeta \in \mu_{l^\infty}(\overline{K})$. Applying the above argument to ϕ mod $\mathfrak{m}_B$ (and replacing both K and B by $B/\mathfrak{m}_B$), we conclude $\phi(\zeta) \in B$.

Let $0 \neq \phi \in \mathcal{O}_{\mathbb{G}_m,1/B}$, with $\phi|U(l) = a\phi$ for $a \in B^\times$. Define $\mathbb{Z}_l(1) = \varprojlim_n \mu_{l^n}(\overline{K})$, and fix a generator $\zeta \in \mathbb{Z}_l(1)$ over $\mathbb{Z}_l$. We define $\zeta^x \in \mu_{l^n}(\overline{K})$ for $x \in \mathbb{Z}/l^n\mathbb{Z}$ by the power ζ_n^x for the image ζ_n of ζ in $\mu_{l^n}(\overline{K})$. We define an l-adic measure $d\phi$ on $\mathbb{Z}_l$ in the following way: For a function $f : \mathbb{Z}/l^n\mathbb{Z} \to B$,

$$\int f d\phi = (la)^{-n} \sum_{x\in\mathbb{Z}/l^n\mathbb{Z}} f(x)\phi(\zeta^x). \tag{1.2.3}$$

This is well defined (by Lemma 1.12) independently of the choice of $\mathbb{Z}/l^n\mathbb{Z}$ through which f factors, because the condition (Dist2) follows from

$$al \cdot \phi(\zeta^y) = \phi|l \cdot U(l)(\zeta^y) = \sum_{x \equiv y \mod l^m} \phi(\zeta^x)$$

for $y \in \mathbb{Z}/l^m\mathbb{Z}$ and x runs over elements in $\mathbb{Z}/l^n\mathbb{Z}$ for $n = m+1$. Formally, we could have defined

$$\int f d\phi(t) = (la)^{-n} \sum_{x \in \mathbb{Z}/l^n\mathbb{Z}} f(x)\phi(\zeta^x t) \in \mathcal{O}_{\mathbb{G}_m, 1/B},$$

and then $\int f d\phi$ is its evaluation at $t = 1$: $\int f d\phi(1)$. We then have for a primitive Dirichlet character $\chi : (\mathbb{Z}/l^n\mathbb{Z})^\times \to B^\times$

$$\begin{aligned}\int \chi d\phi(t) &= (la)^{-n} \sum_x \sum_m \chi(x) a(m,\phi)(\zeta^x t)^m \\ &= (la)^{-n} \sum_m \left(\sum_{x \in \mathbb{Z}/l^n\mathbb{Z}} \chi(x)\zeta^{mx} \right) a(m,\phi) t^m \\ &= G(\chi)(la)^{-n} \sum_m \chi^{-1}(m) a(m,\phi) t^m,\end{aligned}$$

where $G(\chi)$ is the *Gauss sum*: $G(\chi) = \sum_{x \in \mathbb{Z}/l^n\mathbb{Z}} \chi(x)\zeta^x \neq 0$.

Remark 1.13 We may define $\phi|U(l)$ for any function $\phi : \mu_{l^\infty}(\overline{K}) \to B$ by the same formula (U). Then if $\phi|U(l) = a\phi$ for a function $\phi : \mu_{l^\infty}(\overline{K}) \to B$ with $a \in B^\times$, we can define the associated measure $d\phi$ again by the same formula (1.2.3).

1.2.4 Evaluation Formula

Take $B = \mathbb{Z}_p[\mu_N] \subset \overline{\mathbb{Q}}_p$. Applying the computation in Sect. 1.2.3 to $\phi = \Phi_\lambda = \sum_{m=0}^\infty \lambda(m) t^m$, we find $\int \chi d\Phi(t) = G(\chi)(l\lambda(l))^{-n} \Phi_{\chi^{-1}\lambda}(t)$, and evaluating the result at $t = 1$, we get from (1.2.1) (applied to $\chi^{-1}\lambda$ in place of λ)

$$\int \chi d\Phi = (l\lambda(l))^{-n} G(\chi) \Phi_{\chi^{-1}\lambda}(1) = (l\lambda(l))^{-n} G(\chi) L(0, \chi^{-1}\lambda).$$

Since $\chi \neq 1$ is supported on $\mathbb{Z}_l^\times$, we may restrict $d\Phi$ to $\mathbb{Z}_l^\times$.

Since any character $\chi : \mathbb{Z}_l^\times \to \mu_{l^\infty}$ factors through $\Gamma = \mathbb{Z}_l^\times/\mu_{l-1}$, we want to have a measure φ supported on $\Gamma = \mathbb{Z}_l^\times/\mu_{l-1}$ so that we have

$$\int_\Gamma \chi d\varphi = \int_{\mathbb{Z}_l^\times} \chi d\Phi \quad \text{for all characters } \chi \text{ of } \Gamma.$$

The measure φ is not associated with a rational function such as Φ, but if we allow functions on μ_{l^∞}, φ is associated with a function Ψ close to Φ. Note that any $z \in \mathbb{Z}_l$ has l-adic expansion $\sum_{n=0}^\infty a_n l^n$ with integers $0 \leq a_n < l$ and that for $\zeta \in \mu_{l^m}$, $\zeta^z = \zeta^{\sum_n a_n l^n} = \prod_{n=0}^{\infty} \zeta^{l^n a_n} = \prod_{n=0}^{m} \zeta^{l^n a_n}$ is a well-defined root of unity in $\mu_{l^\infty}(\overline{K})$. In particular, we have $\mathbb{Z}_l^\times \subset \mathrm{Aut}_{\mathrm{group}}(\mu_{l^\infty}(\overline{K}))$, and the group $\mu_{l-1} \subset \mathbb{Z}_l^\times$ acts on functions f defined on μ_{l^∞} by $f(\zeta) \mapsto f(\zeta^s)$ $(s \in \mathbb{Z}_l)$. This action commutes with $U(l)$ given in Remark 1.13. Define

$$\Psi(\zeta) = \sum_{\varepsilon \in \mu_{l-1}} \Phi(\zeta^\varepsilon) = \sum_{\varepsilon \in \mu_{l-1}/\{\pm 1\}} (\Phi(\zeta^\varepsilon) + \Phi(\zeta^{-\varepsilon})),$$

which is an eigenfunction of $U(l)$ with eigenvalue $\lambda(l)$. Assuming $l \nmid N$ [i.e., $\lambda(l) \in B^\times$], we get $\varphi = d\Psi$ by construction.

1.2.5 Zariski Density

We now assume that $B = \overline{\mathbb{F}}_p = K$. We prove the Zariski density result in (step 4). We start with the following result in [Sn2]:

Theorem 1.14 (Sinnott) *Let $\Xi \subset \mu_{l^\infty}(\overline{\mathbb{F}}_p)$ be an infinite set. Let $\mathcal{F}$ be the $\overline{\mathbb{F}}_p$-algebra of functions on Ξ with values in $\overline{\mathbb{F}}_p$. If $a_1, \dots, a_r \in \mathbb{Z}_l$ are linearly independent over $\mathbb{Z}$, the algebra homomorphism from the affine ring $R_0 = \overline{\mathbb{F}}_p[y_1, y_1^{-1} \dots, y_r, y_r^{-1}]$ of $\mathbb{G}_m^r$ into $\mathcal{F}$ sending y_j to an element given by $\zeta \mapsto \zeta^{a_i}$ for $i = 1, 2, \dots, r$ is injective.*

We prove this theorem after stating an exercise and a lemma.

Exercise 1.15 *Let K be a field and L/K be a finite field extension. Suppose that $K \supset \mu_l(\overline{K})$. For $\zeta \in \mu_{l^\infty}(L)$, prove that*

$$\mathrm{Tr}_{L/K}(\zeta) = \begin{cases} 0 & \text{if } \zeta \notin K, \\ [L:K]\zeta & \text{if } \zeta \in K. \end{cases}$$

Lemma 1.16 *If $b_1, \dots, b_s \in \mathbb{Z}_l$ are finitely many elements distinct in $\mathbb{Z}/l^N\mathbb{Z}$ for a given integer $N > 0$, then the zero set of the function $f : \mu_{l^\infty} \to \overline{\mathbb{F}}_p$ given by $f(\zeta) = \sum_{j=1}^r c_j \zeta^{b_j}$ for $0 \neq (c_1, \dots, c_s) \in \overline{\mathbb{F}}_p^s$ is a finite set.*

Proof. Let k be the finite field generated by c_j $(j = 1, 2, \dots, r)$ and all elements in $\mu_l(\overline{\mathbb{F}}_p)$ over $\mathbb{F}_p$. We write $l^{N_0} = |k \cap \mu_{l^\infty}(\overline{\mathbb{F}}_p)|$. By Exercise 1.15 applied to $K = k$, we have $\mathrm{Tr}_{k(\zeta)/k}(\zeta) = 0$ if $\zeta \in \mu_{l^\infty}(\overline{\mathbb{F}}_p) - k$. Put $M = N_0 + N$. If $f(\zeta) = 0$ for a root of unity $\zeta \notin \mu_{l^M}(\overline{\mathbb{F}}_p)$, $0 = \mathrm{Tr}_{k(\zeta)/k}(\zeta^{-b_i} f(\zeta)) = [k(\zeta):k]c_i$, because $\zeta^{b_j - b_i} \notin k$ $(i \neq j)$. Since $[k(\zeta):k]$ is an l-power, we find that $c_i = 0$ for all i, a contradiction; hence, the zero set is contained in $\mu_{l^M}(\overline{\mathbb{F}}_p)$.

Here is the proof of Theorem 1.14 by Sinnott: Since $a_1, \dots, a_r$ are linearly independent over $\mathbb{Z}$, for all monomials $y^k = y_1^{k_1} \dots y_r^{k_r}$ with $k = (k_1, \dots, k_r) \in \mathbb{Z}^r$ appearing in a nonzero Laurent polynomial $f(y_1, \dots, y_r)$ in R_0, the l-adic numbers $k_1 a_1 + \dots + k_r a_r$ are distinct. Let us order these finitely many l-adic numbers $k_1 a_1 + \dots + k_r a_r$ as $b_1, b_2, \dots, b_s$. Since they are distinct in $\mathbb{Z}_l$, there exists $0 < N \in \mathbb{Z}$ such that $b_j \mod l^N$ are distinct. Thus, the function $\zeta \mapsto f(\zeta^{a_1}, \dots, \zeta^{a_r})$ has only finitely many zeros in μ_{l^∞}; so, $f \neq 0$ in $\mathcal{F}$. Thus, the map $R_0 \to \mathcal{F}$ is injective. □

To conclude the assertion of (step 4), we need to make the following variable change: Let A be the additive subgroup of $\mathbb{Z}_l$ generated by μ_{l-1} and take a $\mathbb{Z}$-basis $I = \{a_1, \dots, a_r\}$ of A. Take a complete set of representatives $\varepsilon_1, \dots, \varepsilon_n$ for $\mu_{l-1}/\{\pm 1\}$ and write $\varepsilon_j = \sum_i c_{ij} a_i$ $(c_{ij} \in \mathbb{Z})$. Put $T = \mathbb{G}_m^{\mu_{l-1}/\{\pm 1\}} = \mathrm{Spec}(R)$ [resp., $Y = \mathrm{Spec}(R_0) = \mathbb{G}_m^I$] for $R = \overline{\mathbb{F}}_p[t_1, t_1^{-1}, \dots, t_n, t_n^{-1}]$ [resp., $R_0 = \overline{\mathbb{F}}_p[y_1, y_1^{-1}, \dots, y_r, y_r^{-1}]$]. Here t_j is the variable on the factor $\mathbb{G}_m$ indexed by ε_j. Consider the ring homomorphism $\pi^* : R \hookrightarrow R_0$ sending t_j to $\prod_i y_i^{c_{ij}}$. This algebra homomorphism is the pullback of a group homomorphism $\pi : Y \to T$ given as follows: Note $Y = \mathbb{G}_m \otimes_{\mathbb{Z}} \mathbb{Z}^r$ ($A = \mathbb{Z}^r$ via the basis $\{a_i\}$) and $T := \mathrm{Spec}(R) = \mathbb{G}_m \otimes_{\mathbb{Z}} \mathbb{Z}^n$. Regard $\mathbb{Z}^r$ and $\mathbb{Z}^n$ as row vector modules. For the $r \times n$ matrix $C = (c_{ij})$, $C : \mathbb{Z}^r \to \mathbb{Z}^n$ given by $x \mapsto xC$ induces $\pi : Y = \mathrm{Spec}(R_0) \cong \mathbb{G}_m \otimes_{\mathbb{Z}} \mathbb{Z}^r \xrightarrow{\mathrm{id} \otimes C} \mathbb{G}_m \otimes_{\mathbb{Z}} \mathbb{Z}^n \cong \mathrm{Spec}(R) = T$; hence, π is a morphism of algebraic groups. Since $a_1, \dots, a_r$ is a basis of A and $\varepsilon_1, \dots, \varepsilon_n$ generate A, the matrix C has rank r; thus, $C : \mathbb{Z}^r \to \mathbb{Z}^n$ is injective. This shows that $\mathrm{Ker}(\pi : Y(\overline{\mathbb{F}}_p) = (\overline{\mathbb{F}}_p^\times)^r \to (\overline{\mathbb{F}}_p^\times)^n = T(\overline{\mathbb{F}}_p))$ is finite.

Exercise 1.17 *Why is* $\mathrm{Ker}(\pi : Y(\overline{\mathbb{F}}_p) \to T(\overline{\mathbb{F}}_p))$ *finite?*

Lemma 1.18 *For an infinite subset* $\Xi \subset \mu_{l^\infty}$, *the set*

$$\tilde{\Xi} = \{(\zeta^{\varepsilon_1}, \dots, \zeta^{\varepsilon_n}) \in T(\overline{\mathbb{F}}_p) | \zeta \in \Xi\}$$

is the image of $\Xi_Y := \{(\zeta^{a_1}, \dots, \zeta^{a_r}) \in Y(\overline{\mathbb{F}}_p) | \zeta \in \Xi\}$ *under* π *and is Zariski dense in* $\pi(Y)$.

Proof. Note

$$\zeta^{\varepsilon_j} = \zeta^{\sum_i a_i c_{ij}} \Leftrightarrow t_j|_{t_j = \zeta^{\varepsilon_j}} = \prod_i y_i^{c_{ij}}|_{y_i = \zeta^{a_i}} = \pi(t_j)|_{y_i = \zeta^{a_i}}.$$

Thus, we find $\tilde{\Xi} = \pi\left(\{(\zeta^{a_1}, \dots, \zeta^{a_r}) \in Y | \zeta \in \Xi\}\right)$. Take $f \in R$. We need to show that if $f(\pi(\zeta^{\varepsilon_1}, \dots, \zeta^{\varepsilon_n})) = 0$ for all $\zeta \in \Xi$, we have $f|_{\pi(Y)} = 0$. The vanishing $f|_{\tilde{\Xi}} = 0$ implies $f|_{\Xi} = 0$ in $\mathcal{F}$. Then by Theorem 1.14, $f|_{\pi(Y)} = 0$; thus, $\tilde{\Xi}$ is dense in $\pi(Y)$.

In the following section, we conclude the assertion of (step 4) from the following:

Lemma 1.19 *Let the notation be as above. Then a relation of the form*

$$(P_1(t_1) + \dots + P_n(t_n))\,|_{\pi(Y)} = 0 \qquad (*)$$

for $P_j(z) \in \overline{\mathbb{F}}_p[z, z^{-1}]$ *can only occur if* $P_j(z) \in \overline{\mathbb{F}}_p$ *for all* j.

Proof. Since a_j is a basis of A, the equation $\varepsilon_j = \sum_i c_{ij} a_i$ determines integers $(c_{ij})_i$ uniquely. Since $t_j = \prod_i y_i^{c_{ij}}$ on $\pi(Y)$, $P_i(t_i)|_{\pi(Y)}$ and $P_j(t_j)|_{\pi(Y)}$ for $i \neq j$ do not contain common monomials of y_i as an element of R_0. Since monomials of $\{y_i\}_i$ are linearly independent over $\overline{\mathbb{F}}_p$, we find that the relation $(*)$ implies $P_i(z) \in \overline{\mathbb{F}}_p$ for all i.

1.2.6 Proof of Theorem 1.8

We first assume that $\int_\Gamma \chi d\varphi = 0$ [that is, $L(0, \chi^{-1}\lambda) \equiv 0 \mod \mathfrak{P}$] for all $\chi : \Gamma \to \mu_{l^\infty}$. Then, by the orthogonality relation of characters (see [LRF] Sects. 2.3 and 18.2), we find $\tilde{\Psi}(\zeta) = 0$ for all $\zeta \in \mu_{l^\infty}$. Since the image of μ_{l^∞} in $\mathbb{G}_m^{\mu_{l-1}/\{\pm 1\}}$ by $\zeta \mapsto (\zeta^\varepsilon)_\varepsilon$ is Zariski dense in $\pi(Y)$ (over $\overline{\mathbb{F}}_p$), we find that $\Phi(t_\varepsilon) + \Phi(t_\varepsilon^{-1})$ is constant by Lemma 1.19, which is impossible by the t-expansion of Φ. As we show now, from a slightly weaker assumption of infinity of characters χ with vanishing $L(0, \chi^{-1}\lambda) \equiv 0 \mod \mathfrak{P}$, by Zariski density in $Y = \mathbb{G}_m^I$ proven by Sinnott (Theorem 1.14), we can still conclude that a nonconstant linear combination of translations of $\Phi(t^\varepsilon)$ (under the multiplication on $\mathbb{G}_m$) is a constant in $\overline{\mathbb{F}}_p$. From this contradiction, we conclude the generic nonvanishing of the L-values.

Proof of Theorem 1.8. Our way of proof is via contradiction. Thus, we assume that we have an infinite sequence of characters $\{\chi_j\}_j$ of order l^{n_j} with $\int_\Gamma \chi_j d\varphi = 0$, which implies by variable change

$$\int_\Gamma \chi_j(x) d\varphi(ax) \overset{x \mapsto a^{-1}x}{=} \chi(a)^{-1} \int_\Gamma \chi d\varphi = 0.$$

Exercise 1.20 *For any field k of characteristic different from l containing a primitive lth root of unity, prove the following formula for a primitive l^nth root of unity ζ_{l^n} with $n \geq 2$ not in k:*

$$\mathrm{Tr}_{k[\zeta_{l^n}]/k}(\zeta_{l^n}^x) \neq 0 \Leftrightarrow \zeta_{l^n}^x \in k \quad \textit{for } x \in \mathbb{Z}_l.$$

Hint: This follows from Exercise 1.15 and the minimal polynomial of ζ_{l^n}: $X^{l^{n-1}(l-1)} + X^{l^{n-1}(l-2)} + \cdots + 1 = 0$ over $\mathbb{Z}$.

For simplicity, we assume that λ has values in $\mathbb{F}_p^\times$. Then, applying the Frobenius automorphism $F(x) = x^p$, we find for $\chi = \chi_j$

$$\begin{aligned} 0 = \left(\int_\Gamma \chi(x) d\varphi(ax)\right)^{p^n} &= \left(\sum_u \chi(u) \Psi(\zeta_{l^n}^{au})\right)^{p^n} \\ &= \sum_u \chi^{p^n}(u) \Psi(\zeta_{l^n}^{aup^n}) = \chi^{p^n}(p^n)^{-1} \int_\Gamma \chi^{p^n} d\varphi(ax) \end{aligned}$$

for all n. Thus, taking the trace from the field $\mathbb{F}_p[\chi_j]$ generated by the values of χ_j to $\mathbb{F}_p[\mu_l]$, we find that

$$\sum_{u\in\chi_j^{-1}(\mathbb{F}_p[\mu_l])\subset\Gamma/\Gamma^{l^{n_j}}} \chi_j(u)\Psi(\zeta_{l^{n_j}}^{au}) = 0$$

for all $a \in \Gamma$. The above sum only involves $u \in \Gamma^{l^{n_j-m}}/\Gamma^{l^{n_j}}$ for a constant m independent of j. Indeed, writing the order of the l-primary part of $(\mathbb{F}_p[\mu_l])^\times$ as l^m and taking n_j so that $n_j \geq 2m$, we can identify the multiplicative group $\Gamma^{l^{n_j-m}}/\Gamma^{l^{n_j}}$ with the additive one $\mathbb{Z}/l^m\mathbb{Z}$ by $\mathbb{Z}/l^m\mathbb{Z} \ni v \mapsto 1 + l^{n_j-m}v = u$. We can then write $\chi_j(u) = \zeta_{l^m}^{b_j v}$ for $b_j \in (\mathbb{Z}/l^m\mathbb{Z})^\times$ and $\zeta_{l^n}^{au} = \zeta_{l^n}^a\zeta_{l^m}^v$. Since $\{\chi_j\}$ is infinite, replacing $\{\chi_j\}$ by an infinite subset of $\{\chi_j\}$, we may assume that b_j is a constant b. Then we have for any $a \in \mathbb{Z}_l^\times$

$$\sum_{v \mod l^m} \zeta_{l^m}^{bv}\Psi(\zeta_{l^{n_j}}^a\zeta_{l^m}^v) = \sum_{v \mod l^m} \zeta_{l^m}^{bv}\Psi|\zeta_{l^m}^v(\zeta_{l^{n_j}}^a) = 0,$$

where for $f \in \overline{\mathbb{F}}_p(\mathbb{G}_m)$, $f|x \in \overline{\mathbb{F}}_p(\mathbb{G}_m)$ is defined by $f|x(t) = f(tx)$ for $x \in \mathbb{G}_m(\overline{\mathbb{F}}_p)$. Since $\pi(\Xi)$ for $\Xi = \bigcup_j(\mu_{l^{n_j}} \setminus \mu_{l^{n_j-1}})$ is still dense in $\pi(Y)$, applying Lemma 1.19 to $P_j(z) = \Phi_j'(z) + \Phi_j'(z^{-1})$ for $\Phi_j'(z) = \sum_{v \mod l^m} \zeta_{l^m}^{av}\Phi|\zeta_{l^m}(z)$ on $\mathbb{G}_m = \mathrm{Spec}(\overline{\mathbb{F}}_p[z, z^{-1}])$, we conclude $P_j(z) \in \overline{\mathbb{F}}_p$. Let us compute the Taylor expansion at $z = 1$ of $P_j(z)$. We have

$$\begin{aligned}
\Phi_j'(z) &= \sum_{v \mod l^m} \zeta_{l^m}^{bv}\Phi|\zeta_{l^m}(z) \\
&= \sum_v \zeta_{l^m}^{bv}\sum_{n=1}^{\infty}\lambda(n)(\zeta_{l^m}z)^n \\
&= \sum_{n\geq 1}\lambda(n)z^n \sum_{v \mod l^m} \zeta_{l^m}^{(b+n)v} \\
&= l^m \sum_{n>0, n\equiv -b \mod l^m} \lambda(n)z^n,
\end{aligned}$$

which is nonconstant, and P_j is also nonconstant. This finishes the proof of Theorem 1.8.

Exercise 1.21 *Give a proof of* Theorem 1.8 *when* λ *has values in* $\mathbb{F}_q^\times$ *for a general* p*-power* q.

1.3 CM Periods and L-Values

We continue our outline of the principal results of this book; later we will fill in all the details. This section has a twofold purpose:

1. To give a concise proof of the rationality and integrality result (due to Shimura [Sh6] and Katz [K2]) of the special values of (Archimedean and p-adic) arithmetic modular forms at CM points (Sects. 1.3.5 and 1.3.6).
2. To give an outline of the construction of a two-variable p-adic measure interpolating critical Hecke L-values of imaginary quadratic fields (Sect. 1.3.7).

We follow [Sh6] to prove the rationality of the special values of classical arithmetic modular forms and their derivatives (Theorem 1.27).

After some dormant period, the study of the rationality of special values of modular/automorphic forms (particularly at CM points) and L-values at critical integers was revived by many of Shimura's works from the early 1970s on, for example, [Sh6], [Sh7], [Sh8] and [Sh9]. The rationality/algebraicity of modular/automorphic forms at CM points is the base of the construction of Shimura varieties (which Shimura did in the 1960s to early 1970s; see Chap. 7 for an example of a Shimura variety). A principle we found in Shimura's work is that abelian or modular Hecke L-values are often realized as a finite sum of the special values of specific modular/automorphic forms at well-chosen points, hence getting rationality. Invariant differential operators (studied first by Maass and later revived again by Shimura) play a prominent role in such a setting to prove the rationality of nonholomorphic modular/automorphic forms at CM points. As for L-values, an integer is called critical for an L-function if its Γ-factor is finite at the integer and its counterpart of the functional equation. If the L-function is associated with algebro-geometric objects (for example, a projective variety or a motive), in the late 1970s, Deligne made a prediction of rationality of the L-values at critical integers relative to an intrinsic motivic period [D6] (though, strangely, he did not quote much of Shimura's work in [D6]).

Our new proof of p-integrality statements due originally to Katz (Theorem 1.34) is a version of Shimura's argument and can easily be generalized to Hilbert modular forms (and beyond). Again, a p-adic invariant differential operator (constructed by Katz as an analog of Archimedean operators) plays a fundamental role.

An elliptic curve E defined over a base ring B is a specific kind of curve (with a lot of fine structures) defined by the zero set $\Phi(x{:}y{:}z) = 0$ (in the two-dimensional projective space $\mathbf{P}^2$) of a special type of a cubic homogeneous equation $\Phi(x{:}y{:}z)$ with coefficients in B (see Sects. 2.1.3 and 2.1.4 for more of an explanation on such curves). Indeed, for any local B-algebra A, we may regard $E(A) = \{(x{:}y{:}z) \in \mathbf{P}^2(A) | \Phi(x{:}y{:}z) = 0\}$ as a functor $A \mapsto E(A)$ (see Chaps. 2 and 4). A very important property is that $E(A)$ has a canonical group structure (Theorems 2.32 and 6.3).

For any elliptic curve $E_{/B}$, because of the group structure, we have the kernel of $x \mapsto Nx = \overbrace{x + x + \cdots + x}^{N}$ on $E(A)$ for a positive integer N. Write this kernel $E[N](A)$, getting another functor $A \mapsto E[N](A)$ from the category of B-algebras to groups. We put $E[p^\infty](A) = \bigcup_n E[p^n](A)$ for a prime p. A morphism $\mu_{p^n} \to E[p^n]$ is a morphism of group functors. We will study such functor morphisms in Chap. 2 for an algebraically closed field $B = k$ and for a general ring B in Chap. 4. Until then, just take it as a sort of generalization of a usual homomorphism of groups [it is a compatible system of homomorphisms $\mu_{p^n}(A) \to E[p^n](A)$ indexed by algebras A over B].

1.3.1 Elliptic Modular Forms

What are modular forms? In the simplest case of elliptic modular forms, writing $w = {}^t(w_1, w_2)$ for two linearly independent complex numbers [with $\mathrm{Im}(z) > 0$ $(z = w_1/w_2)$], a weight- k modular form is a holomorphic function f of w satisfying $f\left(\left(\begin{smallmatrix} a & b \\ c & d \end{smallmatrix}\right) w\right) = f(w)$ for $\left(\begin{smallmatrix} a & b \\ c & d \end{smallmatrix}\right) \in SL_2(\mathbb{Z})$ and $f(aw) = a^{-k} f(w)$ for $a \in \mathbb{C}^\times$. In other words, f is a "holomorphic" function on the set Lat of lattices in $\mathbb{C}$ with $f(aL) = a^{-k} f(L)$ (see Sect. 2.3.4). If we regard f as a function of $z \in \mathfrak{H} = \{z \in \mathbb{C} | \mathrm{Im}(z) > 0\}$ by setting $f(z) = f\left(2\pi i \left(\begin{smallmatrix} z \\ 1 \end{smallmatrix}\right)\right)$, the two relations can be stated as a single one:

$$f\left(\frac{az+b}{cz+d}\right) = (cz+d)^k f(z)$$

for $\left(\begin{smallmatrix} a & b \\ c & d \end{smallmatrix}\right) \in SL_2(\mathbb{Z})$. This latter definition might be more familiar for some of the readers. Here we put $2\pi i$ to define $f(z)$ to make the q-expansion of f rational if f is algebro-geometrically rational [since $f(2\pi i w) = (2\pi i)^{-k} f(w)$, this is just the division by a power of $(2\pi i)$]. This division has to be done because of $\exp : \mathbb{C}/2\pi i(\mathbb{Z} + \mathbb{Z}z) \cong \mathbb{G}_m(\mathbb{C})/q^{\mathbb{Z}}$ for $q = \exp(2\pi i z)$. We want to prove algebraicity and integrality of the value $f(w)$ when w is a basis of an imaginary quadratic field (up to a canonical complex number, called the *CM period* of the elliptic curve). This rationality/integrality result is due to some mathematicians, notably, Hurwitz, Shimura, and Katz.

We now describe an algebraic interpretation of modular forms. Waiting until Chaps. 2 and 6 for a more thorough treatment of the algebraization, here we give an outline of how to define modular forms algebraically. Pick two linearly independent numbers $w = (w_1, w_2) \in \mathbb{C}^2$. Writing u for the variable on $\mathbb{C}$, the quotient $\mathbb{C}/L_w$ of $\mathbb{C}$ by a lattice $L_w = \mathbb{Z}w_1 + \mathbb{Z}w_2$ gives rise to a pair (E, ω) of elliptic curves and the differential $\omega = du$ of first kind (nowhere vanishing differential). First of all, we note $E(\mathbb{C}) \cong \mathbb{C}/L_w$. Then we embed E into $\mathbf{P}^2$ by $u \mapsto (x(u), y(u), 1) \in \mathbf{P}^2(\mathbb{C})$ for Weierstrass $\wp$-functions (Sect. 2.3.4):

$$x(u) = \wp(u; L_w) = \frac{1}{u^2} + \sum_{0 \neq l \in L_w} \left\{ \frac{1}{(u-l)^2} - \frac{1}{l^2} \right\} = \frac{1}{u^2} + \frac{g_2}{20} u^2 + \frac{g_3}{28} u^4 + \cdots$$

and $y = \frac{dx}{du}$, where $g_2(w) = 60 \sum_{0 \neq l \in L_w} l^{-4}$ and $g_3(w) = 140 \sum_{0 \neq l \in L_w} l^{-6}$. Then the equation satisfied by x and y is $y^2 = 4x^3 - g_2 x - g_3$ and $\omega = du = \frac{dx}{y}$

(see Sect. 2.3.4). The pair w (or lattice L_w) can be recovered by ω so that $w_i = \int_{\gamma_i} \omega$ for a basis (γ_1, γ_2) of the Betti homology group $H_1(E(\mathbb{C}), \mathbb{Z}) = \pi_1^{\mathrm{top}}(E(\mathbb{C}))$ (see Sect. 2.3.3 for the topological fundamental group $\pi_1^{\mathrm{top}}(E)$). For a given $(E,\omega)_{/\mathbb{C}}$, we write L_E for its lattice. Here $\mathbb{C}/L_w$ is an additive group, and this is the canonical group structure of $E(\mathbb{C})$.

Conversely, start with a pair $(E,\omega)_{/A}$ defined over a ring A made of an elliptic curve (a smooth curve of genus 1 with a specific point $0 = 0_E \in E(A)$) and a nowhere vanishing differential ω (see Sect. 2.1.2 for a definition of algebraic differentials over a field and Sect. 4.1.7 over rings/schemes). Then take an algebraic parameter u around 0 so that $\omega = du$. Write $[0]$ for the relative Cartier divisor given by 0 (see Sect. 2.1.5 for divisor and Sect. 6.1.2 for Cartier divisor). Since the line bundle $\mathcal{L}(m[0])$ (made of meromorphic functions having only pole at 0 of order at most m) is free of rank m if $m > 0$ (by the existence of ω), we can find $x \in H^0(E, \mathcal{L}(2[0]))$ having a pole of order 2 whose Laurent expansion has its leading term u^{-2}. If $6^{-1} \in A$, there is a unique way of normalizing x and u so that $y^2 = 4x^3 - g_2x - g_3$ with $y = \frac{dx}{du}$ for a unique pair $(g_2 = g_2(E,\omega), g_3 = g_3(E,\omega)) \in A^2$ (see Sect. 2.2.2 for Weierstrass's computation of the equation). Since $E_{/A}$ is smooth, this pair (g_2, g_3) has to satisfy $\Delta = g_2^3 - 27g_3^2 \in A^\times$ (Proposition 2.37). This shows (see Sect. 2.2.3)

$$[(E,\omega)_{/A}] \overset{\text{1 to 1 and onto}}{\leftrightarrow} \{(g_2,g_3) \in A^2 | \Delta \in A^\times\}$$
$$= \mathrm{Hom}_{ALG}(\mathbb{Z}[\frac{1}{6}, g_2, g_3, \frac{1}{\Delta}], A) = \mathrm{Spec}(\mathbb{Z}[\frac{1}{6}, g_2, g_3, \frac{1}{\Delta}])(A).$$

Here Hom_{ALG} denotes the set of algebra homomorphisms, the ring $\mathbb{Z}[\frac{1}{6}][g_2, g_3]$ is a polynomial ring with variables g_2, g_3, and the straight brackets $[\cdot]$ indicates the set of isomorphism classes of the objects inside; thus, $[(E,\omega)_{/A}] = \{(E,\omega)_{/A}\}/\cong$.

An important fact is that all these functions g_2, g_3 and Δ have Fourier expansions in $\mathbb{Z}[\frac{1}{6}][[q]]$ for $q = \exp(2\pi i z)$. Indeed, g_2 and g_3 are rational multiple of the following Eisenstein series for $k = 4, 6$:

$$\frac{(k-1)!}{2(2\pi i)^k} \sum_{(m,n)\in\mathbb{Z}^2\setminus\{(0,0)\}} \frac{1}{(mz+n)^k}$$
$$= \frac{1}{2}\zeta(1-k) + \sum_{n=1}^{\infty} q^n \sum_{0<d|n} d^{k-1}$$
$$= \frac{1}{2}\zeta(1-k) + \sum_{(m,n)\in\mathbb{Z}^2\setminus\{(0,0)\}/\mathbb{Z}^\times, mn>0} q^{mn} \tag{1.3.1}$$

for even integers $2 < k \in \mathbb{Z}$. This follows from the partial fraction expansion of the cotangent function of $z = \frac{\log q}{2\pi i}$ ([EDM] (2.18) or [EEK] Chap. II):

$$\frac{1}{z}+\sum_{n=1}^{\infty}\left\{\frac{1}{(z+n)}+\frac{1}{(z+n)}\right\}=\pi\cot(\pi z)=\pi i\left\{1-2\sum_{n=1}^{\infty}q^n\right\}, \tag{1.3.2}$$

and its derivatives by $\left((2\pi i)^{-1}\frac{d}{dz}\right)^k=\left(q\frac{d}{dq}\right)^k$:

$$\sum_{n=-\infty}^{\infty}\frac{1}{(z+n)^k}=\frac{(-1)^k(2\pi i)^k}{(k-1)!}\sum_{n=1}^{\infty}n^{k-1}q^n. \tag{1.3.3}$$

In particular, keeping in mind the fact that $g_k(z)=g_k(2\pi i\left(\begin{smallmatrix} z \\ 1\end{smallmatrix}\right))$, we find

$$\begin{aligned}
g_2(q)&=\frac{1}{12}+20\sum_{n=1}^{\infty}\left\{\sum_{0<d|n}d^3\right\}q^n=\frac{1}{12}+20\sum_{n=1}^{\infty}\frac{n^3q^n}{1-q^n},\\
g_3(q)&=-\frac{1}{216}+\frac{7}{3}\sum_{n=1}^{\infty}\left\{\sum_{0<d|n}d^5\right\}q^n=-\frac{1}{216}+\frac{7}{3}\sum_{n=1}^{\infty}\frac{n^5q^n}{1-q^n},\\
\Delta(q)&=q\prod_{n=1}^{\infty}(1-q^n)^{24}.
\end{aligned} \tag{1.3.4}$$

To obtain the product expansion of Δ, one needs to work a little more (see [EEK], IV (36) and [IAT] (4.6.1)).

We can then think of the *Tate curve* Tate(q) defined over the power series ring $\mathbb{Z}[\frac{1}{6}][[q]]$ by the (affine) equation $y^2=4x^3-g_2(q)x-g_3(q)$. As shown by Tate (Sect. 2.4.2), we have $\mathrm{Tate}(q)(\overline{\mathbb{Q}}_p[[q]])\supset(\overline{\mathbb{Q}}_p[[q]])^\times$, and therefore we get a natural inclusion $\phi_{N,can}:\mu_N(\overline{\mathbb{Q}}_p)\hookrightarrow\mathrm{Tate}(q)[N]$, which comes from a unique morphism of finite flat group schemes (see Sect. 4.3 for finite flat group schemes). A Tate curve also has a canonical differential $\omega_{can}=\frac{dx}{y}$, which is a unique (up to scalar multiple) nowhere vanishing differential on Tate(q) (see Sect. 6.1.4). A Tate curve is not an elliptic curve over $\mathbb{Z}[\frac{1}{6}][[q]]$ but is an elliptic curve over $\mathbb{Z}[\frac{1}{6}]((q)):=\mathbb{Z}[\frac{1}{6}][[q]][q^{-1}]$ because the nonunit q is a factor of the discriminant $\Delta(q)$ of the elliptic curve.

Let B be a $\mathbb{Z}[\frac{1}{6}]$-algebra. For general elliptic curve E over a B-algebra R, a *level $\Gamma_1(N)$ structure* is an embedding $\phi_N:\mu_N\hookrightarrow E[N]$ of finite flat group schemes (i.e., a group homomorphism of functors inducing an embedding of group schemes we describe later in Sect. 4.3). For the first reading, if the reader is not familiar with the theory of group schemes, just assume $N=1$ to forget about ϕ_N. If the triple (E,ϕ_N,ω) is all defined over R, we call $(E,\phi_N,\omega)_{/R}$ a *test object* over R. The elements g_2,g_3,Δ in $\mathbb{Z}[g_2,g_3]$ are functions of pairs $(E,\omega)_{/R}$ of an elliptic curve E and a differential ω defined over R with values in R and can be considered as a function of test objects (E,ϕ_N,ω) disregarding ϕ_N. This motivates the following algebraic definition/theorem (see Theorem 2.48) of B-integral elliptic modular forms of level $\Gamma_1(N)$ as functions of test objects $(E,\phi_N:\mu_N\hookrightarrow E[N],\omega)$ satisfying

(G0) f assigns a value $f((E,\phi_N,\omega)_{/R}) \in R$ for each test object

$$(E,\phi_N:\mu_N \hookrightarrow E[N],\omega)_{/R}$$

defined over an B-algebra R. Here R is also a variable (running over the category of B-algebras).

(G1) $f((E,\phi_N,\omega)_{/R}) \in R$ depends only on the isomorphism class of the triple $(E,\phi_N,\omega)_{/R}$.

(G2) If $\varphi: R \to R'$ is a B-algebra homomorphism, we have

$$f((E,\phi_N,\omega)_{/R} \otimes_{R,\varphi} R') = \varphi(f((E,\phi_N,\omega)_{/R})),$$

where $(E,\phi_N,\omega)_{/R} \otimes_{R,\varphi} R'$ means that we regard $(E,\phi_N,\omega)_{/R}$ as defined over R' via φ [so, for example, if $N=1$, $(E,\omega)\otimes_{R,\varphi} R'$ is defined by the equation $y^2 = 4x^3 - \varphi(g_2(E,\omega))x - \varphi(g_3(E,\omega))$].

(G3) $f((E,\phi_N,a\cdot\omega)_{/R}) = a^{-k} f(E,\phi_N,\omega)$ for $a \in R^\times$.

(G4) $f(q) = f((\mathrm{Tate}(q),\phi_N,\omega_{can})_{/R[[q^{1/d}]][q^{-1/d}]}) \in R[[q^{1/d}]]$ for any level N-structures ϕ_N on $\mathrm{Tate}(q)$ defined over $R[[q^{1/d}]][q^{-1/d}]$ for any factor $d|N$. In particular, we have

$$f(q) = f((\mathrm{Tate}(q),\phi_{N,can},\omega_{can})_{/B[[q]][q^{-1}]}) \in B[[q]].$$

Here, for example, if N is invertible in R and R contains all Nth roots of unity, $\mathbb{Z}/N\mathbb{Z} \cong \mu_N$ over R, and we can think of a level $\Gamma_1(N)$-structure $\phi_N^{\mathrm{et}}: \mathbb{Z}/N\mathbb{Z} = \mu_N \cong q^{1/N\mathbb{Z}}/q^{\mathbb{Z}} \subset \mathrm{Tate}(q)[N]$, which is defined over $R[[q^{1/N}]][q^{-1/N}]$. More generally, if $N = dN'$ with d prime to N', assuming $\frac{1}{d} \in R$ and that R contains all dth roots of unity, we can think of

$$\phi_N = \phi_d^{\mathrm{et}} \times \phi_{N',can} : \mu_N = \mathbb{Z}/d\mathbb{Z} \times \mu_{N'} \hookrightarrow \mathrm{Tate}(q),$$

which is defined over $R[[q^{1/d}]][q^{-1/d}]$.

The space of modular forms defined by the conditions (G0–4) will be written as $G_k(N;B) = G_k(\Gamma_1(N);B)$. We confirm $G_k(1;B) = \bigoplus_{4a+6b=k} Bg_2^a g_3^b$ (Theorem 2.48), and $G_k(\Gamma_1(N);\mathbb{Z}[\frac{1}{6}]) \otimes_{\mathbb{Z}[\frac{1}{6}]} \mathbb{C} = G_k(\Gamma_1(N);\mathbb{C})$. Also, if $f \in G_k(\Gamma_1(N);\mathbb{C})$, $f(q)$ with $q = \exp(2\pi i z)$ gives the Fourier expansion of f at the cusp ∞.

In addition to (G1–4), if f satisfies the following condition, we call f a *cusp form*:

(s) $f(q) = f((\mathrm{Tate}(q),\phi_N,\omega_{can})_{/R[[q^{1/d}]][q^{-1/d}]}) \in q^{1/d}R[[q^{1/d}]]$ for any level structures ϕ_N on $\mathrm{Tate}(q)$ defined over $R[[q^{1/d}]][q^{-1/d}]$ with $d|N$. This means that the constant term of f vanishes at all cusps.

We write $S_k(N;B) = S_k(\Gamma_1(N);B) \subset G_k(N;B)$ for the subspace of cusp forms.

1.3.2 A Rationality Theorem, an Application

In 1897, Hurwitz studied in [Hz1] and [Hz2] an analog of the Riemann zeta function:

$$L(4k) = \sum_{a+bi \in \mathbb{Z}[i]-\{0\}} \frac{1}{(a+bi)^{4k}}$$

for Gaussian integers $\mathbb{Z}[i]$ and showed that for a positive integer k,

$$\frac{L(4k)}{\Omega_\infty^{4k}} \in \mathbb{Q} \quad (\Omega_\infty = 2\int_0^1 \frac{dx}{\sqrt{1-x^4}} = \int_\gamma \frac{dx}{y} : \text{CM period of the lemniscate}).$$

Today we regard this value as a special value of a Hecke L-function:

$$L(s,\lambda) = \sum_{\mathfrak{a}} \lambda(\mathfrak{a}) N(\mathfrak{a})^{-s} \quad \text{of the Gaussian field } \mathbb{Q}[i] \ (i = \sqrt{-1}).$$

Here $\lambda = \lambda_{4k}$ is a Hecke ideal character of $\mathbb{Q}[i]$, with $\lambda_{4k}((\alpha)) = \alpha^{-4k}$ and $L(4k) = 4L(0,\lambda)$, and $\mathfrak{a}$ runs over all nonzero ideals of the Gaussian integer ring $\mathbb{Z}[i]$.

The plane affine curve $\mathfrak{G}$ defined by $y^2 = 1 - x^4$ was first studied by Gauss around the same time when he finished writing his first major treatise, *Disquisitiones Arithmeticae*, in 1798 (he seems to have claimed that he could extend his theory of drawing a regular N-gon in a circle to this lemniscate or more general curves, which means, if true, he knew how to compute N section/torsion points of $\mathfrak{G}[N]$). This *Gauss's lemniscate* is equivalent to $Y^2 = 4X^3 - 4X$ by $Y = \frac{(4+4i)y}{((i-1)x+(1-i))^2}$ and $X = \frac{(1+i)x+1+i}{(i-1)x+1-i}$; hence, it is an elliptic curve $(\mathfrak{E}, \omega_{\mathfrak{E}} := \frac{dX}{Y})_{/\mathbb{C}}$ with a given nowhere vanishing differential $\omega_{\mathfrak{E}}$. We have $X = 0 \Leftrightarrow x = -1$ and $X = \infty \Leftrightarrow x = 1$. Since $\int_{-1}^1 \frac{dx}{\sqrt{1-x^4}} = 2\int_0^1 \frac{dx}{\sqrt{1-x^4}}$, we get (cf. [EDM] Sect. 13.9)

$$\Omega_\infty = 2\int_0^1 \frac{dx}{\sqrt{1-x^4}} = \int_0^\infty \frac{dX}{\sqrt{4X^3-4X}}.$$

Here, writing the integral path $\gamma := [0,\infty] = -[0, \mathbf{0}_{\mathfrak{E}}]$ as an element of the topological fundamental group $\pi_1^{\mathrm{top}}(\mathfrak{E}(\mathbb{C}), \mathbf{0}_{\mathfrak{E}})$, we get the "CM period" Ω_∞ in the lattice $L_{\mathfrak{E}} \in Lat$ of $\mathfrak{E}$, which is a part of a $\mathbb{Z}$-basis of $L_{\mathfrak{E}}$.

Gauss's curve has an automorphism $[i] : (x,y) \mapsto (ix,y)$. This corresponds to the automorphism $[i] : (X,Y) \mapsto (iX,-Y)$ of the elliptic curve $\mathfrak{E}$. Note that $[i](\mathbf{0}_{\mathfrak{E}}) = \mathbf{0}_{\mathfrak{E}}$, which implies, by Abel's theorem, that $[i]$ is a group endomorphism of $\mathfrak{E}$ (see Lemma 6.1 and the remark after the lemma).

We can verify this fact differently. Any curve endomorphism $\alpha : \mathfrak{E} \to \mathfrak{E}$ sending $\mathbf{0}_{\mathfrak{E}}$ to $\mathbf{0}_{\mathfrak{E}}$ induces an endomorphism of the universal covering $\alpha : \mathcal{Z} \to \mathcal{Z}$ by sending a path $\gamma : \mathbf{0}_{\mathfrak{E}} \to z$ to $\alpha(\gamma) : \mathbf{0}_{\mathfrak{E}} = \alpha(\mathbf{0}_{\mathfrak{E}}) \to \alpha(\gamma)$ (the image of γ under α). Then, by the definition of composition of paths, we have $\alpha(\gamma\gamma') = \alpha(\gamma)\alpha(\gamma')$. The group structure on $\mathcal{Z} = \mathbb{C}$ is given by this product structure, and $\alpha : \mathbb{C} \to \mathbb{C}$ is an additive group endomorphism. Since

α is holomorphic (as it is a polynomial map in coordinates of $\mathbf{P}^2_{/\mathbb{C}}$ containing $\mathfrak{E}$), α has to be a $\mathbb{C}$-linear map; thus, the original $\alpha : \mathbb{C}/L_{\mathfrak{E}} \to \mathbb{C}/L_{\mathfrak{E}}$ is a group homomorphism. See Sect. 2.3.3 for the covering $\mathcal{Z} = \mathcal{Z}(\mathfrak{E}(\mathbb{C}))$ and $\pi_1^{\mathrm{top}}(\mathfrak{E}(\mathbb{C}), \mathbf{0}_{\mathfrak{E}})$.

Now the vertical line L_{Y_0} defined by $Y = Y_0$ intersects with $\mathfrak{E}(\mathbb{C})$ at $[i]^2(X_0, Y_0) = (-X_0, Y_0)$ and (X_0, Y_0) and the infinity $\mathbf{0}_{\mathfrak{E}}$. Thus, in $\mathrm{End}(\mathfrak{E}_{/\mathbb{C}})$, $[i]$ satisfies $[i]^2 = -1$. In other words, we have an embedding $\mathbb{Z}[i] \hookrightarrow \mathrm{End}(\mathfrak{E}_{/\mathbb{C}})$ given by $a + bi \mapsto a + b[i]$. Therefore, $[i]$ acts on $\pi_1^{\mathrm{top}}(\mathfrak{E}(\mathbb{C}), \mathbf{0}_{\mathfrak{E}}) \cong L_{\mathfrak{E}}$ by $[i]$ on the left-hand side and via multiplication by i on the right-hand side. In short, $\mathfrak{E}$ has complex multiplication by the Gaussian integers $\mathbb{Z}[i]$. Since the path $[-1, 1]$ is one of the generators of $H_1(R, \mathbb{Z})$ for the Riemann surface R associated with $y^2 = 1 - x^4$, we find that $L_{\mathfrak{E}} = \mathbb{Z}[i] \cdot \Omega_\infty \subset \mathbb{C}$.

For a lattice $L = \mathbb{Z}w_1 + \mathbb{Z}w_2 \subset \mathbb{C}$, set

$$E_{2k}(L) = \frac{1}{2} \sum_{aw_1 + bw_2 \in L} \frac{1}{(aw_1 + bw_2)^{2k}} \quad \text{(Eisenstein series)},$$

which is a function of lattices satisfying $E_{2k}(\alpha L) = \alpha^{-2k} E_{2k}(L)$. The quotient $\mathbb{C}/L$ gives rise to an elliptic curve $X(L) \subset \mathbf{P}^2$ by Weierstrass theory. Since we can recover the lattice L as $\{\int_\gamma du | \gamma \in \pi_1^{\mathrm{top}}(X(L))\}$ out of the pair $(X(L), du)$ for the variable u on $\mathbb{C}$, we can think of E_{2k} as a function of the pairs (E, ω) of an elliptic curve E and a nowhere vanishing differential ω satisfying $E_{2k}(E, \alpha\omega) = \alpha^{-2k} E_{2k}(E, \omega)$. We have $E_4(E, \omega) = \frac{1}{120} g_2(E, \omega)$ and $E_6(E, \omega) = \frac{1}{280} g_3(E, \omega)$. Since E_{2k} is a rational isobaric polynomial of g_2 and g_3, E_{2k} is a modular form f of weight $2k$.

Weierstrass's function has the following Laurent expansion:

$$x_L(u) = u^{-2} + 6E_4(E, \omega)u^2 + \cdots + 2(2k+1)E_{2k+2}(E, \omega)u^{2k} + \cdots . \quad (1.3.5)$$

Taking the derivative, we get

$$y_L(u) = -2u^{-3} + 12E_4(E, \omega)u + \cdots + 4k(2k+1)E_{2k}(E, \omega)u^{2k-1} + \cdots . \quad (1.3.6)$$

Plugging these into the relation $y_L^2 = 4x_L^3 - g_2(E, \omega)x_L - g_3(E, \omega)$, we get a relation of E_{2k} (noted by Hurwitz but known to Weierstrass earlier; see [EDM] Sect.10.2). Solving the relation, we can compute the isobaric polynomial $\Phi_{2k}(X, Y) \in \mathbb{Q}[X, Y]$, with $E_{2k} = \Phi_{2k}(g_2, g_3)$. See [EDM] Sect. 10 for more of these relations.

Returning to Gauss's curve $(\mathfrak{E}, \omega_{\mathfrak{E}})$ defined by $y^2 = 4x^3 - 4x$, we have $L_{\mathfrak{E}} = \mathbb{Z}[i] \cdot \Omega_\infty$. Thus,

$$E_{2k}(\mathfrak{E}, \Omega_{\mathfrak{E}}) = E_{2k}(\mathbb{Z}[i] \cdot \Omega_\infty) = \Omega_\infty^{-2k} E_{2k}(\mathbb{Z}i + \mathbb{Z}) = \Omega_\infty^{-2k} E_{2k}(i),$$

where at the extreme right, we regard E_{2k} as a function on $\mathfrak{H}$ by $E_{2k}(z) = E_{2k}(2\pi i(\mathbb{Z}z + \mathbb{Z}))$. On the other hand,

$$E_{2k}(\mathfrak{E}, \omega_{\mathfrak{E}}) = \Phi_{2k}(g_2(\mathfrak{E}, \omega_{\mathfrak{E}}), g_3(\mathfrak{E}, \omega_{\mathfrak{E}})) = \Phi_{2k}(4, 0) \in \mathbb{Q}.$$

Note that $E_{2k}(i) = L(0, \lambda_{2k})$. Thus, we get the following result:

Theorem 1.22 *We have* $\frac{L(0,\lambda_{4k})}{\Omega_\infty^{4k}} \in \mathbb{Q}$ *for all* $0 < k \in \mathbb{Z}$.

See Hurwitz's paper [Hz1] and [Hz2] for numerical values of $\frac{L(0,\lambda_{4k})}{\Omega_\infty^{4k}}$ and a generalization of the von Staut theorem (an integrality result) for these numbers.

We can apply the same technique to the curve defined by $y^2 = x^3 - 1$ that has complex multiplication by the field $\mathbb{Q}[\sqrt{-3}]$, as a cubic root of unity ζ acts on the curve by $(x, y) \mapsto (\zeta x, y)$. See [H81a] Sect. 8 for numerical examples in this cubic root case and [EDM] Sect. 13 for more CM elliptic curves.

This is an obvious analog of the rationality (and recurrence relations) of Riemann zeta values by Euler/Bernoulli (see [EDM] Sect. 11):

$$\frac{\zeta(2k)}{(2\pi i)^{2k}} = \frac{1}{(2\pi i)^{2k}} \sum_{n=1}^{\infty} \frac{1}{n^{2k}} \in \mathbb{Q}.$$

Indeed, $2\pi i$ is the integral $\oint_{|z|=1} \frac{dz}{z}$ of the invariant differential of the multiplicative group $\mathbb{G}_{m/\mathbb{Z}}$ sending a ring A to its multiplicative group $A^\times$, and also note $\mathbb{G}_m(\mathbb{C}) = \mathbb{C}^\times \underset{\exp}{\overset{\sim}{\longleftarrow}} \mathbb{C}/2\pi i\mathbb{Z}$, while $\mathfrak{E}(\mathbb{C}) \cong \mathbb{C}/\Omega_\infty \cdot \mathbb{Z}[i]$.

The rationality/integrality results of Hurwitz is now generalized to all imaginary quadratic field M and L-values of the form $L(0, \lambda_{k,j})$ with integers $j < 0 \leq k$ for characters of the form $\lambda_{k,j}(\alpha) = \alpha^k \overline{\alpha}^j$ by the effort of many mathematicians, notably, Shimura and Katz, and the value is algebraic up to a power of the period $\Omega_\infty(E, \omega)$ and a power of $2\pi i$. Here $\Omega_\infty(E, \omega)$ is given by $\int_\gamma \omega$ for a pair (E, ω) defined over a number field with complex multiplication by the integer ring of M. Elementary treatments of this generalization can be found [EDM] Chap. IV and [EEK].

We can further ask

(Q1) Is there a good expression by an L-value of the value of a non-Eisenstein series $f \in G_k(1; \overline{\mathbb{Q}})$ at $(\mathfrak{E}, \omega_{\mathfrak{E}})$ (or at more general elliptic curves with complex multiplication)?
(Q2) What happens if we evaluate a modular form in $G_k(1; \overline{\mathbb{Q}})$ at a rational elliptic curve $(E, \omega)_{/\mathbb{Q}}$ without complex multiplication?

To get a reasonable answer, we need to assume that the modular form in question is an eigenvector of all Hecke operators (as Eisenstein series are examples of such eigenforms). Under this assumption, as for (Q1), there are good answers by Waldspurger [Wa] (see [H10b] for an elementary treatment) and also by Gross–Zagier [GZ] (which promoted much progress toward solution of the Birch–Swinnerton-Dyer conjecture). As for (Q2), not much is known yet.

1.3.3 p-Adic Elliptic Modular Form

Fix a prime $p \geq 5$ and a positive integer N prime to p. We call a $\mathbb{Z}_p$-algebra R a p-adic algebra if $R = \varprojlim_n R/p^n R$. Thus, $\mathbb{Z}_p$ is a p-adic algebra, but $\mathbb{Q}_p$ is not. Take a p-adic algebra B as a base ring. As we will learn in more detail in Sect. 7.2.4, the space of B-integral p-adic modular form $V(B) = V_{\Gamma_1(N)/B}$ is a collection of rules f assigning a value $f((E, \phi_N, \phi_p : \mu_{p^\infty} \hookrightarrow E[p^\infty])_{/R}) \in R$ for p-adic B-algebras R satisfying the following conditions (V0–3) (where ϕ_p is a group homomorphism of functors inducing an embedding of Barsotti–Tate groups we describe later in Sect. 4.3.7):

(V0) f assigns a value $f((E, \phi_N, \phi_p)_{/R}) \in R$ to each *test object*

$$(E, \phi_N, \phi_p : \mu_{p^\infty} \hookrightarrow E[p^\infty])_{/R}$$

defined over a p-adic B-algebra R. Here R is also a *variable.*

(V1) The value $f((E, \phi_N, \phi_p)_{/R}) \in R$ depends only on the isomorphism class of $(E, \phi_N, \phi_p)_{/R}$.

(V2) If $\varphi : R \to R'$ is an B-algebra homomorphism continuous under the p-adic topology, we have

$$f((E, \phi_N, \phi_p)_R \otimes R') = \varphi(f((E, \phi_N, \phi_p)_{/R})).$$

(V3) $f(q) = f((\mathrm{Tate}(q), z\phi_{p,can}, \phi_N)_{/R[[q^{1/d}]][q^{-1/d}]}) \in R[[q^{1/d}]]$ for any level N-structures ϕ_N on $\mathrm{Tate}(q)$ over $R[[q^{1/d}]][q^{-1/d}]$ for $0 < d|N$ and $z \in \mathbb{Z}_p^\times$. Again, we note $f(q) = f((\mathrm{Tate}(q), \phi_{p,can}, \phi_{N,can})_{/R[[q]][q^{-1}]}) \in R[[q]]$.

Here we have simply written $\phi_{p,can}$ for

$$\phi_{p^\infty,can} := \varinjlim_n \phi_{p^n,can} : \mu_{p^\infty} \hookrightarrow \mathrm{Tate}(q)[p^\infty].$$

We note that $V_{\Gamma_1(N)/B} = V(B)$ is a p-adic B-algebra, as we see in Sect. 7.2.4. Replacing (V3) by the following condition:

(s) $f(q) = f((\mathrm{Tate}(q), \phi_{p,can}, \phi_N)_{/R[[q^{1/d}]][q^{-1/d}]}) \in q^{1/d}R[[q^{1/d}]]$ for any level structures ϕ_N on $\mathrm{Tate}(q)$ defined over $R[[q^{1/d}]][q^{-1/d}]$ for $0 < d|N$,

we define the subspace of p-adic cusp forms $V^{\mathrm{cusp}}_{\Gamma_1(N)/B}$.

As we learn in Sect. 4.4.3, the knowledge of $\mu_{p^\infty/\mathbb{Z}_p} = \varinjlim_n \mu_{p^n/\mathbb{Z}_p}$ is equivalent to the knowledge of the formal group $\widehat{\mathbb{G}}_{m/\mathbb{Z}_p} = \mathrm{Spf}(\widehat{\mathbb{Z}[t,t^{-1}]})$ for $\widehat{\mathbb{Z}[t,t^{-1}]} := \varprojlim_n \mathbb{Z}_p[t,t^{-1}]/(t^{p^n} - 1))$ (see Example 4.79 and Sect. 4.4 for the functor "Spf"). Then $\phi_p : \mu_{p^\infty} \hookrightarrow E$ induces an identification $\widehat{\phi}_p : \widehat{\mathbb{G}}_m \cong \widehat{E} = \varprojlim_n E[p^n]^\circ$ with the formal completion $\widehat{E}$ along the identity of E (in short, the *formal group* of E). Since $\widehat{\mathbb{G}}_m$ has a canonical "invariant" differential $\frac{dt}{t}$, ϕ_p induces a nowhere vanishing differential $\omega_p = \widehat{\phi}_{p,*}\frac{dt}{t}$. Thus, $f \in G_k(\Gamma_1(Np^n); B)$ can be regarded as a p-adic modular form by $f((E, \phi_N, \phi_p : \mu_{p^\infty} \to E[p^\infty])_{/R}) = f(E, \phi_N, \phi_p|_{\mu_{p^n}}, \widehat{\phi}_{p,*}\frac{dt}{t}) \in R$, and we have a canonical B-linear map $G_k(\Gamma_1(Np^n); B) \to V(B) = V_{\Gamma_1(N)/B}$. The following fact is called the *q-expansion principle* based on these two facts (as we will see later): the irreducibility of the Igusa tower over $\mathbb{F}_p$ and the existence of the Tate curve (see Sects. 7.2 and 2.4):

(q0) $f(q) = 0 \Leftrightarrow f = 0$ for any f in $V(B)$ or in $G_k(\Gamma_1(Np^n); B)$. In particular, $G_k(\Gamma_1(Np^n); B) \to V_{\Gamma_1(N)}(B)$ is an injection (see Proposition 7.41), and functions in the image satisfy $f(E, a \cdot \phi_p) = a^{-k} f(E, \phi_N, \phi_p)$ for $a \in \mathbb{Z}_p^\times$.

(q1) Let $f_n \in V_{\Gamma_1(N)}(B)$ be a sequence. Then f_n converges p-adically in $V_{\Gamma_1(N)}(B) \Leftrightarrow f_n(q)$ converges p-adically in $B[[q]] \Leftrightarrow f_n((E, \phi_N, \phi_p)_{/R})$ converges p-adically for all $(E, \phi_N, \phi_p)_{/R}$ and all p-adic B-algebras R (see Remark 7.42).

(q2) If B_0 is a $\mathbb{Z}[\frac{1}{6}]$-algebra p-adically dense in B,

$$G_k(\Gamma_1(Np^\infty); B_0) = \bigcup_n G_k(\Gamma_1(Np^n); B_0)$$

is dense in $V_{\Gamma_1(N)}(B)$ for any $k \geq 2$ (see [H88b] Corollary 5.2).

(q3) If $f \in V_{\Gamma_1(N)}(B) \otimes_{\mathbb{Z}} \mathbb{Q}$ and $f(q) \in B[[q]]$, then $f \in V_{\Gamma_1(N)}(B)$, assuming B is flat over $\mathbb{Z}_p$.

By (q1–2),

> *we may construct a p-adic modular form as a p-adic limit of classical modular forms.* (Limit principle)

This limit principle is used as the definition of p-adic modular forms at the start of the theory by J.-P. Serre (see [Se2]). Because of this limit principle, $V(B)$ is complete under p-adic topology (see Sect. 7.2.4).

1.3.4 CM Elliptic Curve

An elliptic curve $E_{/A}$ is said to have *complex multiplication* if $\mathrm{End}(E_{/A})$ contains an order $\mathfrak{O}'$ of the integer ring $\mathfrak{O}$ of an imaginary quadratic field $M \subset \mathbb{C}$. An order $\mathfrak{O}'$ is a subring of finite index in $\mathfrak{O}$. If $E(\mathbb{C}) = \mathbb{C}/L_w$ has complex multiplication, we have $\mathfrak{O}' \cdot L_w \subset L_w$, and hence $\mathfrak{O}' = \{\alpha \in M | \alpha \cdot L_w \subset L_w\}$. In this way, we have a representation $\rho : M^\times \hookrightarrow GL_2(\mathbb{Q})$ such that $\alpha w = \rho(\alpha) w$ for $\alpha \in M^\times$. Since $\rho(\alpha)(z_0) = \frac{az_0+b}{cz_0+d}$ ($\rho(\alpha) = \left(\begin{smallmatrix} a & b \\ c & d \end{smallmatrix}\right)$) and $z_0 = w_1/w_2$ corresponds to the same elliptic curve, we have $\rho(\alpha)(z_0) = z_0$. If it is necessary to indicate dependence on z_0, we write ρ_{z_0} for ρ.

Suppose that E has complex multiplication by $\mathfrak{O}'$. Take a large complete discrete valuation ring W in $\mathbb{C}_p$ and put $\mathcal{W} = i_p^{-1}(W)$ (we may assume that W has residue field $\overline{\mathbb{F}}_p$). Then, by the finiteness of isomorphism classes over $\mathbb{C}$ of elliptic curves having CM by $\mathfrak{O}'$, E is defined over a number field K, and by a result of Serre–Tate, E is defined over a valuation ring $\mathfrak{V} \subset \mathcal{W}$ of K of residual characteristic p (see Proposition 6.53). For an ideal $\mathfrak{m} \subset \mathfrak{O}'$, we can think of the kernel of multiplication by $\mathfrak{m}$:

$$E[\mathfrak{m}](A) = \{x \in E(A) | \mathfrak{m}x = 0\} = \bigcap_{\alpha \in \mathfrak{m}} E[\alpha](A).$$

We suppose further p splits into a product of two primes $\mathfrak{p}\overline{\mathfrak{p}}$ in $\mathfrak{O}'$ (so, as we will see later, $\mathfrak{O}_p = \mathfrak{O} \otimes_{\mathbb{Z}} \mathbb{Z}_p = \mathfrak{O}' \otimes_{\mathbb{Z}} \mathbb{Z}_p = \mathfrak{O}'_p$). Let $E[\mathfrak{p}^\infty] = \bigcup_n E[\mathfrak{p}^n]$.

Then we may assume that $E[\mathfrak{p}^\infty]_{/\mathfrak{W}}$ is isomorphic to μ_{p^∞} after extending scalars to the strict henselization $\mathcal{W} \subset \overline{\mathbb{Q}}$ of $\mathfrak{W}$ (see Sect. 4.3.9); hence, we have $\phi = \phi_E : \mu_{p^\infty} \cong E[\mathfrak{p}^\infty]$. If $p = \mathfrak{p}\overline{\mathfrak{p}}$ in $\mathfrak{O}$ but not in $\mathfrak{O}'$ (so, $\mathfrak{O}'$ is an order of conductor divisible by p), we just assume that E has an isogeny $E' \xrightarrow{\pi} E$ defined over a finite extension $\mathcal{W}' \subset \overline{\mathbb{Q}}$ of $\mathcal{W}$ such that E' has complex multiplication by $\mathfrak{O}$ and $\mathrm{Ker}(\pi) \cap E'[\mathfrak{p}^\infty] = 0$. Under this setting, $\phi' : \mu_{p^\infty} \cong E'[\mathfrak{p}^\infty]$ induces $\phi = \phi_E = \pi \circ \phi_{E'} : \mu_{p^\infty} \hookrightarrow E[p^\infty]$. We study in deoth later the sheaf $\Omega_{E/\mathcal{W}'}$ of differentials on the scheme E (see Sect. 6.1.4). An important point is that its global sections $H^0(E, \Omega_{E/\mathcal{W}'})$ are a free $\mathcal{W}'$-module of rank 1. Pick $\omega \in H^0(E, \Omega_{E/\mathcal{W}'})$ so that $H^0(E, \Omega_{E/\mathcal{W}'}) = \mathcal{W}'\omega$.

As described above, we fix an embedding $\phi_p : \mu_{p^\infty/\mathcal{W}'} \hookrightarrow E[p^\infty]_{/\mathcal{W}}$ and an isomorphism $E(\mathbb{C}) = \mathbb{C}/i_\infty(\mathfrak{a})$ for a fractional $\mathfrak{O}'$-ideal $\mathfrak{a}$. In this case, we write $E = E(\mathfrak{a})$. Thus, we may assume that $\mathfrak{a} = \mathbb{Z} + \mathbb{Z}z_0$ ($z_0 \in M^\times \cap \mathfrak{H}$). Note $W \cong \varprojlim_n \mathcal{W}/p^n\mathcal{W}$, and set $W' = \varprojlim_n \mathcal{W}'/p^n\mathcal{W}'$, and identify W' with the closure W' of $i_p(\mathcal{W}')$ in $\mathbb{C}_p$. Then we have two numbers $\Omega_\infty \in \mathbb{C}^\times$ and $\Omega_p \in W'^\times$ such that

$$\omega = \Omega_\infty du = \Omega_p \widehat{\phi}_{p,*}\frac{dt}{t}.$$

The following result is plainly a generalization of Theorem 1.22:

Theorem 1.23 *Suppose that $p = \mathfrak{p}\overline{\mathfrak{p}}$ with $\mathfrak{p} \neq \overline{\mathfrak{p}}$ in $\mathfrak{O}$. Let $f = f_\infty \in G_k(\Gamma_1(Np^n))$ be a $\mathcal{W}$-integral form and let $\mathcal{W}'_{/\mathcal{W}}$ be a finite extension of $\mathcal{W}$, which is discrete valuation ring. Write $f_p \in V_{\Gamma_1(N)}(W')$ as the corresponding p-adic modular form coming from f_∞. If $(E(\mathfrak{a}), \omega)_{/\mathcal{W}'}$ has complex multiplication by an order $\mathfrak{O}'$ of $\mathfrak{O}$, we have*

$$\frac{f_\infty(z_0)}{\Omega_\infty^k} \overset{(1)}{=} \frac{f_\infty(E(\mathfrak{a}), \phi_N, du)}{\Omega_\infty^k} = \frac{f_p(E(\mathfrak{a}), \phi_N, \widehat{\phi}_{p,*}\frac{dt}{t})}{\Omega_p^k} \overset{(2)}{=} f(E(\mathfrak{a}), \phi_N, \omega) \in \mathcal{W}'.$$

Here the identity (1) is the base of Shimura's rationality theory of special values of modular/automorphic forms (cf. [EDM] Sect. 12) and (2) is a key to the p-adic version by Katz.

Remark 1.24 Replacing $\mathcal{W}$ by its quotient field, the assertion of the theorem is valid, for an obvious reason, also for $f = \frac{h}{g}$ if $h \in G_{k+k'}(\Gamma_1(Np^n); \mathcal{W})$ and $g \in G_{k'}(\Gamma_1(Np^n); \mathcal{W})$ with $g(E(\mathfrak{a}), \phi_N, du)) \neq 0$.

Remark 1.25 To have a well-defined p-adic period Ω_p, a key point here is to have an embedding $\phi_p : \mu_{p^\infty} \hookrightarrow E[p^\infty]$ of Barsotti–Tate groups for a CM elliptic curve E not E having multiplication by the full integer ring $\mathfrak{O}$. We will scrutinize ϕ_p for $E(\mathbb{C}) = \mathbb{C}/i_\infty(\mathfrak{a})$ for a lattice $\mathfrak{a}$ in M with $\mathfrak{a}_p$ not equal to $\mathfrak{O}_p$ in Sect. 6.4.5. Then the above theorem is valid over a finite extension $\mathcal{W}'/\mathcal{W}$ [over which $E(\mathfrak{a})$ is defined]. The minimum ring $\mathcal{W}'$ is given by $\mathcal{W}[\mu_{p^n}]$ for n depending on how far $\mathfrak{a}_p$ differs from $\mathfrak{O}_p$ (see Sects. 6.3.4 and 6.4). So we state the result hereafter without assuming $\mathfrak{O}'_p = \mathfrak{O}_p$.

1.3.5 Invariant Differential Operators

Shimura studied the effect on modular forms of the following *invariant differential operators* on the upper half-complex plane $\mathfrak{H}$ indexed by $k \in \mathbb{Z}$:

$$\delta_k = \frac{1}{2\pi\sqrt{-1}}\left(\frac{\partial}{\partial z} + \frac{k}{2y\sqrt{-1}}\right) \text{ for } z = x + iy \in \mathfrak{H} \text{ and } \delta_k^r = \delta_{k+2r-2}\cdots\delta_k, \tag{1.3.7}$$

where $r \in \mathbb{Z}$ with $r \geq 0$. For some other properties of these operators not stated here, see [LFE] 10.1. Here are easy identities:

Exercise 1.26 *Show the following formulas:*

1. $\delta_{k+l}(fg) = g\delta_k f + f\delta_l g$.
2. *For a holomorphic function* $f : \mathfrak{H} \to \mathbb{C}$, $\delta_k^r(f|_k\alpha) = (\delta_k^r f)|_{k+2r}\alpha$ *if* $f|_k\alpha(z) = \det(\alpha)^{k/2} f(\alpha(z))(cz+d)^{-k}$ *for* $\alpha = \left(\begin{smallmatrix} a & b \\ c & d \end{smallmatrix}\right) \in GL_2(\mathbb{R})$ *with* $\det(\alpha) > 0$.

By the exercise, if $f \in G_k(\Gamma_1(N); \mathbb{C})$, $\delta_k^r(f)$ satisfies $\delta_k^r(f)|_{k+2r}\gamma = \delta_k^r(f)$ for all $\gamma \in \Gamma_1(N)$. Although $\delta_k^r(f)$ is not a holomorphic function, defining

$$\delta_k^r(f)(w) = w_2^{-k-2r}\delta_k^r(f)(z),$$

we have a well-defined homogeneous modular form. In this sense, $\delta_k^r(f)$ is a real-analytic modular form on $\Gamma_1(N)$ of weight $k + 2r$.

An important fact Shimura found is that the differential operator preserves rationality property at CM points of (arithmetic) modular forms, although it does not preserve holomorphy. Here we call $z_0 \in \mathfrak{H}$ a *CM point* if $z_0 \in \mathfrak{H} \cap i_\infty(M)$ for an imaginary quadratic field M. Then $\mathfrak{a} := i_\infty^{-1}(\mathbb{Z} + \mathbb{Z}z_0) \subset M$ is a lattice, and we have a CM elliptic curve $E(\mathfrak{a})$ with multiplication by $\mathfrak{O}(\mathfrak{a}) = \{\alpha \in M | \alpha\mathfrak{a} \subset \mathfrak{a}\}$ as in Remark 1.25. Here is the rationality result of Shimura [Sh6] (see also [EDM] Theorem 12.2):

Theorem 1.27 (G. Shimura) *Let the notation be as above; in particular, z_0 is a CM point of $\mathfrak{H}$. For $f \in G_k(\Gamma_1(N); \overline{\mathbb{Q}})$, we have*

$$\frac{(\delta_k^r f_\infty)(z_0)}{\Omega_\infty^{k+2r}} = \frac{(\delta_k^r f)(E(\mathfrak{a}), \phi_N, du)}{\Omega_\infty^{k+2r}} \in \overline{\mathbb{Q}}. \tag{Sh}$$

In this theorem, z_0 is just a CM point, and it has nothing to do with the prime p we have chosen. Thus, we do not need to assume that the prime p is split in the imaginary quadratic field $M = \mathbb{Q}[z_0]$.

Proof. We follow the argument of Shimura in [Sh6]. Since

$$(\delta_k^r f)(E, \phi_N, du) = w_2^{-k-2r}(\delta_k^r f)(z) = (\delta_k^r f)(z)$$

for $z = w_1/w_2 \in \mathfrak{H}$ (and $w_2 = 1$), we need to show $\frac{(\delta_k^r f)(z_0)}{\Omega_\infty^{k+2r}} \in \overline{\mathbb{Q}}$. When $r = 0$, the result follows from Theorem 1.23. We have $\rho : M^\times \hookrightarrow GL_2(\mathbb{Q})$

given by $\binom{z_0\alpha}{\alpha} = \rho(\alpha)\binom{z_0}{1}$ for $\alpha \in M \setminus \mathbb{Q}$. Then $\rho(\alpha)(z_0) = z_0$. Writing $\rho(\alpha) = \left(\begin{smallmatrix} a & b \\ c & d \end{smallmatrix}\right)$, we have $cz_0 + d = \alpha$. Apply δ_k to $f|_k\rho(\alpha) = fh$ [putting $h = (f|_k\rho(\alpha))/f$], by Exercise 1.26, we have $(\delta_k f)|_{k+2}\rho(\alpha) = (\delta_k f)h + f(\delta_0 h)$. Specializing this equality at z_0 and choosing $\alpha \in M^\times$ of infinite order with $\det(\rho(\alpha)) = N(\alpha) = 1$, we have

$$\begin{aligned}\alpha^{-k-2}(\delta_k f)(z_0) = (\delta_k f)|_{k+2}\rho(\alpha)(z_0) &= (\delta_k f)(z_0)h(z_0) + f(z_0)(\delta_0 h)(z_0) \\ &= (\delta_k f)(z_0)\alpha^{-k} + f(z_0)(\delta_0 h)(z_0),\end{aligned}$$

because $h(z_0) = (f|_k\rho(\alpha))(z_0)/f(z_0) = \alpha^{-k}$. As $\alpha^2 \neq 1$, we have

$$(\delta_k f)(z_0) = \alpha^k(\alpha^{-2} - 1)^{-1} f(z_0)(\delta_0 h)(z_0).$$

Note that $\delta_0 h$ is a meromorphic modular form of weight 2 defined over $\overline{\mathbb{Q}}$ by the q-expansion principle (see (2.3.3) and Sect. 6.2.8 for the principle). Thus, $\frac{\delta_0 h(z_0)}{\Omega_\infty^2} \in \overline{\mathbb{Q}}$ (Remark 1.24), and this proves the result when $r = 1$. We repeat this process r times. By the Leibnitz formula, $\delta_k^r(fh) = \sum_{0\le s\le r}\binom{r}{s}\delta_k^s f \delta_0^{r-s} h$. From this, we get

$$\delta_k^r(f)|_{k+2r}\rho(\alpha) = (\delta_k^r f)h + \sum_{0<s\le r}\binom{r}{s}(\delta_k^{r-s} f)(\delta_0^s h).$$

Evaluating this at z_0, we finally get

$$(\delta_k^r f)(z_0) = \alpha^k(\alpha^{-2r} - 1)^{-1}\sum_{0<s\le r}\binom{r}{s}(\delta_k^{r-s} f)(z_0)(\delta_0^s h)(z_0). \tag{1.3.8}$$

Note that $\delta_0^s h = \delta_2^{s-1}\delta_0 h$, and as we have already observed, $\delta_0 h$ is a meromorphic $\overline{\mathbb{Q}}$-rational modular form finite at z_0. Then, by the induction hypothesis, we get the desired rationality.

Remark 1.28 Choosing $g \in G_{k+2r}(\Gamma_1(N); \mathcal{W})$ with $g(z_0) \neq 0$ under the notation of the above proof, Shimura actually proved in [Sh6] that $\frac{\delta_k^r f(z_0)}{g(z_0)} \in \overline{\mathbb{Q}}$, which is equivalent to the above theorem by Theorem 1.23.

Remark 1.29 For a given $f \in G_k(\Gamma_1(N); \overline{\mathbb{Q}})$ as above, defining the transformation equation $P(X, f)$ (of Fricke–Klein) by

$$P(X, f) = \prod_{\gamma\in\Gamma_1(N)\backslash SL_2(\mathbb{Z})}(X - f|_k\gamma) = \sum_{j=0}^{d} a_j(z)X^j,$$

we have $a_j \in G_{kd-jd}(1; \mathcal{W})$. Thus, $a_j = Q_j(g_2, g_3)$ for an isobaric polynomial Q_j with coefficients in $\overline{\mathbb{Q}}$. If (E, ω) is defined by the Weierstrass equation $y^2 = 4x^3 - g_2(E,\omega)x - g_3(E,\omega)$, $f(E, \phi_N, \phi_p, \omega)$ satisfies

$$\sum_{j=0}^{d} Q_j(g_2(E,\omega), g_3(E,\omega))X^j = 0.$$

Thus, this gives an algorithm to compute the value $f(E(\mathfrak{a}), \phi_N, \phi_p, \omega)$. Once we know the value $f(E(\mathfrak{a}), \phi_N, \phi_p, \omega) = \frac{f(E(\mathfrak{a}), \phi_N, du)}{\Omega_\infty^k}$, we can then compute $\frac{\delta_k^r(f(z_0))}{\Omega_\infty^{k+2r}}$ following the above proof (in particular, the induction process). See [H81a] Sect. 8 for numerical examples of derived values of Eisenstein series.

1.3.6 p-Adic Differential Operators

On $V(W')$, we have a more standard differential operator $d = \delta_0$ whose effect on q-expansion is $d(\sum_n a_n q^n) = \sum_n n a_n q^n$. An elementary construction of d can be given as follows. Pick $f \in G_k(\Gamma_1(Np^m); \mathcal{W}') \subset G_k(\Gamma_1(Np^m); \mathbb{C})$. Then, for any function $\phi : \mathbb{Z}/p^r\mathbb{Z} \to \mathcal{W}'$, we define its Fourier transform $\phi^* : \mathbb{Z}/p^r\mathbb{Z} \to \mathcal{W}'$ by $\phi^*(x) = \sum_{u \in \mathbb{Z}/p^r\mathbb{Z}} \phi(u)\mathbf{e}(xu/p^r)$, where $\mathbf{e}(x) = \exp(2\pi i x)$.

Exercise 1.30 *Prove* $(\phi^*)^*(x) = p^r\phi(-x)$.

We define

$$f|\phi(z) = p^{-r} \sum_{u \mod \mathbb{Z}/p^r\mathbb{Z}} \phi^*(-u) f(z + \frac{u}{p^r}). \tag{1.3.9}$$

Then we have $(\sum_n a_n q^n)|\phi) = \sum_n \phi(n) a_n q^n \in G_k(\Gamma_1(Np^{2r+m}); \mathcal{W}')$.

Exercise 1.31 *Define* $\phi|x(u) = \phi(xu)$ *for* $x \in (\mathbb{Z}/p^r\mathbb{Z})^\times$ $(r > 0)$. *Prove that*

$$(f|\phi)|_k\gamma = (f|_k\gamma)|(\phi|ad^{-1}) \quad \text{if } \gamma = \left(\begin{smallmatrix} a & b \\ c & d \end{smallmatrix}\right) \in \Gamma_1(Np^m) \cap \Gamma_0(p^{2r})$$

for $f \in G_k(\Gamma_1(Np^m); \mathbb{C})$. *In particular,* $f|\phi \in G_k(\Gamma_1(Np^m); \mathbb{C})$.

Hint: Use the strong approximation theorem of $SL(2)$ and the formula:

$$\left(\begin{smallmatrix} a & b \\ 0 & d \end{smallmatrix}\right)^{-1} \left(\begin{smallmatrix} 1 & u/p^r \\ 0 & 1 \end{smallmatrix}\right) \left(\begin{smallmatrix} a & b \\ 0 & d \end{smallmatrix}\right) = \left(\begin{smallmatrix} 1 & (a^{-1}d)u/p^r \\ 0 & 1 \end{smallmatrix}\right)$$

(see also [IAT] Proposition 3.64).

Choosing $\phi_n : \mathbb{Z}/p^n\mathbb{Z} \to \mathcal{W}'$ so that $\phi_n(u) \equiv u \mod p^n\mathcal{W}'$, the q-expansion $\lim_{n\to\infty}(f|\phi_n)$ converges p-adically to the q-expansion of df. By the limit principle, $G_k(Np^\infty; \mathcal{W}') := \bigcup_m G_k(Np^m; \mathcal{W}')$ is dense in $V(W')$, and we have a unique $df \in V(W')$ (this p-adic approximation process in the space of p-adic modular forms will be fully justified in Remark 7.42). Thus, $d^r f(E, \phi_p) \in W'$ is well defined. The effect of d^r on the q-expansion of a modular form is given by

$$d^r \sum_n a(n)q^n = \sum_n a(n)n^r q^n. \tag{1.3.10}$$

We can let $a \in \mathbb{Z}_p^\times$ act on $f \in V(B)$ by

$$f|a(E,\phi_N,\phi_p) = f|\left(\begin{smallmatrix} a & 0 \\ 0 & a^{-1} \end{smallmatrix}\right)(E,\phi_N,\phi_p) = f(E,\phi_N,a\cdot\phi_p).$$

The operator d has the invariance properties described in the following two lemmas. So it is natural to call it the *p-adic invariant differential operator.*

Lemma 1.32 *If $f \in G_k(\Gamma_1(Np^m);\mathcal{W}')$, then we have*

$$(d^r f)|\left(\begin{smallmatrix} a & 0 \\ 0 & a^{-1} \end{smallmatrix}\right) = a^{-k-2r}(d^r f)$$

for $a \in \mathbb{Z}_p^\times$.

Proof. We can approximate p-adically $g := \left(\begin{smallmatrix} a & 0 \\ 0 & a^{-1} \end{smallmatrix}\right)$ by an element $\gamma_n \in \Gamma_1(Np^m) \cap \Gamma_0(p^{2n})$ so that $\gamma_n \equiv g \mod p^n M_2(\mathbb{Z}_p)$. By Exercise 1.31, we have

$$df|a = a^{-k}\lim_{n\to\infty} f|\phi_n|_k\gamma_n = a^{-k}\lim_{n\to\infty} f|\gamma_n|(\phi_n|a^2) = a^{-k-2}df$$

because $\phi_n(u) \equiv u \mod p^n\mathcal{W}'$. Then, iterating this formula r times, we get the formula in the lemma.

The above action of a [or $g \in T_1(\mathbb{Z}_p))$] is the local one concentrated at p (as ϕ_N is intact), where $T_1(A)$ is the subgroup of $SL_2(A)$ made up of diagonal matrices for any ring A. We have a global action $f \mapsto f|\gamma$ for $\gamma \in GL_2(\mathbb{Z}_{(p)})$. Since γ changes the level structure ϕ_N, the definition is more involved. For that, we actually need to extend the level structure to the prime-to-p Tate module $\eta^{(p)} : (\mathbb{A}^{(p\infty)})^2 \cong V^{(p)}(E) = \left(\varprojlim_{p\nmid m} E[m]\right) \otimes_{\mathbb{Z}} \mathbb{Q}$ so that we can let γ act by $\eta^{(p)} \mapsto \eta^{(p)} \circ \gamma$. This is tantamount to introducing the structure of Shimura variety, and this will be done in Chap. 7. Define

$$(d^r f)|\rho(\alpha)(E,\phi_p,\eta^{(p)}) := (d^r f)(E,\phi_p\circ\alpha_{\mathfrak{p}},\eta^{(p)}\circ\rho^{(p)}(\alpha))$$

for the image $\alpha_{\mathfrak{p}} \in \mathfrak{O}_{\mathfrak{p}}$ of α. Here we just note the following fact:

Lemma 1.33 *Recall the elliptic curve $E(\mathfrak{a})$ with complex multiplication by an imaginary quadratic field M in which p splits into $(p) = \mathfrak{p}\overline{\mathfrak{p}}$. Define $\rho : \mathfrak{O}_{(p)}^\times \hookrightarrow G(\mathbb{Z}_p \times \mathbb{A}^{(p\infty)})$ by $\alpha^{(p)} \circ \eta^{(p)} = \eta^{(p)} \circ \rho(\alpha)$ and $\rho_p(\alpha) = \left(\begin{smallmatrix} \alpha_{\mathfrak{p}} & 0 \\ 0 & \alpha_{\overline{\mathfrak{p}}} \end{smallmatrix}\right)$. Then, for $f \in G_k(\Gamma_1(Np^m);\mathcal{W}')$, if $N(\alpha) = 1$, we have $d(f|\rho(\alpha)) = \alpha_{\mathfrak{p}}^2(df|\rho(\alpha))$ and*

$$((d^r f)|\rho(\alpha))(E(\mathfrak{a}),\phi_p,\eta^{(p)}) = \alpha_{\mathfrak{p}}^{-k-2r}\cdot d^r f(E(\mathfrak{a}),\phi_p,\eta^{(p)}).$$

Proof. By the deformation theory of $E(\mathfrak{a})$ (see Remark 7.49), the differential operator d commutes with the action $\eta^{(p)} \mapsto \eta^{(p)} \circ g$ for $g \in SL_2(\mathbb{A}^{(p\infty)})$. Then the effect of d on the operation $f \mapsto f|\rho(\alpha)$ can be computed by Lemma 1.32. Note that $(E(\mathfrak{a}),\phi_p\circ\alpha_{\mathfrak{p}},\eta^{(p)}\circ\rho^{(p)}(\alpha)) \cong (E(\mathfrak{a}),\phi_p,\eta^{(p)})$ by complex multiplication.

Katz interpreted the differential operator d in terms of the Gauss–Manin connection of the universal elliptic curve over the modular curve $X_1(N)$ and gave a purely geometric definition of the operator d^r acting on $V(R)$ for any p-adic W'-algebra R (see Katz's paper [K2]). Since his definition of d^r is purely geometric, it is valid for classical modular forms and p-adic modular forms at the same time. An important formula given in [K2] (2.6.7) is as follows.

Theorem 1.34 (N. Katz) *Let the notation and the assumption be as in* Theorem 1.23; *in particular, $E(\mathfrak{a})_{/\mathcal{W}'}$ is the CM elliptic curve associated with a lattice $\mathfrak{a} \subset M$ for an imaginary quadratic field M in which p splits, where $\mathcal{W}'$ is a discrete valuation ring finite flat over $\mathcal{W}$ over which $E(\mathfrak{a})$ is defined. For $f \in G_k(p^n;\mathcal{W}')$, we have*

$$\frac{(d^r f)(E(\mathfrak{a}),\phi_N,\phi_p)}{\Omega_p^{k+2r}} = \frac{(\delta_k^r f)(E(\mathfrak{a}),\phi_N,du)}{\Omega_\infty^{k+2r}} \in \mathcal{W}'. \tag{K}$$

If $\mathfrak{a}$ is a fractional ideal of $\mathfrak{O}$, we can take $\mathcal{W}' = \mathcal{W}$ as explained earlier. We can always find a nowhere vanishing differential ω in $\Omega_{E(\mathfrak{a})/\mathcal{W}'}$, as $\mathcal{W}'$ is a discrete valuation ring, and as before $\omega = \Omega_p \widehat{\phi}_{p,*}\frac{dt}{t}$. We give here a proof similar to the argument that proves Theorem 1.27.

Proof. We use the notation introduced in the proof of Theorem 1.27; hence, $\mathfrak{a} = i_\infty^{-1}(\mathbb{Z}+\mathbb{Z}z_0)$ for $z_0 \in \mathfrak{H}$. We take $\alpha \in \mathfrak{O}_{(p)}^\times$ of infinite order, with $\alpha\overline{\alpha} = 1$ and $\alpha \equiv 1 \mod N$.

After identifying geometric forms and analytic ones by q-expansions via the two fixed embeddings $\mathbb{C}_p \xleftarrow{i_p} \mathcal{W}' \xrightarrow{i_\infty} \mathbb{C}$, we see that $d = \frac{1}{2\pi i}\frac{\partial}{\partial z}$. We write $\mathcal{A} = \mathcal{A}(N;\overline{\mathbb{Q}}) = \bigcup_k \{\frac{f}{g} | f,g \in G_k(\Gamma_1(N);\overline{\mathbb{Q}})\}$. For meromorphic functions $h(z) \in \mathcal{A}$, we have

$$\begin{aligned} d(h\circ\rho(\alpha))(z_0) &= \frac{1}{2\pi i}\frac{\partial h(\rho(\alpha)(z_0))}{\partial z} \\ &= \alpha^{-2}\frac{1}{2\pi i}\frac{\partial h}{\partial z}((\rho(\alpha)(z_0)) = \alpha^{-2}(dh)\circ\rho(\alpha)(z_0). \end{aligned}$$

Since $dh = g_1/g_2$ for $g_1 \in G_{k+2}(\Gamma_1(N);\mathcal{W}')$ and $g_2 \in G_k(\Gamma_1(N);\mathcal{W}')$ for sufficiently large k, we have (Remark 1.24)

$$\frac{(dh)(E(\mathfrak{a}),\phi_N,\widehat{\phi}_{p,*}\frac{dt}{t})}{\Omega_p^2} = (dh)(E(\mathfrak{a}),\phi_N,\omega) = (\delta_0 h)(E(\mathfrak{a}),\phi_N,\omega) \in \overline{\mathbb{Q}}. \tag{1.3.11}$$

For sufficiently large k, we will see later in Sect. 6.2.6 that the line bundle generated by weight k forms is very ample; thus, we can write any modular function of level N as a ratio of two modular forms of weight k, and hence, any weight-2 meromorphic modular form is also a ratio of forms, one of weight $k+2$ and another of weight k.

Since $\frac{f(E(\mathfrak{a}),\widehat{\phi}_{p,*}\frac{dt}{t})}{\Omega_p^k} = f(E(\mathfrak{a}),\phi_N,\omega) \in \mathcal{W}'$, we first show

$$\frac{(d^r f)(E(\mathfrak{a}),\phi_N,\phi_p)}{\Omega_p^{k+2r}} = \frac{(\delta_k^r f)(E(\mathfrak{a}),\phi_N,du)}{\Omega_\infty^{k+2r}} \in \overline{\mathbb{Q}}$$

by induction on r. When $r = 0$, this follows from Theorem 1.23. To treat $r > 0$, take $f \in G_k(\Gamma_1(N);\mathcal{W}')$, and define $h \in \mathcal{A}$ by $f|_k\rho(\alpha) = fh$ as in the proof of Theorem 1.27.

Applying d to $f|_k\rho(\alpha) = fh$, we have $(df)|_{k+2}\rho(\alpha) = (df)h + f(dh)$. Specializing this equality at $(E(\mathfrak{a}),\phi_N,\omega)$, we get from Lemma 1.33

$$\begin{aligned}\alpha^{-k-2}(df(E(\mathfrak{a}),\phi_N,\phi_p)) &= ((df)|\rho(\alpha))(E(\mathfrak{a}),\phi_N,\phi_p)\\ &= (df)(E(\mathfrak{a}),\phi_N,\phi_p)h(E(\mathfrak{a})) + f(E(\mathfrak{a}),\phi_N,\phi_p)(dh)(E(\mathfrak{a}),\phi_N,\phi_p).\end{aligned}$$

Since $h(E(\mathfrak{a})) = (f|_k\rho(\alpha))(E(\mathfrak{a}),\phi_N,\phi_p)/f(E(\mathfrak{a}),\phi_N,\phi_p) = \alpha^{-k}$, we have

$$(df)(E(\mathfrak{a}),\phi_N,\phi_p) = \alpha^k(\alpha^{-2}-1)^{-1}f(z_0)(dh)(E(\mathfrak{a}),\phi_N,\phi_p).$$

Thus, we have again proved $\frac{dh(E,\phi_N,\widehat{\phi}_{p,*}\frac{dt}{t}))}{\Omega_p^2} \in \overline{\mathbb{Q}}$, and also this proves the result when $r = 1$. We repeat this process r times. By the Leibnitz formula, we have $d^r(fh) = \sum_{0\le s\le r}\binom{r}{s}d^s f d^{r-s}h$. From this, we get

$$d^r(f)|_{k+2r}\rho(\alpha) = (d^r f)h + \sum_{0<s\le r}\binom{r}{s}(d^{r-s}f)(d^s h).$$

Evaluating this at $(E(\mathfrak{a}),\phi_N,\phi_p)$, we get for $C = \frac{\alpha^k}{(\alpha^{-2r}-1)}$

$$(d^r f)(E(\mathfrak{a}),\phi_N,\phi_p) = C\cdot\sum_{0<s\le r}\binom{r}{s}(d^{r-s}f)(E(\mathfrak{a}),\phi_N,\phi_p)(d^s h)(E(\mathfrak{a}),\phi_N,\phi_p).$$

Dividing by Ω_p^{k+2r}, we finally get

$$\frac{(d^r f)(E(\mathfrak{a}),\phi_N,\phi_p)}{\Omega_p^{k+2r}} = C\cdot\sum_{s>0}\binom{r}{s}\frac{(d^{r-s}f)(E(\mathfrak{a}),\phi_N,\phi_p)}{\Omega_p^{k+2r-2s}}\frac{(d^s h)(E(\mathfrak{a}),\phi_N,\phi_p)}{\Omega_p^{2s}}.$$

By the induction hypothesis, we have, for $s > 0$,

$$\frac{(d^{r-s}f)(E(\mathfrak{a}),\phi_N,\phi_p)}{\Omega_p^{k+2r-2s}} = \frac{(\delta_k^{r-s}f)(E(\mathfrak{a}),\phi_N,du)}{\Omega_\infty^{k+2r-2s}}$$

and

$$\frac{(d^s h)(E(\mathfrak{a}),\phi_N,\phi_p)}{\Omega_p^{2s}} = \frac{(d^{s-1}dh)(E(\mathfrak{a}),\phi_N,\phi_p)}{\Omega_p^{2s}} = \frac{(\delta_2^{s-1}dh)(E(\mathfrak{a}),\phi_N,du)}{\Omega_\infty^{2s}}.$$

Replacing each term as above by the corresponding Archimedean term, we recover the right-hand side of (1.3.8) divided by Ω_∞^{k+2r}. Then, by the induction hypothesis, we get the desired identity:

$$\frac{(d^r f)(E(\mathfrak{a}),\phi_N,\phi_p)}{\Omega_p^{k+2r}} = \frac{(\delta_k^r f)(E(\mathfrak{a}),\phi_N,du)}{\Omega_\infty^{k+2r}}$$

inside $\overline{\mathbb{Q}}$. Since the left-hand side of the above identity is in the completion W' of $i_p(\mathcal{W}')$ in $\mathbb{C}_p$, we conclude the identity in $\mathcal{W}' = i_p^{-1}(W')$.

Remark 1.35 We note that this process of proving algebraicity and p-integrality applies to Hilbert modular forms (and beyond) after an appropriate adjustment (as Shimura's argument proving Theorem 1.27 has been generalized to unitary/symplectic Shimura varieties by himself in [AAF] Chap. III). Therefore, the use of the Gauss–Manin connection to study this type of algebraicity is not necessary (though the use is perhaps more conceptual than our way). In particular, if $\mathfrak{a}_p \neq \mathfrak{O}_p$, the proof via the Gauss–Manin connection is more involved, because one needs to make scalar extension of the ring of definition of $E(\mathfrak{O})$ to split the Hodge filtration on $H^1_{DR}(E(\mathfrak{a}))$ in an appropriate manner [see [K2] (2.4.2)]. We note that [K2] (essentially) covers the case $\mathfrak{a}_p \neq \mathfrak{O}_p$. In any case, the existence of a canonical family of abelian varieties over the Shimura variety is essential in either way one goes.

1.3.7 Katz Measure at a Glance

Recall the binomial polynomial $\binom{x}{m}$ given by

$$\binom{x}{m} = \begin{cases} 1 & \text{if } m = 0, \\ \frac{x(x-1)\cdots(x-n+1)}{m!} & \text{if } m \geq 1. \end{cases}$$

So $\binom{x}{1} = x$, $\binom{x}{2} = \frac{x(x-1)}{2}$, and so on. Consider the differential operator $\binom{d}{m}$ [e.g., $\binom{d}{2}f = \frac{1}{2}(d^2 f - df)$]. An easy property of $\binom{x}{m}$ is $\binom{x}{m}\big|_{x=n} = 0$ for integers $0 \leq n \leq m$, and $\binom{x}{m}\big|_{x=n}$ is the traditional binomial coefficient $\binom{n}{m} \in \mathbb{Z}$. In short, for the set of nonnegative integers $\mathbb{Z}_+$, $x \mapsto \binom{x}{m}$ has values in $\mathbb{Z}_+$. Since a polynomial function is p-adically continuous, we conclude immediately that the polynomial function $x \mapsto \binom{x}{m}$ sends $\mathbb{Z}_p$ into $\mathbb{Z}_p$ as $\mathbb{Z}_+$ is dense in $\mathbb{Z}_p$ (the existence of the p-adic expansion of each p-adic integer). Then we have

$$\binom{d}{m}(\sum_n a_n q^n) = \sum_n \binom{n}{m} a_n q^n.$$

In particular, $\binom{n}{m}a_n \in R$ if $a_n \in R$ for any p-adic algebra R. This shows $\binom{d}{m} : V(R) \to V(R)$. In particular, $\binom{d}{m}f(E,\omega) \in W$ if $f \in G_k(p^n;W)$.

A remarkable fact Mahler found in [Mh] (though elementary) is

Theorem 1.36 *Let $V = \varprojlim_n V/p^nV$ be any p-adic Banach algebra over W with norm $|\cdot|_p$; then the space of continuous functions $\mathcal{C}(\mathbb{Z}_p;V)$ of $\mathbb{Z}_p$ with values in V has a basis $\{\binom{x}{m}\}_{n=0,1,\dots}$ over V. In other words, $\Phi \in \mathcal{C}(\mathbb{Z}_p;V)$ has a unique expansion $\Phi(x) = \sum_{n=0}^{\infty} a_n(\Phi)\binom{x}{n}$ for $a_n(\Phi) \in V$ satisfying $|a_n(\Phi)|_p \to 0$ as $n \to \infty$.*

One can find an easy proof of this theorem in [LFE] Sect. 3.2. Note here that the space $\mathcal{C}(\mathbb{Z}_p;V)$ is again a Banach space under the norm $|\Phi|_p = \mathrm{Sup}_{x\in\mathbb{Z}_p}|\Phi(x)|_p$. Therefore, any bounded V-linear functional $\phi : \mathcal{C}(\mathbb{Z}_p;V) \to V$ (called a bounded measure on $\mathbb{Z}_p$ with values in V) is determined uniquely by the value $\{\phi(\binom{x}{n}) = \int_{\mathbb{Z}_p}\binom{x}{n}d\phi\}_{n=0,1,\dots}$. In short, we have

$$\int_{\mathbb{Z}_p}\Phi(x)d\phi = \lim_{m\to\infty}\left(\sum_{n=0}^{m} a_n(\Phi)\int_{\mathbb{Z}_p}\binom{x}{n}\right).$$

Introducing a formal variable T, we have a formal identity $(1+T)^x = \sum_{n=0}^{\infty}\binom{x}{n}T^n$. Thus, by associating a formal power series $\int(1+T)^x d\phi(x) = \sum_{n=0}^{\infty}\left(\int_{\mathbb{Z}_p}\binom{x}{n}d\phi\right)T^n \in V[[T]]$ to a bounded measure ϕ as above, we have an isomorphism between the power series ring $V[[T]]$ and the space $\mathcal{M}(\mathbb{Z}_p;V)$ of all bounded measures on $\mathbb{Z}_p$ with values in V.

Corollary 1.37 *We have a canonical isomorphism of p-adic Banach spaces: $\mathcal{M}(\mathbb{Z}_p;V) \cong V[[T]]$, where the Banach norm on $\mathcal{M}(\mathbb{Z}_p;V)$ is given by $|\phi|_p = \mathrm{Sup}_{\Phi\neq 0}\left|\int_{\mathbb{Z}_p}\Phi d\phi\right|_p/|\Phi|_p$.*

Exercise 1.38 *Prove $|\phi|_p = \mathrm{Sup}_n\left|\int_{\mathbb{Z}_p}\binom{x}{n}d\phi\right|_p$, and show $\mathcal{M}(\mathbb{Z}_p;V) \cong V[[T]]$ as p-adic Banach spaces.*

By Mahler's theorem, a given continuous function $\Phi : \mathbb{Z}_p \to W$ can be expanded uniquely as $\Phi(x) = \sum_{m\geq 0} a_m(\Phi)\binom{x}{m}$ for $a_m(\Phi) \in W$ tending to 0 as $m \to \infty$. Thus, defining $\int_{\mathbb{Z}_p}\binom{x}{m}df(x) = \binom{d}{m}f(E,\phi_N,\phi_p)$, we have a unique p-adic measure $df : \mathcal{C}(\mathbb{Z}_p;W) \to W$, where $\mathcal{C}(\mathbb{Z}_p;W)$ is the p-adic Banach space of continuous functions on $\mathbb{Z}_p$ with values in W.

Theorem 1.39 *Let the notation and the assumption be as in* Theorem 1.34, *and suppose $f \in G_k(p^n;\mathcal{W})$. If $\Phi(x) = \sum_{m\geq 0} a_m(\Phi)\binom{x}{m} \in \mathcal{C}(\mathbb{Z}_p;W)$, we have*

$$\int_{\mathbb{Z}_p}\Phi(x)df(x) = \Phi(x) = \sum_{m\geq 0} a_m(\Phi)\binom{d}{m}f(E(\mathfrak{a}),\phi_N,\phi_p).$$

In particular, we have

$$\frac{\int_{\mathbb{Z}_p}x^r df(x)}{\Omega_p^{k+2r}} = \frac{d^r f(E(\mathfrak{a}),\phi_N,\phi_p)}{\Omega_p^{k+2r}} = \frac{\delta_k^r f(E(\mathfrak{a}),du)}{\Omega_\infty^{k+2r}} \in \mathcal{W}.$$

Recall the Eisenstein series

$$e_k(w) = \frac{2}{(k-1)!} \sum_{0 \neq l \in L_w} l^{-k}.$$

As seen in (1.3.1), this series e_k has the following q-expansion:

$$e_k(q) = 2^{-1}\zeta(1-k) + \sum_{n>0} q^n \sum_{0<a|n} a^{k-1}.$$

Thus, defining $\mathcal{E}_k = e_k|\mathbf{1}$ for the characteristic function $\mathbf{1}$ of $(\mathbb{Z}/p\mathbb{Z})^\times$ on $(\mathbb{Z}/p\mathbb{Z})$ in order to remove coefficients of q^{mp} for $m = 0, 1, \dots,$ we get

$$\mathcal{E}_k(q) = \sum_{n>0, p\nmid n} q^n \sum_{0<a|n} a^{k-1}.$$

Then, writing $\binom{x}{m} = \sum_{j=0}^m c_j x^j$ and defining $\binom{\mathcal{E}}{m} = \sum_j c_j \mathcal{E}_{j+1}$ as a function of (E, ω), we find its q-expansion in $\mathbb{Z}[[q]]$:

$$\binom{\mathcal{E}}{m} = \sum_{n>0, p\nmid n} q^n \sum_{0<a|n} \binom{a}{m} \in \mathbb{Z}[[q]].$$

Thus, we can create a two-variable measure $dE : \mathcal{C}(\mathbb{Z}_p^2; W) \to W$ as follows.

Theorem 1.40 (N. Katz) *Let the notation and the assumption be as in* Theorem 1.34. *There exists a measure $dE : \mathcal{C}(\mathbb{Z}_p^2; W) \to W$ such that*

$$\int_{\mathbb{Z}_p^2} \binom{x}{m}\binom{y}{n} dE = \binom{d}{m}\binom{\mathcal{E}}{n}(E(\mathfrak{a}), \phi_N, \phi_p) \in W$$

for all integers $n \geq 0$ and $m \geq 0$.

If $E(\mathfrak{a})$ is given by Gauss's lemniscate $\mathfrak{E} = E(\mathfrak{O})$: $y^2 = 1 - x^4 \cong y^2 = 4x^3 - 4x$, we have $\mathfrak{O} = \mathbb{Z}[i]$ for $i = \sqrt{-1}$, $\Omega_\infty = \int_0^1 du = \int_0^\infty \frac{dx}{\sqrt{1-x^4}} dx$, and

$$e_k(E, \omega) = \Omega_\infty^{-k} \frac{2}{(k-1)!} \sum_{0 \neq \alpha \in \mathbb{Z}[i]} \alpha^{-k} = \frac{L(0, \lambda_k)}{(k-1)! \Omega_\infty^k},$$

where $\lambda_k((\alpha)) = \alpha^{-k}$. More generally, the differential operator δ_k^r brings λ_k into $\lambda_{k+r,-r}$, with $\lambda_{k+r,-r}((\alpha)) = \alpha^{-k}(\overline{\alpha}/\alpha)^r$, and we get a p-adic measure interpolating $\frac{L(0, \lambda_{k+r,-r})}{\Omega_\infty^{k+2r}(2\pi i)^{-r}}$ over $\mathbb{Z}_p^2$ [up to an explicit constant like $(k-1)!$]. We may regard $\lambda_{k+r,-r}$ as a continuous character of $\mathbb{Z}_p^\times \times \mathbb{Z}_p^\times$ with values in $\mathbb{Z}_p^\times$ by putting $\widehat{\lambda}_{k+r,-r}(x, y) = x^{-k}(y/x)^r$; thus, $\widehat{\lambda}_{k+r,-r}(\alpha, \overline{\alpha}) = \lambda_{k+r,-r}((\alpha))$ for all $\alpha \in \mathfrak{O}_{(p)}^\times$. We call $\widehat{\lambda}_{k+r,-r}$ the *p-adic avatar* of $\lambda_{k+r,-r}$. See Definition 8.3 for a more theoretical definition of p-adic avatars.

For a general imaginary quadratic field M, one needs to make a combination of the above type of measures to get this type of p-adic interpolation of Hecke L-values. Indeed, writing $Cl(p^n)$ for the ray class group of M modulo p^n, the limit $C = \varprojlim_n Cl(p^n)$ has a decomposition $C = \bigsqcup_{\mathfrak{A} \in Cl} [\mathfrak{A}]\mathfrak{O}_p^\times/\{\pm 1\}$. Each $\mathfrak{A}$ gives rise to an elliptic curve with complex multiplication by $\mathfrak{O}$ with $E_{\mathfrak{A}}(\mathbb{C}) = \mathbb{C}/i_\infty(\mathfrak{A})$. Since we assumed that p splits in $\mathfrak{O}$, we have $\mathfrak{O}_p \cong \mathbb{Z}_p \times \mathbb{Z}_p$ as algebras. By the above construction, we get a measure $dE_{\mathfrak{A}}$ associated with $E_{\mathfrak{A}}$ on the coset $[\mathfrak{A}]\mathfrak{O}_p^\times/\{\pm 1\}$. Since we want to have a measure supported on C, which is the Galois group of the maximal abelian extension of M unramified outside p, we need to modify this construction as the original measure is supported on $\mathbb{Z}_p^2$. We also construct the p-adic avatar $\widehat{\lambda} : C \to \overline{\mathbb{Q}}_p^\times$, so that $\widehat{\lambda}(\mathfrak{a}) = \lambda(\mathfrak{a})$ for all fractional ideals $\mathfrak{a}$ prime to p (see Sect. 9.1.4). After an appropriate (but tricky) adjustment, Katz proved the following theorem from Theorems 1.27 and 1.34 in this way (see Sect. 9.1 for the actual construction):

Theorem 1.41 (N. Katz) *Let M be an imaginary quadratic field in which the prime p splits. Then there exists a measure $d\mu$ on C such that for a Hecke character λ with $\lambda((\alpha)) = \alpha^{-k}\overline{\alpha}^j$ for $\alpha \equiv 1 \mod p^n$ for some $n > 0$, writing its p-adic avatar as $\widehat{\lambda}$,*

$$\frac{\int_C \widehat{\lambda} d\mu}{\Omega_p^{k+j}} = c(\lambda)(1-\lambda(\overline{\mathfrak{p}}))(1-\lambda(\mathfrak{p})^{-1}p^{-1})\frac{L(0,\lambda)}{(2\pi i)^{-j}\Omega_\infty^{k+j}} \in \mathcal{W}[\lambda],$$

where $0 \le j < k$ and $\mathcal{W}[\lambda]$ is the ring generated over $\mathcal{W}$ by $\lambda(\mathfrak{A})$ for $\mathfrak{A}$ running over all fractional ideals of M prime to p. For the simple constant $c(\lambda)$, see Theorem 9.4.

The determination of the μ-invariant and the λ-invariant of the Katz p-adic L-functions is a basic question. We would like to answer this question about μ in the present book.

2

Elliptic Curves and Modular Forms

We now describe basics of elliptic curves and modular curves in three steps:

1. as plane curves over a field;
2. as scheme/group functor over a ring;
3. modular forms as functorial rules on modular curves.

Elliptic curves and modular curves are perhaps the most important objects studied in number theory. The theory gives a base of the proof by Wiles (through Ribet's work) of Fermat's last theorem, it supplies us with the simplest (and perhaps the most beautiful) example of Shimura varieties (cf. [IAT] Chaps. 6 and 7), a fast prime factorization algorithm (cf. [REC] IV), and so on. We give a sketch of the theory of modular curves and modular forms and (a minimal amount of) notation and terminology for us to be able to state main themes in the next chapter. For graduate students just starting to learn elliptic curves and modular forms, this chapter hopefully serves as an introductory illustration of the theory. Experienced readers may skip this chapter, going directly to the Chap. 3.

2.1 Curves over a Field

In this section, we describe basics of plane curves over a fixed field k. We also fix an algebraic closure $\overline{k}$ of k and a sufficiently big algebraically closed field Ω containing $\overline{k}$. Here we suppose that Ω has many transcendental elements over k. An example of this setting is the familiar one: $k = \mathbb{Q} \subset \overline{\mathbb{Q}} \subset \mathbb{C} = \Omega$.

2.1.1 Plane Curves

Let $\mathfrak{a}$ be a principal ideal of the polynomial ring $k[X, Y]$. Note that polynomial rings over a field are a unique factorization domain. We thus have the prime factorization $\mathfrak{a} = \prod_{\mathfrak{p}} \mathfrak{p}^{e(\mathfrak{p})}$ with principal primes $\mathfrak{p}$. We call $\mathfrak{a}$ *square-*

H. Hida, *Elliptic Curves and Arithmetic Invariants*, Springer Monographs in Mathematics, DOI 10.1007/978-1-4614-6657-4_2,

free if $0 \leq e(\mathfrak{p}) \leq 1$ for all principal primes $\mathfrak{p}$. Fix a square-free $\mathfrak{a}$. The set of A-rational points for any k-algebra A of a *plane curve* defined over k is given by the zero set

$$V_{\mathfrak{a}}(A) = \left\{(x,y) \in A^2 \middle| f(x,y) = 0 \text{ for all } f(X,Y) \in \mathfrak{a}\right\}.$$

It is common to take an intermediate field $\Omega/A/k$ classically, but the definition itself works well for any k-algebra A (here a k-algebra is a commutative ring containing k and sharing identity with k). Often in mathematics, if one has more flexibility, proofs become easier; so, we just allow $V_{\mathfrak{a}}(A)$ for any k-algebras A. Obviously, for a generator $f(X,Y)$ of $\mathfrak{a}$, we could have defined

$$V_{\mathfrak{a}}(A) = V_f(A) = \left\{(x,y) \in A^2 \middle| f(x,y) = 0\right\},$$

but this does not depend on the choice f of generators and depends only on the ideal $\mathfrak{a}$; thus, it is more appropriate to write it as $V_{\mathfrak{a}}$. As an exceptional case, we note $V_{(0)}(A) = A^2$. Geometrically, if $\mathfrak{a} \neq \{0\}$, we think of $V_{\mathfrak{a}}(\Omega)$ as a curve in $\Omega^2 = V_{(0)}(\Omega)$. This is intuitively more geometric if we take $k \subset \Omega = \mathbb{C}$ (a "curve" is a two-dimensional "plane" as a real manifold). In this sense, for any algebraically closed field K over k, a point $x \in V_{\mathfrak{a}}(K)$ is called a geometric point with coefficients in K, and $V_{(f)}(K) \subset V_{(0)}(K)$ is called the geometric curve in $V_{(0)}(K) = K^2$ defined by the equation $f(X,Y) = 0$.

By Hilbert's zero theorem (Nullstellensatz; see [CRT] Theorem 5.4 and [ALG] Theorem I.1.3A), writing $\overline{\mathfrak{a}}$ for the principal ideal of $\overline{k}[X,Y]$ generated by $\mathfrak{a}$, we have

$$\overline{\mathfrak{a}} = \left\{g(X,Y) \in \overline{k}[X,Y] \middle| g(x,y) = 0 \text{ for all } (x,y) \in V_{\mathfrak{a}}(\overline{k})\right\}. \tag{2.1.1}$$

Thus we have a bijection

$$\{\text{square-free ideals of } \overline{k}[X,Y]\} \leftrightarrow \{\text{plane curves } V_{\mathfrak{a}}(\overline{k}) \subset V_{(0)}(\overline{k})\}.$$

The association $V_{\mathfrak{a}} : A \mapsto V_{\mathfrak{a}}(A)$ is a covariant functor from the category of k-algebras to the category of sets (denoted by $SETS$; see Sect. 4.1 for category and functor and Example 4.1 for a list of notations for categories). Indeed, for any k-algebra homomorphism $\sigma : A \to A'$, we have an associated map: $V_{\mathfrak{a}}(A) \ni (x,y) \mapsto (\sigma(x),\sigma(y)) \in V_{\mathfrak{a}}(A')$ as $0 = \sigma(0) = \sigma(f(x,y)) = f(\sigma(x),\sigma(y))$. Thus, $\mathfrak{a} = \overline{\mathfrak{a}} \cap k[X,Y]$ is determined uniquely by this functor, but the value $V_{\mathfrak{a}}(A)$ for an individual A may not determine $\mathfrak{a}$. From a number-theoretic viewpoint (Diophantine problems), studying $V_{\mathfrak{a}}(A)$ for a small field (or even a ring, such as $\mathbb{Z}$) is important. Thus, it would often be better to regard $V_{\mathfrak{a}}$ as a functor.

If $\mathfrak{a} = \prod_{\mathfrak{p}} \mathfrak{p}$ for principal prime ideals $\mathfrak{p}$, by definition, we have

$$V_{\mathfrak{a}} = \bigcup_{\mathfrak{p}|\mathfrak{a}} V_{\mathfrak{p}}.$$

The plane curve $V_{\mathfrak{p}}$ (for each prime $\mathfrak{p}|\mathfrak{a}$) is called an *irreducible* component of $V_{\mathfrak{a}}$. Since $\mathfrak{p}$ is a principal prime, we cannot have a finer nontrivial decomposition $V_{\mathfrak{p}} = V \cup W$ with plane curves V and W defined over k. A prime ideal $\mathfrak{p} \subset k[X,Y]$ generates an ideal in $\overline{\mathfrak{p}}$ in $\overline{k}[X,Y]$, which may decompose into a product of primes in $\overline{k}[X,Y]$. If the ideal $\overline{\mathfrak{p}}$ remains prime in $\overline{k}[X,Y]$, we call $V_{\mathfrak{p}}$ *geometrically irreducible.*

Suppose that we have a map $F_A = F(\phi)_A : V_{\mathfrak{a}}(A) \to V_{\mathfrak{b}}(A)$ given by two polynomials $\phi_X(X,Y), \phi_Y(X,Y) \in k[X,Y]$ (independent of A) such that $F_A(x,y) = (\phi_X(x), \phi_Y(y))$ for all $(x,y) \in V_{\mathfrak{a}}(A)$ and all k-algebras A. Such a map is called a *regular k-map* or a *k-morphism* from a plane k-curve $V_{\mathfrak{a}}$ into $V_{\mathfrak{b}}$. Here $V_{\mathfrak{a}}$ and $V_{\mathfrak{b}}$ are plane curves defined over k. If $\mathbf{A}^1 = V_{\mathfrak{b}}$ is the *affine line*, that is, $V_{\mathfrak{b}}(A) \cong A$ for all A (taking, for example, $\mathfrak{b} = (y)$), a regular k-map $V_{\mathfrak{a}} \to \mathbf{A}^1$ is called a *regular* k-function. Regular k-functions are just functions induced by the polynomials in $k[x,y]$ on $V_{\mathfrak{a}}$; hence, $R_{\mathfrak{a}}$ is the ring of regular k-functions of $V_{\mathfrak{a}}$ defined over k. Let $\mathbf{A}^n = (\mathbf{A}^1)^n$.

We write $\mathrm{Hom}_{k\text{-curves}}(V_{\mathfrak{a}}, V_{\mathfrak{b}})$ for the set of regular k-maps from $V_{\mathfrak{a}}$ into $V_{\mathfrak{b}}$. Obviously, only $\phi_? \mod \mathfrak{a}$ $(? = X, Y)$ can possibly be unique. We have a commutative diagram for any k-algebra homomorphism $\sigma : A \to A'$:

$$\begin{array}{ccc} V_{\mathfrak{a}}(A) & \xrightarrow{F_A} & V_{\mathfrak{b}}(A) \\ {\scriptstyle\sigma}\downarrow & & \downarrow{\scriptstyle\sigma} \\ V_{\mathfrak{a}}(A') & \xrightarrow[F_{A'}]{} & V_{\mathfrak{b}}(A'). \end{array}$$

Indeed,

$$\begin{aligned} \sigma(F_A((x,y))) &= (\sigma(\phi_X(x,y)), \sigma(\phi_Y(x,y))) \\ &= (\phi_X(\sigma(x), \sigma(y)), \phi_Y(\sigma(x), \sigma(y)) = F_{A'}(\sigma(x), \sigma(y)). \end{aligned}$$

Thus, the k-morphism is a *natural transformation of functors* (or a *morphism of functors*) from $V_{\mathfrak{a}}$ into $V_{\mathfrak{b}}$. We write $\mathrm{Hom}_{COF}(V_{\mathfrak{a}}, V_{\mathfrak{b}})$ for the set of natural transformations from $V_{\mathfrak{a}}$ into $V_{\mathfrak{b}}$ [we will see later that $\mathrm{Hom}_{COF}(V_{\mathfrak{a}}, V_{\mathfrak{b}})$ is a set].

The polynomials (ϕ_X, ϕ_Y) induce a k-algebra homomorphism $\underline{F}$ from $k[X,Y]$ into itself by pullback; that is, $\underline{F}(\Phi(X,Y)) = \Phi(\phi_X(X,Y), \phi_Y(X,Y))$. Take a class $[\Phi]_{\mathfrak{b}} = \Phi + \mathfrak{b}$ in $B = k[X,Y]/\mathfrak{b}$. Then look at $\underline{F}(\Phi) \in k[X,Y]$ for $\Phi \in \mathfrak{b}$. Since $(\phi_X(x), \phi_Y(y)) \in V_{\mathfrak{b}}(\overline{k})$ for all $(x,y) \in V_{\mathfrak{a}}(\overline{k})$, $\Phi(\phi_X(x,y), \phi_Y(x,y)) = 0$ for all $(x,y) \in V_{\mathfrak{a}}(\overline{k})$. By Nullstellensatz, $\underline{F}(\Phi) \in \overline{\mathfrak{a}} \cap k[X,Y] = \mathfrak{a}$. Thus, $\underline{F}(\mathfrak{b}) \subset \mathfrak{a}$, and $\underline{F}$ induces a (reverse) k-algebra homomorphism

$$\underline{F} : k[X,Y]/\mathfrak{b} \to k[X,Y]/\mathfrak{a},$$

making the following diagram commutative:

$$\begin{array}{ccc} k[X,Y] & \xrightarrow{F} & k[X,Y] \\ \downarrow & & \downarrow \\ k[X,Y]/\mathfrak{b} & \xrightarrow[\underline{F}]{} & k[X,Y]/\mathfrak{a}. \end{array}$$

We write $R_\mathfrak{a} = k[X,Y]/\mathfrak{a}$ and call it the affine ring of $V_\mathfrak{a}$. Here is a useful (but tautological) lemma that is a special case of Yoneda's lemma:

Lemma 2.1 *We have a canonical isomorphism:*

$$\mathrm{Hom}_{COF}(V_\mathfrak{a}, V_\mathfrak{b}) \cong \mathrm{Hom}_{k\text{-}curves}(V_\mathfrak{a}, V_\mathfrak{b}) \cong \mathrm{Hom}_{k-alg}(R_\mathfrak{b}, R_\mathfrak{a}).$$

The first association is covariant, and the second is contravariant.

Here is a sketch of the proof.

Proof. First, we note $V_\mathfrak{a}(A) \cong \mathrm{Hom}_{ALG_{/k}}(R_\mathfrak{a}, A)$ via

$$V_\mathfrak{a}(A) \ni (a,b) \leftrightarrow (\Phi(X,Y) \mapsto \Phi(a,b)) \in \mathrm{Hom}_{ALG_{/k}}(R_\mathfrak{a}, A).$$

Thus, for functors, we have $V_\mathfrak{a}(?) \cong \mathrm{Hom}_{ALG_{/k}}(R_\mathfrak{a}, ?)$. We identify the two functors $A \mapsto V_\mathfrak{a}(A)$ and $A \mapsto \mathrm{Hom}(R_\mathfrak{a}, A)$ in this way. Then the main point of the proof of the lemma is to construct, from a given natural transformation $F \in \mathrm{Hom}_{COF}(V_\mathfrak{a}, V_\mathfrak{b})$, a k-algebra homomorphism $\underline{F} : R_\mathfrak{b} \to R_\mathfrak{a}$, giving F by $V_\mathfrak{a}(A) = \mathrm{Hom}_{ALG_{/k}}(R_\mathfrak{a}, A) \ni \phi \overset{F_A}{\mapsto} \phi \circ \underline{F} \in \mathrm{Hom}_{ALG_{/k}}(R_\mathfrak{b}, A) = V_\mathfrak{b}(A)$. Then the following exercise finishes the proof, as, clearly, if we start with $\underline{F}$, the above association leads to F.

Exercise 2.2 *Let* $\underline{F} = F_{R_\mathfrak{a}}(\mathrm{id}_{R_\mathfrak{a}}) \in V_{R_\mathfrak{b}}(R_\mathfrak{a}) = \mathrm{Hom}_{ALG_{/k}}(R_\mathfrak{b}, R_\mathfrak{a})$, *where* $\mathrm{id}_{R_\mathfrak{a}} \in V_\mathfrak{a}(R_\mathfrak{a}) = \mathrm{Hom}_{ALG_{/k}}(R_\mathfrak{a}, R_\mathfrak{a})$ *is the identity map. Then prove that* $\underline{F}$ *does the required job.*

Recall that $V_\mathfrak{a}$ is *irreducible* (resp., *geometrically irreducible*) if $\mathfrak{a}$ is a prime ideal of $k[x,y]$ (resp., $\overline{\mathfrak{a}} = \mathfrak{a}\overline{k}[X,Y]$ is a prime ideal in $\overline{k}[X,Y]$).

Exercise 2.3 *1. Prove that for any unique factorization domain* R, $R[X]$ *is a unique factorization domain.*
2. *Give an example of two distinct principal prime ideals* $\mathfrak{a}, \mathfrak{b}$ *of* $\mathbb{Q}[X,Y]$ *with* $V_\mathfrak{a}(\mathbb{Q}) = V_\mathfrak{b}(\mathbb{Q})$.
3. *If* $\mathfrak{a}$ *and* $\mathfrak{b}$ *are two distinct principal prime ideals of* $\mathbb{Q}[X,Y]$, *prove* $V_\mathfrak{a}(\overline{\mathbb{Q}}) \neq V_\mathfrak{b}(\overline{\mathbb{Q}})$.
4. *For a principal ideal* $\mathfrak{a} = (f) \subset k[X,Y]$, *prove* $\overline{\mathfrak{a}} \cap k[X,Y] = \mathfrak{a}$.
5. *Show that* $\underline{F} : k[X,Y]/\mathfrak{b} \to k[X,Y]/\mathfrak{a}$ *is uniquely determined by the morphism* $F : V_\mathfrak{a} \to V_\mathfrak{b}$ *of functors.*

An element in the total quotient ring of $R_\mathfrak{a}$ is called a *rational* k-function on $V_\mathfrak{a}$. If $V_\mathfrak{a}$ is irreducible, then rational k-functions form a field. This field is called the *rational function field of* $V_\mathfrak{a}$ *over* k and is written as $k(V_\mathfrak{a})$.

2.1.2 Tangent Space and Local Rings

Suppose $\mathfrak{a} = (f(X,Y))$. Write $V = V_\mathfrak{a}$ and $R = R_\mathfrak{a}$. Let $P = (a,b) \in V_\mathfrak{a}(K)$. We consider partial derivatives

$$\frac{\partial f}{\partial X}(P) := \frac{\partial f}{\partial X}(a,b) \text{ and } \frac{\partial f}{\partial Y}(P) := \frac{\partial f}{\partial Y}(a,b).$$

Then the line tangent to $V_\mathfrak{a}$ at (a,b) has equation

$$\frac{\partial f}{\partial X}(a,b)(X-a) + \frac{\partial f}{\partial Y}(a,b)(Y-b) = 0.$$

We write the corresponding linear space as $T_P = V_\mathfrak{b}$ for the principal ideal $\mathfrak{b}$ generated by $\frac{\partial f}{\partial X}(a,b)X + \frac{\partial f}{\partial Y}(a,b)Y$. We say that $V_\mathfrak{a}$ is *nonsingular* or *smooth* at $P = (a,b) \in V_\mathfrak{a}(K)$ for a subfield $K \subset \Omega$ if this T_P is really a line, in other words, if $(\frac{\partial f}{\partial X}(P), \frac{\partial f}{\partial Y}(P)) \neq (0,0)$.

Example 2.4 Let $\mathfrak{a} = (f)$ for $f(X,Y) = Y^2 - X^3$. Then for $(a,b) \in V_\mathfrak{a}(K)$, we have

$$\frac{\partial f}{\partial X}(a,b)(X-a) + \frac{\partial f}{\partial Y}(a,b)(Y-b) = -3a^2(X-a) + 2b(Y-b).$$

Thus, this curve is singular only at $(0,0)$.

Example 2.5 Suppose that k has characteristic different from 2. Let $\mathfrak{a} = (Y^2 - g(X))$ for a cubic polynomial $g(X) = X^3 + aX + b$. Then the tangent line at (x_0, y_0) is given by $2y_0(X - x_0) - g'(x_0)(Y - y_0)$. This equation vanishes if $0 = y_0^2 = g(x_0)$ and $g'(x_0) = 0$ and hence is singular at only $(x_0, 0)$ for a multiple root x_0 of $g(X)$. Thus, $V_\mathfrak{a}$ is a nonsingular curve if and only if $g(X)$ is separable if and only if its discriminant $-4a^3 - 27b^2 \neq 0$.

Suppose that K/k is an algebraic field extension. Then $K[X,Y]/\mathfrak{a}K[X,Y]$ contains $R_\mathfrak{a}$ as a subring. A maximal ideal $(X-a, Y-b) \subset K[X,Y]/\mathfrak{a}K[X,Y]$ induces a maximal ideal $P = (X-a, Y-b) \cap R_\mathfrak{a}$ of $R_\mathfrak{a}$. Then the *local ring* $\mathcal{O}_{V,P}$ at P is the localization

$$\mathcal{O}_{V,P} = \left\{ \frac{a}{b} \middle| b \in R,\ b \in R \setminus P \right\},$$

where $\frac{a}{b} = \frac{a'}{b'}$ if there exists $s \in R \setminus P$ such that $s(ab' - a'b) = 0$. Write the maximal ideal of $\mathcal{O}_{V,P}$ as $\mathfrak{m}_P$. Then $\mathfrak{m}_P \cap R = P$.

Lemma 2.6 *The linear vector space $T_P(K)$ is the dual vector space of $P/P^2 = \mathfrak{m}_P/\mathfrak{m}_P^2$.*

Proof. Write $\mathfrak{a} = (f)$. Replacing $k[X,Y]/(f)$ by $K[X,Y]/(f)$, we may assume that $K = k$. A K-derivation $\partial : \mathcal{O}_{V,P} \to K$ (at P) is a K-linear map with $\partial(\phi\varphi) = \varphi(P)\partial(\phi) + \phi(P)\partial(\varphi)$. Write $D_{V,P}$ for the space of all K-derivations at P, which is a K-vector space. Clearly, for $\mathbf{A} := V_{(0)}$, $D_{\mathbf{A},P}$ is a two-dimensional vector space generated by $\partial_X : \phi \mapsto \frac{\partial \phi}{\partial X}(P)$ and $\partial_Y : \phi \mapsto \frac{\partial \phi}{\partial Y}(P)$. We have a natural injection $i : D_{V,P} \to D_{\mathbf{A},P}$ given by $i(\partial)(\phi) = \partial(\phi|_V)$. Note that $\Omega_{(a,b)} = (X-a, Y-b)/(X-a, Y-b)^2$ is a two-dimensional vector space over K generated by $X-a$ and $Y-b$. Thus, $D_{\mathbf{A},P}$ and $\Omega_{(a,b)}$ are dual to each other under the pairing $(\alpha(X-a)+\beta(Y-b), \partial) = \partial(\alpha(X-a)+\beta(Y-b))$. The projection $k[X,Y] \twoheadrightarrow R$ induces a surjection $\Omega_{(a,b)} \to \Omega_{V,P} = P/P^2$, whose kernel is spanned by

$$f \mod (X-a, Y-b)^2 = \frac{\partial f}{\partial X}(a,b)(X-a) + \frac{\partial f}{\partial Y}(a,b)(Y-b)$$

if $\mathfrak{a} = (f)$, since we have

$$\phi(X,Y) \equiv \frac{\partial \phi}{\partial X}(a,b)(X-a) + \frac{\partial \phi}{\partial Y}(a,b)(Y-b) \mod (X-a, Y-b)^2.$$

Thus, the above duality between $\Omega_{(a,b)}$ and $D_{\mathbf{A},(a,b)}$ induces the duality $\Omega_{V,P} = P/P^2$ and $T_P(K)$ given by $(\omega, t) = t(\omega)$, where we regard t as a derivation $\mathcal{O}_{V,P} \to K$. □

We call T_P the *tangent* space at P and $\Omega_P = \Omega_{V,P}$ the *cotangent* space at P of V. More generally, a *k-derivation* $\partial : R_{\mathfrak{a}} \to R_{\mathfrak{a}}$ is a k-linear map satisfying the Leibniz condition $\partial(\phi\varphi) = \phi\partial(\varphi) + \varphi\partial(\phi)$ and $\partial(k) = 0$. For a k-derivation as above, $f\partial : \varphi \mapsto f \cdot \partial(\varphi)$ because $f \in R_{\mathfrak{a}}$ is again a k-derivation. The totality of the k-derivation $Der_{V_{\mathfrak{a}}/k}$ is therefore an $R_{\mathfrak{a}}$-module.

To make T_P explicit, first take $\mathfrak{a} = (0)$; thus, $V_{\mathfrak{a}} = \mathbf{A}^2$. We have

$$\partial(X^n) = nX^{n-1}\partial X, \ \partial(Y^m) = mY^{m-1}\partial Y$$
$$\text{and } \partial(X^nY^m) = nX^{n-1}Y^m\partial X + mX^nY^{m-1}\partial Y$$

for $\partial \in Der_{\mathbf{A}^2/k}$; hence, ∂ is determined by its value $\partial(X)$ and $\partial(Y)$. Note that $(\partial X)\frac{\partial}{\partial X} + (\partial Y)\frac{\partial}{\partial Y}$ in $Der_{\mathbf{A}^2/k}$, and the original ∂ has the same value at X and Y; thus, we have

$$\partial = (\partial X)\frac{\partial}{\partial X} + (\partial Y)\frac{\partial}{\partial Y}.$$

Thus, $\{\frac{\partial}{\partial X}, \frac{\partial}{\partial Y}\}$ gives a basis of $Der_{\mathbf{A}^2/k}$.

Assuming $V_{\mathfrak{a}}$ nonsingular [including $\mathbf{A}^2 = V_{(0)}$], we write the $R_{\mathfrak{a}}$-dual as $\Omega_{V_{\mathfrak{a}}/k} := \mathrm{Hom}(Der_{V_{\mathfrak{a}}/k}, R_{\mathfrak{a}})$ (the *space of k-differentials*) with the duality pairing

$$(\cdot,\cdot) : \Omega_{V_{\mathfrak{a}}/k} \times Der_{V_{\mathfrak{a}}/k} \to R_{\mathfrak{a}}.$$

We have a natural map $d : R_{\mathfrak{a}} \to \Omega_{V_{\mathfrak{a}}/k}$ given by $\phi \mapsto (d\phi : \partial \mapsto \partial(\phi)) \in Der_{V_{\mathfrak{a}}/k}$. Note

$$(d(\phi\varphi), \partial) = \partial(\phi\varphi) = \phi\partial(\varphi) + \varphi\partial(\phi) = (\phi d\varphi + \varphi d\phi, \partial)$$

for all $\partial \in Der_{V_{\mathfrak{a}}/k}$. Thus, we have $d(\phi\varphi) = \phi d\varphi + \varphi d\phi$, and d is a k-linear derivation with values in $\Omega_{V_{\mathfrak{a}}/k}$.

Again, let us first look into $\Omega_{\mathbf{A}^2/k}$. Then, by definition, $(dX, \partial) = \partial X$ and $(dY, \partial) = \partial Y$; hence, $\{dX, dY\}$ is the dual basis of $\{\frac{\partial}{\partial X}, \frac{\partial}{\partial Y}\}$. We have $d\Phi = \frac{\partial \Phi}{\partial X} dX + \frac{\partial \Phi}{\partial Y} dY$, as we can check easily that the left-hand and the right-hand sides have the same value on any $\partial \in Der_{\mathbf{A}^2/k}$.

If $\partial : R_{\mathfrak{a}} = k[X,Y]/(f) \to R_{\mathfrak{a}}$ is a k-derivation, we can apply it to any polynomial $\Phi(X,Y) \in k[X,Y]$ and hence regard it as $\partial : k[X,Y] \to R_{\mathfrak{a}}$. By the above argument, $Der_k(k[X,Y], R_{\mathfrak{a}})$ has a basis $\{\frac{\partial}{\partial X}, \frac{\partial}{\partial Y}\}$ now over $R_{\mathfrak{a}}$. Since ∂ factors through the quotient $k[X,Y]/(f)$, it satisfies $\partial(f(X,Y)) = (df, \partial) = 0$. Thus, we have the following:

Lemma 2.7 *We have an inclusion* $Der_{V_{\mathfrak{a}}/k} \hookrightarrow (R_{\mathfrak{a}} \frac{\partial}{\partial X} \oplus R_{\mathfrak{a}} \frac{\partial}{\partial Y})$ *whose image is given by* $\{\partial \in Der_k(k[X,Y], R_{\mathfrak{a}}) | \partial f = 0\}$. *This implies*

$$\Omega_{V_{\mathfrak{a}}/k} = (R_{\mathfrak{a}} dX \oplus R_{\mathfrak{a}} dY)/R_{\mathfrak{a}} df$$

for $df = \frac{\partial f}{\partial X} dX + \frac{\partial}{\partial Y} dY$ *by duality.*

Remark 2.8 If $V_{\mathfrak{a}}$ is irreducible (so that $R_{\mathfrak{a}}$ is an integral domain), the space $k(V_{\mathfrak{a}})\Omega_{V_{\mathfrak{a}}/k} = (k(V_{\mathfrak{a}})dX \oplus k(V_{\mathfrak{a}})dY)/k(V_{\mathfrak{a}})df$ has dimension 1, as $df \neq 0$ in $\Omega_{\mathbf{A}^2/k}$. In particular, if we pick $\psi \in R_{\mathfrak{a}}$ with $d\psi \neq 0$ (i.e., a nonconstant), any differential $\omega \in \Omega_{V_{\mathfrak{a}}/k}$ can be uniquely written as $\omega = \phi d\psi$ for $\phi \in k(V_{\mathfrak{a}})$.

Lemma 2.9 *The four following conditions are equivalent:*

1. *A point* P *of* $V(\overline{k})$ *is a smooth point.*
2. $\mathcal{O}_{V,P}$ *is a local principal ideal domain, not a field.*
3. $\mathcal{O}_{V,P}$ *is a discrete valuation ring with residue field* $\overline{k}$.
4. $\varprojlim_n \mathcal{O}_{V,P}/\mathfrak{m}_P^n \cong \overline{k}[[T]]$ *(a formal power series ring of one variable).*

Proof. Let $K = \overline{k}$. By the above lemma, T_P is a line if and only if $\dim T_P(K) = 1$ if and only if $\dim P/P^2 = \dim \mathfrak{m}_P/\mathfrak{m}_P^2 = 1$. Thus, by Nakayama's lemma (Lemma 10.8), $\mathfrak{m}_P$ is principal. Any nonzero prime ideal of $k[X,Y]$ is either principal or maximal (i.e, the ring $k[X,Y]$ has Krull dimension 2). Thus, any prime ideal of R and $\mathcal{O}_{V,P}$ is maximal. Thus, (1) and (2) are equivalent. The equivalence of (2) and (3) follows from general ring theory (see [CRT] Theorem 11.2). We leave the equivalence (3) $\Leftrightarrow$ (4) as an exercise.

Write x, y for the image of $X, Y \in k[X,Y]$ in $R_{\mathfrak{a}}$. Any $\omega \in \Omega_{V_{\mathfrak{a}}/k}$ can be written as $\phi dx + \varphi dy$. Suppose that $V_{\mathfrak{a}}$ is nonsingular. Since $\mathcal{O}_{V_{\mathfrak{a}},P} \hookrightarrow k[[T]]$ [for $P \in V_{\mathfrak{a}}(k)$] for a local parameter T as above, ϕ, φ, x, y have the "Taylor expansion" as an element of $k[[T]]$, for example, $x(T) = \sum_{n \geq 0} a_n(x) T^n$ with $a_n(x) \in k$. Thus, dx, dy also have a well-defined expansion, say, $dx = d(\sum_{n \geq 0} a_n(x) T^n) = \sum_{n \geq 1} a_n(x) T^{n-1} dT$. Therefore, we may expand

$$\omega = \phi dx + \varphi dy = \sum_{n \geq 0} a_n(\omega) T^n dT$$

once we choose a parameter T at P. This expansion is unique independent of the expression $\phi dx + \varphi dy$. Indeed, if we allow meromorphic functions Φ as coefficients, as we remarked already, we can uniquely write $\omega = \Phi dx$ and the above expansion coincides with the Taylor expansion of Φdx.

Exercise 2.10 *Let $P \in V_{\mathfrak{a}}(K)$ for a finite field extension K/k, and pull back P to a maximal ideal $(X-a, Y-b) \subset K[X,Y]$. Define $(X-a, Y-b) \cap k[X,Y]$, and project it down to a maximal ideal $p \subset R_{\mathfrak{a}} = k[X,Y]/\mathfrak{a}$. Write $\mathcal{O}_{V_{\mathfrak{a}},p}$ for the localization of $R_{\mathfrak{a}}$ at p. Prove the following facts:*

1. *p is a maximal ideal and its residue field is isomorphic to the field $k(a,b)$ generated by a and b over k.*
2. *$(p/p^2) \otimes_{k(a,b)} K \cong P/P^2$ as K-vector space.*
3. *Any maximal ideal of $R_{\mathfrak{a}}$ is the restriction of $P \in V_{\mathfrak{a}}(K)$ for a suitable finite field extension K/k.*
4. *$\mathcal{O}_{V_{\mathfrak{a}},p}$ is a discrete valuation ring if and only if $\mathcal{O}_{V_{\mathfrak{a}},P}$ is a discrete valuation ring.*

Write $\mathrm{Max}(R_{\mathfrak{a}})$ for the set of maximal ideals of $R_{\mathfrak{a}}$. Then, clearly, we have a natural inclusion $V_{\mathfrak{a}}(k) \hookrightarrow \mathrm{Max}(R_{\mathfrak{a}})$ sending (a,b) to $(x-a, y-b)$ for the image x, y in $R_{\mathfrak{a}}$ of $X, Y \in k[X,Y]$. For $P \in \mathrm{Max}(R_{\mathfrak{a}})$, we say P is smooth on $V_{\mathfrak{a}}$ if $\mathcal{O}_{V,P}$ is a discrete valuation ring. By the above exercise, this is consistent with the earlier definition (no more, no less).

For any given affine plane irreducible curve $V_{\mathfrak{a}}$, we call $V_{\mathfrak{a}}$ *normal* if $R_{\mathfrak{a}}$ is integrally closed in its field of fractions.

Corollary 2.11 *Any normal irreducible affine plane curve is smooth everywhere.*

Proof. By ring theory, any localization of a normal domain is normal. Thus, $\mathcal{O}_{V,P}$ is a normal domain. By the exercise below, we may assume that $P \cap k[X,Y] \neq (0)$. Then P is a maximal ideal, and hence $K = k[X,Y]/P$ is an algebraic extension of k. Then $\mathcal{O}_{V,P}$ is a normal local domain with principal maximal ideal, which is a discrete valuation ring (cf. [CRT] Theorem 11.1).

Exercise 2.12 1. *Let $P = k[X,Y] \cap (X-a, Y-b)$ for $(a,b) \in V_{\mathfrak{a}}(\Omega)$, where $(X-a, Y-b)$ is the ideal of $\Omega[X,Y]$. Is it possible to have $P = (0) \subset k[X,Y]$ for a point $(a,b) \in V_{\mathfrak{a}}(\Omega)$?*

2. *If $\mathfrak{a} = (XY)$, is the ring $\mathcal{O}_{V,O}$ for $O = (0,0)$ an integral domain? What is $\dim_k \mathfrak{m}_O/\mathfrak{m}_O^2$?*
3. *For all points $P \in V_{\mathfrak{a}}(\Omega)$ with $R_{\mathfrak{a}} \cap P = (0)$ [regarding $P = (x-a, y-b)$ as an maximal ideal of $\Omega[X,Y]/\mathfrak{a}\Omega[X,Y]$], prove that V is smooth at P.*
4. *If A is a discrete valuation ring containing a field $k \subset A$ that is naturally isomorphic to the residue field of A, prove $\widehat{A} = \varprojlim_n A/\mathfrak{m}_A^n \cong k[[T]]$, where $\mathfrak{m}_A$ is the maximal ideal of A.*

2.1.3 Projective Space

Let A be a commutative ring. Write A_P for the localization at a prime ideal P of A. Thus, $A_P = \{\frac{b}{s}|s \in A \setminus P\}/\sim$, where $\frac{b}{s} \sim \frac{b'}{s'}$ if there exists $s'' \in A \setminus P$ such that $s''(s'b - sb') = 0$. An A-module M is called *locally free* at P if $M_P = \{\frac{m}{s}|s \in A \setminus P\}/\sim = A_P \otimes_A M$ is free over A_P. We call M locally free if M_P is free at all prime ideals P of A. If $\mathrm{rank}_{A_P} M_P$ is constant r independent of P, we write $\mathrm{rank}_A M = r$.

Write $ALG_{/B}$ for the category of B-algebras; hence, $\mathrm{Hom}_{ALG_{/B}}(A, A')$ is made up of B-algebra homomorphisms from A into A', sending the identity 1_A to the identity $1_{A'}$. Here B is a general base ring, and we write ALG for $ALG_{/\mathbb{Z}}$ (and ALG is the category of all commutative rings with identity). We consider a covariant functor $\mathbf{P}^n = \mathbf{P}^n_{/B} : ALG_{/B} \to SETS$ given by

$$\mathbf{P}^n(A)=\left\{L \subset A^{n+1}\middle|L \text{ (resp., } A^{n+1}/L) \text{ is locally } A\text{-free of rank 1 (resp., } n)\right\}.$$

This is a covariant functor (represented by the *projective space* $\mathbf{P}^n$ of dimension n). Indeed, if $\sigma : A \to A'$ is a B-algebra homomorphism, letting it act on A^{n+1} coordinatewise, $L \mapsto \sigma(L) \otimes_A A'$ [the A'-module generated by $\sigma(L)$] induces a map $\mathbf{P}^n(A) \to \mathbf{P}^n(A')$. If A is a field K, then $L \in \mathbf{P}^n(K)$ has to be free of dimension 1 generated by a nonzero vector $x = (x_0, x_1, \ldots, x_n)$. The vector x is unique up to multiplication by nonzero elements of K. Thus, we have proven the first statement (for a field) of the following:

Lemma 2.13 *Suppose that K is a local ring with maximal ideal $\mathfrak{m}$. Then $\mathbf{P}^n(K)$ is canonically in bijection to*

$$\left\{\underline{x} = (x_0, x_1, \ldots, x_n) \in K^{n+1}|\underline{x} \not\equiv (0, \ldots, 0) \mod \mathfrak{m}\right\}/K^\times.$$

Moreover, writing $D_i : ALG_{/B} \to SETS$ for the subfunctor $D_i(A) \subset \mathbf{P}^n(A)$ made up of the classes L whose projection to the ith component $A \subset A^{n+1}$ is surjective, we have $\mathbf{P}^n(K) = \bigcup_i D_i(K)$ and $D_i(A) \cong \mathbf{A}^n$ canonically for all B-algebras A. If A is a local ring K, $D_i \cong \mathbf{A}^n$ is given by sending $(x_0, \ldots, x_n)$ to $(\frac{x_0}{x_i}, \ldots, \frac{x_n}{x_i}) \in K^n$, removing the ith coordinate.

Proof. Since $K = K_\mathfrak{m}$ for its maximal ideal $\mathfrak{m}$, L is free if it is locally free. Thus, we have a generator $\underline{x} = (x_0, \ldots, x_n)$ of L over K. Since K^{n+1}/L is locally free of rank n, it has to be free of rank n over K as K is local. Taking a basis $\overline{v}_1, \ldots, \overline{v}_n$ of K^{n+1}/L, we lift them to $v_i \in K^{n+1}$ so that $\underline{x}, v_1, \ldots, v_n$ form a basis of K^{n+1} over K (by Nakayama's lemma; Lemma 10.8). Thus, $\underline{x} \not\equiv 0 \mod \mathfrak{m}$ for the maximal ideal $\mathfrak{m}$ of K. In particular, for an index i, $x_i \notin \mathfrak{m}$; hence, $x_i \in K^\times$. Since the projection of L to the ith component is generated by $x_i \in K^\times$, it is equal to K, and hence $\underline{x} \in D_i(K)$. Thus, $\mathbf{P}^n(K) = \bigcup_i D_i(K)$.

If $L \in D_i(A)$, we have the following commutative diagram.

$$\begin{array}{ccc} L & \xrightarrow{\hookrightarrow} & A^{n+1} \\ \| \downarrow & & \downarrow i\text{-th proj} \\ L & \xrightarrow{\sim} & A \end{array}$$

Thus, L is free of rank 1 over A and hence has a generator $(x_0, \ldots, x_n)$ with $x_i \in A^\times$. Then $(x_0, \ldots, x_n) \mapsto (\frac{x_0}{x_i}, \ldots, \frac{x_n}{x_i}) \in A^n$ gives rise to a natural transformation of D_i onto $\mathbf{A}^n$ (which is an isomorphism of functors). □

If K is local (in particular, a field), we write $(x_0{:}x_1{:}\cdots{:}x_n)$ for the point of $\mathbf{P}^n(K)$ represented by $(x_0, \ldots, x_n)$, as only the ratio matters.

Exercise 2.14 *Is there any example of a point in $X \in \mathbf{P}^1(A)$ (and a ring A) such that the projections to the first and second coordinates are both not surjective?*

We assume that K is a field for a while. When $n = 1$, we see $\mathbf{P}^1(K) = K^\times \sqcup \{\infty\}$ by $(x{:}y) \mapsto \frac{x}{y} \in K \sqcup \{\infty\}$. Thus, $\mathbf{P}^1(\mathbb{R})$ [resp., $\mathbf{P}^1(\mathbb{C})$] is isomorphic (topologically under Euclidean topology) to a circle (resp., a Riemann sphere).

We now assume that $n = 2$. Set $L = \{(x{:}y{:}0) \in \mathbf{P}^2(K)\}$. Then $\mathbf{P}^1 \cong L$ by $(x{:}y) \mapsto (x{:}y{:}0)$; hence, L is isomorphic to the projective line. We have $\mathbf{P}^2(K) = D(K) \sqcup L$ for fields K, where $D = D_2$. Thus, geometrically (i.e., over algebraically closed fields K), $\mathbf{P}^2(K)$ is the union of the affine plane and a projective line $L \cong \mathbf{P}^1(K)$. We let $L = L_\infty$ (the line at ∞).

2.1.4 Projective Plane Curve

Return to our base field $B = k$. For a plane curve defined by $\mathfrak{a} = (f(x, y))$ for $f(x, y)$ of degree m, we define $F(X, Y, Z) = Z^m f(\frac{X}{Z}, \frac{Y}{Z})$, which is a (square-free) homogeneous polynomial of degree m in $k[X, Y, Z]$. If $L \in \mathbf{P}^2(A)$, we can think of $F(\ell)$ for $\ell \in L$. We write $F(L) = 0$ if $F(\ell) = 0$ for all $\ell \in L$. Thus, for any k-algebra A, we define the functor $\overline{V}_\mathfrak{a} : ALG_{/k} \to SETS$ by

$$\overline{V}_\mathfrak{a}(A) = \{L \in \mathbf{P}^2(A) | F(L) = 0\}.$$

If A is a field K, we have $L \in \mathbf{P}^2(K)$ sent its generator $(a, b, c) \in L$ to identify $\mathbf{P}^2(K)$ with the (classical) projective space with homogeneous coordinate. Since $F(L) = 0$ if and only if $F(a{:}b{:}c) = 0$, we have

$$\overline{V}_\mathfrak{a}(K) = \{(a{:}b{:}c) \in \mathbf{P}^2(K) | F(a, b, c) = 0\},$$

which is called a *projective plane k-curve.*

Since $D_2 \cong \mathbf{A}^2$ canonically via $(x{:}y{:}1) \mapsto (x, y)$ (and this coordinate is well defined even over general A), we have $\overline{V}_\mathfrak{a}(A) \cap D_2(A) = V_\mathfrak{a}(A)$. In this sense, we can think of $\overline{V}_\mathfrak{a}$ as a completion of $V_\mathfrak{a}$, adding the boundary at ∞: $\overline{V}_\mathfrak{a} \cap L_\infty$. Since in $D_j \cong \mathbf{A}^2$ $(j = 0, 1, 2)$, $\overline{V}_\mathfrak{a} \cap D_j$ is a plane affine curve (for example, $\overline{V}_\mathfrak{a} \cap D_0$ is defined by $F(1, y, z) = 0$), $(L_\infty \cap \overline{V}_\mathfrak{a})(\overline{k})$ is a finite set. Thus, $\overline{V}_\mathfrak{a}$ is a completion/compactification of the (open) affine curve $V_\mathfrak{a}$. Of course, we can start with a homogeneous polynomial $F(X, Y, Z)$ [or a homogeneous ideal of $k[X, Y, Z]$ generated by $F(X, Y, Z)$] to define a projective plane curve. Following Lemma 2.1, we define $\mathrm{Hom}_{\text{proj } k\text{-curves}}(\overline{V}_\mathfrak{a}, \overline{V}_\mathfrak{b}) := \mathrm{Hom}_{COF}(\overline{V}_\mathfrak{a}, \overline{V}_\mathfrak{b})$.

A projective plane curve $\overline{V}_\mathfrak{a}$ is nonsingular (or smooth) if $\overline{V}_\mathfrak{a} \cap D_j$ is a nonsingular plane curve for all $j = 0, 1, 2$. The tangent space at $P \in \overline{V}_\mathfrak{a}(K)$ is defined as before since P is in one of $D_j \cap V_\mathfrak{a}$.

Example 2.15 Suppose $\mathfrak{a} = (y^2 - f(x))$ for a cubic $f(x) = x^3 + ax + b$. Then $F(X,Y,Z) = Y^2Z - X^3 - aXZ^2 - bZ^3$. Since L_∞ is defined by $Z = 0$, we find $L_\infty \cap \overline{V}_{\mathfrak{a}} = \{(0{:}1{:}0)\}$ made of a single point (of $\overline{V}_{\mathfrak{a}}$ intersecting with L_∞ with multiplicity 3). We call this point the origin $\mathbf{0}$ of the curve $\overline{V}_{\mathfrak{a}}$. If $\overline{V}_{\mathfrak{a}}$ is smooth, the pair $(\overline{V}_{\mathfrak{a}}, \mathbf{0})$ is called an elliptic curve.

Exercise 2.16 *Suppose $\overline{V}_{\mathfrak{a}}$ is defined by $F(X,Y,Z) = 0$. Let $f(x,y) = F(x,y,1)$ and $g(y,z) = F(1,y,z)$. Then the projective plane curve $\overline{V}_{\mathfrak{a}}$ for $\mathfrak{a} = (f(x,y))$ satisfies $\overline{V}_{\mathfrak{a}} \cap D_0 = V_{(g)}$. Show that $\mathcal{O}_{V_{\mathfrak{a}},P} \cong \mathcal{O}_{V_{(g)},P}$ canonically if $P \in \overline{V}_{\mathfrak{a}} \cap D_0 \cap D_2$.*

By the above exercise, the tangent space (the dual of $\mathfrak{m}_P/\mathfrak{m}_P^2$) at $P \in \overline{V}_{\mathfrak{a}}(K)$ does not depend on the choice of j with $P \in \overline{V}_{\mathfrak{a}} \cap D_j$. If a projective plane curve C is irreducible, the *rational function field* $k(C)$ of C over k is the field of fractions of $\mathcal{O}_{C,P}$ for any $P \in C(\overline{k})$ hence, independent of $C \cap D_j$.

Lemma 2.17 *Take a nonzero $f \in k(C)$. Then there exist homogeneous polynomials $G(X,Y,Z), H(X,Y,Z) \in k[X,Y,Z]$ with $\deg(G) = \deg(H)$ such that $f(x{:}y{:}z) = \frac{H(x,y,z)}{G(x,y,z)}$ for all $(x{:}y{:}z) \in C(\overline{k})$.*

Proof. We may write $C \cap D_2$ $f(x,y,1) = \frac{h(x,y)}{g(x,y)}$. If $m = \deg(h) = \deg(g)$, we just define $H(X,Y,Z) = h(\frac{X}{Z},\frac{Y}{Z})Z^m$ and $G(X,Y,Z) = g(\frac{X}{Z},\frac{Y}{Z})Z^m$. If $\deg(h) > \deg(g)$, we define $H(X,Y,Z) = h(\frac{X}{Z},\frac{Y}{Z})Z^{\deg(h)}$ and $G(X,Y,Z) = g(\frac{X}{Z},\frac{Y}{Z})Z^{\deg(h)}$. If $\deg(h) < \deg(g)$, we define $H(X,Y,Z) = h(\frac{X}{Z},\frac{Y}{Z})Z^{\deg(g)}$ and $G(X,Y,Z) = g(\frac{X}{Z},\frac{Y}{Z})Z^{\deg(g)}$. Multiplying h or g by a power of Z does not change the above identity $f(x,y,1) = \frac{h(x,y)}{g(x,y)}$, because $Z = 1$ on $C \cap D_2$. Thus, adjusting in this way, we get G and H.

Example 2.18 Look at $\phi = cx + dy$ in $k(C)$ for $C = \overline{V}_{\mathfrak{a}}$ for $\mathfrak{a}$ generated by $y^2 - x^3 - ax - b$. Then C is defined by $Y^2Z - X^3 - aXZ^2 - bZ^3 = 0$, and

$$\phi(X{:}Y{:}Z) = c\frac{X}{Z} + d\frac{Y}{Z} = \frac{cX + dY}{Z}.$$

Thus, ϕ has a pole of order 3 at $Z = 0$ (as the infinity on C has multiplicity 3) and three zeros at the intersection of $L := \{cx + dy = 0\}$ and $C \cap D_2 \cap L$.

Take a projective nonsingular plane k-curve $C_{/k}$. Set $C_i = C \cap D_i$, which is an affine nonsingular plane curve. Then we have well-defined global differentials $Der_{C_i/k}$. Since $\partial : Der_{C_i/k}$ induces $\partial_P : \mathcal{O}_{C_i,P} \to K$ for any $P \in C_i(K)$ by $f \mapsto \partial(f)(P)$, we have $\partial_P \in T_P$. If $\partial_i \in Der_{C_i/k}$ given for each $i = 0,1,2$ satisfies $\partial_{i,P} = \partial_{j,P}$ for all (i,j) and all $P \in (D_i \cap D_j)(\overline{k})$, we call $\partial = \{\partial_i\}_i$ a global tangent vector defined on C. Obviously, the totality $T_{C/k}$ of global tangent vectors is a k-vector space. The k-dual of $T_{C/k}$ is called the space of k-differentials over k and written as $\Omega_{C/k}$. We will see that $\Omega_{C/k}$ is finite dimensional over k.

Corollary 2.19 *Suppose that C is nonsingular. Each $\phi \in k(C)$ induces $\phi \in \mathrm{Hom}_{\text{proj } k\text{-curves}}(C, \mathbf{P}^1)$. Indeed, we have $k(C) \sqcup \{\infty\} \cong \mathrm{Hom}_{\text{proj } k\text{-curves}}(C, \mathbf{P}^1)$, where ∞ stands for the constant function sending all $P \in C(A)$ to the image of $\infty \in \mathbf{P}^1(k)$ in $\mathbf{P}^1(A)$.*

Proof. We prove only the first assertion. Suppose $k = \overline{k}$. Write $\phi(x{:}y{:}z) = \frac{h(x,y,z)}{g(x,y,z)}$ as a reduced fraction by the above lemma. For $L \in C(A) \subset \mathbf{P}^2(A)$, we consider the sub-$A$-module $\phi(L)$ of A^2 generated by $\{(h(\ell), g(\ell)) \in A^2 | \ell \in L\}$. We now show that $\phi(L) \in \mathbf{P}^1(A)$; thus, we will show that the map $C(A) \ni L \mapsto \phi(L) \in \mathbf{P}^1(A)$ induces the natural transformation of C into $\mathbf{P}^1$. If A is local, by Lemma 2.13, L is generated by (a, b, c) with at least one unit coordinate. Then any $\ell \in L$ is of the form $\lambda(a, b, c)$ (for a scalar λ), and therefore $\phi(\ell) = \lambda^{\deg(h)}\phi(a, b, c)$. Thus, $\phi(L) = A \cdot \phi(a, b, c)$. Since A is a k-algebra, k is naturally a subalgebra of the residue field $A/\mathfrak{m}$ of A. Since $\phi(P)$ for all $P \in C(k)$ is either a constant in k or ∞, we may assume that $(h(P), g(P)) \neq (0, 0)$ for all $P \in C(k)$. Since $(a, b, c) \not\equiv 0 \mod \mathfrak{m}$ as (a, b, c) generates a direct summand of A^3, thus $(h(a, b, c), g(a, b, c)) \not\equiv (0, 0) \mod \mathfrak{m}$. After tensoring $A/\mathfrak{m}$ over A, $(A/\mathfrak{m})^2/(\phi(L)/\mathfrak{m}\phi(L))$ is one-dimensional. Thus, by Nakayama's lemma (see [CRT] Theorem 2.2–3 and Lemma 10.8 in the text), $A/\phi(L)$ is generated by a single element and has to be a free module of rank 1 as $\phi(L)$ is a free A-module of rank 1. Thus, $\phi(L) \in \mathbf{P}^1(A)$. If k is not algebraically closed, replacing A by $\overline{A} = A \otimes_k \overline{k}$, we find $\phi(L) \otimes_k \overline{k} \in \mathbf{P}^1(\overline{k})$ and hence $\phi(L) \otimes_A A/\mathfrak{m} \in \mathbf{P}^1(k)$, which implies $\phi(L) \in \mathbf{P}^1(A)$.

If A is not necessarily local, applying the above argument to the local ring A_P for any prime ideal P of A, we find that $\phi(L)_P = \phi(L_P)$ and $A_P^2/\phi(L_P)$ are free of rank 1, and so $\phi(L)$ and $A^2/\phi(L)$ are locally free of rank 1. Therefore, $\phi(L) \in \mathbf{P}^1(A)$; hence, $L \mapsto \phi(L)$ induces a natural transformation of functors.

Exercise 2.20 *Prove the following facts:*

1. *If $L_\mathfrak{m}$ is free of finite rank r for a maximal ideal $\mathfrak{m}$ of A, L_P is free of rank r for any prime ideal $P \subset \mathfrak{m}$.*
2. *If $L \subset A^2$ is a free A-submodule of rank 1 and A^2/L is generated by one element over A, A^2/L is A-free of rank 1.*
3. $\mathrm{Hom}_{\text{proj } k\text{-curves}}(C, \mathbf{P}^1) \setminus \infty \cong k(C)$.

2.1.5 Divisors

Let C be a nonsingular projective geometrically irreducible plane curve. Since C is nonsingular, for any point $P \in C(\overline{k})$, $\mathcal{O}_{C,P}$ is a discrete valuation ring, and the rational function field $\overline{k}(C)$ is the quotient field of $\mathcal{O}_{C,P}$ (regarding C as defined over $\overline{k}$). Writing $v_P : \overline{k}(C) \twoheadrightarrow \mathbb{Z} \cup \{\infty\}$ for the additive valuation of $\mathcal{O}_{C,P}$, we have a well-defined $v_P(f) \in \mathbb{Z}$ for any nonzero rational $\overline{k}$-function $f \in \overline{k}(C)$. Since $\mathfrak{m}_P = (t_P)$ and $t_P^{v_P(f)} \| f$ in $\mathcal{O}_{C,P}$, f has a zero of order $v_P(f)$ at P if $v_P(f) > 0$ and a pole of order $|v_P(f)|$ if $v_P(f) < 0$. In other words, the Taylor expansion of f at P is given by $\sum_n a_n(f)t_P^n$ and $v_P(f) = \min(n : a_n(f) \neq 0)$. We start with Bézout's theorem:

Theorem 2.21 *Let C and C' be two plane projective k-curves inside $\mathbf{P}^2$ defined by relatively prime homogeneous equations*

$$F(X,Y,Z) = 0 \quad \text{and} \quad G(X,Y,Z) = 0$$

of degree m and n, respectively. Then, counting with multiplicity, we have $|C(\overline{k}) \cap C'(\overline{k})| = m \cdot n$.

Here we do not assume F and G are square-free. As is clear from the definition of multiplicity given below, for example, if $F = F_0^e$ with $\deg(F_0) = m_0$,

$$|C(\overline{k}) \cap C'(\overline{k})| = e|C_0(\overline{k}) \cap C'(\overline{k})| = em_0 \cdot n = m \cdot n$$

for the curve C_0 defined by $F_0 = 0$. Thus, for the proof of the theorem, we may assume that F and G are square-free.

If C is smooth at $P \in C \cap C'$ in $C \cap D_2$, $\phi = \frac{G(X,Y,Z)}{Z^n}$ is a function vanishing at P. The multiplicity of P in $C \cap C'$ is just $v_P(\phi)$. More generally, if $P = (a,b)$ is not necessarily a smooth point, writing $C \cap D_2 = V_{\mathfrak{a}}$ and $C' \cap D_2 = V_{\mathfrak{b}}$ for principal ideals $\mathfrak{a}, \mathfrak{b}$ in $\overline{k}[X,Y]$ and viewing P as an ideal

$$(X - a, Y - b) \subset \overline{k}[X,Y],$$

the multiplicity is given by the dimension of the localization $(\overline{k}[x,y]/(\mathfrak{a}+\mathfrak{b}))_P$ over $\overline{k}$. The same definition works well for any points in $C \cap D_0$ and $C \cap D_1$. One can find the proof of this theorem with a better definition of multiplicity in good textbooks on algebraic geometry (e.g., [ALG] Theorem I.7.7).

The *divisor group* $\mathrm{Div}(C)$ of a smooth curve C is a formal free $\mathbb{Z}$-module generated by points $P \in C(\overline{k})$. When we consider a point P as a divisor, we write it as $[P]$. For each divisor $D = \sum_P m_P[P]$, we define the *degree* of D by $\deg(D) = \sum_P m_P$.

Consider the space of *meromorphic differentials*:

$$\Omega_{C/k,\eta} = \{f \cdot \omega | f \in k(C), \omega \in \Omega_{C_i/k}\} \ (= k(C) \cdot \Omega_{C/k} \text{ if } \Omega_{C/k} \neq 0).$$

For $\omega \in \Omega_{C/\overline{k},\eta}$, we have its expansion $\sum_n a_n(\omega)t_P^n dt_P$ at each $P \in C(\overline{k})$; thus, we define $v_P(\omega) := \min(n : a_n(\omega) \neq 0)$.

Since there are only finitely many poles and zeros of f, we define divisors $\mathrm{div}(f) = \sum_{P \in C(k)} v_P(f)[P]$, $\mathrm{div}_0(f) = \sum_{P \in C(k), v_P(f)>0} v_P(f)[P]$ (zero divisor) and $\mathrm{div}_\infty(f) = \sum_{P \in C(k), v_P(f)<0} v_P(f)[P]$ (polar divisor) of f. Similarly, for a meromorphic differential ω, we define again $\mathrm{div}(\omega) = \sum_P v_P(\omega)[P]$. By Lemma 2.17, we may write $f(x{:}y{:}z) = \frac{h(x:y:z)}{g(x:y:z)}$ for a homogeneous polynomial h, g in $\overline{k}[x,y,z]$ of equal degree. If the degree of equation defining C is m and C' is defined by $h(X,Y,Z) = 0$, $\deg_0(\mathrm{div}(f)) = |C(\overline{k}) \cap C'(\overline{k})| = m\deg(h) = m\deg(g) = \deg_\infty(\mathrm{div}(f))$. This shows $\deg(\mathrm{div}(f)) = 0$ as $\sum_{P, v_P(f)>0} m_P = m\deg(h)$ and $-\sum_{P, v_P(f)<0} m_P = m\deg(g)$. Thus, we get the following:

Lemma 2.22 *Let C be a smooth projective plane curve. For any $f \in \overline{k}(C)$,* $\deg(\mathrm{div}(f)) = 0$*, and f is a constant in $\overline{k}$ if $f \in \overline{k}(C)$ is regular at all $P \in C$.*

Lemma 2.23 *If $f \in k(C)$ satisfies* $\deg(\mathrm{div}_0(f)) = \deg(\mathrm{div}_\infty(f)) = 1$*, $f : C \to \mathbf{P}^1$ induces an isomorphism of projective plane curve over k.*

Proof. Write $\phi(x{:}y{:}z) = \frac{H(x,y,z)}{G(x,y,z)}$ as a reduced fraction of homogeneous polynomials $G, H \in k[X,Y,Z]$ of degree n. Consider the curve C' defined by $H = 0$. Suppose C is defined by a homogeneous equation of degree m. Then by Bézout's theorem (applied to C and C'), $m \cdot n = \deg(\mathrm{div}_0(\phi)) = 1$. Thus, $m = n = 1$, and it is then clear that $(x{:}y{:}z) \mapsto (G(x,y,z){:}H(x,y,z))$ gives rise to an isomorphism $C \cong \mathbf{P}^1$.

Another proof: By the proof of Corollary 2.19, $\deg(\mathrm{div}_0(f))$ is the number of points over 0 (counting with multiplicity) of the regular map $f : C \to \mathbf{P}^1$. By taking a constant $\alpha \in k \subset \mathbf{P}^1$ away from f, $\deg(\mathrm{div}_0(f - \alpha)) = 1 = \deg(\mathrm{div}_\infty(f - \alpha))$, and $|f^{-1}(\alpha)| = \deg(\mathrm{div}_0(f - \alpha)) = 1$; hence, we find that f is one-to-one and onto. Thus, f is an isomorphism. □

Write $\mathrm{Div}^0(C) = \{D \in \mathrm{Div}(C_{/\overline{k}}) | \deg(D) = 0\}$. Inside $\mathrm{Div}^0(C)$, we have the subgroup $\{\mathrm{div}(f) | f \in \overline{k}(C)^\times\}$. We call two divisors D, D' *linearly equivalent* if $D = \mathrm{div}(f) + D'$ for $f \in \overline{k}(C)$. We say that D and D' are *algebraically equivalent* if $\deg(D) = \deg(D')$. The quotient groups

$$\mathcal{J}(C) = \frac{\mathrm{Div}^0(C)}{\{\mathrm{div}(f) | f \in \overline{k}(C)^\times\}} \quad \text{and} \quad \mathrm{Pic}(C) = \frac{\mathrm{Div}(C)}{\{\mathrm{div}(f) | f \in \overline{k}(C)^\times\}}$$

are called the *jacobian* and the *Picard group* of C, respectively. Sometimes, $\mathcal{J}(C)$ is written as $\mathrm{Pic}^0(C)$ (the degree-0 Picard group).

2.1.6 Riemann–Roch Theorem

We write $D = \sum_P m_P[P] \geq 0$ (resp., $D > 0$) for a divisor D on C if $m_P \geq 0$ for all P (resp., $D \geq 0$ and $D \neq 0$). For a divisor D on $C_{\overline{k}}$,

$$L(D) = \{f \in \overline{k}(C) | \mathrm{div}(f) + D \geq 0\} \cup \{0\}.$$

Clearly, $L(D)$ is a vector space over $\overline{k}$. It is known that $\ell(D) = \dim_{\overline{k}} L(D) < \infty$. For $\phi \in \overline{k}(C)^\times$, $L(D) \ni f \mapsto f\phi \in L(D - \mathrm{div}(\phi))$ is an isomorphism. Thus, $\ell(D)$ depends only on the class of D in $\mathrm{Pic}(C)$.

Example 2.24 Let $C = \mathbf{P}^1$. Take a divisor $D = \sum_{a \in \overline{k}} m_a[a]$ with $m_a \geq 0$ and $m_a > 0$ for some a, regarding $a \in \overline{k}$ as a point $[a] \in \mathbf{P}^1(\overline{k}) = \overline{k} \sqcup \{\infty\}$. On $\mathbf{A}^1(\overline{k}) = \overline{k}$, forgetting about ∞, $\mathrm{div}(f) + D \geq 0$ if $f = \frac{g(x)}{\prod_a (x-a)^{m_a}}$ for a polynomial $g(x)$. If $\deg(D) \geq \deg(g(x))$, the function f does not have a pole at ∞. Thus, $L(D) = \{g(x) | \deg(g(x)) \leq \deg(D)\}$, and we have $\ell(D) =$

$1+\deg(D)$. If C is a plane projective curve, we write $f = \frac{h(X,Y,Z)}{g(X,Y,Z)}$ as a reduced fraction by Lemma 2.17. Take $D = \sum_P m_P[P] \in \mathrm{Div}(C)$, and put

$$|D| = \{P|D = \sum_P m_P[P] \ \text{ with } \ m_P \neq 0\}.$$

If $|D|$ is inside $D_2 \cap C \subset \mathbf{A}^2$ and $D > 0$ (i.e., $m_P > 0$ for some P), we may assume that $V_{(g(X,Y,1))} \cap C$ contains $|D|$. In order not to have a pole in $C \setminus D_2$, $\deg(h)$ has to be bounded; thus, $\ell(D) < \infty$. Since $L(D) \subset L(D_+)$ in general, writing $D = D_+ + D_-$ so that $D_+ \geq 0$ and $-D_- \geq 0$, this shows that $\ell(D) < \infty$.

Exercise 2.25 *Give more details of the proof of* $\ell(D) < \infty$.

Theorem 2.26 (Riemann–Roch) *Let* $C = \overline{V}_{\mathfrak{a}}$ *be a smooth projective irreducible curve over a field* k. *Let* $K = \mathrm{div}(\omega)$ *for meromorphic* $0 \neq \omega \in \Omega_{C/\overline{k},\eta}$. *Then, for* $g = \dim_{\overline{k}} \Omega_{C/\overline{k}}$, *we have* $\ell(D) = 1 - g + \deg(D) + \ell(K - D)$ *for all divisors* D *on* $C(\overline{k})$. *If* $g = 1$, *we have* $K = 0$ *in* $\mathcal{J}(C)$.

The number g in the above theorem is called the *genus* of the curve C and is written as $g(C)$. The divisor K is called a *canonical divisor* K (whose linear equivalence class is unique). Note that

$$L(K) = \{f \in \overline{k}(C)|\mathrm{div}(f\omega) = \mathrm{div}(f) + \mathrm{div}(\omega) \geq 0\} \cong \Omega_{C/\overline{k}}$$

by $f \mapsto f\omega \in \Omega_{C/k}$. Then, by the above theorem,

$$g(C) = \dim \Omega_{C/\overline{k}} = \ell(K) = 1 - g + \deg(K) + \ell(0) = 2 + \deg(K) - g(C),$$

and from this, we conclude $\deg(K) = 2g(C) - 2$. One can find a proof of this theorem in any introductory book of algebraic geometry (e.g., [ALG] IV.1 or [GME] Theorem 2.1.3).

Corollary 2.27 *If* $g(C) = 1$ *and* $\deg(D) > 0$, *then we have* $\ell(D) = \deg(D)$ *and* $\ell(-D) = 0$.

Proof. For a nonconstant $f \in \overline{k}(E)$, $\deg(\mathrm{div}(f)) = 0$ implies that f has a pole somewhere. If $D > 0$, $f \in L(-D)$ does not have pole and hence is constant. Since $D > 0$, f vanishes at $P \in |D|$. Thus, $f = 0$. More generally, if $\deg(D) > 0$ and $\phi \in L(-D)$, then $0 > \deg(-D) = \deg(\mathrm{div}(\phi)) - \deg(D) \geq 0$; thus, $\phi = 0$. Thus, if $\deg(D) > 0$, then $\ell(-D) = 0$. Since $K = 0$ if $K = \mathrm{div}(\omega)$ for $0 \neq \omega \in \Omega_{C/\overline{k}}$, we have by Theorem 2.26 that $\ell(D) = \deg(D) + \ell(0 - D) = \deg(D)$. □

Because $\deg(\mathrm{div}(f)) = 0$, if $D \gg 0$, $\ell(-D) = 0$ [as $f \in L(-D)$ has to vanish over D and regular everywhere outside D]. In particular, $\ell(K - D) = 0$ if $D \gg 0$. Thus, the above theorem implies what Riemann originally proved:

Corollary 2.28 (Riemann) *Let $C = \overline{V}_{\mathfrak{a}}$ be a nonsingular projective curve defined over a field k. Then there exists a nonnegative integer $g = g(C)$ such that $\ell(D) \geq 1 - g + \deg(D)$ for all divisors D on $C(\overline{k})$, and the equality holds for sufficiently positive divisors D.*

By Example 2.24, we conclude $g(\mathbf{P}^1) = 0$ from the corollary.

Exercise 2.29 *Prove $\Omega_{\mathbf{P}^1/\overline{k}} = 0$ (without using the Riemann–Roch theorem).*

2.1.7 Regular Maps from a Curve into a Projective Space

Take a divisor D on a nonsingular projective plane curve C. Suppose $\ell(D) = n > 0$. Take a basis $(f_1, f_2, \ldots, f_n)$ of $L(D)$. Thus, we can write $f_j = \frac{h_j}{g_j}$ with homogeneous polynomials g_j, h_j having $\deg(g_j) = \deg(h_j)$. Replacing (g_j, h_j) by $(g'_0 := g_1 g_2 \cdots g_n, h'_j := h_j g^{(j)})$ for $g^{(j)} = \prod_{i \neq j} g_i$, we may assume $\deg(g'_j) = \deg(h'_j)$ for all j, and further dividing them by the GCD of $(h'_1, \ldots, h'_n, g'_0)$, we may assume that $f_j = \frac{h_j}{g_0}$ with $\deg(h_j) = \deg(g_0)$ for all j and $(g_0, h_1, \ldots, h_n)$ do not have a nontrivial common divisor.

Lemma 2.30 *Let the assumptions on $(g_0, h_1, \ldots, h_n)$ be as above. Suppose that $(g_0(P), h_1(P), \ldots, h_n(P)) \neq (0, 0, \ldots, 0)$ for all $P \in C(\overline{k})$. For $L \in C(A) \subset \mathbf{P}^n(A)$, define $\phi_A(L)$ for an A-submodule of A^{n+1} generated by $\phi(\ell) = (g_0(\ell), h_1(\ell), \ldots, h_n(\ell)) \in A^{n+1}$ for all $\ell \in L$. Then $\phi = \{\phi_A\}_A : C \to \mathbf{P}^n$ is a k-morphism of the projective plane k-curve C into $\mathbf{P}^n_{/k}$.*

The proof of the above lemma is basically the same as that of Corollary 2.19.

Exercise 2.31 *Prove the above lemma.*

2.2 Elliptic Curves

An *elliptic curve* $E_{/k}$ is a nonsingular, projective, geometrically irreducible plane curve of genus 1 with a point $\mathbf{0}_E$ specified. Here we define that the genus $g(E)$ regarding E is defined over $\overline{k}$. We will study elliptic curves in more detail in this section.

2.2.1 Abel's Theorem

When we consider $P \in E(k)$ as a divisor, we write $[P]$. So $3[P]$ is a divisor supported on P with multiplicity 3. We prove

Theorem 2.32 (Abel) *Let $E_{/k}$ be an elliptic curve with origin $\mathbf{0}_E$. The correspondence $P \mapsto [P] - [\mathbf{0}_E]$ induces a bijection $E(\overline{k}) \cong \mathcal{J}(E)$ (the Jacobian of E). In particular, $E(\overline{k})$ is an abelian group.*

Proof. Injectivity: If $[P]-[Q]=[P]-[\mathbf{0}_E]-([Q]-[\mathbf{0}_E])=\mathrm{div}(f)$ with $P\neq Q$ in $E(\overline{k})$, by Lemma 2.23, f is an isomorphism. This is wrong, as $g(\mathbf{P}^1)=0$ while $g(E)=1$. Thus, $P=Q$.

Surjectivity: Pick $D\in \mathrm{Div}^0(E)$. Then $D+[\mathbf{0}_E]$ has degree 1; so, by Corollary 2.27, $\ell(D+[\mathbf{0}_E])=1$, and we have $\phi\in L(D+[\mathbf{0}_E])$. Then $\mathrm{div}(\phi)+D+[\mathbf{0}_E]\geq 0$, and this divisor has degree 1. Any nonnegative divisor with degree 1 is a single point $[P]$. Thus, $D+[\mathbf{0}_E]$ is linearly equivalent to $[P]$; hence, the map is surjective.

Corollary 2.33 *If $0\neq\omega\in\Omega_{E/\overline{k}}$, then* $\mathrm{div}(\omega)=0$.

Proof. Since $E(\overline{k})$ is a group, for each $P\in E(\overline{k})$, $\mathcal{T}_P : Q\mapsto Q+P$ gives an automorphism of the curve E. Thus, $\omega\circ\mathcal{T}_P$ is another element in $\Omega_{E/\overline{k}}$. Since $\dim\Omega_{E/\overline{k}}=1$, we find $\omega\circ\mathcal{T}_P=\lambda(P)\omega$ for $\lambda(P)\in\overline{k}^\times$. Since $\omega\neq 0$, at some point $P\in E(\overline{k})$, $v_P(\omega)=0$. Since $v_Q(\omega\circ\mathcal{T}_P)=v_{P+Q}(\omega)$ and we can bring any point to P by translation, we have $v_P(\omega)=0$ everywhere. Thus, $\mathrm{div}(\omega)=0$.

We can easily show that $\lambda(P)=1$ for all P (see Sect. 6.1.4). Nonzero differentials ω in $\Omega_{E/k}$ are called *nowhere vanishing differentials* as $\mathrm{div}(\omega)=0$. They are unique up to constant multiple.

Exercise 2.34 *Take a line L defined by $aX+bY+cZ$ on $\mathbf{P}^2$ and suppose its intersection with an elliptic curve $E\subset\mathbf{P}^2$ to be $\{P,Q,R\}$. Prove that* $[P]+[Q]+[R]\sim 3[\mathbf{0}_E]$.

A field k is called a *perfect* field if any finite field extension of k is separable (i.e., generated by θ over k, whose minimal equation over k does not have multiple roots). Fields of characteristic 0 and finite fields are perfect.

Exercise 2.35 *Let C be an irreducible plane curve over a perfect field k. Let K be the integral closure of k in $k(C)$. Show*

1. *K/k is a finite field extension;*
2. *$K\otimes_k\overline{k}\cong\overbrace{\overline{k}\times\overline{k}\times\cdots\times\overline{k}}^{d}$ as k-algebras for $d=\dim_k K$;*
3. *C is geometrically irreducible if and only if $K=k$.*

Remark 2.36 If k is perfect, $\overline{k}/k$ is a Galois extension possibly of infinite degree; thus, by Galois theory (see Sect. 4.2.1), we have a bijection between open subgroups G of $\mathrm{Gal}(\overline{k}/k)$ and finite extensions K/k inside $\overline{k}$ by

$$G\mapsto\overline{k}^G=\{x\in\overline{k}|\sigma(x)=x \text{ for all } \sigma\in G\}$$

and $K\mapsto\mathrm{Gal}(\overline{k}/K)$. Since the isomorphism $E(\overline{k})\cong\mathcal{J}(C)$ is Galois equivariant, we have

$$E(K)\cong\mathcal{J}(E)^{\mathrm{Gal}(\overline{k}/K)}=\{D\in\mathcal{J}(E)|\sigma(D)=D \text{ for all } \sigma\in G\},$$

where $\sigma \in \mathrm{Gal}(\overline{k}/k)$ acts on $D = \sum_P m_P[P]$ by $\sigma(D) = \sum_P m_P[\sigma(P)]$. Basically, by definition, we have

$$\mathcal{J}(E)(K) := \mathcal{J}(E)^{\mathrm{Gal}(\overline{k}/K)} = \frac{\{D \in \mathrm{Pic}^0(E) | \sigma(D) = D\}}{\{\mathrm{div}(f) | f \in K(E)^\times\}}.$$

Since any subfield $K \subset \overline{k}$ is a union of finite extensions, the identity $E(K) \cong \mathcal{J}(E)(K)$ is also true for an infinite extension K/k inside $\overline{K}$. Actually, we have a good definition of $\mathrm{Pic}(E)(A)$ for any k-algebra A, and we can generalize the identity $E(K) \cong \mathcal{J}(E)(K)$ to all k-algebras A in place of fields K inside $\overline{k}$ (see Theorem 6.3).

2.2.2 Weierstrass Equations of Elliptic Curves

We now embed $E_{/k}$ into the two-dimensional projective space $\mathbf{P}^2_{/k}$ using a base of $L(3[\mathbf{0}])$ and determine the equation of the image in $\mathbf{P}^2_{/k}$. Choose a parameter $T = t_\mathbf{0}$ at the origin $\mathbf{0} = \mathbf{0}_E$. Consider $L(n[\mathbf{0}])$, which has dimension n if $n > 0$ by Corollary 2.27. We have $L([\mathbf{0}]) = k$ and $L(2[\mathbf{0}]) = k1 + kx$. Since x has to have a pole of order 2 at $\mathbf{0}$, we normalize x so that $x = T^{-2}(1 + \text{higher terms})$ in $k[[T]]$. Here x is unique up to translation: $x \mapsto x + a$ with $a \in k$. Then $L(3[\mathbf{0}]) = k1 + kx + ky$. We then normalize y so that $y = -T^{-3}(1 + \text{higher terms})$. Following the tradition, we later rewrite y for $2y$; thus, the normalization will be $y = -2T^{-3}(1 + \text{higher terms})$ at the end. This y is unique up to the transformation: $y \mapsto y + ax + b$ $(a, b \in k)$.

Proposition 2.37 *Suppose that the characteristic of the base field k is different from 2 and 3. Then, for a given pair (E, ω) of an elliptic curve E and a nowhere vanishing differential ω both defined over k, we can find a unique base $(1, x, y)$ of $L(3[\mathbf{0}])$ such that E is embedded into $\mathbf{P}^2_{/k}$ by $(1, x, y)$ whose image is defined by the affine equation*

$$y^2 = 4x^3 - g_2 x - g_3 \quad \text{with} \quad g_2, g_3 \in k, \tag{2.2.1}$$

and ω on the image is given by $\frac{dx}{y}$. Conversely, a projective algebraic curve defined by the above equation is an elliptic curve with a specific nowhere vanishing differential $\frac{dx}{y}$ if and only if the discriminant $\Delta(E, \omega) = g_2^3 - 27g_3^2$ of $4X^3 - g_2 X - g_3$ does not vanish.

The function $\Delta(E, \omega)$ is called the discriminant function and also *Ramanujan's Δ-function.* An equation of an elliptic curve E as in (2.2.1) is called a *Weierstrass equation* of E, which is determined by the pair (E, ω).

Proof. By the dimension formulas, counting the order of poles at $\mathbf{0}$ of monomials of x and y, we have

$$\begin{aligned} L(4[\mathbf{0}]) &= k + kx + ky + kx^2, \\ L(5[\mathbf{0}]) &= k + kx + ky + kx^2 + kxy \quad \text{and} \\ L(6[\mathbf{0}]) &= k + kx + ky + kx^2 + kxy + kx^3 \\ &= k + kx + ky + kx^2 + kxy + ky^2, \end{aligned}$$

from which the following relation results:

$$y^2 + a_1xy + a_3y = x^3 + a_2x^2 + a_4x + a_6 \quad \text{with } a_j \in k, \tag{2.2.2}$$

because the poles of order 6 of y^2 and x^3 have to be canceled. We homogenize (2.2.2) by putting $x = \frac{X}{Z}$ and $y = \frac{Y}{Z}$ (and multiplying by Z^3). Write C for the projective plane k-curve in $\mathbf{P}^2$ defined by the (homogenized) equation. Thus, we have a k-regular map: $\phi : E \to C \subset \mathbf{P}^2$ given by $P \mapsto (x(P) : y(P) : 1)$. Thus, the function field $k(E)$ contains the function field $k(C)$ by the pullback of ϕ. By definition, $k(C) = k(x, y)$. Since $\mathrm{div}_\infty(x) = 2[\mathbf{0}_E]$ for $x = \frac{X}{Z} : E \to \mathbf{P}^1$, this gives a covering of degree 2; hence, $[k(E) : k(x)] = 2$. Similarly, $[k(E) : k(y)] = 3$. Since $[k(E) : k(C)]$ is a common factor of the two degrees $[k(E) : k(x)] = 2$ and $[k(E) : k(y)] = 3$, we get $k(E) = k(C)$. Thus, if C is smooth, $E \cong C$ by ϕ as a smooth geometrically irreducible curve is determined by its function field. Therefore, assuming C is smooth, $E_{/k}$ can be embedded into $\mathbf{P}^2_{/k}$ via $P \mapsto (x(P), y(P))$. The image is defined by (2.2.2).

Let T be a local parameter at $\mathbf{0}_E$ normalized so that

$$\omega = (1 + \text{higher-degree terms})dT.$$

As $\omega = (a + \text{higher-degree terms})dT$ for $a \in k^\times$, and by replacing T by aT, we achieve this normalization. The parameter T normalized as above is called a *parameter adapted* to ω (and ω is said to be adapted to the parameter T). We normalize x so that $x = T^{-2} + \text{higher-degree terms}$. We now suppose that 2 is invertible in k. Then we further normalize y so that $y = -2T^{-3} +$ higher-degree terms (which we will do soon but not yet; so, for the moment, we still assume $y = T^{-3} + \text{higher-degree terms}$).

The above normalization is not affected by a variable change of the form $y \mapsto y+ax+b$ and $x \mapsto x+a'$. Now, we make a variable change $y \mapsto y+ax+b$ in order to remove the terms of xy and y (i.e., we are going to make $a_1 = a_3 = 0$):

$$\begin{aligned}(y + ax + b)^2 &+ a_1x(y + ax + b) + a_3(y + ax + b)\\ &= y^2 + (2a + a_1)xy + (2b + a_3)y + \text{polynomial in } x.\end{aligned}$$

Assuming that 2 is invertible in k, we take $a = -\frac{a_1}{2}$ and $b = -\frac{a_3}{2}$. The resulting equation is of the form $y^2 = x^3 + b_2x^2 + b_4x + b_6$. We now make the change of variable $x \mapsto x + a'$ to make $b_2 = 0$:

$$y^2 = (x + a')^3 + b_2(x + a')^2 + b_4(x + a') + b_6 = x^3 + (3a' + b_2)x^2 + \cdots .$$

Assuming that 3 is invertible in k, we take $a' = -\frac{b_2}{3}$. We can rewrite the equation as in (2.2.1) (making a variable change $-2y \mapsto y$). By the variable change as above, we have $y = -2T^{-3}(1 + \text{higher terms})$, and from this, we conclude $\omega = \frac{dx}{y}$. The numbers g_2 and g_3 are determined by T adapted to a given nowhere vanishing differential form ω.

If the discriminant $\Delta(E, \omega)$ of $g(x) = 4x^3 - g_2x - g_3$ vanishes, C has only singularity at $(x_0{:}0{:}1)$ for a multiple root x_0 of $g(x) = 0$ (see Example 2.5).

If $g(x)$ has a double zero, C is isomorphic over $\overline{k}$ to the curve defined by $y^2 = x^2(x-a)$ for $a \neq 0$. Let $t = \frac{x}{y}$. Then for $P \in E(\overline{k})$ mapping to $(0,0)$, $v_P(y) = v_P(x)$; hence, P is neither a zero nor a pole of t. The function t never vanishes outside $\mathbf{0}_E$ [having a pole at $(a,0)$]. It has a simple zero at $\mathbf{0}_E$ by the normalization of x and y. Thus, $\deg(\mathrm{div}_0(t)) = 1$, and $\overline{k}(C) = \overline{k}(t)$, which is impossible as $k(C) = k(E)$ and $g(E) = 1$. The case of triple zeros can be excluded similarly. Thus, we conclude $\Delta(E,\omega) \neq 0$ ($\Leftrightarrow C$ is smooth), and we have $E \cong C$ by ϕ.

Conversely, we have seen that any curve defined by (2.2.1) is smooth in Example 2.5 if the cubic polynomial $F(X) = 4X^3 - g_2X - g_3$ has three distinct roots in k. In other words, if the discriminant $\Delta(E,\omega)$ of $F(X)$ does not vanish, E is smooth.

For a given equation, $Y^2 = F(X)$, the algebraic curve E defined by the homogeneous equation $Y^2Z = 4X^3 - g_2XZ^2 - g_3Z^3$ in $\mathbf{P}^2_{/k}$ has a rational point $\mathbf{0} = (0,1,0) \in E(k)$, which is ∞ in $\mathbf{P}^2$ (Example 2.15). Thus, E is smooth over k if and only if $\Delta(E,\omega) \neq 0$ (an exercise following this proof).

We show that there is a canonical nowhere vanishing differential $\omega \in \Omega_{E/k}$ if E is defined by (2.2.1). If such an ω exists, all other holomorphic differentials ω' are of the form $f\omega$ with $\mathrm{div}(f) \geq 0$, which implies $f \in k$; hence, $g = \dim_k \Omega_{E/k} = 1$, and $E_{/k}$ is an elliptic curve. It is an easy exercise to show that $y^{-1}dx$ does not vanish on E (an exercise following this proof).

We summarize what we have seen. Returning to the starting elliptic curve $E_{/k}$, for the parameter T at the origin, we see by definition

$$x = T^{-2}(1 + \text{higher-degree terms}) \ \text{ and } \ y = -2T^{-3}(1 + \text{higher-degree terms}).$$

This shows

$$\frac{dx}{y} = \frac{-2T^{-3}(1+\cdots)}{-2T^{-3}(1+\cdots)}dT = (1 + \text{higher-degree terms})dT = \omega.$$

Thus, the nowhere vanishing differential form ω to which T is adapted is given by $\frac{dx}{y}$. Conversely, if $\Delta \neq 0$, the curve defined by $y^2 = 4x^3 - g_2x - g_3$ is an elliptic curve over k with origin $\mathbf{0} = \infty$ and a standard nowhere vanishing differential form $\omega = \frac{dx}{y}$. This finishes the proof.

Exercise 2.38 *1. If C is defined by $y^2 = x^3$, prove $k(C) = k(t)$ for $t = \frac{x}{y}$.*
2. Compute $v_P(dx/y)$ explicitly at each point P on $E(\overline{k})$.
3. Show that if $\Delta \neq 0$, the curve defined by $y^2 = 4x^3 - g_2x - g_3$ (over a field k of characteristic $\neq 2,3$) is also smooth at $\mathbf{0} = \infty$.

2.2.3 Moduli of Weierstrass Type

We continue to assume that the characteristic of k is different from 2 and 3. Suppose that we are given two elliptic curves $(E,\omega)_{/k}$ and $(E',\omega')_{/k}$ with nowhere vanishing differential forms ω and ω'. We call two pairs (E, ω) and (E', ω') *isomorphic* if we have an isomorphism $\varphi : E \to E'$ with $\varphi^*\omega' = \omega$ and $\varphi(\mathbf{0}_E) = \mathbf{0}_{E'}$. Here, for $\omega' = f dg$, $\varphi^*\omega' = (f \circ \varphi)d(g \circ \varphi)$; in other words, if $\sigma : k(E') \to k(E)$ is the isomorphism of the function fields associated with φ, $\varphi^*\omega' = \sigma(f)d(\sigma(g))$. Let T' be the parameter at the origin $\mathbf{0}_{E'}$ of E' adapted to ω'. If $\varphi : (E, \omega) \cong (E', \omega')$, then the parameter $T = \varphi^*T' \bmod T^2$ is adapted to ω [because $\varphi^*\omega' = \omega$ and $\varphi(\mathbf{0}_E) = \mathbf{0}_{E'}$]. We choose coordinates (x, y) for E and (x', y') for E' relative to T and T' as above. By the uniqueness of the choice of (x, y) and (x', y'), we know $\varphi^*x' = x$ and $\varphi^*y' = y$. Thus, the Weierstrass equations of (E, ω) and (E', ω') coincide. We write $g_2(E, \omega)$ and $g_3(E, \omega)$ for the g_2 and g_3 of the coefficients of the Weierstrass equation of (E, ω). If a field K has characteristic different from 2 and 3, we have

$$\begin{aligned}\wp(K) := \left[(E, \omega)_{/K}\right] &\cong \left\{(g_2, g_3) \in K^2 \middle| \Delta(E, \omega) \neq 0\right\} \\ &\cong \mathrm{Hom}_{ALG}(\mathbb{Z}[\frac{1}{6}, X, Y, \frac{1}{X^3 - 27Y^2}], K),\end{aligned}$$

where $[\cdot]$ indicates the set of isomorphism classes of the objects inside the brackets and $\mathrm{Spec}(R)(K)$ for a ring R is the set of all algebra homomorphisms: $R \to K$. The last isomorphism sends (g_2, g_3) to the algebra homomorphism ϕ with $\phi(X) = g_2$ and $\phi(Y) = g_3$. We will see later this identity is actually valid any algebra A in $ALG_{/\mathbb{Z}[\frac{1}{6}]}$ in place of a field K.

Exercise 2.39 *If k has characteristic 2, show that we cannot have any ring $\mathcal{R}$ such that*

$$\wp(K) = \left[(E, \omega)_{/K}\right] \cong \mathrm{Hom}_{ALG}(\mathcal{R}, K)$$

for all field extensions K/k. Here the isomorphism is a natural transformation between the functors $K \mapsto [(E, \omega)_{/K}]$ and $K \mapsto \mathrm{Hom}_{ALG}(\mathcal{R}, K)$ from the category of fields into $SETS$.

We now classify elliptic curves E, eliminating the contribution of the differential from the pair (E, ω). If $\varphi : E \cong E'$ for (E, ω) and (E', ω'), we have $\varphi^*\omega' = \lambda\omega$ with $\lambda \in K^\times$, because $\varphi^*\omega'$ is another nowhere vanishing differential. Therefore, we study $K^\times$-orbit: $(E, \omega) \bmod K^\times$ under the action of $\lambda \in K^\times$ given by $(E, \omega)_{/K} \mapsto (E, \lambda\omega)_{/K}$, computing the dependence of $g_j(E, \lambda\omega)$ $(j = 2, 3)$ on λ for a given pair $(E, \omega)_{/K}$. Let T be the parameter adapted to ω. Then λT is adapted to $\lambda\omega$. We see

$$\begin{aligned}x(E, \omega) &= \frac{(1 + T\phi(T))}{T^2} \Rightarrow x(E, \lambda\omega) = \frac{(1 + \text{higher terms})}{(\lambda T)^2} = \lambda^{-2}x(E, \omega), \\ y(E, \omega) &= \frac{(-2 + T\psi(T))}{T^3} \Rightarrow y(E, \lambda\omega) = \frac{(-2 + \text{higher terms})}{(\lambda T)^3} = \lambda^{-3}y(E, \omega).\end{aligned}$$

Since $y^2 = 4x^3 - g_2(E, \omega)x - g_3(E, \omega)$, we have

$$(\lambda^{-3}y)^2 = 4\lambda^{-6}x^3 - g_2(E,\omega)\lambda^{-6}x - \lambda^{-6}g_3(E,\omega)$$
$$= 4(\lambda^{-2}x)^3 - \lambda^{-4}g_2(E,\omega)(\lambda^{-2}x) - \lambda^{-6}g_3(E,\omega),$$

$$g_2(E,\lambda\omega) = \lambda^{-4}g_2(E,\omega) \quad \text{and} \quad g_3(E,\lambda\omega) = \lambda^{-6}g_3(E,\omega). \tag{2.2.3}$$

Thus, we have the following result:

Theorem 2.40 *If two elliptic curves $E_{/K}$ and $E'_{/K}$ are isomorphic, then choosing nowhere vanishing differentials $\omega_{/E}$ and $\omega'_{/E'}$, we have $g_j(E',\omega') = \lambda^{-2j}g_j(E,\omega)$ for $\lambda \in K^\times$. The constant λ is given by $\varphi^*\omega' = \lambda\omega$.*

We define the J-invariant of E by $J(E) = \frac{(12g_2(E,\omega))^3}{\Delta(E,\omega)}$. Then J only depends on E (not the chosen differential ω). If $J(E) = J(E')$, then we have

$$\frac{(12g_2(E,\omega))^3}{\Delta(E,\omega)} = \frac{(12g_2(E',\omega'))^3}{\Delta(E',\omega')} \iff g_j(E',\omega') = \lambda^{-2j}g_j(E,\omega)$$

for a 12th root λ of $\Delta(E,\omega)/\Delta(E',\omega')$. Note that the 12th root λ may not be in K if K is not algebraically closed.

Conversely, for a given $j \notin \{0,1\}$, the elliptic curve defined by

$$y^2 = 4x^3 - gx - g \quad \text{for} \quad g = \frac{27j}{j-1}$$

has J-invariant $12^3 j$. If $j = 0$ or 1, we can take the following elliptic curve with $J = 0$ or 12^3. If $J = 0$, then $y^2 = 4x^3 - 1$, and if $J = 12^3$, then $y^2 = 4x^3 - 4x$ (Gauss's lemniscate). Thus, we have the following:

Corollary 2.41 *If K is algebraically closed, then $J(E) = J(E') \Leftrightarrow E \cong E'$ for two elliptic curves over K. Moreover, for any field K, there exists an elliptic curve E with a given $J(E) \in K$.*

Exercise 2.42 *1. Prove that $g_j(E',\omega') = \lambda^{-2j}g_j(E,\omega)$ for suitable ω and ω' and a suitable 12th root λ of $\Delta(E,\omega)/\Delta(E',\omega')$ if $J(E) = J(E')$.*
2. Explain what happens if $J(E) = J(E')$ but $E \ncong E'$ over a field K not necessarily algebraically closed.

2.3 Modular Forms

We give an algebraic definition of modular forms and then relate it to classical definitions.

2.3.1 Elliptic Curves over General Rings

What we have done over fields can also be done over general noetherian rings A. We sketch the theory (see Chap. 6 for detailed proofs). Here is a definition of a plain projective curve over a ring A as a subfunctor $C \subset \mathbf{P}^2$. Recall $L \in \mathbf{P}^2(R)$ for an A-algebra R is a locally free R-submodule of R^3 of rank 1 with locally free quotient R^3/L. For a given homogeneous polynomial $\Phi(X,Y,Z) \in A[X,Y,Z]$, we define $\Phi(L) = 0$ if $\Phi(\ell) = 0$ for all $\ell \in L$. Assume that $F(X,Y,Z)$ is not a zero divisor in $A[X,Y,Z]$. Then a homogeneous polynomial $F(X,Y,Z) \in A[X,Y,Z]$ defines a subfunctor (called a plane projective A-curve) by

$$R \mapsto C(R) = \{L \in \mathbf{P}^2(R) | F(L) = 0\}.$$

Clearly, C is a covariant subfunctor of $\mathbf{P}^2$. If the residue ring $\frac{A[X,Y,Z]}{(F(X,Y,Z))}$ modulo its nilradical is an integral domain, we call C *irreducible*.

Exercise 2.43 *If A is a field k, verify that this definition is equivalent to the definition of irreducibility of the plane k-curve already given earlier.*

We define $\mathrm{Hom}_{A\text{-curves}}(C,C') := \mathrm{Hom}_{COF}(C,C')$, and in this way, we get the category of plane projective A-curves. Fix such a curve $C \subset \mathbf{P}^2_{/A}$. Suppose that A is a local ring with maximal ideal $\mathfrak{m}$. Write k for $A/\mathfrak{m}$. We then define

$$R_0 = \frac{A[Y,Z]}{(F(1,Y,Z))}, \quad R_1 = \frac{A[X,Z]}{(F(X,1,Z))}, \quad R_2 = \frac{A[X,Y]}{(F(X,Y,1))}.$$

Consider a covariant functor $C_i : R \mapsto \mathrm{Hom}_{ALG_{/A}}(R_j, R)$ from $ALG_{/A}$ to $SETS$. This functor can be identified with a subfunctor of C, for example, by

$$C_2(R) \ni \phi \mapsto L = R \cdot (\phi(X), \phi(Y), 1) \in C(R),$$

and C_2 can be identified with the functor sending R to the zero set of $F(X,Y,1)$ in R^2. If R is a local ring, we know $C(R) = C_0(R) \cup C_1(R) \cup C_2(R)$. For any finite field extension K of k, a point $P \in C_i(K)$ gives rise to an A-algebra homomorphism $\phi : R_i \to K$; hence, $\mathrm{Ker}(\phi)$ is a maximal ideal of R_i.

Exercise 2.44 *Under the above setting, prove*

1. $\mathrm{Ker}(\phi)$ *is a maximal ideal of* R_i *if* K/k *is a finite field extension,*
2. *any maximal ideal of* R_i *is given in this way as* $\mathrm{Ker}(\phi)$ *for some* ϕ.

The point $P \in C(\overline{k})$ is called a *maximal point* of C. The *local ring* at P is

$$\mathcal{O}_{C,P} = \{\frac{a}{b} | b \in R_i \setminus \mathrm{Ker}(\phi)\},$$

where $\phi : R_i \to \overline{k}$ is the k-algebra homomorphism inducing the point P. Again, $\mathcal{O}_{C,P}$ is determined independent of the choice of i with $P \in C_i(K)$. Then $\mathcal{O}_{C,P}$ is a local ring with maximal ideal $\mathfrak{m}_P$ with $\mathcal{O}_{C,P}/\mathfrak{m}_P \cong \mathrm{Im}(P) \subset K$.

The cotangent space at P is defined by P/P^2 and the tangent space at P over K is by definition its dual $\mathrm{Hom}_K(P/P^2, K)$. As before, the tangent space is isomorphic to the space of K-derivations $\partial : \mathcal{O}_{C,P} \to K$.

We sketch a general definition of smoothness, but before starting this subtle process of defining smoothness over a ring, we point out that—precise definitions aside—an important point is that we can again prove that an elliptic curve defined by $y^2 = 4x^3 - g_2x - g_3$ is smooth over $A = \mathbb{Z}[\frac{1}{6}, g_2, g_3]$ if and only if $\Delta \in A^\times$. If the reader is not very familiar with the notion of smoothness over rings, he or she can just admit this fact for a while to go through this section and the next (as we learn more about this in Chap. 4).

Here is a formal definition of smoothness. For A-algebras R and R', we define the R'-module of derivations $Der_A(R, R')$ by the R'-module of derivations trivial over A [hence, $(\partial : R \to R') \in Der_A(R, R')$ satisfies $\partial(\varphi\phi) = \varphi\partial(\phi) + \phi\partial(\varphi)$ and $\partial(a) = 0$ for all $a \in A$].

Consider $\mathfrak{m}$-adic *completions*

$$\widehat{A} = \varprojlim_n A/\mathfrak{m}^n \text{ and } \widehat{\mathcal{O}}_{C,P} = \varprojlim_n \mathcal{O}_{C,P}/\mathfrak{m}_P^n.$$

Then $\widehat{\mathcal{O}}_{C,P}$ is naturally an algebra over $\widehat{A}$. Write $\widehat{\mathfrak{m}}_P$ for the maximal ideal of $\widehat{\mathcal{O}}_{C,P}$. We call $P \in C(K)$ *smooth* over A if $\widehat{\mathcal{O}}_{C,P}$ is free of finite positive rank over $\widehat{A}[[T]]$ for a variable $T \in \widehat{\mathcal{O}}_{C,P}$ and any adically continuous derivation of $\widehat{A}[[T]]$ over $\widehat{A}$ with values in any Artinian $\widehat{\mathcal{O}}_{C,P}$-algebra extends uniquely to $\widehat{\mathcal{O}}_{C,P}$; i.e., the ring-theoretic tangent spaces of $\widehat{\mathcal{O}}_{C,P}$ and $\widehat{A}[[T]]$ are equal. This last point means that $\Omega_{\widehat{\mathcal{O}}_{C,P}/\widehat{A}[[T]]} = 0$ (see Sect. 4.1.7 and Lemma 4.32). In short, $\widehat{\mathcal{O}}_{C,P}$ is an étale algebra over $\widehat{A}[[T]]$; i.e., $\mathrm{Spec}(\widehat{\mathcal{O}}_{C,P})$ and $\mathrm{Spec}(\widehat{A}[[T]])$ are locally isomorphic in the sense of algebraic geometry; see Sect. 4.1.8 for a more thorough discussion of étale/smooth morphisms).

If C is smooth over A at all maximal points $P \in C(K)$, we call C *smooth* over A. Assuming that k is algebraically closed, C is smooth over A if and only if $\widehat{\mathcal{O}}_{C,P} \cong \widehat{A}[[T]]$ as $\widehat{A}$-algebras for all maximal points $P \in C$.

For general A not necessarily local, we call C *smooth* over A if C is smooth over the localization of A at every maximal ideal of A.

Exercise 2.45 *Prove that if C is a smooth plane projective curve over an integral local domain A with algebraically closed residue field, C is smooth over the quotient field of A.*

We recall the R_i-module of derivations $Der_{C_i/A} = Der_A(R_i, R_i)$, which is the R_i-module of $\partial : R_i \to R_i$ such that $\partial(\varphi\phi) = \varphi\partial(\phi) + \phi\partial(\varphi)$ and $\partial(a) = 0$ for all $a \in A$. The R_i-dual $\Omega_{C_i/A}$ of $Der_{C_i/A}$ is called the R_i-module of 1-differentials over C_i. Each $\partial \in Der_{C_i/A}$ gives rise to an A_P-derivation $\partial_P : \mathcal{O}_{C,P} \to \mathcal{O}_{C,P}$ given by $\partial_P(\frac{a}{b}) = \frac{\partial(a)b - a\partial(b)}{b^2}$ for a maximal point $P \in C_i$, where A_P is the localization of A at $P \cap A$ (regarding P as a prime ideal of R_i). By duality, $\omega \in \Omega_{C_i/A}$ therefore gives rise to the cotangent vector

$\omega_P \in \Omega_{\mathcal{O}_{C,P}/A_P} := \mathrm{Hom}_{A_P}(Der_{\mathcal{O}_{C,P}/A_P}, \mathcal{O}_{C,P})$. If C_i is smooth over A, then the R_i-module $\Omega_{C_i/A}$ is a locally free R_i-module of rank 1. We define the *relative cotangent module* $\Omega_{C/A}$ to be the collection of all $\omega = (\omega_i \in \Omega_{C_i/A})_i$ such that $\omega_{i,P} = \omega_{j,P}$ for all $P \in (C_i \cap C_j)(\overline{k})$ $((i,j) = (0,1), (1,2), (0,2))$. If C is smooth over A, $\Omega_{C/A}$ is a locally free A-module of some rank g, and this number g is called the *genus* $g(C)$ of a curve C over A.

An elliptic curve over A is a plane projective smooth curve E of genus 1 with a specific point $\mathbf{0}_E \in E(A)$. If $\Omega_{E/A} = A\omega$, the differential ω is called a *nowhere vanishing differential.* If $\phi : E \to E'$ is a morphism of elliptic curve and ∂ is a derivation in $Der_{E/A}$, we define $\phi_*\partial \in Der_{E'/A}$ by $\phi_*\partial_{\phi(P)}(f) = \partial_P(f \circ \phi)$ for all P. By duality, we can pull back a nowhere vanishing differential ω' on E' by ϕ, which is written as $\phi^*\omega'$. Note here that $\phi^*\omega'$ may not be nowhere vanishing (although it is if ϕ is an isomorphism).

Exercise 2.46 *Let $A = \mathbb{F}_p$. Give an example of a nonconstant morphism $\phi : E \to E$ such that $\phi^*\omega = 0$ for a nowhere vanishing differential ω on E.*

If $A \xrightarrow{\sigma} A'$ is an algebra homomorphism and if a plane projective A-curve C is defined by an equation $F(X,Y,Z) = \sum_{i,j,l} c_{i,j,l} X^i Y^j Z^l$, the σ-transform $\sigma(F)(X,Y,Z) = \sum_{i,j,l} \sigma(c_{i,j,l}) X^i Y^j Z^l$ defines a plane projective A'-curve $\sigma(C)$. Note that the affine ring of $\sigma(C_i)$ is the ring $R_i \otimes_{A,\sigma} A'$; hence, often we write $C \otimes_A A'$ for $\sigma(C)$ and call it the base-change $C \otimes_A A'_{/A'}$ of $C_{/A}$. Similarly, if $\partial : R_i \to R_i$ is an A-derivation, $\partial \otimes 1 : R_i \otimes_A A' \to R_i \otimes_A A'$ given by $\partial \otimes 1(\phi \otimes a) = \sigma(a\partial(\phi))$ is an A'-derivation. This shows $Der_{C_i/A} \otimes_A A' = Der_{C_i \otimes A'/A'}$. By duality, we also have $\Omega_{C_i/A} \otimes_A A' = \Omega_{C_i \otimes A'/A'}$. In particular, $\omega \in \Omega_{C/A}$ induces $\sigma_*(\omega) = \omega \otimes 1 \in \Omega_{C \otimes A'/A'}$. We write the pair $(E \otimes_A A', \sigma_*\omega)$ as $(E,\omega) \otimes_A A'$. This makes $\wp : ALG \to SETS$ given by $\wp(A) = \left[(E,\omega)_{/A}\right]$ a covariant functor from ALG into $SETS$. We again have the following result basically in the same way as in the case of fields (see Sect. 6.2.1 for a proof):

Theorem 2.47 *Let $\mathcal{R} = \mathbb{Z}[\frac{1}{6}, g_2, g_3, \frac{1}{\Delta}]$. Then we have a canonical equivalence of functors from $ALG_{/\mathbb{Z}[\frac{1}{6}]}$ to SETS:*

$$\wp(?) \cong \mathrm{Hom}_{ALG_{/\mathbb{Z}[\frac{1}{6}]}}(\mathcal{R}, ?).$$

In other words, for a given pair $(E,\omega)_{/A}$ of an elliptic curve E over A and a nowhere vanishing differential ω, there exists unique $(g_2(E,\omega), g_3(E,\omega)) \in A^2$ such that E is canonically isomorphic to an elliptic curve defined by

$$Y^2Z = 4X^3 - g_2(E,\omega)XZ^2 - g_3(E,\omega)Z^3$$

and ω induces the differential $\frac{dX}{Y}$ on $E_2 = E \cap D_2$ under this isomorphism. Thus, we have the following:

1. If (E,ω) is defined over a $\mathbb{Z}[\frac{1}{6}]$-algebra A, we have $g_j(E,\omega) \in A$, which depends only on the isomorphism class of (E,ω) over A;
2. $g_j((E,\omega) \otimes_A A') = \sigma(g_j(E,\omega))$ for each $\mathbb{Z}[\frac{1}{6}]$-algebra homomorphism $\sigma : A \to A'$;
3. $g_j(E, \lambda\omega) = \lambda^{-2j} g_j(E,\omega)$ for all $\lambda \in A^\times$.

2.3.2 Geometric Modular Forms

Let A be an algebra over $\mathbb{Z}[\frac{1}{6}]$. We restrict the functor $\wp$ to $ALG_{/A}$ and write the restriction $\wp_{/A}$. Then, by Theorem 2.47, for $\mathcal{R}_A := A[g_2, g_3, \frac{1}{\Delta}]$,

$$\wp_{/A}(?) = \mathrm{Hom}_{ALG_{/A}}(\mathcal{R}_A, ?).$$

A morphism of functors $\phi : \wp_{/A} \to \mathbf{A}^1_{/A}$ is, by definition, given by (a collection of) maps $\phi_R : \wp_{/A}(R) \to \mathbf{A}^1(R) = R$ indexed by $R \in ALG_{/A}$ such that for any $\sigma : R \to R'$ in $\mathrm{Hom}_{ALG_{/A}}(R, R')$, $\phi_{R'}((E,\omega) \otimes_R R') = \sigma(f((E,\omega)_{/R}))$. Note that $\mathbf{A}^1_{/A}(?) = \mathrm{Hom}_{ALG_{/A}}(A[X], ?)$ by $R \ni a \leftrightarrow (\varphi : A[X] \to R) \in \mathrm{Hom}_{ALG_{/A}}(A[X], ?)$ with $\varphi(X) = a$. In particular,

$$\phi_{\mathcal{R}_A} : \wp(\mathcal{R}_A) = \mathrm{Hom}_{ALG_{/A}}(\mathcal{R}_A, \mathcal{R}_A) \to \mathbf{A}^1(A[X], \mathcal{R}_A) = \mathcal{R}_A.$$

Thus, $\phi_{\mathcal{R}_A}(\mathrm{id}_{\mathcal{R}_A}) \in \mathcal{R}_A$; hence, write $\phi_{\mathcal{R}_A}(\mathrm{id}_{\mathcal{R}_A}) = \Phi(g_2, g_3)$ for a two-variable rational function $\Phi(x,y) \in A[x, y, \frac{1}{x^3-27y^2}]$. Let $\mathbf{E}_{/\mathcal{R}_A}$ be the universal elliptic curve over $\mathcal{R}_A$ defined by $Y^2Z = 4X^3 - g_2XZ^2 - g_3Z^3$ with the universal differential $\boldsymbol{\omega} = \frac{dX}{Y}$. If we have $(E,\omega)_{/R}$, we have a unique A-algebra homomorphism $\sigma : \mathcal{R}_A \to R$ given by $\sigma(g_j) = g_j(E,\omega)$; in other words, $(E,\omega)_{/R} \cong (\mathbf{E},\omega)_{\mathcal{R}_A} \otimes_{\mathcal{R}_A} R$, and

$$\begin{aligned}\phi_R(E,\omega) &= \phi_R((\mathbf{E},\omega) \otimes_{\mathcal{R}_A} R) = \sigma(\phi_{\mathcal{R}_A}(\mathbf{E},\omega)) \\ &= \sigma(\phi_{\mathcal{R}_A}(\mathrm{id}_{\mathcal{R}_A})) = \Phi(\sigma(g_2), \sigma(g_3)) = \Phi(g_2(E,\omega), g_3(E,\omega)).\end{aligned}$$

Theorem 2.48 *Any functor morphism $\phi : \wp_{/A} \to \mathbf{A}^1_{/A}$ is given by a rational function $\Phi \in \mathcal{R}_A$ of g_2 and g_3 so that $\phi(E,\omega) = \Phi(g_2(E,\omega), g_2(E,\omega))$ for every elliptic curve (E,ω) over an A-algebra.*

Define a weight function $w : A[g_2, g_3] \to \mathbb{Z}$ by $w(g_2^a g_3^b) = 4a + 6b$, and for general polynomials $\Phi = \sum_{a,b} c_{a,b} g_2^a g_3^b$, we put $w(\Phi) = \max(w(g_2^a g_3^b) | c_{a,b} \neq 0)$. A polynomial $\Phi = \sum_{a,b \geq 0} c_{a,b} g_2^a g_3^b$ of g_2 and g_3 is called *isobaric* if $c_{a,b} \neq 0 \Rightarrow 4a + 6b = w$.

A weight w *modular form* defined over A is a morphism of functors $\wp_{/A} \to \mathbf{A}^1_{/A}$ given by an isobaric polynomial of g_2 and g_3 of weight w with coefficients in A. Write $G_w(A) = G_w(\Gamma_0(1); A)$ for the A-module of modular forms of weight w. Then $f \in G_w(A)$ is a functorial rule assigning each isomorphism class of $(E,\omega)_{/R}$ for an A-algebra R an element $f(E,\omega) \in R$ satisfying the following properties:

(G0) $f \in A[g_2, g_3]$;
(G1) if (E,ω) is defined over an A-algebra R, we have $f(E,\omega) \in R$, which depends only on the isomorphism class of (E,ω) over R;
(G2) $f((E,\omega) \otimes_R R') = \sigma(f(E,\omega))$ for each A-algebra homomorphism $\sigma : R \to R'$;

(G3) $f((E, \lambda\omega)_{/R}) = \lambda^{-w} f(E, \omega)$ for any $\lambda \in R^\times$.

Exercise 2.49 *For a field K with $\frac{1}{6} \in K$, prove, for $0 < w \in 2\mathbb{Z}$,*

$$\dim_K G_w(K) = \begin{cases} \left[\frac{w}{12}\right] & \text{if } w \equiv 2 \mod 12, \\ \left[\frac{w}{12}\right] + 1 & \text{otherwise.} \end{cases}$$

2.3.3 Archimedean Uniformization

In the following three sections, we would like to give a sketch of the classical theory of elliptic curves defined over the complex field $\mathbb{C}$ created by Weierstrass. By means of Weierstrass $\wp$-functions, we can identify $E(\mathbb{C})$ (for each elliptic curve $E_{/\mathbb{C}}$) with a quotient of $\mathbb{C}$ by a lattice L. In this way, we can identify $[(E, \omega)_{/\mathbb{C}}]$ with the space of lattices in $\mathbb{C}$. This method is analytic.

We can deduce from the analytic parameterization (combined with geometric technique of Weil–Shimura) many results on the moduli space of elliptic curves, such as the exact field of definition of the moduli, determination of the field of moduli (of each member), and so on (e.g., [IAT] Chap. 6). We have come here in a reverse way: Starting algebraically, mainly by the Riemann–Roch theorem, we have determined a unique Weierstrass equation over A for a given pair $(E, \omega)_{/A}$, and therefore, we know the exact shape of the moduli space before setting out to study the analytic method. After studying analytic theory over $\mathbb{C}$, combining these techniques, we start studying modular forms.

We start with classical homotopy theory for complex manifolds. A *path* γ on a complex manifold M is a piecewise smooth continuous map γ from the closed interval $[0, 1]$ into a complex manifold M. We write the path as $(\gamma : x \to y)$ with $\gamma(0) = x$ and $\gamma(y) = 1$. Morally, we start the point x at time 0 and "walking" to reach y by the unit time 1. Two paths $(\alpha : x \to y)$ and $(\beta : x \to y)$ are homotopy equivalent (for which we write $\alpha \approx \beta$) if we have a bicontinuous map $\phi : [0, 1] \times [0, 1] \to M$ such that $\alpha(t) = \phi(0, t)$ and $\beta(t) = \phi(1, t)$ (i.e., we can fill the area encircled by α and β in M, intuitively). Consider the space $\mathcal{Z} = \mathcal{Z}(M)$ of homotopy classes of paths starting from a fixed point $x \in M$. An open neighborhood U of x is called *simply connected* if we have a bijection: $\mathcal{Z}(U) \cong U$ by projecting $(\gamma : x \to y)$ down to y [i.e., no holes in $\mathcal{Z}(U)$ so no path circling the hole]. Shrinking the neighborhood of x (to avoid holes), we can always find a simply connected open neighborhood of $x \in M$. For example, if U is diffeomorphic to an open disk with center x, it is simply connected (that is, every loop is equivalent to x). If $\gamma : x \to y$ and $\gamma' : y \to z$ are two paths, we define their composed path $\gamma\gamma' : x \to z$ by

$$\gamma\gamma'(t) = \begin{cases} \gamma(2t) & \text{if } 0 \le t \le 1/2 \\ \gamma'(2t-1) & \text{if } 1/2 \le t \le 1. \end{cases}$$

By this composition, $\pi_M = \pi_1^{\mathrm{top}}(M, x) = \{\gamma \in \mathcal{Z}(M) | \gamma : x \to x\}$ becomes a group called *the topological fundamental group* of M. Taking a system of neighborhoods $\mathcal{U}_y$ of $y \in M$ made of simply connected open neighborhoods of y, we equip a topology with $\mathcal{Z}(M)$ so that a fundamental system of neighborhoods of $\gamma : x \to y$ is given by $\{\gamma U | U \in \mathcal{U}_x\}$. Then $\mathcal{Z}(M)$ becomes a complex manifold. Via composition, π_M acts on $\mathcal{Z}(M)$ freely without fixed points. We have a continuous map $\pi : \pi_M \backslash \mathcal{Z}(M) \to M$ given by $\pi(\gamma : x \to y) = y$, which is a local isomorphism, and $\pi : \pi_M \backslash \mathcal{Z}(M) \cong M$ is a homeomorphism. This space $\mathcal{Z}(M)$ is called a *universal covering* space of M.

We apply this general construction to an elliptic curve $(E, \omega)_{/\mathbb{C}}$ defined over $\mathbb{C}$ in the following way. We take as M the following zero set of the equation of E:

$$E(\mathbb{C}) = E(g_2, g_3)(\mathbb{C}) = \{(x{:}y{:}z) \in \mathbf{P}^2(\mathbb{C}) | y^2 z - 4x^3 + g_2 z^2 x + g_3 z^3 = 0\},$$

which is a compact Riemann surface of genus 1. Let $\mathcal{Z} = \mathcal{Z}(E(\mathbb{C}))$ be the set of all equivalence classes of paths emanating from $\mathbf{0}$. Write $\Pi = \pi_1^{\mathrm{top}}(E, \mathbf{0})$. Since $E(\mathbb{C})$ is a commutative group, writing its group multiplication additively, we define the sum $\gamma + \gamma'$ on $\mathcal{Z}$ by, noting that γ and γ' originate at the origin $\mathbf{0}$:

$$(\gamma + \gamma')(t) = \begin{cases} \gamma(2t) & \text{if } 0 \le t \le 1/2 \\ \gamma(1) + \gamma'(2t-1) & \text{if } 1/2 \le t \le 1. \end{cases}$$

Then $(\gamma + \gamma')(1) = \gamma(1) + \gamma'(1)$, and we claim that $\gamma + \gamma' \approx \gamma' + \gamma$. In fact, on the square $[0,1] \times [0,1]$, we consider the path α on the boundary connecting the origin $(0,0)$ and $(1,1)$ passing $(0,1)$, and write β the opposite path from $(0,0)$ to $(1,1)$ passing $(1,0)$. They are obviously homotopy equivalent. Thus, we have a continuous map $\phi : [0,1] \times [0,1] \to [0,1] \times [0,1]$ such that $\phi(0,t) = \alpha(t)$ and $\phi(1,t) = \beta(t)$. Define

$$f : [0,1] \times [0,1] \to E(\mathbb{C}) \text{ by } f(t,t') = \gamma(t) + \gamma'(t').$$

Then it is easy to see that $f \circ \phi(0,t) = (\gamma' + \gamma)(t)$ and $f \circ \phi(1,t) = (\gamma + \gamma')(t)$.

Via the addition induced from the group structure of $E(\mathbb{C})$ as above, $\mathcal{Z}$ becomes an additive complex Lie group. Since $\gamma + \gamma' = \gamma\gamma'$ if $\gamma \in \Pi$ and $\gamma' \in \mathcal{Z}$ by definition, Π is an additive subgroup of $\mathcal{Z}$ and $\Pi \backslash \mathcal{Z} \cong E(\mathbb{C})$. Here we may regard the left-hand side as the quotient group of $\mathcal{Z}$ by the subgroup Π.

We claim to have an isomorphism $\mathcal{Z} \cong \mathbb{C}$ as additive Lie groups. Choose a nowhere vanishing differential form ω on E, and define a map $I : \mathcal{Z} \to \mathbb{C}$ by $\gamma \mapsto \int_\gamma \omega \in \mathbb{C}$. Since ω is holomorphic on $\mathcal{Z}$, the value of I is independent of the representative γ in the homotopy class $[\gamma]$ by Cauchy's integration theorem. Since ω is translation invariant on $E(\mathbb{C})$, it is translation invariant on $\mathcal{Z}$ and $I(\gamma + \gamma') = I(\gamma) + I(\gamma')$. In particular, I is a local homeomorphism because $E(\mathbb{C})$ is one-dimensional and for simply connected U, $\mathcal{Z}(U) \cong I(U)$. Thus, the pair $(E(\mathbb{C}), \omega)$ is isomorphic locally to the pair of the additive group $\mathbb{C}$ and du for the coordinate u on $\mathbb{C}$, because du is the unique translation-invariant differential (up to constant multiple). Since $I^{-1}([\mathbf{0}]) = \{\mathbf{0}\}$, I is a linear isomorphism into $\mathbb{C}$. For an open neighborhood U of $\mathbf{0}$ with $U \cong \mathcal{Z}(U) \ni \gamma \mapsto I(\gamma) = \int_\gamma \omega \in \mathbb{C}$ giving an isomorphism onto a small open disk D in $\mathbb{C}$ centered at 0, we have two $\gamma_1, \gamma_2 \in U$ giving rise to two linearly independent $I(\gamma_j)$ $(j = 1, 2)$. Then $I(m\gamma_1 + n\gamma_2) = mI(\gamma_1) + nI(\gamma_2)$ for all $m, n \in \mathbb{Z}$. Replacing γ_j by $\frac{1}{a}\gamma_j \in \mathcal{Z}(U)$ such that $I(\frac{1}{a}\gamma_j) = \frac{I(\gamma_j)}{a}$ for any positive integer a, by the same argument, we find $I(m\gamma_1 + n\gamma_2) = mI(\gamma_1) + nI(\gamma_2)$ for all $m, n \in \mathbb{Q}$; hence, I is a surjective isomorphism.

In the same way, if $\alpha : E \to E$ is an endomorphism of E with $\alpha(\mathbf{0}_E) = \mathbf{0}_E$, α lifts an endomorphism of $\mathcal{Z}$, sending a path γ from $\mathbf{0}_E$ to $z \in \mathbb{C}$ to a path $\alpha(\gamma)$ from $\alpha(\mathbf{0}_E) = \mathbf{0}_E$ to $\alpha(z)$. In particular, $\alpha(\gamma + \gamma') = \alpha(\gamma) + \alpha(\gamma')$. Thus, α gives rise to a linear map from $\mathbb{C} = \mathcal{Z}$ to $\mathbb{C}$. Since α is holomorphic (as it is a polynomial map of the coordinates of $\mathbf{P}^2_{/\mathbb{C}}$), α is a $\mathbb{C}$-linear map. We thus get a natural inclusion:

$$\mathrm{End}(E_{/\mathbb{C}}) \hookrightarrow \mathbb{C}. \tag{2.3.1}$$

Writing $L = L_E$ for $I(\Pi)$, we can find a base w_1, w_2 of L over $\mathbb{Z}$. Thus, we have a map

$$\wp(\mathbb{C}) \ni (E, \omega) \longmapsto L_E \in \{L | L : \text{lattice in } \mathbb{C}\} =: Lat,$$

and we have $(E(\mathbb{C}), \omega) \cong (\mathbb{C}/L_E, du)$. Therefore, the map $\wp(\mathbb{C}) \to Lat$ is injective. We show its surjectivity in the next subsection.

By the above fact combined with (2.3.1), we get the following:

Proposition 2.50 *We have a ring embedding*

$$\mathrm{End}(E_{/\mathbb{C}}) \hookrightarrow \{u \in \mathbb{C} | u \cdot L_E \subset L_E\},$$

and hence $\mathrm{End}(E_{/\mathbb{C}})$ *is either* $\mathbb{Z}$ *or an order of an imaginary quadratic field.*

Proof. The first assertion follows from (2.3.1). Pick $\alpha \in \mathrm{End}(E_{/\mathbb{C}})$ corresponding to $u \in \mathbb{C}$ as above. Note that $L_E = \mathbb{Z}w_1 + \mathbb{Z}w_2$. Then $uw_1 = aw_1 + bw_2$ and $uw_2 = cw_1 + dw_2$ for integers a, b, c, d. In short, writing $w = \left(\begin{smallmatrix} w_1 \\ w_2 \end{smallmatrix}\right)$ and $\rho(\alpha) = \left(\begin{smallmatrix} a & b \\ c & d \end{smallmatrix}\right)$, we get $uw = \rho(\alpha)w$; hence, $\rho : \mathrm{End}(E_{/\mathbb{C}}) \to M_2(\mathbb{Z})$ is a ring homomorphism. By the first assertion, the image has to be an order of imaginary quadratic field or just $\mathbb{Z}$.

When $\mathrm{End}(E_{/\mathbb{C}}) \neq \mathbb{Z}$, E is said to have *complex multiplication*. We also simply call E a *CM elliptic curve* if it has complex multiplication.

2.3.4 Weierstrass $\wp$-Function

For a given $L \in Lat$, Weierstrass defined his $\wp$-functions by

$$x_L(u) = \wp(u) = \frac{1}{u^2} + \sum_{\ell \in L - \{0\}} \left\{ \frac{1}{(u - \ell)^2} - \frac{1}{\ell^2} \right\} = \frac{1}{u^2} + \frac{g_2}{20} u^2 + \frac{g_3}{28} u^4 + \cdots$$

$$y_L(u) = \wp'(u) = \frac{\partial \wp}{\partial u}(u) = -\frac{2}{u^3} - 2 \sum_{\ell \in L - \{0\}} \frac{1}{(u - \ell)^3} = -2u^{-3} + \cdots,$$

where

$$g_2 = g_2(L) = 60 \sum_{\ell \in L-\{0\}} \frac{1}{\ell^4} \quad \text{and} \quad g_3 = g_3(L) = 140 \sum_{\ell \in L-\{0\}} \frac{1}{\ell^6}.$$

Equating poles, $\varphi = y_L^2 - 4x_L^3 + g_2 x_L + g_3$ is holomorphic everywhere. Since these functions factor through the compact space $\mathbb{C}/L$, φ is bounded and hence must be a constant. Since the constant terms of x_L and y_L both vanish, we conclude $\varphi = 0$. We obtain a holomorphic map $(x_L, y_L) : \mathbb{C}/L - \{0\} \to \mathbf{A}^2_{/\mathbb{C}}$. Counting the order of poles at $\mathbf{0}$, we check that the map (x_L, y_L) has degree 1, that is, an isomorphism onto its image and extends to

$$\Phi = (x_L{:}y_L{:}1) = (u^3 x_L{:}u^3 y_L{:}u^3) : \mathbb{C}/L \to \mathbf{P}^2_{/\mathbb{C}}.$$

Thus, we rediscover the elliptic curve $E_L = E(L) = \Phi(\mathbb{C}/L) = E(g_2(L), g_3(L))$ and the induced differential

$$\omega_L = \frac{dx_L}{y_L} = du.$$

This shows

Theorem 2.51 (Weierstrass) *We have* $\wp(\mathbb{C}) = [(E, \omega)_{/\mathbb{C}}] \cong Lat$.

We now relate the space *Lat* with the upper half-complex plane $\mathfrak{H}$. Two complex numbers $w_1, w_2 \in (\mathbb{C}^\times)^2$ span a lattice if and only if $\mathrm{Im}(w_1/w_2) \neq 0$. Recall that $\mathfrak{H} = \{z \in \mathbb{C} | \mathrm{Im}(z) > 0\}$. By changing the order of w_1 and w_2 without affecting their lattice, we may assume that $\mathrm{Im}(w_1/w_2) > 0$. Thus, we have a natural isomorphism of complex manifolds:

$$\mathcal{B} = \left\{ v = \left(\begin{smallmatrix} w_1 \\ w_2 \end{smallmatrix}\right) \in (\mathbb{C}^\times)^2 \middle| \mathrm{Im}(w_1/w_2) > 0 \right\} \cong \mathbb{C}^\times \times \mathfrak{H}$$

via $\left(\begin{smallmatrix} w_1 \\ w_2 \end{smallmatrix}\right) \mapsto (w_2, w_1/w_2)$.

Two vectors v and v' span the same lattice L if and only if $v' = \alpha v$ for $\alpha \in SL_2(\mathbb{Z})$; hence,

$$Lat \cong SL_2(\mathbb{Z}) \backslash \mathcal{B}.$$

This action of $\alpha = \left(\begin{smallmatrix} a & b \\ c & d \end{smallmatrix}\right) \in SL_2(\mathbb{Z})$ on $\mathcal{B}$ can be interpreted on $\mathbb{C}^\times \times \mathfrak{H}$ as follows:

$$\alpha(u, z) = (cu + d, \alpha(z)) \quad \text{for} \quad \alpha(z) = \frac{az + b}{cz + d}.$$

Exercise 2.52 *Check that the above action is well defined; that is, show that* $(\alpha\beta)(u, z) = \alpha(\beta(u, z))$ *for* $\alpha, \beta \in SL_2(\mathbb{Z})$.

2.3.5 Holomorphic Modular Forms

Since a geometric modular form $f \in G_w(\mathbb{C})$ is a function of $(E,\omega)_{/\mathbb{C}}$, it induces a function of $v \in \mathcal{B}$. Writing $L(v) = L(w_1, w_2)$ for the lattice spanned by $v \in \mathcal{B}$, therefore, we regard f as a holomorphic function on $\mathcal{B}$ by $f(v) = f(E_{L(v)}, \omega_{L(v)})$. Then conditions (G0–3) can be stated as

(G0) $\quad f \in \mathbb{C}[g_2(v), g_3(v)];$
(G1) $\quad f(\alpha v) = f(v)$ for all $\alpha \in SL_2(\mathbb{Z})$;
(G2) $\quad f \in \mathbb{C}[g_2(v), g_3(v), \Delta(v)^{-1}];$
(G3) $\quad f(\lambda v) = \lambda^{-w} f(v)$ $(\lambda \in \mathbb{C}^\times)$.

We may also think of $f \in G_w(\mathbb{C})$ as a function on $\mathfrak{H}$ by $f(z) = f(v(z))$ for $v(z) = 2\pi i \left(\begin{smallmatrix} z \\ 1 \end{smallmatrix}\right)$ $(z \in \mathfrak{H})$. Recall that multiplying $\left(\begin{smallmatrix} z \\ 1 \end{smallmatrix}\right)$ by $2\pi i$ is to adjust the rationality coming from q-expansion to the rationality coming from the universal ring $\mathbb{Z}[\frac{1}{6}][g_2, g_3]$. Indeed, we will see below that $(2\pi i^{-2j} g_j)(\left(\begin{smallmatrix} z \\ 1 \end{smallmatrix}\right))$ has Fourier expansion in $\mathbb{Q}[[q]]$ for $q = \exp(2\pi i z)$. Then we have the following interpretation:

(G0) $\quad f \in \mathbb{C}[g_2(z), g_3(z)];$
(G1,3) $\quad f(\alpha(z)) = f(z)(cz+d)^w$ for all $\alpha = \left(\begin{smallmatrix} a & b \\ c & d \end{smallmatrix}\right) \in SL_2(\mathbb{Z})$;
(G2) $\quad f \in \mathbb{C}[g_2(z), g_3(z), \Delta(z)^{-1}].$

Since $\left(\begin{smallmatrix} 1 & 1 \\ 0 & 1 \end{smallmatrix}\right)(z) = z+1$, any $f \in \mathbb{C}[g_2(z), g_3(z), \Delta^{-1}(z)]$ satisfies $f(z+1) = f(z)$. Defining $\mathbf{e}(z) = \exp(2\pi i z)$ for $i = \sqrt{-1}$, the function $\mathbf{e} : \mathbb{C} \to \mathbb{C}^\times$ induces an analytic isomorphism: $\mathbb{C}/\mathbb{Z} \cong \mathbb{C}^\times$. Let $q = \mathbf{e}(z)$ be the variable on $\mathbb{C}^\times$. Since f is translation invariant, f can be considered as a function of q. Thus, it has a Laurent expansion $f(q) = \sum_{n \gg -\infty} a(n,f) q^n$. We have the following examples (see (1.3.1), the following section, and [LFE] Chap. 5):

$$\begin{aligned} 12g_2 &= 1 + 240 \sum_{n=1}^{\infty} \left\{ \sum_{0<d|n} d^3 \right\} q^n \in \mathbb{Z}[[q]]^\times, \\ -6^3 g_3 &= 1 - 504 \sum_{n=1}^{\infty} \left\{ \sum_{0<d|n} d^5 \right\} q^n \in \mathbb{Z}[[q]]^\times, \\ \Delta &= q \prod_{n=1}^{\infty} (1-q^n)^{24} \in q(\mathbb{Z}[[q]]^\times). \end{aligned} \tag{2.3.2}$$

The last expansion of Ramanujan's function tells us that Δ does not vanish on $\mathfrak{H}$ analytically although this is geometrically clear, as $\Delta(E,\omega) \neq 0$ for any elliptic curve (see [EEK] for a concise proof of the product expansion). In particular, we have

$$J = \frac{(12g_2)^3}{\Delta} = q^{-1} + \cdots \in q^{-1}(1 + \mathbb{Z}[[q]]).$$

By this adjustment (multiplying $2\pi i$), we regard g_2 and g_3 as elements of $\mathbb{Z}[\frac{1}{6}][[q]]$, and we consider the Weierstrass equation as a power series identity with coefficients in $\mathbb{Z}[\frac{1}{6}]$; namely, consider a projective plane curve $E_{\infty/\mathbb{Z}[[q]]}$ (called the Tate curve) defined over the power series ring $\mathbb{Z}[[q]]$ by the equation

$$Y^2Z = 4X^3 - g_2(q)XZ^2 - g_3(q)Z^3$$

and define $\omega_\infty = \frac{dX}{Y}$. Since Δ is a unit in $\mathbb{Z}[\frac{1}{6}]((q)) := \mathbb{Z}[\frac{1}{6}][[q]][\frac{1}{q}]$, we see that $(E_\infty, \omega_\infty)$ gives an elliptic curve over the Laurent series ring $\mathbb{Z}[1/6]((q))$ with nowhere vanishing differential ω_∞. For any $f \in G_w(A)$,

$$f(q) = f((E_\infty, \omega_\infty) \otimes_{\mathbb{Z}[\frac{1}{6}]((q))} A((q))) \in A[[q]]$$

is called the *q-expansion* of f. If $f \in G_w(\mathbb{C})$, the q-expansion $f(q)$ coincides with the analytic Fourier expansion via $q = \mathbf{e}(z)$, because by Theorem 2.48, f is an isobaric polynomial $\Phi(g_2, g_3)$ in g_2 and g_3 and by definition $g_2(q)$ and $g_3(q)$ are their analytic expansions.

Write $\mathbf{P}^1(J)_{/\mathbb{Z}[\frac{1}{6}]}$ for the projective line over $\mathbb{Z}[\frac{1}{6}]$ whose coordinate is given by J [in other words, $\mathbf{P}^1(J) = D_0 \cup D_1$ over local rings with $D_1 = \mathbf{A}^1$ defined by the affine ring $\mathbb{Z}[\frac{1}{6}][J]]$. Since the coordinate at ∞ of $\mathbf{P}^1(J)$ can be given by J^{-1} ($J^{-1} \in q(1 + q\mathbb{Z}[[q]])$), we know that $\mathbb{Z}[[q]] = \mathbb{Z}[[J^{-1}]]$ and

$$\widehat{\mathcal{O}}_{\mathbf{P}^1(J),\infty} \cong \mathbb{Z}[1/6][[q]] \text{ via } q\text{-expansion}, \tag{2.3.3}$$

where $\widehat{\mathcal{O}}_{\mathbf{P}^1(J),\infty}$ is the (q)-adic completion of the local ring $\mathcal{O}_{\mathbf{P}^1(J),\infty}$ at ∞.

We note that

$$Lat/\mathbb{C}^\times = SL_2(\mathbb{Z})\backslash(\mathfrak{H} \times \mathbb{C}^\times)/\mathbb{C}^\times \cong SL_2(\mathbb{Z})\backslash\mathfrak{H},$$

which is isomorphic to $\mathbf{P}^1(J) - \{\infty\}$ by J. Thus, we see that (G0) over $\mathbb{C}$ is equivalent to

(G0′) f is a holomorphic function on $\mathfrak{H}$ satisfying the automorphic property (G1,3), and its analytic q-expansion $f(q)$ is contained in $\mathbb{C}[[q]]$.

More generally, for modular forms $f \in G_w(A)$, we can interpret (G0) as

(G0″) $f : \wp_{/A} \to \mathbf{A}^1_{/A}$ is a morphism of functors satisfying the automorphic property (G3) in Sect. 2.3.2, and its algebraic q-expansion $f(E_\infty, \omega_\infty)$ is contained in $A[[q]]$.

2.4 *p*-Adic Uniformization

We recall Tate's theory of p-adic uniformization of elliptic curves, following Tate's paper [T3] (dating back to 1959). The uniformization has been generalized to higher-dimensional abelian varieties by Mumford and Faltings–Chai [DAV] Chaps. II, III, which is the base of the theory of smooth toroidal compactification of Shimura varieties.

2.4.1 Explicit q-Expansion

As we learned from Weierstrass, every elliptic curve over $\mathbb{C}$ is isomorphic to E with $E(\mathbb{C}) = \mathbb{C}/L$ for a lattice $L = \mathbb{Z}(2\pi i) + \mathbb{Z}\log q$ for an element $q \in \mathbb{C}^\times = \mathbb{G}_m(\mathbb{C})$ with $|q| < 1$. The covering map $\mathbb{C} \to E(\mathbb{C})$ factors through $\exp : \mathbb{C} \to \mathbb{C}^\times$ given by $\exp(x) = e^x = \sum_{n=0}^{\infty} \frac{x^n}{n!}$. Thus, $E(\mathbb{C}) = \mathbb{C}^\times/q^{\mathbb{Z}}$, where $q^{\mathbb{Z}} = \{q^m | m \in \mathbb{Z}\}$, which is a discrete subgroup of $\mathbb{C}^\times$. We see from the definition of Weierstrass functions in Sect. 2.3.4 that

$$\begin{aligned} g_2(L) &= \frac{1}{12} + 20\sum_{n=1}^{\infty}\left\{\sum_{0<d|n} d^3\right\} q^n = \frac{1}{12} + 20\sum_{n=1}^{\infty}\frac{n^3q^n}{1-q^n}, \\ g_3(L) &= -\frac{1}{216} + \frac{7}{3}\sum_{n=1}^{\infty}\left\{\sum_{0<d|n} d^5\right\} q^n = -\frac{1}{216} + \frac{7}{3}\sum_{n=1}^{\infty}\frac{n^5q^n}{1-q^n}, \\ \Delta(L) &= q\prod_{n=1}^{\infty}(1-q^n)^{24}, \\ J(L) &= q^{-1} + j'(q) \quad \text{with } j'(q) \in \mathbb{Z}[[q]]. \end{aligned} \tag{2.4.1}$$

These formulas follow from (1.3.1).

Setting $w = \exp(u) = e^u$ ($u = \log w$), we compute the (q, w)-expansion of $\wp_L(u)$: By (1.3.3), for w with $|q| < |w| < |q|^{-1}$, we obtain

$$\begin{aligned} \wp_L(u) &= \frac{1}{u^2} + \sum_{m=-\infty,\ m\neq 0}^{\infty}\left\{\frac{1}{(u+2\pi i m)^2} - \frac{1}{(2\pi i m)^2}\right\} \\ &\quad + \sum_{n=1}^{\infty}\left\{\sum_{m=-\infty}^{\infty}\frac{1}{(-u+2\pi i m+n\log q)^2} - \frac{1}{(2\pi i m+n\log q)^2}\right\} \\ &\quad + \sum_{n=1}^{\infty}\left\{\frac{1}{(u+2\pi i m+n\log q)^2} - \frac{1}{(2\pi i m+n\log q)^2}\right\} \\ &= \sum_{m=1}^{\infty} m w^m - \frac{2\zeta(2)}{(2\pi i)^2} + \sum_{n=1}^{\infty}\sum_{m=1}^{\infty}\left\{m w^{-m} q^{mn} + m w^m q^{mn} - 2q^{mn}\right\}. \end{aligned}$$

Applying $\frac{d}{dw}$ to $\frac{1}{1-w} = \sum_{m=0}^{\infty} w^m$, we get

$$\frac{w}{(1-w)^2} = \sum_{m=1}^{\infty} m w^m.$$

Then, from the fact $\zeta(2) = \frac{\pi^2}{6}$ (see [LFE] Sect. 2.1), we see

$$x_L(u) = \wp_L(u) = t_L(w) + \frac{1}{12}, \tag{2.4.2}$$

where

$$t(q,w)=t_L(w)=\sum_{m=-\infty}^{\infty}\frac{q^m w}{(1-q^m w)^2}-2\sum_{m=1}^{\infty}\frac{q^m}{(1-q^m)^2}. \tag{2.4.3}$$

The same formula written in a slightly different way is

$$t(q,w)=\frac{w}{(1-w)^2}+\sum_{n=1}^{\infty}\frac{nq^n}{1-q^n}(w^n+w^{-n}-2).$$

This new formula confirms $t(q,w)\in\mathbb{Z}[w,w^{-1},(1-w)^{-1}][[q]]$. Let $A_w=\mathbb{Z}[w,w^{-1},(1-w)^{-1}]$ as a localized polynomial ring, which is plainly a finitely generated $\mathbb{Z}$-algebra. Thus, we have $t(q,w)\in A_w[[q]]$.

Again applying $w\frac{d}{dw}=\frac{d}{du}$ to x_L, we get

$$y_L(u)=\wp'(u)=t_L(w)+2s_L(w), \tag{2.4.4}$$

where

$$s(q,w)=s_L(w)=\sum_{m=-\infty}^{\infty}\frac{(q^m w)^2}{(1-q^m w)^3}+\sum_{m=1}^{\infty}\frac{q^m}{(1-q^m)^2}. \tag{2.4.5}$$

From the identity: $y_L^2=4x_L^3-g_2(L)x_L-g_3(L)$, we get

$$s^2(q,w)+t(q,w)s(q,w)=t(q,w)^3-b_2(q)t(q,w)-b_3(q), \tag{2.4.6}$$

where

$$b_2(q)=b_2(L)=\frac{1}{4}\left(g_2-\frac{1}{12}\right)=5\sum_{n=1}^{\infty}\frac{n^3q}{1-q^n}\in q\mathbb{Z}[[q]]$$

$$b_3(q)=b_3(L)=\frac{1}{4}\left(g_3+\frac{g_2}{12}-\frac{1}{432}\right)=\sum_{n=1}^{\infty}\left(\frac{7n^5+5n^3}{12}\right)\frac{q^n}{1-q^n}\in q\mathbb{Z}[[q]].$$

Note here that the identity is the algebraic identity in the power series ring $A_w[[q]]$, because the identity is valid over the open set $|q|<|w|<|q|^{-1}$ in $\mathbb{C}^2$. We recall the following identity in $A_w[[q]]$:

$$\begin{aligned}\Delta=g_2^3-27g_3^2&=\left(4b_2+\frac{1}{12}\right)^3-27\left(4b_3-\frac{b_2}{3}-\frac{1}{216}\right)^2\\&=b_3+b_2^2+72b_2b_3-432b_3^2+64b_2^3=q\prod_{n=1}^{\infty}(1-q^n)^{24}.\end{aligned} \tag{2.4.7}$$

Since $w\mapsto(t(q,w),s(q,w))$ factors through $\mathbb{C}^\times/q^{\mathbb{Z}}$, we have

$$t(q,qw)=t(q,w)\quad\text{and}\quad s(q,qw)=s(q,w)\quad\text{in }A_w[[q]]. \tag{2.4.8}$$

We check the following identity by power series computation:

$$s(q, w^{-1}) + s(q, w) = -t(q, w) \quad \text{in } A_w[[q]]. \tag{2.4.9}$$

The canonical differential on $E(\mathbb{C})$ is given by

$$\frac{dw}{w} = du = \frac{dx}{\frac{dx}{du}} = \frac{dx}{y}.$$

Exercise 2.53 *1. Show that the projective plane curve C over a field k defined by $X^3 - XYZ - Y^2Z = 0$ is singular at* (0:0:1)*, which is an ordinary double point.*

2. Show the function field of the curve C for $k = \mathbb{F}_p$ as above is isomorphic to $\mathbb{F}_p(w)$ with $x = \frac{X}{Z} = \frac{w}{(1-w)^2}$ and $y = \frac{Y}{Z} = \frac{w^2}{(1-w)^3}$.

2.4.2 Tate Curves

By the computation we have done in the previous section, the projective plane curve E_∞ is defined over $\mathbb{Z}[[q]]$ by the following new equation in $\mathbf{P}^2_{/\mathbb{Z}[[q]]}$ with homogeneous coordinate $(S{:}T{:}U)$:

$$S^2U + TSU - T^3 + b_2(q)TU^2 + b_3(q)U^3 = 0.$$

An important point is that the curve originally defined over $\mathbb{Z}[\frac{1}{6}][[q]]$ is extended to over $\mathbb{Z}[[q]]$. It has an integral point $\mathbf{0}$ given by $(S, T, U) = (1, 0, 0)$. More generally, we can think of a surjective homomorphism:

$$\mathbb{Z}[[q]][S, T, U]/(S^2U + TSU - T^3 + b_2(q)TU^2 + b_3(q)U^3) \to \mathbb{Z}[[q]][S]$$

taking (S, T, U) to $(S, 0, 0)$. To compute the tangent space at $\mathbf{0}$, we use the affine equation of $u = U/S$ and $t = T/S$ given by $u + tu = t^3 - b_2u^2t - b_3u^3$, and we have

$$\Omega_{\widehat{\mathcal{O}}_{E_\infty,\mathbf{0}}/\mathbb{Z}[[q]]} = \mathbb{Z}[[q, t]]dt.$$

This shows that $\widehat{\mathcal{O}}_{E_\infty,\mathbf{0}} = \mathbb{Z}[[q, u]]$ and $\mathbf{0}$ is a **smooth** point of E_∞.

Since $q|\Delta$ in $\mathbb{Z}[[q]]$ (and $\Delta/q \in \mathbb{Z}[[q]]^\times$), after inverting q, that is, over $\mathbb{Z}[[q]][q^{-1}]] = \mathbb{Z}((q))$, the curve $E_{\infty/\mathbb{Z}((q))}$ is an elliptic curve with an invariant differential $\omega_\infty = \frac{dx}{y} = \frac{dt}{t+2s} = \frac{dw}{w}$. The curve E_∞ over $\mathbb{Z}[[q]]$ (without inverting q) has one singular point, that is, $E_\infty \mod q$ is singular only at P with coordinate $(s, t) = (0, 0)$ (which is not the origin of $\overline{E}_\infty = E_\infty \mod q$), and the (completed) stalk $\widehat{\mathcal{O}}_{E_\infty,P}$ is isomorphic to $\mathbb{Z}[[q]][[t, s]]/(ts - q)$, which is a regular ring (cf. [CRT] Sect. 19). Thus, the local ring at every geometric point of $E_{\infty/\mathbb{Z}}$ is a regular ring of relative dimension 1 over $\mathbb{Z}[[q]]$ (we call such a curve a regular curve). The smooth locus of $\overline{E}_\infty$ is isomorphic to $\mathbf{P}^1$ removed two points, say 0 and ∞, which is $\mathbb{G}_m$ (here we may think $\mathbb{G}_m$ as a covariant functor sending a ring A to its multiplicative group $A^\times$). The description of the singularity of $(E_\infty \mod q) = E_\infty \otimes_{\mathbb{Z}[[q]]} \mathbb{Z}[[q]]/(q)$ as above is clear from the equation of $\overline{E}_\infty$: $s^2 + st = t^3$, because $b_2(q) \equiv b_3(q) \equiv 0 \mod q$. Thus, $\overline{E}_{\infty/\mathbb{Z}}$ is a projective regular plane curve with a nowhere vanishing differential ω_∞.

Let K be a complete field with discrete valuation $|\ | = |\ |_K$ (for example, the p-adic field $\mathbb{Q}_p$ and its field extensions finite degree). Write A for the valuation ring of K. We pick $q_E \in K^\times$ with $|q_E| < 1$. The specialization of E_∞ under the algebra homomorphism $q \mapsto q_E$ gives rise to an elliptic curve $E_K = E_\infty \otimes_{\mathbb{Z}[[q]]} K$ defined over K. Let $P, Q, R \in E_K(K)$. By Abel's theorem (Theorem 2.32), we have

$$P + Q + R = \mathbf{0} \iff [P] + [Q] + [R] \sim 3[\mathbf{0}],$$

where "$\sim$" indicates the linear equivalence.

We explore the addition formula in terms of the coordinates s and t in order to understand well the group structure of $E(K)$. In other words, we want to express explicitly the coordinates of the sum $P+Q$ in terms of the coordinates of each P and Q. By the equation defining E_K, we get $3[\mathbf{0}] = E_K \cap L_\infty$ (see Example 2.18), where $L_\infty = \{U = 0\} \subset \mathbf{P}^2$ is the line at infinity. Since any two lines in $\mathbf{P}^2$ are linearly equivalent (that is, $L_\infty - L$ is the divisor of the function U/ϕ_L for the linear form ϕ_L defining L),

$$P + Q + R = \mathbf{0} \iff [P] + [Q] + [R] = L \cap E_K$$

for the line $L \subset \mathbf{P}^2$ passing through two of the three points P, Q, R, because if P and Q are on L (P and Q determine L), we find the third point $R \in L \cap E_K$ by the Bézout theorem. Here the line L is the tangent line at P if $P = Q$.

Write $P = (s, t)$, $Q = (s', t')$, and $R = (s'', t'')$. We suppose that P and Q are different from $\mathbf{0}$; so their coordinates are finite. Suppose that the line L (having P and Q on it) passes through $\mathbf{0}$ (thus, $R = \mathbf{0}$). If a line passes through $\mathbf{0} = (0, 1, 0)$, its equation $\phi(S, T, U) = aT + bS + cU = 0$ ($s = \frac{S}{U}$ and $t = \frac{T}{U}$) satisfies $\phi(1, 0, 0) = b = 0$; hence, L is parallel to the s-axis, and we conclude $t = t'$. Thus, by (2.4.6), assuming $P \neq Q$ (i.e., $s \neq s'$ or equivalently $2P \neq \mathbf{0}$), we have $s^2 + st = {s'}^2 + s't \iff s + s' = -t$. In short,

$$t = t' \quad \text{and} \quad s + s' = -t \iff P + Q = -R = \mathbf{0}. \tag{2.4.10}$$

If $P + Q \neq \mathbf{0}$, then the equation of L can be written as $s = \mu t + \nu$. Again by (2.4.6), we have

$$\mu = \frac{s - s'}{t - t'} = \frac{t^2 + tt' + {t'}^2 - b_2 - s'}{s + s' + t} \tag{2.4.11}$$
$$\nu = s - \mu t = s' - \mu t'.$$

Using the above equations, we find the third point in $L \cap E_\infty$. We get

$$t'' = \mu^2 + \mu - t - t' \ \text{ and } \ s'' = -t'' - \mu t'' - \nu. \tag{2.4.12}$$

Here is a result in [T3] Theorem 1:

Theorem 2.54 (Tate) *Let* $A = \varprojlim_m A/q^m A$ *be a* q*-adically complete local* $\mathbb{Z}[[q]]$*-algebra. Then*

(1) *The map* $w \mapsto (s(q,w), t(q,w), 1) \in \mathbf{P}^2(A)$ *induces an injective homomorphism of* $A^\times$ *into* $E_A(A)$ *for* $E_A = E_\infty \otimes_{\mathbb{Z}[[q]]} A$.
(2) *If* A *is the integer ring of a local field* K *[that is, a finite extension of* $\mathbb{Q}_p$ *or* $\mathbb{F}_p((q))$*], then* π *extends to an isomorphism of* $K^\times/q^{\mathbb{Z}} \cong E_A(K)$.

If an elliptic curve E is isomorphic to E_A over $A[\frac{1}{q}]$, we say that E has *split multiplicative reduction* modulo (q).

Proof. We reproduce the proof [T3]) with some modification as in [GME] Sect. 2.5.2. The power series $t(q,w) - \frac{w}{(1-w)^2}$ and $s(q,w) - \frac{w^2}{(1-w)^3}$ are contained in $\mathbb{Z}[w, w^{-1}][[q]]$. Thus, if $w \in A^\times$, the series $((1-w)^3 s(q,w), (1-w)^3 t(q,w))$ converges in A^2 under the q-adic topology; hence, we get a point

$$\pi(w) = ((1-w)^3 s(q,w) : (1-w)^3 t(q,w) : (1-w)^3) \in E_A(A)$$

as long as one of the coordinates is nonzero. Since $(1-w)^3 s(q,w) \equiv w^2 \mod qA$, we see $(1-w)^3 s(q,w) \in A^\times$ for all $w \in A^\times$. Thus, the map $\pi : A^\times \to E_A(A)$ is well defined. If $\pi(w) = \mathbf{0}_{E_A} = (1{:}0{:}0)$, we have $(1-w)^3 = 0$, and so $w = 1$. Thus, $\pi^{-1}(\mathbf{0}) = \{1\}$.

We do not give a detailed proof of "homomorphy" of π (i.e., π is a group homomorphism) here, but instead we just remark that the assertion (2) implies homomorphy because the addition and the inverse are basically power series identities. More precisely, taking parameters (w, w') on $E_\infty \times E_\infty$, we have a power series $\Phi(W, W') \in \mathbb{Z}[[q]][[W, W']]$ for $W = 1 + w$ and $W' = 1 + w'$ such that if $P \in E_\infty$ has coordinate w and $Q \in E_\infty$ has coordinate w', and then the w-coordinate of $P + Q \in E_\infty$ is given by $\Phi(W, W')$. This fact is valid by (2) after evaluating the variable q of the base ring $\mathbb{Z}[[q]]$ at many different $q_E \in A$, and so it should be valid as a power series identity.

We now prove (2). We first assume that K is of characteristic 0. We can easily check the convergence of

$$\pi(w) = (s(q,w) : t(q,w) : 1) \in \mathbf{P}^2(K) \quad \text{if } |q| < |w| < |q|^{-1}$$

for $q \in K^\times$ with $|q| < 1$. Here $|\cdot|$ is the absolute value of K, and convergence is under the topology of K. We simply put $\pi(w) = \mathbf{0} \in E_A(K)$ if $w \in q^{\mathbb{Z}}$. Thus, $\pi : K^\times/q^{\mathbb{Z}} \to E_A(K)$ is well defined by (2.4.8), and by the first assertion,

$$\pi^{-1}(\mathbf{0}) = q^{\mathbb{Z}}. \tag{2.4.13}$$

We take $u, v, w \in K^\times$ with $w = uv$. Since π depends only on the class modulo $q^{\mathbb{Z}}$ (2.4.8), we may assume $|q| < |u| \le 1$ and $1 \le |v| < |q|^{-1}$. Thus, $|q| < |w| < |q|^{-1}$, and $\pi(u), \pi(v)$ and $\pi(w)$ are well defined [that is, the power series $s(q, ?)$ and $t(q, ?)$ converge at these points]. Since $\pi(1) = \pi(q^0) = \mathbf{0}$ by definition, (2.4.9) and (2.4.10) shows the desired result when $uv = 1$. Thus, we may assume that $\pi(u) = P$, $\pi(v) = Q$, and $\pi(w) = R$ are all different from $\mathbf{0}$ and that $P \neq Q$. Write $P = (s,t)$, $Q = (s',t')$, and $R = (s'',t'')$. By (2.4.10), (2.4.11), and (2.4.12), $\pi(u) + \pi(v) = \pi(w)$ is equivalent to the following simultaneous identities:

$$(t-t')^2 t'' = (s-s')^2 + (s-s')(t-t') - (t-t')^2(t+t'), \qquad (2.4.14)$$
$$(t-t')s'' = -(t-t')(s+t'') + (s-s')(t-t'').$$

Assuming $w = uv$, we want to show that this identity (2.4.14) holds for $\pi(u) = P$, $\pi(v) = Q$, and $\pi(w) = R$. Since $w = uv$, (2.4.14) is the identity in

$$\mathbb{Z}[u, u^{-1}, v, v^{-1}, (1-u)^{-1}, (1-v)^{-1}, (1-uv)^{-1}][[q]].$$

Since $\mathbb{Z}[u, u^{-1}, v, v^{-1}, (1-u)^{-1}, (1-v)^{-1}, (1-uv)^{-1}]$ is finitely generated over $\mathbb{Z}$, we can embed this ring into $\mathbb{C}$. Then the identity holds, by extending this embedding to $K \hookrightarrow \mathbb{C}$; consider $E_{\mathbb{C}} = E_K \otimes_K \mathbb{C}$ over $\mathbb{C}$, since the identities (2.4.14) hold for elliptic curves defined over $\mathbb{C}$. We only verified homomorphy assuming $P \neq \pm Q$, but any map between infinite groups satisfying $\pi(uv) = \pi(u) + \pi(v)$ if $\pi(u) \neq \pm\pi(v)$ can be easily verified to be a homomorphism (see the exercise below and [T3] Lemma 1). This shows that $\pi : K^\times/q^{\mathbb{Z}} \to E_K(K)$ is a homomorphism, and so, as remarked already, the assertion (1) also holds for any q-adically complete A. In particular, π is also a homomorphism for local fields of characteristic p. Then the injectivity follows from (2.4.13).

We only give a sketch of a proof of the surjectivity when A is the integer ring of a finite extension $K/\mathbb{Q}_p$. Since a convergent power series gives an open map on a convergent open disk into an open disk under the p-adic topology (cf. [T3] Corollary 1), $\pi(K^\times)$ is an p-adic open subgroup of $E_A(K)$. Since $\mathbf{P}^2(K)$ is a compact p-adic set, $E_A(K)$ is a compact p-adically closed subset of $\mathbf{P}^2(K)$. Since $E_A(K) = \bigcup_{x \in E_A(K)}(x + \pi(K^\times))$, $E_A(K)$ is covered by finitely many open set of the form $x + \pi(K^\times)$. Thus, $\pi(K^\times)$ is a subgroup of $E_A(K)$ of finite index. Thus, $E_A(K)/\pi(K^\times)$ is a finite group. In other words, for any $x \in E_A(K)$, $Nx \in \pi(K^\times)$. Write $\overline{\mathbb{Q}}_p$ for an algebraic closure of $\mathbb{Q}_p$ containing K. Since $\overline{\mathbb{Q}}_p/q^{\mathbb{Z}}$ is divisible and all torsion points of E_A are contained in $\pi(\overline{\mathbb{Q}}_p^\times)$, we find that $Nx \in \pi(K^\times)$ implies $x \in \pi(\overline{\mathbb{Q}}_p^\times)$. Thus, $\overline{\mathbb{Q}}_p^\times/q^{\mathbb{Z}} \cong E_A(\overline{\mathbb{Q}}_p)$. By definition, $\pi(w^\sigma) = \pi(w)^\sigma$ for $\sigma \in \mathrm{Gal}(\overline{\mathbb{Q}}_p/K)$. If $\pi(w)^\sigma = \pi(w)$, we have $w^\sigma = q^m w$. Since $|q| < 1$ and $|w| = |w^\sigma|$, we find $w^\sigma = w$. Take $\mathfrak{G} := \mathrm{Gal}(\overline{\mathbb{Q}}_p/K)$ invariants of the exact sequence: $1 \to q^{\mathbb{Z}} \to \overline{\mathbb{Q}}_p^\times \to \overline{\mathbb{Q}}_p^\times/q^{\mathbb{Z}} \to 1$. Note that $H^1(\mathfrak{G}, q^{\mathbb{Z}}) \cong \mathrm{Hom}_{\mathrm{cont}}(\mathfrak{G}, \mathbb{Z}) = 0$ (as $\mathbb{Z}$ does not have nontrivial compact subgroups). Thus, we have $H^0(\mathfrak{G}, \overline{\mathbb{Q}}_p^\times/q^{\mathbb{Z}}) = K^\times/q^{\mathbb{Z}}$ and $H^0(\mathfrak{G}, E(\overline{\mathbb{Q}}_p)) = E_A(K)$ by definition. From $\overline{\mathbb{Q}}_p^\times/q^{\mathbb{Z}} \cong E_A(\overline{\mathbb{Q}}_p)$, via a long exact sequence of the continuous cohomology, we get $K^\times/q^{\mathbb{Z}} \cong E_A(K)$ as desired.

Exercise 2.55 *Let $(G, \cdot)$ and $(H, +)$ be infinite abelian groups and let $\pi : G \to H$ be a map with infinite image sending the identity to the identity such that $\pi(v \cdot w) = \pi(v) + \pi(w)$ as long as $\pi(v) \neq \pm\pi(w)$. Show that π is a homomorphism.*

Corollary 2.56 *Let W be a complete discrete valuation ring finite flat over $\mathbb{Z}_p$ inside $\overline{\mathbb{Q}}_p$ and put K for its field of fractions. Let $q \in W$ with $|q|_p < 1$ and E_q be the elliptic curve over W with Tate period q. Then, for a prime l, we have $E[l^n] \cong \mu_{l^n} \times \mathbb{Z}/l^n\mathbb{Z}$ by the map sending $\zeta q^{m/l^n} \in E_q[l^n]$ with $\zeta \in \mu_{l^n}(\overline{\mathbb{Q}}_p)$ and $m \in \mathbb{Z}$ to $(\zeta, (m \mod l^n)) \in \mu_{l^n}(\overline{\mathbb{Q}}_p) \times \mathbb{Z}/l^n\mathbb{Z}$. On $T_l E_q = \varprojlim_n E_q[l^n] \cong \mathbb{Z}_l(1) \times \mathbb{Z}_l$, $\mathrm{Gal}(\overline{\mathbb{Q}}_p/K)$ acts by a nonsemisimple representation isomorphic to $\sigma \mapsto \left(\begin{smallmatrix} \mathcal{N}_l & u \\ 0 & 1 \end{smallmatrix}\right)$, where $\mathcal{N}_l$ is the l-adic cyclotomic character such that $\zeta^\sigma = \zeta^{\mathcal{N}_l(\sigma)}$ for $\zeta \in \mu_{l^\infty}(\overline{\mathbb{Q}}_p)$ and $\mathcal{N}_l(\sigma) \in \mathbb{Z}_l{}^\times$ and for a choice of $q^{1/l^n} \in \overline{\mathbb{Q}}_p$ and the primitive l^nth root $\zeta_{l^n} = i_p(i_\infty^{-1}(\mathbf{e}(\frac{1}{l^n}))) \in \mu_{l^n}(\overline{\mathbb{Q}}_p)$, we have $(q^{1/l^n})^\sigma = \zeta_{l^n}^{u(\sigma)} q^{1/q^n}$ (for all n) with $u(\sigma) \in \mathbb{Z}_l$. In particular, the l^n-torsion points of E_q for n sufficiently large generate a nonabelian soluble extension of K.*

Proof. By Theorem 2.54 (2), the field of rationality of $E_q[l^n]$ is the Kummer extension $K[\mu_{l^n}, q^{1/l^n}]$. Then the above form of representation is given by the Galois action on the basis $\{\zeta_{l^n}, q^{1/l^n}\}$ of $E_q[l^n](\overline{\mathbb{Q}}_p)$.

2.5 What We Study in This Book

We learned in this section that the isomorphism classes of (E, ω) over a field are in bijection to rational points of an affine algebraic variety $\mathrm{Spec}(\mathcal{R})$ for $\mathcal{R} = \mathbb{Z}[\frac{1}{6}, X, Y, \frac{1}{X^3 - 27Y^2}]$. Similarly, the set of isomorphism classes of elliptic curves over an algebraically closed field k is isomorphic to $\mathbf{P}^1(J)(k) - \infty = \mathrm{Spec}(\mathbb{Z}[J])(k) \cong k$. Generalizing this, we may look into the problem classifying pairs $(E, \phi_N)_{/A}$ of an elliptic curve E over a ring A with a level-N-structure for a positive integer N. The level structure ϕ_N can be an embedding of abelian group functors $\phi_N : \mu_N \hookrightarrow E$ defined over the category of A-algebras $ALG_{/A}$. Here we use the group structure of E. We prove later in Chap. 6 that this classification problem gives rise to an affine curve $Y_1(N)_{/\mathbb{Z}[\frac{1}{N}]} = \mathrm{Spec}(\mathcal{R}_0(\Gamma_1(N)))$ defined over the very small ring $\mathbb{Z}[\frac{1}{N}]$. In other words, if $N \geq 4$, the isomorphism classes of $(E, \phi_N)_{/A}$ over an $\mathbb{Z}[\frac{1}{N}]$-algebra A are in bijection with the set $Y_1(N)(A) = \mathrm{Hom}_{\mathrm{alg}}(\mathcal{R}_0(\Gamma_1(N)), A)$ (this is a theorem of Shimura and Igusa; Theorem 6.25). Adding finitely many points (called cusps) to $Y_1(N)$ using the theory of Tate curves, we construct a projective curve $X_1(N)$ canonically containing $Y_1(N)$ as we made $\mathbf{P}^1(J)$ out of $M_1 = \mathbf{P}^1(J) \setminus \{\infty\}$. Since $\phi_N|_{\mu_{N'}}$ for a factor $N'|N$ is a level-N'-structure, we naturally get morphisms $Y_1(N) \to Y_1(N')$ and $X_1(N) \to X_1(N')$. Thus, we get a tower of projective curves $\{X_1(N) \twoheadrightarrow X_1(N')\}_{N'|N}$ indexed by positive integers N under the (reversed) order given by divisibility $N'|N$. The procurve $\varprojlim_N X_1(N)$ and also the proaffine curve $\varprojlim_N Y_1(N)$ are a quotient of the Shimura curve we study in Chap. 7. We study the arithmetic invariant via the theory of Shimura varieties and Hecke operators acting on it. In the following chapter, we present some of the results of this book without going deep into the proof; thus, even if readers do not fully understand the results, we suggest they try to get an overview of the book (the details will be supplied later, after we've given the basics of scheme theory and the relevant theory of modular forms/elliptic curves/Shimura variety).

3

Invariants, Shimura Variety, and Hecke Algebra

In the first chapter, we gave an introductory sketch of a proof of nonvanishing modulo p of Dirichlet L-values. Since we have described basics of elliptic curves in an elementary manner in Chap. 2, at least we could illustrate our main objectives in this book with a rough outline of their proofs. Detailed proofs (for some of them) will be given after we become equipped with a scheme-theoretic description of elliptic curves and their moduli as the simplest example of Shimura varieties in the following chapters. For some others, we give full proofs here, possibly assuming simplifying assumptions (and we refer to quoted research articles about the author's proofs in more general cases).

To prove the nonvanishing/nontriviality of an arithmetic invariant, our principle is to spread out the invariant to a function defined over a larger geometric object and then try to prove the nonvanishing of this function, as we described in the introduction. A prototypical example is the nonvanishing of Dirichlet L-values we studied already, as we somehow extended partial L-values to a rational function on a product of $\mathbb{G}_m$ and get nonvanishing for almost all Dirichlet characters of an l-power conductor for a given prime l. This is an almost optimal result because a given L-value is, through numerical computations, usually a nonzero integer divisible by fairly big primes (hence, a given L-value vanishes modulo p for a prime p appearing sporadically).

The first example of such techniques discussed in this chapter is the Hecke field of a Hecke eigenform (i.e., the number field generated by Hecke eigenvalues of a given eigenform). In this case, it is not a value; thus, in appearance, it is not a nonvanishing problem but a nontriviality problem. We show that the Hecke fields grow bigger when the eigenform is moving/deforming over the (spectrum of) Hecke algebra (except for the cases where the associated Galois representation has an "almost" abelian image). Thus, we could say that this is a solution to a growth problem of fields generated by Hecke eigenvalues (see Theorems 3.2 and 3.25). Then, as an application of the nontrivial growth of Hecke fields, we prove the finiteness of character twist classes of potentially ordinary non-CM abelian varieties of $GL(2)$-type having a good reduction

H. Hida, *Elliptic Curves and Arithmetic Invariants*, Springer Monographs in Mathematics, DOI 10.1007/978-1-4614-6657-4_3,

outside p (Theorem 3.17). As another application, we study the nonvanishing of the $\mathcal{L}$-invariant associated with the exceptional zero of p-adic modular adjoint L-functions (see [HMI] for more about the $\mathcal{L}$-invariant). We prove as a function over Spec($\mathbf{h}$) that Greenberg's $\mathcal{L}$-invariant is nonconstant (except for the cases where the associated Galois representation has an "almost" abelian image; see Theorem 3.28); so it does not vanish at almost all Hecke eigenforms (Greenberg conjectures that it never vanishes). Then we proceed to the vanishing of the Iwasawa μ-invariant of p-adic L-functions, which is actually a nonvanishing result of the power series modulo p of the L-function. In this case, the "bigger" geometric object we use is the Shimura curve/variety as the moduli of elliptic curve (or, more generally, abelian varieties), and roughly, with controlled error terms, the power series of a certain type of p-adic L-functions is expected to be the Taylor expansion of a specific nonzero modular/automorphic form at an appropriate CM point of the variety (so, again, it does not vanish often; Theorem 3.37). Discussions about the μ-invariant and the "big" Hecke algebra in this chapter are only outlines, and details on the invariant and the Hecke algebra will be discussed in later chapters.

3.1 Abelian Component of the "Big" Hecke Algebra

In this section, first, we give an axiomatic definition of the cuspidal "big" ordinary Hecke algebra $\mathbf{h}$ necessary to state our results without proof. Then we give a precise definition of the abelian components. As we mentioned, often Spec($\mathbf{h}$) plays the role of a bigger geometric object when we want to spread a given arithmetic invariant as a schematic function. The construction of the Hecke algebra (logically necessary to show what we state here) will be given in Sects. 7.3 and 7.4. Until all the details are given, we suggest that the reader take axiomatic statements literally to see an overview of the theory. The scheme Spec($\mathbf{h}$) is the universal parameter space of p-adic slope-0 modular forms of a given prime-to-p level. There is a rigid analytic generalization of the parameter space for positive slope forms (the eigencurve; see [Em] for a brief outline of the theory). As described above, the nonvanishing/nontriviality theorems involving the spectrum of the Hecke algebra are often valid outside "abelian" components.

Fix a prime p and, for simplicity, assume $p \geq 3$. We can treat the case of $p = 2$, and all the statements on the big Hecke algebra $\mathbf{h}$ are valid for $p = 2$, replacing p^{r+1} by $4p^r$ (see [GME] Sect. 3.2 for the theory valid for general primes including $p = 2$). Fix a positive integer N prime to p. Take a Dirichlet character ψ modulo Np^{r+1}. Consider the subspace of modular forms $G_{k+1}(\Gamma_0(Np^{r+1}), \psi)$ made up of modular forms f in $G_{k+1}(\Gamma_1(Np^{r+1}))$ satisfying $f(E, \phi_{Np^{r+1}} \circ d, \omega) = \psi(d) f(E, \phi_{Np^{r+1}}, \omega)$ for all $d \in (\mathbb{Z}/Np^{r+1}\mathbb{Z})^\times$. Here $\phi_{Np^{r+1}} \circ d$ is the composition of $\phi_{Np^{r+1}}$ and $\zeta \mapsto \zeta^d$ for $\zeta \in \mu_{Np^{r+1}}$, and we assume $p \nmid N$ and $r \geq 0$. We write the subspace in $G_{k+1}(\Gamma_0(Np^{r+1}), \psi)$ of cusp forms as $S_{k+1}(\Gamma_0(Np^{r+1}), \psi)$ (excluding Eisenstein series). The character ψ

is called the Neben type, or the "Neben" character (taking the terminology of Hecke). These spaces are defined in [IAT] Sect. 3.5 along with the Hecke operators $T(n)$ acting on it, where $S_{k+1}(\Gamma_0(Np^{r+1}),\psi) = S_{k+1}(\Gamma_0(Np^{r+1}),\psi;\mathbb{C})$ is defined. We already defined the spaces of modular forms and cusp forms $G_{k+1}(\Gamma_1(N)) = G_{k+1}(\Gamma_1(N);\mathbb{C})$ and $S_{k+1}(\Gamma_1(N)) = S_{k+1}(\Gamma_1(N);\mathbb{C})$ [see (G1–4) and (s) in Sect. 1.3.1]. We will give a more detailed definition of these spaces and operators later (see Definitions 6.30 and 6.21). In this chapter only, we write the weight of the modular form as $k+1$ since the l-Frobenius eigenvalue has absolute value $l^{k/2}$ for the Galois representation of a cuspidal eigenform f of weight $k+1$. Here we say that f is a Hecke eigenform if $f|T(n) = a_n f$ for all n. Actually, the motive associated with f has Hodge (and Hodge–Tate) weight $(k,0)$, so it might be more natural to call this weight k as the weight of modular form, though we use the traditional weight $k+1$ in this book (see [MTV] for motives and [HMI] Sect. 4.4.1 for Hodge–Tate weight).

Let $\mathbb{Z}[\psi]$ be the subring of $\overline{\mathbb{Q}}$ generated by the values of ψ. The *Hecke algebra* over $\mathbb{Z}[\psi]$ is

$$h = \mathbb{Z}[\psi][T(n)|n = 1,2,\cdots] \subset \mathrm{End}(S_{k+1}(\Gamma_0(Np^{r+1}),\psi)).$$

Thus, h is a subalgebra of the linear endomorphism algebra of the space $S_{k+1}(\Gamma_0(Np^{r+1}),\psi)$ of cusp forms generated by Hecke operators. This algebra is a commutative ring with identity $T(1)$. We put $h_{k+1,\psi} = h_{k+1,\psi/A} = h \otimes_{\mathbb{Z}[\psi]} A$ for a p-adically complete ring $A \subset \mathbb{C}_p$ containing $\mathbb{Z}_p[\psi]$. Often our $T(p)$ is written as $U(p)$ because the level is divisible by p. The *ordinary* part $\mathbf{h}_{k+1,\psi/A} \subset h_{k+1,\psi/A}$ is the maximal ring direct summand on which $U(p)$ is invertible. In other words, $\lim_{n\to\infty} U(p)^{n!}$ converges p-adically in $h_{k+1,\psi/A}$ to the idempotent e (called the *ordinary projector*) of $\mathbf{h}_{k+1,\psi/A} = e \cdot h_{k+1,\psi/A}$.

Exercise 3.1 *Let A be a p-adically complete algebra, and suppose that A is of finite type as a module over $\mathbb{Z}_p$. Prove that the p-adic limit* $\lim_{n\to\infty} a^{n!}$ *for any $a \in A$ exists in A giving an idempotent of A.*

Let $\psi_1 = \psi_N \times$ the tame p-part of ψ, and recall the fixed field embeddings $\mathbb{C} \overset{i_\infty}{\hookleftarrow} \overline{\mathbb{Q}} \overset{i_p}{\hookrightarrow} \mathbb{C}_p$. Fix a complete discrete valuation ring $W \subset \mathbb{C}_p$. Let the ring $W[\psi] \subset \mathbb{C}_p$ be generated by the values ψ over W. Then, as we will prove in Corollary 7.61, we have a unique "big" Hecke algebra $\mathbf{h} = \mathbf{h}_{\psi_1/W}$ characterized by the following two conditions:

1. $\mathbf{h}$ is free of finite rank over the Iwasawa algebra $\Lambda := W[[T]]$ equipped with $T(n) \in \mathbf{h}$ for all n (we write $t = 1 + T \in \Lambda$ as before);
2. if $k \geq 1$, regarding a character $\varepsilon : \mathbb{Z}_p^\times \to \mu_{p^\infty}(\overline{\mathbb{Q}}_p)$ as a ring homomorphism ε_k of Λ onto $W[\varepsilon]$ over W, sending t to $\gamma^k\varepsilon(\gamma)$ $(\gamma = 1+p)$, we have

$$\mathbf{h} \otimes_{\Lambda,\varepsilon_k} W[\varepsilon] \cong \mathbf{h}_{k+1,\varepsilon\psi_k/W[\varepsilon]} \text{ for } \psi_k := \psi_1\omega^{1-k},$$

where ω is the Teichmüller character. This identity sends $T(n)\otimes 1$ to $T(n)$.

As is well known, $G_{k+1}(\Gamma_0(Np^{r+1}),\psi) = 0$ if $\psi(-1) = (-1)^k$; hence, we have $\psi_1(-1) = 1$ (even character) in order to have nontrivial $\mathbf{h}_{\psi_1}$. A (normalized) Hecke eigenform in $G_{k+1}(\Gamma_0(Np^{r+1}),\psi)$ has *slope* $\alpha \in \mathbb{Q}$ if $f|U(p) = a \cdot f$ with $|a|_p = p^{-\alpha}$. We simply put $\alpha = \infty$ if $a = 0$. Since $\lim_{n\to\infty} a^{n!} = 1 \Leftrightarrow \alpha = 0$, the algebra $\mathbf{h}_{k+1,\varepsilon\psi_k}$ acts nontrivially on a Hecke eigen cusp form f in $S_{k+1}(\Gamma_0(N),\varepsilon\psi_k;\overline{\mathbb{Q}}_p)$ if and only if f has slope 0. A slope-0 form is also called an *ordinary form*; hence, we call positive-slope forms *nonordinary*. An important consequence of the above two facts is the *boundedness principle*

(b) *The number of slope-*0 *Hecke eigenforms of level* Np^{r+1}*, of weight* $k+1$*, and of a given character* ψ *is bounded independently of* k*,* r*, and* ψ*.*

If a Hecke eigenform f as in (b) has slope 0, $\lambda : h \to \overline{\mathbb{Q}}_p$ given by $f|T = \lambda(T)f$ for $T \in h$ factors through $\mathbf{h}_{k+1,\psi}$ and $f = \sum_{n=1}^{\infty} a(n,f)q^n = \sum_{n=1}^{\infty} \lambda(T(n))q^n$. Thus, the number of slope-0 forms with Neben character ψ is less than or equal to $\mathrm{rank}_W \mathbf{h}_{k+1,\psi} = \mathrm{rank}_\Lambda \mathbf{h}_{\psi_1}$ independent of r and ε. We define the *Hecke field* of f over a number field K to be $K(f) = K(\lambda(T(n))|n = 1,2,\dots)$ (the subfield of $\overline{\mathbb{Q}}$ generated by Hecke eigenvalues of f). This field $K(f)$ is known to be a finite extension of K (e.g., [IAT] Chap. 3).

The corresponding objects for the modular form are denoted by the corresponding capital characters; so, for example,

$$H = \mathbb{Z}[\psi][T(n)|n = 1,2,\cdots] \subset \mathrm{End}(G_{k+1}(\Gamma_0(Np^{r+1}),\psi)),$$

$H_{k+1,\psi/W} = H \otimes_{\mathbb{Z}[\psi]} W$, the ordinary part $\mathbf{H}_{k+1,\psi} \subset H_{k+1,\psi}$ and the big Hecke algebra $\mathbf{H} = \mathbf{H}_{\psi_1/W}$.

Replacing cusp forms by modular forms (and lowercase symbols by uppercase symbols), we construct the "big" cuspidal Hecke algebra $\mathbf{H}_{\psi_1}$. For this algebra $\mathbf{H}_{\psi_1}$, the corresponding assertions as (1) and (2) hold. We have a surjective Λ-algebra homomorphism $\mathbf{H} \twoheadrightarrow \mathbf{h}$ sending $T(n)$ to $T(n)$. We take an irreducible component $\mathrm{Spec}(\mathbb{I}) \subset \mathrm{Spec}(\mathbf{H})$ (thus $\mathbb{I}$ is a torsion-free integral domain over Λ and is a Λ-module of finite type).

Each point $P \in \mathrm{Spec}(\mathbf{H})$ has a two-dimensional (semisimple) representation ρ_P [of $\mathrm{Gal}(\overline{\mathbb{Q}}/\mathbb{Q})$ unramified outside Np] with coefficients in the residue field $\kappa(P)$ of P such that $\mathrm{Tr}(\rho_P(Frob_l)) = (T(l) \mod P)$ for all primes ℓ outside p and the prime-to-p level N. See [GME] Sects. 4.2 and 4.3.1 for an account of the construction of the Galois representation. Here, as a convention, the terminology "residue field" $\kappa(P)$ means the field of factions of the integral domain $\mathbf{h}/P$. By the semisimplicity of ρ_P, the trace of ρ_P characterizes the representation. Since the Frobenius elements $Frob_l$ for primes l outside a finite set S of primes are dense (in the Galois group of the maximal extension unramified outside S) by Chebotarev's density theorem (e.g., [CFN] Sect. V.6), the trace of $\rho_P(Frob_l)$ determines the isomorphism class of ρ_P uniquely.

In particular, writing $\mathbb{I} = \mathbf{H}/\mathcal{P}$ for a minimal prime $\mathcal{P} \in \mathrm{Spec}(\mathbf{H})$, $\mathbb{I}$ (strictly speaking, its quotient field) carries a Galois representation $\rho_{\mathbb{I}} = \rho_{\mathcal{P}}$ with

$\mathrm{Tr}(\rho_{\mathbb{I}}(Frob_l)) = a(l)$ (for the image $a(l)$ in $\mathbb{I}$ of $T(l)$) for all primes $l \nmid Np$.

For $a \in \mathbb{I}$ and $P \in \mathrm{Spec}(\mathbb{I})(\overline{\mathbb{Q}}_p)$, regarding it as a W-algebra homomorphism $P : \mathbb{I} \to \overline{\mathbb{Q}}_p$, we write $a_P \in \overline{\mathbb{Q}}_p$ for $P(a)$. If a prime P of $\mathrm{Spec}(\mathbb{I})$ with $\mathbb{I}/P \hookrightarrow \overline{\mathbb{Q}}_p$ contains $(t - \varepsilon\psi_k(\gamma)\gamma^k)$ with $k \geq 1$, we have a Hecke eigenform $f_P \in G_{k+1}(\Gamma_0(Np^{r(P)+1}), \varepsilon\psi_k)$ with $f_P|T(n) = a_P(n)f_P$ for $a_P(n) = P(a(n)) \in \overline{\mathbb{Q}}_p$ for all n (see Sect. 7.4.3). Such a P is called *arithmetic* if $k \geq 1$, and we write $\varepsilon_P = \varepsilon$, $\psi_P = \varepsilon\psi_k$, $r(P) = r$, and $k(P) = k$ for such a P. Thus, $\mathbb{I}$ gives rise to a slope 0 *analytic family of modular forms* $\mathcal{F}_{\mathbb{I}} = \{f_P|\text{arithmetic } P \in \mathrm{Spec}(\mathbb{I})(\overline{\mathbb{Q}}_p)\}$ and Galois representations $\{\rho_P\}_{P \in \mathrm{Spec}(\mathbb{I})(\overline{\mathbb{Q}}_p)}$.

We call a Galois representation ρ *abelian* if there exists an open subgroup $G \subset \mathrm{Gal}(\overline{\mathbb{Q}}/\mathbb{Q})$ such that the semisimplification $(\rho|_G)^{ss}$ has an abelian image. We call $\mathbb{I}$ an *abelian component* if $\rho_{\mathbb{I}}$ is abelian. A component $\mathbb{I}$ is called *cuspidal* if $\mathrm{Spec}(\mathbb{I}) \subset \mathrm{Spec}(\mathbf{h})$, and if not we call it an *Eisenstein* component. An abelian cuspidal component is also called a *CM component* for a reason we describe soon. Abelian components are either Eisenstein or CM components (e.g., [GME] Sect. 4.3).

In the description of ρ_P, we have written $\mathrm{Tr}(\rho_P(Frob_l)) = (T(l) \mod P)$. The precise meaning of this congruence formula is

$$\mathrm{Tr}(\rho_P(Frob_l)) = a(l)_P, \quad \det(\rho_P(Frob_l)) = \psi_P(l)l^{k(P)} \text{ for all primes } l \nmid Np. \tag{3.1.1}$$

As for $a(p)$, we have the following fact (see [GME] Theorem 4.3.2). For the decomposition subgroup $D_p \subset \mathrm{Gal}(\overline{\mathbb{Q}}/\mathbb{Q})$,

$$\rho_P|_{D_p} \cong \begin{pmatrix} \epsilon_P & * \\ 0 & \delta_P \end{pmatrix} \tag{3.1.2}$$

for an unramified character δ_P with $\delta_P(Frob_p) = a(p)_P$.

If P is arithmetic, ρ_P is the Galois representation of the Hecke eigenform f_P originally constructed by Shimura for weight-2 forms and by Deligne for higher weight (and by Deligne–Serre for weight 1 if f_P is classical; that is, P with $k(P) = 0$ is not arithmetic, and hence f_P is sometimes classical and sometimes not classical, just p-adic). The classical construction is valid for any classical Hecke eigenform f not necessarily ordinary (i.e., with slope 0), but our construction (for example, in [GME] Sects. 4.2 and 4.3.1) also works for a p-adic modular form f_P at nonarithmetic P.

For each Hecke eigenform $f \in G_{k+1}(\Gamma_0(Np^{r+1}), \psi)$, we have a compatible system of Galois representations $\rho_{f,\mathfrak{p}} : \mathrm{Gal}(\overline{\mathbb{Q}}/\mathbb{Q}) \to GL_2(\mathbb{Q}(f)_{\mathfrak{p}})$ for each prime ideal $\mathfrak{p}$ of $\mathbb{Q}(f)$ with coefficients in the $\mathfrak{p}$-adic completion $\mathbb{Q}(f)_{\mathfrak{p}}$ of $\mathbb{Q}(f)$. Write $V_{\mathfrak{p},I_l} := H_0(I_l, V_{\mathfrak{p}})$ for the maximal quotient of the representation space $V_{\mathfrak{p}} := \mathbb{Q}(f)^2_{\mathfrak{p}}$ of $\rho_{f,\mathfrak{p}}$ on which the inertia group I_l at a prime l acts trivially by $\rho_{f,\mathfrak{p}}$. The collection $\rho_f = \{\rho_{f,\mathfrak{p}}\}_{\mathfrak{p}}$ is a *compatible system* defined in Sect. 1.1.2 because the characteristic polynomial $H_l(X) = \det(1 - X \cdot \rho_{f,\mathfrak{p}}(Frob_l)|_{V_{\mathfrak{p},I_l}})$ of l-Frobenius element $Frob_l$ is independent of $\mathfrak{p}$ (see [C]) as long as $l \notin \mathfrak{p}$ (assuming f is primitive in the sense of [MFM] Chap. 4). By the above identity (3.1.1), we have $L(s,f) = \sum_{n=1}^{\infty} a(n,f)n^{-s} = \prod_l H_l(l^{-s})^{-1}$. If $\mathbb{I}$ is nonabelian, the representation ρ_P is perhaps not in a compatible system if P is not arithmetic [with some exceptions when $k(P) = 0$].

If $P \in \mathrm{Spec}(\mathbb{I})(\overline{\mathbb{Q}}_p)$, as $\mathrm{Gal}(\overline{\mathbb{Q}}/\mathbb{Q})$ is compact, we may assume that ρ_P has coefficients in the p-adic integer ring O of $\kappa(P)$ [see [GME] Remark 4.2.1 and [MFG] Exercise (1) of Sect. 3.2.3]; so ρ_P modulo the maximal ideal of O has its semisimplification $\rho_{\mathfrak{m}_{\mathbb{I}}}$. The representation $\rho_{\mathfrak{m}_{\mathbb{I}}} : \mathrm{Gal}(\overline{\mathbb{Q}}/\mathbb{Q}) \to GL_2(\overline{\mathbb{F}}_p)$ depends only on the maximal ideal $\mathfrak{m}_{\mathbb{I}}$ of $\mathbb{I}$ (because a semisimple representation is determined by its trace; see, [MFG] Sect. 2.1.5).

We let $\rho_{f,\mathfrak{p}}$ act, by conjugation $x \mapsto \rho_{f,\mathfrak{p}}(\sigma)x\rho_{f,\mathfrak{p}}(\sigma)^{-1}$, on the Lie algebra $V = \mathfrak{sl}_2(\mathbb{Q}(f)_{\mathfrak{p}}) = \{x \in M_2(\mathbb{Q}(f)_{\mathfrak{p}}) | \mathrm{Tr}(x) = 0\}$ and write $\mathrm{Ad}(\rho_{f,\mathfrak{p}})$ for this three-dimensional representation. Setting $V_{\mathfrak{p},I_l} = V/\sum_{\sigma \in I_l}(\sigma - 1)V$ to be the maximal quotient of the vector space $\mathfrak{sl}_2(\mathbb{Q}(f)_{\mathfrak{p}})$ on which the inertia group I_l at a prime l acts trivially, we get $Ad_l(X) = \det(1 - X \cdot \mathrm{Ad}(\rho_{f,\mathfrak{p}})(Frob_l)|_{\mathrm{Ad}(V)_{\mathfrak{p},I_l}})(\mathfrak{p} \nmid l)$ in $\mathbb{Q}(f)[X]$ independently of $\mathfrak{p}$. We get the adjoint compatible system $\mathrm{Ad}(\rho_f) = \{\mathrm{Ad}(\rho_{f,\mathfrak{p}})\}_{\mathfrak{p}}$ and the *adjoint L-function* by

$$L(s, \mathrm{Ad}(f)) = L(s, \mathrm{Ad}(\rho_f)) = \prod_l Ad_l(l^{-s})^{-1}. \tag{3.1.3}$$

This L-function converges on a right half-plane and can be continued to an entire function on the complex plane (see [Sh5] and [GeJ]) if f is a cusp form. We write $L(s, \mathrm{Ad}(\rho_P))$ for $L(s, \mathrm{Ad}(f_P))$ (if P is arithmetic).

Hereafter we assume $\mathbb{I}$ to be cuspidal. As seen in [MFG] Sect. 5.3.6 and [H88a] (see also [H81a] and [H81b]), we have a p-*adic adjoint L-function*

$$L_p = L_p(\mathrm{Ad}(\rho_{\mathbb{I}})) := L_p(1, \mathrm{Ad}(\rho_{\mathbb{I}})) = L_p(1, \rho_{\mathbb{I}}^{sym\otimes 2} \otimes \det(\rho_{\mathbb{I}})^{-1}) \in \mathbb{I}$$

interpolating

$$L_p(P) := P(L_p) = (L_p \mod P) = \frac{L(1, \mathrm{Ad}(\rho_P))}{\text{period}} \quad \text{for all arithmetic } P.$$

Writing $\mathrm{Spec}(\mathbf{h}) = \mathrm{Spec}(\mathbb{I}) \cup \mathrm{Spec}(\mathbb{X})$ for the complement $\mathbb{X}$, we have (under mild assumptions)

$$\mathrm{Spec}(\mathbb{I}) \cap \mathrm{Spec}(\mathbb{X}) = \mathrm{Spec}(\mathbb{I} \otimes_{\mathbf{h}} \mathbb{X}) \cong \mathrm{Spec}(\mathbb{I}/(L_p)) \text{ (a congruence criterion).}$$

The assumptions are that for the connected component $\mathrm{Spec}(\mathbb{T})$ of $\mathrm{Spec}(\mathbf{h})$ containing $\mathrm{Spec}(\mathbb{I})$, $\mathbb{T}$ is a Gorenstein ring (see [MFG] Sect. 5.3.4 for Gorenstein ring) and that $\mathrm{Spec}(\mathbb{I})$ is normal (see [H11c] Sect. 3.1 for more explanation). Actually, so far, all known computed examples of the normalization of $\mathbb{I}$ are regular isomorphic to a one-variable power series ring over a valuation ring finite over $\mathbb{Z}_p$, and as for $\mathbb{T}$, non-Gorenstein examples are rare, but they exist (see [KiW]).

Writing the image of $T(n)$ in $\mathbb{I}$ (resp., $\mathbb{X}$) as $a(n)$ [resp., $b(n)$] and regarding them as functions on $\mathrm{Spec}(\mathbb{I})$ and $\mathrm{Spec}(\mathbb{X})$, we have $a(n)(P) = b(n)(P)$ for P in the intersection $\mathrm{Spec}(\mathbb{I}) \cap \mathrm{Spec}(\mathbb{X})$; in other words, we have $a(n) \equiv b(n) \mod P$ (congruence between two Hecke eigenvalues). This is a reason for our calling this criterion a "congruence" criterion. The study relating congruence primes and adjoint L-values was started in [H81a]. See also [DoHI] for a historical account of the early research on congruence between Hecke eigenforms in the late 1970s and some open conjectures (in particular, in the mid-1970s, Doi found computationally highly nontrivial numerical examples of congruence primes between primitive forms of the same conductor, the same Neben character, and the same weight).

Congruence modulo a prime l of a fixed Hecke eigenform f in a fixed space $S_{k+1}(\Gamma_0(Np^{r+1}), \psi)$ with another Hecke eigenform g in the same space is controlled by the canonical integer part $c(f)$ of the adjoint L-value $L(1, \mathrm{Ad}(f))$ [i.e., basically, $l|c(f) \Leftrightarrow$ the existence of g with $g \equiv f \mod \mathfrak{l}$ for a prime $\mathfrak{l}|l$]. Thus, $c(f)$ is a product of the prime factors of the discriminant/conductor of the Hecke order $\mathbb{Z}[f] = \mathbb{Z}[\lambda(T(n)]_n$ inside $\mathbb{Q}(f)$ and more if we have g nonconjugate (but congruent) to f. For this reason, the number $c(f)$ is called the *congruence number* of f. This implication from the adjoint L-value to congruences of a cusp form was first discovered by the author in [H81a] and [H81b] and developed into the criterion in the above form in [H86a] (0.7). See [H12b] Sects. 5 and 6 and [MFG] Sect. 5.3.6 for an up-to-date treatment of the criterion and the construction of L_p. The criterion characterizes L_p, and at this moment, L_p is well defined up to units in $\mathbb{I}$ (see [H11c] Conjecture 3.1 on how to normalize L_p without ambiguity).

If we interpolate the adjoint L-values including the cyclotomic variable, that is, adding a variable s that interpolates $L(s, \mathrm{Ad}(\rho_P))$, moving s, we need to multiply the L-value by a nontrivial modifying Euler p-factor as described in Sect. 1.1.2. For this enlarged two-variable adjoint L-function, the modifying factor vanishes at $s = 1$ (see Example 1.6); so $L_p(s, \mathrm{Ad}(\rho_{\mathbb{I}}))$ has an exceptional zero at $s = 1$, and for an $\mathcal{L}$-invariant $0 \neq \mathcal{L}^{an}(\mathrm{Ad}(\rho_{\mathbb{I}})) \in \mathbb{I}[\frac{1}{p}]$, we expect to have $L'_p(1, \mathrm{Ad}(\rho_{\mathbb{I}})) \stackrel{?}{=} \mathcal{L}^{an}(\mathrm{Ad}(\rho_{\mathbb{I}}))L_p$ (in the style of Mazur–Tate–Teitelbaum). Greenberg proposed a definition of a number $\mathcal{L}(\mathrm{Ad}(\rho_P))$ conjectured to be equal to $\mathcal{L}^{an}(\mathrm{Ad}(\rho_P))$ for arithmetic P. We interpolate Greenberg's $\mathcal{L}$-invariant $\mathcal{L}(\mathrm{Ad}(\rho_P))$ over arithmetic P to get a function $\mathcal{L}(\mathrm{Ad}(\rho_{\mathbb{I}})) \neq 0$ in $\mathbb{I}[\frac{1}{p}]$ so that $\mathcal{L}(\mathrm{Ad}(\rho_{\mathbb{I}}))(P) = \mathcal{L}(\mathrm{Ad}(\rho_P))$ for all arithmetic P.

3.1.1 Is Characterizing Abelian Components Important?

Here is a list of such characterizations (possibly conjectural):

- A cuspidal $\mathbb{I}$ is abelian $\Leftrightarrow$ the cuspidal $\mathbb{I}$ is a CM component $\Leftrightarrow$ there exist an imaginary quadratic field $M = \mathbb{Q}[\sqrt{-D}]$ in which p splits into $\mathfrak{p}\overline{\mathfrak{p}}$ and a character $\Psi = \Psi_{\mathbb{I}} : G_M = \mathrm{Gal}(\overline{\mathbb{Q}}/M) \to \tilde{\mathbb{I}}^{\times}$ of conductor $\mathfrak{c}\mathfrak{p}^{\infty}$ for an ideal $\mathfrak{c}$ with $\mathfrak{c}\overline{\mathfrak{c}}D_M|N$ such that $\rho_{\mathbb{I}} \cong \mathrm{Ind}_M^{\mathbb{Q}} \Psi$, where D_M is the discriminant of M

and $\mathrm{Spec}(\tilde{\mathbb{I}})$ is the normalization of $\mathrm{Spec}(\mathbb{I})$. This should be well known; see (cm1–4) in Sect. 3.1.5 in the text and [GME] Theorem 4.3.18. Thus, we called a cuspidal abelian component a *CM component*. Since one can check $\mathrm{Ad}(\mathrm{Ind}_M^{\mathbb{Q}} \Psi) = \left(\frac{M/\mathbb{Q}}{}\right) \oplus \mathrm{Ind}_M^{\mathbb{Q}} \Psi^-$, this implies $L_p = L_p(\Psi^-)L(0, \left(\frac{M/\mathbb{Q}}{}\right))$, where $\Psi^-(\sigma) = \Psi(c\sigma c^{-1}\sigma^{-1})$ for complex conjugation c, and $L_p(\Psi^-)$ is the *anticyclotomic* Katz p-adic L-function associated with Ψ^-. This is a key to the proof by Mazur/Tilouine (e.g., [Ti], [MzT] and also [H82] as a precursor of their result) of the anticyclotomic main conjecture, different from those by Rubin [Ru1] and [Ru2] via an Euler system. See [H06a] and [H09b] for a version taking care of general CM fields.

- $\mathbb{I}$ is abelian $\Leftrightarrow$ ρ_P is abelian for a single arithmetic prime P. By Ribet [Ri3], if ρ_P is abelian, ρ_P has complex multiplication or Eisenstein. Then P has to be on a CM component or on an Eisenstein component(see [H12b] Sects. 3 and 4).
- $\mathbb{I}$ is abelian $\Leftrightarrow$ $\rho_{\mathbb{I}} \mod p$ is abelian? This is almost equivalent to the vanishing of the Iwasawa μ-invariant for $L_p(\Psi^-)$ (see Sect. 3.5 for the vanishing). This is a main result in [H12b].
- (Vertical conjecture in [H11a]) Consider the field $\mathcal{V}_r(\mathbb{I}) \subset \overline{\mathbb{Q}}$ generated by $a_P(p)$ for all arithmetic P with level $\leq Np^{r+1}$ for a fixed $r \geq 0$. Then we conjecture that $\mathbb{I}$ is abelian $\Leftrightarrow$ $[\mathcal{V}_r(\mathbb{I}) : \mathbb{Q}] < \infty$ (see Conjecture 3.24 for a more precise statement). This was a question that L. Clozel asked me in the early 1990s. This holds true if the family contains some weight-2 cusp form whose abelian variety has good ordinary reduction modulo p or more generally a weight-$k \geq 2$ cusp form whose motive is potentially crystalline ordinary at p (see Theorem 3.25). Here a crystalline motive is ordinary if its Newton polygon of the crystalline Frobenius coincides with the Hodge polygon. By applying this crystalline-ordinary criterion, the family $\mathcal{F}_\Delta$ containing Ramanujan's Δ-function has $\mathcal{V}_0(\mathbb{I})$ of infinite degree over $\mathbb{Q}$. In the 1970s, Y. Maeda made a conjecture asserting that $\mathbb{Q}(a(p, f))$ for any normalized Hecke eigenform f in $S_k(SL_2(\mathbb{Z}))$ has degree equal to $d := \dim S_k(SL_2(\mathbb{Z}))$, with its Galois closure having Galois group isomorphic to the symmetric group $\mathfrak{S}_d$ of d letters (cf. [HM]). This conjecture is numerically verified up to large weight and large p (e.g., [A]) and implies our conjecture if $N = 1$.
- (Horizontal theorem in [H11a]) Fix $k \geq 1$ and consider the field $\mathcal{H}_k(\mathbb{I})$ generated by $a_P(p)$ over $\mathbb{Q}(\mu_{p^\infty})$ for all arithmetic P with a fixed weight $k \geq 1$. Then $\mathbb{I}$ is abelian $\Leftrightarrow$ $[\mathcal{H}_k(\mathbb{I}) : \mathbb{Q}(\mu_{p^\infty})] < \infty$ (see Theorem 3.2).
- The representation $\rho_{\mathbb{I}}$ restricted to the decomposition group at p is completely reducible $\Leftrightarrow$ $\mathbb{I}$ is abelian. This is the result of Ghate–Vatsal in [GhV]. See also [BGV], [Zh], and [H13a] for Hilbert modular versions.
- For cuspidal $\mathbb{I}$, $\mathcal{L}(\mathrm{Ad}(\rho_{\mathbb{I}}))$ is a constant function over $\mathrm{Spf}(\mathbb{I})$ if and only if $\mathbb{I}$ is a CM component. This is a corollary of the horizontal theorem given in [H11c], and we will give an outline of the proof in Sect. 3.4.

- (Conjecture/Question) Does a cuspidal component $\mathbb{I}$ have CM by an imaginary quadratic field M if

$$\mathcal{L}(\mathrm{Ad}(f_P)) = \log_p(\mathfrak{p}) \text{ (up to algebraic numbers) for one arithmetic } P$$

  taking a prime factor $\mathfrak{p}$ of p in M? Here $\log_p(\mathfrak{p}) = -\log_p(\overline{\mathfrak{p}})$ is as in Example 1.3. This is a wild guess, but a main point is that if the $\mathcal{L}$-invariant is closely related to an imaginary quadratic field, the form f_P has to come from the imaginary quadratic field.

All statements seem to have good arithmetic consequences, and these examples convinced the author of the importance of giving as many characterizations of abelian components as possible. Another remark is that rigidity lemmas (asserting that a formal subscheme of a formal torus stable under an open subgroup of the automorphism group of the formal torus is essentially a formal subtorus) play an important role in the proof of these characterizations, in particular, in the proof of the horizontal/vertical theorems (Theorems 3.2 and 3.25) and also the vanishing of the μ-invariant. This type of lemma has been studied by Chai [Ch3] over a characteristic p base (see Theorems 10.1 and 10.6), but we also present some characteristic 0 versions (for example, Lemmas 3.12, 3.30, 3.45, and 4.85).

3.1.2 Horizontal Theorem

Let $\mathrm{Spec}(\mathbb{I})$ be a component of $\mathrm{Spec}(\mathbf{h})$. Here is what we prove in this section:

Theorem 3.2 *Pick an infinite set $\mathcal{A} \subset \mathrm{Spec}(\mathbb{I})(\overline{\mathbb{Q}}_p)$ of arithmetic points P with fixed weight $k(P) = k \geq 1$. Write $\mathcal{H}_{\mathcal{A}}(\mathbb{I}) \subset \mathcal{H}_k(\mathbb{I})$ for the field generated over $K := \mathbb{Q}(\mu_{p^\infty})$ by $\{a_P(p)\}_{P\in\mathcal{A}}$. Then the Hecke field $\mathcal{H}_{\mathcal{A}}(\mathbb{I})$ is a finite extension of K if and only if $\mathbb{I}$ is abelian. Moreover, if $\mathbb{I}$ is not abelian,*

$$\limsup_{P\in\mathcal{A}} [K(a_P(p)) : K] = \infty.$$

We may conjecture that $\mathbb{I}$ is abelian if and only if

$$\limsup_{P\in\mathcal{A}} [K(a_P(l)) : K] = \infty$$

for any given prime $l \nmid Np$.

We state a corollary of the boundedness principle (b) for Hecke fields:

Lemma 3.3 *Let $\mathcal{F}$ be a slope-0 p-adic analytic family of Hecke eigenforms with coefficients in $\mathbb{I}$. Then we have*

1. *Fix $0 \leq r < \infty$. Let $K = \mathbb{Q}$. Then the degree $[K(f_P) : K(a_P(p))]$ for arithmetic P with $r(P) \leq r$ is bounded independently of P.*
2. *Let $K = \mathbb{Q}(\mu_{p^\infty})$ and fix $k \geq 1$. Then the degree $[K(f_P) : K(a_P(p))]$ for arithmetic P with $k(P) = k$ is bounded independently of P.*

Proof. If $f \in G_{k+1}(\Gamma_0(Np^{r+1}), \psi; \overline{\mathbb{Q}})$, by the rationality (see Theorem 2.48 and its generalization in Sect. 6.2), the Galois conjugate f^σ given by the q-expansion $\sum_{n=0}^{\infty} a(n,f)^\sigma q^n$ for $\sigma \in \mathrm{Gal}(\overline{\mathbb{Q}}/\mathbb{Q})$ is again a modular form in $G_{k+1}(\Gamma_0(Np^{r+1}), \psi^\sigma; \overline{\mathbb{Q}})$ [where $\psi^\sigma(n) = \psi(n)^\sigma$ for $n \in \mathbb{Z}$]. If $\sigma \in \mathrm{Gal}(\overline{\mathbb{Q}}/K[\psi_1, \omega])$ fixes $a_P(p)$, f_P^σ is still ordinary (i.e., slope 0) Hecke eigenforms of the same level and the same Neben character. The number of such forms is bounded by $\mathrm{rank}_{\mathbb{Z}_p[[T]]}\,\mathbf{h}$ [by principle (b)]. Thus,

$$[K(f_P) : K(a_P(p))] \le [K[\psi_1, \omega] : K]\,\mathrm{rank}_{\mathbb{Z}_p[[T]]}\,\mathbf{h}.$$

This finishes the proof.

Hereafter in this section, we fix $\mathcal{A}$ (as in Theorem 3.2) and assume that $[\mathcal{H}_{\mathcal{A}}(\mathbb{I}) : K] < \infty$ for $K := \mathbb{Q}(\mu_{p^\infty})$. We try to prove that $\mathbb{I}$ is abelian. For a prime l outside Np, let $A(l)$ be a root of $\det(X - \rho_{\mathbb{I}}(Frob_l)) = 0$ in an algebraic closure $\overline{Q}$ of the field of fractions of $\mathbb{I}$. Extending $\mathbb{I}$, we assume that $A(l) \in \mathbb{I}$. Then the value $\alpha_{l,P} := A_P(l)$ is a root of $X^2 - a_P(l)X + \psi_k(l)l^{k(P)} = 0$. If $l = p$, we put $A(l) = a(l)$. Fix l. By Lemma 3.3, $L_P = K[\alpha_{l,P}]$ has bounded degree over K independently of l and P for all $P \in \mathcal{A}$; so l is tamely ramified in L_P/K for sufficiently large primes $l \gg 0$.

3.1.3 Weil Numbers

We gather here some results on Weil numbers for the proof of Theorem 3.2. For a prime l, a *Weil l-number* $\alpha \in \mathbb{C}$ of integer weight $k \ge 0$ satisfies

$$(1)\ \alpha \text{ is an algebraic integer; } (2)\ |\alpha^\sigma| = l^{k/2} \text{ for all } \sigma \in \mathrm{Gal}(\overline{\mathbb{Q}}/\mathbb{Q}).$$

Weil numbers of weight 0 are roots of unity (see Lemma 3.6).

Weil numbers are related to Diophantine problems over finite fields. Let $X_{/\mathbb{F}_q}$ be a smooth geometrically connected projective variety of dimension d defined over a finite field $\mathbb{F}_q$. Write $q = l^f$ for a prime l. In elementary terms, fixing an algebraic closure $\mathbb{F} = \overline{\mathbb{F}}_q$, the set of $\mathbb{F}$-points $X(\mathbb{F})$ of X is the zero set $(\Phi(x), \ldots, \Phi_m(x))$ of finitely many homogeneous equations in $\mathbb{F}_q[x]$ in a projective space

$$\mathbf{P}^N(\mathbb{F}) = \left\{x := (x_0, x_1, \ldots, x_N) \in \mathbb{F}^{N+1} \middle| x \neq 0 := (0, 0, \ldots, 0)\right\} / \mathbb{F}^\times$$

in Lemma 2.13. Thus, $X(\mathbb{F}) = \left\{x \in \mathbf{P}^N(\mathbb{F}) | \Phi_1(x) = \cdots = \Phi_m(x) = 0\right\}$. See Chaps. 2 and 4 for smoothness and connectedness. Let $N_n = |X(\mathbb{F}_{q^n})| = |X(\mathbb{F}) \cap \mathbf{P}^N(\mathbb{F}_{q^n})|$ (the number of $\mathbb{F}_{q^n}$-rational points of X). Then the zeta function $\zeta(s, X)$ of $X_{/\mathbb{F}_q}$ is defined by a Dirichlet series

$$\zeta(s,X) = \exp(\sum_{n=1}^{\infty} \frac{N_n}{n} q^{-ns})$$
$$= 1 + \sum_{n=1}^{\infty} \frac{N_n}{n} q^{-ns} + \frac{1}{2!}\left(\sum_{n=1}^{\infty} \frac{N_n}{n} q^{-ns}\right)^2 + \frac{1}{3!}\left(\sum_{n=1}^{\infty} \frac{N_n}{n} q^{-ns}\right)^3 + \cdots \tag{3.1.4}$$

convergent over a right half-plane of $s \in \mathbb{C}$. Let $P_0(T) = (1-T)$ and $P_{2d}(T) = (1 - q^d T)$.

Exercise 3.4 *Prove that $\zeta(s,X)$ is absolutely convergent over a right half-plane, using $|\mathbf{P}^N(\mathbb{F}_q)| = \frac{q^{N+1}-1}{q-1}$.*

In 1949, Weil conjectured in [Wel, p. 507] the existence of Weil l-numbers $\{\alpha_{ij}\}$ of weight $f \cdot i$ for $j = 1, \ldots, 2g(i)$ for a nonnegative integer $g(i)$ for each integer i with $1 \le i \le 2d-1$ such that

$$\zeta(s,X) = \prod_{i=0}^{2d} P_i(q^{-s})^{(-1)^{i+1}} \text{ for } P_i(T) = \prod_{j=1}^{g(i)} (1 - \alpha_{ij} T).$$

This expression, as a rational function of q^{-s}, implies that $\zeta(s,X)$ can be continued to a meromorphic function on the entire complex plane with zero and pole on the line $\mathrm{Re}(s) = \frac{i}{2}$ $(i = 0, 1, \ldots, 2d)$ and satisfies a certain functional equation. See [Wel] for more details on the conjecture [particularly, his prediction of existence of cohomology theory over X with degree i cohomology group having rank $g(i)$] and [ECH] VI.12 for Grothendieck's proof of rationality via his algebro-geometric cohomology theory (note also an earlier proof by Dwork via p-adic analysis). The conjecture (including the statement for α_{ij}) was proven by Deligne in the above form in [D4]. Weil originally proved his conjecture in the case of curves and abelian varieties, and we will see this when the curve is an elliptic curve. The case of elliptic curves is a theorem of Hasse (Theorem 6.6) earlier than Weil's.

If we start with a smooth projective variety $X \subset \mathbf{P}^n$ defined over $\mathbb{Z}[\frac{1}{N}]$, we can think of the base change $X_{/\mathbb{F}_p}$ taking the zero set in $\mathbf{P}^n(\overline{\mathbb{F}}_p)$ given by reduction modulo (p) for a prime $p \nmid N$ of the polynomials defining $X_{/\mathbb{Z}[\frac{1}{N}]}$. Thus, we have a well-defined $P_i(T)$ for each $0 \le i \le 2d$ and $p \nmid N$, which we write as $P_{i,p}(T)$. The degree-i part of the (imprimitive) L-function of X is defined by an Euler product converging a right half-plane:

$$L_i(s,X) = \prod_{p \nmid N} P_{i,p}(p^{-s})^{-1}. \tag{3.1.5}$$

The conjecture of Hasse-Weil (e.g., [IAT] Sect. 7.1) tells us that this L-function has meromorphic continuation to the entire complex plane and satisfies a functional equation after supplementing with missing Euler factors and Γ-factors appropriately.

Here is one more example of the natural appearance of Weil numbers. For any Hecke eigenform $f \in S_{k+1}(\Gamma_0(Np^{r+1}), \psi)$ with $f|T(l) = a_l f$, if l is prime to Np, then the roots α, β of $X^2 - a_l X + \psi(l)l^k = 0$ are Weil numbers of weight k. When f is Ramanujan's Δ-function, which spans $S_{12}(SL_2(\mathbb{Z}))$, this fact is conjectured by Ramanujan; for general k, it was conjectured by Petersson [cf. [IAT] Sect. 3.5 (Z) and (Z')]. If $f = f_P$ for an arithmetic point P of $\mathrm{Spec}(\mathbf{h})(\overline{\mathbb{Q}}_p)$, the numbers α, β are the eigenvalues of $\rho_P(Frob_l)$ for the Frobenius element $Frob_l$ at the prime l in $\mathrm{Gal}(\overline{\mathbb{Q}}/\mathbb{Q})$, because $\mathrm{Tr}(\rho_P(Frob_l)) = a_l$ and $\det(\rho_P(Frob_l)) = \psi(l)l^k$ by (3.1.1). We define the modular Hecke L-function of f by $L(s, f) = \prod_l (1 - a_l l^{-s} + \psi(l)l^{k-s})^{-1} = \sum_{n=1}^{\infty} a(n, f)n^{-s}$.

As we described in the earlier chapters (and we go deeper into this subject in Sect. 6.2.4), each modular form is a function classifying elliptic curves with additional structures, for example, (E, ω) for the level-1 case and (E, ϕ_N, ω) (with a level-N structure ϕ_N) for level-N cases. Moreover, we remarked that $X_1(N)$ is defined over $\mathbb{Z}[\frac{1}{N}]$ [i.e., $X_1(N)(\overline{\mathbb{Q}})$ is a zero set in a projective space $\mathbf{P}^n(\overline{\mathbb{Q}})$ of homogeneous polynomials defined over $\mathbb{Z}[\frac{1}{N}]$]. We prove in Theorem 6.25 that $X_1(N)$ is (at least) defined over $\mathbb{Z}[\frac{1}{6N}]$, throwing away primes 2 and 3. Thus, we can think of the reduction $X_1(N)_{/\mathbb{F}_p}$ modulo p and degree-1 part of the zeta function $L(s, X_1(N)) = L_1(s, X_1(N))$.

When $k = 1$ (that is, for weight-2 cusp forms), the Hasse–Weil conjecture for $X_1(N)$ and the Ramanujan–Petersson conjecture were proven by Shimura by computing $L(s, X_1(N))$ as a product of $L(s, f)$ for Hecke eigenforms of weight 2 on $X_1(N)$ via identifying the roots α, β of $X^2 - a_l X + \psi(l)l = 0$ with l-Frobenius eigenvalues (see [Sh1] and [IAT] Sect. 7.5) and then reducing the proof of $|\alpha| = |\beta| = \sqrt{p}$ to Weil's work for curves. For higher-weight modular forms, Deligne went down a similar path to prove the Ramanujan–Petersson conjecture, reducing it to his proof of Weil's conjecture taking the fibered product $\overbrace{\mathbf{E} \times_X \mathbf{E} \times_X \cdots \times_X \mathbf{E}}^{k-1}$ (see [D1] and also [Sch]). For weight-1 cusp forms (i.e., the case for $k = 0$), the problem is reduced analytically by Deligne and Serre to higher-weight cases (see [DS]). Thus, one might expect to have a Weil number of weight k even for Hecke eigenvalues of $U(p)$ for a cusp form of weight $k + 1 \geq 2$, as "old" Hecke eigenforms of level Np coming from a form of the original level $p \nmid N$ have α or β as their $U(p)$-eigenvalue. This is not always true for eigenforms properly of level Np^{r+1} ($r \geq 0$), called *primitive* forms (see [MFM] Sect. 4.6 for primitive forms). For example, if $p^{r+1} = C_p(\psi)p^e$ for the p-conductor $C_p(\psi)$ of ψ, then a_p is equal to 0 if $e \geq 2$ or $\pm\sqrt{\psi_0(p)}p^{(k-1)/2}$ if $e = 1$ (Weil number of weight $k - 1$). Here ψ_0 is the primitive character associated with ψ if ψ is imprimitive at p. See [MFM] Corollary 4.6.18 for these facts. The case where we have Weil number of weight $k - 1$ for $k \geq 1$ is called the Steinberg (or special) case, as the automorphic representation generated by the cusp form has p-component isomorphic to the Steinberg representation (see [AAG] about Steinberg/special representations and automorphic representations).

On the other hand, if ψ has p-conductor p^{r+1} (with $r \geq 0$), writing $f|U(p) = a_p \cdot f$, a_p is a Weil number of weight k. This fact follows from [MFM] Theorem 4.6.17. In this case, the proof is more elementary without recourse to arithmetic geometry.

By these facts, Weil numbers have an intimate relation to modular forms and Diophantine geometry; thus, it is natural to ask how often we find such numbers in a given algebraic number field of finite or infinite degree over $\mathbb{Q}$. This is what we study now.

Lemma 3.5 *Let $K/\mathbb{Q}$ be a finite extension of $\mathbb{Q}$ inside $\mathbb{C}$ stable under the "complex conjugation" c [so write $c \in \mathrm{Aut}(K)$ for the field automorphism induced by the complex conjugation]. If, for any field embedding $\sigma : K \hookrightarrow \mathbb{C}$, we have $c \circ \sigma = \sigma \circ c$, K is a subfield of a CM field. In particular, if α is a Weil number, $\mathbb{Q}(\alpha)$ is contained in a CM field.*

Here a *CM field* is a totally imaginary quadratic extension of a totally real field of finite degree (such a field appears as the endomorphism field of an abelian variety with complex multiplication; see [ACM]). A *totally real field F* is a subfield of $\overline{\mathbb{Q}}$ such that $\sigma(F)$ remains a subfield of $\mathbb{R}$ for every $\sigma \in \mathrm{Gal}(\overline{\mathbb{Q}}/\mathbb{Q})$.

Proof. Let K^+ be the subfield of K fixed by c; hence, $K^+ = K \cap \mathbb{R}$ in $\mathbb{C}$. Suppose $c = \mathrm{id}_K$. Then $K = K^+$ is real. Since $\sigma(K) \subset \mathbb{C}$ is also fixed by c by $c \circ \sigma = \sigma \circ c$, $\sigma(K) \subset \mathbb{R}$; hence, all complex embeddings of K send K into $\mathbb{R}$. Thus, $K = K^+$ is totally real, and hence in a CM field. Suppose $c \neq \mathrm{id}_K$. Then K/K^+ is fixed by order-2 automorphism, and K/K^+ is a quadratic extension. By the above argument applied to K^+, K^+ is totally real. By the commutativity $c \circ \sigma = \sigma \circ c$, c is nontrivial on $\sigma(K)$; hence, K/K^+ is a totally imaginary quadratic extension of the totally real field K^+ (i.e., a CM field).

Let $K = \mathbb{Q}(\alpha)$. Then for any embedding $\sigma : K \hookrightarrow \mathbb{C}$, $\alpha^\sigma \alpha^{\sigma c} = |\alpha|^2 = l^k = l^{k\sigma} = (\alpha\alpha^c)^\sigma = \alpha^\sigma \alpha^{c\sigma}$. Thus, we get $c \circ \sigma = \sigma \circ c$, and so $\mathbb{Q}(\alpha)$ is in a CM field.

Here is another lemma due to Kronecker:

Lemma 3.6 *If ζ is a Weil number of weight 0, then there exists a positive integer N such that $\zeta^N = 1$.*

Proof. Let $K = \mathbb{Q}(\zeta)$ and $K_\infty = K \otimes_{\mathbb{Q}} \mathbb{R}$. Then we have $\dim_{\mathbb{Q}} K = \dim_{\mathbb{R}} K_\infty < \infty$, and K_∞ is a product of copies of $\mathbb{C}$ and $\mathbb{R}$. Write $K_\infty = \prod_F F$ for fields F and $|\cdot|_F$ for the Archimedean absolute value of F. Put

$$K_\infty^1 = \left\{ x \in K_\infty^\times : |x|_F = 1 \text{ for all factors } F \text{ of } K_\infty \right\},$$

which is a compact group. By definition, $\zeta \otimes 1 \in K_\infty$ is an element in the multiplicative group $K_\infty^1 \cap (\mathfrak{O}_K^\times \otimes 1)$. By Dirichlet's unit theorem (and its proof in [LFE] Sect. 1.2 Theorem 3), $\mathfrak{O}_K^\times \otimes 1 \subset K_\infty^\times$ is a discrete subgroup. Thus, $K_\infty^1 \cap (\mathfrak{O}_K^\times \otimes 1)$ is compact and discrete. Thus, this group is finite and has order N. In particular, $\zeta^N = 1$.

We call two nonzero algebraic numbers a and b equivalent (written as $a \sim b$) if a/b is a root of unity.

Lemma 3.7 *Let K be a finite field extension of $\mathbb{Q}(\mu_{p^\infty})$ inside $\overline{\mathbb{Q}}$. Then for a given prime l and weight $k \geq 0$, there are only finitely many Weil l-numbers of weight k in K up to equivalence. If $K = \mathbb{Q}(\mu_{p^\infty})$ and there is only one prime in $\mathbb{Z}[\mu_{p^\infty}]$ above (l) (for example, if $l = p$), any Weil l-number of weight k is equivalent to $(l^*)^{k/2}$ in $\overline{\mathbb{Q}}$, where $l^* = (-1)^{(l-1)/2}l$ if l is odd, and $l^* = 2$ if $l = 2$.*

An analytic result of Loxton confirms that, up to equivalence, there are only finitely many Weil l-numbers of a given weight in the maximal abelian extension $\mathbb{Q}^{ab}$ of $\mathbb{Q}$ (see [Lx] Lemma 7). Thus, the finiteness result of the lemma follows from this analytic result of Loxton. We give an algebraic proof.

Proof. The decomposition group of each prime l is of finite index in $\mathrm{Gal}(K/\mathbb{Q})$ [identifying $\mathrm{Gal}(K/\mathbb{Q}) \cong \mathbb{Z}_p^\times$ by the p-adic cyclotomic character, the decomposition group is generated by l if $l \neq p$, and otherwise $l = p$, p is fully ramified in $\mathbb{Q}[\mu_{p^\infty}]$; see [ICF] Chap. 2]; hence, there are only finitely many primes $\mathfrak{L}$ of $\mathbb{Z}[\mu_{p^\infty}]$ above (l). Thus, for a Weil l-number α of weight k, there are only finitely many possibilities of prime factorization of (α) if $l \neq p$. If $(\alpha) = (\beta)$ for two Weil l-numbers α, β, then α/β is a Weil number of weight 0; so $\alpha \sim \beta$ by Kronecker's theorem (Lemma 3.6). If there is only one prime over l in $\mathbb{Z}[\mu_{p^\infty}]$, any Weil l-number of weight k is equivalent to $(l^*)^{k/2}$. Thus, the result follows from this if $K = \mathbb{Q}(\mu_{p^\infty})$.

Let $W(l,k)$ [resp., $W_K(l,k)$] be a complete set of representatives of Weil l-numbers in $\mathbb{Q}(\mu_{p^\infty})$ (resp., in K) of weight k under the equivalence. By the above argument, $W(l,k)$ is a finite set, and we want to prove that $W_K(l,k)$ is finite. Write $d = [K : \mathbb{Q}(\mu_{p^\infty})]$. If $\alpha \in K$ is such a Weil l-number, then $N_{K/\mathbb{Q}(\mu_{p^\infty})}(\alpha)$ is equivalent to a number in $W(l,kd)$. Thus, $N_{K/\mathbb{Q}(\mu_{p^\infty})}$ induces a map $N : W_K(l,k) \to W(l,dk)$. Write L for the field generated by elements in $W(l,dk)$. Then $L/\mathbb{Q}$ is a finite abelian extension in $\mathbb{Q}(\mu_{p^\infty})$. Since no prime completely splits in $\mathbb{Q}(\mu_{p^\infty})$, the decomposition subgroup D of l in $\mathrm{Gal}(K/\mathbb{Q})$ is an open subgroup of finite index. Thus, there are only finitely many valuations v of K with $v(l) = 1$. Let $\mathcal{V}$ be the set of valuations v of K with $v(l) = 1$, which is a finite set. For $v \in \mathcal{V}$ and $\alpha \in W_K(l,k)$, $v(\alpha) \in [0,k] \cap d^{-1}v(L)$, because $N_{K/\mathbb{Q}(\mu_{p^\infty})}(\alpha)$ is in $W(l,dk)$ up to roots of unity. Let $V := \prod_{v \in \mathcal{V}} \left([0,k] \cap d^{-1}v(L)\right)$, which is a finite set. We have a map $\mathrm{ord}_l : W_K(l,k) \to V$ sending α to $\mathrm{ord}_l(\alpha) = (v(\alpha))_{v \in \mathcal{V}}$. If $\mathrm{ord}_l(\alpha) = \mathrm{ord}_l(\beta)$ $(\alpha, \beta \in W_K(l,k))$, then α/β is an algebraic integer with complex absolute value $|(\alpha/\beta)^\sigma| = 1$ for all $\sigma \in \mathrm{Gal}(\overline{\mathbb{Q}}/\mathbb{Q})$; hence, by Kronecker's theorem (Lemma 3.6), $\alpha \sim \beta$. Thus, ord_l is an injection, proving the finiteness of $W_K(l,k)$.

A prime l is *tamely ramified* in a finite extension K/F inside $\overline{\mathbb{Q}}_l$ if the ramification index over F of all primes above l in K is prime to l. It is easy to see that the composite of two tamely ramified extensions of F is still tamely ramified. Thus, we think of the composite of all tamely ramified extensions of F inside $\overline{\mathbb{Q}}_l$, which is called the maximal tamely ramified extension of F. To prove an improvement of Lemma 3.7, we first remark

Lemma 3.8 *Let $K = \mathbb{Q}_l[\mu_{p^\infty}]$ inside $\overline{\mathbb{Q}}_l$, and let K^t/K (resp., K^{ur}/K) be the maximal tamely l-ramified extension (resp., the maximal unramified extension inside K^t) in $\overline{\mathbb{Q}}_l$. Then $K^t_{/K}$ has the l-inertia group isomorphic to $\widehat{\mathbb{Z}}^{(l)}(1)$, where $\widehat{\mathbb{Z}}^{(l)}(1) = \varprojlim_{l \nmid N} \mu_N(\overline{\mathbb{Q}}_l)$ as $\mathrm{Gal}(K^{ur}/\mathbb{Q}_l)$-modules, and we have*

$$\mathrm{Gal}(K^t/K) \cong \begin{cases} \widehat{\mathbb{Z}}^{(p)} \ltimes \widehat{\mathbb{Z}}^{(l)}(1) & \text{if } l \neq p, \\ \widehat{\mathbb{Z}} \ltimes \widehat{\mathbb{Z}}^{(l)}(1) & \text{if } l = p \end{cases} \tag{3.1.6}$$

a semidirect product with $\mathrm{Gal}(K^t/K) \rhd \widehat{\mathbb{Z}}^{(l)}(1)$, and

$$\mathrm{Gal}(K^{ur}/K) \cong \begin{cases} \widehat{\mathbb{Z}}^{(p)} & \text{if } l \neq p, \\ \widehat{\mathbb{Z}} & \text{if } l = p. \end{cases}$$

In particular, for a given $d > 0$, there are finitely many extensions in K^t/K of degree $\leq d$.

Proof. Let L/F be a finite Galois extension for a finite extension F of $\mathbb{Q}_l$. Write $\mathfrak{l}$ for the maximal ideal of the l-adic integer ring O_F of F. We introduce a filtration on the l-adic inertia group $I(L/F)$ of $\mathrm{Gal}(L/F)$. Let $\mathfrak{L}$ be the maximal ideal of the l-adic integer ring O_L of L. By definition,

$$I(L/F) = I_0(L/F) = \left\{\sigma \in \mathrm{Gal}(L/F) \middle| x^\sigma \equiv x \mod \mathfrak{L} \text{ for all } x \in O_L\right\}.$$

Thus, for each integer $j > 0$, we define

$$I_j(L/F) = \left\{\sigma \in \mathrm{Gal}(L/F) \middle| x^\sigma \equiv x \mod \mathfrak{L}^{j+1} \text{ for all } x \in O_L\right\}.$$

Then, by definition, the subgroups $I_j(L/F)$ of $\mathrm{Gal}(L/F)$ are normal. Let $U_0 = O_L^\times$ and $U_j = 1 + \mathfrak{L}^j \subset U_0$ for $j > 0$. Then we claim that the map

$$I_j(L/F)/I_{j+1}(L/F) \to U_j/U_{j+1}$$

given by $\sigma \mapsto \varpi^{\sigma-1}$ for a generator ϖ of $\mathfrak{L}$ is an injective homomorphism. The injectivity immediately follows from the existence of an $\mathfrak{L}$-adic expansion of any $x \in O_L^\times$: $\sum_{j \geq 0} a_j \varpi^j$ with $a_j^\sigma = a_j$ for all $\sigma \in I(L/F)$, which implies

$$\varpi^{\sigma-1} \in 1 + \mathfrak{L}^{j+1} \iff \sigma \in I_{j+1}(L/F).$$

Note that $(\sigma-1)(\tau-1) = (\sigma\tau-1)-(\sigma-1)-(\tau-1)$. Then $\varpi^{(\sigma-1)(\tau-1)} \in U_{j+1}$ because, again by $\mathfrak{L}$-adic expansion, $(\sigma-1)$ for $\sigma \in I_j(L/F)$ takes U_j into U_{j+1} for all j. Thus, it is a homomorphism.

The above claim shows that $(I_j : I_{j+1})$ is a l-power for $j > 0$ and for $r = \dim_{\mathbb{F}_F} \mathbb{F}_L$ with $\mathbb{F}_F := O_F/\mathfrak{l}$ and $\mathbb{F}_L := O_L/\mathfrak{L}$,

$$(I_0 : I_1)\big||\mathbb{F}_L^\times| = |\mathbb{F}_F|^r - 1,$$

which is prime to l. Since $I_\infty(L/F) = \{1\}$, $I_1(L/F)$ is the l-Sylow subgroup of $I(L/F)$. In particular, $I(L/F)/I_1(L/F) \hookrightarrow \mathbb{F}_L^\times$, which is an abelian group. By the theory of Lubin–Tate formal groups (see [CFN] Sect. III.7), there is an explicit way of constructing such L fully ramified and abelian over F with given Galois group isomorphic to $\mathbb{F}_F^\times$, and if $F = \mathbb{Q}_l$, $\mathbb{Q}_l(\mu_l)$ does the job. Thus, $D(L/F)$ is a soluble group, and $I_1(L/F)$ is nilpotent, because $(I_j, I_j) \subset I_{j+1}$.

Let $I = I_F := \mathrm{Gal}(\overline{\mathbb{Q}}_l/F^{ur})$ and $D = \mathrm{Gal}(\overline{\mathbb{Q}}_l/F)$. By class field theory, there always exists a totally ramified extension of F with Galois group isomorphic to $\mathbb{F}_F^\times$. Let $I_l \subset I$ be the maximal l-profinite subgroup. Since $I = \mathrm{Gal}(\overline{\mathbb{Q}}_l/F^{ur})$,

$$I^{(l)} = I/I_l = \varprojlim_{L \subset F^{ur}} I(L/F)/I_1(L/F) \cong \varprojlim_{L \subset F^{ur}} \mathbb{F}_L^\times \cong \widehat{\mathbb{Z}}^{(l)},$$

where $\widehat{\mathbb{Z}}^{(l)} = \prod_{\ell \neq l} \mathbb{Z}_\ell$. In particular, if we write $I_{(q)}$ for the maximal q-profinite quotient of $I^{(l)}$ (and hence of I), it is isomorphic to $\mathbb{Z}_q$. Thus, we have a unique l-totally ramified $\mathbb{Z}_q$-extension $F^{(q)}/F^{ur}$. Let $F_n^{(q)} \subset F^{(q)}$ be the subfield such that $\mathrm{Gal}(F_n^{(q)}/F^{ur}) \cong \mathbb{Z}/q^n\mathbb{Z}$. Since F^{ur} contains all q-power roots of unity, by Kummer theory ([CFN] 1.5), $F_n^{(q)} = F^{ur}(\sqrt[q^n]{\varpi})$ for a prime element ϖ in F^{ur}. Thus, if $\phi \in \mathrm{Gal}(F^{(q)}/F)$ induces the Frobenius and $\sigma \in \mathrm{Gal}(F^{(q)}/F^{ur})$, we see from this that

$$\phi\sigma\phi^{-1} = \sigma^{|\mathbb{F}_F|} = \sigma^{\mathcal{N}(\phi)},$$

for the q-adic cyclotomic character $\mathcal{N} : \mathrm{Gal}(\overline{\mathbb{Q}}_l/\mathbb{Q}_l) \to \mathbb{Z}_q^\times$. This shows that

$$I_{(q)} \cong \mathbb{Z}_q(1) \quad \text{as } \mathrm{Gal}(F^{ur}/F)\text{-module}, \tag{3.1.7}$$

where $\mathbb{Z}_q(m)$ is a Galois module isomorphic to $\mathbb{Z}_q$ as (plain) modules on which the Galois group acts via $\mathcal{N}^m$ for the q-adic cyclotomic character $\mathcal{N}$ realized on $\mathbb{Z}_q(1) = \varprojlim_n \mu_{q^n}$. In particular, $D/I^{(q)} \cong \mathrm{Gal}(\overline{\mathbb{F}}_p/\mathbb{F}_F) \ltimes \mathbb{Z}_q(1) = \phi^{\widehat{\mathbb{Z}}} \ltimes \mathbb{Z}_q(1)$, where $I^{(q)}$ is the maximal prime-to-q-profinite subgroup of I. Thus, we have $\mathrm{Gal}(F^t/F) \cong \mathrm{Gal}(\overline{\mathbb{F}}_p/\mathbb{F}_F) \ltimes \widehat{\mathbb{Z}}^{(l)}(1) = \phi^{\widehat{\mathbb{Z}}} \ltimes \widehat{\mathbb{Z}}^{(l)}(1)$, moving $q \neq l$.

Now we assume that $F \subset K = \mathbb{Q}_l(\mu_{p^\infty})$, and pass to the projective limit $\mathrm{Gal}(K^t/K) = \varprojlim_{F \subset K} \mathrm{Gal}(F^t/F)$. If $l = p$, $\mathbb{F}_F = \mathbb{F}_p$ for any F, so $\mathrm{Gal}(F^t/F)$ is independent of F, and we get the desired result. If $l \neq p$, $\mathrm{Gal}(\mathbb{F}_K/\mathbb{F}_l) \cong \mathbb{Z}_p^\times$, which contains a $\mathbb{Z}_p$-module free of rank 1 (i.e., $1 + p\mathbb{Z}_p$ for p odd and $1 + 4\mathbb{Z}_2$ if $p = 2$). Thus, after passing to the limit, we lose one $\mathbb{Z}_p$-free component from $\widehat{\mathbb{Z}} \cong \phi^{\widehat{\mathbb{Z}}} = \mathrm{Gal}(\overline{\mathbb{F}}_l/\mathbb{F}_l)$ for the Frobenius element ϕ with $\phi(x) = x^l$; hence, $\mathrm{Gal}(K^t/K) \cong \widehat{\mathbb{Z}}^{(p)} \ltimes \widehat{\mathbb{Z}}^{(l)}(1)$ if $l \neq p$.

By this expression, there are finitely many open subgroups of $\mathrm{Gal}(K^t/K)$ of index $\leq d$, and so there are only finitely many extensions of K inside K^t of degree $\leq d$ (see Sect. 4.2.1). This finishes the proof.

We call two numbers $\alpha, \beta \in \overline{\mathbb{Q}}^{\times}$ *p-equivalent* if $\alpha/\beta \in \mu_{p^\infty}(\overline{\mathbb{Q}})$. We write $\alpha \sim_p \beta$ if they are p-equivalent.

Proposition 3.9 *Let $\mathcal{K}_d$ be the set of all finite extensions of $\mathbb{Q}[\mu_{p^\infty}]$ of fixed degree d inside $\overline{\mathbb{Q}}$ whose ramification at l is tame. Then there are only finitely many Weil l-numbers of a given weight, up to p-equivalence, in the set-theoretic union $\bigcup_{L \in \mathcal{K}_d} L$ in $\overline{\mathbb{Q}}$.*

The point of the proof is as follows. Writing $K = \mathbb{Q}[\mu_{p^\infty}]$ and $K_l = K \otimes_{\mathbb{Q}} \mathbb{Q}_l$, by tameness, there are only finitely many isomorphism classes of $K \otimes_{\mathbb{Q}} \mathbb{Q}_l$-algebras $L_l = L \otimes_{\mathbb{Q}} \mathbb{Q}_l$ for $L \in \mathcal{K}_d$. Thus, we only need to prove finiteness for Weil numbers of a given weight contained in a fixed isomorphism class of L_l. We look at the universal composite $L_l \otimes_{K_l} L_l$, which is a product of fields indexed by l-adic nonequivalent normalized valuations $v_1, \ldots, v_n$. Indeed, for any composite X of two copies of L_l embedded as K_l-algebras inside a commutative semisimple K_l-algebra, $a \otimes b \mapsto a \cdot b \in X$ extends to a surjective algebra homomorphism $L_l \otimes_{K_l} L_l \to X$ by the universality of the tensor product. Considering $L_l \otimes_{K_l} L_l$, we can think of any possible composite containing α and β. Another important point is that the simple components of $L_l \otimes_{K_l} L_l$ are indexed by equivalence classes of valuations v_is (see [BCM] VI.8). With these facts in mind, consider a tuple

$$V(\alpha) = (v_1(\alpha \otimes 1), \ldots, v_n(\alpha \otimes 1), v_1(1 \otimes \alpha), \ldots, v_n(1 \otimes \alpha)).$$

If $\alpha \sim \beta$, we have $V(\alpha) = V(\beta)$. The tuple $V(\alpha)$ determines the prime factorization of (α) in any possible composite $K(\alpha, \beta)$; hence, if $V(\alpha) = V(\beta)$, $(\alpha) = (\beta)$ in $K(\alpha, \beta)$; thus, by Kronecker's theorem (Lemma 3.6), $\alpha \sim \beta$. Since there are only finitely many possibilities of $V(\alpha)$, there are only finitely many classes. Here are the details of the proof.

Proof. For each $L_{/K}$ in $\mathcal{K}_d$, consider the Galois closure L^{gal} of L over K (see Sect. 4.2.1 in the text and [TCF] Chap. 2 for Galois theory). Then, by the permutation representation, $\mathrm{Gal}(L^{gal}/K)$ is embedded into the symmetric group of d letters $\mathfrak{S}_d$, hence, writing $\mathfrak{G}$ for the set of all subgroups of $\mathfrak{S}_d$ and $\mathcal{K}_G$ (for a finite group G) for the set of Galois extensions of K with Galois group isomorphic to G inside $\overline{\mathbb{Q}}$ whose ramification at l is tame. Then $\bigcup_{L \in \mathcal{K}_d} L \subset \bigcup_{G \in \mathfrak{G}} \bigcup_{L \in \mathcal{K}_G} L$. Since $\mathfrak{G}$ is a finite set, we only need to prove the proposition, replacing $\mathcal{K}_d$ by $\mathcal{K}_G$.

Let $\mathcal{V}$ be the set of normalized l-adic valuations of $K := \mathbb{Q}[\mu_{p^\infty}]$. Write $\mathbb{F}_v$ for the residue field of $v \in \mathcal{V}$ inside a fixed algebraic closure $\overline{\mathbb{F}}_l$. If $l \neq p$, $\mathbb{F}_v$ is an infinite extension of $\mathbb{F}_l$ generated by all p-power roots of unity; otherwise, $\mathbb{F}_v = \mathbb{F}_p$. As we have remarked, $\mathcal{V}$ is a finite set. Let $L \in \mathcal{K}_G$; hence, L/K is a finite extension of degree $d = |G|$ tamely ramified at l. We claim that there are only finitely many isomorphism classes of K_l-algebras $L_l = L \otimes_{\mathbb{Q}} \mathbb{Q}_l$ for L running through $\mathcal{K}_G$. To see this, write $K_l = K \otimes_{\mathbb{Q}} \mathbb{Q}_l = \prod_{v \in \mathcal{V}} K_v$ for $K_v = \mathbb{Q}_l[\mu_{p^\infty}]$ inside the v-adic completion of K. By Lemma 3.8, there are finitely many open subgroups of $\mathrm{Gal}(K_v^t/K_v)$ with index $\leq d$, and hence there are finitely many isomorphism classes of tamely ramified extensions of K_v of degree $\leq d$. Thus, the claim follows.

Consider the set Ξ of all isomorphism classes of K_l-algebras $L \otimes_{\mathbb{Q}} \mathbb{Q}_l$, where L runs over $\mathcal{K}_G$. Thus, Ξ is a finite set. Pick $\mathcal{L} \in \Xi$ and consider the set $\mathcal{K}_{\mathcal{L}}$ of all $L \in \mathcal{K}_G$ such that $L \otimes_{\mathbb{Q}} \mathbb{Q}_l \cong \mathcal{L}$ as K-algebras. Then $\mathcal{K}_G = \bigsqcup_{\mathcal{L} \in \Xi} \mathcal{K}_{\mathcal{L}}$. The set $\mathcal{K}_{\mathcal{L}}$ is a countable set. Assume $\mathcal{K}_{\mathcal{L}} \neq \emptyset$. We just order its elements as $L_1, L_2, \ldots$. We fix a K-algebra isomorphism $\iota_j : L_j \otimes_{\mathbb{Q}} \mathbb{Q}_l \cong \mathcal{L}$. Thus, $L_j \subset \mathcal{L}$ for every j. We take the (set-theoretic) union $\Omega_{\mathcal{L}} := \bigcup_{j=1}^{\infty} \iota_j(L_j)$ inside $\mathcal{L}$. Since L_j/K is a Galois extension, another choice of ι_j is equal to $\iota_j \circ \sigma$ for $\sigma \in \mathrm{Gal}(L_j/K)$. Thus, $\mathrm{Im}(\iota_j : L_j \to \mathcal{L})$ is independent of the choice of ι_j, and we therefore forget about ι_j. Note that $K_l = K \otimes_{\mathbb{Q}} \mathbb{Q}_l$ is a product of field K_v indexed by $v \in \mathcal{V}$ such that K_v has a valuation inducing v on K. Similarly, $\mathcal{L}$ is a product of finitely many fields with l-adic valuation. Write $\mathcal{L} = \prod_{j=1}^{r} \mathbb{L}_j$ for a simple component $\mathbb{L}_j$, and write V_j for the valuation giving the l-adic topology of $\mathbb{L}_j$ extending one of the members of $\mathcal{V}$. Write $\mathcal{V}_{\mathcal{L}} = \{V_1, \ldots, V_r\}$, which is a finite set.

Let $\mathcal{W}(l,k)$ [resp., $\mathcal{W}(l,k,\mathcal{L})$] be the set of Weil l-numbers of weight k in $\bigcup_{L \in \mathcal{K}_G} L$ (resp., in $\Omega_{\mathcal{L}}$). Thus, for each $\alpha \in \mathcal{W}(l,k,\mathcal{L})$, we find a Galois extension L_α/K containing $L(\alpha)$ over K inside $\mathcal{K}_G$ with Galois group G such that $L_\alpha \otimes_{\mathbb{Q}} \mathbb{Q}_l \cong \mathcal{L}$. Thus, we have

$$\mathcal{W}(l,k) = \bigcup_{\mathcal{L} \in \Xi} \mathcal{W}(l,k,\mathcal{L}).$$

The intersection $\mathcal{W}(l,k,\mathcal{L}) \cap \mathcal{W}(l,k,\mathcal{L}')$ for $\mathcal{L}, \mathcal{L}' \in \Xi$ obviously contains all Weil l-numbers of weight k in K (and possibly more). For each $\alpha \in \mathcal{W}(l,k,\mathcal{L})$, we define a tuple of nonnegative positive rational numbers $S_1(\alpha) \in (\mathbb{Q}_{\geq 0})^{\mathcal{V}_{\mathcal{L}}}$ given by $S_1(\alpha) = (V(\alpha))_{V \in \mathcal{V}_{\mathcal{L}}}$.

If $l \neq p$, l is unramified in $K/\mathbb{Q}$; hence, $V(\alpha) \in |G|^{-1}\mathbb{Z}$ and $0 \leq V(\alpha) \leq k$. Thus,

$$\mathcal{S} := \{S_1(\alpha) | \alpha \in \mathcal{W}(l,k,\mathcal{L})\}$$

is a finite set. If $l = p$, by the above lemma, there are only finitely many Weil p-numbers (up to equivalence) in K of a given weight. Thus, the set $\mathbb{V} \subset \mathbb{Q}$ of valuations of Weil numbers in K of weight $k|G|$ is a finite set. For each $\alpha \in \mathcal{W}(p,k,\mathcal{L})$, take a Galois extension L_α/K (inside $\mathcal{K}_G$) with $\mathrm{Gal}(L_\alpha/K) = G$ containing α with $L_\alpha \otimes_{\mathbb{Q}} \mathbb{Q}_l \cong \mathcal{L}$. Then $\sum_{V \in \mathcal{V}_{\mathcal{L}}} e_V V(\alpha) = v(N_{L_\alpha/K}(\alpha)) \in \mathbb{V}$ for the (tame) ramification index e_V of V over K and $V(\alpha) \in |G|^{-1}\mathbb{V}$. Again,

$$\mathcal{S} := \{S_1(\alpha) | \alpha \in \mathcal{W}(p,k,\mathcal{L})\}$$

is a finite set. If $\alpha/\beta \in \mu_{p^\infty}(K)$, we have $K(\alpha) = K(\beta)$ inside $\mathcal{L}$ and hence $S_1(\alpha) = S_1(\beta)$. By Lemma 3.11 following this proposition, the group of roots of unity in the composite $\mathbf{L}$ of L for all $L \in \mathcal{K}_{\mathcal{L}}$ in $\overline{\mathbb{Q}}$ contains $\mu_{p^\infty}(K)$ as a subgroup of finite index; hence, $S_1(\alpha) = S_1(\beta)$ up to a permutation of valuations in $\mathcal{V}_{\mathcal{L}}$ if $\alpha \sim \beta$.

We consider for each $s \in \mathcal{S}$ the set $\mathcal{W}(s) = \{\alpha \in \mathcal{W}(l,k,\mathcal{L}) | S_1(\alpha) = s\}$. By definition, we have a decomposition

$$\mathcal{W}(l,k,\mathcal{L}) = \bigsqcup_{s \in \mathcal{S}} \mathcal{W}(s).$$

We want to show that $\mathcal{W}(s)/\mu_{p^\infty}(K)$ is a finite set. Write $\mathcal{O}$ for the integral closure of $\mathbb{Z}_l$ in $\mathcal{L}$ (the l-adic integer ring of $\mathcal{L}$). By definition, for $\alpha, \beta \in \mathcal{W}(s)$, the principal $\mathcal{O}$-ideal (α) is equal to (β), because $S_1(\alpha) = S_1(\beta) = s$ determines the prime factorization of $(\alpha) = (\beta)$ in $\mathcal{L}$. Write the tensor product $\mathcal{L} \otimes_{K_l} \mathcal{L} = \mathcal{L}_1 \oplus \mathcal{L}_2 \oplus \cdots \oplus \mathcal{L}_n$ for fields $\mathcal{L}_i$ and write $\mathcal{V}_i$ for the valuation of $\mathcal{L}_i$ normalized so that it extends a member of $\mathcal{V}$. Then the possibility of $\nu(\alpha) = (\mathcal{V}_i((\alpha \otimes 1), \mathcal{V}_i(1 \otimes \alpha))_i$ for α running over $\mathcal{W}(s)$ is finitely many, and if $\nu(\alpha) = \nu(\beta)$, α/β is a Weil l-number of weight 0, because the prime factorization of (α) and (β) in $L_\alpha \otimes_K L_\beta$ is equal. Thus, α/β is a root of unity by Kronecker's theorem (Lemma 3.6), and hence $\alpha \sim \beta$. As already remarked (see Lemma 3.11), {roots of unity in $\mathbf{L}$} $= \bigcup_{\varsigma \in Z} \zeta \cdot \mu_{p^\infty}(K)$ for a finite set Z of roots of unity. Thus, $\mathcal{W}(s)/\mu_{p^\infty}(K)$ is finite, which implies $\mathcal{W}(l,k)/\{p\text{-power roots of unity}\}$ is finite.

Exercise 3.10 *Explain in details why in the above proof we can replace the set $\mathcal{K}_d$ by $\mathcal{K}_G$.*

It is not very difficult to prove the following result.

Lemma 3.11 *Let $\mathcal{K}_?$ be one of $\mathcal{K}_\mathcal{L}$, $\mathcal{K}_G$, and $\mathcal{K}_d$. Suppose $\mathcal{K}_? \neq \emptyset$. Then the group of roots of unity in the composite $\mathbf{L}$ of L for $L \in \mathcal{K}_?$ in $\overline{\mathbb{Q}}$ contains $\mu_{p^\infty}(K)$ as a subgroup with a finite index.*

By this lemma, we can replace the equivalence $\alpha \sim \beta$ by the finer p-equivalence $\alpha \sim_p \beta$ requiring $\alpha/\beta \in \mu_{p^\infty}$, and still the finer equivalence classes in the union $\bigcup_{L \in \mathcal{K}_d} L$ of Weil l-numbers of a given weight are finite. Though we have stated the lemma for $\mathcal{K}_d$, as is clear from the proof below, the assertion of Lemma 3.11 remains true even if we replace $\mathcal{K}_d$ by the set of all extensions of degree $\leq d$ over $\mathbb{Q}[\mu_{p^\infty}]$ (without the tame ramification condition).

Proof. We first take $\mathcal{K}_? = \mathcal{K}_\mathcal{L}$ or $\mathcal{K}_G$. Let $\mu^{(p)} \subset \overline{\mathbb{Q}}$ be the group of roots of unity of order prime to p. We need to prove that the subgroup $\mu^{(p)} \cap \mathbf{L} \subset \mu^{(p)}$ is finite. For any finite subset $\mathcal{K} \subset \mathcal{K}_?$, the composite of $L \in \mathcal{K}$ in $\overline{\mathbb{Q}}$ is a Galois extension of K whose Galois group is a subgroup of $\prod_{L \in \mathcal{K}} G$ contained in

$$\{(\sigma_L)_{L \in \mathcal{K}} | \sigma_L \in \mathrm{Gal}(L/K),\ \sigma_L|_{L \cap L'} = \sigma_{L'}|_{L \cap L'} \text{ for all pairs } (L, L') \in \mathcal{K}^2\}.$$

Thus, the order of any $\sigma \in \mathrm{Gal}(\mathbf{L}/K)$ is a factor of $|G|$ and, hence, bounded. Pick any infinite subgroup $\nu \subset \mu^{(p)}$, and decompose it into a product of ℓ-primary subgroups ν_ℓ for primes $\ell \neq p$. Let $\Xi = \{\ell | \nu_\ell \neq 1\}$. If Ξ is finite, we have $\nu \supset \mu_{\ell^\infty}$ for a prime $\ell \in \Xi$, and $\mathrm{Gal}(K(\nu)/K)$ surjects down to

$\mathrm{Gal}(\mathbb{Q}(\mu_{\ell^\infty})/\mathbb{Q}) = \mathbb{Z}_\ell^\times$; hence, the elements of $\mathrm{Gal}(K(\nu)/K)$ have unbounded order, and hence $\mathbb{Q}(\nu) \not\subset \mathbf{L}$. If $|\Xi| = \infty$, $\mathrm{Gal}(K(\nu_\ell)/K)$ for odd ℓ is cyclic of order $\geq \ell - 1$ and, therefore, elements of $\mathrm{Gal}(K(\nu)/K)$ have unbounded order. Thus, the subgroup $\mu^{(p)} \cap \mathbf{L}$ has to be finite. As for $\mathcal{K}_? = \mathcal{K}_d$, the Galois closure of $L \in \mathcal{K}_d$ over K has Galois group G realized inside $\mathfrak{S}_d$, and the same proof works as what we used isonly the boundedness of $|G| \leq d!$.

3.1.4 A Rigidity Lemma

Recall the base ring W, which is a p-adically complete discrete valuation ring inside $\mathbb{C}_p$. We start with a rigidity lemma:

Lemma 3.12 *Let $\Phi(T) \in W[[T]]$. If there is an infinite subset $\Omega \subset \mu_{p^\infty}(\overline{K})$ such that $\Phi(\zeta - 1) \in \mu_{p^\infty}(\overline{\mathbb{Q}}_p)$ for all $\zeta \in \Omega$, then there exist $\zeta_0 \in \mu_{p^\infty}(W)$ and $s \in \mathbb{Z}_p$ such that $\zeta_0^{-1}\Phi(T) = t^s = \sum_{n=0}^{\infty} \binom{s}{n} T^n$ $(t = 1 + T)$.*

By the assumption, for $s \in \mathbb{Z}_p^\times$ sufficiently close to 1, $\zeta \mapsto \zeta^s$ is an automorphism of $W[[\mu_{p^\infty}]]$ over W; hence, $\Phi(\zeta^s - 1) = \Phi(\zeta - 1)^s \Leftrightarrow \Phi(t^s - 1) = \Phi(t-1)^s$, and the power series is the desired form as stated in a remark of Chai's [Ch1] Remark 6.6.1 (iv). Here is an elementary proof supplied to the author by Kiran Kedlaya (a proof following Chai can be found in [H11a] Sect. 5; see also Lemma 3.30 in the text).

Proof. Making the variable change $T \mapsto \zeta_1^{-1}(T+1) - 1$ for a $\zeta_1 \in \Omega$ (replacing W by its finite extension if necessary), we may replace Ω by $\zeta_1^{-1}\Omega \ni 1$; hence, rewriting $\zeta_1^{-1}\Omega$ as Ω, we may assume that $1 \in \Omega$. Note that $t = 1 \Leftrightarrow T = 0$.

Write the valuation of W as v (and use the same symbol v for an extension of v to $W[\mu_{p^\infty}]$). Normalize v so that $v(p) = 1$. We are trying to show that $\Phi(T) = (1+T)^s \zeta'$ for some $s \in \mathbb{Z}_p$ and some p-power root of unity ζ'. Anyway, we write $\Phi(0) = \zeta' \in \mu_{p^\infty}(\overline{\mathbb{Q}}_p)$. Replacing Φ by $\zeta'^{-1}\Phi$ (and extending the scalars to a finite extension of W if necessary), we may assume that $\Phi(0) = 1$.

Suppose that $\Phi(T) \notin W$ (nonconstant). Write $\Phi(T) - 1 = \sum_{i=1}^{\infty} a_i T^i$. Since W is a discrete valuation ring, there is a least index $j > 0$ for which $v(a_j)$ is minimized. For ϵ sufficiently small, if $v(\tau) = \epsilon$, then $v(\Phi(\tau) - 1) = v(a_j) + j\epsilon$. In particular, for ζ a p-power root of unity, taking $\tau = \zeta - 1$, we have $v(\zeta - 1) = p^{-m}/(p-1)$ for some nonnegative integer m, so we have infinitely many relations of the form $jp^{-m}/(p-1) + v(a_j) = p^{-n}/(p-1)$. Then we may assume $m \to \infty \Rightarrow n \to \infty$, replacing Ω by its infinite subset. Indeed, by the nonconstancy of $\tau \mapsto \Phi(\tau)$, the image of Ω under Φ is an infinite set, and hence over Ω, the order of $\Phi(\zeta - 1)$ is unbounded. Thus, taking limits under $m \to \infty$ yields $v(a_j) = 0$. Also, j must be a power of p, say $j = p^h$, and for m large, we have $n = m - h$.

Since $v(a_j) = 0$, $a_j \mod \mathfrak{m}_W$ is in $\mathbb{F}^\times$. For the moment, assume $\mathbb{F} = \mathbb{F}_p$. That is, a_j reduces to an integer b_0 coprime to p in the residue field of W. We can thus replace $\Phi(T)$ by $\Phi_1(T)$ defined by $\Phi(T) = \Phi_1(T) \times (1+T)^s$ for some s (namely, $s = b_0 j = b_0 p^{h_0}$ for $h_0 := h$) so as to increase the least index

j for which $v(a_j) = 0$. Indeed, writing $\Phi(T) = \sum_{n=0}^{j} a_n T^n + T^{j+1} f(T)$ with $f(T) \in W[[T]]$, we have

$$\sum_{n=0}^{j} a_n T^n \equiv 1 + b_0 T^{p^{h_0}} \equiv (1 + T^{p^{h_0}})^{b_0} \equiv (1+T)^s \mod (\mathfrak{m}_W + (T^{j+1})).$$

We also have $\Phi_1(T) \equiv 1 + T^{j+1} f(T)(1+T)^{-s} \equiv 1 \mod (\mathfrak{m}_W + (T^{j+1}))$. Thus, if we write j_1 for the j for this new Φ_1, $j_1 > j$, and $j_1 = p^{h_1}$ with $h_1 > h_0$ and $a_{j_1} \equiv b_1 \mod \mathfrak{m}_W$ for $b_1 \in \mathbb{Z}$. Repeating this, for $s = \sum_{k=0}^{\infty} b_k p^{h_k} \in \mathbb{Z}_p$, $\Phi(T)/(1+T)^s - 1 = \sum_{n=1} a_n T^n$ no longer has a least j with minimal $v(a_j)$; hence, $\Phi(T)/(1+T)^s = 1$, and we get $\Phi(T) = (1+T)^s$.

Suppose now that $\mathbb{F} \neq \mathbb{F}_p$. We have the Frobenius automorphism ϕ fixing $\mathbb{Z}_p[\mu_{p^\infty}] \subset W[\mu_{p^\infty}]$. Letting ϕ act on power series by $(\sum_n a_n T^n)^\phi = \sum_n a_n^\phi T^n$, we find $\Phi^\phi(t^\phi) = \Phi(t)^\phi$. Since $\Phi(\zeta - 1)$ is a p-power root of unity for ζ in an infinite set $\Omega \subset \mu_{p^\infty}$, we have $\Phi^\phi(\zeta - 1) = \Phi^\phi(\zeta^\phi - 1) = \Phi(\zeta - 1)^\phi = \Phi(\zeta - 1)$. Since $\Omega \subset \widehat{\mathbb{G}}_m$ is Zariski dense, we find that $\Phi^\phi = \Phi$, which shows $\Phi \in W^\phi[[T]]$ for the subring $W^\phi \subset W$ fixed by ϕ. Note that the residue field of W^ϕ is $\mathbb{F}_p$, and the earlier argument applies to $\Phi \in W^\phi[[T]]$.

3.1.5 CM Components

The Galois representation $\rho_{\mathbb{I}}$ restricted to the p-decomposition group D_p is reducible by (3.1.2). We write $\rho_{\mathbb{I}}^{ss}$ for its semisimplification over D_p. More precisely than (3.1.2), we have (see, for example, [GME] Sect. 4.3.1–2)

$$\rho_{\mathbb{I}}^{ss}([\gamma^s, \mathbb{Q}_p]) \sim \begin{pmatrix} (1+T)^{-s} & 0 \\ 0 & 1 \end{pmatrix}, \quad \rho_{\mathbb{I}}^{ss}([p, \mathbb{Q}_p]) \sim \begin{pmatrix} * & 0 \\ 0 & a(p) \end{pmatrix}, \qquad \text{(Gal)}$$

where $\gamma^s = (1+p)^s \in \mathbb{Z}_p^\times$ for $s \in \mathbb{Z}_p$ and $[x, \mathbb{Q}_p]$ is the local Artin symbol.

We say that $\mathcal{F}_{\mathbb{I}} = \{f_P\}$ is a *CM family* if one of the following four equivalent conditions is satisfied (see [H12b] Sect. 3 or [H11a] (CM1–3) in Sect. 1):

(cm1) There exist an arithmetic point $P \in \mathrm{Spec}(\mathbb{I})$ and a nontrivial Galois character χ such that $\rho_P \otimes \chi \cong \rho_P$ (which, after extending scalars, implies $\rho_P \cong \mathrm{Ind}_{\mathrm{Ker}(\chi)}^{\mathrm{Gal}(\overline{\mathbb{Q}}/\mathbb{Q})} \lambda$ for a Galois character λ; see [MFG] Lemma 2.15);
(cm2) for all arithmetic points $P \in \mathrm{Spec}(\mathbb{I})$ and a nontrivial Galois character χ, we have $\rho_P \otimes \chi \cong \rho_P$;
(cm3) for some arithmetic points $P \in \mathrm{Spec}(\mathbb{I})$, f_P is a theta series of a binary quadratic form [i.e., $f_P = \sum_{\mathfrak{a}} \lambda(\mathfrak{a}) q^{N(\mathfrak{a})}$ for a Hecke character λ of an imaginary quadratic field M, where $\mathfrak{a}$ runs over integral ideals of M];
(cm4) for all arithmetic points $P \in \mathrm{Spec}(\mathbb{I})$, f_P is a binary theta series.

If $\mathcal{F}$ has CM, χ as in (cm1) cuts out an imaginary quadratic field M, and $\rho_{\mathbb{I}}$ is isomorphic to the induced representation $\mathrm{Ind}_M^{\mathbb{Q}} \Psi$ for a character

$\Psi : \mathrm{Gal}(\overline{\mathbb{Q}}/M) \to \mathbb{I}^{\times}$. The decomposition $(\rho_{\mathbb{I}}|_D)^{ss} = \epsilon \oplus \delta$ can only happen if p splits into $\mathfrak{p}\overline{\mathfrak{p}}$ in M so that Ψ ramifies at $\mathfrak{p}$ and $\Psi^c(\sigma) = \Psi(c\sigma c^{-1})$ is unramified at $\mathfrak{p}$. Then D is the decomposition group of $\mathfrak{p}$, $\epsilon = \Psi$, and $\delta = \Psi^c$. Write $\mathfrak{O}$ for the integer ring of M. At an arithmetic point, f_P is the theta series of a Hecke character Ψ_P whose p-adic avatar $\Psi_P = P \circ \Psi$ has p-type $\Psi_P([x, M_{\mathfrak{p}}]) = \psi_1 \varepsilon_P(x)\langle x\rangle^{-k(P)}$ $(x \in \mathfrak{O}_{\mathfrak{p}}^{\times})$ identifying $\mathfrak{O}_{\mathfrak{p}}$ with $\mathbb{Z}_p$, and $\Psi_P^c([\mathfrak{p}, M_{\mathfrak{p}}]) = \Psi_P^c([p, M_{\mathfrak{p}}]) = a(p)$; hence, we can verify

$$a(p) = \zeta_0(1+T)^{\log_p(\overline{\mathfrak{p}})/\log_p(\gamma)}$$

for a root of unity ζ_0, where $\log_p(\overline{\mathfrak{p}}) = \frac{1}{h}\log_p(\alpha)$, taking h such that $\overline{\mathfrak{p}}^h = (\alpha)$ with $\alpha \in M$. Thus, in this case, the derivative $\frac{da(p)}{dT}|_{T=0}$ is essentially $\log_p(\mathfrak{p})$, which is closely related to the $\mathcal{L}$-invariant of $\chi = \left(\frac{M/\mathbb{Q}}{}\right)$ mentioned in Theorem 1.5 (2) and also in the statement "Conjecture/Question" in Sect. 3.1.1. This is not a coincidence, and we relate the adjoint $\mathcal{L}$-invariant to the derivative $\frac{da(p)}{dT}$ for the non-CM family in Theorem 3.31.

By a similar computation, we confirm, for any prime $\mathfrak{l}$ of M outside Np,

$$\Psi(Frob_{\mathfrak{l}}) = \zeta_{\mathfrak{l}}(1+T)^{\log_p(\mathfrak{l})/\log_p(\gamma)}$$

with a root of unity $\zeta_{\mathfrak{l}}$ (see [LFE] Sect. 7.6).

3.1.6 An Eigenvalue Formula

Let Q be the field of fractions of $\mathbb{I}$ and fix an algebraic closure $\overline{Q}$ of Q. Recall the root $A(l) \in \overline{Q}$ of $\det(X - \rho_{\mathbb{I}}(Frob_l)) = 0$ for a prime $l \nmid Np$. Extend $\mathbb{I}$ so that $A(l) \in \mathbb{I}$. Further extending $\mathbb{I}$ to its integral closure in its quotient field, we assume that $\mathbb{I}$ is normal. We have written $\alpha_{l,P} = P(A(l))$ for $P \in \mathrm{Spec}(\mathbb{I})(\overline{\mathbb{Q}}_p)$. For $l = p$, we simply put $\alpha_{p,P} = P(a(p))$ for the image $a(p)$ in $\mathbb{I}$ of $U(p)$.

Proposition 3.13 (Frobenius eigenvalue formula) *Fix a rational prime $l \nmid N$ and put $A = A(l)$. Write $K := \mathbb{Q}[\mu_{p^\infty}]$ and $L_P = K(A_P)$ for each arithmetic point P with $k(P) = k$. Suppose that $W = \overline{\mathbb{Q}}_p \cap \mathbb{I}$ and that there exists an infinite set $\mathcal{A}$ of arithmetic points with $k(P) = k \geq 1$ satisfying the following condition:*

(BT) *L_P/K is a finite extension of bounded degree independent of $P \in \mathcal{A}$ and in L_P/K; the prime l is at worst tamely ramified for all $P \in \mathcal{A}$.*

Then we have $A \in W[\mu_{p^n}][[T]][t^{1/p^n}] \cap \mathbb{I}$ $(t = 1+T)$ in $\overline{Q}$ for $0 \leq n \in \mathbb{Z}$, and there exist two Weil l-numbers α_0 and α_1 both of weight 1 such that

$$A(X) = c_0(1+X)^s \quad \textit{with } c_0 = \alpha_1\langle\alpha_0\rangle^{-1} \textit{ and } s = \frac{\log_p(\alpha_0)}{\log_p(\gamma)}.$$

Here $\langle x\rangle = \exp_p(\log_p(x))$ is the p-torsion-free p-profinite projection of $x \in \mathbb{C}_p$.

Under the assumption of Theorem 3.2, the constant c_0 turns out to be a root of unity, as we prove later (see Theorem 3.15). In any case, it is close to the form of the eigenvalues in the CM case discussed in Sect. 3.1.5.

Proof. In this book, we give a proof assuming $\mathbb{I} = \Lambda = W[[T]]$, referring to [H13c] Proposition 5.2 for a proof dealing with the general case. By Proposition 3.9 (and a remark after Lemma 3.11), we have only a finite number of Weil l-numbers of weight k in $\bigcup_{P \in \mathcal{A}} L_P$ up to multiplication by roots of unity in $\mu_{p^\infty}(K)$, and hence A_P for $P \in \mathcal{A}$ hits one such Weil l-number α of weight k infinitely many times, up to p-power order roots of unity, except for the case where $\alpha_{l,P}$ is not a Weil l-number of weight k. If $\alpha_{l,P}$ is not a Weil l-number of weight k, then f_P is Steinberg at l and l is a factor of N; thus, this case is excluded by our assumption $l \nmid N$.

The automorphic representation generated by f_P for arithmetic P has l-component in the principal series if $l \nmid N$ or $k(P) > 1$. Indeed, in the Steinberg case at p, we have $a_P(p) = p^{(k(P)-1)/2}$ up to roots of unity, which has to have slope 0. Hence, it is limited to $k(P) = 1$. This follows from the fact that in the Steinberg case, the central character (i.e., the Neben character) is trivial at p, and we have $a_P(p) = p^{(k(P)-1)/2}$ up to roots of unity [e.g., [MFM] Theorem 4.6.17 (2)]. If $\varepsilon_P \neq 1$, the p-conductor and the p-level match, and as already explained, the $U(p)$-eigenvalue is a Weil number of weight $k = k(P)$.

Note that $\mathrm{Spf}(W[[T]]) = \widehat{\mathbb{G}}_m$ (the formal multiplicative group; see Example 4.79) as $\widehat{\mathbb{G}}_m = \mathrm{Spf}(\widehat{W[t,t^{-1}]})$ for $\widehat{W[t,t^{-1}]} = \varprojlim_n W[t,t^{-1}]/(t^n - 1) \cong W[[T]]$ by $t = 1+T$. As a functor, $\widehat{\mathbb{G}}_m$ sends each local p-profinite W-algebra R to the following multiplicative group:

$$\{x \in R^\times | x \equiv 1 \mod \mathfrak{m}_R\}$$

for the maximal ideal $\mathfrak{m}_R$ of R. Thus, an arithmetic point P as a point of $\widehat{\mathbb{G}}_m(W[\zeta])$ is given by $\gamma^k \zeta$ for a p-power root $\zeta \in \mu_{p^\infty}$ of unity [$k = k(P)$ and $\zeta = \varepsilon_P(\gamma)$], and we can make the variable change $t \mapsto \gamma^{-k} t$ to bring arithmetic points to p-power roots of unity in $\widehat{\mathbb{G}}_m$. This translation operation is very useful. In terms of $T = t - 1$, this variable change is given by $T \mapsto Y = \gamma^{-k}(1+T) - 1$. Then we have $A(Y)|_{Y=0} = A(T)|_{T=\gamma^k - 1}$.

Let $\Phi(Y) := \alpha^{-1} A(Y) = \alpha^{-1} A(\gamma^{-k}(1+T) - 1) \in W[[Y]]$. Here note that $|\alpha|_p = 1$, hence, $\alpha^{-1} \in W$. Let $\Omega_1 = \{\varepsilon_P(\gamma) | P \in \mathcal{A}\}$, which is an infinite set in $\mu_{p^\infty}(K)$. The subset Ω of Ω_1 made up of $\zeta \in \Omega_1$ such that $\Phi(\zeta - 1) \in \mu_{p^\infty}(K)$ is again an infinite set. Then Φ satisfies the assumption of Lemma 3.12, and for a root of unity ζ, we have $A(Y) = \alpha_k (1+Y)^{s_1}$ for $s_1 \in \mathbb{Z}_p$ and $\alpha_k = \zeta \alpha$ $(k \geq 1)$, and $A(T) = \alpha_k (\gamma^{-k}(1+T))^{s_1}$. Let $T = \zeta' \gamma^\kappa - 1$ for $\zeta' \in \mu_{p^\infty}(K)$. Then $A(\zeta' \gamma^\kappa - 1) = \alpha_k (\zeta' \gamma^{-k+\kappa})^{s_1}$, which is equal to a Weil l-number of weight κ. Thus, $c_0 = \alpha_k \gamma^{-k s_1} \in W^\times$.

To get the expression of s as in the proposition, take $\kappa > 1$. Then

$$\alpha_0 := \frac{A(\gamma^\kappa - 1)}{A(\gamma^{\kappa-1} - 1)} = \frac{\alpha_k (\zeta' \gamma^{-k+\kappa})^{s_1}}{\alpha_k (\zeta' \gamma^{-k+\kappa-1})^{s_1}} = \gamma^{s_1},$$

which is an algebraic number α_0 independent of κ. Note that for $\kappa > 1$, α_0 is a ratio of Weil l-numbers of weight $\kappa - 1$ and κ, and hence α_0 is not a root of unity. Thus, we have $s := s_1 = \frac{\log_p(\alpha_0)}{\log_p(\gamma)}$.

We now show that α_0 is a Weil l-number of weight 1. Regard A as a function of $t = 1 + T$, and write it as $A(t)$. Since we have

$$A(t) = \alpha_k(\gamma^{-k}t)^{s_1} \quad \text{for} \quad s_1 = \log_p(\alpha_0)/\log_p(\gamma),$$

we get a Weil number of weight κ $A(\gamma^\kappa) = \alpha_k\gamma^{(\kappa-k)s_1}t^{s_1}|_{t=\gamma^\kappa} = \alpha_k\alpha_0^{\kappa-k}$. Take any normalized l-adic valuation v of $\overline{\mathbb{Q}}$. Since $A(\gamma^\kappa)$ is a Weil l-number of weight κ, we have

$$0 \le v(A(\gamma^\kappa)) = v(\alpha_k) + (\kappa - k)v(\alpha_0) \le \kappa;$$

hence, we get

$$0 \le \frac{v(\alpha_k)}{\kappa - k} + v(\alpha_0) \le \frac{\kappa}{\kappa - k} \quad \text{if } \kappa > k.$$

Taking the limit $\kappa \to \infty$, we get $0 \le v(\alpha_0) \le 1$, and hence α_0 is a Weil l-number of weight 1. Thus, we get $A(t) = \alpha_1\gamma^{-s_1}t^{s_1}$ for $s_1 = \log_p(\alpha_0)/\log_p(\gamma)$ and $c_0 = \alpha_1/\langle\alpha_0\rangle \in i_p(\overline{\mathbb{Q}}^\times)$.

3.1.7 Polynomial Representations of $GL(2)$

We prepare one more preliminary result. We consider the space of homogeneous polynomials $L_n(A) = AX^n + AX^{n-1}Y + \cdots + AY^n$ of degree $n > 0$ with coefficients in a commutative ring A. If $n = 0$, we simply put $L_0(A) = A$. Then we let $\alpha \in GL_2(A)$ act on $P(X,Y) \in L_n(A)$ by $\alpha \cdot P(X,Y) = P((X,Y)\alpha)$ by the right matrix multiplication $(X,Y) \mapsto (X,Y)\alpha$.

For a representation (i.e., a group homomorphism) $\rho : G \to GL_2(A)$ of a group G, we write as $\rho^{sym\otimes n}$ the representation ρ composed with the group homomorphism $GL_2(A) \to \mathrm{Aut}(L_n(A)) \cong GL_{n+1}(A)$ we constructed and call it the symmetric nth tensor representation of ρ. Note that $\rho^{sym\otimes 0}$ is the trivial representation. Simply, taking the basis of $L_n(A)$ given by monomials of degree n, we get the following simple formula if $\rho(g) = (\begin{smallmatrix}\alpha & * \\ 0 & \beta\end{smallmatrix}))$ for $g \in G$ is upper triangular:

$$\mathrm{Tr}(\rho^{sym\otimes n}(g)) = \alpha^n + \alpha^{n-1}\beta + \cdots + \beta^n = \sum_{j=0}^{n} \beta^j\alpha^{n-j}. \tag{3.1.8}$$

More generally, if $\rho(g)$ has eigenvalues α and β, we have

$$\mathrm{Tr}(\rho^{sym\otimes n}(g)) = \alpha^n + \alpha^{n-1}\beta + \cdots + \beta^n = \sum_{j=0}^{n} \beta^j\alpha^{n-j}$$
$$\mathrm{Tr}(\rho(g)^n) = \alpha^n + \beta^n = \mathrm{Tr}(\rho^{sym\otimes n}(g)) - \det(\rho(g))\mathrm{Tr}(\rho^{sym\otimes(n-2)}(g)). \tag{3.1.9}$$

The last formula is valid if $n \ge 2$.

Lemma 3.14 *Let ρ, φ be representations of G into $GL_2(A)$. Suppose that A is a field and that $\rho^{sym\otimes n}$, $\rho^{sym\otimes(n-2)}$, $\varphi^{sym\otimes n}$, and $\varphi^{sym\otimes(n-2)}$ are all absolutely irreducible. If $\mathrm{Tr}(\rho^n) = \mathrm{Tr}(\varphi^n)$ as a function on G, then we have*

$$\rho^{sym\otimes n} \cong \varphi^{sym\otimes n} \text{ and } \det(\rho) \otimes \rho^{sym\otimes(n-2)} \cong \det(\varphi) \otimes \varphi^{sym\otimes(n-2)},$$

where $\det(\rho) \otimes \rho^{sym\otimes(n-2)}(g) = \det(\rho(g)) \cdot \rho^{sym\otimes(n-2)}(g) \in GL_{n+1}(A)$.

Here $\mathrm{Tr}(\rho^n) : G \to A$ is just a function on G whose value at g is given by the trace of the nth matrix power $\rho(g)^n$.

Proof. By (3.1.9), we find

$$\begin{aligned}\mathrm{Tr}(\rho^{sym\otimes n}(g)) - \det(\rho(g))\mathrm{Tr}(\rho^{sym\otimes(n-2)}(g)) &= \mathrm{Tr}(\rho(g)^n) = \mathrm{Tr}(\varphi(g)^n)\\ &= \mathrm{Tr}(\varphi^{sym\otimes n}(g)) - \det(\varphi(g))\mathrm{Tr}(\varphi^{sym\otimes(n-2)}(g)).\end{aligned}$$

This implies

$$\mathrm{Tr}(\rho^{sym\otimes n} \oplus (\det(\varphi) \otimes \varphi^{sym\otimes(n-2)})) = \mathrm{Tr}(\varphi^{sym\otimes n} \oplus (\det(\rho) \otimes \rho^{sym\otimes(n-2)})).$$

By absolute irreducibility, the direct sums $\rho^{sym\otimes n} \oplus (\det(\varphi) \otimes \varphi^{sym\otimes(n-2)})$ and $\varphi^{sym\otimes n} \oplus (\det(\rho) \otimes \rho^{sym\otimes(n-2)})$ are semisimple representations. Since the isomorphism class of semisimple representations is determined by the trace (see [MFG] Sect. 2.1 or [LRF]), we conclude that

$$\rho^{sym\otimes n} \oplus (\det(\varphi) \otimes \varphi^{sym\otimes(n-2)} \cong \varphi^{sym\otimes n} \oplus (\det(\rho) \otimes \rho^{sym\otimes(n-2)}.$$

Since $\dim \rho^{sym\otimes n} = \dim \varphi^{sym\otimes n} = n+1$ and $\dim(\det(\rho) \otimes \rho^{sym\otimes(n-2)}) = \dim(\det(\varphi) \otimes \varphi^{sym\otimes(n-2)}) = n-1$, the factors on the left- and right-hand sides are nonisomorphic. By comparing the dimension, we conclude the assertion of the lemma.

3.1.8 Proof of Theorem 3.2

Recall the root $A(l) \in \overline{\mathbb{Q}}$ (we fixed in Sect. 3.1.6) of $\det(X - \rho_{\mathbb{I}}(Frob_l)) = 0$ for a prime $l \nmid Np$. Extend $\mathbb{I}$ so that $A(l) \in \mathbb{I}$. Extending $\mathbb{I}$ further to its integral closure in its quotient field, we assume that $\mathbb{I}$ is normal. We have written $\alpha_{l,P} = P(A(l))$ for $P \in \mathrm{Spec}(\mathbb{I})(\overline{\mathbb{Q}}_p)$. For $l|Np$, we simply put $\alpha_{l,P} = P(a(l))$ for the image $a(l)$ in $\mathbb{I}$ of $U(l)$.

The horizontal theorem follows from the following theorem:

Theorem 3.15 *Let the notation be as above, and write $K := \mathbb{Q}[\mu_{p^\infty}]$ and $E_P = K(\alpha_{p,P})$ for each arithmetic point P with $k(P) = k$. Suppose that there exists an infinite set $\mathcal{A}$ of arithmetic points with $k(P) = k \geq 1$ satisfying*

(bd) *E_P/K is a finite extension of bounded degree independent of $P \in \mathcal{A}$.*

Then we have an imaginary quadratic field M in which p splits such that the component $\mathbb{I}$ has complex multiplication by M. In particular, for α_0, α_1 given in Proposition 3.13, *we conclude that $\alpha_1 = \alpha_0$ up to roots of unity.*

In this book, as before, we give a proof under the assumption that $\mathbb{I} = W[[T]]$ (see [H13c] for the treatment when $\mathbb{I} \neq W[[T]]$).

Proof. By (bd) and Lemma 3.3 (2) combined, $[K(A(l)_P) : K]$ is bounded by a constant B independent of $P \in \mathcal{A}$ and $l \nmid Np$, and hence the condition (BT) of Proposition 3.13 is satisfied if $l > B!$. Assume that $l > B!$ and write $A = A_l = A(l)$. We use the symbols introduced in the proof of Proposition 3.13. As shown in Proposition 3.13, slightly modifying the two Weil l-numbers of weight 1 there (by multiplying roots of unity if necessary) and writing them again as α_0 and α_1, we have $A_l(\zeta\gamma^\kappa) = \zeta^\kappa \alpha_1 \alpha_0^{\kappa-1}$ (for all $\zeta \in \mu_{p^\infty}$). We now vary $l > B!$.

Pick a p-power root of unity $\zeta \neq 1$ of order $m = p^e$. Write $f = f_P$ for $P = (t - \gamma^k)$ and g for the cusp form $f_{P'}$ for $P' = (t - \zeta\gamma^k)$. Put $\alpha_{f,l} = \alpha_l = \alpha_1 = A_l(\gamma^k)$ and $\alpha_{g,l} = \beta_l = A_l(\zeta\gamma^k)$. Thus, $\alpha_{f,l} = \zeta_l \alpha_{g,l}$ with $\zeta_l^m = 1$. Consider the compatible system of Galois representations associated with f and g. Write ρ_f (resp., ρ_g) for the p-adic member $\rho_{f,\mathfrak{p}}$ (resp., $\rho_{g,\mathfrak{p}}$) of the system associated with f (resp., g).

For a continuous representation $\rho : \mathrm{Gal}(\overline{\mathbb{Q}}/F) \to GL_2(\overline{\mathbb{Q}}_p)$, let $\rho^{sym\otimes j}$ denote the jth *symmetric* tensor representation into $GL_{j+1}(\overline{\mathbb{Q}}_p)$ introduced in the previous section. Suppose that f [and hence g by the equivalence of (cm1–4)] does not have complex multiplication. Then by [Ri3] (see [GME] Theorem 4.3.18), $\mathrm{Im}(\rho_{f,\mathfrak{p}})$ and $\mathrm{Im}(\rho_{g,\mathfrak{p}})$ contain an open subgroup of $SL_2(\mathbb{Z}_p)$. Then $\rho_{?,\mathfrak{p}}^{sym\otimes j}$ $(? = f, g)$ is absolutely irreducible for all $j \geq 0$. Since $\beta_l = \zeta_l \alpha_l$ for an mth root of unity ζ_l, we conclude that $\beta_l^m = \alpha_l^m$ for the integer $m = p^e$. Thus, $\mathrm{Tr}(\rho_{f,\mathfrak{p}}^m(Frob_l)) = \mathrm{Tr}(\rho_{g,\mathfrak{p}}^m(Frob_l))$ for all prime $l \nmid Np$, where $\mathrm{Tr}(\rho_{?,\mathfrak{p}}^m)(g)$ is just the trace of the mth matrix power $\rho_{?,\mathfrak{p}}^m(g)$. Since the continuous functions $\mathrm{Tr}(\rho_{f,\mathfrak{p}}^m)$ and $\mathrm{Tr}(\rho_{g,\mathfrak{p}}^m)$ match on the dense subset of Frobenius conjugacy classes of all primes outside Np, we find that $\mathrm{Tr}(\rho_{f,\mathfrak{p}}^m) = \mathrm{Tr}(\rho_{g,\mathfrak{p}}^m)$ on $\mathrm{Gal}(\overline{\mathbb{Q}}/\mathbb{Q})$. From Lemma 3.14, we get, by the absolute irreducibility of the representations involved,

$$\rho_{f,\mathfrak{p}}^{sym\otimes m} \cong \rho_{g,\mathfrak{p}}^{sym\otimes m}.$$

Write, by (3.1.2), $\rho_{?,\mathfrak{p}}|_{\mathrm{Gal}(\overline{\mathbb{Q}}_p/\mathbb{Q}_p)} \cong \left(\begin{smallmatrix} \epsilon_? & * \\ 0 & \delta_? \end{smallmatrix}\right)$ with unramified $\delta_?$ $(? = f, g)$. Then we conclude $\{\epsilon_f^j \delta_f^{m-j} | j = 0, \ldots, m\} = \{\epsilon_g^j \delta_g^{m-j} | j = 0, \ldots, m\}$ from $\rho_{f,\mathfrak{p}}^{sym\otimes m} \cong \rho_{g,\mathfrak{p}}^{sym\otimes m}$. By (Gal), we have $\delta_?([\gamma, \mathbb{Q}_p]) = 1$ and $\epsilon_?([\gamma, \mathbb{Q}_p]) = \zeta_? \gamma^k$ for $\gamma \in \mathbb{Z}_p^\times$ and a root of unity $\zeta_?$. Thus, we conclude that $\epsilon_f^j \delta_f^{m-j} = \epsilon_g^j \delta_g^{m-j} \Leftrightarrow (\epsilon_f/\epsilon_g)^j = (\delta_g/\delta_f)^{m-j}$ for all j. From (3.1.2), we conclude the triviality $\epsilon_f/\epsilon_g = 1 = \epsilon_P/\epsilon_{P'}$ on the inertia group at p as $\delta_?$ is unramified. This is a contradiction, as we have chosen $1 \neq \zeta = \epsilon_P/\epsilon_{P'}([\gamma, \mathbb{Q}_p])$. Thus, f and hence g have complex multiplication by the same imaginary quadratic field, and hence α_1/α_0 is a root of unity by the explicit form of $A(l)$ described in Sect. 3.1.5.

3.2 Finiteness of Abelian Varieties

We insert here a geometric application of Theorem 3.2 (in order to give an example of geometric consequences of the theorem). Our proof is purely arithmetic (no analytic techniques are used). The result presented here is more advanced than the other topics in this book in the sense that we study abelian schemes (not just elliptic curves) and motives associated with Hecke eigenforms without much pedagogical description (see Sect. 4.3.6 for the definition of abelian schemes). In addition to this, some familiarity with the theory of automorphic representations is assumed here (including the local Langlands correspondence between admissible representations and local Galois representations). Since the results here will not be used later, this section is not fully self-contained (quoting results from some research articles); so, for the first reading, the reader may skip this section. For abelian varieties attached by Shimura to a weight-2 Hecke eigenform, we refer to [IAT] Sect. 7.5; for automorphic representations, see [AAG].

A g-dimensional abelian variety $A_{/F}$ over a number field F is of $GL(2)$-*type* if $\mathrm{End}^0(A_{/F}) := \mathrm{End}(A_{/F}) \otimes_{\mathbb{Z}} \mathbb{Q}$ contains a field E of degree g (see [GME] Sect. 5.3.1). We call such a field E an endomorphism field of A. We say an F-simple abelian variety $A_{/F}$ over a number field F of dimension g has CM if $\mathrm{End}^0(A_{/F} \times_F \overline{\mathbb{Q}})$ contains a semisimple commutative algebra of degree $2g$ over $\mathbb{Q}$ (cf. [ACM]). An abelian variety $A_{/F}$ is F-*simple* if A does not have an abelian subvariety $B \subsetneq A$ defined over F (e.g., [GME] Sect. 4.1.6).

The Galois action on the Tate module of an F-simple abelian variety $A_{/F}$ of $GL(2)$-type with endomorphism field E produces a two-dimensional strictly compatible system of Galois representations (as in the case of elliptic curves)

$$\rho_A = \{\rho_\lambda : \mathrm{Gal}(\overline{\mathbb{Q}}/F) \to GL_2(E_\lambda)\}_\lambda$$

indexed by primes λ of E. Thus, we have its L-function $L(s, \rho_A)$.

Two F-simple abelian varieties A and $B = A_\chi$ [of $GL(2)$-type] are *twist equivalent* if $L(s, \rho_B) = L(s, \rho_A \otimes \chi)$ for a finite order character $\chi : \mathrm{Gal}(\overline{\mathbb{Q}}/F) \to \overline{\mathbb{Q}}^\times$. Here the identity of two L-functions means the identity of the coefficients of Dirichlet series after choosing suitable embeddings of the endomorphism fields of A and B, respectively, into $\overline{\mathbb{Q}}$.

Note that dimension is possibly unbounded over a twist-equivalent class. Since Tate's conjecture has been proven by Faltings for abelian varieties, one could formulate this equivalence by insisting that the two abelian varieties share a simple component over an abelian extension of F. In such a geometric context, $\mathbb{Q}$-simple abelian factors of modular jacobians have been studied in depth by Ribet (for example, see [Ri4] and his papers quoted there). However, we adopt an analytic definition of twist equivalence using their L-function, as it can also be applied to rank-2 $\mathbb{Q}$-motives. The Tate conjecture is not completely known for two given geometric rank-2 motives, though through the modularity of two-dimensional compatible systems of $\mathrm{Gal}(\overline{\mathbb{Q}}/\mathbb{Q})$ by Khare–Wintenberger [KhW] I Theorem 10.1 combined with the Tate conjecture for "modular" rank-2 motives proven by Brown and Ghate [BrG] Corollary 2.4.2, once two rank-2 Grothendieck motives are (both) realized in the cohomology of modular curves (not just as compatible systems), Galois endomorphisms come from motivic endomorphisms (as Tate's conjecture predicts).

An abelian scheme A of dimension g over a field κ of characteristic p is called *ordinary* if we can embed the finite flat group scheme μ_p^g into A over an algebraic closure of κ. An abelian scheme over a number field F has a *good reduction* at a prime $\mathfrak{p}$ if A extends to an abelian scheme over the valuation ring $O_\mathfrak{p}$ of $\mathfrak{p}$. It has *potentially good reduction* if there exists a valuation ring V finite flat over $O_\mathfrak{p}$ over which A has good reduction. We say that A has ordinary potential good reduction if the reduction modulo $\mathfrak{m}_V$ of the extended abelian scheme over V is ordinary (for the maximal ideal $\mathfrak{m}_V$ of V). In the following conjecture, the term *number field* means a finite extension of $\mathbb{Q}$.

Conjecture 3.16 *For a given base number field F, fixing a prime ideal $\mathfrak{p}$ of F, there are only finitely many twist-equivalence classes of non-CM F-simple abelian varieties of $GL(2)$-type with good reduction everywhere outside $\mathfrak{p}$ and potentially ordinary good reduction modulo $\mathfrak{p}$.*

As a result related to the conjecture, we have Faltings' theorem (see [ARG] Chap. II, Sect. 6) asserting the finiteness of the number of abelian varieties of fixed dimension with good reduction outside a finite set S of primes of a number field. As we do not bound the dimension of abelian varieties we consider, Faltings' theorem does not immediately imply the conjecture.

Assuming $F \neq \mathbb{Q}$, if $F_\mathfrak{p} = \mathbb{Q}_p$ and F is not an imaginary quadratic field, there are only finitely many characters of $\mathfrak{p}$-power conductor; hence, one may replace twist equivalence by isomorphism in the conjecture. If $F_\mathfrak{p} \neq \mathbb{Q}_p$, there could be infinitely many characters of $\mathfrak{p}$-power conductor (this is a problem related to the Leopoldt conjecture). Anyway, in the extreme case of $[F_\mathfrak{p} : \mathbb{Q}_p] = [F : \mathbb{Q}]$, we have infinitely many characters of p-power conductor, and starting from an F-simple A, the twist A_χ by a character χ with p-power conductor satisfies an inequality $\dim A_\chi \geq [\mathbb{Q}(\chi) : \mathbb{Q}]$ if the order of χ is sufficiently large [since $\mathrm{End}^0(A) \supset \mathbb{Q}(\chi)$]. Thus, without bounding the dimension, we cannot replace twist equivalence by isomorphism. We have formulated the conjecture in the above way since by doing this, there is no need for a bounding dimension.

For some primes p, it is easy to see that there are infinitely many twist-equivalence classes of non-CM supersingular $\mathbb{Q}$-simple abelian varieties A of $GL(2)$-type having potentially good reduction at p and good reduction everywhere else (see [H12a] Remark 3.2). The abelian variety is *supersingular* if it has no nontrivial p-torsion $\overline{\mathbb{F}}_p$-points. If we fix a $\mathbb{Q}$-simple non-CM abelian variety A with $\dim A \leq 2$ and vary primes, A is ordinary at each prime (of a well-chosen number field possibly different from $\mathbb{Q}$) in a set with Dirichlet density 1 (see [O] 2.7 and [H13a] Sect. 7). The density 1 result is only known up to abelian surfaces but is expected to be true in general. On the other hand, if we vary such abelian varieties, even allowing potential ordinarity at $\mathfrak{p}$ (but requiring good reduction outside $\mathfrak{p}$), we believe that its twist-equivalence classes are finitely many in number. What we prove as an application of Theorem 3.2 follows:

Theorem 3.17 *If $F = \mathbb{Q}$, then the above conjecture holds.*

Actually, for small primes $p = 2, 3, 5$ and 7, there is no such abelian variety defined over $\mathbb{Q}$ (see the argument after Corollary 3.19). There is also a motivic version of this theorem (see Remark 3.23). This theorem is given in [H12a] under the assumption $p > 2$. The assumption $p > 2$ can be removed as it is now eliminated in [H13c] from the horizontal theorem. We prove a stronger version of this theorem as Theorem 3.22, assuming $p > 2$ as before.

If we take for granted the principle that the L-function of the compatible system ρ_A of an F-simple abelian variety A of $GL(2)$-type should be given by the L-function of a cohomological automorphic representation π of $GL_2(F_{\mathbb{A}})$ of weight corresponding to Hodge weights $(1, 0)$ and $(0, 1)$, we would expect that there are finitely many isogeny classes of F-simple abelian varieties with everywhere good reduction, since such a π must have level 1 and there are only finitely many such π of level 1. Since the modularity of ρ_A has been proven at least potentially if F is totally real by Taylor (cf. [Ta]), this finiteness (under the bounding *conductor* of ρ_A) could be accessible now, but some serious problems remain. See [GME] Theorem 5.1.9 for the conductor of ρ_A.

More precisely, for each ρ_A over a totally real field F, Taylor finds a totally real extension $F'_{/F}$ dependent on A over which ρ_A is modular. A conceivable idea is to bound the number of isogeny classes of modular abelian varieties A' of $GL(2)$-type having good reduction at every prime of F' that descends to F. If we find such a finite extension F' independent of A, the desired finiteness follows, as there are finitely many automorphic forms of level 1 on $GL(2)_{/F'}$ of a given weight. The difficulty is to find F' independent of A without assuming the desired finiteness over F outright. Clearly, there is some serious work to be done to make the field F' explicit (at present, we do not know even the finiteness of isomorphism classes of two-dimensional mod p odd Galois representations of a given prime-to-p conductor for a general $F \neq \mathbb{Q}$).

More geometrically, it is a celebrated theorem of Fontaine's that there is no abelian scheme over $\mathbb{Z}$ (see [Fo]). Fontaine's theorem is a starting point of the induction process found by Khare–Wintenberger for proving the Serre's mod-p modularity conjecture (see [KhW] and [Kh1]). If we ease the reduction property to potentially good ordinary reduction, one has to allow all π of the p-power level; hence, even for twist-equivalence classes, it is not evident that the number of classes is finite. By [KhW] I Theorem 10.1, if $F = \mathbb{Q}$, ρ_A is known to be associated with an elliptic Hecke eigenform f and A is isogenous to Shimura's abelian factor A_f in [IAT] Theorem 7.14 of the Jacobian of $X_1(N)$ for the *conductor* N of ρ_A. In other words, the association is given by $L(s, f) = L(s, \rho_{A_f})$. We use this fact and Theorem 3.2 to prove the above theorem. If $f = f_P$ is in an analytic family $\mathcal{F}(\mathbb{I})$ [for $P \in \mathrm{Spec}(\mathbb{I})(\overline{\mathbb{Q}}_p)$], we often write A_P for A_f.

3.2.1 CM Type

For a set of field embeddings Σ of a number field into $\overline{\mathbb{Q}}$, we define $\Sigma^c = \{c \circ \sigma | \sigma \in \Sigma\}$ for complex conjugation c induced by i_∞. For a CM field F, a *CM type* Σ is such a set satisfying $\Sigma \cap \Sigma^c = \emptyset$ and that $\Sigma \cup \Sigma^c$ is the total set of embeddings of F into $\overline{\mathbb{Q}}$. By Lemma 3.5, this definition is independent of the choice of $i_\infty : \overline{\mathbb{Q}} \hookrightarrow \mathbb{C}$. By the theory of complex multiplication, CM types of F are in bijection to the isogeny classes of abelian varieties over $\overline{\mathbb{Q}}$ with complex multiplication by F (see [IAT] Chap. 5).

We define $\mathrm{Inf}_{L/F} \Sigma = \{\sigma : L \hookrightarrow \overline{\mathbb{Q}} : \sigma|_F \in \Sigma\}$ for a finite extension L/F. Assume that F and L are CM fields. Then $\mathrm{Inf}_{L/F} \Sigma$ is again a CM type of L if Σ is a CM type of F. For a CM type Σ of F, write Σ_p for the p-adic places of F induced by $i_p \circ \sigma$ for $\sigma \in \Sigma$. If $\Sigma_p \cap \Sigma_p^c = \emptyset$, we call Σ_p a *p-adic CM type* (of F). We prove the following result:

Theorem 3.18 *Let $\mathcal{A}$ be an infinite set of arithmetic points in* Spec($\mathbb{I}$) *outside $((1+T) - \gamma)$. Suppose that $k(P)$ for all $P \in \mathcal{A}$ is a constant $k \geq 1$. Then the family $\mathcal{F}$ has CM if and only if*

$$\Sigma_P = \{\sigma : \mathbb{Q}(a_P(p)) \hookrightarrow \overline{\mathbb{Q}} : |i_p(\sigma(a_P(p)))|_p = 1\}$$

is a CM type of $\mathbb{Q}(a_P(p))$ for all $P \in \mathcal{A}$.

A CM abelian variety of type (F, Σ) over a number field has potentially *ordinary* good reduction at the place induced by i_p if and only if Σ_p is a p-adic CM type (see a remark after the proof and [H11a] Corollary 2.2).

Proof. Since the "only if" part is easier, we just refer the proof of the "only if" part to [H12a]. Let $F_P = \mathbb{Q}(a_P(p))$. To prove the "if part," let $K_P = \mathbb{Q}(\varepsilon_P)$ and $L_P = F_P(\varepsilon_P)$. Then $\mathrm{Inf}_{L_P/F_P} \Sigma_P$ is a CM type of L_P. Thus, we have $2|\mathrm{Inf}_{L_P/F_P} \Sigma_P| = [L_P : \mathbb{Q}]$. Let $K = \mathbb{Q}[\mu_{p^\infty}]$. By the definition of Σ_P, we have $\Sigma_{P,p} \cap \Sigma_{P,p}^c = \emptyset$; i.e., Σ_P gives rise to a p-adic CM type. Since K has only one p-adic place and $[K_P : \mathbb{Q}] = p^{r(P)-1}(p-1)$, for a fixed embedding $\sigma_0 : K \hookrightarrow \overline{\mathbb{Q}}$, we have

$$\begin{aligned} \left|\mathrm{Inf}_{L_P/F_P} \Sigma_P\right| &= \left|\{\sigma \in \mathrm{Inf}_{L_P/F_P} \Sigma_P : \sigma|_{K_P} = \sigma_0|_{K_P}\}\right| \cdot [K_P : \mathbb{Q}] \\ &= C p^{r(P)} [L_P : K_P] \qquad (3.2.1) \end{aligned}$$

for a constant C independent of P [which is equal to $(p-1)/p$ if $p > 2$]. Take W sufficiently large so that all characters of $(\mathbb{Z}/Np\mathbb{Z})^\times$ have values in $W^\times$.

Write $H = \bigoplus_\psi \mathbf{h}_\psi$, where ψ runs over all (even) characters of $(\mathbb{Z}/Np\mathbb{Z})^\times$. Since $\left|\mathrm{Inf}_{L_P/F_P} \Sigma_P\right|$ is at most the number of conjugate slope-0 forms f_P^σ indexed by $\sigma \in \Sigma_P$ (which is bounded by the rank of the Hecke algebra $H/(1+T)^{p^r} - \gamma^k)H$ acting on them), we have, for $r = r(P)$,

$$|\operatorname{Inf}_{L_P/F_P} \Sigma_P| \le \operatorname{rank}_W H/((1+T)^{p^r} - \gamma^k)H = p^{r(P)} \cdot \operatorname{rank}_\Lambda H. \quad (3.2.2)$$

Since $[K(a_P(p)) : K] = [L_P : L_P \cap K] \le [L_P : K_P]$ as $L_P \cap K \supset K_P$, (3.2.1) and (3.2.2) combined tells us $[K(a_P(p)) : K] \le C^{-1} \operatorname{rank}_\Lambda H$. This is impossible if $\mathcal{F}$ does not have CM, since $\sup_{P \in \mathcal{A}}[K(a_P(p)) : K] = \infty$ by Theorem 3.2. □

If $k(P) = 1$ (i.e., the weight of modular forms is 2) with nontrivial Neben character, Σ_P is a CM type if and only if the abelian variety A_P associated with f_P in [IAT] Theorem 7.14 has ordinary good reduction over $\mathbb{Z}_p[\mu_{p^{r(P)+1}}]$. To see this, we note that the Frobenius endomorphism Φ of $\tilde{A}_P := A_P \otimes_{\mathbb{Z}_p[\mu_{p^{r(P)+1}}]} \mathbb{F}_p$ coincides with $U(p)$ on the étale Barsotti–Tate group $\tilde{A}_P[p^\infty]^{et}$ [e.g.,[GME] Theorem 4.2.6 (1)]; hence, the characteristic polynomial $\mathcal{P}(X)$ of Φ over $\mathbb{Q}$ coincides with a power of the characteristic polynomial of $\alpha := a(p, f_P)$. Then, decomposing the total set of field embeddings of $\mathbb{Q}(f_P)$ into

$$\operatorname{Inf}_{\mathbb{Q}(f_P)/F_P} \Sigma_P \sqcup \Xi_P \sqcup \operatorname{Inf}_{\mathbb{Q}(f_P)/F_P} \Sigma_P^c,$$

the set Ξ_P is characterized by

$$\left\{\sigma : \mathbb{Q}(f_P) \hookrightarrow \overline{\mathbb{Q}}_p \mid p^{-1} < |i_p(\alpha^\sigma)|_p < 1\right\}.$$

Thus, $\Xi_P \neq \emptyset$ ($\Leftrightarrow \Sigma_P$ is not a CM type) implies that the Newton polygon of $\mathcal{P}(X)$ has a middle positive slope between 0 and 1; hence, $\tilde{A}_P$ is ordinary if and only if Σ_P is a CM type. See Corollary 6.39 for an elliptic curve version of this fact. Thus, we get the following fact from the above theorem.

Corollary 3.19 *Let $\mathcal{A}$ be an infinite set of arithmetic points in* Spec($\mathbb{I}$) *outside $((1+T)-\gamma)$. Suppose $k(P) = 1$ for all $P \in \mathcal{A}$. The family $\mathcal{F}$ has CM if and only if the abelian variety A_P associated with f_P by Shimura has ordinary good reduction over $\mathbb{Z}_p[\mu_{p^{r(P)+1}}]$ for all $P \in \mathcal{A}$. In particular, if $\mathcal{F}$ does not have CM, the set made up of arithmetic points P with A_P having ordinary good reduction over $\mathbb{Z}_p[\mu_{p^{r(P)+1}}]$ is a finite set.*

We would ask the following questions:

For a given slope-0 family $\mathcal{F}$ without CM indexed by Spec($\mathbb{I}$), *let*

$$Ord_2(\mathbb{I}) = \{P | A_P \text{ has ordinary good reduction over } \mathbb{Z}_p[\mu_{p^{r(P)+1}}]\}.$$

By the corollary above, this set is a finite set. What are the numbers

$$|Ord_2(\mathbb{I})| \text{ and } R = \sup_{P \in Ord_2(\mathbb{I})} r(P)?$$

Could $Ord_2(\mathbb{I})$ be empty?

For $N = 1$ and for very small primes $p = 2, 3, 5$, and 7, because of the nonexistence of slope-0 analytic families of prime-to-p level 1, we conclude the nonexistence of non-CM $\mathbb{Q}$-simple abelian variety of $GL(2)$-type defined over $\mathbb{Z}[\frac{1}{p}]$ with potentially ordinary good reduction at p. Similarly, it is known that $X_1(Np)$ has genus 0 if $1 \le Np \le 10$ and $Np = 12$; hence, we can extend our list of (N, p) a bit without potentially p-ordinary abelian variety of $GL(2)$-type of prime-to-p conductor N. Here an abelian variety A of $GL(2)$-type has *conductor* N if the compatible system ρ_A has conductor N. On the other hand, the prime $p = 11$ is the smallest to have a nontrivial family (of prime-to-p level 1), but even for this simplest family at $p = 11$ containing Ramanujan's Δ-function, we do not know if $Ord_2(\mathbb{I}) = \emptyset$. Thus, it would be interesting to make a specific computation targeted to find potentially ordinary factors of the modular jacobians for a fixed small level N.

Eknath Ghate informed me of the following example of A_P with nonordinary good reduction at $p = 13$: Take weight 2, level $221 = 13 \cdot 17$, with quadratic nebentypus. Then the Hecke algebra breaks into 4 Galois orbits (each of dimension $4+4+6+6$-dimensional over $\mathbb{Q}$). The two six-dimensional factors have equal Hecke polynomial for $T(13)$ given by

$$x^6 + 8x^5 + 31x^4 + 104x^3 + 403x^2 + 1352x + 2197 \equiv x^4(x^2 + 8x + 5) \mod 13.$$

Thus, we have slopes 0, 1/2, and 1, each of multiplicity 2, for the above polynomial. These abelian varieties are associated with slope-0 forms but do not have ordinary good reduction over $\mathbb{Z}_{13}$. See [BrG] Table I for some other information on this example. Numerically, we can find A_P with nonordinary reduction modulo p, but they are rather rare for small-level primes p; so $Ord_2(\mathbb{I})$ could be large although it is a finite set.

3.2.2 Supercuspidality Implies Supersingularity

Let f be an elliptic Hecke eigenform of weight 2 that generates an automorphic representation $\pi = \otimes_l \pi_l$ with local representation π_l of $GL_2(\mathbb{Q}_l)$ at primes l. See [AAG] for the terminology of representations of $GL_2(\mathbb{Q}_l)$. Assume $p > 2$ for simplicity in this book as before. Let $\rho_\lambda : \mathrm{Gal}(\overline{\mathbb{Q}}/\mathbb{Q}) \to GL_2(\overline{\mathbb{Q}}_l)$ be the λ-adic Galois representation associated with f. Then we know by [C] Théorème A and [We2] that if π_p is supercuspidal and $p > 2$, for $\lambda \nmid p$, there exist a quadratic extension $M/\mathbb{Q}_p$ and an infinite order character $\varphi : \mathrm{Gal}(\overline{\mathbb{Q}}_p/M) \to \overline{\mathbb{Q}}_l^\times$ such that $\rho_\lambda|_{D_p}$ is irreducible, $\rho_\lambda|_{D_p} \cong \mathrm{Ind}_M^{\mathbb{Q}} \varphi$, and $\varphi|_{I_p}$ has finite order [for the decomposition group $D_p = \mathrm{Gal}(\overline{\mathbb{Q}}_p/\mathbb{Q}_p)$ and its inertia subgroup I_p]. Since any Galois conjugate f^σ has its automorphic representation supercuspidal at p, writing l for the rational prime below λ, the l-adic Tate module of Shimura's abelian variety A_f attached to f (cf. [IAT] Theorem 7.14) becomes unramified at p over a finite extension of $\mathbb{Q}_p$; so A_f has potentially good reduction at p by the Néron–Ogg–Shafarevich theorem in [ST].

Proposition 3.20 *If π is supercuspidal at p, A_f has potentially good reduction $\tilde{A}_{f/\overline{\mathbb{F}}_p}$ at p but can never be ordinary [i.e., $\tilde{A}_f[p](\overline{\mathbb{F}}_p) = \{0\}$].*

Proof. Though the result is valid even for $p = 2$, we assume $p > 2$ here for simplicity. We use the notation introduced above the proposition. In particular, take $K = \mathbb{Q}_p$, and $M_{/K}$ is the quadratic extension specified above the proposition. Let E be the subfield of $\overline{\mathbb{Q}}$ generated over $\mathbb{Q}$ by the Hecke eigenvalues of f; hence, we identify E with the E-linear endomorphism algebra $\mathrm{End}^0_E(A_{f/\mathbb{Q}})$.

As we noted, A_f has potentially good reduction modulo p. Take a finite extension $L \subset \overline{\mathbb{Q}}_p$ of $\mathbb{Q}_p$ such that A_f extends to an abelian scheme A over the integer ring of L. By extending L, we may assume that L contains the field M and is a Galois extension of K. Let $\tilde{A}$ be the special fiber of A. Define another character $\psi : \mathrm{Gal}(\overline{\mathbb{Q}}_p/M) \to \overline{\mathbb{Q}}_l^{\times}$ by $\psi(g) = \varphi(\tau g \tau^{-1})$ for an element $\tau \in \mathrm{Gal}(\overline{\mathbb{Q}}_p/K)$ inducing a nontrivial automorphism on M. Write ξ_L for the restriction of ξ to $\mathrm{Gal}(\overline{\mathbb{Q}}_p/L) \subset \mathrm{Gal}(\overline{\mathbb{Q}}_p/M)$ for any character ξ of $\mathrm{Gal}(\overline{\mathbb{Q}}_p/M)$. Then the action of $\mathrm{Gal}(\overline{\mathbb{Q}}_p/L)$ on the λ-adic Tate module $T_\lambda \tilde{A}$ is isomorphic to $\varphi_L \oplus \psi_L$. The two characters φ_L and ψ_L are unramified. Thus, for the prime element ϖ of L and its Artin symbol $\phi := [\varpi, L]$, $\varphi_L(\phi) + \psi_L(\phi) \in E$, and $\varphi_L(\phi)\psi_L(\phi)$ is equal to the value at ϕ of the l-adic cyclotomic character (after extending L further if necessary); hence, we have $\varphi(\phi)\psi(\phi) = |N_{L/\mathbb{Q}_p}(\varpi)|_p^{-1}$. Thus, $P(X) = (X - \varphi(\phi))(X - \psi(\phi))$ is in $E[X]$. Then the characteristic polynomial over $\mathbb{Q}$ of the Frobenius endomorphism Φ of $\tilde{A}$ over the residue field $\mathbb{F}$ of L is given by $\prod_{\sigma:E\hookrightarrow\overline{\mathbb{Q}}} P^\sigma(X)$ for conjugates $P^\sigma(X)$ of $P(X)$ under field embeddings $\sigma : E \to \overline{\mathbb{Q}}$. If $(\varpi)^e = (p)$ in the integer ring of L, we get

$$\psi^\sigma(\phi^e) = \psi^\sigma([p, L]) = \varphi^\sigma(\tau[p, L]\tau^{-1}) = \varphi^\sigma([p^\tau, L]) = \varphi^\sigma([p, L]) = \varphi^\sigma(\phi^e).$$

This implies $|\varphi^\sigma(\phi)|_p = |\psi^\sigma(\phi)|_p$, and by $\varphi(\phi)\psi(\phi) = |N_{L/\mathbb{Q}_p}(\varpi)|_p^{-1}$ already remarked, we have $|\varphi^\sigma(\phi^e)\psi^\sigma(\phi^e)|_p = |\varphi(\phi)|_p^{2e} = |\psi(\phi)|_p^{2e} = |N_{L/\mathbb{Q}_p}(p)|_p$. So we get $|\varphi_L^\sigma(\phi)|_p = |\psi^\sigma(\phi)|_p = (N_{L/\mathbb{Q}}(p))^{-1/2e}$. Thus, $0 < |\psi^\sigma(\phi)|_p < 1$, and hence, all eigenvalues of Φ have a positive slope between 0 and 1. Thus, $\tilde{A}$ cannot be ordinary, and, moreover, $\tilde{A}[p](\overline{\mathbb{F}}_p) = \{0\}$.

Proposition 3.21 *If A_f has potentially ordinary good reduction modulo p, then π_p is in principal series of the form $\pi(\alpha, \beta)$ for two characters $\alpha, \beta : \mathbb{Q}_p^\times \to \overline{\mathbb{Q}}_p^\times$, and one of $\alpha(p)$ and $\beta(p)$ is a p-adic unit; hence, f is nearly p-ordinary, and its twist by a character modulo p power has slope* 0.

Proof. By the above proposition, π_p has to be either in principal series or Steinberg. By [C] Théorème A, Steinberg cases correspond to potentially multiplicative reduction at p; hence, π_p has to be in principal series, and so $\pi_p = \pi(\alpha, \beta)$. Since $|\alpha\beta(p)|_p = |p|_p$, we may assume that $1 \geq |\alpha(p)|_p \geq |\beta(p)|_p > 0$. We have canonical isomorphisms $\mathbb{Z}_p^\times \cong \mathrm{Gal}(\mathbb{Q}_p[\mu_{p^\infty}]/\mathbb{Q}_p) \cong \mathbb{Z}_p^\times \cong \mathrm{Gal}(\mathbb{Q}[\mu_{p^\infty}]/\mathbb{Q})$ via the cyclotomic character $\mathcal{N} : \mathrm{Gal}(\mathbb{Q}[\mu_{p^\infty}]/\mathbb{Q}) \to \mathbb{Z}_p^\times$. Thus, regarding α, β as Galois characters, we can lift $\alpha|_{I_p}$ and $\beta|_{I_p}$ to a unique global character $\tilde{\alpha}, \tilde{\beta} : \mathrm{Gal}(\mathbb{Q}[\mu_{p^\infty}]/\mathbb{Q}) \to \overline{\mathbb{Q}}^\times$ of finite order. Regarding these characters as adelic Hecke characters, we can think of $\pi \otimes \tilde{\alpha}^{-1}$ and $\pi \otimes \tilde{\beta}^{-1}$. We write A' for Shimura's abelian variety associated with $\pi \otimes \tilde{\alpha}^{-1}$. Again by [C] Théorème A or [AME] Sect. 14.5, A' has good reduction modulo p over the fixed field of $\mathrm{Ker}(\tilde{\alpha}^{-1}\beta)$, and the Frobenius endomorphism of A' shares the same characteristic polynomial with the $U(p)$ operator (e.g., [GME] Theorem 4.2.6) and is given by $\tilde{\alpha}^{-1}\alpha(p)$ (up to Galois conjugation). Thus, if A' has ordinary good reduction at p, we get $|\alpha(p)| = |\tilde{\alpha}^{-1}\alpha(p)|_p = 1$, and the primitive form $f \otimes \tilde{\alpha}^{-1}$ associated with $\pi \otimes \tilde{\alpha}^{-1}$ has slope 0.

3.2.3 Twist Classes of Abelian Varieties of $GL(2)$-Type

Theorem 3.17 is a special case $N = 1$ of the following theorem:

Theorem 3.22 *For a given integer $N > 0$ prime to p, there are finitely many twist classes of non-CM $\mathbb{Q}$-simple abelian varieties of $GL(2)$-type with potentially good ordinary reduction modulu p and of prime-to-p conductor N.*

Proof. If A is such an abelian variety as in the theorem, by the Khare–Wintenberger theorem ([KhW] Theorem 10.1) combined with a theorem of Faltings' (e.g., [ARG]), A is isogenous to A_f for an elliptic Hecke eigenform f. Twisting f by a Dirichlet character of p-power conductor, we may assume that f is a p-ordinary form of slope 0 (see Propositions 3.20 and 3.21 and their proofs). The prime-to-p conductor of f remains N after twist by a character of p-power conductor. Then f belongs to a p-adic analytic family of slope-0 forms of prime-to-p conductor N. Since there are only finitely many such families and each family can contain only finitely many A_f of potentially ordinary good reduction by Corollary 3.19, the desired result follows.

Remark 3.23 Fix an integer $k \geq 2$. We look into a rank-2 $\mathbb{Q}$-simple motive of Hodge weights $(k-1, 0)$ and $(0, k-1)$, which has its compatible system of Galois representations. Though we do not know if the identity of L-functions of two Grothendieck motives implies that the two motives are isomorphic (assuming their coefficient fields are the same), we can anyway think of twist-equivalence classes of such motives in an obvious sense. If two motives can be realized in cohomology of modular curves, the Tate conjecture is basically known by [BrG] Corollary 2.4.2. Again by [Kh2] Theorem 10.1, such a motive is twist equivalent to a modular motive (constructed in [Sch]) belonging to a slope-0 family. Then we can prove the finiteness of twist classes of a potentially ordinary crystalline $\mathbb{Q}$-simple motive of rank 2 of a given prime-to-p conductor. For the same small pairs (N, p) we listed above, there are no such motives (independent of weight k).

3.3 Vertical Version

Let $\mathcal{F} = \mathcal{F}_{\mathbb{I}} = \{f_P = \sum_{n=1}^{\infty} a(n, f_P)q^n\}_{P \in \mathrm{Spec}(\mathbb{I})(\overline{\mathbb{Q}}_p)}$ be a cuspidal p-adic analytic family of p-ordinary Hecke eigencusp forms of slope 0. Let $\mathcal{V}_r(\mathbb{I})$

be the subfield of $\overline{\mathbb{Q}}$ generated by $a(n, f_P)$ for all n and all arithmetic $P \in \mathrm{Spec}(\mathbb{I})(\overline{\mathbb{Q}}_p)$ with $r(P) \leq r$. In the early 1990s, L. Clozel asked the author if (or when) the Hecke field $\mathcal{V}_r(\mathbb{I})$ for a finite r is a finite extension of $\mathbb{Q}$. At the time, due to a scarcity of examples, we felt that it "probably" is finite if and only if the family contains a CM theta series (i.e., a binary theta series) of weight $k(P) \geq 2$. If the family contains a binary theta series of weight $k \geq 2$ of an imaginary quadratic field M, all forms have CM by the same (fixed) imaginary quadratic field M, and we can prove via Galois deformation theory that $\mathcal{V}_r(\mathbb{I})$ ($r < \infty$) is contained in a finite extension of the imaginary quadratic field M (see [H11a] Sect. 4 for a proof). The following "vertical" conjecture is a slightly stronger version of our thought at the time:

Conjecture 3.24 *Let $\mathcal{A}$ be an infinite set of arithmetic points with bounded level $r(P) \leq r$ for a fixed $r \geq 0$. Let $\mathcal{V}_{\mathcal{A}}(\mathbb{I})$ be the field generated over $\mathbb{Q}$ by $\{\alpha_{p,P}\}_{P\in\mathcal{A}}$, where P runs over all arithmetic points with $\mathrm{Im}(\varepsilon_P) \subset \mu_{p^r}$ for a fixed r. Then the field $\mathcal{V}_{\mathcal{A}}(\mathbb{I})$ is a finite extension of $\mathbb{Q}$ for a fixed $r < \infty$ if and only if $\mathbb{I}$ is abelian. Moreover, if $\mathbb{I}$ is a nonabelian component, then* $\limsup_{P\in\mathcal{A}}[\mathbb{Q}(f_P) : \mathbb{Q}] = \infty$.

Pick a prime l different from p, and write $\mathcal{V}^{(l)}_{\mathcal{A}}(\mathbb{I})$ for the field generated by $\{\alpha_{l,P}, \beta_{l,P}\}$ for all $P \in \mathcal{A}$, where P runs over all points in $\mathcal{A}$. We could formulate a similar conjecture, replacing $\mathcal{V}_{\mathcal{A}}(\mathbb{I})$ by $\mathcal{V}^{(l)}_{\mathcal{A}}(\mathbb{I})$. However, there is a tricky point when, for some arithmetic P and $l|N$, f_P generates an automorphic representation square integrable at l (then actually, for all arithmetic points, this assertion holds). It is known that even if $\mathbb{I}$ is nonabelian, in such a case of "square-integrable" l, $\mathcal{V}^{(l)}_{\mathcal{A}}(\mathbb{I})$ is a finite extension of $\mathbb{Q}$. Thus, we just refer to Conjecture 3.1 of [H11a] for a description of what happens for $\mathcal{V}^{(l)}_{\mathcal{A}}(\mathbb{I})$.

We give an outline of a proof in [H11a] of the vertical theorem.

Theorem 3.25 (Vertical theorem) *Let r be a nonnegative integer, and assume $p > 2$. For an infinite set $\mathcal{A}$ of arithmetic points P with bounded level $r(P) \leq r$ for a finite $r \geq 0$, assume that $\mathcal{V}_{\mathcal{A}}(\mathbb{I})$ is a finite extension of $\mathbb{Q}$. If there exists an arithmetic point $P_0 \in \mathcal{A}$ with $k(P_0) \geq 1$ such that*

1. *$\alpha_0 = a_{P_0}(p)$ is a Weil number,*
2. *$\Sigma_{\alpha_0} = \{\sigma : \mathbb{Q}(\alpha_0) \hookrightarrow \overline{\mathbb{Q}} : |i_p(\alpha_0^\sigma)|_p = 1\}$ is a CM type of $\mathbb{Q}(\alpha_0)$.*

Then $\mathbb{I}$ has complex multiplication.

As explained after Theorem 3.2 of [H11a], condition (2) is equivalent to the following two assertions:

(C) The motive M attached to f_{P_0} by Scholl [Sch] is potentially crystalline at p (see [MTV] for motives).

(O) Regarding M as a motive with coefficients in $\mathbb{Q}$ [not in $\mathbb{Q}(f_{P_0})$], M is ordinary, i.e., the p-adic Newton polygon of α_0 over $\mathbb{Q}$ coincides with the Hodge polygon with three vertices

$$(0,0), ([\mathbb{Q}(f_{P_0}) : \mathbb{Q}], 0), (2[\mathbb{Q}(f_{P_0}) : \mathbb{Q}], k(P_0) - 1).$$

In particular, if M has coefficients in $\mathbb{Q}$ (for example, if M is associated with Ramanujan's Δ function), it satisfies the requirements of the theorem (assuming ordinarity of M). Note here that making the weight higher and higher, it becomes increasingly difficult to find rational Hecke eigenforms. It might be the case that all rational Hecke eigencusp forms have complex multiplication if the weight ≥ 27 (another conjectural characterization of CM forms).

3.3.1 Results Toward the Vertical Theorem

We prepare

Proposition 3.26 *Let $\mathcal{F} = \{f_P\}_{P \in \mathrm{Spec}(\mathbb{I})(\overline{\mathbb{Q}}_p)}$ be a p-adic analytic family of classical p-ordinary Hecke eigenforms and $\mathcal{A} \subset \mathrm{Spec}(\mathbb{I})(\overline{\mathbb{Q}}_p)$ be an infinite set of arithmetic points P with $r(P) \leq r$ for a fixed $r \geq 0$ $\mathcal{V}_{\mathcal{A}}(\mathbb{I})$ is a finite extension of $\mathbb{Q}$. Assume for $P_0 \in \mathcal{A}$ that*

1. *$\alpha_0 = a_{P_0}(p)$ is a Weil number,*
2. *$\Sigma_{\alpha_0} = \{\sigma : \mathbb{Q}(\alpha_0) \hookrightarrow \overline{\mathbb{Q}} : |i_p(\alpha_0^\sigma)|_p = 1\}$ is a CM type of $\mathbb{Q}(\alpha_0)$.*

Then there exists a Weil p-number α of weight 1 with $|i_p(\alpha)|_p = 1$ such that $a(p, f_P) = \zeta(\varepsilon_P(\gamma))^{\log_p(\alpha)/\log_p(\gamma)} \langle\alpha\rangle^{k(P)}$ for a root of unity ζ for all arithmetic P with $k(P) \geq 1$, where $\langle\alpha\rangle = \exp_p(\log_p(i_p(\alpha)))$.

Proof. In this book, to simplify the argument, we only deal with the case where $M := \mathcal{V}_{\mathcal{A}}(\mathbb{I})$ is an imaginary quadratic field and $r = 0$; hence, $M = \mathbb{Q}(\alpha_0)$ (see [H11a] Proposition 7.2 for the general case). Since we assumed $r = 0$, we can forget about $\varepsilon_P(\gamma) = 1$. Take $P \in \mathcal{A}$ with $k(P) > 1$. Then $\alpha_{p,P}$ is a Weil number of weight $k(P) > 1$ with $|\alpha_{p,P}|_p = 1$. Thus, (p) has to split in M, and so $(p) = \mathfrak{p}\overline{\mathfrak{p}}$ in M. Thus, $\Sigma_{\alpha_{p,P}}$ consists of the single element $\iota = i_p|_M$, and for each k, there exists at most one Weil number $\alpha_k \in M$ of weight k (up to roots of unity in M) such that $|\alpha_k|_p = 1$. In M, $(\alpha_k) = \overline{\mathfrak{p}}^k$ for the prime ideal $\mathfrak{p}$ of M corresponding to $i_p|_M$. Fix such a k. Taking a kth root $\alpha = \sqrt[k]{\alpha_k}$, we have $\alpha_l = \alpha^l$ up to roots of unity for all l as $(\alpha_l) = \overline{\mathfrak{p}}^l$.

Since $\mathcal{A}$ is an infinite set, we find an infinite sequence, $P_1, P_2, \ldots, P_n, \ldots$ in $\mathcal{A}$, with increasing weight $k(P_1) < k(P_2) < \cdots$ such that $(a_{P_j}(p)) = \overline{\mathfrak{p}}^{k(P_j)}$ for all $j > 0$. Put

$$\langle\alpha\rangle = \exp(\frac{1}{k(P_0)} \log_p(a(p, f_{P_0})) = \exp(\log_p(\alpha)).$$

Since $(a_{P_j}(p)) = \overline{\mathfrak{p}}^{(k(P_j))}$, $a_{P_j}(p)/\langle\alpha\rangle^{k(P_j)}$ is a Weil number of weight 0; that is, it is an algebraic integer with all its conjugates having absolute value 1. Then, by Kronecker's theorem, we find $a_{P_j}(p) = \zeta_{P_j}\langle\alpha\rangle^{k(P_j)}$ for a root of unity ζ_{P_j}. Note that $\langle\alpha\rangle$ is contained in a finite extension M'/M. Since there are finitely many roots of unity in M', we have only finitely many possibilities of ζ_{P_j}. Therefore, replacing $\{P_j\}_j$ by its subsequence, we find an infinite sequence $P_1, P_2, \cdots, P_n, \cdots$ of increasing weights such that $a_{P_j}(p) = \zeta\langle\alpha\rangle^{k(P_j)}$ for all $j = 1, 2, \ldots$ for a fixed root of unity ζ. We have a power series $\Phi_\alpha(T) \in W[[T]]$ with coefficients in a discrete valuation ring W finite flat over $\mathbb{Z}_p$ such that $\Phi_\alpha(\gamma^k - 1) = \zeta\langle\alpha\rangle^k$ for all integers k. Since $\mathcal{F}$ is an ordinary family, there exists an element $A \in \mathbb{I}$ such that $a(p, f_P) = (A \mod P)$ for all primes P of $\mathrm{Spec}(\mathbb{I})(\overline{\mathbb{Q}}_p)$ over $(1 + T - \gamma^{k(P)}) \in \mathrm{Spec}(\Lambda)$. Thus, we find $A \equiv \Phi_\alpha \mod P_j$ for infinitely many distinct primes P_j; hence, $A = \Phi_\alpha$, as desired.

3.3.2 Proof of Theorem 3.25

Suppose that $\mathcal{V}_\mathcal{A}(\mathbb{I})$ is a finite extension and assume the existence of an arithmetic point P_0 as in the theorem. Therefore, the assumptions of Proposition 3.26 are met. By Proposition 3.26, we find a Weil number α of weight 1 and a power series $\Phi_\alpha(T) \in W[[T]]$ such that $a(p, f_P) = \Phi_\alpha(\varepsilon_P(\gamma)\gamma^{k(P)} - 1) = \zeta(\varepsilon_P(\gamma))^{\log_p(\alpha)/\log_p(\gamma)}\langle\alpha\rangle^{k(P)}$ for all arithmetic P, where ζ is a root of unity independent of P; in short, $a(p) = \Phi_\alpha \in W[[T]] \subset \mathbb{I}$. Then, for the entire set $\mathcal{B}$ of arithmetic points P with $k(P) = 1$, we find $\mathcal{H}_\mathcal{B}(\mathbb{I}) \subset \mathbb{Q}(\mu_{p^\infty(p-1)})(\zeta, \alpha)$, which is a finite extension of $\mathbb{Q}(\mu_{p^\infty})$. By the horizontal theorem, $\mathbb{I}$ has complex multiplication. The converse is easier (in the same manner as in the proof of Theorem 3.2; see [H11a] Sect. 4 for more details). □

We could make the following conjecture stronger than Conjecture 3.24:

Conjecture 3.27 *Let $\mathcal{A} \subset \mathrm{Spec}(\mathbb{I})(\overline{\mathbb{Q}}_p)$ be an infinite set of arithmetic points P with bounded level $r(P) \le r$. Suppose that $\mathbb{I}$ does not have complex multiplication. Then we have*

$$\limsup_{P \in \mathcal{A}} [\mathbb{Q}(a(p, f_P)) : \mathbb{Q}] = \infty.$$

3.4 Nonconstancy of Adjoint $\mathcal{L}$-Invariant

Consider a cuspidal slope-0 family of Hecke eigenforms

$$\mathcal{F}_\mathbb{I} = \mathcal{F} = \{f_P | P \in \mathrm{Spec}(\mathbb{I})(\overline{\mathbb{Q}}_p)\}$$

indexed by points of $\mathrm{Spec}(\mathbb{I})(\overline{\mathbb{Q}}_p)$ and its family of Galois representations $\{\rho_P\}_P$. For each p-decomposition subgroup $D \subset \mathrm{Gal}(\overline{\mathbb{Q}}/\mathbb{Q})$, we have $\rho_P|_D \cong \left(\begin{smallmatrix} \epsilon_P & * \\ 0 & \delta_P \end{smallmatrix}\right)$ with unramified quotient character δ_P (e.g., [GME] Theorem 4.2.6). Here, for each $P \in \mathrm{Spec}(\mathbb{I})$, f_P is a p-adic modular form of slope 0 of level $Np^{r(P)+1}$ for a fixed prime to p-level N $(p \nmid N)$. Consider the adjoint representation $\mathrm{Ad}(\rho_P)$ realized in the trace-zero subspace in $\mathfrak{sl}_2(\kappa(P)) \subset M_2(\kappa(P))$ by conjugation action. Thus, $\mathrm{Ad}(\rho_P)(Frob_p)$ has an eigenvalue 1; hence, $L_p(s, \mathrm{Ad}(\rho_P))$ has an exceptional zero of order 1 at $s = 1$. For the $\mathcal{L}$-invariant

$\mathcal{L}(\mathrm{Ad}(\rho_P))$ defined by Greenberg [Gr] (which we will recall in Sect. 3.4.5), his conjecture $\mathcal{L}^{an}(\mathrm{Ad}(\rho_P)) \stackrel{?}{=} \mathcal{L}(\mathrm{Ad}(\rho_P))$ is still an open question (except in some cases). In any case, we get a function $P \mapsto \mathcal{L}(\mathrm{Ad}(\rho_P))$ defined on the set of arithmetic points of $\mathrm{Spec}(\mathbb{I})$. As we will see, this function is interpolated analytically on $\mathrm{Spec}(\mathbb{I})$. Thus, we may write $P \mapsto \mathcal{L}(\mathrm{Ad}(\rho_P))$ for general $P \in \mathrm{Spec}(\mathbb{I})(\overline{\mathbb{Q}}_p)$. Either supposing an almost known conjecture on the smoothness at P of the ordinary deformation space of ρ_P (see Conjecture 3.32 below) or defining the $\mathbb{I}$-adic $\mathcal{L}$-invariant as in Sect. 3.4.7, we prove the following result in this section:

Theorem 3.28 *The analytic function $P \mapsto \mathcal{L}(\mathrm{Ad}(\rho_P))$ is constant if and only if the family $\mathcal{F}$ has CM.*

We prove this theorem assuming a formula relating $\frac{da(p)}{dT}$ with the $\mathcal{L}$-invariant (Theorem 3.31); after the proof, we revisit Greenberg's definition of his $\mathcal{L}$-invariant in Sects. 3.4.2–3.4.5 and prove the derivative formula in Sect. 3.4.6.

By this theorem, if $\mathcal{F}$ is a non-CM family, $P \mapsto \mathcal{L}(\mathrm{Ad}(\rho_P))$ is a nonconstant function; hence, except for finitely many Galois representations in the family, Greenberg's conjecture (see [Gr]) predicting the nonvanishing of $\mathcal{L}(\mathrm{Ad}(V))$ is true. We have a Hilbert modular slope-0 family as described in [HMI]. We can make the following conjecture for Hilbert modular families.

Conjecture 3.29 *For a slope-0 parallel weight family of Hilbert modular Galois representations $\{\rho_P\}_{P\in\mathrm{Spec}(\mathbb{I})}$ of $\mathrm{Gal}(\overline{\mathbb{Q}}/F)$ for a totally real field F, $P \mapsto \mathcal{L}(\mathrm{Ad}(\rho_P))$ is constant if and only if the family has CM.*

If the conjecture holds, the $\mathcal{L}$-invariant $\mathcal{L}(\mathrm{Ind}_F^{\mathbb{Q}} \mathrm{Ad}(\rho_P))$ vanishes only over a thin Zariski closed subset of $\mathrm{Spec}(\mathbb{I})$.

Before starting the proof of the theorem, we prepare to state another form of the rigidity lemma. Identify Λ with $\widehat{W[t,t^{-1}]} = \varprojlim_n W[t]/(t^{p^n}-1)$ by $t = 1+T$. Then $t \mapsto t^z$ for $z \in \mathbb{Z}_p$ extends linearly to an endomorphism ι_n^z of the ring $W[t]/(t^{p^n}-1)$. Indeed, in $W[t]/(t^{p^n}-1)$, $t^{p^n}=1$; hence, cutting the p-adic expansion $z = \sum_{m=0}^{\infty} a_m p^m$ (with integers $0 \le a_m \le p-1$) to $z_0 = \sum_{m-0}^{n-1} a_m p^m$, we legitimately define t^z by $t^z = t^{z_0}$ in $W[t]/(t^{p^n}-1)$, and we then define $\iota_n^z(\sum_{j=0}^{p^n-1} c_j t^j) = \sum_{j=0}^{p^n-1} c_j t^{zj}$. Clearly, ι_n^z is a W-algebra endomorphism of $W[t]/(t^{p^n}-1)$. Then passing to the limit, $z \in \mathbb{Z}_p$ gives rise to an algebra endomorphism $\iota^z = \varprojlim_n \iota_n^z$ of Λ. In this way, $\mathbb{Z}_p^\times$ acts on Λ through $z \mapsto \iota^z \in \mathrm{Aut}(\lambda_{/W})$. Note that $t^z = (1+T)^z = \sum_{n=0}^{\infty}\binom{z}{n}T^n \in W[[T]]$.

Write $\Lambda\widehat{\otimes}_W\Lambda$ as $\widehat{W[t,t^{-1},t',t'^{-1}]} = \varprojlim_n W[t,t^{-1},t',t'^{-1}]/(t^{p^n}-1, t'^{p^n}-1)$ for the variable t of the left factor and t' of the right factor, and let $z \in \mathbb{Z}_p^\times$ act on $\widehat{W[t,t^{-1},t',t'^{-1}]}$ by $(t,t') \mapsto (t^z, t'^z)$ for $z \in \mathbb{Z}_p^\times$. Here is a version of the rigidity lemma (Lemma 3.12), which we prove later as Lemma 4.85 after some preparation:

Lemma 3.30 *Let $\Phi(t,t') = t'^d + a_1(t)t'^{d-1} + \cdots + a_d(t) \in \Lambda[t']$ be an irreducible polynomial over the quotient field Q of Λ such that $\Phi(1,1) = 0$. Suppose that the integral closure of $\mathbb{Z}_p$ in $\mathbb{J} := W[\widehat{t,t^{-1},t',t'^{-1}}]/(\Phi(t,t'))$ is equal to W. If there exists an open subgroup U of $\mathbb{Z}_p^\times$ such that $(\Phi(t^z,t'^z)) = (\Phi(t,t'))$ as principal ideals of $W[\widehat{t,t^{-1},t',t'^{-1}}]$ for all $z \in U$, we have $\Phi(t,t') = t'^d - t^b$ for a p-power d and $b \in \mathbb{Z}_p$ with $d\mathbb{Z}_p + b\mathbb{Z}_p = \mathbb{Z}_p$.*

3.4.1 Proof of Theorem 3.28

We assume $p > 2$ in the rest of this section. We may assume that the integral closure of $\mathbb{Z}_p$ in $\mathbb{I}$ is equal to W (by extending scalars if necessary). As we will prove in the following sections as Theorem 3.31 (for which we need to assume Conjecture 3.32), $\mathcal{L}(\mathrm{Ad}(\rho_P))$ is a constant multiple of

$$\left(a(p)^{-1}\frac{da(p)}{dX}\right)\Big|_{X=0},$$

where if$P \cap \Lambda = (X)$ for $X = \gamma^{-k}\zeta^{-1}t - 1$ for $t = 1 + T$. After proving the theorem assuming this formula, we recall the proof of the formula (from [H04a]).

We rewrite

$$\left(a(p)^{-1}\frac{da(p)}{dX}\right)\Big|_{X=0} = \left(a(p)^{-1}t\frac{da(p)}{dt}\right)\Big|_{t=\zeta\gamma^k}.$$

The constancy of $\mathcal{L}(\mathrm{Ad}(\rho_P))$ implies the constancy of

$$a(p)^{-1}(1+T)\frac{da(p)}{dT} = a(p)^{-1}t\frac{da(p)}{dt}.$$

Then $t\frac{da}{dt} = s \cdot a$ for $a(t) = a(p)(t)$ for $s \in W \otimes_{\mathbb{Z}_p} \mathbb{Q}_p$. Setting $b(x) = \log_p \circ a(\exp_p(x))$ (for $x = \log_p(t)$), as $dx = \frac{dt}{t}$, we get

$$\frac{db}{dx} = \frac{da}{dx}\frac{db}{da} = \frac{da}{dx}\frac{d\log_p(a)}{da} = s \cdot a \cdot \frac{1}{a} = s$$

from the chain rule. Then b must be a linear function of x with slope s:

$$\log_p(a) = sx + c \Leftrightarrow a = C\exp_p(s \cdot \log_p(t)) = Ct^s \ (C = \exp_p(c)).$$

Now, we want to show $s \in \mathbb{Q}_p$. We get $a(p) = Ct^s \in F[[T]]$ for $t^s = \exp_p(s \cdot \log_p(t))$, where F is the quotient field of W. Since $a(p)$ is a unit in $\mathbb{I}$, C is a p-adic unit in $W^\times$; hence, $\Phi(t) := t^s = C^{-1}a(p) \in \mathbb{I}$. Consider the subalgebra $\mathbb{J} = \Lambda[\Phi(t)] \subset \mathbb{I}$ topologically generated by $\Phi(t)$. We have a surjective W-algebra homomorphism $W[\widehat{t,t^{-1},t',t'^{-1}}] \twoheadrightarrow \mathbb{J}$ sending (t,t') to $(1+T,\Phi(t))$. This surjection gives rise to an integral formal closed subscheme

$Z \cong \mathrm{Spf}(\mathbb{J})$ in $\widehat{\mathbb{G}}_m \times \widehat{\mathbb{G}}_m$ (see Sect. 4.4 for formal scheme and the functor "Spf"). Since $\Phi(t)$ is a power series $t^s = \sum_{n=0}^{\infty} \binom{s}{n} T^n \in F[[T]]$ convergent over a closed disk $D_\epsilon = \{x \in F : |x|_p \le \epsilon\}$ for some $\epsilon \in |F^\times|_p$, it is in an affinoid ring $F\{T\}$, which is the convergent power series ring over D_ϵ (see [Cd] for rigid geometry and general affinoid rings). Since the action $t \mapsto t^z$ for $z \in \mathbb{Z}_p^\times$ extends to $F\{T\}$, we have $\Phi(t^z) = \Phi(t)^z$ in $F\{T\}$ as $\Phi(t) = t^s$. Thus, $\Phi(t^z) = \Phi(t)^z$ in $\mathbb{J}$, as we may regard $\mathbb{J}$ as a subring of $F\{T\}$. Then the formal subscheme Z is stable under the action $(t,t') \mapsto (t^z, t'^z)$ of $\mathbb{Z}_p^\times$ on $\widehat{\mathbb{G}}_m \times \widehat{\mathbb{G}}_m$, and we conclude that $s \in \mathbb{Q}_p$ from Lemma 3.30.

To finish, let us write $f_{P,\zeta} = f_P$ for the eigenform in the family corresponding to a prime P above $(X) = (\gamma^{-k}\zeta^{-1}t - 1)$ with $\zeta \in \mu_{p^r}$ to emphasize the dependence on ζ. The cusp form $f_{P,\zeta}$ is a Hecke eigenform in $S_k(\Gamma_1(Np^{r+1}))$, and we have $a(p, f_{P,\zeta}) = C\gamma^{ks}\zeta^s$. Take $\zeta = 1$. Then $a(p, f_{P,1}) = C\gamma^{ks}$ is an algebraic number α. This shows that for any $\zeta \in \mu_{p^\infty}$, $a(p, f_{P,\zeta}) = \alpha$ up to p-power roots of unity. Thus, the field generated by $a(p, f_{P,\zeta})$ for all $\zeta \in \mu_{p^\infty}$ is a finite extension of $\mathbb{Q}[\mu_{p^\infty}]$. Then by Theorem 3.2, we conclude that $\mathcal{F}$ is a CM family.

Conversely, we suppose that $\mathcal{F}$ is a CM family. Then we find a Galois character $\Psi : \mathrm{Gal}(\overline{\mathbb{Q}}/M) \to \tilde{\mathbb{I}}^\times$ for an imaginary quadratic field M such that $\rho_{\mathbb{I}} \cong \mathrm{Ind}_M^{\mathbb{Q}} \Psi$ [and hence $\rho_P = \mathrm{Ind}_M^{\mathbb{Q}} \Psi \bmod P$ for all $P \in \mathrm{Spec}(\tilde{\mathbb{I}})$] and Ψ is unramified at the factor $\overline{\mathfrak{p}}|p$ in M [where $\mathfrak{p} = \{\alpha \in \mathfrak{O} : |i_p(\alpha)|_p < 1\}$]. Then $a(p)$ is the value of the character $\Psi(Frob_{\overline{\mathfrak{p}}})$ at the Frobenius element $Frob_{\overline{\mathfrak{p}}}$ at $\overline{\mathfrak{p}}$. By computation, we find that $\Psi(Frob_{\overline{\mathfrak{p}}}) = t^{\log_p(\overline{\mathfrak{p}})/\log_p(\gamma)}$ up to a root of unity. This shows the constancy of the $\mathcal{L}$-invariant for the CM family [the constant is equal to the one $\mathcal{L}^{an}(\chi)$ in Theorem 1.5 (1); see also [HMI] Sect. 5.3.3]. □

3.4.2 Review of $\mathcal{L}$-Invariants

After Mazur–Tate–Teitelbaum [MzTT], many number theorists proposed diverse definitions of the $\mathcal{L}$-invariant that are expected to give the correction factor relating the conjectural arithmetic part of the leading term of the Taylor expansion of a given p-adic motivic L-function at an exceptional zero to its Archimedean counterpart (see [Cz] for a description of other $\mathcal{L}$-invariants we do not touch upon). For an elliptic curve $E_{/\mathbb{Q}}$ with multiplicative or ordinary good reduction modulo p, its p-adic L-function $L_p(s, E)$ has the following evaluation formula at $s = 1$ (as we have seen in Example 1.4):

$$L_p(1, E) = (1 - a_p^{-1}) \frac{L_\infty(1, E)}{\text{period}},$$

where $L_\infty(s, E)$ is the Archimedean L-function of E, and a_p is the eigenvalue of the arithmetic Frobenius element at p on the unramified quotient of the p-adic Tate module $T(E)$ of E. If E has *split* multiplicative reduction, $a_p = 1$, and $L_p(s, E)$ has zero at $s = 1$. This type of zero of a p-adic L-function resulting from the modification Euler p-factor is called an *exceptional zero*, and it is believed that if the Archimedean L-values doesnot vanish, the order of the zero is the number e of such Euler p-factors (see Conjecture 1.2); so, in this case, $e = 1$. Then $L'_p(1, E) = \frac{dL_p(s,E)}{ds}|_{s=1}$ was expected to be equal to the Archimedean value $\frac{L_\infty(1,E)}{\text{period}}$ times a theoretically explicit error factor $\mathcal{L}^{an}(E)$, the so-called $\mathcal{L}$-invariant:

$$L'_p(1, E) = \mathcal{L}^{an}(E)\frac{L_\infty(1, E)}{\text{period}}.$$

The problem regarding $\mathcal{L}$-invariants of motives is to find an explicit formula in terms of their p-adic realization without recourse to p-adic L-functions. Writing $E(\overline{\mathbb{Q}}_p) = \overline{\mathbb{Q}}_p^\times/q^{\mathbb{Z}}$ for the Tate period $q \in p\mathbb{Z}_p$, the solution [Theorem 1.5 (2)] conjectured by [MzTT] and proved by Greenberg–Stevens [GS] is

$$\mathcal{L}^{an}(E) = \frac{\log_p(q)}{\text{ord}_p(q)}.$$

This value is shown to be nonzero by [BDGP].

Since E is modular, $L(s, E) = L(s, f_E)$ for an elliptic Hecke eigenform f_E of weight 2. In particular, $a(p, f_E) = a_p = 1$ and $a(1, f_E) = 1$. We may now have $\mathcal{F}_{\mathbb{I}}$ normalized so that f_E is a specialization of $\mathcal{F}$ at $X = 0$. Then one of the key ingredients of their proof in [GS] is the following formula:

$$\mathcal{L}^{an}(E) = -2\log_p(\gamma)\frac{da(p)}{dX}\Big|_{X=0},$$

where γ is the generator of $\Gamma = 1 + p\mathbb{Z}_p$ corresponding to $1 + X$ under the identification $\mathbb{Z}_p[[\Gamma]] = \mathbb{Z}_p[[X]]$. These formulas are actually proven in [GS] for abelian varieties A split multiplicative at p associated with weight-2 cusp forms in place of elliptic curve E.

Here is an analogous formula for Greenberg's $\mathcal{L}$-invariant of $\text{Ad}(\rho_P)$ in [H04a] whose proof we give in Sect. 3.4.6 after recalling Galois deformation theory and Greenberg's definition of his $\mathcal{L}$-invariant:

Theorem 3.31 *Let p be an odd prime. Then we have*

$$\mathcal{L}(\text{Ad}(\rho_P)) = -2\log_p(\gamma)a_P(p)^{-1}\frac{da(p)}{dX}\Big|_{X=0}.$$

In particular, if $f_P = f_E$ as above, we have

$$\mathcal{L}(\text{Ad}(\rho_P)) = \mathcal{L}^{an}(E) = \frac{\log_p(q)}{\text{ord}_p(q)}.$$

The last formula follows from the first and the Greenberg–Stevens' formula we described before stating the theorem. We prove the first assertion in the rest of this section.

3.4.3 Galois Deformation

A main ingredient in the proof of Theorem 3.31 is Galois deformation theory developed by Mazur in [Mz2]. In this section, we assume $p > 2$. Since ρ_P (for an arithmetic P) is irreducible (by a result of Ribet's; see [Ri2], [Ri3] and [GME] Theorem 4.3.18) and $\mathrm{Tr}(\rho_{\mathbb{I}}) \in \mathbb{I}$, via the method of pseudo-representation (due to Wiles) in [MFG] Sect. 2.2.1, we may arrange $\rho_{\mathbb{I}}$ to have values in $\mathbb{I}_P$ for arithmetic P by the irreducibility of ρ_P (see [GME] Theorem 4.3.1).

Fix an arithmetic P, and write ρ for ρ_P for simplicity. Let $\widehat{\mathbb{I}}_P$ be the P-adic localization-completion $\widehat{\mathbb{I}}_P := \varprojlim_n \mathbb{I}_P/P^n\mathbb{I}_P$. It is known that $\widehat{\mathbb{I}}_P \cong \kappa(P)[[X]]$ (see [HMI] Proposition 3.78). The character $\det(\rho_{\mathbb{I}})^{-1}\det(\rho)$ has values in the p-profinite group $1 + \mathfrak{m}_{\mathbb{I}}$ for the maximal ideal $\mathfrak{m}_{\mathbb{I}}$ of $\mathbb{I}$, and hence we have its unique square root ψ with values in $1 + \mathfrak{m}_{\mathbb{I}}$. Define a representation $\boldsymbol{\rho} : \mathrm{Gal}(\overline{\mathbb{Q}}/\mathbb{Q}) \to GL_2(\widehat{\mathbb{I}}_P)$ with $\det(\boldsymbol{\rho}) = \det(\rho)$ by $\boldsymbol{\rho}(\sigma) = \psi(\sigma)\rho_{\mathbb{I}}(\sigma)$. Note that $\boldsymbol{\rho} \equiv \rho_{\mathbb{I}} \mod P$. Fix a decomposition subgroup $D_p = \mathrm{Gal}(\overline{\mathbb{Q}}_p/\mathbb{Q}_p) \subset \mathrm{Gal}(\overline{\mathbb{Q}}/\mathbb{Q})$ at p (corresponding to i_p). Normalize ρ_P so that $\rho_P|_{D_p} = \left(\begin{smallmatrix}\epsilon_P & * \\ 0 & \delta_P\end{smallmatrix}\right)$ with unramified δ_P (by (3.1.2)). Then $\epsilon_P \neq \delta_P$ and ϵ_P is ramified by (3.1.1). If, further, $\epsilon_P \not\equiv \delta_P \mod \mathfrak{m}_{\mathbb{I}}$, we call $\rho_{\mathbb{I}}$ *p-distinguished* (this definition is independent of the choice of P, because this is equivalent to $\rho_{\mathfrak{m}_{\mathbb{I}}}|_{D_p} \cong \left(\begin{smallmatrix}\overline{\epsilon} & * \\ 0 & \overline{\delta}\end{smallmatrix}\right)$ with $\overline{\epsilon} \neq \overline{\delta}$).

Simply write $\kappa := \kappa(P)$. Let S be the set of places of $\mathbb{Q}$ made up of all prime factors of Np and ∞. Consider the *Galois deformation functor* D^{gal} into $SETS$ from the category of local Artinian κ-algebras with residue field κ. See [Mz2] and Sect. 4.4.2 for the generality of deformation functors. The value $D^{gal}(A)$ at a local Artinian κ-algebra A with maximal ideal $\mathfrak{m}_A$ is given by the set of isomorphism classes of two-dimensional continuous Galois representations $\rho_A : \mathrm{Gal}(\overline{\mathbb{Q}}/\mathbb{Q}) \to GL_2(A)$ unramified outside S such that

(D1) $(\rho_A \mod \mathfrak{m}_A) \cong \rho_P$;

(D2) writing $\iota : \kappa \to A$ for the structure homomorphism of κ-algebras, we have the identity of the determinant characters:

$$\iota \circ \det(\rho) = \det(\rho_A);$$

(D3) we have an exact sequence $\rho_A|_{D_p} \cong \left(\begin{smallmatrix}\epsilon_A & * \\ 0 & \delta_A\end{smallmatrix}\right)$ with $\delta_A \equiv \delta_P \mod \mathfrak{m}_A$.

Condition (D3) is the near-ordinarity, and we call the character δ_A of D_p the *nearly ordinary character* of ρ. By the work started by Wiles/Taylor (and practically ended by Kisin), we know the following conjecture for most cases:

Conjecture 3.32 *Suppose that P is an arithmetic point of* $\mathrm{Spec}(\mathbb{I})(\overline{\mathbb{Q}}_p)$. *The above functor is pro-represented by the pair $(\widehat{\mathbb{I}}_P, \boldsymbol{\rho})$. In other words, if we have a deformation ρ_A as above, there exists a unique κ-algebra homomorphism $\varphi : \widehat{\mathbb{I}}_P \to A$ such that ρ_A is equivalent to $\varphi \circ \boldsymbol{\rho}$.*

Note here that $\widehat{\mathbb{I}}_P \cong \kappa(P)[[X]]$. Though we have taken the field κ as a base ring, we could have taken $W = W(\mathbb{F})$ in place of κ and the deformation functor from the category of Artinian W-algebras similarly defined. This type of deformation over W was studied by Taylor–Wiles, Kisin, and others, and from their results, we can conclude the above conjecture in the following cases:

1. ρ_P is p-distinguished [as defined above (D1)] and $\overline{\rho} := \rho_{\mathfrak{m}_{\mathbb{I}}}$ is absolutely irreducible over $\mathbb{Q}[\sqrt{p^*}]$ for $p^* = (-1)^{(p-1)/2}p$ (e.g., essentially the original $R = T$ theorem of Wiles/Taylor; see [DiFG], [MFG] Theorem 5.29, and [HMI] Proposition 3.78).
2. This case is via the theory of eigenvariety (cf. [Em]). Supposing that $\overline{\rho}$ is absolutely irreducible, the conjecture holds if the slope is less than $k(P)/2$, since the eigenvariety is étale around P over the arithmetic point below P on the weight space; see also [Ki1] Theorem 11.10). The result includes the identification of the rigid-analytic full (dimension-3) Galois deformation space with the maximal spectrum of the full Hecke algebra (p-inverted) by Gouvêa/Mazur, Coleman/Mazur, Kisin, and Chenevier (see [Ki2] Theorem 8.2, [Cn1], and [Cn2]). Here the word "full" means we do not impose reducibility on the restriction of the deformation to D_p.
3. $\overline{\rho} \cong \chi_1 \oplus \chi_2$ for two Galois characters with $\chi_1|_{\mathrm{Gal}(\overline{\mathbb{Q}}/K)} \neq \chi_2|_{\mathrm{Gal}(\overline{\mathbb{Q}}/K)}$ for $K = \mathbb{Q}[\mu_{p^\infty}]$ ([Ki1] Theorem 11.10 and [Ki2] Theorem 8.2).

The only remaining case is when $\overline{\rho}$ is reducible and $\chi_1|_{\mathrm{Gal}(\overline{\mathbb{Q}}/\mathbb{Q}[\mu_{p^\infty}])} = \chi_2|_{\mathrm{Gal}(\overline{\mathbb{Q}}/\mathbb{Q}[\mu_{p^\infty}])}$ (which means p-distinguishedness as the representation is odd). In this exceptional case, the Skinner–Wiles result [SW] applies (so at least $\widehat{\mathbb{I}}_P$ is equal to the reduced part of the universal ring).

Case (1) is the combination of the result in [DiFG] and the method in [HMI]. Indeed, if we have an $R = T$ theorem for one arithmetic point P, it extends by the method of [HMI] Chap. 3 to the entire connected component $\mathrm{Spec}(\mathbb{T})$ of $\mathrm{Spec}(\mathbf{h})$ containing $\mathrm{Spec}(\mathbb{I})$. The specific one point P with $R = T$ is essentially covered by either [DiFG] or [HMI] Chap. 3 (especially Sect. 3.2.4).

In cases (2) and (3), the full Galois deformation ring $\mathcal{R}$ [without imposing (D3)] around an arithmetic point is smooth over the weight space by the work quoted there. The full deformation ring often does not have information on $a(p)$ [or $U(p)$], because we often have a deformation whose restriction on D_p does not satisfy (D3). Then the natural projection $\mathrm{Spec}(R) \to \mathrm{Spec}(\mathcal{R})$ (for the ordinary deformation ring R) has an image on one of the two leaves (of the infinite fern) crossing at P if the slope $\mathrm{ord}_p(a_P(p))$ is 0 [or more generally, $\mathrm{ord}_p(a_P(p))$ is less than $k(P)/2$]; hence, it is étale around the point $\tilde{P} \in \mathrm{Spec}(R)$ coming from the arithmetic point $P \in \mathrm{Spec}(\mathbb{I})$. Thus, we have $\mathrm{Spec}(R_{\tilde{P}}) \xrightarrow{\text{étale}} \mathrm{Spec}(\mathcal{R}_P) \xrightarrow{\text{smooth}} \mathrm{Spec}(W[[T]]_P)$, which shows that the composite is étale, and hence $\widehat{R}_{\tilde{P}} = \widehat{\mathbb{I}}_P = \kappa(P)[[X]]$.

In the following sections, we start with a brief review of Greenberg's definition of the Selmer group and his $\mathcal{L}$-invariant.

3.4.4 Adjoint Selmer Groups

We describe Greenberg's definition of his Selmer group associated with the adjoint square Galois representation. For simplicity, we assume that $S = \{p, \infty\}$ (so $N = 1$). One can find the definition in the general case in [Gr] and in [HMI] Sect. 1.2.3. We may assume that κ has a p-adic integer ring W. Let $\mathbb{Q}^S$ be the maximal extension over $\mathbb{Q}$ unramified outside S. All Galois cohomology groups are continuous cohomology groups described in [MFG] 4.3.3. Write $\mathfrak{G}^S = \mathrm{Gal}(\mathbb{Q}^S/\mathbb{Q})$ and I_p for the inertia subgroup of the decomposition subgroup $D_p \subset \mathfrak{G}^S$. The group D_p is expected to be isomorphic to $\mathrm{Gal}(\overline{\mathbb{Q}}_p/\mathbb{Q})$ (and this fact is close to be true; see [CnC]).

Write V for the space of ρ_P. Let $\mathfrak{G}^S$ act on $\mathrm{End}_\kappa(V)$ by conjugation and put $\mathrm{Ad}(V) \subset \mathrm{End}_\kappa(V)$ (the trace 0 subspace of dimension 3). We have a filtration:

$$V \supsetneq F^+V \supsetneq \{0\} \tag{ord}$$

stable under the decomposition group D_p such that D_p acts on the quotient V/F^+V by δ_P. Then $\mathrm{Ad}(V)$ has the following three-step filtration stable under D_p:

$$\mathrm{Ad}(V) \supset F^- \supset F^+ \supset \{0\}, \tag{F}$$

where

$$F^- = F^-\mathrm{Ad}(V) = \{\phi \in \mathrm{Ad}(V) | \phi(F^+V) \subset F^+V\} \ \text{(upper triangular)},$$
$$F^+ = F^+\mathrm{Ad}(V) = \{\phi \in \mathrm{Ad}(V) | \phi(F^+V) = 0\} \ \text{(upper nilpotent)}.$$

Note that D_p acts trivially on F^-/F^+; so $F^-/F^+ \cong \kappa$; hence, the p-adic L-function of $\mathrm{Ad}(V)$ has an exceptional zero at $s = 1$. Put

$$U_p(\mathrm{Ad}(V)) = \mathrm{Ker}(\mathrm{Res} : H^1(D_p, \mathrm{Ad}(V)) \to H^1(I_p, \frac{\mathrm{Ad}(V)}{F^+(\mathrm{Ad}(V))})).$$

Then we define

$$\mathrm{Sel}(\mathrm{Ad}(V)) = \mathrm{Ker}(H^1(\mathfrak{G}^S, \mathrm{Ad}(V)) \to \frac{H^1(D_p, V)}{U_p(V)}). \tag{3.4.1}$$

Replacing $U_p(\mathrm{Ad}(V))$ by the bigger

$$U_p^-(\mathrm{Ad}(V)) = \mathrm{Ker}(\mathrm{Res} : H^1(D_p, \mathrm{Ad}(V)) \to H^1(I_p, \frac{\mathrm{Ad}(V)}{F^-(\mathrm{Ad}(V))}))$$

for $\mathfrak{p}|p$, we define a bigger "$-$" Selmer group $\mathrm{Sel}^-(\mathrm{Ad}(V)) \supset \mathrm{Sel}(\mathrm{Ad}(V))$.

In the above definition, replacing $\mathfrak{G}^S$ by the stabilizer $\mathfrak{G}^S_\infty := \mathrm{Gal}(\mathbb{Q}^S/\mathbb{Q}_\infty)$ of the cyclotomic $\mathbb{Z}_p$-extension $\mathbb{Q}_{\infty/\mathbb{Q}}$ and $\mathrm{Ad}(V)$ by $A = \mathrm{Ad}(V)/L$ for a Galois stable lattice L of $\mathrm{Ad}(V)$, one can define the Selmer group $\mathrm{Sel}_{\mathbb{Q}_\infty}(A)$. The characteristic power series $\Phi(T)$ of the Pontryagin dual of $\mathrm{Sel}_{\mathbb{Q}_\infty}(A)$ is conjectured to be the adjoint p-adic L-function $L_p(s, \mathrm{Ad}(\rho_P))$ (the adjoint main conjecture). Here the cyclotomic $\mathbb{Z}_p$-extension means that a unique subextension $\mathbb{Q}_\infty/\mathbb{Q}$ of $\mathbb{Q}(\mu_{p^\infty})$ such that the cyclotomic character induces $\mathrm{Gal}(\mathbb{Q}_\infty/\mathbb{Q}) \cong 1 + \mathbf{p}\mathbb{Z}_p \cong \mathbb{Z}_p$ for $\mathbf{p} = 4$ if $p = 2$ and $\mathbf{p} = p$ if $p > 2$.

It is easy to see $\mathrm{Sel}^-(A)$ is sent (with finite kernel) into $\mathrm{Sel}_{\mathbb{Q}_\infty}(A)$ [as p-ramification of cocycles giving $\mathrm{Sel}^-(A)$ projected to F^-A/F^+A is absorbed by the wild ramification of $\mathbb{Q}_\infty/\mathbb{Q}$]. Therefore, we have the nontrivial image on which $\mathrm{Gal}(\mathbb{Q}_\infty/\mathbb{Q})$ acting trivially, and so $T|\Phi(T)$; that is, the image produces the exceptional zero at $s = 1$ of the algebraic p-adic L-function

$$s \mapsto \Phi(\gamma^s - 1) =: L_p^{alg}(s, \mathrm{Ad}(\rho_P)).$$

Greenberg's principle (behind his way of defining the $\mathcal{L}$-invariant) is that the $\mathcal{L}$-invariant must be produced out of cocycles in $\mathrm{Sel}^-(A)$. Assuming $\mathcal{L}(\mathrm{Ad}(\rho_P)) \neq 0$, $L_p^{alg}(s, \mathrm{Ad}(\rho_P))$ has order-1 zero at $s = 1$ and $L_p'(1, \mathrm{Ad}(\rho_P)) = \mathcal{L}(\mathrm{Ad}(\rho_P))L(1, \mathrm{Ad}(\rho_P))/\text{period}$ up to units under mild conditions (see [Gr] Proposition 4, [H07b] Theorem 3.1, and [MFG] Theorem 5.20).

Taking the dual $\mathrm{Ad}(V)^*(1) = \mathrm{Hom}_\kappa(\mathrm{Ad}(V), \kappa)(1)$ with a single Tate twist, and the filtration dual to (F), we define the dual Selmer group $\mathrm{Sel}(\mathrm{Ad}(V)^*(1))$.

Lemma 3.33 *Suppose* Conjecture 3.32. *Then we have*

$$\dim \mathrm{Sel}^-(\mathrm{Ad}(V)) = 1 \quad \textit{and} \quad \mathrm{Sel}(\mathrm{Ad}(V)) = \mathrm{Sel}(\mathrm{Ad}(V)^*(1)) = 0. \tag{V}$$

In the earlier article [H04a], the balanced Selmer group $\overline{\mathrm{Sel}}_{\mathbb{Q}}$ [see [Gr] (16) and [HMI] Sect. 1.5.1] is used to prove this type of result. This technicality is not actually necessary. To see this, we note that by definition, $\mathrm{Sel}_{\mathbb{Q}}(\mathrm{Ad}(V)) \supset \overline{\mathrm{Sel}}_{\mathbb{Q}}(\mathrm{Ad}(V))$ and, by duality, $\mathrm{Sel}_{\mathbb{Q}}(\mathrm{Ad}(V)^*(1)) \subset \overline{\mathrm{Sel}}_{\mathbb{Q}}(\mathrm{Ad}(V)^*(1))$. Then, by Greenberg (see [Gr] Proposition 2 or [HMI] Proposition 3.82), we have

$$\dim \overline{\mathrm{Sel}}_{\mathbb{Q}}(\mathrm{Ad}(V)) = \dim \overline{\mathrm{Sel}}_{\mathbb{Q}}(\mathrm{Ad}(V)^*(1)),$$

and therefore, to prove the vanishing of all such Selmer groups, we only need to show $\mathrm{Sel}_{\mathbb{Q}}(\mathrm{Ad}(V)) = 0$.

Proof. Here is a sketch of the proof (whose details can be found in the articles quoted above). For any derivation $\partial : \widehat{\mathbb{I}}_P \to \kappa$, consider $c_\partial := (\partial\boldsymbol{\rho})\boldsymbol{\rho}_P^{-1} : \mathfrak{G}^S \to \mathrm{End}(V)$. Applying ∂ to $\boldsymbol{\rho}(\sigma)\boldsymbol{\rho}(\tau) = \boldsymbol{\rho}(\sigma\tau)$, we verify that c_∂ is a cocycle. Since $\det(\boldsymbol{\rho})$ is constant, c_∂ has values in $\mathrm{Ad}(V)$. Since $\boldsymbol{\rho}|_{D_p}$ is upper triangular, $[c_\partial] \in \mathrm{Sel}^-(\mathrm{Ad}(V))$. By universality, any such cocycle is of the form c_∂. Thus, the tangent space $\mathcal{T}_P \cong \kappa$ of $\mathrm{Spec}(\widehat{\mathbb{I}}_P)$ at P is isomorphic to $\mathrm{Sel}^-(\mathrm{Ad}(V))$; hence, $\dim_\kappa \mathrm{Sel}^-(\mathrm{Ad}(V)) = 1$. Since the diagonal entry of c_∂ is nontrivial, $\mathrm{Sel}(\mathrm{Ad}(V))$ is a proper subspace of of $\mathrm{Sel}^-(\mathrm{Ad}(V))$, and so it vanishes. The size of the dual Selmer group must be closely related to the value at the counterpart of the functional equation of $L(s, \mathrm{Ad}(\rho_P))$ of the value corresponding to the original Selmer group. Indeed, according to Greenberg, $\dim_\kappa \mathrm{Sel}(\mathrm{Ad}(V)) = \dim_\kappa \mathrm{Sel}(\mathrm{Ad}(V)^*(1))$ [strictly speaking, $\dim_\kappa \overline{\mathrm{Sel}}(\mathrm{Ad}(V)) = \dim_\kappa \overline{\mathrm{Sel}}(\mathrm{Ad}(V)^*(1))$ as remarked; see [HMI] Lemma 1.84]; hence, the desired vanishing also follows for the dual.

We have the Poitou–Tate exact sequence (e.g., [MFG] Theorem 4.50 (5)):

$$0 \to \mathrm{Sel}(\mathrm{Ad}(V)) \to H^1(\mathfrak{G}^S, \mathrm{Ad}(V)) \to \frac{H^1(D_p, \mathrm{Ad}(V))}{U_p(\mathrm{Ad}(V))} \to \mathrm{Sel}(\mathrm{Ad}(V)^*(1))^*.$$

Thus, by (V), we have

$$H^1(\mathfrak{G}^S, \mathrm{Ad}(V)) \cong \frac{H^1(D_p, \mathrm{Ad}(V))}{U_p(\mathrm{Ad}(V))}. \tag{I}$$

3.4.5 Greenberg's $\mathcal{L}$-Invariant

Greenberg defined in [Gr] his invariant $\mathcal{L}(\mathrm{Ad}(V))$ in the following way. Write $F^-H^1(D_p, \mathrm{Ad}(V))$ for the image of $H^1(D_p, F^-\mathrm{Ad}(V))$ in $H^1(D_p, \mathrm{Ad}(V))$. By the definition of $U_p(\mathrm{Ad}(V))$, the subspace $\frac{F^-H^1(D_p,\mathrm{Ad}(V))}{U_p(\mathrm{Ad}(V))}$ inside the right-hand side of (I) is isomorphic to $\mathrm{Sel}^-(\mathrm{Ad}(V)) \cong \kappa$. Namely, we have

$$\mathrm{Sel}^-(\mathrm{Ad}(V)) \xrightarrow[\mathrm{Res}]{\sim} \frac{F^-H^1(D_p, \mathrm{Ad}(V))}{U_p(\mathrm{Ad}(V))} \subset \frac{H^1(D_p, \mathrm{Ad}(V))}{U_q(\mathrm{Ad}(V))}.$$

By projecting down to $F^-\mathrm{Ad}(V)/F^+\mathrm{Ad}(V) \cong \kappa$ with trivial D_p-action, cocycles in $\mathrm{Sel}^-(\mathrm{Ad}(V))$ give rise to a subspace L of

$$\mathrm{Hom}(D_p^{ab}, F^-\mathrm{Ad}(V)/F^+\mathrm{Ad}(V)) = \mathrm{Hom}(D_p^{ab}, \kappa).$$

Note that

$$\mathrm{Hom}(D_p^{ab}, \kappa) \cong \kappa \times \kappa$$

canonically by $\phi \mapsto (\frac{\phi([u,\mathbb{Q}_p])}{\log_p(u)}, \phi([p, \mathbb{Q}_p]))$ for any $u \in \mathbb{Z}_p^\times$ of infinite order. Here $[x, \mathbb{Q}_p]$ is the local Artin symbol (suitably normalized).

If a cocycle c representing an element in $\mathrm{Sel}^-(\mathrm{Ad}(V))$ is unramified, it falls in $\mathrm{Sel}(\mathrm{Ad}(V))$. By the vanishing (V) of $\mathrm{Sel}(\mathrm{Ad}(V))$, this implies $c = 0$; hence, the projection of L to the first factor κ [via $\phi \mapsto \phi([u, \mathbb{Q}_p])/\log_p(u)$] is surjective. Thus, this subspace L is a graph of a κ-linear map

$$\mathcal{L} : \kappa \to \kappa,$$

which is given by the multiplication by an element $\mathcal{L}(\mathrm{Ad}(V))=\mathcal{L}(Ad(\rho_P))\in\kappa$.

3.4.6 Proof of Theorem 3.31

Write $\boldsymbol{\rho}|_{D_p} \cong \left(\begin{smallmatrix}\epsilon & * \\ 0 & \boldsymbol{\delta}\end{smallmatrix}\right)$ with nearly ordinary character $\boldsymbol{\delta}$. We know that $c := c_\partial$ for $\partial = \frac{d}{dX}$ gives a nontrivial element in $\mathrm{Sel}^-(\mathrm{Ad}(V))$. The image of c in $\mathrm{Hom}(D_p^{ab}, \kappa)$ is $\delta_P^{-1}\partial\boldsymbol{\delta}|_{X=0}$. We know that $\delta_P^{-1}\boldsymbol{\delta}([p, \mathbb{Q}_p]) = a_P(p)^{-1}a(p)$ and

$$\delta_P^{-1}\boldsymbol{\delta}([u, \mathbb{Q}_p]) = (\zeta\gamma^k)^{-\log_p(u)/2\log_p(\gamma)} t^{\log_p(u)/2\log_p(\gamma)} = (1 + X)^{\log_p(u)/2\log_p(\gamma)}$$

by our construction [and $t = \gamma^{-k}\zeta(1+X)$]. More precisely, we have

$$c([p,\mathbb{Q}_p]) = a_P(p)^{-1}\frac{da(p)}{dX}\Big|_{X=0},$$
$$c([u,\mathbb{Q}_p]) = \frac{d(1+X)^{\log_p(u)/2\log_p(\gamma)}}{dX}\Big|_{X=0}. \tag{3.4.2}$$

Taking $u = \gamma^s$, we have

$$\frac{d\delta([\gamma^s,\mathbb{Q}_p])}{dX}\Big|_{X=0} = -\frac{s}{2} \text{ and } \frac{d\delta([\gamma^s,\mathbb{Q}_p])}{dX}\Big|_{X=0}\log_p(\gamma^s)^{-1} = -\frac{1}{2\log_p(\gamma)}.$$

Thus, we get the desired result from the definition of $\mathcal{L}(\mathrm{Ad}(V))$. □

3.4.7 $\mathbb{I}$-adic $\mathcal{L}$-Invariant

We can go through a similar argument to define the $\mathcal{L}$-invariant of $\mathrm{Ad}(\rho_{\mathbb{I}})$, which interpolate $\mathcal{L}(\mathrm{Ad}(\rho_P))$. Let Q be the quotient field of $\mathbb{I}$, and regard $\boldsymbol{\rho}$ as a representation into $GL_2(Q)$, writing $\mathbb{V}$ for its space. Taking $\partial = t\frac{d}{dt}$, we can think of $\mathbf{c} = \boldsymbol{\rho}^{-1}\partial\boldsymbol{\rho} : \mathfrak{G}^S \to \mathrm{Ad}(\mathbb{V})$. Fixing a basis of $\mathbb{V}$ so that $\boldsymbol{\rho}|_{D_p} = \left(\begin{smallmatrix}\epsilon & *\\ 0 & \delta\end{smallmatrix}\right)$, again we have three-step filtration

$$\mathrm{Ad}(\mathbb{V}) \supset F^-\mathrm{Ad}(\mathbb{V}) \supset F^+\mathrm{Ad}(\mathbb{V}) \supset 0$$

in exactly the same manner as in (F). Note that $\mathbf{c}$ is a 1-cocycle of $\mathfrak{G}^S$ with values in $\mathrm{Ad}(\mathbb{V})$ whose restriction to D_p has values in $F^-\mathrm{Ad}(\mathbb{V})$. Note that the cohomology class of $\mathbf{c}$ depends only on $\rho_{\mathbb{I}}$, not the character twist $\boldsymbol{\rho}$, which is dependent on P. Then considering $\mathbf{c}_{ab} = \mathbf{c}|_{D_p} \mod F^+\mathrm{Ad}(\mathbb{V})$, we get $\mathbf{c}_{ab} \in \mathrm{Hom}(D_p^{ab}, Q) \cong Q \times Q$. Then we define an $\mathbb{I}$-adic version of $\mathcal{L}$-invariant by

$$\mathcal{L}(\mathrm{Ad}(\rho_{\mathbb{I}})) = \mathcal{L}(\mathrm{Ad}(\mathbb{V})) = \frac{\mathbf{c}_{ab}([p,\mathbb{Q}_p])}{\frac{\mathbf{c}_{ab}([\gamma,\mathbb{Q}_p])}{\log_p(\gamma)}}. \tag{L}$$

If $P \in \mathrm{Spec}(\mathbb{I})(\overline{\mathbb{Q}}_p)$ is arithmetic, we have $\mathcal{L}(\mathrm{Ad}(\mathbb{V}))$ inside $\mathbb{I}_P$, since $\boldsymbol{\rho}$ has values in $GL_2(\mathbb{I}_P)$. Thus, $\mathcal{L}(\mathrm{Ad}(\mathbb{V})) \in \mathbb{I}_P$, and hence $\mathcal{L}(\mathrm{Ad}(\mathbb{V}))(P) := P(\mathcal{L}(\mathrm{Ad}(\mathbb{V})))$ is a well-defined number in the residue field $\kappa(P) \subset \overline{\mathbb{Q}}_p$ of P. If, further, $(\widehat{\mathbb{I}}_P, \boldsymbol{\rho})$ is universal, any cocycle class in $\mathrm{Sel}^-(\mathrm{Ad}(\rho_P))$ is of the form c_∂ as in the proof of Lemma 3.33. We get $\mathcal{L}(\mathrm{Ad}(\mathbb{V}))(P) = \mathcal{L}(\mathrm{Ad}(\rho_P))$. In any case, we can take (L) to be the definition of $\mathcal{L}(\mathrm{Ad}(\rho_{\mathbb{I}}))$ (which does not require any assumption except for the existence of $\rho_{\mathbb{I}}$), and from the proof of Theorem 3.31 without assuming any conjecture, we get the following:

Theorem 3.34 *We have* $\mathcal{L}(\mathrm{Ad}(\rho_{\mathbb{I}})) = -2\log_p(\gamma)a^{-1}t\frac{da}{dt}$ *for* $a = a(p) \in \mathbb{I}$.

This is the meaning of the analytic continuation of the adjoint $\mathcal{L}$-invariant mentioned just before Theorem 3.28.

3.5 Vanishing of the μ-Invariant of Katz L-Functions

The last two sections in this chapter are an introductory discussion of problems concerning vanishing of the Iwasawa μ-invariant of p-adic L-functions. Full proofs will be given in Chap. 9. This type of results has found applications in divisibility problems of class numbers, as seen in Sect. 1.1.1, in proofs of the main conjectures in Iwasawa's theory in [H06a] and [H09b], and in proving the open image property of mod p Λ-adic modular Galois representations in [H12b] (and [H13b]). In the past 10 years, some new methods of proving the vanishing have emerged in the work of Vatsal, Cornut [V1], [V2], [CV1], [CV2], Finis [Fi1], [Fi2] and this author. See [V3] for an overview.

We describe a geometric method which was started by Sinnott in [Sn1] and [Sn2] and has been generalized in [H04b], [H07b], [H10a], [H11b], [Hs2], [BuH] and [Bu] via the theory of Shimura varieties. We rely on a philosophical *rigidity principle* (proposed by Chai, Oort, and others): "A Hecke invariant subvariety of a Shimura variety is a Shimura subvariety." For any power series $\Phi(x_1,\dots,x_d) \in W[[x_1,\dots,x_d]]$, define $\mu(\Phi)$ by the exact power $p^{\mu(\Phi)} \| \Phi(X)$ in $W[[x_1,\dots,x_d]]$. The W-valued measure space on $\mathbb{Z}_p$ can be identified with a one-variable power series ring $W[[T]]$ by $\varphi \mapsto \Phi(T) = \int_{\mathbb{Z}_p}(1+T)^s d\varphi(s) \in W[[T]]$. Thus, we define *$\mu$-invariant* of φ by $\mu(\varphi) = \mu(\Phi)$. Introducing the supremum norm $\|\phi\| = \mathrm{Sup}_{x\in\mathbb{Z}_p} |\phi(x)|_p$ for continuous functions $\phi : \mathbb{Z}_p \to W$, we have $p^{-\mu(\varphi)} = \mathrm{Sup}_{\phi:\|\phi\|=1} |\int_{\mathbb{Z}_p} \phi d\varphi|_p = \mathrm{Sup}_{\phi:\|\phi\|\neq 0} |\int_{\mathbb{Z}_p} \phi d\varphi|_p / \|\phi\|$.

Exercise 3.35 *Prove the above formula for $\mu(\varphi)$ using* Theorem 1.36.

A measure φ with values in W on an abelian almost p-profinite group $G = \Delta \times \boldsymbol{\Gamma}$ with torsion-free $\boldsymbol{\Gamma} \cong \mathbb{Z}_p^m$ and a finite group Δ can be regarded as an element of the group algebra $W[[G]] = \varprojlim_H W[G/H]$ for H running over open subgroups of G. A character $\psi : \Delta \to W^\times$ gives rise to a projector $\psi_* : W[[G]] = W[[\boldsymbol{\Gamma}]][\Delta] \to W[[\boldsymbol{\Gamma}]]$ sending group element δ to $\psi(\delta) \in W$. The projection $\psi_*\varphi$ is called a *branch* of φ with *branch character* ψ.

For a number field $K \subset \overline{\mathbb{Q}}$, a Galois extension K_∞/K is called a $\mathbb{Z}_p$-*extension* if $\mathrm{Gal}(K_\infty/K) \cong \mathbb{Z}_p$. If K is a finite extension of $\mathbb{Q}$, the composite of all $\mathbb{Z}_p$-extension of K (inside $\overline{\mathbb{Q}}$) has Galois group isomorphic to $\mathbb{Z}_p^m$ for some $m > 0$.

Let $p > 2$ be a prime. Let M be a CM field of degree $2d$ in which p is *unramified.* This means that M is a totally imaginary quadratic extension of a totally real field F (and each prime factor of p in F splits in M). In this setting, we take $\boldsymbol{\Gamma}$ to be the Galois group of the composite of all $\mathbb{Z}_p$-extensions of M. The Leopoldt conjecture holds for p and M if and only if $\mathrm{rank}_{\mathbb{Z}_p} \boldsymbol{\Gamma} = d+1$. Though it is not necessary, to describe the result concisely, we just assume $\mathrm{rank}_{\mathbb{Z}_p} \boldsymbol{\Gamma} = d+1$.

We assume to be able to split primes of M over p into a disjoint union $\Sigma_p \sqcup \Sigma_p^c$ for complex conjugation c on M. Katz associated to Σ_p and each finite order branch character ψ of p-power conductor a p-adic L-function $L_p = L_p(\psi)$.

Recall fixed embeddings $\mathbb{C} \overset{i_\infty}{\leftrightarrow} \overline{\mathbb{Q}} \overset{i_p}{\hookrightarrow} \overline{\mathbb{Q}}_p$. We have a CM type associated $\Sigma = \{\sigma : M \hookrightarrow \mathbb{C}\}$ to Σ_p [so $\mathrm{Hom}_{\mathrm{field}}(M, \mathbb{C}) = \Sigma \sqcup \Sigma^c$]. We may view the p-adic L-function as a power series $L_p(x_\sigma, y)_{\sigma\in\Sigma} \in W[[x_\sigma, y]]$ of $d+1$-variables for the p-adic big unramified complete discrete valuation ring $W \subset \mathbb{C}_p$ with algebraic closed residue field $\mathbb{F} = \overline{\mathbb{F}}_p$. For each fractional ideal $\mathfrak{a}$ of M prime to p, its power $\mathfrak{a}^h$ becomes principal generated by $\alpha \in M^\times$. Define $\langle \mathfrak{a}^\sigma \rangle \in \overline{\mathbb{Q}}_p^\times$ by $\exp_p(\frac{1}{h}\log_p(\alpha^\sigma))$ for the p-adic logarithm $\log_p$. Then

$$\widehat{\lambda}_{\kappa,k} : \mathfrak{a} \mapsto \langle \mathfrak{a}^{-k\Sigma-\kappa(1-c)} \rangle := \prod_{\sigma\in\Sigma} \langle \mathfrak{a}^{-k\sigma-\kappa_\sigma\sigma(1-c)} \rangle$$

is the *p-adic avatar* of an arithmetic *Hecke character* $\lambda_{\kappa,k}$ of conductor at most p with infinity type $\sum_{\sigma\in\Sigma} k\sigma + \kappa_\sigma(1-c)\sigma$.

Exercise 3.36 *Prove that the product $\prod_{\sigma\in\Sigma}\langle \mathfrak{a}^{-k\sigma-\kappa_\sigma\sigma(1-c)}\rangle$ does not depend on the choice of the generator α of $\mathfrak{a}^h$, though the individual factor $\langle \mathfrak{a}^{-k\sigma-\kappa_\sigma\sigma(1-c)}\rangle$ may depend on it.*

For $\kappa \geq 0 (\Leftrightarrow \kappa_\sigma \geq 0\ \forall \sigma \in \Sigma)$ and $k > 0$, we have

$$\frac{L_p(\widehat{\lambda}_{\kappa,k})}{\Omega_p^{k\Sigma+2\kappa}} := \frac{L_p(\gamma_\sigma^{\kappa_\sigma}-1, \gamma^k-1)}{\Omega_p^{k\Sigma+2\kappa}} = *E(\psi\lambda_{\kappa,k})\frac{\pi^\kappa L(0,\psi\lambda_{\kappa,k})}{\Omega_\infty^{k\Sigma+2\kappa}} \quad \text{for } \gamma_\sigma = 1+p.$$

This is a consequence of Theorem 1.41 when $F = \mathbb{Q}$ (and the general case is shown in [K2]). Here we put $\Omega_?^{k\Sigma+2\kappa} = \prod_\sigma \Omega_{?,\sigma}^{k+2\kappa_\sigma}$ for $\Omega_? = (\Omega_{?,\sigma})_{\sigma\in\Sigma}$, which is the p-adic/complex Néron period of CM abelian variety of CM type Σ (with ordinary good reduction at p), $*$ is a simple constant with $|*|_p = 1$ including the Γ/ϵ-factor, and $E(\lambda) = \prod_{\mathfrak{p}\in\Sigma_p}(1-\lambda(\mathfrak{p}^c))(1-N(\mathfrak{p})^{-1}\lambda(\mathfrak{p})^{-1})$.

Assume hereafter in this chapter that $\mathbb{F} = \overline{\mathbb{F}}_p$ (the algebraic closure of $\mathbb{F}_p$). Limiting ourselves to the case of imaginary quadratic M, we describe a sketch of the proof of the following result:

Theorem 3.37 *If p is unramified in F, we have $\mu(L_p(\psi)) = 0$; that is,*

$$p \nmid L_p(x_\sigma, y) \ \ \text{in} \ \ W[[x_\sigma, y]].$$

In the case of an imaginary quadratic field M, the composite of all $\mathbb{Z}_p$-extensions of M has Galois group $\boldsymbol{\Gamma} \cong \mathbb{Z}_p^2$ over M by class field theory, and $W[[x, y]]$ is canonically identified with $W[[\boldsymbol{\Gamma}]]$ (see the construction of Katz's measure in Chap. 9). Then we have other ways of proving the vanishing. Indeed, if one can find a $\mathbb{Z}_p$-extension M_∞/M such that the projected image of $L_p(\psi) \in W[[\boldsymbol{\Gamma}]]$ to $W[[\mathrm{Gal}(M_\infty/M)]]$ has vanishing μ, the original $\mu(L_p(\psi))$ must vanish. There are some choices when M is quadratic:

1. If ψ factors through the Galois group of abelian extension of $\mathbb{Q}$, one can take the cyclotomic $\mathbb{Z}_p$-extension M_∞/M and the original result of Ferrero–Washington gives the vanishing (either via Gross's factorization formula in [Gs] or via the proof of the main conjectures in the cyclotomic case [MzW] and in the imaginary quadratic case [Ru1] and [Ru2]).
2. One can take the unique $\mathbb{Z}_p$-extension unramified at the factor $\overline{\mathfrak{p}}$ of p. This case is dealt with by Gillard [Gil] and Schneps [Scn].
3. A line in $G = \mathbb{Z}_p^2$ parallel to the anticyclotomic $\mathbb{Z}_p$-extension by Finis [Fi2].

In this book, we take *anticyclotomic* $\mathbb{Z}_p$-extension M_∞^-/M and lines parallel to M_∞^-/M inside the $\mathbb{Z}_p^2$-extension of M to prove the above theorem. The extension M_∞^-/M is the unique $\mathbb{Z}_p$-extension such that $c\sigma c^{-1} = \sigma^{-1}$ for all $\sigma \in \mathrm{Gal}(M_\infty^-/M)$, where c is a complex conjugation. This method works well also for CM fields (see [H10a], [H11b], [Hs2], and [BuH] for general cases).

For a weight $k > 0$, we prove $\inf_{\zeta\in\mu_{p^\infty}} \mu(L_p(\psi)(x_\sigma, \zeta\gamma^k - 1)) = 0$, which implies $\mu(L_p(\psi)(x_\sigma, y)) = 0$. If we forget about notational complication coming from dealing with a general CM field, the proof is in essence the same for any choice of F, in this book, we assume

1. $F = \mathbb{Q}$; hence, M is an imaginary quadratic field with integer ring $\mathfrak{O}$;
2. the prime p splits in M; that is, $(p) = \mathfrak{p}\overline{\mathfrak{p}}$ and $\Sigma_p = \{\mathfrak{p}\}$;
3. $p \geq 5$, $\psi = 1$, any ring to have $\frac{1}{6}$ and $|\mathfrak{O}^\times| = 2$.

In addition to these three hypotheses, to make the exposition less technical in this section, we assume an extra assumption:

(Ex) The class number of M is prime to p.

In the detailed proof of Theorem 3.37 we will give later in Sect. 9.2, we do not assume assumption (Ex).

3.5.1 Eisenstein Series

As we have seen in (1.3.1), for each lattice $L = \mathbb{Z}w_1 + \mathbb{Z}w_2 \subset \mathbb{C}$, we define an Eisenstein series

$$\frac{(2\pi i)^k}{(k-1)!}G_k(L) = \frac{1}{2}\sum_{\ell\in L-\{0\}} \frac{1}{\ell^k} = \frac{1}{2}\zeta(1-k) + \sum_{n=1}^{\infty}\Big(\sum_{0<d|n} d^{k-1}\Big)q^n,$$

which is a function of lattices satisfying $G_k(\alpha L) = \alpha^{-k}G_k(L)$. The quotient $\mathbb{C}/L$ gives rise to an elliptic curve $E(L) \subset \mathbf{P}^2$ by Weierstrass theory (which we recalled in Sect. 2.3.4).

Since $\Omega_{E(L)/\mathbb{C}}$ is generated by du for the variable u of $\mathbb{C}$ and we can recover from $(E(L), du)$ the lattice L as $\{\int_\gamma du | \gamma \in \pi_1^{\mathrm{top}}(E)\}$, we regard G_k as a function of the pairs (E, ω) of an elliptic curve E with a generator ω of $H^0(E, \Omega_{E/\mathbb{C}})$ satisfying $G_k(E, \alpha\omega) = \alpha^{-k}G_k(E, \omega)$. For a given base ring $B_{/\mathbb{Z}[\frac{1}{6}]}$, a modular form f defined over B of weight k and of level 1 can be interpreted as a functorial rule assigning a number in A to the isomorphism class of a pair $(E,\omega)_{/A}$ of an elliptic curve E over a B-algebra A and a differential with $H^0(E,\Omega_{E/A}) = A\omega$ such that [see (G0–2) in Sect. 6.2.4 and (G3) in Sect. 6.2.8]

1. $f((E,\omega)\times_{A,\rho} A') = \rho(f(E,\omega))$ for each B-algebra homomorphism $\rho : A \to A'$;
2. $f(E,a\omega) = a^{-k} f(E,\omega)$ for $a \in A^\times$;
3. f is finite at cusps (the value at the Tate curve at each cusp lands in $B[[q]]$).

If a modular form f defined over $\mathbb{C}$ has q-expansion in $B[[q]]$ at the infinity cusp, f is actually defined over B (assuming $B \subset \mathbb{C}$). Indeed, f is then an isobaric polynomial $\Phi(g_2,g_3)$ in $B[g_2,g_3]$, and if (E,ω) is defined over A by $y^2 = 4x^3 - g_2(E,\omega)x - g_3(E,\omega)$ with $\omega = \frac{dx}{y}$, $f(E,\omega) = \Phi(g_2(E,\omega), g_3(E,\omega)) \in A$. We take $B := \mathcal{W} = W \cap \overline{\mathbb{Q}}$.

Removing p-coefficients, $\mathcal{G}_k(z) = \sum_{n>0,p\nmid n}(\sum_{0<d|n} d^{-1}\langle d\rangle^k)q^n$ gives rise to a p-adic analytic family with $\psi_1 = \omega^{-1}$. It is a part of the family $\{\mathcal{G}_P\}_{P\in\mathrm{Spec}(\Lambda)}$ such that $a(n,\mathcal{G}_P) = \sum_{0<d|n}\epsilon_P(d)d^{-1}\langle d\rangle^k$ for arithmetic $P = (1+T-\epsilon_P(\gamma)\gamma^{k-1})$. Often we write this $\mathcal{G}_P$ as $\mathcal{G}_{k,\zeta}$ for $\zeta = \varepsilon_P(\gamma) \in \mu_{p^\infty}(\overline{\mathbb{Q}}_p)$ and also $\varepsilon_\zeta = \varepsilon_P$. The form $\mathcal{G}_{k,\zeta}$ is also an Eisenstein series with possibly nontrivial Nebentypus.

Let $C = M_{\mathbb{A}}^\times / M^\times(\widehat{\mathfrak{O}}^{(\mathfrak{p})})^\times M_\infty^\times$ be the mod $\mathfrak{p}^\infty$ ray class group of M, which fits into the exact sequence

$$1 \to \mathfrak{O}_{\mathfrak{p}}^\times/\{\pm 1\} \to C \to Cl \to 1$$

for the class group Cl of M. Consider $\Gamma = 1 + p\mathbb{Z}_p$ a subgroup of $\mathfrak{O}_{\mathfrak{p}}^\times/\{\pm1\}$ by the inclusion map $\mathbb{Z}_p^\times \cong \mathfrak{O}_{\mathfrak{p}}^\times$. Since $p \nmid |Cl|$ by the extra assumption (Ex), we have the splitting $C = \Delta \times \Gamma$, and we may regard ε_ζ as a character $\tilde{\varepsilon}_\zeta$ of C projecting down C to Γ. Here $\Delta \subset C$ is the maximal finite subgroup of C, and in the proof of Theorem 3.37 in Sect. 9.2, if p divides the class number (though we may not have splitting), we compute the μ-invariant of each branch of the Katz measure restricted to Γ (as the original μ vanishes if the "restricted" μ vanishes). This is how we eliminate assumption (Ex) later.

The CM curve $E(\mathfrak{a})$ is defined over $\mathcal{W}$ and has a differential $\omega(\mathfrak{a})$ with $\omega(\mathfrak{a}) = \pi^*\omega(\mathfrak{O})$ for a fixed $\omega(\mathfrak{O})$, where $\pi : E(\mathfrak{a}) \to E(\mathfrak{O})$ is an étale isogeny of degree $[R:\mathfrak{a}]$. Fix a generator $\alpha \in \pi_1^{\mathrm{top}}(E(\mathfrak{O}))$, and set $\Omega_\infty = \int_\alpha \omega(\mathfrak{O})$. We find $\frac{\mathcal{G}_{k,\zeta}(\mathfrak{a})}{\Omega_\infty^k} = \mathcal{G}_{k,\zeta}(\Omega_\infty\mathfrak{a}) = \mathcal{G}_{k,\zeta}(E(\mathfrak{a}),\omega(\mathfrak{a})) \in \mathcal{W}[\zeta]$,

$$\frac{\mathcal{G}_k(\mathfrak{a})}{\lambda_{0,k}(\mathfrak{a})} = \frac{1}{2\lambda_{0,k}(\mathfrak{a})}\sum_{\alpha\in\mathfrak{a},(a)+(p)=\mathfrak{O}} \langle\alpha\rangle^{-k} \doteqdot \sum_{\alpha\in\mathfrak{a}} \frac{\lambda_{0,k}(\alpha\mathfrak{a}^{-1})}{N(\alpha\mathfrak{a}^{-1})^s}\bigg|_{s=0} \doteqdot L_{\mathfrak{a}^{-1}}(0,\lambda_{0,k})$$

and $\dfrac{\mathcal{G}_{k,\zeta}(\mathfrak{a})}{\tilde{\varepsilon}_\zeta\lambda_{0,k}(\mathfrak{a})} \doteqdot L_{\mathfrak{a}^{-1}}(0,\tilde{\varepsilon}_\zeta\lambda_{0,k})$,

where "$\doteqdot$" indicates that we need to multiply the Euler-like factor $E(?)$.

Applying the invariant differential operator (of Maass–Shimura)

$$\delta_k = \frac{1}{2\pi i}\left(\frac{k}{2iy} + \frac{\partial}{\partial z}\right) \quad \text{and} \quad \delta_k^\kappa = \overbrace{\delta_{k+2\kappa-2}\cdots\delta_k}^{\kappa},$$

we have, by Shimura,

$$\frac{\delta_k^\kappa \mathcal{G}_{k,\zeta}(\mathfrak{a})}{\tilde{\varepsilon}_\zeta \lambda_{\kappa,k}(\mathfrak{a})} \doteqdot L_{\mathfrak{a}^{-1}}(0, \tilde{\varepsilon}_\zeta \lambda_{\kappa,k}) \;\; (\textit{only dependent on the class of } \mathfrak{a}).$$

This can be seen as follows: For $z_0 = z_0(\mathfrak{a})$ with $\mathfrak{a} = \mathbb{Z}z_0 + \mathbb{Z}$ (and $\mathrm{Im}(z_0) > 0$), define $\rho = \rho_{\mathfrak{a}} : M \to M_2(\mathbb{Q})$ by $\rho(\alpha)\binom{z_0}{1} = \binom{z_0}{1}\alpha$. Then $\rho(\alpha)(z_0) = z_0$. We take a local parameter τ around z_0 so that $\rho(\alpha)(\tau) = \alpha^{1-c}\tau$ and $\tau = 0 \leftrightarrow z = z_0$ (for example, if $z_0 = i = \sqrt{-1}$, $\tau = \frac{z-i}{z+i}$). Then we find, regarding $\langle \mathfrak{a}^{1-c}\rangle \in \mathbb{C}^\times$

$$\frac{\delta_k^\kappa \mathcal{G}_k(\langle \mathfrak{a}^{1-c}\rangle \tau)|_{\tau=0}}{\lambda_{0,k}(\mathfrak{a})} = \frac{\langle \mathfrak{a}^{1-c}\rangle^\kappa \delta_k^\kappa \mathcal{G}_k(z_0(\mathfrak{a}))}{\lambda_{0,k}(\mathfrak{a})} = \frac{\delta_k^\kappa \mathcal{G}_k(z_0(\mathfrak{a}))}{\lambda_{\kappa,k}(\mathfrak{a})} \doteqdot L_{\mathfrak{a}^{-1}}(0, \lambda_{\kappa,k}).$$

There is a canonical p-adic Serre–Tate parameter t around z_0 (as a point of a modular curve). Heuristically, $\log_p(t)$ behaves like τ: $\tau = 0 \Leftrightarrow t = 1$ and $t \circ \rho(\alpha) = t^{\alpha^{1-c}}$ (see Proposition 7.44 for a proof). For $\theta := t\frac{d}{dt}$, by Katz, with a specific p-adic period $\Omega_p \in W^\times$ of $E(\mathfrak{O})$ (we recall later),

$$\frac{L_p(\widehat{\lambda}_{\kappa,k})}{\Omega_p^{k+2\kappa}} = \sum_{\mathfrak{a}} \frac{\theta^\kappa(\mathcal{G}_k(\tau^{\langle \mathfrak{a}^{1-c}\rangle}))|_{\tau=1}}{\lambda_{0,k}(\mathfrak{a})} = \sum_{\mathfrak{a}} \frac{\delta_k^\kappa(\mathcal{G}_k(\Omega_\infty \mathfrak{a}))}{\lambda_{\kappa,k}(\mathfrak{a})} \doteqdot \frac{\pi^\kappa L(0, \lambda_{\kappa,k})}{\Omega_\infty^{k+2\kappa}}$$

for $\mathfrak{a}$ running through ideal classes. Thus, we can compute the Taylor expansion of $E = \sum_{\mathfrak{a}} \frac{\mathcal{G}_k(t^{\langle \mathfrak{a}^{1-c}\rangle})}{\lambda_{0,k}(\mathfrak{a})}$ with respect to $x' = \log_p(t)$ by computing the derivative with respect to θ.

Since E is defined over W, out of this identification of the Taylor expansion, we conclude that $L_p(x, \gamma^k - 1)$ is almost the expansion of E with respect to $x \doteqdot T = t - 1$ (which we call *t-expansion* by an abuse of symbol instead of T-expansion, as this T has nothing to do with the weight variable "T" of the Hecke algebra $\mathbf{h}$). Strictly speaking, first, the t-expansion is the expansion of the measure given by E as a measure on $\mathbb{Z}_p$, not on $1 + p\mathbb{Z}_p$. Second, we want to know the nonvanishing of the t-expansion modulo p of the restriction of the measure on $Cl(p^\infty)$ to $1 + p\mathfrak{O}_p \cong \Gamma^2$. Thus, we need to replace $\mathcal{G} := \{\mathcal{G}_{k,\zeta}\}$ by a family $\{\mathcal{G}'_{k,\zeta,\mathfrak{b}}\}$ of Eisenstein series of level p^2 (somehow restricting the measure to the torsion-free part $1 + p\mathfrak{O}_p$ increases the level; see Sect. 9.2, as this is subtle). Since $L_{\mathfrak{a}^{-1}}(s, \lambda)$ (resp. $\mathcal{G}$) is a sum of partial L-functions for a class modulo p (resp., a sum of Eisenstein series of level p^2), we have $L_{\mathfrak{a}^{-1}}(s, \lambda) = \sum_{\mathfrak{b} \equiv \mathfrak{a} \bmod p, [\mathfrak{b}] \in Cl_M(p)} L_{\mathfrak{b}^{-1}}(s, \lambda)$, and

$$E_\zeta = \sum_{\mathfrak{b} \in Cl_M(p)} \frac{\mathcal{G}'_{k,\zeta,\mathfrak{b}}(t^{\langle \mathfrak{b}^{1-c}\rangle})}{\tilde{\varepsilon}_\zeta \lambda_{0,k}(\mathfrak{b})}$$

gives rise to the exact power series $L_p(t-1, \gamma^k\zeta - 1)$ as a measure on $\mathbb{Z}_p$.

Note that $t-1$ is the local parameter around $z_0(\mathfrak{O})$. Take the following fact for granted (we will give an outline of the proof at the end of this section and give details later in Sect. 11.1.5):

Theorem 3.38 (Linear Independence Theorem) *For any nonzero, non-constant mod p-modular form $f_\mathfrak{b}$ of weight k indexed by ideal classes, the set $\{f_\mathfrak{b}(t^{\langle \mathfrak{b}^{1-c}\rangle})\}_{[\mathfrak{b}]\in Cl_M(p)/\sim}$ is linearly independent over $\mathbb{F}$ in $\mathbb{F}[[(t-1)]]$, where $\{[\mathfrak{b}]\}$ is a representative set of ray classes modulo p under the equivalence: $[\mathfrak{b}] \sim [\mathfrak{c}] \Leftrightarrow \langle \mathfrak{b}^{1-c}\rangle = \alpha(\mathfrak{c})^{1-c}\langle \mathfrak{c}^{1-c}\rangle$ for $\alpha(\mathfrak{c}) \in M^\times$.*

We prove this later showing that $\{t^{\langle \mathfrak{b}^{1-c}\rangle}\}_\mathfrak{b}$ is algebraically independent in $\mathbb{F}[[(t-1)]]$ over $\mathbb{F}$, which enables us to compute the $\mu(L_p(x, \zeta\gamma^k - 1))$ [via the almost identity $L_p(x, \zeta\gamma^k - 1) \doteqdot$ the t-expansion of E_ζ under $x \doteqdot t-1$] by q-expansion of $f_{\zeta,\mathfrak{b}} = \sum_{\mathfrak{c}\sim\mathfrak{b}} \mathcal{G}'_{k,\zeta,\mathfrak{b}}(\tau^{\alpha(\mathfrak{c})^{c-1}})$. This is because E_ζ is a linear combination of $f_{\zeta,\mathfrak{b}}((t^{\langle \mathfrak{b}^{1-c}\rangle})$. Our conclusion is

$$\mu(L_p(x, \gamma^k\zeta - 1)) = \min(\mathrm{ord}_p(f_{\zeta,\mathfrak{b}}))_\mathfrak{b},$$

where $\mathrm{ord}_p(f) = \min_n(\mathrm{ord}_p(a(n,f)))$. This is because

$$p^{\mu(\mathfrak{b})} \parallel f_{\zeta,\mathfrak{b}}(t) \in W[[t-1]] \Leftrightarrow p^{\mu(\mathfrak{b})} \parallel f_{\zeta,\mathfrak{b}}(q) \in W[[q]].$$

Thus, dividing E_ζ by p^μ for $\mu = \min_\mathfrak{b} \mu(\mathfrak{b})$, and applying Theorem 3.38 to $p^{-\mu} f_{\zeta,\mathfrak{b}}$, we find that $\mu(L_p(x, \zeta\gamma^k - 1)) = \mu(L_p(t-1, \zeta\gamma^k - 1)) = \mu$ and that $L_p(t-1, \zeta\gamma^k - 1)$ is the t-expansion of E_ζ.

For $\zeta \in \mu_{p^\infty}(\overline{\mathbb{Q}}_p)$, we prove $\inf_{\zeta,k} \mathrm{ord}_p(f_{\zeta,\mathfrak{b}}) = 0$, and $p \nmid L_p(x,y)$ follows as $\mu(L_p(x,y)) \le \mu(L_p(x, \zeta\gamma^k - 1))$. If $\mathfrak{b} = \rho^{-1}\mathfrak{a}$ for $\rho \in (\mathfrak{a}/p\mathfrak{a})^\times \cong (\mathfrak{O}/p\mathfrak{O})^\times$, $\mathcal{G}'_{k,\zeta,\mathfrak{b}}$ is "something like" the sum over $\alpha \in \mathfrak{a}$ with $\alpha \equiv \rho \mod p$. Thus, for a well-chosen prime l, the q-expansion coefficient $a(l, f_{\zeta,\mathfrak{b}})$ is $1 + \varphi_{k,\zeta}(l) l^{k-1}$ for a suitable character $\varphi_{k,\zeta}(l) \doteqdot \zeta^{\log_p(l)/\log_p(\gamma)}$ up to p-adic units. Thus, $\min(\mathrm{ord}_p(f_{\zeta,\mathfrak{b}}))_\mathfrak{b} \le \min_\zeta(\mathrm{ord}_p(a(l, f_{\zeta,\mathfrak{b}}))) = 0$.

3.5.2 Modular Curves as Shimura Variety

To prove the linear independence theorem (Theorem 3.38), we study a subvariety of the self-product of modular curves stable under the diagonal "toric" action by $\rho(\alpha)$.

We define the modular curve by a classification problem of elliptic curves $E_{/A}$ over a ring $A_{/B}$ for $B = \mathbb{Z}[\frac{1}{N}, \mu_N]$ (with specific primitive root $\zeta \in \mu_N$), looking into the following moduli functor of level $\Gamma(N)$ and writing "$[\cdot]$" for "$\{\cdot\}/\cong$":

$$\mathcal{E}_{\Gamma(N),\zeta}(A) = \left[(E, \phi_N : (\mathbb{Z}/N\mathbb{Z})^2 \cong E[N])_{/A} \middle| \langle \phi_N(1,0), \phi_N(0,1)\rangle = \zeta\right],$$

which is represented by a geometrically irreducible curve $Y_\zeta(N)$ (see Sect. 6.2.6 for a proof). Here $\langle \cdot, \cdot \rangle$ is the Weil pairing (see Sect. 6.1.6 for the pairing).

We know classically that $\mathcal{E}_{\Gamma(N),\zeta}(\mathbb{C}) \cong \Gamma(N)\backslash\mathfrak{H} = Y_\zeta(N)(\mathbb{C})$. If we remove the contribution upon ζ and consider the functor $\mathcal{E}_{\Gamma(N)}(A) = [(E,\phi_N)_{/A}]$ defined on the category of $\mathbb{Z}[\frac{1}{N}]$-algebras, we have $\mathcal{E}_{\Gamma(N)} = \bigsqcup_\zeta \mathcal{E}_{\Gamma(N),\zeta}$, and this functor is represented by a geometrically nonconnected curve $Y(N) = \bigsqcup_\zeta Y_\zeta(N)$ defined over $\mathbb{Z}[\frac{1}{N}]$ if $N \geq 3$.

Write $G = GL(2)_{/\mathbb{Z}}$. We can let $\alpha \in G(\mathbb{Z}/N\mathbb{Z})$ act on $Y(N)$ by $(E,\phi) \mapsto (E,\phi\circ\alpha)$. Thus, the group $G(\widehat{\mathbb{Z}}) = \varprojlim_N G(\mathbb{Z}/N\mathbb{Z})$ acts on the limit $Y = \varprojlim_N Y(N)$ (which is a proscheme defined over $\mathbb{Q}$), and $SL_2(\widehat{\mathbb{Z}})$ preserves the connected component $Y_{\zeta_\infty} = \varprojlim_N Y_{\zeta_N}(N)$.

A remarkable fact Shimura found is that this action of $G(\widehat{\mathbb{Z}})$ can be extended to the finite adele group $G(\mathbb{A}^{(\infty)}) = G(\mathbb{A})/G(\mathbb{R})$ (see [IAT] Chap. 6 and Theorem 7.18 in the text). In other words, the action of $g \in G(\mathbb{A}^{(\infty)})$ on the limit Y can be interpreted as a family of morphisms/correspondences (Hecke action/Hecke correspondence) on finite-level modular curves. A pro-variety equipped with such adelic action is rare, but all Shimura varieties have such a large symmetry (we study the simplest Shimura curve Y in details in Chap. 7 and refer to [PAF] for general Shimura varieties).

Deligne's interpretation of the symmetry is equally useful. To explain Deligne's idea, consider the Tate module $T(E) = \varprojlim_N E[N]$ for an elliptic curve $E_{/A}$ for a $\mathbb{Q}$-algebra A. This means, taking a geometric point $s = \mathrm{Spec}(k) \in \mathrm{Spec}(A)$ (for an algebraically closed field k) in each connected component of $\mathrm{Spec}(A)$, we think of the $\widehat{\mathbb{Z}}$-module $T(E_s) = \varprojlim_N E_s[N](k)$ (the choice of s does not matter; see Lemma 7.10). Then $T(E) \cong \widehat{\mathbb{Z}}^2$ and $V(E) = T(E)\otimes_{\mathbb{Z}}\mathbb{Q} \cong (\mathbb{A}^{(\infty)})^2$. Deligne realized that Y represents the following functor defined over $\mathbb{Q}$-algebras (see Theorem 7.11 in Sect. 7.1.3 for a proof):

$$\mathcal{E}^{\mathbb{Q}}(A) = \{(E,\eta : (\mathbb{A}^{(\infty)})^2 \cong V(E))_{/A}\}/\text{isogenies}.$$

Here $\mathbb{A}^{(\infty)}$ is the finite adele ring. Then $g \in G(\mathbb{A})$ sends a point $(E,\eta)_{/A} \in \mathcal{E}^{\mathbb{Q}}(A)$ to $(E,\eta\circ g^{(\infty)})_{/A}$ for the projection $g^{(\infty)}$ of g to $\mathbb{A}^{(\infty)}$.

Take the quotient $Y^{(p)} = \varprojlim_{p\nmid N} Y(N) = Y/G(\mathbb{Z}_p)$. Put $V^{(p)}(E) = T(E)\otimes_{\widehat{\mathbb{Z}}}\mathbb{A}^{(p\infty)}$, and consider the prime-to-p-level structure $\eta^{(p)} : (\mathbb{A}^{(p\infty)})^2 \cong V^{(p)}(E)$. By Theorem 7.13, $Y^{(p)}$ over $\mathbb{Z}_{(p)}$ represents the following functor defined over $\mathbb{Z}_{(p)}$-algebras:

$$\mathcal{E}^{(p)}(A) = \{(E,\eta^{(p)} : (\mathbb{A}^{(p\infty)})^2 \cong V^{(p)}(E))_{/A}\}/\text{prime-to-}p\text{ isogenies},$$

where an isogeny ϕ is prime to p if $\deg(\phi)$ is prime to p. On $Y^{(p)}$ and its p-fiber $Y^{(p)}_{/\mathbb{F}}$ over $\mathrm{Spec}(\mathbb{F})$, again $g \in G(\mathbb{A})$ acts by $\eta^{(p)} \mapsto \eta^{(p)}\circ g^{(p\infty)}$.

An endomorphism $\phi : E \to E$ of elliptic curves is called *central* if $\phi(x) = mx$ for an integer m for all x. If we have a prime-to-p **noncentral** endomorphism $\alpha : E \to E$, then E has complex multiplication by an imaginary quadratic field $M = \mathbb{Q}[\alpha] \subset \mathrm{End}(E)\otimes_{\mathbb{Z}}\mathbb{Q}$, and we can write $\alpha \circ \eta^{(p)} = \eta^{(p)} \circ \rho^{(p)}(\alpha)$ for $\rho^{(p)}(\alpha) \in G(\mathbb{A}^{(p\infty)})$. Since α is "non-central" (i.e., $\alpha \notin \mathbb{Z}$), $\rho^{(p)}(\alpha)$ is not in the center of $G(\mathbb{A}^{(p\infty)})$. Thus, if $z_0 = (E, \eta^{(p)}) \in Y^{(p)}(A)$ ($A = \mathcal{W}$ and $\mathbb{F}$), we find that $\rho^{(p)}(\alpha)(z_0) = z_0$, and

$$\mathfrak{O}_{(p)}^{\times}/\mathbb{Z}_{(p)}^{\times} \underset{\rho}{\xrightarrow{\sim}} \{g \in \mathrm{Aut}(Y^{(p)}) | g(z_0) = z_0\}.$$

Pick the elliptic curve $E := E(\mathfrak{O})_{/\mathcal{W}}$ with CM by the integer ring $\mathfrak{O}$ of M. Since $\Sigma_p^c = \{\overline{\mathfrak{p}}\}$, we have $\mathfrak{p} = \mathfrak{O} \cap \mathfrak{m}_{\mathcal{W}}$ and $\mathcal{W}/\mathfrak{m}_{\mathcal{W}} = \overline{\mathbb{F}}_p$, and $E[\overline{\mathfrak{p}}^\infty]$ is étale constant and $E[\mathfrak{p}^\infty] \cong \mu_{p^\infty}$ over $\mathcal{W}$ (see Sect. 4.3 particularly Sect. 4.3.7 for such group schemes). We fix a level-p structure $\eta_p^\circ : \mu_{p^\infty} \cong E[\mathfrak{p}^\infty]$ and $\eta_p^{et} : \mathbb{Q}_p/\mathbb{Z}_p \cong E[\overline{\mathfrak{p}}^\infty]$. Then η_p° induces an isomorphism of formal groups: $\widehat{\eta}_p^\circ : \widehat{\mathbb{G}}_m = \mathrm{Spf}(\widehat{W[t, t^{-1}]}) \cong \widehat{E}$ (see Lemma 4.82 in Sect. 4.4.3); so we define the p-adic Néron period Ω_p by

$$\omega(\mathfrak{O}) = \Omega_p \cdot (\widehat{\eta}_{p,*}^\circ \frac{dt}{t}) \text{ for } \Omega_p \in W^\times. \tag{3.5.1}$$

This is the p-adic period we discussed earlier.

Write $\eta_p = (\eta^\circ, \eta_p^{et}) : \mu_{p^\infty} \times \mathbb{Q}_p/\mathbb{Z}_p \cong E[\mathfrak{p}^\infty] \times E[\overline{\mathfrak{p}}^\infty]$, and define a homomorphism ρ_p of $\mathfrak{O}_{(p)}^\times$ into the diagonal torus of $G(\mathbb{Z}_p)$ by $\alpha \circ \eta_p = \eta_p \circ \rho_p(\alpha)$ for $\alpha \in \mathfrak{O}_{(p)}$. Thus, $\eta_p^\circ \circ \rho_p(\alpha) = \alpha \eta_p^\circ$, identifying $\mathfrak{O}_\mathfrak{p}$ with $\mathbb{Z}_p$ and $\eta_p^{et} \circ \rho_p(\alpha) = \alpha^c \eta_p^{et}$.

Fix a base w_1, w_2 of $\widehat{\mathfrak{O}}^{(p)} \cong T^{(p)}(E)$ over $\widehat{\mathbb{Z}}^{(p)}$, and identify $M_\mathbb{A}^{(p\infty)}$ with $(\mathbb{A}^{(p\infty)})^2$. The choice induces a prime-to-p-level structure $\eta^{(p)} : (\mathbb{A}^{(p\infty)})^2 \cong \mathfrak{O} \otimes_\mathbb{Z} \mathbb{A}^{(p\infty)} = V^{(p)}(E)$. We put $\eta = \eta_p \times \eta^{(p)}$. Define $\rho : \mathfrak{O}_{(p)}^\times \to G(\mathbb{Z}_p \times \mathbb{A}^{(p\infty)})$ by $\eta \circ \rho(\alpha) = \alpha \circ \eta$. Since $\alpha \in \mathfrak{O}_{(p)}^\times$ induces an isogeny $\alpha : E \to E$ sending $\alpha \eta^{(p)} = \eta^{(p)} \rho^{(p)}(\alpha)$, the point $z_0(\mathfrak{O}) = (E(\mathfrak{O}), \eta^{(p)}) \in Y^{(p)} = Y/G(\mathbb{Z}_p)$ is fixed by $\rho(\alpha)$. Pick a fractional ideal $\mathfrak{a} \subset M$ prime to p; hence, $\mathfrak{a} = (a\widehat{\mathfrak{O}}) \cap M$ for an idele $a \in M_\mathbb{A}^\times$ with $a_p = a_\infty = 1$. Then we have $z_0(\mathfrak{a}) = (E(\mathfrak{a}), \eta(\mathfrak{a})) = \rho(a)^{-1}(z_0(\mathfrak{O}))$.

Consider the formal completion $\widehat{Y} = \widehat{Y}_{z/W}$ of $Y_{/\mathcal{W}}^{(p)}$ along $z = z_0(\mathfrak{a}) \in Y^{(p)}(\mathbb{F})$. See Sect. 4.4 for completing the process of a scheme. Then, by the universality of $Y^{(p)}$, $\widehat{Y}$ satisfies

$$\widehat{Y}(A) \cong \widehat{\mathcal{E}}(A) := \left\{X_{/A}\text{: elliptic curves over } A \middle| X \otimes_A \mathbb{F} = E(\mathfrak{a})_{/\mathbb{F}}\right\} / \cong,$$

where A runs through p-profinite local W-algebras with $A/\mathfrak{m}_A = W/\mathfrak{m}_W = \mathbb{F} = \overline{\mathbb{F}}_p$.

By Serre–Tate's deformation theory (see Sect. 6.3), we have $\widehat{Y} \cong \widehat{\mathbb{G}}_m$ canonically. Indeed, first $X_{/A} \in \widehat{\mathcal{E}}(A)$ is determined by the extension $X[p^\infty]^\circ \hookrightarrow X[p^\infty] \twoheadrightarrow X[p^\infty]^{et}$ of the Barsotti–Tate groups. By Theorem 6.47, such an extension over A is classified by

$$\mathrm{Ext}(X[p^\infty]^{et}, X[p^\infty]^\circ) \cong \mathrm{Hom}(\mathbb{Q}_p/\mathbb{Z}_{p/A}, \mu_{p^\infty/A}) = \varprojlim_n \mu_{p^n}(A) = \widehat{\mathbb{G}}_m(A).$$

For this identification, we used $\eta_p^\circ : \mu_{p^\infty} \cong E(\mathfrak{a})[\mathfrak{p}^\infty]$ and its dual inverse $\eta_p^{et} : \mathbb{Q}_p/\mathbb{Z}_p \cong E(\mathfrak{a})[\overline{\mathfrak{p}}^\infty]$. The above identification only depends on $\eta_p = (\eta_p^\circ, \eta_p^{et})$, and multiplication by an idele a with $a_p = 1$ does not affect η_p. Thus the above identification is independent of a and $\mathfrak{a}$.

Since $\rho(\alpha)$ fix $z_0(\mathfrak{a})$, it acts on $\widehat{Y}$. Since $\alpha \in \mathfrak{O}$ acts on $X[p^\infty]^{\mathrm{et}} = \mathbb{Q}_p/\mathbb{Z}_p$ by $\alpha \mapsto \alpha^c$ and on $X[p^\infty]^\circ = \mu_{p^\infty}$ by $\alpha \mapsto \alpha$, only reasonable action of $M^\times$ on the deformation space $t \in \mathrm{Hom}(\mathbb{Q}_p/\mathbb{Z}_{p/A}, \mu_{p^\infty/A}) = \widehat{\mathbb{G}}_m$ is by $t \mapsto \alpha \circ t \circ \alpha^{-c} = t^{\alpha^{1-c}}$. This turns out to be true, as we verify as Proposition 7.44 in Sect. 7.2.5 (see also, [H10a] Proposition 3.4):

Lemma 3.39 *Identifying $\widehat{Y}$ with $\widehat{\mathbb{G}}_m = \mathrm{Spf}(\varprojlim_n W[t, t^{-1}]/(t-1)^{p^n})$, if $\alpha \in M^\times$, we have $\rho(\alpha)(t) = t^{\alpha^{1-c}}$ for complex conjugation c.*

3.5.3 Subvarieties Stable under Hecke Action

We write $I_{\mathfrak{a}}$ for the irreducible component of $Y^{(p)}_{/\mathbb{F}} = Y^{(p)} \times_{\mathcal{W}} \mathbb{F}$ containing $z_0(\mathfrak{a})$. Let $\mathfrak{a}$ be a fractional ideal prime to p of M with $\widehat{\mathfrak{a}} = a\widehat{\mathfrak{O}}$ for $a \in M_{\mathbb{A}}^\times$ with $a_p = a_\infty = 1$. Then $\rho(a)$ gives an isomorphism of $I_{\mathfrak{a}}$ onto $I := I_{\mathfrak{O}}$ sending $z_0(\mathfrak{O})$ to $z_0(\mathfrak{a})$. We identify $I = I_{\mathfrak{a}}$ for any $\mathfrak{a}$. For any $\alpha, \beta \in \mathfrak{O}_{(p)} \cap M^\times$, we have a skew diagonal $\Delta_{\alpha,\beta} = \mathrm{Im}(\rho(\alpha) \times \rho(\beta) : I \to I \times_{\mathbb{F}} I)$ in $I \times_{\mathbb{F}} I$ for $\alpha, \beta \in \mathfrak{O}_{(p)} \cap M^\times$. Here is an example of the rigidity principle:

Theorem 3.40 *Let $H \subsetneq \overbrace{I \times_{\mathbb{F}} \cdots \times_{\mathbb{F}} I}^{n}$ (with $n \geq 1$) be a proper closed irreducible subscheme with a dominant projection to the product I^{n-1} of the first $n-1$ factors and to the last factor I. If $z_0(\mathfrak{O})^n \in H$ and H is stable under the diagonal action of a p-adic open subgroup of $\mathfrak{O}_{(p)}^\times/\mathbb{Z}_{(p)}^\times$, up to permutations of the first $(n-1)$ factors, we have*

$$H = \overbrace{I \times \cdots \times I}^{n-2} \times \Delta_{\alpha,\beta}.$$

This can be proven via Chai's theory of Hecke invariant subvariety of the Shimura variety. We sketch the proof in the following section (and give full details in Chap. 11).

3.5.4 Conclusion

First, we prove Theorem 3.38 from Theorem 3.40: Let $a_j \in \mathbb{Z}_p^\times$ $(j = 1, 2, \ldots, h)$. Regard $a = a_j \in \mathrm{Aut}(\widehat{Y}) = \mathrm{Aut}_{gp}(\widehat{\mathbb{G}}_m)$ given by $t \mapsto t^a$. Let $z = z_0(\mathfrak{O})$ and $\mathcal{O}_z$ for the stalk of $z \in Y^{(p)} \mod p$. Suppose that the algebra homomorphism $\mathcal{O}_z^{\otimes h} := \overbrace{\mathcal{O}_z \otimes_{\mathbb{F}} \mathcal{O}_z \otimes \cdots \otimes_{\mathbb{F}} \mathcal{O}_z}^{h} \to \mathbb{F}[[T]] = \mathcal{O}_{\widehat{\mathbb{G}}_m/\mathbb{F}}$ given by

$f_1(t) \otimes \cdots \otimes f_h(t)$ to $\prod_j f_j(t^{a_j})$ has a nontrivial kernel $\mathfrak{K}$. The schematic closure H of $\mathrm{Spec}(\mathcal{O}_z^{\otimes h}/\mathfrak{K})$ in I^h is stable under the action of $\rho(\mathfrak{O}_{(p)}^\times)$. Thus, by Theorem 3.40, there exist $i \neq j$ such that $\mathfrak{O}_p^\times/\mathbb{Z}_p^\times \ni a_i/a_j \in (O_{(p)}^\times \cap M^\times)/\mathbb{Z}_{(p)}^\times$. Let $\mathfrak{a}_1, \ldots, \mathfrak{a}_h$ be the representatives of $Cl_M(p)/\sim$. Let $a_j = \langle \mathfrak{a}_j^{1-c} \rangle$. Then $a_i/a_j \notin (\mathfrak{O}_{(p)} \cap M^\times)/\mathbb{Z}_{(p)}^\times$ for all $i \neq j$. Therefore, $\mathcal{O}_z^{\otimes h}$ must inject into $\mathbb{F}[[T]]$; that is, $f_j(t^{a_j})$ $(j = 1, \ldots, h)$ are algebraically independent if each $f_j \in \mathcal{O}_z$ is nonconstant. This finishes the sketch of the proof of the linear independence theorem. □

We have $L_p(t-1, \gamma^k\zeta - 1) = \sum_j f_{\zeta,\mathfrak{a}_j}(t^{a_j})$ for the sum of the Eisenstein series $f_{\zeta,\mathfrak{a}_j}$ of weight $k = p-1$ with q-expansion $\sum_{n=0}^\infty a(n, f_{\zeta,\mathfrak{a}_j})q^n$. Dividing $f_{\zeta,\mathfrak{a}_j}$ by the Hasse invariant h does not change the q-expansion and the value of $f_{\zeta,\mathfrak{a}_j}$. Thus, we have

$$\mu(L_p(t-1, \gamma^k\zeta - 1)) \overset{\text{Theorem 3.38}}{=} \min_{n,j}(\mathrm{ord}_p(a(n, f_{\zeta,\mathfrak{a}_j})),$$

as $T = t - 1$ is a local parameter at $z = z_0(\mathfrak{O})$ and q is a local parameter at the cusp ∞ of the irreducible modular curve I. By computation, we can find a prime ℓ and index j such that $a(\ell, f_{\zeta,\mathfrak{a}_j}) = 1 + \zeta^{\log_p(\ell)/\log_p(\gamma)}\ell^{k-1}$ independent of the choice of $\zeta \in \mu_{p^\infty}(\overline{\mathbb{Q}})$. Thus,

$$0 \le \mu(L_p) \le \inf_\zeta \mu(L_p(t-1, \zeta\gamma^k - 1)) \le \inf_\zeta(\mathrm{ord}_p(1 + \zeta^{\log_p(\ell)/\log_p(\gamma)}\ell^{k-1})) = 0,$$

as the p-power order of the root of unity $\zeta^{\log_p(\ell)/\log_p(\gamma)}$ grows indefinitely. This concludes the sketch of the proof of Theorem 3.37 in this chapter. □

More generally, as Theorem 9.15, we will prove the following:

Corollary 3.41 *Suppose $F = \mathbb{Q}$. Then the μ-invariant of the anticyclotomic Katz p-adic L-function $L_p^-(x) = L_p(x, 0)$ also vanishes.*

When $F \neq \mathbb{Q}$, writing $L_p^-(x_\sigma) = L_p(x_\sigma, 0)$, $\mu(L_p^-)$ could be positive though $\mu(L_p(x_\sigma, y)) = 0$ always (see [H11b]). This rare positivity $\mu(L_p^-) > 0$ only occurs if $[F : F(\mu_p)] = 2$ and M/F is unramified everywhere at finite places [see (M1–M3) in [H10a] for a precise set of conditions for $\mu(L_p^-) > 0$].

3.6 Hecke Stable Subvariety

In this section, we provide a sketch of the proof of the specific case (we used) of the conjecture asserting that *a Hecke invariant subvariety of modulo p Shimura variety is a Shimura subvariety.* We can also prove this (conjectural) principle for the Hilbert modular variety and its self-products, but in this book, we only deal with modular curves and their self-products (and full details will be given in Chap. 11; see [H10a] Sect. 3 for the proof dealing with

Hilbert modular Shimura varieties). Any essential ingredients for the proof of the general case show up in the simplest case of modular curves. Write $G = GL(2)_{/\mathbb{Z}}$ with center $Z \cong \mathbb{G}_{m/\mathbb{Z}}$. In this section, the word "variety" means a finite-type reduced scheme over $\mathbb{F} = \overline{\mathbb{F}}_p$.

We recall the following lemma we mentioned already in Sect. 3.5.1:

Lemma 3.42 *Identifying $\widehat{Y}$ with $\widehat{\mathbb{G}}_m = \mathrm{Spf}(\varprojlim_n W[t,t^{-1}]/(t-1)^{p^n})$, if $\alpha \in M^\times$, we have $\rho(\alpha)(t) = t^{\alpha^{1-c}}$ for complex conjugation c.*

Note that if $\alpha \in M^\times$ is not prime to p, the action of $\rho(\alpha)$ is an endomorphism of $Y^{(p)}$, not an automorphism. We prove this as Proposition 7.44 (more general cases are treated in [H10a] as Proposition 3.4). The action of $\rho(\alpha)$ on the Serre–Tate coordinate is given by $t \mapsto t^{\alpha^{1-c}}$ factoring through $G(\mathbb{A}^{(p\infty)})/Z(\mathbb{Q})$, since $Z(\mathbb{Q})$ acts trivially on the Shimura variety $Y^{(p)}$.

3.6.1 Hecke Invariant Subvariety Is a Shimura Subvariety

We write I for the irreducible component of $Y^{(p)}_{/\mathbb{F}} = Y^{(p)} \times_{\mathcal{W}} \mathbb{F}$ containing $z_0 = z_0(\mathfrak{O})$; hence, the formal completion along z_0 is $\widehat{Y} = \widehat{I}$. Recall that $\mathbb{F} = \overline{\mathbb{F}}_p$.

We want to give a sketch of a proof of the following two theorems (see Corollaries 11.8 and 11.11 for detailed proofs, and also [H10a] Corollaries 3.16 and 3.19 for Hilbert modular versions):

Theorem 3.43 *Suppose that $H \subsetneq I \times_{\mathbb{F}} I$ is a closed irreducible subvariety of codimension 1 containing $(z_0, z_0) \in I \times_{\mathbb{F}} I$ stable under the action of a p-adic open subgroup of $\mathfrak{O}^\times_{(p)}/\mathbb{Z}^\times_{(p)} \overset{1-c}{\hookrightarrow} \mathbb{Z}_p^\times$. Then either $H = z_0 \times I$ or $H = I \times z_0$ or $H = \Delta_{\alpha,\beta}$ for $\alpha, \beta \in \mathfrak{O}_{(p)} \cap M^\times$.*

Theorem 3.44 *Let $H \subset \overbrace{I \times_{\mathbb{F}} \cdots \times_{\mathbb{F}} I}^{n}$ $(n \geq 2)$ containing z_0^n be a closed irreducible subvariety with a dominant projection to the product of the first $n-1$ factors and to the last factor. If H is of codimension 1 stable under the diagonal action of a p-adic open subgroup of $\mathfrak{O}^\times_{(p)}/\mathbb{Z}^\times_{(p)} \overset{1-c}{\hookrightarrow} \mathbb{Z}_p^\times$, up to permutations of the first $(n-1)$ factors, we have $H = \overbrace{I \times \cdots \times I}^{n-2} \times \Delta_{\alpha,\beta}$.*

Note here that $\Delta_{\alpha,\beta}$ is isomorphic to I by the first or second projection; hence, H is isomorphic to the self-product of the modular pro-curve I.

3.6.2 Rigidity Lemma and a Sketch of Proofs

We start with general lemmas. Let $T \subset \mathfrak{O}^\times_{(p)}/\mathbb{Z}^\times_{(p)}$ be the open subgroup (under p-adic topology) fixing H by the diagonal action of $\rho(\alpha) \times \cdots \times \rho(\alpha)$

($\alpha \in T$). Then the formal completion $\widehat{H}$ along z_0^n is also stable under T, since z_0^n is fixed by T. By the Serre–Tate theory, $\widehat{H} \subset \widehat{I}^n \cong \widehat{\mathbb{G}}^n_{m/\mathbb{F}}$.

As we have seen, if a power series $\Phi(T) = \Phi(t)$ ($T = t - 1$) satisfies $\Phi(t^z) = \Phi(t)^z$ for all z in an open subgroup of $\mathbb{Z}_p^\times$, then $\Phi(t) = t^s$ for $s \in \mathbb{Z}_p$ (Lemma 3.30). Note that

$$\widehat{\mathbb{G}}_m = \mathrm{Spf}(\widehat{W[t,t^{-1}]}) = \mathrm{Spf}(W[[T]]).$$

The cocharacter group of $\mathbb{G}_m^n$ is isomorphic to $\mathbb{Z}^n$, which we write as $X_*(\mathbb{G}_m^n)$. Then, by tautology, $\mathbb{G}_m^n = \mathbb{G}_m \otimes_{\mathbb{Z}} X_*(\mathbb{G}_m^n)$. Similarly, in the formal setting, putting $X_*(\widehat{\mathbb{G}}_m^n) := \mathrm{Hom}_{\text{formal gp}}(\widehat{\mathbb{G}}_m, \widehat{\mathbb{G}}_m^n) = X_*(\mathbb{G}_m^n) \otimes_{\mathbb{Z}} \mathbb{Z}_p = \mathbb{Z}_p^n$ (the formal cocharacter group), we have $\widehat{\mathbb{G}}_m^n = \widehat{\mathbb{G}}_m \otimes_{\mathbb{Z}_p} X_*(\widehat{\mathbb{G}}_m^n)$. A slightly more general version of Chai's rigidity lemma can be stated as follows (e.g., [H10a] Lemma 3.7):

Lemma 3.45 (C.-L. Chai) *If $Z \subset \widehat{\mathbb{G}}^n_{m/\mathbb{F}}$ is a reduced equidimensional formal subscheme of dimension r stable under the diagonal action of an open subgroup of $\mathrm{Aut}_{gp}(\widehat{\mathbb{G}}_m) = \mathbb{Z}_p^\times$, then*

$$Z = \bigcup_L \widehat{\mathbb{G}}_m \otimes_{\mathbb{Z}_p} L \subset \widehat{\mathbb{G}}^n_{m/\mathbb{F}},$$

where L runs over the (finitely many) $\mathbb{Z}_p$-direct summand of $X_*(\widehat{\mathbb{G}}_m^n)$ of rank r.

The proof is given in [Ch3] but is technical and long, so we postpone the proof of this lemma until Chap. 10.

We apply this to the formal completion $\widehat{H}$ along $z_0^n = (z_0, z_0, \ldots, z_0) \in I^n$ inside $\widehat{I}^n = \widehat{\mathbb{G}}_m^n$. Since H is stable under $t \mapsto t^{\alpha^{1-c}}$ for $\alpha \in \mathfrak{O}_{(p)}^\times/\mathbb{Z}_{(p)}^\times \overset{1-c}{\hookrightarrow} \mathbb{Z}_p^\times$, by continuity, $\widehat{H}$ is stable under the closure $\mathbb{Z}_p^\times$ of $\{\alpha^{1-c} | \alpha \in \mathfrak{O}_{(p)}^\times\}$. Since H is an excellent irreducible scheme, $\widehat{H}$ is reduced equidimensionally of dimension $n-1$. Thus, by the above lemma,

$$\widehat{H} = \bigcup_L \widehat{\mathbb{G}}_m \otimes_{\mathbb{Z}_p} L \subset \widehat{\mathbb{G}}_m^n = \widehat{I}^n, \tag{3.6.1}$$

where L runs over the (finitely many) $\mathbb{Z}_p$-direct summand of $X_*(\widehat{I}^n)$ of rank $n-1$. Let $\mathcal{H} \to H$ be the normalization of H. Since H is irreducible, $\mathcal{H}$ is irreducible. By (3.6.1), each point over z_0^n of $\mathcal{H}$ is indexed by $\{L\}$, and for the point $y_L \in \mathcal{H}$ over z_0^n corresponding to L, $\widehat{\mathcal{H}}_{y_L}$ is étale over $\widehat{\mathbb{G}}_m \otimes L$. Write $I^n = I' \times I''$ for $I' = I^{n-1}$ and $I'' = I$ for the last component.

Lemma 3.46 *The scheme $\mathcal{H}$ is finite flat over I' around z_0^{n-1}. In particular, each L is of rank $n-1$ and projects down to an open subgroup of $X_*(\widehat{\mathbb{G}}_m^{n-1}) \cong \mathbb{Z}_p^{n-1}$. If one of L surjects down onto $X_*(\widehat{I}')$ ($\widehat{I}' = \widehat{\mathbb{G}}_m^{n-1}$), all of L surjects down onto $X_*(\widehat{I}')$, and the projection $\mathcal{H} \to I'$ is étale finite around z_0^{n-1}.*

Proof. Since the projection of the first $(n-1)$-factor $I' = I^{n-1}$ is dominant, at least one of L, call it L_0, projects down to an open $\mathbb{Z}_p$-submodule of $X_*(\widehat{\mathbb{G}}_m^{n-1})$. If there is L with image in $X_*(\widehat{I'})$ of rank $< n-1$, the nonflat locus $\mathcal{H}^{nf} \subset \mathcal{H}$ of $\mathcal{H} \to I'$ is a nonempty proper closed subscheme of $\mathcal{H}$. Since $\dim \widehat{\mathbb{G}}_m \otimes L = \operatorname{rank} L = n-1$, $\mathcal{H}^{nf}$ has dimension $n-1$ equal to $\dim \mathcal{H}$, and so $\mathcal{H}$ has to be reducible, a contradiction. Thus, $\mathcal{H} \to I'$ is finite flat around z_0^{n-1} via a faithfully flat descent from $\widehat{\mathcal{H}}/\widehat{I'}$ to $\mathcal{H}/I'$.

If one of L, call it L_0, surjects down to $X_*(\widehat{I'})$ and another L_1 has image smaller than $X_*(\widehat{I'})$, the ramified locus $\mathcal{H}^{ram}$ of $\mathcal{H} \to I'$ is a nontrivial proper closed subscheme of dimension $n-1$, again a contradiction to the irreducibility of $\mathcal{H}$; so $\mathcal{H} \to I'$ is étale finite around z_0^{n-1}, again via a faithfully flat descent from $\widehat{\mathcal{H}}/\widehat{I'}$ to $\mathcal{H}/I'$. □

When $n = 2$, by applying a power of the p-power Frobenius or its dual to H [that is, applying $\rho(\alpha)$ for α generating $\mathfrak{p}\mathfrak{O}_{(p)}$ or its dual $\rho(\overline{\alpha})$)], we may assume that at least one L surjects down to $X_*(\widehat{I'})$; hence, by the above lemma, all instances of L surject down to $X_*(\widehat{I'})$. Thus, we may assume that $\mathcal{H} \to I'$ is étale finite around z_0. Now assume $n = 2$. Then, over an open dense subscheme $U \subset \mathcal{H}$ containing all points above z_0^2, the two projections $\pi_L : U \to I' = I$ and $\pi_R : U \to I'' = I$ are étale finite.

We consider the universal elliptic curve $(\mathbf{E}, \boldsymbol{\eta})_{/I}$. We pull it back to $\mathcal{H}$: $(\mathbf{A}, \eta_A) = \pi_L^*(\mathbf{E}, \boldsymbol{\eta})$ and $(\mathbf{B}, \eta_B) = \pi_R^*(\mathbf{E}, \boldsymbol{\eta})$. For a point $y \in \mathcal{H}$ over z_0^2, $\widehat{\mathcal{H}} := \widehat{\mathcal{H}}_y = \{(t^b, t^a) | t \in \widehat{I} = \widehat{\mathbb{G}}_m\} \subset \widehat{I} \times \widehat{I}$. Since $\pi_j : \mathcal{H} \to I$ is étale finite around y, we may assume that $a, b \in \mathbb{Z}_p^\times$,and so we may assume that $b = 1$. Let $E = E(\mathfrak{O})$. The map $\widehat{I} \ni t \mapsto t^a \in \widehat{I}$ is induced on $t \in \widehat{I} = \operatorname{Hom}_{gp}(E[\overline{\mathfrak{p}}^\infty], E[\mathfrak{p}^\infty])$ by regarding a as an endomorphism of $E[\mathfrak{p}^\infty]$. Thus identifying $E = E(\mathfrak{O})_{/\mathbb{F}}$ with the fibers $\mathbf{A}_y = \mathbf{B}_y$ of $\mathbf{A}$ and $\mathbf{B}$ at y, we regard the unit $a \in \operatorname{End}(E[\mathfrak{p}^\infty])$ as a $\mathfrak{O}_p$-linear map $a : \mathbf{A}_y[p^\infty] = E[p^\infty] \to E[p^\infty] = \mathbf{B}_y[p^\infty]$ inducing identity on $E[\overline{\mathfrak{p}}^\infty]$. We note the following fact (see [H10a] Proposition 3.15):

Lemma 3.47 (C.-L. Chai) *Further shrinking the open neighborhood U of y in $\mathcal{H}$, we may assume that the isomorphism $a : \mathbf{A}_y[p^\infty] = E[p^\infty] \to E[p^\infty] = \mathbf{B}_y[p^\infty]$ extends to $\tilde{a} : \mathbf{A}_{/U}[p^\infty] \to \mathbf{B}_{/U}[p^\infty]$. This implies that $\widehat{\mathcal{H}}_u \cong \widehat{\mathbb{G}}_m$ by $(t, t^{\tilde{a}}) \leftrightarrow t$ at any point $u \in U(\mathbb{F})$.*

Here is a sketch of a proof of the above lemma. Since a can be approximated p-adically by $\alpha_n \in R_{(p)}$ modulo p^n, a can be extended to $\rho(\alpha_n) : \mathbf{A}[p^n]_{\widehat{\mathcal{H}}} \to \mathbf{B}[p^n]_{\widehat{\mathcal{H}}}$. Passing to a limit, we have an extension $\widehat{a} : \mathbf{A}[p^\infty]_{\widehat{\mathcal{H}}} \to \mathbf{B}[p^\infty]_{\widehat{\mathcal{H}}}$. Write $\mathcal{O}$ for the stalk of $\mathcal{O}_{\mathcal{H}}$ at y; thus, $\widehat{\mathcal{H}} = \operatorname{Spf}(\widehat{\mathcal{O}})$. Since $\widehat{a}$ is determined by its restriction a to $\mathbf{A}_y[p^\infty]$, it is a unique extension of a. Since $\widehat{\mathcal{O}} \otimes_{\mathcal{O}} \widehat{\mathcal{O}}$ is reduced (because of the excellence of $\mathcal{O}$; see Sects. 4.1.11 and 4.4.4, particularly, Lemma 4.87), the pullbacks of $\widehat{a}$ by two projections $\widehat{\mathcal{H}} \times_{\mathcal{H}} \widehat{\mathcal{H}} \to \widehat{\mathcal{H}}$ and three projections $\widehat{\mathcal{H}} \times_{\mathcal{H}} \widehat{\mathcal{H}} \times_{\mathcal{H}} \widehat{\mathcal{H}} \to \widehat{\mathcal{H}} \times_{\mathcal{H}} \widehat{\mathcal{H}}$ coincide; hence, $\widehat{a}$ satisfies the descent

datum (cf., [GME] Sect. 1.11) with respect to $\widehat{\mathcal{O}}/\mathcal{O}$, obtaining the desired U.
□

Proof of Theorem 3.43. Since the two projections $\pi_j : \mathcal{H} \to I$ ($j = L, R$) are dominant, we have $\mathrm{End}(\mathbf{A}) \otimes \mathbb{Q} = \mathrm{End}(\mathbf{B}) \otimes \mathbb{Q} = \mathbb{Q}$. Let $\mathbb{Y}_{/\mathcal{H}} = \mathbf{A} \times_{\mathcal{H}} \mathbf{B}$. Thus, there are only two possibilities for $\mathrm{End}^{\mathbb{Q}}(\mathbb{Y}) = \mathrm{End}(\mathbb{Y}_{/\mathcal{H}}) \otimes \mathbb{Q}$: Either $\mathrm{End}^{\mathbb{Q}}(\mathbb{Y}) = \mathbb{Q} \times \mathbb{Q}$ or $\mathrm{End}^{\mathbb{Q}}(\mathbb{Y}) = M_2(\mathbb{Q})$.

Suppose that $\mathrm{End}^{\mathbb{Q}}(\mathbb{Y}) = M_2(\mathbb{Q})$. By the semisimplicity of the category of abelian schemes, we have two commuting idempotent $e_? \in \mathrm{End}^{\mathbb{Q}}(\mathbb{Y})$ such that $e_A(\mathbb{Y}) = \mathbf{A}$ and $e_B(\mathbb{Y}) = \mathbf{B}$. Since $\mathrm{End}^{\mathbb{Q}}(\mathbb{Y}) = M_2(\mathbb{Q})$, we can find an invertible element $\tilde{\beta}$ in $G(\mathbb{Z}_{(p)}) \subset M_2(\mathbb{Q})$ such that $\tilde{\beta} \circ e_A = e_B$; hence, $\tilde{\beta} : \mathbf{A} \to \mathbf{B}$ is an isogeny with $\tilde{\beta} \circ \eta_A = \eta_B$, whose specialization to the fiber of $\mathbf{A}$ and $\mathbf{B}$ at y gives rise to an endomorphism $\beta \in \mathrm{End}(E(\mathfrak{O})) \otimes \mathbb{Q}$. Thus, the isogeny $\tilde{\beta}$ is induced by $\rho(\beta)$, and we conclude that $\Delta_{1,\beta} = H$.

We suppose that $\mathrm{End}^{\mathbb{Q}}(\mathbb{Y}) = \mathbb{Q} \times \mathbb{Q}$ and try to get a contradiction [in order to prove that $\mathrm{End}^{\mathbb{Q}}(\mathbb{Y}) = M_2(\mathbb{Q})$]. We pick a sufficiently small open compact subgroup $K \subset G(\mathbb{A}^{(p\infty)})$ maximal at p so that the normalization $\mathcal{H}_K$ of $H_K \subset Y_K \times Y_K$ is smooth at the image of y. The variety Y_K is naturally defined over a finite extension $\mathbb{F}_q/\mathbb{F}_p$ as the solution of the moduli problem $\mathcal{E}^{(p)}/K$. The universal elliptic curve $\mathbf{E}_K$ is therefore defined over $I_{K/\mathbb{F}_q}$, and $\mathcal{H}_K$ is a finite-type variety over $\mathbb{F}_q$. Let η be the generic point of $\mathcal{H}_{K/\mathbb{F}_q}$, and write $\overline{\eta}$ for the geometric point over η and $\mathbb{F}_q(\overline{\eta})^{sep}$ for the separable algebraic closure $\mathbb{F}_q(\overline{\eta})^{sep}$ of $\mathbb{F}_q(\eta)$ in $\mathbb{F}_q(\overline{\eta})$.

Take an odd prime $\ell \neq p$, and consider the ℓ-adic Tate module $T_\ell(\mathbb{Y}_{\overline{\eta}})$ for the generic fiber $\mathbb{Y}_{\overline{\eta}}$ of $\mathbb{Y}$. We consider the image of the Galois action $\mathfrak{G} := \mathrm{Im}(\mathrm{Gal}(\mathbb{F}_q(\overline{\eta})^{sep}/\mathbb{F}_q(\eta)))$ in $GL_{\mathfrak{O}_\ell \times \mathfrak{O}_\ell}(T_\ell(\mathbb{Y}_{\overline{\eta}}))$. Then, by Zarhin's result ([DAV] Theorem V.4.7), the Zariski closure over $\mathbb{Q}$ of $\mathfrak{G}$ is a reductive subgroup $\mathcal{G}$ of $GL_{\mathbb{Q}_\ell \times \mathbb{Q}_\ell}(T_\ell(\mathbb{Y}_{\overline{\eta}}) \otimes \mathbb{Q})$, and $\mathfrak{G}$ is an open subgroup of $\mathcal{G}(\mathbb{Q}_\ell)$. Moreover, by Zarhin's theorem, the centralizer of $\mathcal{G}$ in $\mathrm{End}_{\mathbb{Q}_\ell \times \mathbb{Q}_\ell}(T_\ell(\mathbb{Y}_{\overline{\eta}}) \otimes \mathbb{Q})$ is $\mathrm{End}(\mathbb{Y}) \otimes \mathbb{Q}_\ell$. Since the reductive subgroups of $GL(2)$ either are tori or contain $SL(2)$ (cf. [GME] Corollary 4.3.14 in Sect. 4.3.5), the derived group $\mathcal{G}_1(\mathbb{Q}_\ell)$ of $\mathcal{G}(\mathbb{Q}_\ell)$ has to be $SL_2(\mathbb{Q}_\ell \times \mathbb{Q}_\ell)$.

By a geometric version of Chebotarev's density theorem for function fields (cf. [Se1] Theorem 7; see also [L]), we can find a set of closed points $u \in \mathcal{H}_K(\mathbb{F})$ with positive density such that the Zariski closure in $\mathcal{G}$ of the subgroup generated by the Frobenius element $Frob_u \in \mathrm{Im}(\mathrm{Gal}(\mathbb{F}_q(\overline{\eta})^{sep}/\mathbb{F}_q(\eta)))$ at u with $\pi_j(u) = u_j$ [$u_j \in I_K(\mathbb{F})$] is a torus containing a maximal torus $T_u = (T_{u_1} \times T_{u_2}) \cap \mathcal{G}_1$ of the derived group $\mathcal{G}_1$ of $\mathcal{G}$. In particular, the centralizer of T_u in $\mathcal{G}_1$ is itself. Thus, $\mathbb{Y}_u$ is isogenous to a product of two nonisogenous elliptic curves $Y_1 = E_{u_1}$ and $Y_2 = E_{u_2}$ defined over a finite field.

The endomorphism algebra $M_j = \mathrm{End}^{\mathbb{Q}}(Y_j)$ is an imaginary quadratic field of $\mathbb{Q}$ generated over $\mathbb{Q}$ by the relative Frobenius map ϕ_j induced by $Frob_u$, and $M_1 \neq M_2$. The relative Frobenius map $Frob_u$ acting on $X_*(\widehat{I}_{u_1}) \cong \mathbb{Z}_p$ has one eigenvalue: $\phi_1^{(1-c)\sigma}$ for the CM type $\Sigma_1 = \{\sigma\}$ of Y_1, which differs from the eigenvalues of $\phi_2 \in \mathrm{End}(Y_2)$ on $X_*(\widehat{I}_{u_2}) \cong \mathbb{Z}_p$. Since we have proven that over the open dense subscheme U of $\mathcal{H}$, the formal completion of U at $u \in U$ with $u = (u_1, u_2) \in X \subset V^2$ is canonically isomorphic to a formal subtorus $\widehat{Z} \subset \widehat{I}_{u_1} \times \widehat{I}_{u_2}$ with cocharacter group $X_*(\widehat{Z}) \cong \mathbb{Z}_p$, we may assume that our point $u = (u_1, u_2)$ as above is in the (open dense) image U_K of U in H_K. Projecting $X_*(\widehat{Z})$ down to the left and right factors I_K, the projection map $X_*(\widehat{Z}) \to X_*(\widehat{I}_{u_j})$ is actually an injection commuting with the action of $Frob_u$. Thus, $Frob_u$ has more than one distinct eigenvalue on the $\mathbb{Z}_p$-module $X_*(\widehat{Z})$ of rank 1, which is a contradiction. Thus, we conclude that $\mathrm{End}^{\mathbb{Q}}(\mathbb{Y}) = M_2(\mathbb{Q})$ for any choice of small open compact subgroups K maximal at p.

As we have remarked at the beginning, $\widehat{\mathcal{H}}_y \subset \widehat{H}_{z_0^2} \subset \widehat{I} \times \widehat{I}$ is given by $\{t, t^{\beta^{1-c}})|t \in \widehat{\mathbb{G}}_m\}$ for nonzero $\beta \in \mathfrak{O}_{(p)}$. Suppose that y corresponds to L, and so $\widehat{\mathcal{H}}_y \subset \widehat{I} \times \widehat{I}$ coincides with $\widehat{\mathbb{G}}_m \otimes L$. On the other hand, we have the skew-diagonal $\Delta_\beta = \Delta_{1,\beta} = \{(z, \rho(\beta)(z))|z \in I\} \subset I \times I$. The formal completion $\widehat{\Delta}_\beta$ along (z_0, z_0) therefore coincides with $\widehat{\mathcal{H}}_y$ and $\widehat{\Delta}_\beta = \widehat{\mathbb{G}}_m \otimes L \subset \widehat{H}_{(z_0,z_0)}$ inside $\widehat{I}^2$. Thus, $\Delta_\beta \subset H$. By the irreducibility of H, we conclude that $H = \Delta_\beta$. □

There are two ways of proving Theorem 3.44. One is an induction reducing things to Theorem 3.43, whose details we give in Chap. 11. Here is a brief sketch of the second proof, which is based on the argument in [Ch1] Sect. 8. By using Chai's globalization of the Serre–Tate coordinate in [Ch1] Sect. 2, we find a dense open subscheme $Y^{lin} \subset Y$ $[y \in Y^{lin}]$ such that Y is locally linear at every closed point of Y^{lin} (see Sect. 11.1.1 for local linearity). The existence of $Y^{lin} \subset Y$ follows basically from Proposition 5.3 in [Ch1] and its proof (applied to Y not X). Consider the abelian scheme $\mathbb{Y} = \mathbf{E}^m \times_{V^m} Y$. Then $\mathrm{End}^{\mathbb{Q}}(\mathbb{Y}_{/Y})$ is isomorphic to either $\mathbb{Q}^m$ or $\mathbb{Q}^{m-2} \times M_2(\mathbb{Q})$, because Y is dominant over V^{m-1} and V by the two projections. Then in a manner similar to the above sketch of the proof of Theorem 3.43, we can prove the impossibility of $\mathrm{End}^{\mathbb{Q}}(\mathbb{Y}_{/Y}) \cong \mathbb{Q}^m$. Thus, we have $\mathrm{End}^{\mathbb{Q}}(\mathbb{Y}_{/Y}) \cong \mathbb{Q}^{m-2} \times M_2(\mathbb{Q})$; in other words, for an index $i < m$, the ith factor $\mathbb{Y}_i$ obtained by pulling back the ith factor $\mathbf{E}$ of $\mathbf{E}^m$ to Y is isogenous to the last factor $\mathbb{Y}_m$. This isogeny is induced by a nonzero $\alpha/\beta \in M \cong \mathrm{End}^{\mathbb{Q}}(E_x)$ with $\alpha, \beta \in \mathfrak{O}$. Then we conclude that $\widehat{\mathcal{X}} \supset \widehat{S}^{m-1} \times \widehat{\Delta}_{(\alpha,\beta)}$ for $\Delta_{(\alpha,\beta)}$ plugged in the product of the ith and mth copies of V in V^m, and hence X has the desired form.

4

Review of Scheme Theory

Elliptic curves are algebraic groups in addition to being algebraic curves; a modular form classifies elliptic curves by assigning a value in a ring A to an elliptic curve with some extra structures defined over A. From a different viewpoint, automorphic forms are generally defined on the adele points of a linear algebraic group; in particular, modular forms are functions on $GL_2(\mathbb{A})$ (e.g., [AAG]). Thus, a minimal amount of knowledge of group schemes is necessary to describe elliptic curves and modular forms. Group schemes are best understood from the functorial viewpoint of scheme theory, regarding a scheme as a functor from the category of algebras to sets taking each algebra R to the set of R-rational points $S(R)$ of the scheme S.

In a standard text, one first introduces the category of affine schemes as the category of local ringed spaces constructed from the spectrums of commutative rings with identity. Then, gluing affine schemes, again as a local ringed space, one constructs the category of schemes. After these basics, usually one identifies (by tautology; see Lemma 4.17) the schemes with the associated covariant functor from the category of commutative rings (with identity) into the category of sets. This functorial point of view is natural because a closed algebraic subvariety X in a projective space over a ring A associates with each A-algebra R the set $X(R)$ of points with coordinates in R (which is the covariant functor associated with X).

Since there are many textbooks on the geometric treatment of algebraic geometry (such as [ALG] and [GME] Chap. 1), in this chapter, assuming some knowledge of the standard geometric theory of schemes, we try to facilitate a smooth transition to our functorial viewpoint. If the reader has good experience in dealing with schemes as functors, he or she can skip this chapter (and use it as a dictionary between geometric theory and functorial theory).

H. Hida, *Elliptic Curves and Arithmetic Invariants*, Springer Monographs in Mathematics, DOI 10.1007/978-1-4614-6657-4_4,

4.1 Functorial Algebraic Geometry

In this section, we briefly review definitions and results in functorial algebraic geometry as succinctly as possible in a style suitable to deal with group schemes (cf. [RAG] I.1). To make our exposition short and to quickly reach the core of the theory necessary to our later treatment, we define schemes as covariant functors outright, and in this sense our exposition is unconventional. Since the spectrum $\mathrm{Spec}(A)$ over the base ring B [or equivalently, over the base scheme $\mathrm{Spec}(B)$] is traditionally first defined as a local ringed space made up of prime ideals of A with Zariski topology, we often write S_A for the associated covariant functor defined on the category of B-algebras $R \mapsto S_A(B) = \mathrm{Hom}_{B\text{-alg}}(A, R)$ only in this chapter and the next [later we write this functor as $R \mapsto \mathrm{Spec}_B(A)(R)$, identifying the functor and the scheme]. All rings R we consider are commutative and have the identity element 1_R, and we denote by 0_R the zero element of R.

4.1.1 Affine Variety

Let k be an algebraically closed field. A Zariski closed subset V in k^n is defined by the zero set of finitely many polynomials $f_1, \dots, f_m \in k[X] = k[X_1, \dots, X_n]$; hence, $V = \{(x = (x_1, \dots, x_n) \in k^n | f_j(x) = 0 \ (j = 1, \dots, m)\}$. The polynomials $f_1, \dots, f_m$ generate an ideal $I = \sum_{j=1}^m k[X]f_j \subset k[X]$, and

$$V = V(I) = \{x \in k^n | f(x) = 0 \text{ for all } f \in I\}.$$

Thus, we write this closed set as $V = V(I)$ for the ideal I. Since $k[X]$ is noetherian, an ideal I is generated by finitely many polynomials; hence, any Zariski closed subset has the form $V(I)$ with an ideal I.

It is easy to verify $V(k[X]) = \emptyset$, $V((0)) = k^n$, $V(I \cap J) = V(I) \cup V(J)$, and $V(\sum_{i \in I} I_i) = \bigcap_{i \in I} V(I_i)$. Thus, the family of subsets of the form $V(I)$ for ideals I satisfies the axiom of closed subsets of k^n, giving the Zariski topology on k^n. By Hilbert's zero theorem (cf. [CRT] Sect. 5), $V(I) = V(\sqrt{I})$ for the radical $\sqrt{I} = \{f \in k[X] | f^n \in I \text{ for some } 0 < n \in \mathbb{Z}\}$. Thus, the association $\{\text{ideals of } k[X]\} \to \{\text{Zariski closed subsets of } k^n\}$ given by $I \mapsto V(I)$ is surjective but not injective.

For any k-algebra R, define

$$\mathcal{V}_I(R) = \{x = (x_1, \dots, x_n) \in R^n | f(x) = 0 \text{ for all } f \in I\}.$$

Then $R \mapsto \mathcal{V}_I(R)$ is a covariant functor from the category of k-algebras into sets (see the following section for the definition of functors), as any k-algebra homomorphism $\phi \in \mathrm{Hom}_{k\text{-alg}}(R, R')$ induces a map $\mathcal{V}_I(\phi) : \mathcal{V}_I(R) \to \mathcal{V}_I(R')$ given by $(x_1, \dots, x_n) \mapsto (\phi(x_1), \dots, \phi(x_n))$, which satisfies $\mathcal{V}_I(\phi' \circ \phi) = \mathcal{V}_I(\phi') \circ \mathcal{V}_I(\phi)$. If $\alpha \in \mathcal{V}_I(R)$, we have a k-algebra homomorphism $\phi_\alpha : k[X] \to R$ given by $\phi_\alpha(f(X)) = f(\alpha)$ for $f(X) \in k[X] = k[X_1, \dots, X_n]$. Since $f(\alpha) = 0$ for all $f \in I$, ϕ_α factors through $k[X]/I$. Thus, we get a map $\phi_R : \mathcal{V}_I(R) \to \mathrm{Hom}_{k\text{-alg}}(k[X]/I, R)$, sending $\alpha \in \mathcal{V}_I(R)$ to ϕ_α. For any given $\varphi \in \mathrm{Hom}_{k\text{-alg}}(k[X]/I, R)$, it is plain that $\varphi_R = \phi_\alpha$ for $\alpha = (\varphi(X_1), \dots, \varphi(X_n))$. Therefore,

$$\phi : \mathcal{V}_I(?) \to \mathrm{Hom}_{k\text{-alg}}(k[X]/I, ?) \text{ is an isomorphism of functors.} \tag{4.1.1}$$

Even if $V(I) = V(\sqrt{I})$, it is obvious that $\mathcal{V}_I \neq \mathcal{V}_{\sqrt{I}}$ as functors, as easily checked $\mathrm{Hom}_{k\text{-alg}}(k[X]/\sqrt{I}, k[X]/I) \neq \mathrm{Hom}_{k\text{-alg}}(k[X]/I, k[X]/I)$ if $I \neq \sqrt{I}$. We call the functor $\mathcal{V}_{\sqrt{I}}$ an affine variety, which is a reduced affine k-scheme (a general definition of k-schemes will be given in a following section).

If $W \subset k^r$ is another Zariski closed subset defined by an ideal, $J \subset k[Y] = k[Y_1, \dots, Y_r]$. A map $\varphi : V \to W$ is a regular map if there exists a set of polynomials $\varphi_1, \dots, \varphi_r$ in $k[X]$ such that $\varphi(x) = (\varphi_1(x), \dots, \varphi_r(x)) \in k^r$ for all $x \in V$. This induces a morphism of functors $\mathcal{V}_I \to \mathcal{V}_J$ given by $\varphi_R : \mathcal{V}_I(R) \to \mathcal{V}_J(R)$ with $\varphi_R(x) = (\varphi_1(x), \dots, \varphi_r(x))$ for all $x \in \mathcal{V}_I(R) \subset R^n$. The set of polynomials $\varphi = (\varphi_1, \dots, \varphi_r)$ gives rise to a k-algebra homomorphism $\underline{\varphi} : k[Y] \to k[X]$ given by $\underline{\varphi}(Y_j) = \varphi_j(X)$. It induces, therefore, a k-algebra homomorphism $\underline{\varphi} : k[Y] \to k[X]/I$ by composing the projection $k[X] \twoheadrightarrow k[X]/I$. If $f(Y) \in J$, then $\underline{\varphi}(f(Y)) = f(\varphi_1(X), \dots, \varphi_r(X)) \mod I$. Since $\varphi(x) = (\varphi_1(x), \dots, \varphi_r(x)) \in W$ for all $x \in V$, we get $\underline{\varphi}(f(Y))(x) = 0$; thus, by Hilbert's zero theorem, we conclude that $\underline{\varphi}(f(Y)) \in \sqrt{I}$. Thus, φ induces $\underline{\varphi} : k[Y]/\sqrt{J} \to k[X]/\sqrt{I}$. In particular, if $\sqrt{I} = I$ and $\sqrt{J} = J$, we get

$$\mathrm{Hom}_{COF}(\mathcal{V}_I, \mathcal{V}_J) \ni \varphi \mapsto \underline{\varphi} \in \mathrm{Hom}_{k\text{-alg}}(k[Y]/J, k[X]/I).$$

As we will see in (4.1.5), this is an isomorphism, so the left-hand side is a set. In essence, knowing the affine variety $\mathcal{V}_I$ is equivalent to knowing the algebra $k[X]/I$ (cf. Yoneda's lemma; see Lemma 4.17).

4.1.2 Categories

We give a brief outline of the definition of categories. A category $\mathcal{C}$ consists of two data: objects of $\mathcal{C}$ and morphisms of $\mathcal{C}$. For two objects X and Y of $\mathcal{C}$, we have a **set** $\mathrm{Hom}_{\mathcal{C}}(X, Y)$ of morphisms satisfying the following three rules:

(Ct1) *For three objects X, Y, Z, there is a composition map:*

$$\mathrm{Hom}_{\mathcal{C}}(Y, Z) \times \mathrm{Hom}_{\mathcal{C}}(X, Y) \to \mathrm{Hom}_{\mathcal{C}}(X, Z) : (g, f) \mapsto g \circ f.$$

(Ct2) *(Associativity). For three morphisms $X \xrightarrow{f} Y \xrightarrow{g} Z \xrightarrow{h} W$, we have $h \circ (g \circ f) = (h \circ g) \circ f$.*

(Ct3) *For each object X, there is a specific element $1_X \in \mathrm{Hom}_{\mathcal{C}}(X, X)$ such that $1_X \circ f = f$ and $g \circ 1_X = g$ for all $f : Y \to X$ and $g : X \to Z$.*

For two objects X and Y in $\mathcal{C}$, we write $X \cong Y$ if there exist morphisms $f : X \to Y$ and $g : Y \to X$ such that $f \circ g = 1_Y$ and $g \circ f = 1_X$. Often $\mathcal{C}$ is a subcategory of $SETS$, and in that case, 1_X is given by the identity map id_X; so, in such a case, we write id_X for 1_X.

Example 4.1 A list of examples of categories: In the table below, X denotes a given topological space (it is often the underlying topological space of a scheme). See [GME] Chap. 1 for a brief outline of the traditional theory of schemes and its language (see [EGA] for real details). We will recall a functorial definition of scheme in a couple of sections ahead.

Category	*Objects*	*Morphisms*
$SETS$	*Sets*	*Maps between sets*
AB	*Abelian groups*	*Group homomorphisms*
ALG	*Algebras*	*Algebra homomorphisms*
$ALG_{/B}$	*B-algebras*	*B-algebra homomorphisms*
$\widehat{ALG}_{/B}$	*Complete B-algebras*	*Continuous B-algebra homomorphisms*
$O(X)$	*Open subsets in X*	*Inclusions*
$PS(X)$	*Presheaves on X*	*Morphisms of presheaves*
$S(X)$	*Sheaves on X*	*Morphisms of presheaves*
$QS(X)$	*Quasicoherent sheaves*	*Morphisms of presheaves*
$SCH_{/S}$	*S-schemes*	*Morphisms of functors$_{/S}$*
$GSCH_{/S}$	*Group S-schemes*	*Group homomorphisms of $SCH_{/S}$*
$BT_{/S}$	*Barsotti–Tate groups$_{/S}$*	*Morphisms of $GSCH_{/S}$*
$ELL_{/SCH}$	*Elliptic curves $E_{/S}$*	*Morphisms of $GSCH_{/S}$*
$MOD_{/B}$	*B-modules*	*B-linear maps*

In the above table, $ELL_{/SCH}$ is actually a *fiber category* over the category SCH of schemes. This means that each elliptic curve E is defined over a scheme S, and $\mathrm{Hom}_{ELL}(E_{/S}, E'_{/S'})$ is made up of a commutative diagram

$$\begin{array}{ccc} E & \longrightarrow & E' \\ \downarrow & & \downarrow \\ S & \longrightarrow & S' \end{array}$$

with the top arrow sending the origin of E to that of E' (see Sect. 6.1.1 for a more precise description of this).

A category $\mathcal{C}'$ is a *subcategory* of $\mathcal{C}$ if the following two conditions are satisfied:

(i) Each object of $\mathcal{C}'$ is an object of $\mathcal{C}$ and $\mathrm{Hom}_{\mathcal{C}'}(X,Y) \subset \mathrm{Hom}_{\mathcal{C}}(X,Y)$.
(ii) The composition of morphisms is the same in $\mathcal{C}$ and $\mathcal{C}'$.

A subcategory $\mathcal{C}'$ is called a *full subcategory* of $\mathcal{C}$ if the sets of homomorphisms are equal: $\mathrm{Hom}_{\mathcal{C}'}(X,Y) = \mathrm{Hom}_{\mathcal{C}}(X,Y)$ for any two objects X and Y in $\mathcal{C}'$.

4.1.3 Functors

A *covariant* (resp., *contravariant*) *functor* $F : \mathcal{C} \to \mathcal{C}'$ is a rule associating an object $F(X)$ of $\mathcal{C}'$ and a morphism $F(f) \in \mathrm{Hom}_{\mathcal{C}'}(F(X), F(Y))$ [resp.,

$F(f) \in \mathrm{Hom}_{\mathcal{C}'}(F(Y), F(X))]$ with each morphism $f : X \to Y$ of $\mathcal{C}$. We require $F(1_X) = 1_{F(X)}$ and

$$F(f \circ h) = F(f) \circ F(h) \text{ [resp., } F(f \circ h) = F(h) \circ F(f)]. \tag{4.1.2}$$

Example 4.2 Let X be a topological space. Then the category $O(X)$ of open sets consists of open subsets and inclusions:

$$\mathrm{Hom}_{O(X)}(U, V) = \begin{cases} Inc : U \hookrightarrow V & \text{if } U \subset V \\ \emptyset & \text{otherwise.} \end{cases}$$

Then a presheaf $\mathcal{F}$ is a contravariant functor from $O(X)$ into AB.

A *morphism* f between two contravariant functors F, $G : \mathcal{C} \to \mathcal{C}'$ is a system of morphisms $\{\phi_X \in \mathrm{Hom}_{\mathcal{C}'}(F(X), G(X))\}_{X \in \mathcal{C}}$, making the following diagram commutative for all $u \in \mathrm{Hom}_{\mathcal{C}}(X, Y)$:

$$\begin{array}{ccc} F(Y) & \xrightarrow{F(u)} & F(X) \\ {\scriptstyle \phi_Y}\downarrow & & \downarrow{\scriptstyle \phi_X} \\ G(Y) & \xrightarrow[G(u)]{} & G(X). \end{array} \tag{4.1.3}$$

In general, the totality of morphisms between two functors may not be a set; thus, we cannot define the category of contravariant functors outright by the collection of all functors from $\mathcal{C}$ to $\mathcal{C}'$. However, we pretend to have the "category" of contravariant functors $CTF(\mathcal{C}, \mathcal{C}')$ using the above definition of morphisms between functors. Similarly, we have the "category" $COF(\mathcal{C}, \mathcal{C}')$ of covariant functors by reversing the direction of morphisms $F(u)$ and $G(u)$. In practice, we impose some conditions on functors in these "categories" to have a well-defined full subcategory (this does not pose any real problems as long as we are careful).

Under the notation introduced in Example 4.1, by definition, the category of presheaves $PS(X)$ on a topological space X is given by $CTF(O(X), AB)$, and $S(X)$ and $QS(X)$ are full subcategories of $PS(X)$. Following the tradition, for a presheaf $F \in PS(X)$, we write $\Gamma(X, F) = H^0(X, F)$ for $F(X)$. Heuristically, a sheaf is a presheaf that is determined by local data (whose exact meaning will be given later in different disguises). If one believes this prosaic definition, one might imagine that collecting somehow locally determined "sections" in the image of a given presheaf, we would be able to make a sheaf out of a presheaf in a universal way. Indeed, the *sheafication* $\mathcal{F} \mapsto \mathcal{F}^{\#}$ (defined in [GME] Sect. 1.1.2) is a covariant functor from $PS(X)$ into $S(X)$, satisfying the following universal property: We have a morphism of presheaves $\iota : \mathcal{F} \to \mathcal{F}^{\#}$ such that for any morphism $\phi : \mathcal{F} \to \mathcal{S}$ for a sheaf $\mathcal{S}$, we have a unique morphism $\phi^{\#} : \mathcal{F}^{\#} \to \mathcal{S}$ such that $\phi^{\#} \circ \iota = \phi$. In particular, $\mathcal{F} \mapsto \mathcal{F}^{\#}$ is the identity on $S(X)$. See Sect. 4.1.10 for more details.

The association Spec : $A \mapsto \mathrm{Spec}(A)$ for each algebra A is a contravariant functor from ALG into AFF. For each A-module M, the association $M \mapsto \widetilde{M}$ is a functor from $MOD_{/A}$ into $QS(\mathrm{Spec}(A))$. These are standard constructions that can be found in any introductory book on algebraic geometry; for example, see [GME] Sect. 1.2 (and Remark 4.20 in the text for a slightly nonstandard construction).

The functor Spec : $ALG \to AFF$ and the above functor, $MOD_{/A} \to QS(\mathrm{Spec}(A))$, have inverses given by $AFF \ni X \mapsto \Gamma(X, \mathcal{O}_X) \in ALG$ and $QS(\mathrm{Spec}(A)) \ni \mathcal{F} \mapsto \Gamma(\mathrm{Spec}(A), \mathcal{F})$. We see $\mathrm{Spec}(\Gamma(X, \mathcal{O}_X)) \cong X$ and $\Gamma(\mathrm{Spec}(A), \mathcal{O}_{\mathrm{Spec}(A)}) = A$, canonically. When we have two functors $F : \mathcal{C} \to \mathcal{C}'$ and $G : \mathcal{C}' \to \mathcal{C}$ such that $F(G(X)) \cong X$ and $G(F(Y)) \cong Y$ for each object $X \in \mathcal{C}'$ and $Y \in \mathcal{C}$, we say that $\mathcal{C}$ is equivalent to $\mathcal{C}'$ and write $\mathcal{C} \approx \mathcal{C}'$. Thus, $AFF \approx ALG$ and $QS(\mathrm{Spec}(A)) \approx MOD_{/A}$. When a functor $F : \mathcal{C} \to \mathcal{C}'$ gives an equivalence of $\mathcal{C}$ with a full subcategory of $\mathcal{C}'$, we call F *fully faithful.*

If $F, G : \mathcal{C} \to \mathcal{C}'$ for a subcategory $\mathcal{C}'$ of $SETS$ are functors, and $F(A) \subset G(A)$ for all $A \in Ob(\mathcal{C})$, we call F a *subfunctor* of G. Here, for $\phi \in \mathrm{Hom}_{\mathcal{C}}(A, A')$, $G(\phi)$ induces $F(\phi)$ [in other words, the inclusion $i_A : F(A) \hookrightarrow G(A)$ is a functor morphism]. If $G_i \subset G$ are subfunctors indexed by an index set I, assuming intersection exists in $\mathcal{C}'$, clearly $A \mapsto \bigcap_{i \in I} G_i(A) \subset G(A)$ is a subfunctor of G. We call this functor the intersection of all G_i in G.

We can also define the product of functors $F, G : \mathcal{C} \to SETS$ by putting $(F \times G)(A) := F(A) \times G(A)$ (the set-theoretic product) with $(F \times G)(\phi) = F(\phi) \times G(\phi)$. If $f : F \to S$ and $g : G \to S$ are morphisms of functors $F, G, S : \mathcal{C} \to SETS$, we define the *fibered product* $F \times_S G : \mathcal{C} \to SETS$ by

$$A \mapsto (F \times_S G)(A) = \{(x, y) \in F(A) \times G(A) | f_A(x) = g_A(y)\} \qquad (4.1.4)$$

with $(F \times_S G)(\phi)(x, y) = (F(\phi)(x), G(\phi)(y))$. If, further, $f : F \hookrightarrow S$ and $g : G \hookrightarrow S$ are inclusions of functors, identifying F and G with subfunctors of S by f and g, it is plain that $F \times_S G(R) = F(R) \cap G(R) = (F \cap G)(R)$ in S.

If $f : X \to Y$ is a morphism in $COF(\mathcal{C}, SETS)$ and $Y' \subset Y$ is a subfunctor, we define its *pullback functor* by $f^{-1}(Y') \in COF(\mathcal{C}, SETS)$ by $f^{-1}(Y')(A) = f_A^{-1}(Y'(A))$. Clearly, $f^{-1}(Y')$ is a subfunctor of X as $f^{-1}(Y') = X \times_Y Y'$.

4.1.4 Affine Schemes

As we remarked, knowing the affine variety $\mathcal{V}_I$ is equivalent to knowing the algebra $A = k[X]/I$. So we may start with a k-algebra A in place of the zero set V of the ideal I to make a theory. Slightly more generally, writing B for a fixed-base commutative ring B, we consider a B-algebra A. The *affine B-scheme* $S = S_A = \mathrm{Spec}_B(A)$ associated with A is a functorial rule of assigning to each B-algebra R the set given by $S(R) = \mathrm{Hom}_{B\text{-alg}}(A, R)$. When the underlying base algebra B is evident in the context, we simply write

Spec(A) for Spec$_B(A)$. Although Spec(A) has another aspect of a topological local ringed space as described in, for example, [GME] Sect. 1.2, here we concentrate on its functorial property. Since the category $SCH_{/B}$ of B-schemes is a full subcategory of $COF(ALG_{/B}, SETS)$, this is legitimate (see [GME] Sect. 1.4.3 and Lemma 4.17 in the text). Hereafter we simply write $\mathcal{C}_B$ for $COF(ALG_{/B}, SETS)$, which is the biggest "category" containing our geometric objects. As we remarked already, $COF(ALG_{/B}, SETS)$ is not really a category in the true sense [i.e., $\mathrm{Hom}_{COF}(F, G)$ may not be a set], but our category of schemes $SCH_{/B}$ is defined later, imposing a sheaf condition to ensure that it is a category by Yoneda's lemma (cf. Lemma 4.17).

The set $S_A(R)$ is called the set of *R-rational points* (or *R-points*) of S_A. A *B-morphism* $\phi : S_A \to S_C$ is given by $\phi(P) = P \circ \underline{\phi}$ for an underlying B-algebra homomorphism $\underline{\phi} : C \to A$ satisfying the following commutative diagram:

$$\begin{array}{ccc} A & \xrightarrow{P} & R \\ {\scriptstyle \underline{\phi}}\uparrow & & \uparrow{\scriptstyle P\circ\underline{\phi}} \\ C & = & C. \end{array}$$

In other words, $\mathrm{Hom}_{ALG_{/B}}(C, A) \hookrightarrow \mathrm{Hom}_{AFF}(S_A, S_C)$ (by $\underline{\phi} \mapsto \phi$). By definition, we have the following properties of the functor S_A:

(F1) If $R \xrightarrow{f} R' \xrightarrow{g} R''$ are B-algebra homomorphisms, then we have maps $f_* : S_A(R) \to S_A(R')$ and $g_* : S_A(R') \to S_A(R'')$ given by $f_*(P) = f \circ P$ and $g_*(Q) = g \circ Q$, and we have $(g \circ f)_* = g_* \circ f_*$.

(F2) If $R' = R''$ and g as above is the identity map $\mathrm{id}_{R'} : R' \to R'$, we have $\mathrm{id}_{R',*} \circ f_* = f_*$. If $R = R'$ and f as above is the identity map $\mathrm{id}_R : R \to R$, we have $g_* \circ \mathrm{id}_{R,*} = g_*$.

(F3) For the identity map $\mathrm{id}_R : R \to R$, $\mathrm{id}_{R,*} : S_A(R) \to S_A(R)$ is the identity map of the set $S_A(R)$.

In short, $R \mapsto S_A(R)$ is a covariant functor of $ALG_{/B}$ into $SETS$, hence an object in $\mathcal{C}_B$. For two affine schemes S and T over B, a morphism $\phi : S \to T$ is a family of maps $\phi_R : S(R) \to T(R)$ indexed by B-algebras R such that the following diagram commutes for any B-algebra homomorphism $\alpha : R \to R'$:

$$\begin{array}{ccc} S(R) & \xrightarrow{\phi_R} & T(R) \\ {\scriptstyle \alpha_*}\downarrow & & \downarrow{\scriptstyle \alpha_*} \\ S(R') & \xrightarrow[\phi_{R'}]{} & T(R'). \end{array}$$

We often write $\mathrm{Hom}_B(S, T) = \mathrm{Hom}_{\mathcal{C}_B}(S, T)$ for the set of all morphisms from S into T. If $S = S_A$ and $T = S_C$, we have $\mathrm{Hom}_B(S, T) = \mathrm{Hom}_{ALG_{/B}}(C, A)$ by Yoneda's lemma (see (4.1.5), Exercise 2.2, and Lemma 4.17).

By definition, we also have the following properties of affine schemes:

(cf1) If $A \xrightarrow{\phi} C \xrightarrow{\psi} D$ are B-algebra homomorphisms, then we have morphisms of schemes $S_D \xrightarrow{\underline{\psi}} S_C \xrightarrow{\underline{\phi}} S_A$ such that $\phi \circ \psi$ is associated with $\underline{\psi} \circ \underline{\phi}$.
(cf2) If $A = C$ and ϕ in (cf1) is the identity map $\underline{\mathrm{id}}_A$ of A, we have $\mathrm{id}_A \circ \underline{\psi} = \underline{\psi}$. If $C = D$ and ψ in (cf1) is the identity map $\underline{\mathrm{id}}_C$ of C, we have $\underline{\phi} \circ \mathrm{id}_C = \underline{\phi}$.
(cf3) For the identity map $\underline{\mathrm{id}}_A : A \to A$, $\mathrm{id}_A : S_A(R) \to S_A(R)$ is the identity map for all B-algebras R.

Thus, the functor $A \mapsto S_A$ is a contravariant functor from $ALG_{/B}$ into $\mathcal{C}_B$. One of the most basic facts in functorial algebraic geometry is the full-faithfulness of this functor (e.g., [GME] Sect. 1.4.3):

$$\mathrm{Hom}_{ALG_{/B}}(A', A) \cong \mathrm{Hom}_B(S_A, S_{A'}) \text{ via } \underline{\alpha} \leftrightarrow \alpha. \tag{4.1.5}$$

A main point of the proof of this fact is to construct from a given morphism $\phi \in \mathrm{Hom}_B(S_A, S_{A'})$ a B-algebra homomorphism $\varphi : A' \to A$ such that $\underline{\varphi} = \phi$.
Exercise 4.3 *Let* $\varphi = \phi_A(\mathrm{id}_A) \in S_{A'}(A) = \mathrm{Hom}_{ALG_{/B}}(A', A)$, *where* $\mathrm{id}_A \in S_A(A) = \mathrm{Hom}_{ALG_{/B}}(A, A)$ *is the identity map. Then prove that* $\underline{\varphi} = \phi$.

Here are some examples of affine schemes:

Example 4.4 Take $f(X,Y,Z) = X^p + Y^p - Z^p$ for a prime p, and let $B = \mathbb{Z}$. Then consider $A = \mathbb{Z}[X,Y,Z]/(f(X,Y,Z))$. For each algebra R, we claim

$$S_A(R) \cong \{(x,y,z) \in R^3 | x^p + y^p = z^p\}.$$

Indeed, for each solution $P = (x,y,z)$ of Fermat's equation in R, we define an algebra homomorphism $\underline{\phi} : B[X,Y,Z] \to R$ by sending polynomials $\Phi(X,Y,Z)$ to its value $\Phi(x,y,z) =: \underline{\phi}(\Phi) \in R$. Since $\Phi \in (f(X,Y,Z)) \Leftrightarrow \Phi = \Psi f$, we find that $\underline{\phi}(\Psi f) = \Psi(x,y,z) f(x,y,z) = 0$; so $\underline{\phi}$ factors through the quotient A, getting $\phi \in S_A(R)$. In this way, we get an injection from the right-hand side to $S_A(R)$. If we start from $\underline{\phi} : A \to R$ in $S_A(R)$, we find

$$0 = \underline{\phi}(0) = \underline{\phi}(X^p + Y^p - Z^p) = \underline{\phi}(X)^p + \underline{\phi}(Y)^p - \underline{\phi}(Z)^p.$$

Thus, $(x,y,z) := (\underline{\phi}(X), \underline{\phi}(Y), \underline{\phi}(Z))$ is an element in the right-hand side, getting the isomorphism. By Fermat's last theorem, we have

$$S_A(\mathbb{Z}) \cong \{(a,0,a), (0,b,b), (c,-c,0) | a,b,c \in \mathbb{Z}\} \text{ if } p \text{ is a prime} \geq 3.$$

There is a simpler example: We have

$$S_{\mathbb{Z}[X_1,\dots,X_n]}(R) = R^n \text{ via } \phi \mapsto (\underline{\phi}(X_1), \dots, \underline{\phi}(X_n)).$$

Often $S_{\mathbb{Z}[X_1,\dots,X_n]}$ is written as $\mathbf{A}^n$ and is called the affine space of dimension n. When we regard $\mathbf{A}^n$ as a functor having values in AB, we write $\mathbb{G}_a^n$ (the additive group) for $\mathbf{A}^n$ [and $\mathbb{G}_a = \mathbb{G}_a^1$]. We have an algebra homomorphism $B[X,Y,Z] \to A$ for A in Example 4.4 sending Φ to $(\Phi \mod f(X,Y,Z))$. This in turn induces a morphism $i : S_A \to \mathbf{A}^3$, which is obviously injective.

An A-algebra A' is of *finite type* over A if $A' \cong A[X_1,\dots,X_n]/\mathfrak{a}$ for the polynomial ring $A[X_1,\dots,X_n]$ of finitely many variables and an ideal $\mathfrak{a}$. If, further, $\mathfrak{a}$ is generated by finitely many elements of $A[X_1, \dots, X_n]$, we call A' *finitely presented* over A (cf. [EGA] IV.1.4). If A is noetherian, $A[X_1, \dots, X_n]$ is noetherian, and so a finite presentation of $A[X]/\mathfrak{a}$ is automatic. However, when we treat nonnoetherian rings (for example, if we want to show that the classification functor of elliptic curves $E_{/R}$ given by $R \mapsto \{E_{/R}\}/\cong$ is a scheme), the finite-presentation property guarantees that the number of equations defining affine scheme is finite (and hence the property is important technically). The corresponding morphism $S_{A'} \to S_A$ of affine schemes is also called *of finite type* (resp., *finitely presented*) if A'/A is of finite type (resp., finitely presented).

A morphism $\phi : S_A \to S_{A'}$ is *flat* if $M \mapsto M \otimes_{A',\underline{\phi}} A$ is a left exact functor from the category of A'-modules to the category of A-modules. Here the functor $M \mapsto M \otimes_{A',\underline{\phi}} A$ is *left exact* if it preserves injective morphisms of A-modules; that is,

$$\mathrm{Ker}(\iota \otimes 1 : M \otimes_{A',\underline{\phi}} A \to M' \otimes_{A',\underline{\phi}} A) = 0 \ \text{ if } \ \mathrm{Ker}(\iota : M \to M') = 0.$$

The morphism ϕ is *faithfully flat* if (i) it is flat and (ii) $0 \to M \to M' \to M'' \to 0$ is exact if $0 \to M \otimes_{A',\underline{\phi}} A \to M' \otimes_{A',\underline{\phi}} A \to M'' \otimes_{A',\underline{\phi}} A \to 0$ is exact for any sequence $M \to M' \to M''$ of A'-modules. We quote the following fact.

Proposition 4.5 *Suppose that A is a noetherian B-algebra and A' is an A-algebra of finite type. Let M be an A'-module of finite type.*

1. *$\{P \in \mathrm{Spec}(A') | M_P$ is flat over $A_P\}$ is an open subscheme of $S_{A'}$.*
2. *If A' is A-flat, the morphism $S_{A'} \to S_A$ is an open map.*

We refer the proof to [CRT] Theorem 24.3 for (1), [CMA] page 48 or [EGA] IV.2.4.6 for (2).

A morphism $\phi : S_{A'} \to S_A$ is called *finite* if the A-module A' (via $\underline{\phi}$) is of finite type (that is, there exist a positive integer n and a surjective A-linear map $A^n \twoheadrightarrow A'$). If $S_{A'}$ is finite flat, localizations $A'_{\mathfrak{m}}$ at each prime ideal $\mathfrak{m}$ of A are free of finite rank over $A_{\mathfrak{m}}$, and therefore, $S_{A'}$ is also called locally free of finite rank over S_A if ϕ is finite flat.

For a maximal ideal $\mathfrak{m}$ of A', ϕ is *smooth* at $\mathfrak{m}$ if $A'_{\mathfrak{m}}$ is flat over $A_{\mathfrak{m}}$ and for any A-algebra R with a square-zero ideal I, $\mathrm{Hom}_{ALG_{/A}}(A'/\mathfrak{m}^n A', R) \to \mathrm{Hom}_{ALG_{/A}}(A'/\mathfrak{m}^n A', R/I)$ is surjective for all positive integers n (that is, any A-algebra homomorphism $\phi_0 : A'/\mathfrak{m}^n A' \to R/I$ can be lifted to an A-algebra homomorphism $\phi_1 : A'/\mathfrak{m}^n A' \to R$ such that $\phi_1 \mod I = \phi_0$). Here $A'_{\mathfrak{m}}$ is the localization of A' at $\mathfrak{m}$ and $A_{\mathfrak{m}}$ is the localization of A at $\underline{\phi}^{-1}(\mathfrak{m})$. We call $\phi : S_{A'} \to S_A$ *smooth* if ϕ is smooth at all maximal ideals $\mathfrak{m}$ of A. Since smoothness at $\mathfrak{m}$ is an "infinitesimal" property only depending modulo a power of the maximal ideal, it only depends on the adic completion with respect to the maximal ideal.

Suppose that B is a perfect field k and that S_A is of finite type over S_k (so $K = A/\mathfrak{m}$ is a finite separable extension of k for each maximal ideal $\mathfrak{m}$

of A). Then $S_A \to S_k$ is smooth at $\mathfrak{m}$ if and only if the $\mathfrak{m}$-adic completion $\widehat{A} = \varprojlim_n A/\mathfrak{m}^n$ is isomorphic to a power series ring $K[[X_1, \dots, X_n]]$ for $n = \dim A$ (cf. [BCM] IX.3.3). Thus, if B is a perfect field k and A is noetherian of (Krull) dimension n, S_A is smooth over S_k if and only if the local ring of A at every maximal ideal $\mathfrak{m}$ has completion isomorphic to $(A/\mathfrak{m})[[X_1, \dots, X_n]]$ (i.e., any $f \in A$ has a Taylor expansion at $\mathfrak{m}$ in $(A/\mathfrak{m})[[X_1, \dots, X_n]]$).

Exercise 4.6 *Prove that* $\mathbb{G}_a^n$ *is smooth over* $S_{\mathbb{Z}} = \mathrm{Spec}(\mathbb{Z})$.

For a maximal ideal $\mathfrak{m}$ of A', $\phi : S_{A'} \to S_A$ is *étale* at $\mathfrak{m}$ if $S_{A'_\mathfrak{m}}$ is flat over $S_{A_\mathfrak{m}}$ and for any A-algebra R with a square-zero ideal I, $\mathrm{Hom}_{ALG_{/A}}(A'/\mathfrak{m}^n, R) \to \mathrm{Hom}_{ALG_{/A}}(A'/\mathfrak{m}^n, R/I)$ is bijective for all positive integers n.

Lemma 4.7 *If* A' *is a local ring étale finite over a local ring* A *sharing the same residue field* k, *then* $A/\mathfrak{m}_A^n = A'/\mathfrak{m}_{A'}^n$ *for all* n; *in particular, we have*

$$\widehat{A} = \varprojlim_n A/\mathfrak{m}_A^n = \varprojlim_n A'/\mathfrak{m}_{A'}^n = \widehat{A}'.$$

Proof. By assumption, $A/\mathfrak{m}_A = A'/\mathfrak{m}_{A'} = k$ for the maximal ideals $\mathfrak{m}_A$ and $\mathfrak{m}_{A'}$. Since $\mathfrak{m}_A$ is square-zero ideal of $A/\mathfrak{m}_A^2$, taking R to be $A/\mathfrak{m}_A^2$, we have

$$\mathrm{Hom}_{ALG_{/A}}(A'/\mathfrak{m}_{A'}^2, A/\mathfrak{m}_A) \cong \mathrm{Hom}_{ALG_{/A}}(A'/\mathfrak{m}_{A'}^2, A/\mathfrak{m}_A^2),$$

whose left-hand side is a singleton given by the projection $A'/\mathfrak{m}_{A'}^2 \twoheadrightarrow k$. Similarly, taking R to be $A'/\mathfrak{m}_{A'}^2$,

$$\mathrm{Hom}_{ALG_{/A}}(A'/\mathfrak{m}_{A'}^2, A'/\mathfrak{m}_{A'}) \cong \mathrm{End}_{ALG_{/A}}(A'/\mathfrak{m}_{A'}^2),$$

whose left-hand side is a singleton. Thus,

$$\mathrm{Hom}_{ALG_{/A}}(A'/\mathfrak{m}_{A'}^2, A/\mathfrak{m}_A^2) = \{\varphi_2\} \text{ and } \mathrm{End}_{ALG_{/A}}(A'/\mathfrak{m}_{A'}^2) = \{\mathrm{id}_{A'/\mathfrak{m}_{A'}^2}\}$$

are singletons. Since $A'/\mathfrak{m}_{A'}^2$ is an A-algebra, $\mathrm{Hom}_{ALG_{/A}}(A/\mathfrak{m}_A^2, A'/\mathfrak{m}_{A'}^2) = \{\phi_2 = \iota_A \mod \mathfrak{m}_{A'}^2\}$. Taking $A' = A$, we find $\mathrm{End}_{ALG_{/A}}(A/\mathfrak{m}_A^2) = \{\mathrm{id}_{A/\mathfrak{m}_A^2}\}$. Thus, $\varphi_2 \circ \phi_2 = \mathrm{id}_{A/\mathfrak{m}_A^2}$ and $\phi_2 \circ \varphi_2 = \mathrm{id}_{A'/\mathfrak{m}_{A'}^2}$. So $A'/\mathfrak{m}_{A'}^2 = A/\mathfrak{m}_A^2$.

We now proceed by induction on n assuming $A/\mathfrak{m}_A^{n-1} = A'/\mathfrak{m}_{A'}^{n-1}$ and prove that $A/\mathfrak{m}_A^n = A'/\mathfrak{m}_{A'}^n$. Since $\mathfrak{m}_{A'}^{n-1} \subset A'/\mathfrak{m}_{A'}^n$ and $\mathfrak{m}_A^{n-1} \subset A/\mathfrak{m}_A^n$ are square-zero ideals, taking (R, I) to be $(A'/\mathfrak{m}_{A'}^n, \mathfrak{m}_{A'}^{n-1})$ and $(A'/\mathfrak{m}_{A'}^n, \mathfrak{m}_{A'}^{n-1})$, by the same argument as above, we find

$$\mathrm{Hom}_{ALG_{/A}}(A'/\mathfrak{m}_{A'}^n, A/\mathfrak{m}_A^n) \cong \mathrm{Hom}_{ALG_{/A}}(A'/\mathfrak{m}_{A'}^n, A/\mathfrak{m}_A^{n-1}) = \{\varphi_n\}$$
$$\mathrm{Hom}_{ALG_{/A}}(A'/\mathfrak{m}_{A'}^n, A'/\mathfrak{m}_{A'}^{n-1}) \cong \mathrm{End}_{ALG_{/A}}(A'/\mathfrak{m}_{A'}^n) = \{\mathrm{id}_{A'/\mathfrak{m}_{A'}^2}\}.$$

Then for a unique A-algebra homomorphism $\phi_n = (\iota_A \mod \mathfrak{m}_{A'}^n) : A/\mathfrak{m}_A^n \to A'/\mathfrak{m}_{A'}^n$, we find $\varphi_n \circ \phi_n = \mathrm{id}_{A/\mathfrak{m}_A^n}$ and $\phi_n \circ \varphi_n = \mathrm{id}_{A'/\mathfrak{m}_{A'}^n}$, and we are done.

Corollary 4.8 *Suppose that $B = k$ is an algebraically closed field and that A is a local k-algebra with residue field k. Then $A'_{/A}$ is étale finite over A at every maximal ideal $\mathfrak{m}'$ of A' if and only if $\widehat{A}' = A' \otimes_A \widehat{A}_{\mathfrak{m}} \cong \widehat{A}_{\mathfrak{m}} \times \cdots \times \widehat{A}_{\mathfrak{m}}$ (a finite product) for the $\mathfrak{m}$-adic completion $\widehat{A}_{\mathfrak{m}}$ of A.*

Proof. Write $\mathfrak{m} = \mathfrak{m}_A$ (the maximal ideal of A). Since $A'/\mathfrak{m}A'$ is a finite-dimensional vector space over $k = A/\mathfrak{m}$, $A'/\mathfrak{m}A'$ and A' has only finitely many maximal ideals, say, $\mathfrak{m}_i$ $(i = 1, \dots, m)$. Since $A'/\mathfrak{m}^n A' = \prod_i A'/\mathfrak{m}_i^{e_i}$ for sufficiently large $e_i \geq n$, we have $\widehat{A}' = \prod_i \widehat{A}'_{\mathfrak{m}_i}$. Because of this, A' is étale over A at every $\mathfrak{m}_i$ if and only if $\widehat{A}' = \varprojlim_n A'/\mathfrak{m}^n A'$ is étale over $\widehat{A}_{\mathfrak{m}}$ for every maximal ideal of $\widehat{A}'$. Since $A_{\mathfrak{m}_i}$ is étale over $A = A_{\mathfrak{m}}$, we have $\widehat{A}_{\mathfrak{m}_i} = \widehat{A}_{\mathfrak{m}}$, and the desired equivalence.

We call $\phi : S_A \to S_{A'}$ *étale* if ϕ is étale at all maximal ideals of A'.

Exercise 4.9 *Suppose we have morphisms of B-schemes $S_{A''} \to S_{A'} \to S_A$. If $S_{A''}$ is étale over $S_{A'}$ and $S_{A'}$ is smooth over S_A, prove that $S_{A''}$ is smooth over S_A. But if $S_{A''}$ is smooth over S_A and $S_{A''}$ is étale over $S_{A'}$, is $S_{A'}$ smooth over S_A?*

4.1.5 Zariski Open Covering

When we have a morphism of affine schemes $\phi : S_A \to S_C$, and if $\underline{\phi} : C \to A$ is a surjective ring homomorphism, we call ϕ a *closed immersion*. Then ϕ_R is injective (for any B-algebra R), and we identify S_A with its image all the time, so $S_A \subset S_C$. In this case, S_A, regarded as a subfunctor of S_C, is called *B-closed* in S_C. If $S_i \subset S_C$ is closed for affine schemes $\{S_i\}_{i \in I}$, the intersection $R \mapsto \bigcap_{i \in I} S_i(R)$ is again an affine closed subscheme [e.g., Exercise 4.16 (2)]. Thus, we can give a topology on $S_C(R)$ for each R so that closed sets are given either by the empty set $\emptyset$ or by those of the form $S_A(R)$ for a closed immersion $S_A \hookrightarrow S_C$. This topology is called the *Zariski B-topology* of S_C. When B is obvious in the context, we call it the Zariski topology on S_C.

If B' is a B-algebra, we may regard $A' = B' \otimes_B A$ as a B'-algebra by $b' \mapsto b' \otimes 1$. Then we get a new scheme $S_{A'}$ over the ring B', which is sometimes written as $S_{B'} \times_B S_A$ and is called the *fibered product* of S_A and $S_{B'}$ over B. Indeed as a functor, it is identical to $S_{B'} \times_{S_B} S_A$ [Exercise 4.16 (3–4)]. If we have a point $\phi \in S_A(R)$ for a B'-algebra R, we can extend $\underline{\phi} : A \to R$ to $\underline{\phi}' : A' = B' \otimes_B A \to R$ by $\underline{\phi}'(b \otimes a) = b\underline{\phi}(a)$. Thus, $\phi \mapsto \phi'$ gives the natural map $S_A(R) \to S_{A'}(R)$ for all B'-algebras R. This map is an isomorphism, because for any given $\phi' \in S_{A'}(R)$, $\underline{\phi}(a) = \underline{\phi}'(1 \otimes a)$ gives a point $\phi \in S_A(R)$ as long as R is a B'-algebra [Exercise 4.16 (3)]. However, a B'-closed subset of $S_{A'}$ may not be B-closed; hence, the Zariski topology depends on the base ring B.

Exercise 4.10 *Let $f : S_{A'} \to S_A$ be a morphism of affine scheme. If $S_{\overline{A}} \subset S_A$ is a closed subscheme, prove that $f^{-1}(S_{\overline{A}})$ is a closed subscheme (so a morphism of affine schemes is "continuous" with respect to the Zariski topology).*

Pick a non-nilpotent element $f \in A$. Then $\overline{A} = A/(f)$ is a surjective image of A. Thus, $S_{\overline{A}} \subset S_A$ is a closed subscheme. For each point $\phi \in S_A(R)$, $f : \phi \mapsto \phi(f)$ gives rise to a map $f_R : S_A(R) \to \mathbb{G}_a(R) = R$. This collection of maps $\{f_R\}_R$ can be easily checked to be a morphism $S_A \to \mathbb{G}_a$ of functors, which we again call $f : S_A \to \mathbb{G}_a$. In this way, we regard $f \in A$ as a function (or more precisely, a functor morphism) defined on S_A with values in $\mathbb{G}_a$. Any B-algebra homomorphism $(P : A \to R) \in S_A(R)$ factoring through $\overline{A}$ satisfies $f(P) = P(f) = 0$. Thus, $S_{\overline{A}}(R) = \{P \in S_A(R) | f(P) = 0\}$ (which is the zero set of f). We consider the ring of fractions $A_f = A[\frac{1}{f}] = \{\frac{a}{f^n} | 0 \leq n \in \mathbb{Z}\}$. Then the natural map $b \mapsto \frac{b}{1}$ is a B-algebra homomorphism of A into A_f, so $S_{A_f} \subset S_A$. If $P \in S_A(R)$ factors through A_f, we have $f(P) := P(f) \in R^\times$, as $f \in A$ is invertible in A_f. Thus, $S_A \supset S_{\overline{A}} \sqcup S_{A_f}$, and $S_A(k) = S_{\overline{A}}(k) \sqcup S_{A_f}(k)$ for any field k (which is a B-algebra). Hence, we call S_{A_f} an open subscheme of S_A, giving outside the zero set of the function $f : P \mapsto f(P)$.

Exercise 4.11 *Prove that $S_A(k) = S_{\overline{A}}(k) \sqcup S_{A_f}(k)$ for any field k, and give an example of a B-algebra R such that $S_A(R) \supsetneq S_{\overline{A}}(R) \sqcup S_{A_f}(R)$.*

More generally, a subfunctor $U \subset S_A$ is called an *open subscheme* if U is a subfunctor and there exists a subset $I \subset A$ such that

$$\begin{aligned} U(R) = U_I(R) &= \{P \in S_A(R) | \sum_{f \in I} f(P)R = R\} \\ &= \{P \in \mathrm{Hom}_{B\text{-alg}}(A, R) | \sum_{f \in I} P(f)R = R\}. \end{aligned} \tag{4.1.6}$$

Obviously, for the ideal generated (I) by I, $U_{(I)} = U_I$, and we may assume that I is an ideal. For $I = (0_A)$, $U_{(0)}(R) = \emptyset$ for any B-algebra R, and so $\emptyset$ is an open subscheme. Similarly, $U_A(R) = S_A(R)$ for all R, so S_A itself is an open subscheme. If $\{I_i\}_{i \in \mathcal{I}}$ is a family of ideals, we have $U_{\sum_{i \in \mathcal{I}} I_i} \supset \bigcup_{i \in \mathcal{I}} U_{I_i}$ and $U_{\sum_{i \in \mathcal{I}} I_i}(k) = \bigcup_{i \in \mathcal{I}} U_{I_i}(k)$ if k is a field.

Exercise 4.12 *Check that $R \mapsto U_I(R)$ is a subfunctor of S_A, and verify the following facts:*

(i) $S_{A_f} = U_I$ if $I = \{f\}$ for non-nilpotent $f \in A$;
(ii) $U_I \cap U_J = U_{I \cdot J}$ for the ideal product $I \cdot J$.

Exercise 4.13 *If a nonempty open subscheme $U \subset S_A$ is isomorphic to an affine scheme $S_{A'}$, prove that there exists a multiplicative set $S \subset A$ such that A' is isomorphic as B-algebras to the ring of fractions $S^{-1}A$.*

Exercise 4.14 *Let $B = \mathbb{C}$ and $A = \mathbb{C}[X,Y]$ (the polynomial ring of the indeterminates X and Y). Define a closed subscheme S_C of S_A for $C = A/(X,Y)$. Prove that there exists an open subscheme U of S_A such that $U(k) = S_A(k) - S_C(k)$ for any field extension $k/\mathbb{C}$, but U is not isomorphic to any affine $\mathbb{C}$-scheme.*

Lemma 4.15 *Let U, U' be open subschemes of S_A. If $U(k) = U'(k)$ for any field k over B, we have $U = U'$.*

Proof. Let $U = U_I$ and $U' = U_J$ for ideals I and J. If $U \neq U'$, we can find a B-algebra R such that $P \in U(R)$, but $P \notin U'(R)$. Concerning $P \in S_A(R) = \mathrm{Hom}_{B\text{-alg}}(A,R)$, we find that $P(I)R = R$ and $P(J)R \subsetneq R$. Since $P(J)R$ is a proper ideal of R, we can find a maximal ideal $\mathfrak{m}$ of R such that $\mathfrak{m} \supset P(J)R$ (cf. [CRT] Theorem 1.1). Let $k = R/\mathfrak{m}$ and define $\overline{P} \in \mathrm{Hom}_{B\text{-alg}}(A,k)$ by composing P with the projection $R \twoheadrightarrow k$. Then $\overline{P} \in U(k)$ and $\overline{P} \notin U'(k)$.

Exercise 4.16 *Prove the following facts:*

1. *A closed immersion $i : S_A \hookrightarrow S_C$ gives rise to an injection $i_R : S_A(R) \hookrightarrow S_C(R)$ for any B-algebras$R$.*
2. *If $i : S_A \subset S_C$ and $j : S_D \subset S_C$ are closed, then $R \mapsto S_A(R) \cap S_D(R)$ is closed in S_C and is isomorphic to S_E for $E = A \otimes_C D$, where the tensor product is taken with respect to the associated algebra homomorphisms $\underline{i} : C \to A$ and $\underline{j} : C \to D$. Moreover, $\mathrm{Ker}(\underline{i} \otimes \underline{j}) = \mathrm{Ker}(\underline{i}) + \mathrm{Ker}(\underline{j})$.*
3. *$S_A(R) \cong S_{A'}(R)$ if $A' = B' \otimes_B A$ and R is a B'-algebra, where B' is another B-algebra. Here the left-hand side is considered an affine B-scheme and the right-hand side is considered an affine B'-scheme.*
4. *For two B-algebras A and C, $S_{A \otimes_B C}(R) = S_A(R) \times S_C(R)$ for any B-algebra R. Hint: $\phi \in S_A(R)$ and $\psi \in S_C(R)$; we associate $\phi \otimes \psi \in S_{A \otimes_B C}(R)$ given by $(\phi \otimes \psi)(b \otimes c) = \phi(b)\psi(c)$. Thus, a product of affine schemes is again an affine B-scheme. The scheme $S_{A \otimes_B C}$ is called the fibered product of S_A and S_C over B.*

Recall the simplified notation $\mathcal{C}_B = COF(ALG_{/B}, SETS)$. We can generalize open/closed subschemes to open/closed subfunctors in the following way. A subfunctor $U \subset X$ is called an *open* (resp., *closed*) *subfunctor* if, for any affine B-scheme S_A and any morphism $f : S_A \to X$, the pullback $f^{-1}(U)$ is an open (resp., closed) subscheme. So $f : S_A \to X$ is "continuous" (and X has the strongest "topology," making f continuous). By Exercise 4.10, if $X \cong S_A$ is affine, the definition of the closedness of U is equivalent to the definition of closed affine subschemes. An open subscheme of an affine scheme is determined by its value over fields (Lemma 4.15), and the value of an open subscheme at a field is the complement of a closed subscheme. This shows that if $X \cong S_A$ is affine, the definition of being open for U in X is equivalent to the definition of an open affine subscheme.

Here Yoneda's lemma is more general than Exercise 2.2:

Lemma 4.17 *For any $X \in \mathcal{C}_B$, we have a canonical isomorphism*

$$\iota : \mathrm{Hom}_B(S_A, X) \cong X(A)$$

given by $\iota(f) = f_A(\mathrm{id}_A)$ for the identity map $\mathrm{id}_A : A \cong A$.

Taking $X = S_{A'}$, this proves (4.1.5).

Proof. We have ι well defined as above. Take a (variable) B-algebra R. Then, for any $\phi \in \mathrm{Hom}_{B\text{-alg}}(A, R) = S_A(R)$, we have $X(\phi) \circ f_A = f_R \circ S_A(\phi)$ as f is a morphism of functors. Then, writing $I_f = f_A(\mathrm{id}_A)$, we get $f_R(\phi) = X(\phi)(I_f)$ as $S_A(\phi)(\mathrm{id}_A) = \phi \circ \mathrm{id}_A = \phi$. Since R is a variable, f is uniquely determined by I_f. Thus, ι is injective. Conversely, we create the inverse map π of ι in the following way. For any $x \in X(A)$, we define $\pi(x) \in \mathrm{Hom}_B(S_A, X)$ by $\pi(x)(\phi) = X(\phi)(x)$ for $\phi \in S_A(R) = \mathrm{Hom}_{B\text{-alg}}(A, R)$. We leave the reader to verify $\iota(\pi(x)) = x$. In this way, we can recover $f = \pi(x)$ from x, and hence ι is surjective.

Lemma 4.18 *If $U, U' \subset X \in \mathcal{C}_B$ are open subfunctors, then $U = U'$ if $U(k) = U'(k)$ for all fields k over B.*

Proof. Suppose that $U \neq U'$. Then we find a B-algebra A such that $U(A) \neq U'(A)$. Thus, we find $P \in U(A)$ but $P \notin U'(A)$. By Lemma 4.17, we may regard $P \in \mathrm{Hom}_B(S_A, U) \subset \mathrm{Hom}_B(S_A, X)$. Then $P : S_A(A) \to X(A)$ and $\mathrm{id}_A \in P^{-1}(U(A))$, but $\mathrm{id}_A \notin P^{-1}(U'(A))$; hence, $P^{-1}(U) \neq P^{-1}(U')$. Then by Lemma 4.15, we find a field k with $P^{-1}(U)(k) \neq P^{-1}(U')(k)$, and so $U(k) \neq U'(k)$. □

A family $\{U_i\}_{i \in I}$ of open subfunctors of X is called an *open covering* if $X(k) = \bigcup_{i \in I} U_i(k)$ for all fields k over B.

For a given $X \in \mathcal{C}_B$, we can think of the category $O(X)$ of open subfunctors of X, whose objects are made up of open subfunctors of X and $\mathrm{Hom}_{O(X)}(U, U')$ is either the inclusion $U \subset U'$ or the empty set according as U is a subfunctor of U' or not. By Lemma 4.18, the category $O(X)$ looks very close to the category of open subsets of a topological space.

4.1.6 Zariski Sheaves

A functor $X \in \mathcal{C}_B$ is called *local* if it satisfies the following set-theoretic exact sequence for any open covering $\{U_i\}_{i \in I}$ of any object Y in $\mathcal{C}_B$:

$$\mathrm{Hom}_{\mathcal{C}_B}(Y, X) \to \prod_{i \in I} \mathrm{Hom}_{\mathcal{C}_B}(U_i, X) \rightrightarrows \prod_{i,j \in I} \mathrm{Hom}_{\mathcal{C}_B}(U_i \cap U_j, X), \qquad (4.1.7)$$

in other words, writing $\mathrm{Res}_i : \mathrm{Hom}_{\mathcal{C}_B}(Y, X) \to \mathrm{Hom}_{\mathcal{C}_B}(U_i, X)$ for the restriction to U_i and $\mathrm{Res}_{i,j} : \mathrm{Hom}_{\mathcal{C}_B}(U_i, X) \to \mathrm{Hom}_{\mathcal{C}_B}(U_i \cap U_j, X)$ for the restriction to $U_i \cap U_j$, if $\mathrm{Res}_{i,j}(\alpha_i) = \mathrm{Res}_{j,i}(\alpha_j)$ for all (i, j) with

$(\alpha_i) \in \prod_{i\in I} \mathrm{Hom}_{\mathcal{C}_B}(U_i, X)$, then $\alpha_i = \mathrm{Res}_i(\alpha)$ for all i for a *unique* $\alpha \in \mathrm{Hom}_{\mathcal{C}_B}(Y, X)$. Thus, any morphism from Y to X is determined by the local data α_i. This is the least requirement for X being a geometric object, as a morphism from a geometric object must be determined by local data.

A *B-scheme* is a local functor in $\mathcal{C}_B$ that has an open covering by affine B-schemes. For a B-scheme S, an S-scheme is a pair consisting of a morphism $\iota_X : X \to S$ and a scheme X. We write $SCH_{/S}$ for the category of S-schemes. A morphism of S-schemes $\phi : X \to Y$ by definition satisfies $\iota_Y \circ \phi = \iota_X$. If $X_{/S}$ and $Y_{/S}$ are S-schemes, the fibered product functor $X \times_S Y$ is clearly an S-scheme (so $SCH_{/B}$ has a fibered product; see [GME] Sect. 1.5.2).

A morphism $X \to Y$ of a B-scheme is *separated* if the diagonal subfunctor $\Delta_{X/Y}(R) = \{(x, x) \in X(R) \times_{Y(R)} X(R) | x \in X(R)\}$ is a closed subscheme of $X \times_Y X$. We define a morphism of scheme $f : X \to S$ to be *proper* if the following two conditions are satisfied by f:

1. f is separated and of finite type.
2. For any S-scheme T, the projection $f_T : X_T = X \times_S T \to T$ is a closed map (of the associated topological spaces).

A morphism $X \xrightarrow{f} Y$ of a B-scheme is *quasi-compact* if, for any open subscheme $U \subset Y$, the pullback functor $f^{-1}(U)$ is quasi-compact [i.e., any open covering of $f^{-1}(U)$ has a finite subcovering].

As expected, we have (cf. [RAG] 1.8)

Lemma 4.19 *An affine B-scheme $X = S_R$ is a B-scheme.*

Here is a sketch of a proof.

Proof. We have $X(A) = \mathrm{Hom}_{\mathcal{C}_B}(S_A, S_R) = \mathrm{Hom}_{ALG_{/B}}(R, A)$ by Lemma 4.17 and (2.2). Thus, for an open affine covering $\{S_{A_i}\}_{i\in I}$ (for fraction rings A_i) of S_A, we need to prove exactness for $X = S_R$ of

$$\mathrm{Hom}_{\mathcal{C}_B}(S_A, X) \to \prod_{i\in I} \mathrm{Hom}_{\mathcal{C}_B}(S_{A_i}, X) \rightrightarrows \prod_{i,j\in I} \mathrm{Hom}_{\mathcal{C}_B}(S_{A_i} \cap S_{A_j}, X),$$

which turns out to be

$$\mathrm{Hom}_{ALG_{/B}}(R, A) \to \prod_{i\in I} \mathrm{Hom}_{ALG_{/B}}(R, A_i) \rightrightarrows \prod_{i,j\in I} \mathrm{Hom}_{ALG_{/B}}(R, A_i \otimes_A A_j),$$

since $S_{A_i} \cap S_{A_j} = S_{A_i \otimes A_j}$ with $\mathrm{Res}_{i,j}$ gives rise to $A_i \ni a \mapsto a \otimes 1 \in A_i \otimes_A A_j$. By Yoneda's lemma, the exactness of the above sequence is equivalent to the exactness of

$$A \to \prod_{i\in I} A_i \rightrightarrows \prod_{i,j\in I} A_i \otimes_A A_j.$$

Since $O(S_A)$ has a base of open subsets of the form S_{A_f}, we may assume $A_i = A_{f_i}$ for $\{f_i\}_{i\in I}$. Then $A_i \otimes_A A_j = A_{f_i f_j}$ by $a \otimes b \mapsto ab$, and the exactness of the above sequence follows from standard commutative ring theory (cf. [ALG] II.2.2 or [GME] Sect. 1.2.1). □

For $A \in ALG_{/B}$, we have the structure morphism $\underline{\iota}_A : B \to A$ given the B-algebra structure on A, and hence $\iota_A : S_A \to S_B$. Let X be a B-scheme; so $X = \bigcup_{i \in I} S_{A_i}$ for open affine subschemes S_{A_i}. Since X is local, we have the exact sequence

$$\mathrm{Hom}_{\mathcal{C}_B}(X, S_B) \to \prod_{i \in I} \mathrm{Hom}_{\mathcal{C}_B}(S_{A_i}, S_B) \rightrightarrows \prod_{i,j \in I} \mathrm{Hom}_{\mathcal{C}_B}(S_{A_i} \cap S_{A_j}, S_B)$$

and $\{\iota_{A_i} : S_{A_i} \to S_B\}_i$ is in the kernel of the second double arrows, and so we have a unique *structure morphism* $\iota_X : X \to S_B$.

Let X be a B-scheme. A *presheaf* F on X is a contravariant functor from $O(X)$ to AB. A morphism of presheaves $F \to G$ is a morphism of contravariant group functors. A *sheaf* F is a presheaf that satisfies the following exact sequence (of abelian groups) for any open covering $\{U_i\}_{i \in I}$ of any object U in $O(X)$:

$$0 \to F(U) \xrightarrow{\prod_i \mathrm{Res}_i} \prod_{i \in I} F(U_i) \xrightarrow{\prod_{i,j}(\mathrm{Res}_i - \mathrm{Res}_j)} \prod_{i,j \in I} F(U_i \cap U_j),$$

where Res_i indicates the restriction map $F(U) \to F(U_i)$, and $\mathrm{Res}_i - \mathrm{Res}_j$ is taken in $F(U_i \cap U_j)$ after restricting to $U_i \cap U_j$. The category of sheaves $S(X)$ is a full subcategory of the category $PS(X)$ of presheaves. For a presheaf F and an open subscheme U, often we write $H^0(U, F)$ for $F(U)$ following the tradition of cohomology theory. For any open subscheme $U \subset X$ of a scheme and a sheaf F on X, F induces a sheaf on U as any open subscheme of U is open in X. We write $F|_U$ for the induced sheaf.

The *stalk* at $P \in X(k)$ (for a field k) of a sheaf F is an abelian group given by $F_P = \varinjlim_U H^0(U, F)$, where U runs over affine open subschemes such that $P \in U(k)$ (such an open subscheme is called an open neighborhood of P). Since the stalk at P is determined by open affine neighborhoods of P, any property of affine schemes determined by stalks can be extended to general schemes; so, for example, flatness, smoothness, étaleness, and so on are well defined for the general scheme morphism $f : X \to Y$.

We call P a *maximal point* if it has an affine open neighborhood S_A with $P \in S_A(k) = \mathrm{Hom}_{ALG_{/B}}(A, k)$ surjecting down to k. A point $P \in X(k)$ is called a *geometric point* if k is an algebraically closed field. A geometric point may not be maximal as $P : A \to k$ may not be surjective for any open neighborhood S_A of P. For a geometric point $P \in X(k)$, we sometimes write $k(P)$ for the field of fraction of the image of P in k. For any integral domain R, $P \in X(R)$ induces a geometric point $\overline{P} \in X(k)$ for an algebraic closure k of the field of fraction of R as we have a natural map $X(R) \to X(k)$.

Since a sheaf has values in abelian groups, it has a natural structure of an additive category. We call a sequence of sheaves $F \to G \to H$ exact if $F_P \to G_P \to H_P$ is exact for stalks at all maximal points P. Then the category of sheaves on X becomes an abelian category (see [GME] Sect. 1.4.4

for abelian/additive categories). If $f : X \to Y$ is a morphism of B-schemes and F is a sheaf on X, the *direct image* sheaf on Y, written as f_*F, is defined by $f_*F(U) = F(f^{-1}(U))$ for open subschemes U of Y.

We have a unique sheaf $\mathcal{O}_S$ of "schematic functions" on S such that $\mathcal{O}_S(S_A) = H^0(S_A, \mathcal{O}_S) = A$ for any affine open subscheme $S_A \hookrightarrow S$. Indeed, for any open subscheme $U \subset S$, we have $H^0(U, \mathcal{O}_S) = \mathrm{Hom}_B(U, \mathbb{G}_a)$ as we have $\mathrm{Hom}_B(S_A, \mathbb{G}_a) = \mathrm{Hom}_{ALG_{/B}}(B[X], A) \cong A$ by $\phi \mapsto \phi(X)$. This sheaf is called the *structure sheaf* of S. Since $\mathbb{G}_a$ is a ring scheme, $\mathcal{O}_S(U) = H^0(U, \mathcal{O}_S)$ is a ring. Thus, $\mathcal{O}_S$ is a sheaf of rings (i.e., as a presheaf, $\mathcal{O}_S$ has values in $ALG_{/B}$). A scheme S is called *reduced* if $H^0(\mathcal{O}_S, U)$ does not have nontrivial nilpotent elements for all open subschemes $U \subset S$.

A sheaf F over a B-scheme S is called an $\mathcal{O}_S$-module if we have a functor morphism $\mathcal{O}_S \times F \to F$, which gives rise to an $\mathcal{O}_S(U)$-module structure on $F(U)$ for every open subscheme $U \subset S$. Consider a sheaf of rings $\mathcal{A}_{/S}$ that is at the same time an $\mathcal{O}_S$-module under the structure $\mathcal{O}_S$-algebra homomorphism $\iota_{\mathcal{A}} : \mathcal{O}_S \to \mathcal{A}$. Such a sheaf is called a sheaf of $\mathcal{O}_S$-algebras. We write the category of $\mathcal{O}_S$-algebras as $ALG_{/\mathcal{O}_S}$ whose morphism f is a sheaf $\mathcal{O}_S$-algebra homomorphism; that is, $f : \mathcal{A} \to \mathcal{A}'$ is a morphism of functors from $O(S)$ to AB, and, at the same time, $f_U : \mathcal{A}(U) \to \mathcal{A}'(U)$ is an $\mathcal{O}_S(U)$-algebra homomorphism.

We may replace S_B by a general B-scheme S in the construction of B-schemes. Consider the category $\mathcal{C}_S = COF(ALG_{/\mathcal{O}_S}, SETS)$. Since we have a natural functor $ALG_{/\mathcal{O}_S} \to ALG_{/B}$ given by $\mathcal{A} \mapsto H^0(S, \mathcal{A})$, we may restrict S_A for any B-algebra A to $ALG_{/\mathcal{O}_S}$, getting $S_A \in \mathcal{C}_S$. A subfunctor U of $X \in \mathcal{C}_S$ is called open/closedif $f^{-1}(U)$ is open for any morphism $S_A \xrightarrow{f} X$ from an affine scheme S_A. Here we define the pullback functor $f^{-1}(U)$ by $f^{-1}(U)(R) = f_R^{-1}(U(R))$. An open covering $\{U_i\}_i$ of $Y \in \mathcal{C}_S$ is made up of open subschemes $U_i \subset Y$ such that $Y(k) = \bigcup_i U_i(k)$ for any field k over the algebra $\mathcal{O}_S(U)$. Here U runs over open affine subschemes of S. Using this definition of open coverings, we call $X \in \mathcal{C}_S$ *local* if it satisfies the sheaf exact sequence (4.1.7) (replacing $\mathcal{C}_B$ by $\mathcal{C}_S$) for open coverings $\{U_i\}_i$ of Y. A functor $X \in \mathcal{C}_S$ is called *S-schemes* if it is local and covered by affine open subschemes.

Pick $\mathcal{A} \in ALG_{/\mathcal{O}_S}$, and consider $F \in \mathcal{C}_S = COF(SCH_{/S}, SETS)$ given by $F(f : X \to S) = \mathrm{Hom}_{\mathcal{O}_S\text{-algebra}}(\mathcal{A}, f_*\mathcal{O}_X)$. Here $f_*\mathcal{O}_X$ is a sheaf of $\mathcal{O}_S$-algebras given by $f_*\mathcal{O}_X(U) = \mathcal{O}_X(f^{-1}(U))$. Since $\mathcal{A}$ and $f_*\mathcal{O}_X$ are local as they are sheaves over S, the functor F is also local. Thus, F is an S-scheme. We often write $F = Spec_S(\mathcal{A})$ and call it the *relative spectrum* of $\mathcal{A}$. A morphism $X \xrightarrow{f} S$ is called *affine relative to* S if $X = Spec_S(\mathcal{A})$ for a sheaf of $\mathcal{O}_S$-algebras. By definition, for any affine open subscheme $S_A \subset S$, $f^{-1}(S_A)$ is an affine scheme over A if and only if $X = Spec_S(\mathcal{A})$ (see [GME] Sect. 1.5.4).

For a scheme S, the presheaf $\mathfrak{n}$ of nilradicals of $\mathcal{O}_S$ sending an open subscheme $U \subset S$ to the nilradical of $H^0(U, \mathcal{O}_S)$ is actually a sheaf. Its sheaf quotient $\mathcal{O}_S/\mathfrak{n}$ gives rise to a maximal reduced closed subscheme

$S^{red} = Spec_S(\mathcal{O}/\mathfrak{n})$ called the *reduced part* of S. Clearly, the topological spaces of S and S^{red} are identical. A scheme S is *integral* if $H^0(U, \mathcal{O}_S)$ is an integral domain for all open subschemes $U \subset S$. The reduced part S^{red} is a union of maximal integral closed subschemes. Each maximal integral closed subscheme of S is called an *irreducible component* of S. If $S = \mathrm{Spec}(A)$, for each minimal prime ideal $\mathfrak{p}$ of A, $\mathrm{Spec}(A/\mathfrak{p})$ gives an irreducible component; thus, minimal prime ideals are in bijection with irreducible components. Similarly, irreducible components of a general scheme S are in bijection to subsheaves of minimal prime ideals of $\mathcal{O}_S$. We call S an *irreducible* scheme if S^{red} is integral. See [ALG] Chap. 1 for the topological definition of irreducibility.

An $\mathcal{O}_X$-module F on a scheme X is called *invertible* or a *line bundle* if there exists an open covering $\{U_i\}_{i \in I}$ of X such that $F|_{U_i} \cong \mathcal{O}_{U_i}$ for each i. An $\mathcal{O}_X$-module F is called *quasi-coherent* if we have a sheaf exact sequence $\mathcal{O}_X^I \to \mathcal{O}_X^J \to F \to 0$, where $\mathcal{O}_X^I$ is the product of copies of $\mathcal{O}_X$ indexed by any set I. If we can take I and J as finite, the $\mathcal{O}_X$-module F is called *coherent*.

Remark 4.20 If $X = S_A$ is affine, for a quasi-coherent $\mathcal{O}_X$-module F, putting $M = F(X)$, we have $F(S_{A_f}) = M_f = A_f \otimes_A M$; thus, F is determined by M and sometimes written as $\widetilde{M}$. For any given A-module, we have a unique sheaf $\widetilde{M}$ such that $H^0(S_{A_f}, \widetilde{M}) = M_f$ (cf. [GME] Sect. 1.2.3). Indeed, if we write $\mathbb{G}_a \otimes_A M$ for the A-module functor $R \mapsto R \otimes_A M$ in $\mathcal{C}_A$, we have $\widetilde{M}(U) = \mathrm{Hom}_{\mathcal{C}_A}(U, \mathbb{G}_a \otimes_A M)$.

We write $QS(X)$ for the full subcategory in $S(X)$ of quasi-coherent $\mathcal{O}_X$-modules. The morphisms of this category are defined to be $\mathcal{O}_X$-linear morphisms in the category of (abelian) presheaves on X.

For a quasi-coherent $\mathcal{O}_S$-module $\mathcal{M}$ of a scheme S,

$$\mathfrak{a}(U) = \{f \in \mathcal{O}_S(U) | f \cdot \mathcal{M}(U) = 0\} = \bigcap_{m \in \mathcal{M}(U)} \mathrm{Ker}(f \mapsto f \cdot m)$$

is a subsheaf of ideals of $\mathcal{O}_S$ (the *annihilator ideal* of $\mathcal{M}$). Write its sheaf quotient as $\mathcal{O}_S/\mathfrak{a}$, which is a sheaf of $\mathcal{O}_S$-algebras. We define $\mathrm{Supp}(\mathcal{M}) = Spec_S(\mathcal{O}_S/\mathfrak{a})$, which is a closed subscheme of S called the *support* of $\mathcal{M}$. If k is a field, we can check $\mathrm{Supp}(\mathcal{M})(k) = \{x \in S(k) | \mathcal{F}_x \neq 0\}$, and this is the reason we call it a "support" of $\mathcal{M}$ (see [CRT] Sect. 4 for this fact).

Exercise 4.21 *Let S be a scheme. Prove the following facts:*

1. *For a morphism of $\mathcal{O}_S$-modules $f : \mathcal{M} \to \mathcal{M}'$,*

$$\mathrm{Ker}(f)(U) = \mathrm{Ker}(f : \mathcal{M}(U) \to \mathcal{M}'(U))$$

   is a $\mathcal{O}_S$-submodule of $\mathcal{M}$ [in particular, $U \mapsto \mathrm{Ker}(f)(U)$ is a sheaf].
2. *For a set I of morphisms of $\mathcal{O}_S$-modules $f : \mathcal{M} \to \mathcal{M}'$, $\bigcap_{f \in I} \mathrm{Ker}(f)$ is an $\mathcal{O}_S$-submodule of $\mathcal{M}$ (in particular, it is a sheaf).*
3. *The presheaf $\mathfrak{a}$ as above is a sheaf of $\mathcal{O}_S$-ideals.*

If $X \xrightarrow{f} Y$ is a morphism of S-schemes, the direct image $F \mapsto f_*F$ gives a covariant functor from $QS(X)$ to $QS(Y)$. It has a left adjoint (contravariant) functor $f^* : QS(Y) \to QS(X)$. In other words, the contravariant functor $G \mapsto \mathrm{Hom}_{QS(Y)}(G, f_*F)$ [for $F \in QS(X)$ and $G \in QS(Y))$] is represented by a quasi-coherent sheaf f^*G over X; hence, we have a canonical isomorphism (of bifunctors) $\mathrm{Hom}_{QS(Y)}(G, f_*F) \cong \mathrm{Hom}_{QS(X)}(f^*G, F)$. The sheaf f^*G is called the *inverse image* of G (see [GME] Sect. 1.5.3).

Exercise 4.22 *Using the functoriality of "$\otimes$" and "Hom_A", check that if* $X = S_{A'}$ *and* $Y = S_A$ *and* $G = \widetilde{M}$, *we then have* $f^*G = \widetilde{(M \otimes_A A')}$.

A morphism $X \xrightarrow{f} Y$ of B-schemes is a *finite* morphism if $f_*\mathcal{O}_X$ is a coherent $\mathcal{O}_Y$-module. If f is finite, for all fields k over B, the fiber of the map $f_k : X(k) \to Y(k)$ is finite. If this property of finiteness of the fiber is satisfied for all fields k over B, f is called a *quasi-finite morphism.*

Exercise 4.23 *Give an example of a quasi-finite morphism that is not finite.*

Lemma 4.24 *If a B-morphism $X \xrightarrow{f} Y$ is finite, we have $X \cong Spec_Y(f_*\mathcal{O}_X)$; hence, f is an affine morphism.*

Proof. Let $\mathcal{A} = f_*\mathcal{O}_X$. Then $S_{\mathcal{A}}(T) = \mathrm{Hom}_{ALG_{/\mathcal{O}_Y}}(\mathcal{A}, g_*\mathcal{O}_T)$ for any Y-scheme $T \xrightarrow{g} Y$. In particular, $S_{\mathcal{A}}(X) = \mathrm{Hom}_{ALG_{/\mathcal{O}_Y}}(\mathcal{A}, \mathcal{A})$ has a canonical element $\mathrm{id}_{\mathcal{A}}$, which gives rise to a morphism $f' : X \to Spec_Y(\mathcal{A})$. For any maximal point $y \in Y(k)$, $\mathcal{A}_y$ is a finite-type $\mathcal{O}_{Y,y}$-module with $\mathcal{A}_y/\mathfrak{m}_y\mathcal{A}_y$, giving rise to $\mathcal{O}_{X,f^{-1}(y)}/\mathfrak{m}_y\mathcal{O}_{X,f^{-1}(y)}$. By Nakayama's lemma ([CRT] Theorem 2.2 and Lemma 10.8 in the text), we have $\mathcal{O}_{X,f^{-1}(y)} = \mathcal{A}_y$. This shows that f' gives rise to an isomorphism of all stalks, and hence f' is an isomorphism.

A scheme $X \in SCH_{/B}$ is *locally noetherian* if, for any open affine subscheme S_A of X, A is a noetherian ring. If, further, the topological space of X is quasi-compact, X is called *noetherian.* A morphism $X \xrightarrow{f} Y$ of B-schemes is *of finite type* if f is quasi-compact and there exists an open covering $\{U_i\}_i$ of Y such that $f^{-1}(U_i)$ is covered with (finitely many) affine open subschemes S_{A_j} with A_j finitely generated over $\mathcal{O}_Y(U_i)$. A B-morphism $X \xrightarrow{f} Y$ is called *proper* if (i) it is separated of finite type and (ii) for any Y-scheme T, the base change $f_T : X \times_Y T \to T = Y \times_Y T$ is a closed map of the corresponding topological space. The valuative criterion of separatedness/properness is given as follows:

Proposition 4.25 *Suppose $f : X \to Y$ is a finite-type morphism with X noetherian in $SCH_{/B}$. Suppose we have a commutative diagram:*

$$\begin{array}{ccc} S_K & \xrightarrow{i_X} & X \\ \downarrow & & \downarrow \\ S_V & \xrightarrow[i_Y]{} & Y \end{array}$$

for a valuation ring V with quotient field K in $ALG_{/B}$. Then f is separated (resp., proper) if and only if there exists at most one (resp., there exists a unique) morphism $i : S_V \to X$ making the above diagram commutative (after inserting the diagonal arrow given by i).

See, for example, [ALG] II.4.7 for a proof.

Corollary 4.26 *Let the notation be as in* Proposition 4.25. *If the morphism $f : X \to Y$ is finite, it is proper.*

Proof. We apply the above proposition. Since Spec(V) is irreducible, reduced, and affine, we may assume that $X = \mathrm{Spec}(A')$ and $Y = \mathrm{Spec}(A)$ are irreducible and reduced. Since X is finite over Y, A' is integral over A. Then, having the diagram

$$\begin{array}{ccc} S_K & \xrightarrow{i_X} & X \\ \downarrow & & \downarrow \\ S_V & \xrightarrow[i_Y]{} & Y \end{array}$$

is equivalent to having

$$\begin{array}{ccc} K & \xleftarrow{\underline{i}_X} & A' \\ \cup\uparrow & & \uparrow\cup \\ V & \xleftarrow[\underline{i}_Y]{} & A. \end{array}$$

Since the valuation ring is integrally closed in K, $\underline{i}_X$ maps A' into V and hence induces $S_V \to X$ with desired commutativity.

4.1.7 Sheaf of Differential Forms

Let $S = S_A$ for a B-algebra A and let R be an A-algebra. We have the structure morphism $X = S_R \to S$. For each R-module M, an A-derivation $\delta : R \to M$ is an A-linear map satisfying $\delta(fg) = f\delta(g) + g\delta(f)$ for all $f, g \in R$. We write $Der_A(R, M)$ for the R-module of A-derivations. Note that $\delta(1) = \delta(1 \times 1) = \delta(1) + \delta(1)$, and $\delta(1) = 0$. For $a \in A$, $\delta(a) = \delta(a1) = a\delta(1) = 0$ by the A-linearity of δ. Thus, each A-derivation kills A. Conversely, if a derivation δ kills A, we see $\delta(ar) = a\delta(r) + r\delta(a) = a\delta(r)$, and δ is A-linear.

In algebraic geometry, there is no naive definition of differentiation. However, we can think of derivations of the structure sheaf $\mathcal{O}_X = \widetilde{R}$ for each affine scheme $X = Spec(R)$. Thus, if the covariant functor $M \mapsto Der_A(R, M)$ from R–MOD to AB is representable by an R-module $\Omega_{X/S} = \Omega_{R/A}$, i.e., $Der_A(R, M) \cong \mathrm{Hom}_R(\Omega_{R/A}, M)$, we may think of $\Omega_{R/A}$ as an analog of the cotangent bundle in differential geometry. Indeed, we have the following:

Proposition 4.27 *Let R be an A-algebra. The functor $R\text{–}MOD \ni M \mapsto Der_A(R, M) \in AB$ is representable by an R-module $\Omega_{R/A}$ with a universal derivation $d : R \to \Omega_{R/A}$, which is unique up to automorphisms of $\Omega_{R/A}$ over A. In other words, if $\delta : R \to M$ is an A-derivation, then there is a unique R-linear map $\phi : \Omega_{R/A} \to M$ such that $\delta = \phi \circ d$.*

Proof. Take the free R-module F generated by symbols $\{dr | r \in R\}$, and make a quotient by the R-submodule generated by $d(ar + a'r') - adr - a'dr'$ and $drr' - rdr' - r'dr$ for all $r, r' \in R$ and $a, a' \in A$. Then the resulting R-module $\Omega_{R/A}$ represents the functor.

There is another construction of $\Omega_{R/A}$. The multiplication $a \otimes b \mapsto ab$ induces an A-algebra homomorphism $m : R \otimes_A R \to R$ taking $a \otimes b$ to ab. We set $I = \mathrm{Ker}(m)$, which is an ideal of $R \otimes_A R$. Then we define $\Omega_{R/A} = I/I^2$. It is easy to check that the map $d : R \to \Omega_{R/A}$ given by $d(a) = a \otimes 1 - 1 \otimes a \mod I^2$ is an A-derivation. Thus, we have a morphism of functors: $\mathrm{Hom}_R(\Omega_{R/A}, ?) \to Der_A(R, ?)$ given by $\phi \mapsto \phi \circ d$. Since $\Omega_{R/A}$ is generated by $d(A)$ as A-modules, the above map is injective.

To show that $\Omega_{R/A}$ represents the functor, we need to show the surjectivity. Define $\phi : R \times R \to M$ by $(a, b) \mapsto a\delta(b)$ for $\delta \in Der_A(R, M)$. Then $\phi(ab, c) = ab\delta(c) = \phi(b, c)$ and $\phi(a, bc) = a\delta(bc) = ab\delta(c) = b\phi(a, c)$ for $a, c \in R$ and $b \in A$, and ϕ gives a continuous A-bilinear map. By the universality of the tensor product, $\phi : R \times R \to M$ extends to an A-linear map $\phi : R \otimes_A R \to M$. Now we see that $\phi(a \otimes 1 - 1 \otimes a) = a\delta(1) - \delta(a) = -\delta(a)$ and

$$\phi((a \otimes 1 - 1 \otimes a)(b \otimes 1 - 1 \otimes b)) = -a\delta(b) - b\delta(a) + \delta(ab) = 0.$$

Thus, $\phi|_I$-factors through $I/I^2 = \Omega_{R/A}$, and we have $\delta = \phi \circ d$, as desired.

Remark 4.28 From the first proof of the above proposition, we get

$$\Omega_{R_f/A_f} = \Omega_{R/A,f} = \Omega_{R/A} \otimes_R R_f$$

for a non-nilpotent element $f \in A$; hence, the sheaf $\widetilde{\Omega}_{R/A}$ over S_R satisfies $H^0(S_{R_f}, \widetilde{\Omega}_{R/A}) = \Omega_{R_f/A_f}$. Thus, the modules of differential glue well into a sheaf of differentials $\Omega_{S_R/S_A} := \widetilde{\Omega}_{R/A}$, as in Remark 4.20.

More generally, let A' and R be A-algebras. Write

$$I = \mathrm{Ker}(R \otimes_A R \xrightarrow{a \otimes b \mapsto ab} R),\ R' = A' \otimes_A A',\ I' = \mathrm{Ker}(R' \otimes_{A'} R' \xrightarrow{a \otimes b \mapsto ab} R').$$

From the second construction of $\Omega_{R/A}$, we see that

$$\Omega_{R/A} \otimes_R (A' \otimes_A R) \cong I/I^2 \otimes_R R' \cong I'/I'^2 \cong \Omega_{R'/A'}$$

canonically. Thus, we get the following:

Corollary 4.29 *Let $R \leftarrow A \to A'$ be morphisms of algebras. Then we have $\Omega_{R/A} \otimes_R (A' \otimes_A R) \cong \Omega_{A' \otimes_A R/A'}$.*

By the first proof of Proposition 4.27, any element of $\Omega_{R/A}$ can be written as $\sum_i f_i dr_i$ for some r_i and f_i in R. By Corollary 4.29, taking $A' = S^{-1}A$ for a multiplicative subset S in A, we get

$$S^{-1}\Omega_{R/A} = \Omega_{R/A} \otimes_R S^{-1}R = \Omega_{S^{-1}A \otimes_A R/S^{-1}A'} = \Omega_{S^{-1}R/S^{-1}A}. \quad (4.1.8)$$

Let $f : X \to S$ be a morphism of schemes. Let $S = \bigcup_i S_i$ $[S_i = Spec(A_i)]$ be an open affine covering and $X = \bigcup_\alpha X_\alpha$ $[X_\alpha = Spec(R_\alpha)]$ be an open affine covering of X so that f induces $f_\alpha : X_\alpha \to S_{i(\alpha)}$. Let Ω_α be the sheaf $\widetilde{\Omega}_{R_\alpha/A_{i(\alpha)}}$ on X_α. Then by (4.1.8) and the universality, we have

$$\varphi_{\alpha\beta} : \Omega_\alpha|_{X_\alpha \cap X_\beta} \cong \Omega_\beta|_{X_\beta \cap X_\alpha} \quad \text{and} \quad \varphi_{\beta\gamma} \circ \varphi_{\alpha\beta} = \varphi_{\alpha\gamma} \quad \text{on} \quad X_\alpha \cap X_\beta \cap X_\gamma.$$

The sheaves Ω_α's glue together into a quasi-coherent sheaf $\Omega_{X/S}$. It is coherent if X is noetherian. This sheaf is called the sheaf of *differential* 1*-forms* on X over S. By this construction, Corollary 4.29 implies

Corollary 4.30 *Let $X \to Y \leftarrow Y'$ be morphism of schemes. Then we have $\Omega_{X/Y} \otimes_{\mathcal{O}_X} \mathcal{O}_{X \times_Y Y'} \cong \Omega_{X \times_Y Y'/Y'}$ canonically. In particular, for a point $y \in Y$, writing the fiber of X at y as X_y, we have $\Omega_{X/Y} \otimes_{\mathcal{O}_X} \mathcal{O}_{X_y} \cong \Omega_{X_y/y}$.*

Here the fiber X_y of $X \xrightarrow{f} Y$ at $y \in Y$ as a functor is the pullback functor $f^{-1}(y)$, where $y \subset Y$ is the subfunctor giving the point y. As a scheme, X_y is given by the fibered product $X_y = X \times_Y y$.

Remark 4.31 Write C_X for the category of functors either from $S(X)$ or from $SCH_{/X}$ into $SETS$. Here $SCH_{/X}$ is the category whose object is a pair (Y, f) of a scheme Y. The set of morphisms is given by

$$\mathrm{Hom}_{SCH/X}((Y,f),(Z,g)) = \left\{h \in \mathrm{Hom}_{SCH}(Y,Z) \middle| g \circ h = f\right\}.$$

We slightly generalize the notion of local functor from schemes to functors in C_X. A functor $F \in C_X$ is called *local* if it satisfies the following set-theoretic exact sequence for any open covering $\{U_i\}_{i \in I}$ of any object Y in C_X:

$$\mathrm{Hom}_{C_X}(Y,F) \to \prod_{i \in I} \mathrm{Hom}_{C_X}(U_i, F) \rightrightarrows \prod_{i,j \in I} \mathrm{Hom}_{C_X}(U_i \cap U_j, F).$$

The functor F is local if F is determined by local data depending only on a (whatever) small neighborhood of each point of X. If F is local and F restricted to $SCH_{/Y}$ or $S(Y)$ is a representable functor whenever the X-scheme Y is affine, then F is representable over $SCH_{/X}$ for the starting scheme X. Indeed, taking an affine open covering $X = \bigcup_\alpha X_\alpha$, if W_α represents F restricted over X_α, we have a canonical isomorphism $\varphi_{\alpha\beta} : W_\alpha|_{X_\alpha \cap X_\beta} \cong W_\beta|_{X_\alpha \cap X_\beta}$ with $\varphi_{\beta\gamma} \circ \varphi_{\alpha\beta} = \varphi_{\alpha\gamma}$ on $X_\alpha \cap X_\beta \cap X_\gamma$ by the universality. Thus, W_α glues together well into an object W, which represents F over X.

By the above remark, the sheaf $\Omega_{X/S}$ represents a functor: $QS(X) \ni \mathcal{F} \mapsto Der_{\mathcal{O}_S}(\mathcal{O}_X, \mathcal{F})$, where $Der_{\mathcal{O}_S}(\mathcal{O}_X, \mathcal{F})$ is a sheaf generated by the presheaf assigning the module $Der_{\mathcal{O}_S(f(U))}(\mathcal{O}_X(U), \mathcal{F}(U))$ to an open set U. Here $\mathcal{O}_S(f(U)) = \varinjlim_{V \supset f(U)} \mathcal{O}_S(V)$. The sheaf $\mathcal{O}_X$-dual $\mathcal{T}_{X/Y}$ whose section over U is given by $\mathrm{Hom}_{\mathcal{O}_U}(\Omega_{X/Y}|_U, \mathcal{O}_U)$ is the *tangent sheaf* relative to X/Y. See [ALG] II.8 or [GME] Sect. 1.5.1 for geometric details.

Lemma 4.32 *For a finite flat S-morphism $f : X \to Y_{/S}$, f is étale if and only if $\Omega_{X/Y} = \mathrm{Coker}(f^* : f^*\Omega_{Y/S} \to \Omega_{X/S}) = 0$.*

Proof. Since f is affine by Lemma 4.24, we may assume that $S = \mathrm{Spec}(B)$, $X = \mathrm{Spec}(A')$ and $Y = \mathrm{Spec}(A)$. Suppose f is étale. Consider the exact sequence defining $\Omega_{A'/A}$: $0 \to I/I^2 \to A' \otimes_A A'/I^2 \xrightarrow{a\otimes b \mapsto ab} A' \to 0$. The map $A' \ni a \mapsto a \otimes 1 \in A' \otimes_A A'/I^2$ gives the section of multiplication $a \otimes b \mapsto ab$, so we have $A' \otimes_A A'/I = A' \oplus I/I^2 = A' \oplus \Omega_{A'/A}$ with projection π to $\Omega_{A'/A}$. Note that $\mathrm{Hom}_{ALG_{/A}}(A', A' \otimes_A A'/I^2) \cong Der_A(A', \Omega_{A'/A})$ by $\phi \mapsto \pi \circ \phi$. Since I is a square-zero ideal of $A' \otimes_A A'/I^2$, by étaleness, $Der_A(A', \Omega_{A'/A}) = \{0\}$. Thus, $\Omega_{A'/A} = 0$. The fact $\Omega_{X/Y} = \mathrm{Coker}(f^* : f^*\Omega_{Y/S} \to \Omega_{X/S})$ is a restatement of the fundamental exact sequence (e.g., [CRT] Theorem 25.1 or [GME] Proposition 1.5.4).

Here is an outline of the proof of the converse. Assume $\Omega_{A'/A} = 0$ and that we have A-algebra homomorphisms $\phi, \varphi : A' \to E$ extending an A-algebra homomorphism $A' \to E/J$ for a square-zero ideal J of an A-algebra E. Define an A-algebra homomorphism $\Phi : A' \otimes_A A' \to E$ by $\Phi(a \otimes b) = \phi(a)\varphi(b)$. Since $\phi \mod J = \varphi \mod J$, we have $\Phi(I) \subset J$ and $\Phi(d(a) \cdot d(a')) = 0$ for $d(a) = a \otimes 1 - 1 \otimes a$ by computation. Thus, Φ induces $\Omega_{A'/A} = I/I^2 \to J$. Since $\Omega_{A'/A} = 0$, $\Phi|_I = 0$. Since $\phi(a) - \varphi(a) = \Phi(d(a)) = 0$, we get $\varphi = \phi$, and so A'/A is unramified; that is, the lift $A' \xrightarrow{\phi} E$ of a given $A' \xrightarrow{\phi_0} E/J$ is unique.

The liftability of ϕ_0 to ϕ follows by descent from the natural map $A' = A' \otimes_A A' \to A' \otimes_A A' \otimes_A E = A' \otimes_A E$. By $\Omega_{A'/A} = 0$, this map satisfies descent data for getting $A' \to E$ (see [CRT] Sect. 28 or [EGA] IV.22, and [GME] Sect. 1.11 for descent).

4.1.8 Scheme and Variety

In Sect. 4.1.6, we defined a B-scheme as a local functor in $\mathcal{C}_B$ covered by open affine subschemes. Here is a more down-to-earth definition. Let $f : S_1 \to S$ and $g : S_2 \to S$ be morphisms in $\mathcal{C}_B$. Recall

$$S_1 \times_S S_2(R) = \{(x, y) \in S_1(R) \times S_2(R) | f_R(x) = g_R(y)\} \qquad (*)$$

as an object in $\mathcal{C}_B$. If T is the fourth functor with morphisms $f' : T \to S_1$ and $g' : T \to S_2$ such that $g \circ g' = f \circ f'$, we have a unique morphism $h : T \to S_1 \times_S S_2$ given by $h(x) = (f'(x), g'(x))$ such that $f \circ h = g' \circ h$. In this sense, $S_1 \times_S S_2$ satisfies the universality of the fibered product over S. Now a finitely presented quasi-compact B-*scheme* $S_{/B}$ is a covariant functor from the category $ALG_{/B}$ of B-algebras into $SETS$ such that

1. We have finitely many affine schemes $S_i = \mathrm{Spec}(B_i)$ for finitely presented B-algebras B_i with inclusion $S_i \hookrightarrow S$ such that $S = \bigcup_i S_i$ [i.e., $S(k) = \bigcup_i S_i(k)$ for all fields k over B].
2. $S_i \cap S_j = S_i \times_S S_j$ is Zariski open in S_i and S_j for all pairs (i, j) of indices (so $\{S_i\}$ is an open covering of S).

In practice, the scheme S is often constructed by gluing the affine schemes S_i, and each $\phi \in S(R)$ is determined by its "restriction" ϕ_i to S_i in the following way: For each B-algebra R, noting that $S_R \times_S S_i = \phi^{-1}(S_i)$ by $(*)$, we require the local property that $\phi \in S(R)$ be determined by the tuple

$$\{\phi_i \in \mathrm{Hom}(S_R \times_S S_i, S_i)\}$$

such that ϕ_i and ϕ_j induce an identical morphism of $S_R \times_S S_i \times_S S_j \to S_i \times_S S_j$ (ignoring the indices i with $S_R \times_S S_i = \emptyset$); then we may define [again noting that $S_R \times_S S_i \times_S S_j = \phi^{-1}(S_i \cap S_j)$]

$$S(R) = \bigcup_i S_i := \left\{ (\phi_i)_i \middle| \begin{matrix} \phi_i \in \mathrm{Hom}(S_R \times_S S_i, S_i), \quad \text{and} \\ \phi_i \text{ and } \phi_j \text{ coincide on } S_R \times_S S_i \times_S S_j \text{ for all } i, j \end{matrix} \right\}.$$

Thus, S is a local functor covered with "finitely" many affine open subschemes (this finiteness is "quasi-compactness," as all affine schemes are quasi-compact; see [GME] Sect. 1.2.2).

A B-*morphism* $\phi : S \to T$ of schemes is a morphism in $\mathcal{C}_B$; hence, the category of B-schemes $SCH_{/B}$ is a full subcategory of $\mathcal{C}_B$. The category $SCH_{/B}$ contains the category of affine B-schemes as a full subcategory (by Lemma 4.17).

Let $\phi : S \to T$ be a morphism of B-schemes. Covering T by affine schemes $\{S_{A'_i}\}_i$ for B-algebras A'_i, that is, $T = \bigcup_i S_{A'_i}$, then $\phi^{-1}(S_{A'_i})$ can be covered again by suitable affine schemes $S_{A_{ij}}$ for A'_i-algebras A_{ij}. If $S_{A_{ij}}$ is smooth (resp., étale) over $S_{A'_i}$ for all i and j, we call S is *smooth* (resp., *étale*) over T.

A separated geometrically reduced scheme V of finite type over a field k is called an algebraic *variety* in this book. Here the term "geometrically reduced" means that $\overline{V} := V \times_k \overline{k}$ for an algebraic closure $\overline{k}$ of k is reduced (i.e., $\mathcal{O}_{\overline{V}}$ does not have nontrivial nilpotent elements). We do not include connectedness in the notion of "varieties." If $V_{/k}$ is an irreducible variety, for any dense affine open subscheme $S_A \subset V$, the Krull dimension $\dim A$ is independent of the choice of the affine open subscheme. We set $\dim V = \dim A$. A variety is equidimensional if all irreducible components have equal dimension, written as $\dim V$.

A variety $V_{/k}$ (of equidimension n) over a perfect field k is smooth over k if and only if its stalk at all maximal points has completion isomorphic to

$(\mathcal{O}_{V,x}/\mathfrak{m}_x)[[X_1,\dots,X_n]]$ (hence, also if and only if $\mathcal{O}_{V,x}$ is regular for all maximal points x). A *normal* variety is a variety V whose stalk $\mathcal{O}_{V,x}$ is integrally closed in its total quotient ring at any closed point (hence at any point) of V. In particular, a smooth variety is normal (as power series rings over a field are normal).

A variety having dimension 1 is called a *curve*. Since any normal noetherian domain of dimension 1 is a discrete valuation ring (cf. [CRT] Theorem 11.2), a curve is smooth over a perfect field k if and only if it is normal (cf. Corollary 2.11). A maximal point of a smooth curve C over a perfect field k gives rise to a discrete valuation of the function field $k(C)$ as seen in Sect. 2.1.5. Conversely, each discrete valuation ring associated with a valuation trivial over k gives rise to a unique maximal point of C (see [PAF] Sect. 2.4). In short, a smooth curve C over a perfect field is determined by its function field $k(C)$. This type of exact relation between discrete valuations and points of a variety is only valid for curves.

4.1.9 Projective Schemes

In order to give a typical example of nonaffine schemes, we introduce graded algebras over a commutative algebra B. A graded algebra R over an algebra B is a direct sum $\bigoplus_d R_d$ for a B-subalgebra R_0 and R_0-modules R_d for integers d such that $R_nR_m \subset R_{m+n}$. A B-morphism $\phi : R \to R'$ of graded B-algebras R and R' is a B-algebra homomorphism with $\phi(R_d) \subset R'_d$ for all d. An element x of a graded algebra R is called *homogeneous* of degree d if $x \in R_d$. The polynomial ring $B[T]$ is a graded algebra with $B[T]_d = T^dB$ for $d \geq 0$ and $B_d = 0$ for negative d. More generally, the polynomial ring $B[T_1,\dots,T_n]$ of n variables is a graded algebra such that each monomial of degree d is homogeneous of degree d.

If $x \in R$ is homogeneous of degree $d \neq 0$, then $R[\frac{1}{x}]$ is a graded ring by putting $R[\frac{1}{x}]_n = \sum_{jd+m=n} x^j R_m$ [in other words, for a homogeneous element $a \in R$, $\deg(\frac{a}{x^j}) = \deg(a) - jd$]. Suppose that the base ring B is noetherian. If a graded algebra R is noetherian, then there are finitely many homogeneous elements $x_0,\dots,x_n$ (of degree $d_0,\dots,d_n$, respectively) in R that generate the B-algebra R. Thus, the algebra homomorphism $B[T_0,\dots,T_n] \to R$ sending a polynomial $P(T_0,\dots,T_n)$ to $P(x_1,\dots,x_n) \in R$ is a surjective algebra homomorphism.

Write $D_i = \mathrm{Spec}(B[T_0,\dots,T_n][\frac{1}{T_i}]_0)$. Since we have

$$\begin{aligned}(B[T_0,\dots,T_n][T_i^{-1}] \otimes_{B[T_0,\dots,T_n]} B[T_0,\dots,T_n][T_j^{-1}])_0 \\ \cong B[T_0,\dots,T_n][T_i^{-1}]_0 \cdot B[T_0,\dots,T_n][T_j^{-1}]_0 = B[T_0,\dots,T_n][(T_iT_j)^{-1}]_0\end{aligned}$$

for $i \neq j$ (in $B[T_0,\dots,T_n][\frac{1}{T_0},\frac{1}{T_1},\dots,\frac{1}{T_n}]_0$), we may identify $D_i \cap D_j$ with $\mathrm{Spec}(B[T_0,\dots,T_n][(T_iT_j)^{-1}]_0)$ canonically. In this way, we can define the projective space of dimension n by $\mathbf{P}^n = \bigcup_{j=0}^n D_j$, which is not affine. If A is a B-algebra that is either a field or a valuation ring, by definition, we have

$$\mathbf{P}^n(A) \cong \{(x_0, \ldots, x_n) | x_j \in A^\times \text{ for at least one } j\}/A^\times.$$

Let R be a general noetherian graded B-algebra. Suppose we have a finite set of homogeneous generators $x_0, \ldots, x_n$ of degree 1. Define a surjective B-algebra homomorphism $B[T_0, \ldots, T_n]$ sending T_j to x_j. Then the algebra homomorphism induces a surjection $B[T_0, \ldots, T_n][T_i^{-1}]_0 \to R[x_i^{-1}]_0$, which in turn gives rise to a closed subscheme $V_i = \mathrm{Spec}(R[x_i^{-1}]_0) \subset D_i$. Then we define $\mathrm{Proj}(R) = \bigcup_i V_i \subset \bigcup_i D_i = \mathbf{P}^n$.

We can generalize this definition to any graded B-algebra R generated by finitely many homogeneous elements (not necessarily of degree 1; see [GME] Sect. 1.3), because $\mathrm{Proj}(R) = \mathrm{Proj}(R^{(n)})$ for $R^{(n)} = \bigoplus_d R_{nd}$. If $R^\times$ contains a homogeneous element of nonzero degree, we have $\mathrm{Proj}(R) \cong \mathrm{Spec}(R_0)$ by definition (compare with [GME] Lemma 1.3.1); however, if R has no negative-degree elements, $\mathrm{Proj}(R)$ is not affine.

If we can realize a B-scheme X as a closed subscheme of $\mathbf{P}^n$ (i.e., if we have a closed immersion $X \hookrightarrow \mathbf{P}^n_{/B}$), X is called B-projective. More generally, if we can realize a morphism $f : X \to Y$ so that it factors through a closed immersion $X \hookrightarrow \mathbf{P}^n_Y = \mathbf{P}^n_{/B} \times_{S_B} Y$, f is called a *projective morphism.* This is equivalent to having an invertible sheaf $\mathcal{L}$ with generators $x_0, \ldots, x_n$ on X whose graded algebra $R = \bigoplus_{n \geq 0} H^0(X, \mathcal{L}^{\otimes n})$ has $\mathrm{Proj}(R)$ isomorphic to X. Here $\mathcal{L}^{\otimes n} = \mathcal{L} \otimes_{\mathcal{O}_X} \mathcal{L} \otimes_{\mathcal{O}_X} \cdots \otimes_{\mathcal{O}_X} \mathcal{L}$ (n times). The module R is a ring under the tensor product; that is, for $x \in H^0(X, \mathcal{L}^{\otimes n})$ and $y \in H^0(X, \mathcal{L}^{\otimes m})$, the product $x \cdot y$ in R is given by $x \otimes y \in H^0(X, \mathcal{L}^{\otimes m+n})$. Since $\mathcal{L}_x \cong \mathcal{O}_{X,x}$, the tensor product commutes: $a \otimes b = b \otimes a$; hence, R is a commutative ring. Then $x_0, \ldots, x_n$ are regarded as homogeneous generators of R. The morphism $X \to \mathrm{Proj}(R)$ is the canonical one described in [GME] Sect. 1.5.5.

If $(X, f : X \to Y)$ is isomorphic in $SCH_{/Y}$ to an open subscheme $(U, g : U \to Y)$ of a closed subscheme $(C, h : C \to Y)$ of $\mathbf{P}^n_{/Y}$ (i.e., $U \xrightarrow[\subset]{\text{open}} C$ as Y-schemes), f is called *quasi-projective.*

Exercise 4.33 *Prove that any projective morphism $X \to Y$ of B-schemes is a proper morphism* (hint: use the valuative criterion: Proposition 4.25).

Exercise 4.34 *Suppose that the base ring B is an algebraically closed field k. Prove that for any k-morphism f from $\mathbf{P}^n_{/k}$ into an affine scheme S, $f_K : \mathbf{P}^n(K) \to S(K)$ has one point image for any field K/k. Using this fact, show further that $\mathbf{P}^n$ is not affine if $n \geq 1$.*

Similar to the relative spectrum $Spec_S(\mathcal{A})$ for a quasi-coherent $\mathcal{O}_S$-algebra, we can construct relative "Proj" for a quasi-coherent graded $\mathcal{O}_S$-algebras $\mathcal{A} = \bigoplus_{n \geq 0} \mathcal{A}_n$. We assume that $\mathcal{A}$ is finitely generated over $\mathcal{A}_0$ and $\mathcal{A}_0$ is a coherent $\mathcal{O}_S$-algebras (so $\mathcal{A}_n$ is also coherent). Then for any affine open $\mathrm{Spec}(R) \subset S$, $\mathcal{A}(\mathrm{Spec}(R))$ is a graded algebra of finite type over R, and we have $\mathrm{Proj}(\mathcal{A}(\mathrm{Spec}(R))$. Clearly, if we cover S by an affine open subscheme $\mathrm{Spec}(R_i)$, $\{\mathrm{Proj}(\mathcal{A}(\mathrm{Spec}(R_i)))\}_i$ glue naturally over S, giving rise to an S-scheme $Proj_S(\mathcal{A}) \to S$ (which is locally projective).

4.1.10 Grothendieck Topology on Schemes

We briefly recall the sheaf/cohomology theory under Grothendieck topology. See [ECH] for more details. The theory was originally introduced by Grothendieck to supply schemes with good cohomology theory (perhaps motivated in part to solve Weil's conjecture on L-functions of varieties discussed briefly in Sect. 3.1.3).

Let S be a scheme and τ be a class of morphisms of the category of schemes over S such that

(T1) All isomorphisms are in τ.
(T2) The composite of two morphisms in τ is in τ.
(T3) Any base change of a morphism in τ is in τ.

Here for an S-morphism $X \xrightarrow{f} Y$ and an S-scheme T, the *base change* of f is $X_T \xrightarrow{f \times \mathrm{id}_T} Y_T$ for $X_T = X \times_S T$ and $Y_T = Y \times_S T$. Here are examples of the class τ we use:

(Z) $\tau = (Zar)$ is made of all open immersions;
(E) $\tau = (et)$ is made of all étale morphisms of finite type;
(F) $\tau = (fppf)$ is made of all separated finitely presented flat morphisms (i.e., fppf: "fidèlement plat de presentation finie" in French).

Since we usually assume that our scheme is locally noetherian (i.e., covered by affine open noetherian subschemes) and a morphism is quasi-compact, an fppf morphism is a finitely presented morphism as defined above in Sect. 4.1.8. If we do not assume these simplifying conditions, see [EGA] IV.1.6 for a more general definition of a morphism of finite presentation.

Let S_τ be the full subcategory of S-schemes whose structure morphisms are in τ. Recall that "full subcategory" means that

$$\mathrm{Hom}_{S_\tau}(X, Y) = \mathrm{Hom}_{SCH_{/S}}(X, Y).$$

Write $g(X)$ for the topological image of a morphism of schemes $X \xrightarrow{g} Y$. A τ-covering of an object U in S_τ is a family $(U_i \xrightarrow{g_i} U)_{i \in I}$ of morphisms in τ such that $U = \bigcup_i g(U_i)$. Thus, if $\tau = (Zar)$, (et), or $(fppf)$, $g : \bigsqcup_{i \in I} U_i \to U$ induced from a τ-covering is faithfully flat. Thus, τ gives rise to a notion of τ-covering on S_τ, and this notion is called the *τ-topology* on S_τ. The category S_τ with τ-topology is called the *small τ-site* over S.

A presheaf on S_τ is a contravariant functor from S_τ into the category AB of abelian groups. A morphism $F \to G$ of preserves on S_τ is a morphism of functors (a natural transformation). A presheaf G is a sheaf if the following two conditions are satisfied:

(s1) If there exists a τ-covering $\{U_i \xrightarrow{g_i} U\}_{i \in I}$ such that $u_i = G(g_i)(u) = 0$ for all i, then $u = 0$ in $G(U)$.

(s2) For a τ-covering $\{U_i \xrightarrow{g_i} U\}_{i\in I}$, write $g_{ij,i} : U_i \times_U U_j \to U_i$ for the morphism induced by g_i and g_j. For a given $u_i \in G(U_i)$, if $G(g_{ij,i})(u_i) = G(g_{ij,j})(u_j)$ in $G(U_i \times_U U_j)$ for all possible (i,j), then there exists $u \in G(U)$ such that $u_i = G(g_i)(u)$.

In short,

$$0 \to G(U) \xrightarrow{\prod_i G(g_i)} \prod_i G(U_i) \xrightarrow{\prod_{i,j}(G(g_{ij,i})-G(g_{ij,j}))} \prod_{i,j} G(U_i \times_U U_j)$$

is exact (a usual sheaf condition). Thus, we can form the category of τ-sheaves $\mathcal{S}(S_\tau)$ on S_τ.

Recall that we have a *sheafication* functor $P \mapsto P^\#$ from the category of presheaves into $\mathcal{S}(S_\tau)$ that is universal with respect to morphisms of presheaves P into abelian τ-sheaves F; more precisely, we have a canonical morphism $\iota : P \to P^\#$, and if we have a morphism $a : P \to F$ for a sheaf F, there is a unique morphism $a^\# : P^\# \to F$ such that $a^\# \circ \iota = a$ (see [ECH] Theorem II.2.11). Hence, in the category of abelian τ-sheaves, we have a quotient sheaf. So $\mathcal{S}(S_\tau)$ is an abelian category (see [GME] Sect. 1.4.4 for abelian category); that is, for a monomorphism $i : \mathcal{H} \to \mathcal{G}$ of abelian τ-sheaves, first making the naive presheaf $(\mathcal{G}/\mathcal{H})^P : U \mapsto \mathcal{G}/\mathcal{H}(U)$, its sheafication $\mathcal{G}/\mathcal{H} := ((\mathcal{G}/\mathcal{H})^P)^\#$ gives the sheaf quotient. This is why the category of abelian τ-sheaves is abelian. See Sect. 4.2.3 for monomorphisms.

This category $\mathcal{S}(S_\tau)$ has enough injectives (that is, any sheaf G can be embedded into an injective sheaf I on S_τ; see [ECH] Proposition III.1.1). Taking an injective resolution $G \hookrightarrow I^\bullet$, for any left exact functor F from $\mathcal{S}(S_\tau)$ into an abelian category $\mathcal{C}$, we can define the *right-derived (cohomology) functor* $R^iF : \mathcal{S}(S_\tau) \to \mathcal{C}$ given by $R^iF(G) = \frac{\mathrm{Ker}(I^i \to I^{i+1})}{\mathrm{Im}(I^{i-1}\to I^i)}$ for $i > 0$ [and we put $R^0F(G) = F(G)$]. The object $R^iF(G)$ is independent of the choice of $I^\bullet$ up to canonical isomorphisms. If $F : G \mapsto G(S) \in AB$, we write $H^i(S_\tau, G) = H^i_\tau(S, G)$ for $R^iF(G)$. By a standard argument, an exact sequence $0 \to H \to G \to K \to 0$ of sheaves over S_τ produces a *long exact sequence* with connection maps δ_q (see [GME] Sect. 1.10.1):

$$\begin{aligned} 0 \to R^0F(H) \to R^0F(G) \to R^0F(K) \xrightarrow{\delta_0} R^1F(H) \to R^1F(G) \\ \to R^1F(K) \to \cdots \to R^qF(K) \xrightarrow{\delta_q} R^{q+1}F(H) \to \cdots . \end{aligned} \tag{4.1.9}$$

If S is a smooth projective variety over $\mathrm{Spec}(\overline{k})$ for an algebraically closed field $\overline{k}$, we can take $F(G) = G(S) =: H^0(S, G)$. In this case, we write $H^q_\tau(S, G) := R^qF(G)$. For example, if $G = G_n$ given by $G_n(U) = \mathrm{Hom}_S(U, (\mathbb{Z}/l^n\mathbb{Z})_{/U})$ for the constant group scheme $(\mathbb{Z}/l^n\mathbb{Z})_{/U} = (\mathbb{Z}/l^n\mathbb{Z})_{/S} \times_S U$ for a prime $l \neq \mathrm{char}(\overline{k})$, $H_{\mathrm{et}}(S, \mathbb{Q}_l) := (\varprojlim_n H_{\mathrm{et}}(S, G_n)) \otimes_{\mathbb{Z}_l} \mathbb{Q}_l$ gives a cohomology group with coefficients in $\mathbb{Q}_l$ Weil predicted (see Sect. 3.1.3 for Weil's conjecture on algebro geometric cohomology). If, further, $\overline{k} = \overline{\mathbb{F}}_p$ or $\overline{\mathbb{Q}}$ and S is defined over a subfield $k \subset \overline{k}$, for $\overline{S} = S \times_{\mathrm{Spec}(k)} \mathrm{Spec}(\overline{k})$, $\mathrm{Gal}(\overline{k}/k)$ naturally acts on $\overline{S}_{\mathrm{et}}$ and hence on $H_{\mathrm{et}}(\overline{S}, \mathbb{Q}_l)$. This gives a natural generalization of the Galois module $T_l E$ for an elliptic curve $E_{/k}$ as $T_l E \otimes_{\mathbb{Z}_l} \mathbb{Q}_l$ is canonically a dual of $H^1_{\mathrm{et}}(\overline{E}, \mathbb{Q}_l)$. Further, by Deligne, if $k \subset \overline{\mathbb{Q}}$ is a number field, this Galois representation on $H^q_{\mathrm{et}}(\overline{S}, \mathbb{Q}_l)$ indexed by primes l gives rise to a compatible system of Galois representations whose $\mathfrak{p}$-Frobenius eigenvalues (for primes $\mathfrak{p} \nmid l$ of k) are Weil p-numbers of weight qf (if $|O_k/\mathfrak{p}| = p^f$) as long as S has smooth reduction modulo $\mathfrak{p}$ (i.e., S extends to a smooth scheme over the valuation ring of $\mathfrak{p}$). As in the case of Betti cohomology, one can identify $H^1_{\mathrm{et}}(S, G_n)$ with continuous group cohomology $H^1(\pi_1(S,s), \mathbb{Z}/l^n\mathbb{Z})$ [for a point $s \in S(\overline{k})$; see Sect. 4.2 for $\pi_1(S,s)$], and from this, it is easy to see $T_l E$ is the dual of $H^1(\overline{E}, \mathbb{Z}_l)$ (cf. (7.1.8)). If G is a coherent sheaf, $H^q_{\mathrm{Zar}}(S, G) := R^q F(G)$ is more classical coherent cohomology (Serre studied) with respect to Zariski topology. This does not give a Weil cohomology, as its cohomological dimension is known to be $\dim S$ (see [ALG] III.7), while the cohomology Weil predicted must have cohomological dimension $2 \dim S$.

4.1.11 Excellent Rings and Schemes

We recall some properties of excellent rings from [EGA] IV.7 without much proof. All rings in this section are commutative with multiplicative identity.

Though the definition itself is not very practical, we recall the definition of excellent rings for completeness. A ring is called *catenary* if, for any two prime ideals $\mathfrak{p} \subset \mathfrak{p}'$, we can find a sequence $\mathfrak{p} = \mathfrak{p}_0 \subsetneq \mathfrak{p}_1 \subsetneq \cdots \subsetneq \mathfrak{p}_{n-1} \subsetneq \mathfrak{p}_n = \mathfrak{p}'$ of prime ideals without any more subdivision (i.e., saturated) and any such sequence has constant (finite) length (that is, n). For example, a field is obviously catenary, and any polynomial ring $k[X_1, \ldots, X_n]$ $(0 \leq n < \infty)$ over a field is catenary (cf. [CRT] Exercise 5.1 and Theorem 31.4). By definition, if A is catenary, $A/\mathfrak{a}$ for any ideal $\mathfrak{a}$ of A is catenary, as we can pull back a prime ideal saturated sequence of $A/\mathfrak{a}$ to a prime ideal saturated sequence of A. Similarly, any localization of a catenary ring is catenary. The *Krull dimension* of a ring is the maximal length of saturated prime ideal chain. Any noetherian ring has finite dimension (cf. [CRT] Sect. 15).

A noetherian ring A is called *universally catenary* if every finitely generated A-algebra is catenary (so $k[X_1, \ldots, X_n]$ is universally catenary). Since any residue ring of a catenary ring is catenary (and any finitely generated ring over A is a residue ring of $A[X_1, \ldots, X_n]$), A is universally catenary if $A[X_1, \ldots, X_n]$ is catenary for all $n > 0$. Any localization of universally catenary ring is universally catenary.

A noetherian ring is called *regular* if, for the localization $A_\mathfrak{m}$ at any maximal ideal $\mathfrak{m}$ of A, $\dim A_\mathfrak{m} = \dim_{A/\mathfrak{m}} \mathfrak{m}/\mathfrak{m}^2$, where $\dim A_\mathfrak{m}$ is the Krull dimension of $A_\mathfrak{m}$. If A is regular, any localization is regular; in particular, $A_\mathfrak{p}$ is regular for any prime ideal $\mathfrak{p}$ ([CRT] Theorem 19.3). A regular ring A over a field k is called a *geometrically regular ring* over k if, for any finite extension field k'/k, $A \otimes_k k'$ is regular. A regular "local" ring is an integral domain ([CRT] Theorem 14.3). Since $\mathbb{F}_p(x) \otimes_{\mathbb{F}_p(x^p)} \mathbb{F}_p(x)$ for the rational function field $\mathbb{F}_p(x)$ has nilpotent element $x \otimes 1 - 1 \otimes x$, $\mathbb{F}_p(x)$ is not geometrically regular over $\mathbb{F}_p(x^p)$, though it is geometrically regular over itself.

For a prime ideal $\mathfrak{p}$ of a ring A, we write $k(\mathfrak{p})$ for the quotient field of the integral domain $A/\mathfrak{p}$. A morphism of rings $A \to A'$ is called a *regular morphism* if it is A-flat and $A' \otimes_A k(\mathfrak{p})$ is geometrically regular over $k(\mathfrak{p})$ for any prime ideal $\mathfrak{p}$ of A.

A ring A is called *excellent* if the following three conditions are satisfied:

(e1) A is universally catenary.
(e2) For any prime ideal $\mathfrak{p}$ of A, $A_{\mathfrak{p}} \to \widehat{A}_{\mathfrak{p}}$ is regular, where $\widehat{A}_{\mathfrak{p}}$ is the $\mathfrak{p}$-adic completion of $A_{\mathfrak{p}}$.
(e3) For any integral domain A' of finite type over A, there exists $0 \neq a \in A'$ such that $A'[\frac{1}{a}]$ is regular.

Remark 4.35 Clearly, the field K is regular and excellent. Since properties (e1–3) are stable under localization, any localization of excellent ring is excellent. It is known that a Dedekind domain with characteristic 0 quotient field is excellent. Any algebra of finite type over an excellent ring is excellent.

See [EGA] IV.7.8 for (the location of) a proof of nontrivial facts in the above remark.

A ring is called *semilocal* if it has finitely many maximal ideals. If A is semilocal, $\widehat{A} = \varprojlim_n A/J(R)^n$ denotes the $J(R)$-adic completion of A [here $J(R)$ is the *Jacobson radical*, which is the intersection of all maximal ideals of A]. We have $\widehat{A} = \prod_i \widehat{A}_{\mathfrak{m}_i}$, writing its maximal ideals $\mathfrak{m}_1, \dots, \mathfrak{m}_n$ as $J(R) = \bigcap_i \mathfrak{m}_i$ (cf. [CRT] Theorem 8.15). We need the following fact later (whose proof can be found in [EGA] IV.7.6.4).

Theorem 4.36 (Zariski–Nagata) *Let A be a semilocal noetherian ring. Then the following three conditions are equivalent:*

1. *For any reduced finite A-algebra A', $\widehat{A'} = A' \otimes_A \widehat{A}$ is reduced.*
2. *For any prime ideal $\mathfrak{p}$ of A, $\widehat{A/\mathfrak{p}} = A/\mathfrak{p} \otimes_A \widehat{A}$ is reduced and is a separable algebra over $k(\mathfrak{p})$.*
3. *For any prime ideal $\mathfrak{p}$ of A, $\widehat{A/\mathfrak{p}} \otimes_A k(\mathfrak{p})$ is a separable algebra over $k(\mathfrak{p})$.*

For a field K, a K-algebra A is separable if $A \otimes_K L$ is reduced for any finite extension L/K.

Corollary 4.37 *If A is an excellent reduced noetherian local ring with maximal ideal $\mathfrak{m}$, then $\widehat{A} = \varprojlim_n A/\mathfrak{m}^n$ is reduced.*

By condition (e2), $\widehat{A/\mathfrak{p}} \otimes_A k(\mathfrak{p})$ for all prime ideals $\mathfrak{p}$ of A is geometrically regular over $k(\mathfrak{p})$; in particular, it is reduced and separable over $k(\mathfrak{p})$. Thus, by the theorem, $\widehat{A}$ is reduced.

4.2 Fundamental Groups

This section gives a sketch of the theory of algebraic fundamental groups. The simplest example of algebraic fundamental groups is the Galois group; so we will start with a summary of Galois theory, which is a fundamental group of (the spectrum of) a field. Then we go on to define general fundamental groups of a scheme.

4.2.1 Galois Extensions of Infinite Degree

We study the Galois theory when $\dim_K L$ is infinite for a Galois extension L/K. First, we recall Galois' fundamental theorem when L/K is finite.

Theorem 4.38 (E. Galois) *Suppose that L/K is a finite Galois extension. For an intermediate field $L/M/K$, L/M is a Galois extension, and we have*

(1) *There is one-to-one onto correspondence*

$$\{M : \textit{intermediate field} | L/M/K\} \leftrightarrow \{H : \textit{subgroup of } \mathrm{Gal}(L/K)\}$$

given by $M \mapsto \mathrm{Gal}(L/M)$ and $H \mapsto L^H = \{x \in L | x = \sigma(x) \forall \sigma \in H\}$.

(2) *For two intermediate fields M, M', we have*
- $M \supset M' \iff \mathrm{Gal}(L/M) \subset \mathrm{Gal}(L/M')$.
- $\mathrm{Gal}(M \cap M')$ *is the subgroup of* $\mathrm{Gal}(L/K)$ *generated by the two subgroups* $\mathrm{Gal}(L/M)$ *and* $\mathrm{Gal}(L/M')$.
- $\mathrm{Gal}(L/MM') = \mathrm{Gal}(L/M) \cap \mathrm{Gal}(L/M')$.
- *For* $\sigma \in \mathrm{Gal}(L/K)$, $\mathrm{Gal}(L/\sigma(M)) = \sigma \mathrm{Gal}(L/M)\sigma^{-1}$.

(3) *M/K is a Galois extension if and only if* $\mathrm{Gal}(L/M)$ *is a normal subgroup in the Galois group* $\mathrm{Gal}(L/K)$. *In particular, by $\sigma \mapsto \sigma|_M$, we have an isomorphism:* $\mathrm{Gal}(L/K)/\mathrm{Gal}(L/M) \cong \mathrm{Gal}(M/K)$.

We admit this theorem, just referring the reader to any book on elementary Galois theory (for example, [TCF]).

What we want to do in this subsection is to generalize the above theorem to an infinite Galois extension. Hereafter we suppose that L/K is an infinite Galois extension. For an intermediate extension $L/M/K$, the *Galois closure* M^{gal} of M over K is the intersection of all Galois extensions of K inside L containing M; so M^{gal}/K is a Galois extension. Since each element $\xi \in L$ satisfies a polynomial equation of finite degree with coefficients in K, there are only finitely many conjugates of ξ. Thus, the Galois closure $K[\xi]^{gal}$ of $K[\xi]$ over K is a finite Galois extension of K, and from Theorem 4.38 (2), we get

$$L = \bigcup_{\xi \in L} K[\xi]^{gal} \Leftrightarrow \bigcap_{\xi \in L} \mathrm{Gal}(L/K[\xi]^{gal}) = \{\mathrm{id}\}. \tag{4.2.1}$$

We first generalize the third assertion of the above theorem.

Lemma 4.39 *Suppose M is an intermediate Galois extension of a Galois extension L/K. Then the group* $\mathrm{Gal}(L/M)$ *is a normal subgroup of* $\mathrm{Gal}(L/K)$, *and $\sigma \mapsto \sigma|_M$ induces an isomorphism* $\mathrm{Gal}(L/K)/\mathrm{Gal}(L/M) \cong \mathrm{Gal}(M/K)$. *In particular, if M/K is a finite extension, then we have the equality* $[\mathrm{Gal}(L/K) : \mathrm{Gal}(L/M)] = [M : K] < \infty$.

Proof. If M/K is a finite extension, for any finite Galois extension L' with $L \supset L' \supset M$, we can extend $\sigma \in \mathrm{Gal}(M/K)$ to $\tau \in \mathrm{Gal}(L'/K)$ [assertion (3) of Theorem 4.38]. Since L is a union of finite Galois extensions of L', $\sigma \in \mathrm{Gal}(L'/K)$ extends to $\tau \in \mathrm{Gal}(L/K)$. If M/K is an infinite extension, writing M as a union of finite Galois extensions M'/K and applying the above fact to M'/K, we find that any $\sigma \in \mathrm{Gal}(M/K)$ can be extended to $\tau \in \mathrm{Gal}(L/K)$. Thus, $\sigma \mapsto \sigma|_M$ is a surjective homomorphism of $\mathrm{Gal}(L/K)$ onto $\mathrm{Gal}(M/K)$. Its kernel is given by $\mathrm{Gal}(L/M)$, which is normal.

Basically, the same proof as above yields the following result, which appears stronger than the lemma.

Corollary 4.40 *Let M be an intermediate extension of L/K. Then $\sigma \mapsto \sigma|_M$ induces an isomorphism* $\mathrm{Gal}(L/K)/\mathrm{Gal}(L/M) \cong \mathrm{Hom}_K(M, L)$, *where the set* $\mathrm{Hom}_K(M, L)$ *is the collection of K-linear field homomorphisms of M into L.*

Proving this corollary is left to the reader as an exercise (Exercise 4.47 (1)).

We now give a topology on the group $\mathrm{Gal}(L/K)$ in the following way. Define a fundamental system of open neighborhoods of the identity by the collection of subgroups of the form $\mathrm{Gal}(L/M)$ for a finite extension M/K. If $H = \mathrm{Gal}(L/M)$ for a finite extension M/K, the Galois closure M^{gal} is still a finite Galois extension of K; so we find a normal open subgroup $\mathrm{Gal}(L/M^{gal})$ inside any open subgroup H. Thus, we may define the system to be the set $\mathfrak{U}$ of all normal subgroups $N = \mathrm{Gal}(L/M)$ for finite Galois extensions M/K inside L. For each $\sigma \in \mathrm{Gal}(L/K)$, we define the system of neighborhoods of σ to be $\sigma\mathfrak{U} = \mathfrak{U}\sigma = \{\sigma N = N\sigma | N \in \mathfrak{U}\}$. By this, if $\mathfrak{U}$ satisfies the axiom of a fundamental system of open neighborhoods of a point, G becomes a topological group, that is, G is a group with a topology for which the multiplication $(a, b) \mapsto ab$ and inverse $a \mapsto a^{-1}$ are continuous. The axiom is verified by the next result:

Lemma 4.41 *We have*

(1) *For $\sigma \neq 1$, there exists $N \in \mathfrak{U}$ with $\sigma \notin N$ (so, G is a Hausdorff group).*
(2) $N, H \in \mathfrak{U} \Rightarrow N \cap H \in \mathfrak{U}$.

Proof. By (4.2.1), $L = \bigcup_M M$ for finite Galois extensions M/K. If $\sigma \in \bigcap_M \mathrm{Gal}(L/M)$, σ is the identity over all M, and so $\sigma = 1$ in $\mathrm{Gal}(L/K)$. Thus, there exists a finite Galois extension M/K inside L such that $\sigma \notin N = \mathrm{Gal}(L/M)$ if $\sigma \neq 1$. By definition, we have $N \in \mathfrak{U}$, and so assertion (1) follows.

Write $N = \mathrm{Gal}(L/M)$ and $H = \mathrm{Gal}(L/M')$. Then MM' is a composite of two finite Galois extensions and so is a finite Galois extension. By definition, σ is trivial over $MM' \Leftrightarrow \sigma$ is trivial over M and M'. Then by Lemma 4.39, we have $\mathrm{Gal}(L/MM') = N \cap H$, and hence $N \cap H \in \mathfrak{U}$, which proves (2).

The topology we have defined was first introduced by Krull and is therefore called the *Krull topology* on $\mathrm{Gal}(L/K)$.

Proposition 4.42 *For an infinite Galois extension L/K, $\mathrm{Gal}(L/K)$ is a compact group under the Krull topology.*

Proof. Since $\mathrm{Gal}(L/K)$ is Hausdorff, we need to show that for any given infinite subset Σ in $\mathrm{Gal}(L/K)$, there exists an element $\sigma \in \mathrm{Gal}(L/K)$ such that $\Sigma \cap H\sigma$ is infinite for all $H \in \mathfrak{U}$; that is, Σ has a limit point $\sigma \in \mathrm{Gal}(L/K)$. Since $H = \mathrm{Gal}(L/M)$ for a finite Galois extension M/K, there are finitely many automorphisms of M/K. Thus, there exists an infinite subset Σ_H of Σ that induces a single automorphism σ_M on M. For any extension $\sigma' \in \Sigma_H$ of σ_M to L, the coset $H\sigma'$ is the collection of $\tau \in \mathrm{Gal}(L/K)$ (by Lemma 4.39) with $\tau|_M = \sigma_M$. We make the construction of σ_M compatible with an increasing sequence of Galois extensions M_j of K inside L. In other words, we find a sequence of finite Galois extensions M_j/K $(j = 1, 2, \ldots)$ with $K \subset M_j \subset M_{j+1} \subset L$ and $\sigma_j = \sigma_{M_j} \in \mathrm{Gal}(M_j/K)$ such that $L = \bigcup_j M_j$ and $\sigma_{j+1}|_{M_j} = \sigma_j$ for all j. We have found an element $\sigma \in \mathrm{Gal}(L/K)$ such that $\sigma|_{M_j} = \sigma_{M_j}$ for all j. This σ is a limit point of Σ.

Corollary 4.43 *If M is an intermediate field of an infinite Galois extension L/K, $\mathrm{Gal}(L/M)$ is a closed subgroup of $\mathrm{Gal}(L/K)$.*

Proof. For any finite Galois extension M'/K inside L, we have by definition the following identity: $i^{-1}(\mathrm{Gal}(L/MM')) = \mathrm{Gal}(L/MM')$ for the inclusion map $i : \mathrm{Gal}(L/M) \hookrightarrow \mathrm{Gal}(L/K)$. For any $N \in \mathfrak{U}$, we find a finite Galois extension M'/K inside L such that $N = \mathrm{Gal}(L/M')$. Since MM'/M is a finite Galois extension, $i^{-1}(N) = \mathrm{Gal}(L/MM')$ is open in $\mathrm{Gal}(L/M)$, and so i is continuous. The image of a compact set by a continuous map is compact; in particular, it is closed.

We now prove a generalization of Theorem 4.38.

Theorem 4.44 *Let L/K be an infinite Galois extension. We have the following canonical one-to-one onto correspondence:*

$$\{\textit{intermediate fields } M \textit{ of } L/K\} \leftrightarrow \{\textit{closed subgroups } H \textit{ of } \mathrm{Gal}(L/K)\}$$

induced by $M \mapsto \mathrm{Gal}(L/M)$ and $H \mapsto L^H$. Moreover,

(1) *Every open subgroup of $\mathrm{Gal}(L/K)$ is closed, and open subgroups correspond to finite extensions L^H/K.*

(2) *Let* $\{N_i\}_{i\in I}$ *for an index set* I *be a collection of closed subgroups of the Galois group* $\mathrm{Gal}(L/K)$. *Writing* $N_i = \mathrm{Gal}(L/K_i)$, *for the composite* M *of all* $\{K_i\}_{i\in I}$, *we have* $\bigcap_{i\in I} N_i = \mathrm{Gal}(L/M)$.
(3) *If* $\{K_i\}_{i\in I}$ *is a collection of intermediate fields of* L/K, *then the group* $\mathrm{Gal}(L/\bigcap_{i\in I} K_i)$ *is the closure of the subgroup in* $\mathrm{Gal}(L/K)$ *generated by* $\mathrm{Gal}(L/K_i)$ *for all* i.
(4) *For* $\sigma \in \mathrm{Gal}(L/K)$, *we have* $\mathrm{Gal}(L/\sigma(M)) = \sigma \cdot \mathrm{Gal}(L/M)\sigma^{-1}$.
(5) *If* M/K *is a Galois extension inside* L, *then* $\mathrm{Gal}(L/M)$ *is a normal closed subgroup and* $\mathrm{Gal}(L/K)/\mathrm{Gal}(L/M) \cong \mathrm{Gal}(M/K)$ *induced by* $\sigma \mapsto \sigma|_M$ *is an isomorphism of topological groups.*

Proof. Let $\mathfrak{M}$ be the set of all intermediate fields of L/K and let $\mathfrak{H}$ be the set of all closed subgroups of $\mathrm{Gal}(L/K)$. For any $M \in \mathfrak{M}$, $\mathrm{Gal}(L/M)$ is a closed subgroup of $\mathrm{Gal}(L/K)$ by Corollary 4.43; hence, $M \mapsto \mathrm{Gal}(L/M)$ defines a map $\mathfrak{M} \to \mathfrak{H}$. We define a reverse map $\mathfrak{H} \to \mathfrak{M}$ by $H \mapsto L^H$.

We first prove that $\mathrm{Gal}(L/L^H) = H$. Since $\mathrm{Gal}(L/L^H) \subset \mathrm{Gal}(L/K)$ is the collection of all automorphisms of L fixing L^H, we confirm that $H \subset \mathrm{Gal}(L/L^H)$. Conversely, for each finite Galois extension M/L^H inside L, H acts on M nontrivially if $M \neq L^H$, and the image under $\sigma \mapsto \sigma|_M$ in $\mathrm{Gal}(M/L^H)$ is a subgroup H' of $\mathrm{Gal}(M/L^H)$ and $L^H = M^{H'}$. By Theorem 4.38 (1), $H' = \mathrm{Gal}(M/L^H)$. In other words, for each $\sigma \in \mathrm{Gal}(L/L^H)$, we find $h_M \in H$ such that $h_M|_M = \sigma|_M$. We consider the infinite set $\Sigma = \{h_M\}_M \subset H$ with M running through all finite Galois extensions M/L^H. Since H is a closed subgroup of a compact group $\mathrm{Gal}(L/K)$, H itself is compact. Then Σ has a unique accumulation point h. Then for each finite Galois extension M/L^H, $\sigma|_M = h|_M$. Since $L = \bigcup_M M$ for finite Galois extensions of M/L^H, we find that $h = \sigma$ and $H = \mathrm{Gal}(L/L^H)$.

We now prove $L^{\mathrm{Gal}(L/M)} = M$. By Corollary 4.43, $H = \mathrm{Gal}(L/M) \in \mathfrak{H}$. By definition, $M \subset L^H$. Supposing that $L^H \neq M$, we try to get a contradiction. Pick $\xi \in L^H - M$. Then $M[\xi]/M$ is a finite extension. Thus, $M[\xi]^{gal} \subset L$ because L/K is normal. Since $\xi \notin M$, $\mathrm{Gal}(M[\xi]^{gal}/M) \neq \{1\}$. Pick $\sigma \in \mathrm{Gal}(M[\xi]^{gal}/M)$ with $\sigma(\xi) \neq \xi$. Then by Lemma 4.39, we find $\tau \in \mathrm{Gal}(L/M) = H$ such that $\tau(\xi) = \sigma(\xi) \neq \xi$. This is wrong since $\sigma(\xi)$ has to be ξ because $\xi \in L^H$. Thus, we find that $L^{\mathrm{Gal}(L/M)} = M$, and the correspondence is one-to-one and onto.

For an open subgroup H, $\mathrm{Gal}(L/K)/H$ is discrete and compact; thus, it is finite [see Exercise 4.47 (4)]. Since H is the kernel of the continuous map $\mathrm{Gal}(L/K) \twoheadrightarrow \mathrm{Gal}(L/K)/H$, H is closed. By $H = \mathrm{Gal}(L/L^H)$, we see from Lemma 4.39 and Corollary 4.40 that $\mathrm{Gal}(L/K)/H = \mathrm{Hom}_K(L^H, L)$, and hence L^H/K is a finite extension by Corollary 4.40. This shows (1).

Assertion (2) follows from the fact that $\sigma \in \mathrm{Gal}(L/K)$ is the identity on every K_i if and only if σ is the identity on the composite of all K_i [Exercise 4.47 (2)].

Let $\mathrm{Gal}(L/\bigcap_{i\in I} K_i) = H'$. Then, by definition, H' is a closed subgroup containing all $H_i = \mathrm{Gal}(L/K_i)$, and it contains the closure H of the subgroup generated by H_i for all i. Since H is a closed subgroup (so in $\mathfrak{H}$) containing H_i, we see $L^H \subset K_i$ and hence $L^H \subset \bigcap_{i \in I} K_i$. Thus, $H \supset H' = \mathrm{Gal}(L/\cap_{i \in I} K_i)$. This shows $H = H'$.

Assertion (4) is obvious from the definition of the Galois group, and the last assertion follows from Lemma 4.39.

4.2.2 Automorphism Group of a Field

We study $\mathrm{Aut}(L/K)$ for an arbitrary extension L/K of fields. Such a generalization of Galois' theory was given in [IAT] 6.3. An intermediate extension M of L/K is called *finite type* over K if there are finitely many elements $x_1, \ldots, x_n$ in M such that M is the quotient field of the ring $K[x_1, \ldots, x_n] \subset M$. We define a fundamental system $\mathfrak{U}$ of open neighborhoods of 1 on $\mathrm{Aut}(L/K)$ by

$$\mathfrak{U} = \{H = \mathrm{Aut}(L/M) | M/K \text{ is of finite type}\}. \tag{4.2.2}$$

Since a composite of two intermediate fields of finite type is again of finite type, $\mathfrak{U}$ satisfies the axiom of the fundamental system of neighborhoods [see the proof of Lemma 4.41 and Exercise 4.47 (3)]. Then define a system of neighborhoods of $\sigma \in \mathrm{Aut}(L/K)$ by $\sigma\mathfrak{U} \cup \mathfrak{U}\sigma$. This topology gives a topological group structure on $\mathrm{Aut}(L/K)$. Let $\mathfrak{H}$ be the set of all compact subgroups of $\mathrm{Aut}(L/K)$, and let $\mathfrak{M}$ be the set of all intermediate fields M in L/K such that L/M is a Galois extension. Then we have the following:

Proposition 4.45 (Jacobson) *We have a canonical bijective correspondence $\mathfrak{M} \cong \mathfrak{H}$ given by $M \mapsto \mathrm{Gal}(L/M)$ and $H \mapsto L^H$.*

Proof. Pick a compact subgroup H of $\mathrm{Aut}(L/K)$. Then for each $\xi \in L$, the topological group $\mathrm{Aut}(L/K(\xi))$ is an open subgroup of $\mathrm{Aut}(L/K)$. Thus, $H' = H \cap \mathrm{Gal}(L/K(\xi))$ is an open subgroup of H. In particular, H/H' is compact and discrete, and so it is finite [Exercise 4.47 (4)]. Since the polynomial $f(X) = \prod_{h \in H/H'}(X - h(\xi))$ has coefficients in L^H, ξ is algebraic over L^H. Since $h(\xi) = h'(\xi) \Leftrightarrow hH' = h'H'$, the roots of $f(X)$ are all distinct, and so $L^H(\xi)/L^H$ is a separable extension. Since $h(\xi) \in L$ for all $h \in H$, all the roots of $f(X)$ are in L. The Galois closure M of $L^H(\xi)$ over L^H is a Galois extension of L^H inside L. Since $L = \bigcup_{\xi \in L} L^H(\xi)$, we find that L/L^H is a Galois extension. Then by Theorem 4.44, we find that $H = \mathrm{Gal}(L/L^H)$. Starting from $M \in \mathfrak{M}$, we find that $\mathrm{Gal}(L/M) \in \mathfrak{H}$ and $M = L^{\mathrm{Gal}(L/M)}$ by Theorem 4.44.

A field extension L/K is of finite type if L is generated over K by finitely many elements (as a field). Let $\mathfrak{H}^o$ be the set of all open compact subgroups of $\mathrm{Aut}(L/K)$ and $\mathfrak{M}^o$ be the subset of $\mathfrak{M}$ made up of fields M that are of finite type over K.

Corollary 4.46 *If $\mathfrak{M}^o$ is nonempty, then* $\mathrm{Aut}(L/K)$ *is locally compact, and the one-to-one onto correspondence in* Proposition 4.45 *induces $\mathfrak{M}^o \cong \mathfrak{H}^o$.*

Proof. If $\mathfrak{M}^o$ is nonempty, pick $M \in \mathfrak{M}^o$. Then the identity of $\mathrm{Aut}(L/K)$ has an open neighborhood $\mathrm{Gal}(L/M)$, which is also compact by Proposition 4.42. Translating this neighborhood to any point σ on $\mathrm{Aut}(L/K)$ through multiplication by σ, every point has an open-compact neighborhood, and so $\mathrm{Aut}(L/K)$ is locally compact. We need to show that L^H/K is of finite type over K if $H \in \mathfrak{H}^o$. Consider $F = ML^H$. Then L/F is a Galois extension; hence, $H' = \mathrm{Gal}(L/F)$ is a compact subgroup of $\mathrm{Gal}(L/M)$. Since F/M is a finite extension, F is of finite type over M and hence over K. Thus, the subfield L^H of a field F is of finite type over K.

Exercise 4.47 *1. Prove Corollary 4.40.*
2. Give a detailed proof of assertions (2) and (4) of Theorem 4.44.
3. Prove that the Krull topology on $\mathrm{Aut}(L/K)$ *given by* (4.2.2) *is well defined.*
4. Prove that a compact and discrete set is a finite set.

4.2.3 Galois Theory in Categorical Setting

Following closely the exposition in [PAF] Sect. 4.4, we give a short description of how one can axiomatize Galois theory (the idea is due to Grothendieck, and the main reference is [SGA] V.4). This subsection interprets the result in Sect. 4.2.1 in terms of category theory.

We recall some terminology from category theory. Let $\mathcal{C}$ be a category. A morphism $X \xrightarrow{f} Y$ in this category is called an *epimorphism* if, for any third object Z and morphisms $Y \underset{h}{\overset{g}{\rightrightarrows}} Z$, $g \circ f = h \circ f \Leftrightarrow g = h$ (so f is a right unit). In other words, for any other object Z in $\mathcal{C}$, the pullback map $\mathrm{Hom}_{\mathcal{C}}(Y, Z) \xrightarrow{f^*} \mathrm{Hom}_{\mathcal{C}}(X, Z)$ is injective if f is an epimorphism. Similarly, a morphism $X \xrightarrow{f} Y$ in this category is called a *monomorphism* if, for any third object Z and morphisms $Z \underset{h}{\overset{g}{\rightrightarrows}} X$, $f \circ g = f \circ h \Leftrightarrow g = h$ (so f is a left unit). In many categories, for example, $SETS$ and AB, monomorphisms are injections, and epimorphisms are surjections, but, for example, in the abelian category of sheaves, say, $S(X)$ and $\mathcal{S}(X_\tau)$, an epimorphism may not be a surjection of sections of sheaves.

Recall that a covariant (resp., contravariant) functor $F : \mathcal{C} \to SETS$ is called *representable* if there is an object $\mathcal{F}$ in $\mathcal{C}$ such that we have a system of functorial bijections $\phi_X : F(X) \to \mathrm{Hom}_{\mathcal{C}}(\mathcal{F}, X)$ [resp., $\phi_X : F(X) \to \mathrm{Hom}_{\mathcal{C}}(X, \mathcal{F})$]. Here the word "functorial" means that $f_* \circ \phi_X = \phi_Y \circ F(f)$ in the covariant case and $f^* \circ \phi_Y = \phi_X \circ F(f)$ in the contravariant case for any morphism $X \xrightarrow{f} Y$ in $\mathcal{C}$, where $f_*(g) = f \circ g$ for $g \in \mathrm{Hom}_{\mathcal{C}}(\mathcal{F}, X)$ and $f^*(h) = h \circ f$ for $h \in \mathrm{Hom}_{\mathcal{C}}(Y, \mathcal{F})$. The object in $\mathcal{C}$ representing a functor is unique up to isomorphisms (a characterization by a universal property; see Lemma 4.17).

If a covariant functor $F_{\mathbf{0}}$ sending every object in $\mathcal{C}$ to a singleton $\{\mathbf{0}\}$ is representable in $\mathcal{C}$, the object representing the functor is called an *initial* object in $\mathcal{C}$. Similarly, if a contravariant functor $F^{\mathbf{0}}$ sending every object in $\mathcal{C}$ to a singleton $\{\mathbf{0}\}$ is representable in $\mathcal{C}$, the object representing the functor is called a *final* object in $\mathcal{C}$.

If the covariant functor $T \mapsto \mathrm{Hom}_{\mathcal{C}}(X,T) \times \mathrm{Hom}_{\mathcal{C}}(Y,T)$ for two given objects X, Y of $\mathcal{C}$ is representable by an object of $\mathcal{C}$, the object is written as $X \oplus Y$ and is called the *direct sum*, and we have inclusions $i_X : X \to X \oplus Y$ and $i_Y : Y \to X \oplus Y$ such that $\mathrm{Hom}_{\mathcal{C}}(X \oplus Y, T) \cong \mathrm{Hom}_{\mathcal{C}}(X,T) \times \mathrm{Hom}_{\mathcal{C}}(Y,T)$ is given by $\phi \mapsto (\phi \circ i_X, \phi \circ i_Y)$ (see [MFG] Sect. 4.4). If the contravariant functor $T \mapsto \mathrm{Hom}_{\mathcal{C}}(T,X) \times \mathrm{Hom}_{\mathcal{C}}(T,Y)$ for two given objects X, Y of $\mathcal{C}$ is representable by an object of $\mathcal{C}$, the object is written as $X \times Y$ and is called a *direct product*, and we have projections $p_X : X \times Y \to X$ and $p_Y : X \times Y \to Y$ such that the isomorphism $\mathrm{Hom}_{\mathcal{C}}(T, X \times Y) \cong \mathrm{Hom}_{\mathcal{C}}(T,X) \times \mathrm{Hom}_{\mathcal{C}}(T,Y)$ is given by $\phi \mapsto (p_X \circ \phi, p_Y \circ \phi)$ (see [MFG] Sect. 4.4).

For a given object S in $\mathcal{C}$, we write $\mathcal{C}_{/S}$ for the category of arrows $X \xrightarrow{f} S$. Morphisms from $X \xrightarrow{f} S$ into $Y \xrightarrow{g} S$ are given by morphisms $h : X \to Y$ with $g \circ h = f$. If we apply the above construction to $\mathcal{C}_{/S}$, the direct product in $\mathcal{C}_{/S}$ (if it exists) is the *fibered product* $X \times_S Y$ of X and Y over S (see (4.1.4)). So in this case, the contravariant functor $T \mapsto \mathrm{Hom}_{\mathcal{C}_{/S}}(T,X) \times \mathrm{Hom}_{\mathcal{C}_{/S}}(T,Y)$ is representable by $X \times_S Y$ in $\mathcal{C}_{/S}$. If $\mathcal{C} = SETS$, we see

$$X \times_S Y = \left\{(x,y) \in X \times Y \middle| f(x) = g(y)\right\}.$$

Dualizing the above construction, for an object S of $\mathcal{C}$, we consider a category $\mathcal{C}^S$ whose objects are arrows $S \xrightarrow{f} X$ and $\phi \in \mathrm{Hom}(S \xrightarrow{f} X, S \xrightarrow{g} Y)$ is a morphism $\phi : X \to Y$ with $\phi \circ f = g$. Then $X \oplus^S Y$ in $\mathcal{C}^S$ if it exists represents the covariant functor $T \mapsto \mathrm{Hom}_{\mathcal{C}^S}(X,T) \times \mathrm{Hom}_{\mathcal{C}^S}(Y,T)$. The object $X \oplus^S Y$ is called the *fibered sum* or *push-out* or *contracted product* of X and Y under S. If $\mathcal{C} = SETS$, we see easily that $X \oplus^S Y$ is the quotient of the disjoint union $X \sqcup Y$ under the relation that $x \sim y \Leftrightarrow x = f(s)$ and $y = g(s)$ (see [MFG] Sect. 4.2.1).

An epimorphism $X \xrightarrow{f} Y$ in $\mathcal{C}$ is called a *strict epimorphism* if, for any other object Z of $\mathcal{C}$,

$$\mathrm{Im}(f^* : \mathrm{Hom}_{\mathcal{C}}(Y,Z) \to \mathrm{Hom}_{\mathcal{C}}(X,Z)) = \left\{g \in \mathrm{Hom}_{\mathcal{C}}(X,Z) \middle| p_1^* g = p_2^* g\right\},$$

where $p_j : X \times_Y X \to X$ are the two projections. If this is the case, we call the following sequence *exact*: $X \times_Y X \overset{p_1,p_2}{\rightrightarrows} X \xrightarrow{f} Y$. The notion of strict epimorphisms is stronger than that of epimorphisms. For example, $i : \mathrm{Spec}(\mathbb{Q}) \to \mathrm{Spec}(\mathbb{Z})$ is an epimorphism in SCH (because it is dominant). Since $\mathbb{Q} \otimes_{\mathbb{Z}} \mathbb{Q} = \mathbb{Q}$, if i were a strict epimorphism, the empty set $\mathrm{Hom}_{\mathrm{alg}}(\mathbb{Q}, \mathbb{Z})$ would have to be identical to the singleton $\mathrm{Hom}_{\mathrm{alg}}(\mathbb{Q}, \mathbb{Q})$. This is a contradiction, and hence i is not a strict epimorphism. A faithfully flat morphism $X \to Y$ of finite type is a strict epimorphism (see [ECH] Theorem I.2.17) in the category of schemes. In the category $SETS$, the notion of strict epimorphisms and that of epimorphisms coincide and are equivalent to surjectivity.

Let $G \subset \mathrm{Aut}_{\mathcal{C}}(X)$ be a finite subgroup for an object X of $\mathcal{C}$. Write $\mathrm{Hom}_{\mathcal{C}}(X,T)^G$ for the set of morphisms $X \xrightarrow{f} T$ with $f \circ g = f$ for all $g \in G$. If the covariant functor $T \mapsto \mathrm{Hom}_{\mathcal{C}}(X,T)^G$ is representable by an object Y of $\mathcal{C}$ with a unique morphism $\pi : X \to Y$ inducing $\pi^* : \mathrm{Hom}_{\mathcal{C}}(Y,T) \cong \mathrm{Hom}_{\mathcal{C}}(X,T)^G$, we call Y the *categorical quotient* of X by G and the morphism $\pi : X \to Y$ the quotient morphism (see [GME] Sect. 1.8.2).

Here is an example of $\mathcal{C}$ with fibered products and fibered sums. We start with a profinite compact group Π. A typical example for Π is the Galois group $\mathrm{Gal}(K/k)$ of a (possibly infinite) Galois extension K/k. We denote $\mathcal{C}(\Pi)$ for the category of finite sets on which Π acts continuously (from the left). Thus, for each object X in $\mathcal{C}(\Pi)$, the stabilizer in Π of each point $x \in X$ is an open subgroup of Π. For X, Y in $\mathcal{C}(\Pi)$, a map $f : X \to Y$ is a morphism of $\mathcal{C}(\Pi)$ if $f(\pi(x)) = \pi(f(x))$ for all $x \in X$ and $\pi \in \Pi$. The collection of all such maps is written as $\mathrm{Hom}_{\Pi}(X,Y)$.

Let $\mathbf{0}$ be the set of a single point on which Π acts trivially. Then $\mathrm{Hom}_{\Pi}(X,\mathbf{0})$ is made of one map taking everything in X to $\mathbf{0}$. In other words, $\mathbf{0}$ is the final object. The empty set $\emptyset$ is considered to be an object in $\mathcal{C}(\Pi)$, which gives an initial object in $\mathcal{C}(\Pi)$. Since $\mathcal{C}(\Pi)$ is a subcategory of $SETS$, the fibered product $X \times_S Y$ for $X \xrightarrow{f} S$ and $Y \xrightarrow{g} S$ is given by $X \times_S Y = \{(x,y) \in X \times Y \,|\, f(x) = g(y)\}$ with two projections p_X and p_Y induced by two projections $X \times Y$ onto X and Y. The fibered sum of $S \xrightarrow{f} X$ and $S \xrightarrow{g} Y$ is given by the set-theoretic fibered sum $X \oplus^S Y$ already described, and in particular, the direct sum is the fibered sum for an initial object S and is $X \sqcup Y$ with two obvious inclusions i_X and i_Y. If $G \subset \mathrm{Aut}_{\mathcal{C}(\Pi)}(X)$ is a finite group, the set-theoretic quotient X/G gives the categorical quotient in $\mathcal{C}(\Pi)$. If $X \xrightarrow{f} Y$ is a morphism in $\mathcal{C}(\Pi)$, it factors as $X \twoheadrightarrow \mathrm{Im}(f) \hookrightarrow Y$. Here $p : X \twoheadrightarrow \mathrm{Im}(f)$ is a surjection by definition and hence is a strict epimorphism, and $i : \mathrm{Im}(f) \hookrightarrow Y$ is a monomorphism inducing a direct sum decomposition $Y = (Y - \mathrm{Im}(f)) \sqcup \mathrm{Im}(f)$. In summary, we proved the following three properties for $\mathcal{C} = \mathcal{C}(\Pi)$.

(C1) $\mathcal{C}$ has a final object and a fibered product of any two morphisms $X \to S$ and $Y \to S$ in $\mathcal{C}$ over S.

(C2) $\mathcal{C}$ has an initial object and a direct sum of any two objects in $\mathcal{C}$. For each object X of $\mathcal{C}$ and each finite subgroup G of $\mathrm{Aut}_{\mathcal{C}}(X)$, the categorical quotient X/G exists.

(C3) Each morphism $X \xrightarrow{f} Y$ in $\mathcal{C}$ factors as $X \xrightarrow{f'} X' \xrightarrow{f''} Y$ so that f' is a strict epimorphism and f'' is a monomorphism.

A category $\mathcal{C}$ satisfying the three conditions (C1–3) is called a *Galois category*. A covariant functor $F : \mathcal{C} \to \mathcal{C}'$ is called *left exact* if it takes monomorphisms of $\mathcal{C}$ to monomorphisms of $\mathcal{C}'$ and $F(X \times_S Y) = F(X) \times_{F(S)} F(Y)$. We note that $F(X) \times_{F(S)} F(Y) = \{(x,y) \in F(X) \times F(Y) \,|\, F(f)(x) = F(g)(y)\}$ for $X \xrightarrow{f} S$ and $Y \xrightarrow{g} S$ if $\mathcal{C}' = SETS$. Similarly, a covariant functor $F : \mathcal{C} \to \mathcal{C}'$ is called *right exact* if it takes epimorphisms of $\mathcal{C}$ to epimorphisms of $\mathcal{C}'$ and commutes with fibered sums. A left exact covariant functor $F : \mathcal{C} \to \mathcal{C}'$ is called *exact* if it preserves monomorphisms, epimorphisms, and exact sequences. The inclusion functor $F : \mathcal{C}(\Pi) \hookrightarrow SETS$ obviously satisfies the following three conditions:

(f1) F is left exact and has values in the full subcategory of finite sets in $SETS$.

(f2) $F(X \oplus Y) = F(X) \sqcup F(Y)$ for objects X, Y in $\mathcal{C}$; $F(f)$ is a surjection if $F : X \to Y$ is a strict epimorphism; and if $X \xrightarrow{\pi} Y$ is the categorical quotient in $\mathcal{C}$ by a finite group $G \subset \mathrm{Aut}_{\mathcal{C}}(X)$, $F(X) \xrightarrow{F(\pi)} F(Y)$ is a categorical quotient in $SETS$ by the same group G sitting inside $\mathrm{Aut}(F(X))$, which is just a set-theoretic quotient.

(f3) A morphism $X \xrightarrow{f} Y$ in $\mathcal{C}$ is an isomorphism if $F(f) : F(X) \to F(Y)$ is an isomorphism.

For a Galois category $\mathcal{C}$, a covariant functor $F : \mathcal{C} \to SETS$ satisfying the three conditions (f1-3) is called a *fundamental functor.*

We quote the following theorem of Grothendieck from [SGA] V.4.

Theorem 4.48 *If a pair* $(\mathcal{C}, F)$ *of a category* $\mathcal{C}$ *and a covariant functor* $F : \mathcal{C} \to SETS$ *satisfies the six conditions* (C1–3) *and* (f1–3), *there exists a unique profinite group* Π *(up to isomorphisms) such that* $\mathcal{C}$ *is equivalent to the category* $\mathcal{C}(\Pi)$.

The group Π is called the fundamental group of $(\mathcal{C}, F)$ and is sometimes written as $\pi_1(\mathcal{C}, F)$. We refer to [SGA] V.4 for the proof. We call an object X in $\mathcal{C}$ *connected* if it is not a direct sum of two noninitial objects. Here we list steps of the proof and give a brief description of each step.

1. A morphism $f \in \mathrm{Hom}_{\mathcal{C}}(X, Y)$ is a monomorphism if and only if $F(f)$ is an injection.

   In fact, $p_X : X \times_Y X \to X$ is an isomorphism if and only if $f : X \to Y$ is a monomorphism. Indeed, if $h, g \in \mathrm{Hom}_{\mathcal{C}}(Z, X)$ satisfy $f \circ g = f \circ h = f'$, regarding Z as an object in $\mathcal{C}_{/Y}$ by f', we have $g \times_Y h \in \mathrm{Hom}_{\mathcal{C}}(Z, X \times_Y X)$. Thus, f, being a monomorphism, is equivalent to the fact that the two functors $Z \mapsto \mathrm{Hom}_{\mathcal{C}}(Z, X)$ and $Z \mapsto \mathrm{Hom}_{\mathcal{C}}(Z, X \times_Y X)$ are isomorphic. This implies by the uniqueness of the object representing a functor (see Lemma 4.17) the above statement that $X \times_Y X \cong X$. Then by (C1), (f1), and (f3) combined, we conclude the assertion.

2. A sequence $X_1 \xrightarrow{f_1} X_2 \xrightarrow{f_2} X_3 \to \cdots \to X_j \to X$ of monomorphisms f_j for an object X in $\mathcal{C}$ stabilizes after the finite steps; that is, for sufficiently large j, f_j are all isomorphisms.

   This assertion is obvious in the category of finite sets. Since F preserves monomorphisms and brings $\mathcal{C}$ into the category of finite sets, it follows by (f3) from the result for the category of finite sets.

3. There exist a connected partially ordered set I and a projective system $\{P_i, \pi_{ij} : P_i \to P_j\ (i > j)\}_{i\in I}$ in $\mathcal{C}$ such that
   a) For each object X in $\mathcal{C}$, there exists $i_0 \in I$ such that we have a unique morphism $\phi_i : P_i \to X$ for each $i_0 < i \in I$ compatible with the projective system.
   b) There exist a universal element $\varphi = \varprojlim_i \varphi_i \in \varprojlim_{i\in I} F(P_i)$ and an isomorphism $\iota_X : F(X) \cong \varinjlim_{i\in I} \mathrm{Hom}_{\mathcal{C}}(P_i, X)$ induced by $\iota_X(f) = \varprojlim_i F(\phi_i)(\varphi_i)$.

   A partially ordered set I is *connected* if. for any pair of elements (i_0, j) in I, we can find a finite sequence $i_0, i_1, \cdots, i_m = j$ such that $i_k > i_{k+1}$ or $i_k < i_{k+1}$ for $k = 0, \ldots, m-1$. We hereafter write $\mathrm{Hom}(P, X)$ for $\varinjlim_{i\in I} \mathrm{Hom}_{\mathcal{C}}(P_i, X)$, and consider $P = \varprojlim_{i\in I} P_i$ in the procategory of $\mathcal{C}$. The object P is called the *fundamental pro-object* of $\mathcal{C}$ with respect to the fundamental functor F.

4. P_i can be chosen so that
   a) P_i for all i is connected;
   b) P_i is not an initial object for all i;
   c) π_{ij} is an epimorphism for all $i > j$.

   For an axiomatic proof of (3) and (4) that is based on assertions (1) and (2) besides the axioms (C1–3) and (f1–3), see [FGA] Exposé 195.

   We describe what is going on via examples. Suppose that $\mathcal{C}$ is the opposite category of the category of separable finite-dimensional semisimple commutative algebras over a given field k. In other words, $\mathcal{C}$ is made of $\mathrm{Spec}(K)$ for separable semisimple extensions K/k with $\dim_k K < \infty$. Let Ω (resp., Ω^{sep}) be an algebraic (resp., separable) closure of k [$\mathrm{Spec}(\Omega^{sep})$ is a fundamental pro-object]. Then $F : \mathrm{Spec}(K) \mapsto \mathrm{Hom}_k(K, \Omega^{sep}) = \mathrm{Hom}_{\mathcal{C}}(\mathrm{Spec}(\Omega^{sep}), \mathrm{Spec}(K))$ gives the covariant functor. One can easily check that $(\mathcal{C}, F)$ is a Galois category. In this case, we may take I to be the set of all finite Galois extensions in the fixed algebraically closed field Ω containing k. The order of I is induced by the inclusion of two fields.

   Let $\mathcal{C}$ be the full subcategory of S_{et} made up of finite étale coverings of an irreducible variety S of finite type over a field k. Fix a geometric point $\overline{s} \in S(\Omega)$ and take $F(X) = \mathrm{Hom}_{SCH_{/S}}(\overline{s}, X)$. We often write this category as $ET_{/S} \subset S_{\mathrm{et}}$. In this case, I is the category of all étale irreducible Galois coverings with ordering $X \to Y$ on I given by S-morphisms. For each $X \in I$, its function field is a finite separable extension of the function field of S. Since the function field determines X over S (because epimorphisms in the category of schemes are dominant morphisms), the category I is actually a set.

5. If X is not an initial object and Y is connected, any morphism $f : X \to Y$ in $\mathcal{C}$ is a strict epimorphism.

In the above example of the two categories (of semisimple separable finite-dimensional commutative algebras over k and of étale coverings over S), this assertion is obvious. However, this is just a consequence of axioms (C3) and (f2–3). An axiomatic proof is in [SGA] V.4.e.

6. The following conditions are equivalent.
   a) The natural injection $\mathrm{Hom}_{\mathcal{C}}(P_i, P_i) \to \mathrm{Hom}(P, P_i) \xrightarrow{\iota_{P_i}} F(P_i)$ is surjective.
   b) $\mathrm{Aut}(P_i)$ acts on $F(P_i)$ transitively.
   c) $\mathrm{Aut}(P_i)$ acts on $F(P_i)$ simply transitively.

   We say that P_i is *of Galois type* if one of the above equivalent properties is satisfied by P_i.

   This is again obvious for our two examples in item 4 of $\mathcal{C}$, because $\mathrm{End}(P_i)$ is the monoid of field endomorphisms of a finite extension over a given base field k. In general, we apply the functor F and prove the fact in the image of F by using the previous assertions (1), (4), and (5).

7. Every object X in $\mathcal{C}$ is covered by an object P_i of Galois type so that the unique map $P \to X$ in 3.a)factors through P_i.

8. Let I^{gal} be the subset of I made of all indices i with P_i of Galois type. Then we have $\mathrm{Hom}(P, P) = \mathrm{Aut}(P) = \varprojlim_{i \in I^{gal}} \mathrm{Aut}_{\mathcal{C}}(P_i)$, which is set-theoretically in bijection with $F(P) = \varprojlim_i F(P_i)$. We set $\Pi = \mathrm{Aut}(P)$.

9. Define a functor $G : \mathcal{C}(\Pi) \to \mathcal{C}$ so that

$$\mathrm{Hom}_{\mathcal{C}}(G(Z), ?) \cong \mathrm{Hom}_{\Pi}(Z, \mathrm{Hom}(P, ?))$$

   as functors with variable X in $\mathcal{C}$.

   For each connected X in $\mathcal{C}(\Pi)$, the group Π acts transitively on X by its connectedness; hence, $X \cong \Pi/H$ for a stabilizer of a point $x \in X$. Since H acts faithfully on P, for sufficiently large j in I^{gal}, if $i > j$, π_{ij} induces an isomorphism $P_i/H \cong P_j/H$ of the categorical quotients, which gives P/H (independently of $i > j$). We set $G(X) = P/H$. Here the existence of the categorical quotients P_i/H in $\mathcal{C}$ is guaranteed by (C2). If X is not connected, we can write X uniquely as $X = \bigsqcup_Z Z$ for connected components Z. Then we define $G(X) = \bigoplus_Z G(Z)$, which gives the inverse G of F. See [SGA] V.4.h for more details of the argument.

   In our example of Galois extensions, one can check that $\mathrm{Spec}(K)/H = \mathrm{Spec}(K^H)$ gives the categorical quotient of $\mathrm{Spec}(K)$ by a finite group $H \subset \mathrm{Gal}(K/k)$. For the category of étale coverings of a base scheme S, the existence of categorical quotients is explained in [GME] Proposition 8.4.

10. The two functors F and G give the equivalence of categories.

    In our two examples in item 4, this point is obvious.

In the first example of item 4 (that is the example of finite extensions K/k inside Ω^{sep}), we write $\pi_1(\mathrm{Spec}(k), \overline{s})$ for $\pi_1(\mathcal{C}, F)$. Then the following lemma is clear from the above steps:

Lemma 4.49 *Let k be a field with separable closure Ω^{sep} whose algebraic closure we denote by Ω; thus, $\overline{s} = \mathrm{Spec}(\Omega) \hookrightarrow \mathrm{Spec}(k)$ is a geometric point of $\mathrm{Spec}(k)$. Then we have $\pi_1(\mathrm{Spec}(k), \overline{s}) = \mathrm{Gal}(\Omega^{sep}/k)$.*

We now quote two more propositions from [SGA] V Propositions 5.5–6.

Proposition 4.50 *Let $\{Q_i\}_{i \in I}$ be a pro-object in $\mathcal{C} = \mathcal{C}(\Pi)$ (i.e., a projective system of objects in $\mathcal{C}$). Suppose that Q_i is not an initial object for any i and that Q_i is connected for all i. Define a covariant functor $G : \mathcal{C} \to SETS$ by $G(X) = \mathrm{Hom}_{\mathcal{C}}(Q, X)$. Then Q is isomorphic to Π/H for a closed subgroup H of Π, and the functor G is isomorphic to the functor $E \mapsto E^H$ for the fixed-point subset E^H of E under H.*

The subgroup H is given by the stabilizer of one point $q \in Q$. Since Q_i is connected, the action of Π is transitive, and $Q \cong \Pi/H$ via $\pi q \leftrightarrow \pi H$.

Proposition 4.51 *Let $(\mathcal{C}, F, P)$ be as above (a Galois category, a fundamental functor, and the fundamental pro-object with respect to F). Let $P' = \varprojlim_{j \in J} P'_j$ be a pro-object in the procategory of $\mathcal{C}$, and define $F' : \mathcal{C} \to SETS$ by $F'(X) = \mathrm{Hom}(P', X) = \varinjlim_{j \in J} \mathrm{Hom}_{\mathcal{C}}(P'_j, X)$. Then the following three conditions are equivalent:*

1. *$P' \cong P$ and also $F' \cong F$;*
2. *P' is fundamental and also F' is fundamental;*
3. *F' brings direct sums to direct sums, and $F'(X) \neq \emptyset$ if X is not an initial object.*

We may identify $\mathcal{C}$ with $\mathcal{C}(\Pi)$. For $\mathcal{C}(\Pi)$, the fundamental pro-object P is isomorphic to Π as a profinite set with continuous Π-action. Thus, if P' is another fundamental pro-object, we find the stabilizer of a single point p of P' has to be trivial. Since P' is connected, the action of Π is transitive; hence, $\pi \mapsto \pi p$ gives an isomorphism $\Pi \cong P' \cong \Pi'$. This basically shows the equivalence as above.

Remark 4.52 If $\mathcal{C}$ is the category of étale coverings of a connected variety S of finite type over a field k and $F(X) = F_{\overline{s}}(X) = \mathrm{Hom}_{SCH_{/S}}(\overline{s}, X)$ for a geometric point $\overline{s}$ of $S(\Omega)$, we write the fundamental group for $(\mathcal{C}, F)$ as $\pi_1(S, \overline{s})$. If one changes a point $\overline{s}$ to another geometric point $\overline{t}$, $F_{\overline{t}}$ is still a fundamental functor (see [SGA] V.7); hence, we have an isomorphism $\sigma : F_{\overline{t}} \cong F_{\overline{s}}$, which induces an isomorphism again denoted by σ of the fundamental pro-objects $P_{\overline{t}} \cong P_{\overline{s}}$ associated with $F_{\overline{t}}$ and $F_{\overline{s}}$, respectively. Then $\pi_1(S, \overline{t}) \cong \pi_1(S, \overline{s})$ by $\tau \mapsto \sigma\tau\sigma^{-1}$. We call an isomorphism $\sigma : F_{\overline{t}} \cong F_{\overline{s}}$ a *path* from $\overline{s}$ to $\overline{t}$.

Remark 4.53 Here is a slight generalization of Lemma 4.49. Suppose that S is reduced and irreducible. If we choose t to be the generic point η of S, a connected étale Galois covering $X \to S$ gives an irreducible Galois covering of S; hence, it is determined by its function field $k(X)$ as a Galois extension of $k(\eta) = k(S)$. Thus, $\pi_1(S, \overline{\eta})$ is canonically isomorphic to $\mathrm{Gal}(k(P)/k(S))$ for a fundamental pro-object P in $ET_{/S}$.

4.2.4 Algebraization of Fundamental Groups

Let X be a locally Noetherian connected scheme. We fix a geometric point $\overline{x}$ in $X(\Omega)$. Then the category $ET_{/X}$ of étale faithfully flat coverings of X and the fundamental functor $F_{\overline{x}} : ET_{/X} \to SETS$ given by $F_{\overline{x}}(Y) = \mathrm{Hom}_{SCH_{/X}}(\overline{x}, Y)$ supply us with a Galois category and give the profinite group $\pi_1(X, \overline{x})$ so that $F_{\overline{x}}$ gives the identification of $ET_{/X}$ with $\mathcal{C}(\pi_1(X, \overline{x}))$. The group $\pi_1(X, \overline{x})$ is called the algebraic fundamental group of X with base point $\overline{x}$. We briefly describe functorial properties of the algebraic fundamental group.

When X is a smooth variety over $\mathbb{C}$, we can think of the topological fundamental group $\pi_1^{\mathrm{top}}(X, x)$ for $x \in X(\mathbb{C})$ defined as the homotopy classes of paths starting from x ending up at x. Then, taking the universal cover $p : U \to X$, any étale covering Y of X is canonically isomorphic to U/Γ for a subgroup $\Gamma = \pi_1(Y, y)$ [for a point $y \in Y(\mathbb{C})$ over x] of finite index in $\pi_1^{\mathrm{top}}(X, x)$ (see Sect. 2.3.3). Then $F_x(Y) \cong \pi_1^{\mathrm{top}}(X, x)/\Gamma$, and thus we get

$$\pi_1(X, x) \cong \varprojlim_{\Gamma} \pi_1^{\mathrm{top}}(X, x)/\Gamma = \widehat{\pi}_1^{\mathrm{top}}(X, x), \tag{4.2.3}$$

where Γ runs over all normal subgroups of finite index in $\pi_1^{\mathrm{top}}(X, x)$; in other words, the algebraic fundamental group $\pi_1(X, x)$ is the profinite completion of the topological fundamental group $\pi_1^{\mathrm{top}}(X, x)$. If we pick another point $x' \in X(\mathbb{C})$ and choose a path σ from x to x', we find $\pi_1^{\mathrm{top}}(X, x) \cong \pi_1^{\mathrm{top}}(X, x')$ by $\tau \mapsto \sigma\tau\sigma^{-1}$, and σ induces a unique isomorphism of functors $F_x \cong F_{x'}$.

For a morphism $f : X \to Y$, the pullback $f^*T = X \times_Y T$ of an étale covering $T_{/Y}$ is an étale covering of X, because a pullback of a faithful flat morphism is faithfully flat (see [ECH] Corollary I.3.6). If X and Y are connected, obviously we have $F_{\overline{x}} \circ f^* \cong F_{f(\overline{x})}$ for any geometric point $\overline{x} \in X(\Omega)$, because an étale covering of a connected scheme X is determined by its fiber [with the action of $\pi_1(X, \overline{x})$] at any given geometric point $\overline{x}$ ([ECH] I.3.12 and [SGA] I.5.5). Then f^* is an exact functor from $ET_{/Y}$ into $ET_{/X}$.

Suppose now that X and Y are connected locally Noetherian. Then $F_{\overline{x}} \circ f^*$ gives a covariant functor from $ET_{/Y} \to SETS$. By definition, $\pi_1(X, \overline{x}) = \mathrm{Aut}(F_{\overline{x}})$, and each $\sigma \in \pi_1(X, \overline{x})$ induces an automorphism of $F_{\overline{x}} \circ f^* \cong F_{f(\overline{x})}$. Thus, $\mathrm{Aut}(F_{\overline{x}} \circ f^*) = \mathrm{Aut}(F_{f(\overline{x})}) = \pi_1(Y, f(\overline{x}))$, and we have a homomorphism of profinite groups, $\pi(f) : \pi_1(X, \overline{x}) \to \pi_1(Y, f(\overline{x}))$. In other words, the association $(X, \overline{x}) \mapsto \pi_1(X, \overline{x})$ is a covariant functor from the category of pointed schemes into the category of profinite groups.

Using this, we compare fundamental groups of the fibers of $X \xrightarrow{f} Y$ and those of X and Y. We always suppose that X and Y are connected locally Noetherian and that f is proper, surjective, and separable with $f_*(\mathcal{O}_X) = \mathcal{O}_Y$ so that each geometric fiber of f is connected by Theorem 5.7 (see also [ALG] III.11.3 and [SGA] X.2). A scheme over a field k is called *separable* if, for any field extension K/k, its base change to K is a reduced scheme (no nontrivial nilpotents in the structure sheaf). A morphism $X \xrightarrow{f} Y$ is called *separable* if f is flat and the fiber $X_y = X \otimes_Y Spec(k(y))$ is separable for all points y with residue field $k(y)$ (see [SGA] X.1).

We choose a point y in the image of f in Y and a geometric point $\overline{y}$ over y so that its field $k(\overline{y})$ is an algebraic closure of $k(y)$. Take a point x in X with $f(x) = y$ and a geometric point $\overline{x}$ over x so that $k(\overline{x})$ is an algebraic closure of $k(x)$ and $f(\overline{x}) = \overline{y}$. Then we take a geometric point $\overline{a} \in \overline{X}_y = X \otimes_{k(y)} k(\overline{y})$. We write $i : \overline{X}_y \hookrightarrow X$ for the inclusion morphism. By the functoriality of π_1, we have the following sequence of profinite groups:

$$\pi_1(\overline{X}_y, \overline{a}) \xrightarrow{i_*} \pi_1(X, \overline{x}) \xrightarrow{f_*} \pi_1(Y, \overline{y}) \to 1, \tag{4.2.4}$$

which is induced by the sequence of functors $ET_{/Y} \xrightarrow{f^*} ET_{/X} \xrightarrow{i^*} ET_{/\overline{X}_y}$.

Theorem 4.54 *Suppose that X and Y are connected locally Noetherian and that f is proper, surjective, and separable with $f_*(\mathcal{O}_X) = \mathcal{O}_Y$. Then the sequence* (4.2.4) *is an exact sequence of profinite groups.*

Here is a sketch of the proof (see [SGA] X.1.3 for more details).

Proof. If $Y' \xrightarrow{g} Y$ is a connected étale finite covering, then f^*Y'/X is a finite étale covering and has to be connected because $f_*(\mathcal{O}_X) = \mathcal{O}_Y$. This shows that $F_{\overline{x}}(f^*Y) = F_{\overline{y}}(Y)$, and hence f^* induces a surjection of the universal pro-object of $ET_{/X}$ onto that of $ET_{/Y}$. Since the universal pro-object is in bijection with the fundamental group, we find that $\pi_1(X, \overline{x}) \xrightarrow{f_*} \pi_1(Y, \overline{y})$ is a surjection.

Since the fiber $g^{-1}(\overline{y})$ of Y' over $\overline{y}$ is a finite set of closed points, $i^* \circ f^*(Y')$ is a disjoint union of $\overline{X}_y$ indexed by points of $g^{-1}(\overline{y})$. Since $\overline{X}_y$ is the final object of $ET_{/\overline{X}_y}$, we find that $\mathrm{Im}(i_*) \subset \mathrm{Ker}(f_*)$.

To prove the reverse inclusion, $\mathrm{Im}(i_*) \supset \mathrm{Ker}(f_*)$, we need to show that for any finite étale Galois covering $X' \xrightarrow{h} X$, we have an étale Galois covering Y'/Y so that $X' \cong Y' \times_Y X$ if and only if $\overline{X}'_y/\overline{X}_y$ has a section (i.e., a disjoint union of copies of $\overline{X}_y$). The only nontrivial point is the "if" part. By the properness of f and the finiteness of h, $\mathcal{A} = (f \circ h)_*(\mathcal{O}_{X'})$ is a sheaf of $\mathcal{O}_Y$-algebras finite over $\mathcal{O}_Y$ (see Lemma 5.5 and also [EGA] III.3.2.1). Let Y' be the relative spectrum $Spec_Y(\mathcal{A})$ defined in Sect. 4.1.6. Since $f_*(\mathcal{O}_X) = \mathcal{O}_Y$, f is a surjection onto the topological space of Y. By the separability of f, Y'_y/y is flat for all $y \in Y$, and hence Y'/Y is flat (by the fiber-by-fiber criterion of flatness;

see [EGA] IV.11.3.11). Again by the separability of f, $\Omega_{Y'_y/y} = 0$ for all $y \in Y$, and basically by Corollary 4.30, we have $\Omega_{Y'/Y} \otimes_{\mathcal{O}_{Y'}} \mathcal{O}_{Y'_y} = \Omega_{Y'_y/y} = 0$ for all y and $\Omega_{Y'/Y} = 0$. Hence, Y' is étale at all $y \in Y$ by Lemma 4.32. Thus, Y'/Y is an étale covering. Then the assertion is plain.

Applying the above proposition to a proper separable, geometrically connected morphism $X \to \mathrm{Spec}(k)$ for a field k with separable closure Ω^{sep} inside an algebraic closure Ω, Lemma 4.49 tell us that for $\overline{y} : \mathrm{Spec}(\Omega) \to \mathrm{Spec}(k)$ with $f(\overline{x}) = \overline{y}$

Corollary 4.55 *Let the notation and assumption be as above. We have the following exact sequence:*

$$\pi_1(\overline{X}_y, \overline{a}) \xrightarrow{i_*} \pi_1(X, \overline{x}) \xrightarrow{f_*} \mathrm{Gal}(\Omega^{sep}/k) \to 1.$$

By Hensel's lemma, if A is a Henselian local ring with residue field κ, the category of finite étale algebras over κ is equivalent to that over A by taking fibers over κ (see [ECH] I.4.2 and I.4.4). We have the following generalization of this fact to proper smooth schemes by A. Grothendieck and M. Artin:

Theorem 4.56 *Let A be a Henselian local ring and $X_{/A}$ be a smooth proper scheme with special fiber $X_{0/\kappa}$. Then we have $\pi_1(X, \overline{x}) \cong \pi_1(X_0, \overline{x}_0)$ canonically, where x_0 is a geometric point in the special fiber, and $\overline{x}$ is the geometric point of X whose schematic closure contains the image of $\overline{x}_0$.*

See [ECH] Remark I.4.6 for an indication of references containing the proof.

4.3 Group Schemes

If a scheme $G_{/B}$ as a functor induces a covariant functor from $ALG_{/B}$ into the category GP of groups, G is called a *group scheme*. If it has values in abelian groups AB, we call it a commutative group scheme. We present here a functorial viewpoint of group schemes. A main reference is [RAG].

4.3.1 Affine Algebraic Groups

We start with examples. Let G be an affine scheme over a ring B. Thus, G is a covariant functor from B-algebras $ALG_{/B}$ to $SETS$. If the functor $R \mapsto G(R)$ for all B-algebras R factors through the subcategory GP of groups in $SETS$, [i.e., $G(R)$ is a group and $\phi_* : G(R) \to G(R')$ for any B-algebra homomorphism $\phi : R \to R'$ is a group homomorphism], G is called an *affine group scheme* or an affine algebraic group defined over B. The group functor μ_N sending each B-algebra R to its Nth root of unity $\mu_N(R)$ is given by S_A for $A = B[X]/((1+X)^N - 1) - 1)$ and is an example of finite flat (equivalently, locally free of finite rank) affine group schemes.

Exercise 4.57 *Prove that* $\mu_N(R) = \mathrm{Hom}_{ALG_{/B}}(B[X]/((1+X)^N-1)-1), R)$ *is in bijection to* $\{\zeta \in R^\times | \zeta^N = 1\}$ *by sending* $\phi : B[X]/((1+X)^N-1)-1) \to R$ *to* $\phi(X) = \zeta$.

Similarly, if an affine scheme $\mathcal{R}_{/B}$ is a covariant functor from the category of B-algebras into the category of rings, $\mathcal{R}$ is called an *affine ring scheme.* For two affine algebraic group G, G' defined over B, we define

$$\begin{aligned}\mathrm{Hom}_{B\text{-alg gp}}(G, G') &= \mathrm{Hom}_{GSCH_{/B}}(G, G') \\ &:= \left\{\phi \in \mathrm{Hom}_{SCH_{/B}}(G, G') \middle| \phi_R \text{ is a group homomorphism for all } R\right\}.\end{aligned} \tag{4.3.1}$$

Example 4.58

1. Let $A = B[X_1, \dots, X_n]$. Then $\mathbb{G}_a^n(R) := S_A(R) = R^n$ (as already remarked), which is an additive group. Since
$$\phi_*(r_1, \dots, r_n) = (\phi(r_1), \dots, \phi(r_n))$$
for each algebra homomorphism $\phi : R \to R'$, ϕ_* is a homomorphism of additive groups/rings. Thus, $\mathbb{G}_a^n$ is an additive group/ring scheme.
2. More generally, we can think of $C = B[X_{ij}]$ for n^2 variables. Then $S_C(R) = M_n(R)$, and S_C is not just a group scheme but is a ring scheme. This ring scheme is written often as M_n. As additive group schemes (ignoring ring structure), M_n is isomorphic to $\mathbb{G}_a^{n^2}$.
3. Consider $A = B[t, t^{-1}]$. Then $S_A(R) = \mathrm{Hom}_{ALG_{/B}}(A, R) = R^\times$ by sending $\phi \in S_A(R)$ to $\phi(t) \in R$. Thus, this is a group scheme, denoted by $\mathbb{G}_m$ and called the *multiplicative group.* Note that if $\phi : \mathbb{G}_m \to \mathbb{G}_m$ is a scheme morphism, then $\phi^*(t) = b \cdot t^n$ for $b \in B^\times$ as $B[t, t^{-1}]^\times = B^\times \times t^{\mathbb{Z}}$. If, further, ϕ induces a group homomorphism $\mathbb{G}_m \to \mathbb{G}_m$, the constant b has to be 1. Thus, $\mathrm{Hom}_{GSCH_{/B}}(\mathbb{G}_m, \mathbb{G}_m) = \mathbb{Z}$ by $\phi \mapsto n$ if $\phi^*(t) = t^n$. Consider the group algebra $B[\mathbb{Z}]$ of the additive group $\mathbb{Z}$. Then $B[t, t^{-1}] \cong B[\mathbb{Z}]$ by $t^n \leftrightarrow [n] \in B[\mathbb{Z}]$, so $\mathbb{G}_m = \mathrm{Spec}(B[\mathrm{End}_{GSCH_{/B}}(\mathbb{G}_m)])$.
4. Let L be a free $\mathbb{Z}$-module of rank n with basis $e_1, \dots, e_n$. Consider the functor $R \mapsto R^\times \otimes_{\mathbb{Z}} L$, where $R^\times$ is considered to be an abelian group and $R^\times \otimes_{\mathbb{Z}} L$ is a usual tensor product of two abelian groups. Write this functor as $\mathbb{G}_m \otimes_{\mathbb{Z}} L$. Then $R^\times \otimes_{\mathbb{Z}} L \cong (R^\times)^n$ by sending $\sum_i a_i \otimes e_i$ to $(a_1, \dots, a_n)$. Thus, we have $\mathbb{G}_m \otimes_{\mathbb{Z}} L \cong \mathbb{G}_m^n$. Since $\mathrm{End}_{GSCH_{/B}}(\mathbb{G}_m) = \mathbb{Z}$, we have $L^* = \mathrm{Hom}_{\mathbb{Z}}(L, \mathbb{Z}) \cong \mathrm{Hom}_{GSCH_{/B}}(\mathbb{G}_m \otimes_{\mathbb{Z}} L, \mathbb{G}_m) =: X^*(L)$ (the *character group* of $\mathbb{G}_m \otimes L$) by sending $t \otimes \ell$ to $t^{\ell^*(\ell)}$ for $\ell^* \in L^*$. In other words, we have $\mathbb{G}_m \otimes_{\mathbb{Z}} L = \mathrm{Spec}(\mathbb{Z}[L^*])$ for the group algebra $\mathbb{Z}[L^*]$ of the additive group L^*. Put $X_*(\mathbb{G}_m \otimes_{\mathbb{Z}} L) = \mathrm{Hom}_{GSCH_{/B}}(\mathbb{G}_m, \mathbb{G}_m \otimes_{\mathbb{Z}} L)$. We call it the *cocharacter group* of $\mathbb{G}_m \otimes_{\mathbb{Z}} L$. We have a pairing $(\cdot, \cdot) : X_*(\mathbb{G}_m \otimes_{\mathbb{Z}} L) \times X^*(\mathbb{G}_m \otimes_{\mathbb{Z}} L) \to \mathrm{End}_{GSCH_{/B}}(\mathbb{G}_m) = \mathbb{Z}$ by $(\phi, \chi) = \chi \circ \phi$. Clearly, this pairing is perfect. The group of the form $\mathbb{G}_m \otimes_{\mathbb{Z}} L$ is often called a *B-split torus.*

5. Consider the ring $D = B[X_{ij}, \frac{1}{\det(X)}]$ for n^2 variables X_{ij} and the variable matrix $X = (X_{ij})$. Then $S_D(R) = GL_n(R)$ and S_D is a group scheme under matrix multiplication, which is a subscheme of S_C because $GL_n(R) \subset M_n(R)$ for all R. This group scheme S_D is written as $GL(n)$. In particular, $S_{B[t,t^{-1}]} = GL(1)$ is equal to $\mathbb{G}_m$.
6. For a given B-module X free of rank n, we define $X_R = X \otimes_B R$ (which is R-free of the same rank n) and

$$GL_X(R) = \left\{\alpha \in \mathrm{End}_R(X_R) \middle| \text{there exists } \alpha^{-1} \in \mathrm{End}_R(X_R)\right\}.$$

Then GL_X is isomorphic to $GL(n)_{/B}$ by choosing a basis of X; so GL_X is an affine group scheme defined over a ring B. We can generalize this to a locally free B-module X, but if X is not free, it is slightly more demanding to prove that GL_X is an affine scheme.
7. We can then think of $E = B[X_{ij}]/(\det(X) - 1)$. Then

$$S_E(R) = \{x \in GL_n(R) | \det(x) = 1\}.$$

This closed subscheme of $GL(n)$ is written as $SL(n)$ and is a group scheme (under matrix multiplication) defined over B.
8. Let X is a free B-module of finite rank. We fix a nondegenerate bilinear form $S : X \times X \to B$. Then we consider

$$G(R) = \{\alpha \in GL_X(R) | S_R(x\alpha, y\alpha) = S_R(x, y) \ \text{ for all } x, y \in X_R\},$$

where $S_R(r \otimes x, s \otimes y) = rsS(x, y)$ for $r, s \in R$ and $x, y \in X$.
To see that this G is an affine algebraic group defined over B, we fix a base $x_1, \ldots, x_n$ of X over B and define a matrix S by $S = (S(x_i, x_j)) \in M_n(B)$. Then every (ij) entry $s_{ij}(X)$ of the matrix $XS \cdot {}^tX - S$ $(X = (X_{ij}))$ is a quadratic polynomial with coefficients in B. Then we consider $L = B[X_{ij}, \det(X)^{-1}]/(s_{ij}(X))$. By definition,

$$S_L(R) = \left\{\alpha \in GL_X(R) \middle| \alpha S^t\alpha = S\right\} \cong G(R).$$

We find $\alpha S^t\alpha = S \Rightarrow S = \alpha^{-1}S \cdot {}^t\alpha^{-1}$; so the inverse exists, and G is an affine algebraic group. If $X = B^n$ and $S(x, y) = xS^ty$ for a nondegenerate symmetric matrix S, G as above is written as $O_{S/B}$ and is called the *orthogonal group* of S. If $X = Y \times Y$ and S is nondegenerate and skew symmetric of the form $S((y, y'), (z, z')) = T(y, z') - T(z, y')$ for a symmetric bilinear form $T : Y \times Y \to B$, we write $G = Sp_{T/B}$. In particular, if $S(x, y) = x \left(\begin{smallmatrix} 0 & -1_n \\ 1_n & 0 \end{smallmatrix}\right) {}^ty$, the group G is written as $Sp_{n/B}$ and is called the *symplectic group* of genus n.
9. We consider a quadratic polynomial $f(T) = T^2 + aT + b \in \mathbb{Z}[T]$. Then define $S_f(R) = \mathbb{G}_a(R)[T]/(f(T))$. As a scheme, $S_f \cong \mathbb{G}_a^2$, but its value is a ring all the time. If $\phi : R \to R'$ is an algebra homomorphism, $\phi_*(r + sT) = \phi(r) + \phi(s)T$, and so it is a ring homomorphism of $S_f(R) = R[T]/(f(T))$

into $S_f(R') = R'[T]/(f(T))$. Thus, S_f is a ring scheme, and writing O for the order of the quadratic field $\mathbb{Q}[\sqrt{a^2-4b}]$ generated by the root of $f(T)$, we have $S_f(R) \cong R \otimes_{\mathbb{Z}} O$.

10. Since any given number field F is generated by one element, we know $F = \mathbb{Q}[T]/(f(T))$ for an irreducible monic polynomial $f(T)$. For any $\mathbb{Q}$-algebra R, define $S_f(R) = R[T]/(f(T))$. Then, in the same way as above, S_f is a ring scheme defined over $\mathbb{Q}$ such that $S_f(R) = F \otimes_{\mathbb{Q}} R$.
11. Let G be an affine algebraic group defined over a number field F. Then we define a new functor G' defined over $\mathbb{Q}$-algebras R by $G'(R) = G(S_f(R)) = G(F \otimes_{\mathbb{Q}} R)$. We can prove that G' is an affine group scheme defined over $\mathbb{Q}$, which we write as $G' = \mathrm{Res}_{F/\mathbb{Q}} G$ [see Exercise 4.59 (3)].
12. Assume that f is a quadratic polynomial in $\mathbb{Q}[T]$. Then $S_f(\mathbb{Q}) = F$ is a quadratic extension with $\mathrm{Gal}(F/\mathbb{Q}) = \{1, \sigma\}$. Let X be a finite-dimensional vector space over $\mathbb{Q}$, and let $\mathrm{Gal}(F/\mathbb{Q})$ act on $X_F = F \otimes_{\mathbb{Q}} X$ through F. Suppose we have a Hermitian form $H : X_F \times X_F \to F$ such that $H(x,y) = \sigma(H(y,x))$. Then for a $\mathbb{Q}$-algebra R,

$$U_H(R) = \left\{\alpha \in GL_X(S_f(R)) \middle| H_{S_f(R)}(x\alpha, y\alpha) = H_{S_f(R)}(x,y)\right\}$$

is an affine algebraic group, which is called the unitary group of H. Note that U_H is defined over $\mathbb{Q}$ (not over F).

Exercise 4.59

1. *Prove that if $\phi \in \mathrm{Hom}_{GSCH_{/B}}(\mathbb{G}_m, \mathbb{G}_m) =: \mathrm{End}(\mathbb{G}_m)$, the corresponding algebra homomorphism $\phi^* : B[t,t^{-1}] \to B[t,t^{-1}]$ satisfies $\phi(t) = t^n$ for an integer n [so $\mathrm{End}(\mathbb{G}_m) \cong \mathbb{Z}$].*
2. *Let F be a number field with the integer ring O. Is there any affine ring scheme S defined over $\mathbb{Z}$ such that $S(R) = O \otimes_{\mathbb{Z}} R$?*
3. *Let $S : X \times X \to B$ be a bilinear form for a B-free module X of finite rank n, and suppose that $X \cong \mathrm{Hom}_B(X, B)$ by S. Prove that the matrix of S is in $GL_n(B)$ for any choice of basis of X over B.*
4. *For an affine algebraic group G over a number field F (that is, a finite extension of $\mathbb{Q}$), prove that $\mathrm{Res}_{F/\mathbb{Q}} G$ is an affine algebraic group defined over $\mathbb{Q}$.*
5. *Show that the unitary group U_H over $\mathbb{Q}$ as above is an affine algebraic group.*

More generally than the above Exercise 4.59 (4), we start with an affine group scheme H over a ring R'. For a subalgebra R of R', if the covariant functor $C \mapsto H(C \otimes_R R')$ defined on the category of R-algebras is isomorphic to a scheme $H'_{/R}$, we write $H'_{/R} = \mathrm{Res}_{R'/R} H$ and call it the *Weil restriction* of H with respect to R'/R (this is not changing the base ring of $H_{/R'}$ to the subalgebra R).

Theorem 4.60 *Let the notation and the assumption be as above. If R'/R is locally R-free of finite rank, the group functor $\mathrm{Res}_{R'/R} H$ is an affine group scheme over R.*

For a proof, see [NMD] 7.6, Theorem 4.

4.3.2 Basic Diagrams

The group structure of a group scheme gives rise to morphisms of schemes by Lemma 4.17; for example, the group multiplication induces the multiplication morphism $m : G \times G \to G$ and the existence of identity can be formulated to be the existence of a closed immersion $\mathrm{Spec}(B) \to G$, which satisfies the group law. For example, associativity is equivalent to the commutativity of the following diagram:

$$\begin{array}{ccc} G \times_{S_B} G \times_{S_B} G & \xrightarrow{(x,y,z)\mapsto(xy,z)} & G \times_{S_B} G \\ {\scriptstyle (x,y,z)\mapsto(x,yz)}\downarrow & & \downarrow{\scriptstyle m} \\ G \times_{S_B} G & \xrightarrow[m]{} & G. \end{array}$$

If $G = S_A$ is affine, the dual of this commutative diagram is

$$\begin{array}{ccc} A \otimes_B A \otimes_B A & \xleftarrow{\underline{m}\otimes\mathrm{id}} & A \otimes_B A \\ {\scriptstyle \mathrm{id}\otimes\underline{m}}\uparrow & \circlearrowleft & \uparrow{\scriptstyle \underline{m}} \\ A \otimes_B A & \xleftarrow[\underline{m}]{} & A. \end{array}$$

The B-algebra homomorphism $\underline{m}$ is called *comultiplication*. Similarly, the identity e_R of the group $G(R)$ induces the functor morphism

$$e_R : S_B(R) = \{\text{the structure morphism } \iota_R : B \to R\} \ni \iota_R \mapsto e_R \in G(R).$$

If $G = S_A$ is affine, the dual B-algebra homomorphism $\underline{e} : A \to B$ is called the *co-identity*. The group inverse map $i : G(R) \to G(R)$ induces an involution $\underline{i}$ of $\mathcal{O}_G$ (or A if $G = S_A$) called the *co-inverse*. These maps make the following diagram commutative:

$$\begin{array}{ccc} A & \xrightarrow{\underline{e}} & B \\ {\scriptstyle \underline{m}}\downarrow & & \downarrow{\scriptstyle \iota_A} \\ A \otimes_B A & \xrightarrow[\mathrm{id}_A\otimes\underline{i}]{} & A, \end{array} \quad \text{and} \quad \begin{array}{ccc} A & \xrightarrow{\underline{e}} & B \\ {\scriptstyle \underline{m}}\downarrow & & \downarrow{\scriptstyle \iota_A} \\ A \otimes_B A & \xrightarrow[\underline{e}\otimes\mathrm{id}_A]{} & B \otimes_B A = A, \end{array}$$

A B-algebra A with comultiplication, co-inverse, and co-identity (satisfying the above commutative diagrams) is called a *B-bialgebra* (or Hopf B-algebra). Once we take a dual $A^* = \mathrm{Hom}_B(A, B)$, A^* is also a bialgebra under the dual maps, as long as A is locally free of finite rank over B. The bialgebra A^* is called the *dual B-bialgebra* of a B-bialgebra A. We write $BIALG_{/B}$ for the category of B-bialgebras whose morphisms are B-algebra homomorphisms compatible with comultiplication, co-inverse, and co-identity. The category of group schemes $GSCH_{/B}$ is a subcategory of $SCH_{/B} \subset \mathcal{C}_B$ made up of B-schemes having values in the category of groups, whose morphisms are B-scheme morphisms preserving the group structure. The association $A \mapsto S_A$ induces a contravariant functor of $BIALG_{/B}$ into $GSCH_{/B}$, which gives rise to an (anti-)equivalence of categories between $BIALG_{/B}$ and the full subcategory of affine group schemes in $GSCH_{/B}$.

Exercise 4.61 *Consider* $\mathbb{G}_m = \mathrm{Spec}(B[t, t^{-1}])$. *Let*

$$\underline{m} \in \mathrm{Hom}_{ALG_{/B}}(B[t, t^{-1}], B[t, t^{-1}] \otimes_B B[t, t^{-1}])$$

with $\underline{m}(t) = t \otimes t$. *Show that the corresponding morphism* $m : \mathbb{G}_m \times \mathbb{G}_m \to \mathbb{G}_m$ *is given by* $m(a, b) = ab$ *for* $a, b \in \mathbb{G}_m(R) = R^\times$.

4.3.3 Functorial Representations

Consider a group scheme $G_{/B}$. We have a notion of a module over a given group. We generalize this to a notion of an action of a group scheme on a functor having values in $MOD_{/B}$. We give only a brief outline of the theory (see [RAG] I.2) for general theory). For any B-module M, we define a group functor $\underline{M}$ from $ALG_{/B}$ into $MOD_{/B}$ by $\underline{M}(R) = M \otimes_B R$. Let $G_{/B}$ be a group scheme. A B-module M is called a *functorial* G*-module* or *schematic* G*-module* if we have a functorial action $G(R) \times \underline{M}(R) \to \underline{M}(R)$ that is R-linear. This means that we have a morphism of covariant functors $G \times \underline{M} \to \underline{M}$ such that for each $R \in ALG_{/B}$, the induced map $G(R) \times \underline{M}(R) \to \underline{M}(R)$ is an action of the group $G(R)$ on the R-module $\underline{M}(R)$. In particular, if $G = \mathrm{Spec}(A)$ for an B-bialgebra A, $G(A) = \mathrm{Hom}_S(G, G)$ acts on $\underline{M}(A) = M \otimes_B A$. Thus, $\mathrm{id}_G \in G(A)$ acts on $\underline{M}(A)$ [here id_G is not the identity e_A of the group $G(A)$ but the identity map in $\mathrm{Hom}_S(G, G) = G(A)$].

Define a map $\Delta : M \to M \otimes_B A \xrightarrow{\mathrm{id}_G} \underline{M}(A)$ by $\Delta(\mu) = \mathrm{id}_G(\mu \otimes 1)$ $(\mu \in M)$, where 1 is the identity element of the ring A. Then we claim that Δ determines the G-module structure on $\underline{M}$. In fact, for $g \in G(R) = \mathrm{Hom}_{ALG_{/B}}(A, R)$, we have $G(g)(\mathrm{id}_G) = g$ and a commutative diagram:

$$\begin{array}{cccccc}
(\mathrm{id}_G, \mu \otimes 1) \in & G(A) \times \underline{M}(A) & \longrightarrow & \underline{M}(A) \\
\downarrow & \downarrow G(g) \times (\mathrm{id}_M \otimes g) & & \downarrow \mathrm{id}_M \otimes g \\
(g, \mu \otimes 1) \ \in & G(R) \times \underline{M}(R) & \longrightarrow & \underline{M}(R).
\end{array}$$

We see that $g(\mu \otimes 1) = (\mathrm{id}_M \otimes g) \circ \mathrm{id}_G(\mu \otimes 1) = (\mathrm{id}_M \otimes g) \circ \Delta(\mu)$, and Δ actually determines the destination of $\mu \otimes 1$ under g. By the associativity of the action, for the multiplication $m : G \times_S G \to G$ and the identity $e : S_B \to G$ [sending $S_B(R) = \{\iota_R\}$ to the identity e_R of the group $G(R)$], we get

$$(\Delta \otimes \mathrm{id}_A) \circ \Delta = (\mathrm{id}_M \otimes m) \circ \Delta \quad \text{and} \quad (\mathrm{id}_M \otimes e) \circ \Delta = \mathrm{id}_M \,.$$

Example 4.62 Suppose $G = \mathbb{G}_m = \mathrm{Spec}(B[t, t^{-1}])$ for a variable t. Then we have $\Delta(\mu) = \sum_{n\in\mathbb{Z}} p_n(\mu)t^n$, where $p_n \in \mathrm{End}_B(M)$. Then we see from $\underline{e}(t) = 1$ and $\underline{m}(t) = t \otimes t$ that

$$\mathrm{id}_M = (\mathrm{id}_M \otimes \underline{e}) \circ \Delta = \sum_n p_n \quad \text{and}$$

$$\sum_{k,j\in\mathbb{Z}} p_k \circ p_j t^k \otimes t^j = (\Delta \otimes \mathrm{id}_A) \circ \Delta = (\mathrm{id}_M \otimes \underline{m}) \circ \Delta = \sum_n p_n t^n \otimes t^n.$$

This implies $p_k \circ p_j = \delta_{k,j} p_j$ for Kronecker's δ. Writing $M_n = p_n(M)$, we have $M = \oplus_{k\in\mathbb{Z}} M_k$. It is easy to see that

$$M_k(R) = \left\{ x \in \underline{M}(R) \middle| g(x) = g^k x \quad \text{for all } g \in \mathbb{G}_m(R) \right\},$$

and we have

$$\text{If } G = \mathbb{G}_m \text{ and } M \text{ is a } G\text{-module, then } M = \bigoplus_{k\in\mathbb{Z}} M_k. \tag{4.3.2}$$

A group scheme $G_{/B}$ is called *diagonalizable* if any functorial G-module is a direct sum of eigenspaces under the action of G. In particular, $\mathbb{G}_m$ and $\mathbb{G}_m^n$ are diagonalizable.

4.3.4 Duality of Hopf Algebras

A finite flat group scheme $f : G \to S$ over S is sometimes called a *locally free group* scheme since $G = Spec_S(f_*\mathcal{O}_G)$ with locally free $\mathcal{O}_S$-module $f_*\mathcal{O}_G$ of finite rank. Let $G_{/B}$ be a locally free commutative group scheme. By definition, $G = \mathrm{Spec}(A)$ for a B-bialgebra A. We suppose that G has constant rank N [that is, the rank $\mathrm{rank}_B\, \mathcal{O}_G = N$ is independent of the connected component of $\mathrm{Spec}(B)$]. We have the following correspondence of morphisms:

Morphisms of S-schemes	Homomorphisms of B-algebras
$m : G \times_S G \to G$ (multiplication)	$\underline{m} : A \to A \otimes_B A$
$i : G \to G$ (inverse)	$\underline{i} : A \to A$
$e : S \to G$ (identity)	$\underline{e} : A \to B$

Let $A^* = \mathrm{Hom}_B(A, B)$ be the B-linear dual of A. Then A^* is a free B-module of rank N. Using the morphisms of B-algebras in the left column of the above table, we define the product

$$m^* : A^* \otimes_B A^* \to A^* \quad \text{by} \quad (\phi, \eta) \mapsto \phi \cdot \eta = (\phi \otimes_B \eta) \circ \underline{m}.$$

Recall the commutative diagrams defining (commutative) bialgebras:

1. The commutativity of comultiplication $\underline{m}$:

$$\begin{array}{ccc} A & \xrightarrow{\underline{m}} & A \otimes_B A \\ {\scriptstyle \underline{m}}\downarrow & & \downarrow {\scriptstyle \underline{m}\otimes \mathrm{id}_A} \\ A \otimes_B A & \xrightarrow[\mathrm{id}_A \otimes \underline{m}]{} & A \otimes_B A \otimes_B A \end{array}$$

implies the associativity:

$$\begin{aligned}(\phi \cdot \eta) \cdot \xi = (\phi \otimes \eta \otimes \xi) \circ (\underline{m} \otimes \mathrm{id}_A) \circ \underline{m} \\ = (\phi \otimes \eta \otimes \xi) \circ (\mathrm{id}_A \otimes \underline{m}) \circ \underline{m} = \phi \cdot (\eta \cdot \xi)\end{aligned}$$

for $\phi, \eta, \xi \in A^*$.

2. The commutativity defining co-inverse $\underline{e}$:

$$\begin{array}{ccc} A & \xrightarrow{\underline{m}} & A \otimes_B A \\ \| \downarrow & & \downarrow {\scriptstyle \underline{e}\otimes \mathrm{id}_A} \\ A & = & A \end{array}$$

implies the existence of the identity:

$$e \cdot \phi = (e \otimes \phi) \circ \underline{m} = \phi \circ (\underline{e} \otimes \mathrm{id}_A) \circ \underline{m} = \phi \circ \mathrm{id}_A = \phi.$$

3. The commutativity of the following diagram:

$$\begin{array}{ccc} A & \xrightarrow{\underline{m}} & A \otimes_B A \\ {\scriptstyle \underline{m}}\downarrow & & \downarrow {\scriptstyle sw} \\ A \otimes_B A & = & A \otimes_B A \end{array}$$

implies the commutativity:

$$\phi \cdot \eta = (\phi \otimes \eta) \circ \underline{m} = (\eta \otimes \phi) \circ \underline{m} = \eta \cdot \phi,$$

where $\mathrm{id}_A : A \to A$ is the identity map of the set A and $sw : A \otimes_B A \to A \otimes_B A$ is switching the coordinates: $sw(a \otimes b) = b \otimes a$. We can check the distributive law for A^* as follows:

$$(\phi + \eta) \cdot \xi = ((\phi + \eta) \otimes \xi) \circ \underline{m} = (\phi \otimes \xi + \eta \otimes \xi) \circ \underline{m} = \phi \cdot \xi + \eta \cdot \xi.$$

Therefore, A^* becomes a commutative B-algebra.

We now see what kind of morphism the multiplication $\mu : A \otimes_B A \to A$ induces on A^*. The commutativity of the left diagram in the following diagrams implies the commutativity of the right:

$$\begin{array}{ccc} A\otimes A\otimes A & \xrightarrow[\mu\otimes \mathrm{id}_A]{} & A\otimes A \\ {\scriptstyle \mathrm{id}_A\otimes\mu}\downarrow & & \downarrow{\scriptstyle\mu} \\ A\otimes A & \xrightarrow[\mu]{} & A, \end{array} \quad\Rightarrow\quad \begin{array}{ccc} A^* & \xrightarrow{\mu^*} & A^*\otimes A^* \\ {\scriptstyle \mu^*}\downarrow & & \downarrow{\scriptstyle\mu^*\otimes\mathrm{id}_{A^*}} \\ A^*\otimes A^* & \xrightarrow[\mathrm{id}_{A^*}\otimes\mu^*]{} & A^*\otimes A^*\otimes A^*. \end{array}$$

Writing $G^* = S_{A^*}$, the above commutative diagram implies the associativity of multiplication m^* induced by μ^*. Recall that $\iota_A : B \to A$ denotes the structure B-algebra homomorphism. We see easily again that ι induces the identity $e^* \in G^*$. and that $i^* = (\underline{i})^*$ gives the inverse of G^*. Thus, the affine scheme G^* is a locally free commutative group scheme of constant rank N. By definition, $(G^*)^* \cong G$ canonically. The group scheme G^* is called the *Cartier dual* of G. The association $A \mapsto A^*$ is a contravariant involutive functor from locally free B-bialgebras into itself, and at the same time, $G \mapsto G^*$ is a contravariant involutive functor from locally free group schemes over B into itself.

4.3.5 Duality of Finite Flat Groups

We define the category of group schemes $GSCH_{/S}$ over a scheme S as a full subcategory of $CTF(SCH_{/S}, GP)$ made of group schemes over S. Thus, for two group schemes $G, F : SCH_{/S} \to GP$, we are going to write $\mathrm{Hom}_{GSCH_{/S}}(G, F)$ for the set of morphisms of two contravariant functors *preserving the group structure*. We now extend the definition of the duality to locally free group schemes:

Theorem 4.63 *If $G_{/S}$ is a locally free commutative group scheme of rank N, the functor $G^* : SCH_{/S} \to AB$ given by*

$$X \mapsto \mathrm{Hom}_{GSCH_{/X}}(G \times_S X, \mathbb{G}_m \times_S X)$$

is representable by (i.e., isomorphic to) a locally free group scheme $G^*_{/S}$ of rank N, which is called the Cartier dual *of G. The correspondence $G \mapsto G^*$ is an antiequivalence of the category of locally free groups schemes over S.*

Here *antiequivalence* means that it gives the equivalence of the source category to the opposite category of the target category. The opposite category $\mathcal{C}^\circ$ of $\mathcal{C}$ is obtained by reversing the direction of arrows, that is, we define $\mathrm{Hom}_{\mathcal{C}^\circ}(X, Y) := \mathrm{Hom}_{\mathcal{C}}(Y, X)$ without changing objects. As is clear from the proof, this duality is induced locally by the duality of bialgebras discussed in the previous subsection.

Proof. We follow Mumford's proof in [ABV] III.14. Since G^* is local, we may assume that $S = \mathrm{Spec}(B)$, $G = \mathrm{Spec}(A)$ for an B-bialgebra A, where A is B-free of rank N. We show that G^* is isomorphic to $G^* = S_{A^*}$. We have for any B-algebra R: $\mathrm{Hom}_{ALG_{/B}}(A^*, R) \cong \mathrm{Hom}_{ALG_{/R}}(A^* \otimes_B R, R)$ via $\phi \leftrightarrow \phi \otimes \mathrm{id}_R$ for the identity map $\mathrm{id}_R : R \to R$. Then we see

$$\begin{aligned}
S_{A^*}(R) &= \mathrm{Hom}_{ALG_{/B}}(A^*, R) \cong \mathrm{Hom}_{ALG_{/R}}(A^* \otimes_B R, R) \\
&\cong \mathrm{Hom}_{ALG_{/R}}(\mathrm{Hom}_R(A \otimes_B R, R), R) \quad (A^* \otimes R \cong (A \otimes R)^* \Leftarrow A \text{ is } B\text{-free}) \\
&\cong \mathrm{Hom}_{ALG_{/R}}((A \otimes_B R)^*, R) \\
&= \left\{ f \in \mathrm{Hom}_R((A \otimes_B R)^*, R) \middle| f(\phi \cdot \eta) = f(\phi)f(\eta),\ f(e) = 1 \right\} \\
&\cong \left\{ \xi \in A \otimes_B R \middle| (\phi \otimes \eta)(\underline{m}(\xi)) = \phi(\xi)\eta(\xi) = (\phi \otimes \eta)(\xi \otimes \xi),\ \underline{e}(\xi) = 1 \right\} \\
&= \left\{ \xi \in A \otimes_B R \middle| \underline{e}(\xi) = 1 \text{ and } \underline{m}(\xi) = \xi \otimes \xi \right\} =: \varXi,
\end{aligned}$$

where ϕ and η run over all elements in $(A \otimes_B R)^*$. We write $\varXi$ for the last set of ξ in the above formula. Here the second identity from the last follows from the fact

$$(A \otimes_B R)^* \otimes_R (A \otimes_B R)^* \cong (A \otimes_B A \otimes_B R)^*.$$

If $\xi \in \varXi$, then

$$\underline{m}(\xi) = \xi \otimes \xi \Rightarrow \xi \cdot \underline{i}(\xi) = \underline{m}((1_A \otimes \underline{i})(\underline{m}(\xi))) = \underline{e}(\xi) = 1 \Rightarrow \xi \in (A \otimes_B R)^\times.$$

This shows

$$S_{A^*}(R) = \left\{ \xi \in (A \otimes_B R)^\times \middle| \underline{e}(\xi) = 1 \text{ and } \underline{m}(\xi) = \xi \otimes \xi \right\}.$$

We now compute $\mathrm{Hom}_{GSCH_{/R}}(G_{/R}, \mathbb{G}_{m/R})$ and show that this is isomorphic to the above set: Note that the comultiplication m_0, the co-inverse i_0, and the co-identity e_0 of $\mathbb{G}_{m/R} = Spec(R[t, t^{-1}])$ are given by $m_0(t) = t \otimes t$, $e_0(t) = 1$, and $i_0(t) = t^{-1}$. Then we see

$$\mathrm{Hom}_{GSCH_{/R}}(G_{/R}, \mathbb{G}_{m/R}) = \mathrm{Hom}_{BIALG_{/R}}(R[t, t^{-1}], A \otimes_B R) \subset (A \otimes_B R)^\times$$

by $\lambda \mapsto \lambda(t)$. Computing the image of $\lambda \in \mathrm{Hom}_{BIALG_{/R}}(R[t, t^{-1}], A \otimes_B R)$ in $(A \otimes_B R)^\times$ and writing $\xi = \lambda(t)$, we get

$$\begin{aligned}
&\underline{e}(\xi) = \underline{e}(\lambda(t)) = \lambda(e_0(t)) = 1, \\
&\qquad \xi \otimes \xi = \lambda(t) \otimes \lambda(t) = \lambda \otimes \lambda(m_0(t)) = \underline{m}(\lambda(t)) = \underline{m}(\xi),
\end{aligned}$$

which shows the assertion.

Exercise 4.64 *Let the notation and assumption be as in the theorem. Prove* $(\mathbb{Z}/N\mathbb{Z})^* = \mu_N$ *and* $\mu_N^* = \mathbb{Z}/N\mathbb{Z}$.

Here $(\mathbb{Z}/N\mathbb{Z}) = \mathrm{Spec}(\bigoplus_{n \in \mathbb{Z}/N\mathbb{Z}} B)$; so, for a B-scheme X, writing $\pi_0(X)$ for the set of connected components of the topological space of X, we have $(\mathbb{Z}/N\mathbb{Z})(X) = (\mathbb{Z}/N\mathbb{Z})^{\pi_0(X)}$, which is a locally free group.

4.3.6 Abelian Schemes

Up to now, we have studied affine group schemes (except for elliptic curves). A group scheme $A_{/S}$ is called an *abelian scheme* over a base B-scheme S if $A \to S$ is smooth, proper, and geometrically connected. Here the phrase *geometrically connected* means that the fiber $A_k = A \times_{S,\phi} S_k$ over $\phi : \mathrm{Spec}(k) \to S$ for any algebraically closed field k over B is connected (i.e., irreducible). Since A is proper and geometrically connected, it cannot be affine unless $A = S$.

Lemma 4.65 *If $A_{/S}$ is an abelian scheme, then A as a group functor has values in the category of abelian groups.*

We give a sketch of a proof (see [ABV] Sect. 4 for details).

Proof. For simplicity, we assume that $S = \mathrm{Spec}(k)$ for a field k. Then we consider the stalk $\mathcal{O}_{A,e}$ at the group identity e. Consider $\mathcal{O}_{A,e}$ and $M_n = \mathcal{O}_{A,e}/\mathfrak{m}_e^n$ a schematic representation of A on which $x \in A(R)$ acts by conjugation (as $y \mapsto xyx^{-1}$ preserves the identity e, hence induces an action on the stalk). Thus, we have a morphism of functors $A(R) \to \mathrm{End}_R(M_n \otimes_B R)$. Since $R \mapsto \mathrm{End}_R(M_n \otimes_B R)$ is an affine ring scheme $\mathrm{Spec}(C_n)$, we get a morphism of schemes $A \to S_{C_n}$. Since A is proper and geometrically connected, the image of this morphism is a point (as there is no proper geometrically connected subscheme inside an affine scheme except for a maximal point). Since e induces the identity of S_{C_n}, we find that the entire A fixes M_n for all n, and hence fixes the stalk $\mathcal{O}_{A,e}$. Since A is smooth and geometrically connected, A is irreducible; hence, $\mathcal{O}_A(U) \hookrightarrow \mathcal{O}_{A,e}$ if U is an open subscheme containing e, and the conjugation has to be trivial on A. Thus, A has values in abelian groups. The above proof works over any local ring B and hence over any scheme, as schemes are local ringed spaces.

Lemma 4.66 *Let $A_{/S}$ be an abelian scheme. Then multiplication by an integer $N > 0$: $x \mapsto Nx$ induces a finite flat morphism $N : A \to A$.*

Here is a brief outline of a proof.

Proof. When $S = \mathrm{Spec}(k)$ for an algebraically closed field k, the finiteness of $A[N] = \mathrm{Ker}(N : A \to A)$ follows from the properness of abelian schemes (see Corollary 5.10 in the text and [ABV] Sect. 6). In general, by the finiteness of $A[N](k)$, $N : A \to A$ is quasi-finite. Since A is proper, $N : A \to A$ is proper; hence, N is finite (see Corollary 5.10). If k is a field, N induces an endomorphism $\underline{N}$ of the function field $k(A)$, hence, $k(A)/\underline{N}(k(A))$ is a finite extension of fields, and so N is generically flat. The nonflat locus of N is therefore a proper closed subset F (see Proposition 5.3). Since $A(\overline{k})$ acts transitively on $A(\overline{k})$ (by an automorphism of the scheme $A_{/\overline{k}}$) for an algebraic closure $\overline{k}/k$, this proper closed set has to be empty. Thus, $N : A \to A_{/k}$ is flat. In general, by the above argument, fiber by fiber, $N : A \to A$ is flat, and hence $N : A \to A$ is flat (see [GME] Theorem 1.9.7 and [EGA] IV.11.3.11).

4.3.7 Barsotti–Tate Groups

For a prime p, the notion of a Barsotti–Tate group is technically important in many aspects of arithmetic algebraic geometry, in particular, in the classification problem of abelian schemes, as we can often reduce the study of abelian schemes to that of affine finite flat group schemes via the Barsotti–Tate groups.

Let $G : ALG_{/B} \to AB$ be a covariant functor into the category AB of abelian groups. If we regard G as a functor from the category of affine B-schemes, it is contravariant. Suppose that, for any faithfully flat extension of finite presentation $\iota : R \hookrightarrow R_1$ of B-algebras (such an extension is called an *fppf extension*),

1. The group $G(R)$ injects into $G(R_1)$; that is, $G(\iota) : G(R) \hookrightarrow G(R_1)$ is injective.
2. Let $R_2 = R_1 \otimes_R R_1$ and $R_3 = R_1 \otimes_R R_1 \otimes_R R_1$. Write $\iota_i : R_1 \hookrightarrow R_2$ $(i = 1, 2)$ for the two natural inclusions [with $\iota_1(r) = r \otimes 1$ and $\iota_2(r) = 1 \otimes r$] and $\iota_{ij} : R_2 \hookrightarrow R_3$ for the three natural inclusions [i.e., $\iota_{12}(r \otimes s) = r \otimes s \otimes 1$, and so on]. If $x \in G(R_1)$ satisfies $y = G(\iota_1)(x) = G(\iota_2)(x)$ and $G(\iota_{12})(y) = G(\iota_{23})(y) = G(\iota_{13})(y)$, then x is in the image of $G(R)$ under $G(\iota)$.

In short, the above two conditions are equivalent to requiring that for any fppf covering $\{U_i\}_{i \in I}$ of U in $SCH_{/B}$, the following (now familiar) sequence is exact:

$$0 \to G(U) \to \prod_{i \in I} G(U_i) \to \prod_{i,j \in I} G(U_i \times_U U_j).$$

Thus, G is an abelian sheaf on $SCH_{/B}$ under the $fppf$-topology (so we call it an *abelian $fppf$-sheaf*). As before, we denote by $\mathcal{S}(B_{fppf})$ the category of abelian fppf sheaves over B. More generally, for a general scheme S, we call an object in $\mathcal{S}(S_{fppf})$ an abelian fppf sheaf over S. As we remarked in Sect. 4.1.10, the category $\mathcal{S}(S_{fppf})$ is an abelian category.

Since a finite flat group scheme G over B is affine represented by a B-bialgebra A finite flat over B, A is finitely presented faithfully flat; thus, G gives rise to an object in $\mathcal{S}(B_{fppf})$. Thus, the category $FFG_{/B}$ of finite flat group schemes is a full subcategory of $\mathcal{S}(B_{fppf})$. If H is a finite flat subgroup scheme of a flat group scheme (giving rise to an fppf abelian sheaf), the quotient G/H originally in $\mathcal{S}(B_{fppf})$ is actually represented by a flat group scheme (see, for example, [GME] Proposition 1.12.2). The multiplication of H induces a functorial representation of H on $\mathcal{O}_G$ and, therefore, we have a sheaf of rings $\mathcal{O}_G^H$ made up of H-invariant sections. If G is finite flat, we indeed have $Spec_B(\mathcal{O}_G^H) = G/H$. Though any morphism $f : G \to G'$ in $FFG_{/B}$ may not have cokernel in $FFG_{/B}$, if f is a closed immersion, we have the quotient $\mathrm{Coker}(f)$ in $FFG_{/B}$. If f is an epimorphism in $\mathcal{S}(B_{fppf})$, we have $\mathrm{Ker}(f) \in FFG_{/B}$ (see [GME] Lemma 1.12.1). In short, $FFG_{/B}$ is an exact category (but may not be an abelian category).

Here is a definition of p-divisibility for fppf sheaves (which is *weaker* than Tate's notion of p-divisible groups). We call an abelian fppf sheaf G a *p-divisible* fppf sheaf if, for any $x \in G(R)$, there exists an fppf extension R_1 of R and a point $y \in G(R_1)$ such that $x = py$.

We call a p-divisible fppf sheaf $G_{/B}$ a *p-divisible* group or a *Barsotti–Tate* group if

(i) $G = \varinjlim_n G[p^n]$ for $G[p^n] = \mathrm{Ker}(p^n : G \to G)$;
(ii) $G[p^n]_{/B}$ is finite flat group schemes over B with closed immersions $G[p^n] \hookrightarrow G[p^m]$ for $m > n$;
(iii) the multiplication $[p^{m-n}] : G[p^m] \to G[p^n]$ is an epimorphism in the category of finite flat group schemes.

Lemma 4.67 *Let $G_{/B}$ be a smooth, geometrically connected group scheme of finite dimension with $x \mapsto px$ inducing a finite morphism $G \to G$. Then G is a p-divisible abelian fppf sheaf, and $G[p^\infty] = \bigcup_n G[p^n]$ for $G[p^n]$ given by the kernel* $\mathrm{Ker}(p^n : G \to G)$ *is a Barsotti–Tate group.*

Proof. Write $f : G \to G$ for the multiplication by p. For any point $s \in S(K)$ for a field K, the fiber $G_{s/K}$ is a connected group scheme. Thus, as in the proof of Lemma 4.66, $f : G_s \to G_s$ is generically flat; so the nonflat locus G_s^{nfl} of f is a proper closed subset of G_s (see Sect. 5.1.1). Since $G_s(\overline{K})$ is a group (for an algebraic closure $\overline{K}/K$), $G_s(\overline{K})$ acts transitively on $G_s(\overline{K})$, and so G_s^{nfl} has to be empty. Thus, f is flat on each fiber G_s. By the flatness criterion fiber by fiber (cf. [EGA] IV.11.3.11), $f : G \to G$ is flat. Since G is geometrically connected, f is dominant; hence, f and multiplication by p^n given by f^n are finite faithfully flat of finite presentation, proving the first assertion. Thus, $G[p^n] = G \times_{G,[p^n],e} S_B$ is finite flat over S_B, as the pullback of a finite flat morphism is finite flat (cf. [ALG] III.9.2). By definition, $G[p^n] \hookrightarrow G[p^m]$ for $m > n$ is a closed immersion. Since G is an fppf abelian sheaf, $[p^{m-n}] : G[p^m] \to G[p^n]$ is an epimorphism in the category of fppf abelian sheaves; hence, it is an epimorphism in the category of finite flat group schemes. Thus, $G[p^\infty]$ is a Barsotti–Tate group.

By Lemmas 4.67 and 4.66 combined, if $A_{/S}$ is an abelian scheme, then $R \mapsto A(R) = \mathrm{Hom}_{\mathcal{C}_B}(S_R, A)$ is an fppf abelian sheaf, and $A[p^\infty] = \bigcup_n A[p^n]$ for $A[p^n] = \mathrm{Ker}(p^n : A \to A)$ is a Barsotti–Tate p-divisible group. In the same way, $\mu_{p^\infty} = \bigcup_n \mathbb{G}_m[p^n]$ is a Barsotti–Tate group (over any base, as $\mathbb{G}_{m/\mathbb{Z}}$ is smooth).

4.3.8 Connected–étale Exact Sequence

For a given locally free group scheme $G_{/S}$ over a connected scheme S, intuitively, we expect that the connected component G° would be a locally free group scheme with étale quotient G/G°. This is not generally true, but we can prove the following fact:

Proposition 4.68 *For a locally free group scheme $G_{/S}$ of finite rank over $S = \mathrm{Spec}(A)$ for a Henselian local ring A, we have a canonical exact sequence of locally free group schemes:*

$$0 \to G^{\circ} \to G \to G^{\mathrm{et}} \to 0,$$

where G° is connected and G^{et} is étale.

Proof. Since $G \to S$ is finite, it is affine; hence, $G = \mathrm{Spec}(\mathcal{A})$ for a sheaf $\mathcal{A}$ of bialgebras finite flat over A. Since $ET_{/S}$ is a Galois category (see Sect. 4.2.3), for any two étale A-subalgebras $\mathcal{A}', \mathcal{A}'' \subset \mathcal{A}$, the composite $\mathcal{A}' \cdot \mathcal{A}''$ is still étale over A. Thus, there exists a unique maximal étale $\mathcal{O}$-subalgebra $\mathcal{A}^{\mathrm{et}} \subset \mathcal{A}$. Consider the comultiplication $m^* : \mathcal{A} \to \mathcal{A} \otimes_A \mathcal{A}$.

If $\mathcal{E} \subset \mathcal{A} \otimes_A \mathcal{A}$ is an étale algebra over A, then $\mathcal{E} \cdot (1 \otimes \mathcal{A}^{\mathrm{et}}) \subset \mathcal{A} \otimes_A \mathcal{A}$ is étale over $\mathcal{A}^{\mathrm{et}}$. Thus, by replacing $\mathcal{E}$ by $\mathcal{E} \cdot (\mathcal{A}^{\mathrm{et}} \otimes_A \mathcal{A}^{\mathrm{et}}) \subset \mathcal{A} \otimes_A \mathcal{A}$, we may assume that $\mathcal{E} \supset \mathcal{A}^{\mathrm{et}} \otimes_A \mathcal{A}^{\mathrm{et}}$. Writing $(\mathcal{A} \otimes_A \mathcal{A})^{\mathrm{et}}$ for the maximal étale extension of A inside $\mathcal{A} \otimes_A \mathcal{A}$, we thus have

$$(\mathcal{A} \otimes_A \mathcal{A})^{\mathrm{et}} \supset \mathcal{E} \supset \mathcal{A}^{\mathrm{et}} \otimes_A \mathcal{A}^{\mathrm{et}}.$$

For the moment, assume that $S = \mathrm{Spec}(k)$ for a field k. Then $\mathcal{A}^{\mathrm{et}}/k$ is the maximal separable extension of k inside $\mathcal{A}$. Thus, $\mathcal{A}^{\mathrm{et}} \otimes_k \mathcal{A}^{\mathrm{et}}$ is the maximal separable extension of k inside $\mathcal{A} \otimes_k \mathcal{A}$ (cf. [TCF] II.2.5.3), and

$$(\mathcal{A} \otimes_k \mathcal{A})^{\mathrm{et}} = \mathcal{E} = \mathcal{A}^{\mathrm{et}} \otimes_k \mathcal{A}^{\mathrm{et}}.$$

Taking the unique closed point $x \in S$ and looking into stalks at x, we thus have

$$(\mathcal{A}_x \otimes_A \mathcal{A}_x)^{\mathrm{et}} = \mathcal{E}_x = \mathcal{A}_x^{\mathrm{et}} \otimes_A \mathcal{A}_x^{\mathrm{et}} \Rightarrow (\mathcal{A} \otimes_k \mathcal{A})^{\mathrm{et}} = \mathcal{E} = \mathcal{A}^{\mathrm{et}} \otimes_k \mathcal{A}^{\mathrm{et}}$$

as $ET_{\mathcal{O}_{S,x}} \cong ET_{/k(x)}$ by Theorem 4.56. Since an injective A-algebra homomorphism from an étale extension of A into an A-algebra brings it into an étale extension of A, we find that m^* sends $\mathcal{A}^{\mathrm{et}}$ into $\mathcal{A}^{\mathrm{et}} \otimes_A \mathcal{A}^{\mathrm{et}}$. Similarly, a co-inverse sends $\mathcal{A}^{\mathrm{et}}$ into $\mathcal{A}^{\mathrm{et}}$. In short, $\mathcal{A}^{\mathrm{et}}$ is an A-bialgebra. Let $G^{\mathrm{et}} = \mathrm{Spec}(\mathcal{A}^{\mathrm{et}})$; thus, we have a morphism of finite flat group schemes $\pi : G \to G^{\mathrm{et}}$. We define $\mathrm{Ker}(\pi) = G^{\circ}$ and call it the connected component of G. Since A is local ring, over the separable closure $\overline{k}(x)$ of the residue field $k(x)$, G° has only one physical point; hence, $G^{\circ}(\overline{k}(x)) = \{\mathbf{0}_E\}$. Therefore, G° is indeed connected, and $G^{\circ} = \mathrm{Spec}(\mathcal{A}^{\circ})$ for the maximal local residue ring $\mathcal{A}^{\circ} \twoheadleftarrow \mathcal{A}$ of $\mathcal{A}$ through which the identity section $S \to G$ factors. In particular, $\mathcal{A}^{\circ}$ is finite flat over $\mathcal{O}_S$. $\square$

If G is a Barsotti–Tate group over A, $G^{\circ} := \bigcup G[p^n]^{\circ}$ and $G^{\mathrm{et}} := \bigcup G[p^n]^{\mathrm{et}}$ are obviously Barsotti–Tate groups, as connectedness and étaleness are kept under the inclusion maps and also multiplication by p.

Corollary 4.69 *For a Barsotti–Tate group $G_{/S}$ over $S = \mathrm{Spec}(A)$ with a Henselian local ring A, we have a canonical exact sequence of Barsotti–Tate groups:*

$$0 \to G^{\circ} \to G \to G^{\mathrm{et}} \to 0,$$

where G° is connected and G^{et} is étale.

4.3.9 Ordinary Barsotti–Tate Group

Let S be a scheme over which the fixed prime p is p-adically nilpotent.

Lemma 4.70 *A locally free group scheme $G_{/S}$ of a finite p-power rank is étale if and only if after base change to an étale finite extension, it is a direct sum of the finitely many constant group schemes of p-power order.*

See [GME] Lemma 1.12.5 for a proof.

A morphism of schemes $f : S' \to S$ (for S with p-adically nilpotent p) is called *quasi-compact* if $f^{-1}(U) = S' \times_S U$ for any open affine quasi-compact subscheme $U \subset S$ is quasi-compact. A Barsotti–Tate group G over a scheme S is called *ordinary* if there exists a faithfully flat quasi-compact covering $S' \to S$ such that $G_{/S'} = G \times_S S'$ fits into the exact sequence of Barsotti–Tate group of the following type:

$$0 \to \mu^s_{p^\infty} \to G_{/S'} \to (\mathbb{Q}_p/\mathbb{Z}_p)^r \to 0,$$

where $\mathbb{Q}_p/\mathbb{Z}_p = \bigcup_n p^{-n}\mathbb{Z}/\mathbb{Z}$ for the étale group scheme $p^{-n}\mathbb{Z}/\mathbb{Z}$ over S'. The integer r is called the *p-rank of* G and written as $r(G) := r$.

For the moment, we assume that S' is connected. Since S'/S is faithfully flat, $S' \to S$ is a surjection of the topological space (the going up/down theorem of ring theory); thus, S is connected. The topological space of μ_{p^n} is connected (as S is connected and on S, p is not invertible). If we write $G^{mult}_{/S'}$ for the image of $\mu^r_{p^\infty}$ in G, $G^{mult}_{/S'}$ is the maximal connected subgroup over each connected component of S' of $G_{/S'}$. Thus, defining $G^{mult}_{/S}$ by the maximal connected subgroup over S, we have $G^{mult}_{/S} \times_S S' \cong G^{mult}_{/S'}$, canonically. In other words, we get a canonical extension

$$0 \to G^{mult} \to G \to G^{\mathrm{et}} \to 0$$

such that G^{et} is the maximal étale quotient and G^{mult} is the maximal multiplicative subgroup.

Lemma 4.71 *Suppose that S is a reduced $\mathbb{Z}_{(p)}$-scheme and either that $p = 0$ all over S or that $\mu_{p^\infty}(k(\eta))$ is a finite group of bounded order for all generic points η of S. If the exact sequence $G^{mult} \hookrightarrow G \twoheadrightarrow G^{\mathrm{et}}$ of ordinary Barsotti–Tate groups splits over S, the splitting is unique.*

Proof. We may assume that S is irreducible and that G^{mult} is nontrivial.

We first assume that S is affine given by $\mathrm{Spec}(R)$. Consider the reduced part G^{red}, which is a closed subscheme of G. Since the reduced part is uniquely determined by G, the group structure of G preserves G^{red}, and hence G^{red} is a group. Since $G^{red}[p^n]$, $G[p^n]$, $G^{mult}[p^n]$ and $G^{\mathrm{et}}[p^n]$ are locally free of finite rank over S, and so they are affine. Write $G^?[p^n] = \mathrm{Spec}(R_n^?)$ for a locally free R-algebra $R_n^?$ of finite rank. Then R_n^{et} is the maximal subalgebra of R_n étale over R. Thus, if R is reduced, R_n^{et} is reduced. In particular, for the nilradical $nil(R_n)$, $nil(R_n) \cap R_n^{\mathrm{et}} = 0$. Thus, we get an injective R-algebra homomorphism $R_n^{\mathrm{et}} \hookrightarrow R_n^{red}$.

First suppose that $p = 0$ on S. Since $\mu_{p^n/\mathbb{F}_p} \cong \mathrm{Spec}(\mathbb{F}_p[X]/(X^{p^n}))$ as schemes, and $G^{mult}[p^n]$ is isomorphic to $\mu_{p^n}^r$ étale locally, G^{mult} is not reduced (if $G^{mult} \neq 0$). Thus, $G^{red}[p^n]$ is a proper subgroup of $G[p^n]$. Pick a geometric point $\mathrm{Spec}(k) \to S$ for an algebraically closed field k. Then $R_n^{\mathrm{et}} \otimes_R k \cong k^{G^{\mathrm{et}}[p^n](k)}$ as R-algebra, and $G^{mult}[p^n] \times_R k \cong \mathrm{Spec}(k[T]/((1+T)^{p^n} - 1))^r$. Thus, we have

$$G[p^n] \times_{\mathrm{Spec}(R)} \mathrm{Spec}(k) \cong \bigsqcup_{x \in G^{\mathrm{et}}[p^n](k)} \mathrm{Spec}(k[T]/((1+T)^{p^n} - 1))^r$$

as schemes. This shows that $R_n^{\mathrm{et}} \otimes_R k \cong R_n^{red} \otimes_R k \cong k^{G^{\mathrm{et}}[p^n](k)}$ for any geometric point of S; hence, $R_n^{\mathrm{et}} \cong R_n^{red}$. Thus, $G^{red}[p^n] \subset G[p^n]$ gives a unique splitting of $G[p^n] \twoheadrightarrow G^{\mathrm{et}}[p^n]$. This shows the uniqueness of the splitting $G = G^{mult} \times_S G^{\mathrm{et}}$ (cf. [ABV] page 136).

Now assume that $p \neq 0$ in S. Since S is irreducible and reduced, it is integral, and so p is not a zero divisor in R. Choose an integral domain R'/R faithfully flat étale over R so that G^{et} is constant over R' and $G^{mult} \cong \mu_{p^\infty}^r$ over R'. Such an R' exists by Lemma 4.70 combined with the duality theory of locally free groups (see Sect. 4.3.5). Suppose we have two splittings $G = G^{\mathrm{et}} \oplus G^{mult} = E \oplus T$ for multiplicative Barsotti–Tate groups G^{mult} and T and étale Barsotti–Tate groups G^{et} and E. Then consider the composed map $i : E \hookrightarrow G \twoheadrightarrow G^{mult}$. If the two splittings are not identical, the morphism i is nontrivial. Write k for the field of fraction of R'. Then $k[\mu_{p^\infty}]/k$ is an infinite Galois extension, since $\mu_{p^\infty}(R)$ is finite and R'/R is an étale extension. Write $T_pX = \varprojlim_n X[p^n](k)$ for the Tate module of X for a Barsotti–Tate group over an algebraically closed field k of characteristic 0. Then the morphism $i : G^{mult} \to E$ induces a morphism of $\mathrm{Gal}(k[\mu_{p^\infty}]/k)$-modules $i_* := T_pE \to T_pG^{mult}$. Since the Galois action on T_pE is trivial and it is via the cyclotomic character on T_pG^{mult} because $G^{mult} \cong \mu_{p^\infty}^r$ over R', we find that $i_* = 0$. Since $\mathrm{Hom}(E_{/R}, G^{mult}_{/R}) \hookrightarrow \mathrm{Hom}(E_{/R'}, G^{mult}_{/R'}) \hookrightarrow \mathrm{Hom}_{\mathrm{Gal}(k[\mu_{p^\infty}]/k)}(T_pE_{/k}, T_pG^{mult}_{/k})$ (e.g. [T1] Theorem 4), we find that $i = 0$, a contradiction. Thus, the splitting is unique.

In general, if S is not affine, cover S by affine open subschemes $\mathrm{Spec}(R)$, and apply the above argument to $G \times_S \mathrm{Spec}(R)$. Then the splitting (if it exists) must be unique. □

Let $T_{/S}$ and $E_{/S}$ be a multiplicative and an étale Barsotti–Tate group, respectively. Consider the sheafication

$$\underline{\mathrm{Hom}}_{\mathbb{Z}/p^n\mathbb{Z},S_{\mathrm{fppf}}}(E[p^n],T[p^n]) \ \ (\text{resp., } \underline{\mathrm{Ext}}^1_{\mathbb{Z}/p^n\mathbb{Z},S_{\mathrm{fppf}}}(E[p^n],T[p^n]))$$

of the following presheaf

$$U \mapsto \mathrm{Hom}_{\mathbb{Z}/p^n\mathbb{Z},U}(E[p^n]_{/U},T[p^n]_{/U})$$
$$(\text{resp., } U \mapsto \mathrm{Ext}^1_{\mathbb{Z}/p^n\mathbb{Z},U}(E[p^n]_{/U},T[p^n]_{/U}))$$

over the small fppf site S_{fppf} over S.

Lemma 4.72 *Any extension $T[p^n] \to X \twoheadrightarrow E[p^n]$ in the category of finite flat $\mathbb{Z}/p^n\mathbb{Z}$-modules over S splits over an fppf extension S'/S; thus, we have*

$$\underline{\mathrm{Ext}}^1_{S_{fppf}}(E[p^n],T[p^n]) = 0.$$

If S is reduced, satisfying the assumption of Lemma 4.71, *the splitting $X = E[p^n] \oplus T[p^n]$ (if it exists) is canonical.*

Proof. The uniqueness follows from the previous lemma. To prove the existence, first suppose that $E[p^n]$ is constant over R. Since $X \to E[p^n]$ is the epimorphism as abelian fppf R-sheaves, we can find an fppf extension R'/R such that the set-theoretic group homomorphism $X(R') \to E[p^n](R')$ is surjective. We may assume that $\mathrm{Spec}(R')$ is connected, replacing $\mathrm{Spec}(R')$ by its connected component. Choosing a basis $\overline{x}_1,\ldots,\overline{x}_r$ of $E[p^n](R')$ over $\mathbb{Z}/p^n\mathbb{Z}$, we can lift $\overline{x}_j$ to a point x_j in $X(R')$. Then we have $p^n x_j = 0$ because $X(R')$ is a $\mathbb{Z}/p^n\mathbb{Z}$-module. Then the group subscheme generated by $x_1,\ldots,x_r$ is a constant scheme over R', splitting $X \twoheadrightarrow E[p^n]$ over R'. By extending scalars to a finite étale extension, we may always make $E[p^n]$ constant, and thus we can find an fppf extension of R over which a splitting of $X \twoheadrightarrow E[p^n]$ takes place.

4.4 Completing a Scheme

We now assume that the base ring B is a local noetherian ring complete under the adic topology of the maximal ideal $\mathfrak{m} = \mathfrak{m}_B$. Set $B_n = B/\mathfrak{m}^n$, and define $ALG^{(\infty)}_{/B}$ as a full subcategory of $ALG_{/B}$ whose objects are rings killed by a power of $\mathfrak{m}_B$ (i.e., it is an object of $ALG_{/B_n}$ for sufficiently large n); thus, heuristically, $ALG^{(\infty)}_{/B} = \bigcup_{n>0} ALG_{/B_n}$. We also consider the category $\widehat{ALG}_{/B}$ of B-algebras complete under the $\mathfrak{m}_B$-adic topology. Thus, $\widehat{ALG}_{/B}$ is made of projective limits $A = \varprojlim_n A_n$ for $A_n \in ALG_{/B_n}$ with projection making the following diagram commutative for all $m > n > 0$:

$$\begin{array}{ccc} A_m & \longrightarrow & A_n \\ \uparrow & & \uparrow \\ B_m & \xrightarrow[\text{mod } \mathfrak{m}^n]{} & B_n. \end{array}$$

The morphism set $\mathrm{Hom}_{\widehat{ALG}_{/B}}(A, A')$ is made up of continuous B-algebra homomorphisms. Since any B_n-algebra is $\mathfrak{m}_B$-adically complete and any B_n-algebra homomorphism is $\mathfrak{m}_B$-adically continuous, $ALG_{/B}^{(\infty)}$ is a full subcategory of $\widehat{ALG}_{/B}$. Then we put

$$\widehat{\mathcal{C}}_B = COF(\widehat{ALG}_{/B}, SETS) = COF(ALG_{/B}^{(\infty)}, SETS).$$

Replacing the category $ALG_{/B}$ by $\widehat{ALG}_{/B}$ or $ALG_{/B}^{(\infty)}$, we reconstruct the theory described in the earlier sections.

4.4.1 Formal Schemes

For $A \in \widehat{ALG}_{/B}$, we define $\widehat{S}_A = \mathrm{Spf}_B(A)$ as a functor in $\widehat{\mathcal{C}}_B$ given by $\widehat{S}_A(R) = \mathrm{Hom}_{\widehat{ALG}_{/B}}(A, R)$. We call $\widehat{S}_A$ an *affine formal scheme* over B. Since $\mathrm{Hom}_{\widehat{ALG}_{/B}}(A, A') = \varprojlim_n \mathrm{Hom}_{ALG_{/B_n}}(A/\mathfrak{m}_B^n A, A'/\mathfrak{m}_B^n A')$, any functor $F \in COF(ALG_{/B}^{(\infty)}, SETS)$ extends uniquely to $\widehat{F} \in \widehat{\mathcal{C}}_B$. So, from time to time, we may identify $\widehat{\mathcal{C}}_B$ with $COF(ALG_{/B}^{(\infty)}, SETS)$. If $\mathfrak{a}$ is a closed ideal of A, we define $\overline{A} = A/\mathfrak{a}$, and the projection $A \twoheadrightarrow \overline{A}$ induces an inclusion of functors $\widehat{S}_{\overline{A}} \hookrightarrow \widehat{S}_A$. Such a formal subscheme is called a closed formal subscheme, and the morphism $\widehat{S}_{\overline{A}} \hookrightarrow \widehat{S}_A$ is called a closed immersion. For a subset $\{f_i\}_{i \in I}$ of A, define a subfunctor $\widehat{U}_I(R) = \{P \in \widehat{S}_A(R) | \sum_{i \in I} P(f_i)R = R\}$ of $\widehat{S}_A$. Then, by Nakayama's lemma (Lemma 10.8 in the text, [CRT] Theorem 2.2 and Theorem 8.4, or [BCM] II.3.2),

$$\sum_{i \in I} P(f_i)R = R \Leftrightarrow \sum_{i \in I} P(f_i)R/\mathfrak{m}_B R = R/\mathfrak{m}_B R.$$

Thus, $\widehat{U}_I$ only depends on $\{f_i \mod \mathfrak{m}_B A\}_{i \in I}$, and for this reason, the topological space associated with $\mathrm{Spf}_B(A)$ is given by the space of $\mathrm{Spec}_{B_1}(A/\mathfrak{m}_B A)$.

Example 4.73 *Formal additive group.* Let $A = B[X] = B[X_1, \ldots, X_n]$ for $X = (X_1, \ldots, X_n)$ (the n-variable polynomial ring), and set $\widehat{A} = \varprojlim_n B/\mathfrak{m}_B^n[X]$. Each element $\Phi(X) \in \widehat{A}$ is a power series $\Phi(X) = \sum_\alpha a_\alpha X^\alpha$ (with coefficients in B) for $\alpha = (\alpha_1, \ldots, \alpha_n)$ with $X^\alpha = \prod_j X_j^{\alpha_j}$ such that $\Phi \mod \mathfrak{m}_B^n \in B/\mathfrak{m}_B^n[X]$ for all $n > 0$; so, writing $|\alpha| = \sum_j \alpha_j$, we have

$$\mathrm{ord}_B(a_\alpha) = \min\{n | \mathfrak{m}_B^n \ni a_\alpha\} \to \infty \text{ as } |\alpha| \to \infty. \qquad (*)$$

Then we write $\widehat{S}_{\widehat{A}} \in \widehat{\mathcal{C}}_B$ as $\widehat{G}^n_{a/B}$. Note that $\widehat{G}^n_{a/B}(R) = R^n$. Indeed, a continuous B-algebra homomorphism $\phi : \widehat{A} \to R$ is determined by the value at X, and for any $x \in R^n$, $\Phi(x) = \lim_{n\to\infty} \sum_{|\alpha|<n} a_\alpha x^\alpha$ converges in $\mathfrak{m}_B$-adically complete R, giving rise to a continuous B-algebra homomorphism with $\phi(\Phi(X)) = \Phi(x)$ [so $\phi(X_j) = x_j$ for all j]. Thus, $\widehat{S}_{\widehat{A}}(R) \cong R^n$ by sending ϕ to $(\phi(X_1), \ldots, \phi(X_n)) \in R^n$. Since $\widehat{G}_{a/B}$ has values actually in the category of commutative algebras, $\widehat{G}_{a/B}$ is a ring formal scheme and called a formal additive group over B.

Example 4.74 Let $\widehat{A} = B[[X]] = B[[X_1, \ldots, X_n]]$ (n-variable formal power series ring) for $X = (X_1, \ldots, X_n)$. Then $B[[X]]$ is $\mathfrak{m}_B$-adically complete. Then we write $\widehat{S}_{B[[X]]} \in \widehat{\mathcal{C}}_B$ as $\widehat{\mathbb{G}}^n_a$. Note that $\widehat{\mathbb{G}}^n_a(R) = J(R)^n$ for the Jacobson radical $J(R)$ of R [here $J(R)$ is the intersection of all maximal ideals of R]. By definition, $J(R) \supset \mathfrak{m}_B R$ and $\mathfrak{m}_B J(R) \subset \mathfrak{m}_B R$. Thus, R is not only $\mathfrak{m}_B$-adically complete but also $J(R)$-adically complete. A continuous B-algebra homomorphism $\phi : \widehat{A} \to R$ is determined by the value at X, and for any $x \in J(R)^n$, $\Phi(x) = \lim_{n\to\infty} \sum_{|\alpha|<n} a_\alpha x^\alpha$ converges in the $J(R)$-adically complete R, giving rise to a continuous B-algebra homomorphism with $\phi(\Phi(X)) = \Phi(x)$ [so $\phi(X_j) = x_j$ for all j]. Thus, $\widehat{S}_{\widehat{A}}(R) \cong J(R)^n$ by sending ϕ to $(\phi(X_1), \ldots, \phi(X_n)) \in J(R)^n$. The formal scheme $\widehat{\mathbb{G}}_a$ is a group functor; so it is a formal group, which is also called the *formal additive group*.

Exercise 4.75 *Explain the difference between $\widehat{G}_a$ and $\widehat{\mathbb{G}}_a$.*

Almost all definitions of formal schemes are verbatim translations of the corresponding ones of schemes, replacing $\mathcal{C}_B$ by $\widehat{\mathcal{C}}_B$. An *open formal subfunctor* (resp., *closed formal subfunctor*) $U \subset X$ for $X \in \widehat{\mathcal{C}}_B$ is defined to be a functor such that $f^{-1}(U)$ is an open (resp., closed) affine formal scheme for any morphism $f : \widehat{S}_A \to X$ in $\widehat{\mathcal{C}}_B$. Since morphisms of $ALG^{(\infty)}_{/B}$ are just B-algebra homomorphisms [i.e., $ALG^{(\infty)}_{/B}$ is a full subcategory of $ALG_{/B}$], we can still apply Lemma 4.18 to the functor $X \in \widehat{\mathcal{C}}_B$ and obtain

Lemma 4.76 *If $U, U' \subset X$ are open subfunctors of $X \in \widehat{\mathcal{C}}_B$, then $U = U'$ if and only if $U(k) = U'(k)$ for all fields k in $ALG_{B/\mathfrak{m}_B}$.*

Note here any field in $ALG^{(\infty)}_{/B}$ actually is an object of $ALG_{B/\mathfrak{m}_B}$; so, in the above statement, we could have replaced $ALG_{B/\mathfrak{m}_B}$ by the (possibly) bigger category $ALG^{(\infty)}_{/B}$. This also supports the idea of associating the topological space $\mathrm{Spec}(A/\mathfrak{m}_B A)$ with $\widehat{S}_A$.

Again, an open (formal) covering $\{U_i\}_{i\in I}$ of a functor X in $\widehat{\mathcal{C}}_B$ is made of open subfunctors U_i in $\widehat{\mathcal{C}}_B$ such that $X(k) = \bigcup_{i\in I} U_i(k)$ for all fields k in $ALG_{B/\mathfrak{m}_B}$. A functor $X \in \widehat{\mathcal{C}}_B$ is called *local* if it satisfies the following set-theoretic exact sequence for any open covering $\{Y_i\}_{i\in I}$ of any object Y in $\widehat{\mathcal{C}}_B$:

$$\mathrm{Hom}_{\widehat{\mathcal{C}}_B}(Y,X) \to \prod_{i\in I} \mathrm{Hom}_{\widehat{\mathcal{C}}_B}(Y_i,X) \rightrightarrows \prod_{i,j\in I} \mathrm{Hom}_{\widehat{\mathcal{C}}_B}(Y_i \cap Y_j, X).$$

A *formal B-scheme* is a local formal functor in $\widehat{\mathcal{C}}_B$ that has an open covering made up of affine formal schemes. The structure sheaf $\mathcal{O}_X$ is sending each open formal B-subscheme $U \subset X$ to $\mathrm{Hom}_{\widehat{\mathcal{C}}_B}(U, \widehat{G}_{a/B})$. For each maximal point, we have completed stalk $\widehat{\mathcal{O}}_{X,x} \in \widehat{ALG}_{/B}$ given by $\varprojlim_n \mathcal{O}_{X,x}/\mathfrak{m}_x^n$ for $\mathcal{O}_{X,x} = \varinjlim_{U\ni x} \mathcal{O}_X(U)$ for U running over an open neighborhood under the order given by inclusion.

Example 4.77 *Formal completion along the special fiber.* Let X in $\mathcal{C}_B$ be a B-scheme with structure morphism $\iota_X : X \to S_B$. Take an integer $n > 0$, and consider $X_n \in \mathcal{C}_{B/\mathfrak{m}_B^n}$ given by $X_n(R) = X(R)$ for any $B/\mathfrak{m}_B^n$-algebra R (i.e., X_n is the restriction of X to the full subcategory $\mathcal{C}_{B/\mathfrak{m}B^n}$ of $\mathcal{C}_B$). Note that, for $B/\mathfrak{m}^n$-algebra R,

$$\begin{aligned}(X \times_{S_B} S_{B/\mathfrak{m}^n})(R) &= \{(x,y) \in X(R) \times S_{B/\mathfrak{m}^n}(R) | \iota_X(x) = \iota_{B/\mathfrak{m}^n}(y)\} \\ &= X_n(R/\mathfrak{m}^n R) = X_n(R).\end{aligned}$$

In particular, $S_{A,n} = S_{A/\mathfrak{m}^n A}$. Since X is local, its restriction X_n is local. Since $X = \bigcup_{i\in I} S_{A_i}$ for an affine open subscheme S_{A_i}, $X_n = \bigcup_{i\in I} S_{A_i,n} = \bigcup_{i\in I} S_{A_i/\mathfrak{m}^n A_i}$. Thus, the topological local ringed space associated with X_n is the one associated with the special fiber $X \times_{S_B} S_{B/\mathfrak{m}^n}$. We now restrict X to $ALG_{/B}^{(\infty)}$, getting $\widehat{X} \in \widehat{\mathcal{C}}_B$. Clearly, $\widehat{X}(R) = \varprojlim_n X_n(R/\mathfrak{m}^n R) = \varprojlim_n X(R/\mathfrak{m}^n R)$. The topological space of $\widehat{X}$ is just the one associated with X_1; however, the local ring at the point $P \in \widehat{X}$ is the $\mathfrak{m}$-adic completion of the local ring of X at P (so their ringed space structures are different). The formal scheme $\widehat{X}$ is called the formal completion of X along the special fiber at the maximal point of S_B. Note that $\widehat{G}_a$ is a formal completion of $\mathbb{G}_{a/B}$ along its special fiber.

Example 4.78 *Formal completion along a maximal point.* Take a maximal point $x \in X(k)$ for a B-scheme X. Then the stalk $\mathcal{O}_{X,x}$ is a local B-algebra with maximal ideal $\mathfrak{m}_x$. We can think of the completion $\widehat{\mathcal{O}}_{X,x} = \varprojlim_n \mathcal{O}_{X,x}/\mathfrak{m}_x^n$ and its formal affine scheme $\widehat{X}_x := \mathrm{Spf}_B(\widehat{\mathcal{O}}_{X,x})$ whose topological space is $\mathrm{Spec}(k)$. Since we can find an affine open neighborhood $U = S_A$ of x, x gives rise to a B-algebra homomorphism $A \to \mathcal{O}_{X,x}$ coming from the definition of stalks. Thus, this map gives rise to a maximal ideal $\mathfrak{m}_x \subset A$ with $A_{\mathfrak{m}_x} = \mathcal{O}_{X,x}$. Thus, $\mathrm{Spec}(\mathcal{O}_{X,x})$ is a subscheme of X, and its formal completion along x is the formal scheme $\widehat{X}_x = \mathrm{Spf}(\widehat{\mathcal{O}}_{X,x})$. The formal group $\widehat{\mathbb{G}}_a$ is a formal completion along the identity element 0 of $\mathbb{G}_a$. Since $\widehat{\mathcal{O}}_{X,x}$ is a complete local ring with residue field k, without losing generality, we can restrict the functor $\widehat{X}_x$ to the category $CL_{/B}$ of local B-algebras with residue field k complete under the adic topology of the maximal ideal. The morphism in $CL_{/B}$ is a B-algebra homomorphism whose pullback of the maximal ideal is maximal in the source (i.e., local B-algebra homomorphisms).

Example 4.79 *Formal multiplicative group.* We have already defined the additive formal groups $\widehat{G}_{a/B}$ and $\widehat{\mathbb{G}}_a$. If we start with the multiplicative group $\mathbb{G}_m = \mathrm{Spec}(B[t,t^{-1}])$ for a variable t, we can make a formal completion along the identity element $\overline{1} \in \mathbb{G}_m(B/\mathfrak{m}_B) = (B/\mathfrak{m}_B)^\times$. We write $\widehat{\mathbb{G}}_m$ for the resulting formal scheme. For the point $1_A \in \mathbb{G}_m(A) = A^\times$, its kernel as a B-algebra homomorphism contains $T := t - 1$. Thus, the kernel of $\overline{1}$ is $\mathfrak{m}_1 = (T, \mathfrak{m}_B)$, and $\mathcal{O}_{\mathbb{G}_m,1} = B[t,t^{-1}]_{\mathfrak{m}_1}$ (the localization at 1). This shows that

$$\widehat{\mathcal{O}}_{\mathbb{G}_m,\overline{1}} = \varprojlim_n B[t,t^{-1}]_{\mathfrak{m}_1}/\mathfrak{m}_1^n = \varprojlim_n B[T]/(T^n) \cong B[[T]].$$

Thus, as a scheme, $\widehat{\mathbb{G}}_m$ is isomorphic to $\widehat{\mathbb{G}}_a$, and $\widehat{\mathbb{G}}_m(R) \cong J(R)$ as sets by sending $\phi \in \mathrm{Hom}_{B\text{-alg, cont}}(B[t,t^{-1}], R)$ to $\phi(T) = \phi(t) - 1_R$. Since t is the variable compatible with group structure of $\mathbb{G}_m$ (the multiplication of the group scheme $\mathbb{G}_m$ is induced by the B-algebra homomorphism, sending $t \in B[t,t^{-1}]$ to $t \otimes t \in B[t,t^{-1}] \otimes_B B[t,t^{-1}]$; see Exercise 4.61), we have $\widehat{\mathbb{G}}_m(R) = 1 + J(R)$. Since the multiplicative group $1 + J(R)$ is not always isomorphic to the additive group $J(R)$, $\widehat{\mathbb{G}}_m \not\cong \widehat{\mathbb{G}}_a$ as formal groups over B. Similarly, we can take $\widehat{G}_{m/B}$, which is a formal completion of $\mathbb{G}_{m/B}$ along its special fiber. Then $\widehat{G}_m(R) = \{x \in R | x \mod \mathfrak{m}_B R \in (R/\mathfrak{m}_B R)^\times\}$.

Exercise 4.80 *Give an example of a local ring R such that the multiplicative group $1 + J(R)$ is not isomorphic to the additive group $J(R)$.*

4.4.2 Deformation Functors

Assume that B is a local ring with residue field k. Recall the category $CL_{/B}$ of local complete B-algebras with residue field k (introduced in Definition 4.78). A covariant functor $D \in COF(CL_{/B}, SETS)$ is called a *deformation functor* if $D(k)$ is a singleton (i.e., $|D(k)| = 1$). A deformation functor is representable if $D \cong \widehat{S}_A$ over $CL_{/B}$ for an object A in $CL_{/B}$. Let $CL_{/B}^{(\infty)}$ be the full subcategory of $CL_{/B}$ whose object is killed by a power of $\mathfrak{m}_B$. As before, we do not lose generality by restricting a deformation functor to $CL_{/B}^{(\infty)}$, though the ring A representing D may not be in $CL_{/B}^{(\infty)}$.

For any R-scheme $X_{/R}$ ($R \in CL_{/B}$), we write $X \times_R k$ for the restriction of X to $ALG_{/k}$ (the special fiber of X). Give oneself a property P of schemes relative to the base compatible to base change (i.e., if $T_{/S}$ satisfies P relative to S, then $T \times_S S'$ satisfies P relative to S' for all S-schemes S'). The property

P can be flatness, smoothness, finiteness, or étaleness. Give oneself a group scheme $G_{/k}$ with property P over k. Consider a functor $D_G^P : CL_{/B} \to SETS$ given by

$$D_G^P(R) = \left\{ (\mathcal{G}_{/R}, \iota) \middle| \mathcal{G} \in GSCH_{/R} \text{ satisfies } P \text{ over } R,\ \iota : \mathcal{G} \times_R k \cong G \right\} / \cong .$$

Here two pairs, $(\mathcal{G}, \iota)$ and $(\mathcal{G}', \iota')$, are isomorphic if we have an isomorphism $I : \mathcal{G} \to \mathcal{G}'$, making the following diagram commutative:

$$\begin{array}{ccc} \mathcal{G} \times_B k & \xrightarrow{\iota} & G \\ \wr\downarrow I & & \downarrow \| \\ \mathcal{G}' \times_B k & \xrightarrow{\iota'} & G. \end{array}$$

Obviously, D_G^P is a deformation functor.

Lemma 4.81 *Take $P = ffl$ to be the property of $\mathcal{G}_{/R}$ being finite flat over R, and take $G = \mathbb{Z}/N\mathbb{Z}$ or μ_N. Then D_G^{ffl} is represented by B.*

Proof. We need to prove that $|D_G^{ffl}(R)| = 1$ for all R in $CL_{/B}$. Since $\mathbb{Z}/N\mathbb{Z}_{/R}$ always gives a deformation [i.e., an element of $D_G^{ffl}(R)$], we need to prove that any deformation $\mathcal{G}_{/R}$ is isomorphic to $\mathbb{Z}/N\mathbb{Z}_{/R}$. Since $\mathcal{G}_{/R}$ is finite, it is affine by Lemma 4.24. Thus, $\mathcal{G} = S_A$ for a B-algebra A finite flat over R; hence, $A \cong R^n$ as an R-module because R is a local ring (cf. [CRT] Theorem 7.10). Since $A/\mathfrak{m}_R A \cong \bigoplus_{i \in \mathbb{Z}/N\mathbb{Z}} k$, by Hensel's lemma (cf. [CRT] Theorem 8.3), the completeness of R tells us that each idempotent $\overline{e}_i$ corresponding to the ith factor of $A/\mathfrak{m}_R A$ can be uniquely lifted to an idempotent e_i of A ([BCM] III.4.6); so $A = \bigoplus_{i \in \mathbb{Z}/N\mathbb{Z}} R$. Thus, for any R-algebra R' in $CL_{/R}$, $S_A(\pi) : S_A(R') \cong S_A(k)$ as groups by the unique projection $\pi : R' \twoheadrightarrow k$. Thus, S_A is a constant functor having value $\mathbb{Z}/N\mathbb{Z}$ all over $CL_{/R}$; so $\mathcal{G} \cong \mathbb{Z}/N\mathbb{Z}_{/R}$. If $G = \mu_N$, pick $\mathcal{G}_{/R} \in D_{\mu_N}^{ffl}(R)$. Then the Cartier dual $\mathcal{G}^*_{/R}$ is a deformation in $D_{\mathbb{Z}/N\mathbb{Z}}^{ffl}(R)$; so $\mathcal{G}^* \cong \mathbb{Z}/N\mathbb{Z}_{/R}$. Taking the dual back, we find $\mathcal{G} \cong \mu_{N/R}$, as desired.

4.4.3 Connected Formal Groups

Consider the kernel $\mathbb{G}_m[N]$ of the multiplication $[N] : \mathbb{G}_m(R) \ni q \mapsto q^N \in \mathbb{G}_m(R)$ by an integer $N > 1$. Since $\mathbb{G}_m = \mathrm{Spec}(B[t, t^{-1}])$, we have for the group identity $e \in \mathbb{G}_m$ [with $\underline{e}(t) = 1$]

$$\begin{aligned} \mathbb{G}_m[N] = \mathbb{G}_m \times_{\mathbb{G}_m, [N]} e &= \mathrm{Spec}(B[t, t^{-1}] \otimes_{B[t,t^{-1}], t \mapsto t^N} B) \\ &= \mathrm{Spec}(B[T]/((1+T)^N - 1) = \mu_N, \end{aligned}$$

where $T = t - 1$. Suppose that B is p-profinite noetherian local ring with characteristic p residue field. Then $B/\mathfrak{m}_B^n$ is a finite ring with p-power order, so, for any $n > 1$, $\widehat{\mathbb{G}}_m(B/\mathfrak{m}_B^n) = 1 + J(B/\mathfrak{m}_B^n) = 1 + \mathfrak{m}_B/\mathfrak{m}_B^n$ is made of a p-power root of unity. Thus, $\widehat{\mathbb{G}}_m(B/\mathfrak{m}_B^n) = \mu_{p^N}(B/\mathfrak{m}_B^n)$ for sufficiently large N. For any Artinian local ring A over B, by the same token, $\widehat{\mathbb{G}}_m(A) = \mu_{p^N}(A)$ for sufficiently large N.

Write $ART_{/B}$ for the full subcategory of $CL_{/B}$ made up of Artinian local B-algebras with residue field $k = B/\mathfrak{m}_B$. Note that

$$COF(CL_{/B}, SETS) \xrightarrow{F \mapsto F|_{ALT}} COF(ART_{/B}, SETS)$$

is an equivalence of category, as $F(\varprojlim_n R/\mathfrak{m}_R^n) = \varprojlim_n F(R/\mathfrak{m}_R^n)$. This shows that $\widehat{\mathbb{G}}_m$ and $\varinjlim_n \mu_{p^n}$ induces the same functor on $ALT_{/B}$. Since $\widehat{\mathbb{G}}_m = \mathrm{Spf}(B[[T]])$ and $\mu_N = \mathrm{Spf}(B[T]/((1+T)^{p^n} - 1)$, we can also check this fact from the identity $B[[T]] = \varprojlim_n B[T]/((1+T)^{p^n} - 1)$. In other words, $\varinjlim_n \mu_{p^n}$ over $ALT_{/B}$ is prorepresented by $\widehat{\mathbb{G}}_m$ in $\widehat{\mathcal{C}}_B$.

A group functor $\widehat{G}$ in $COF(CL_{/B}, AB)$ is called a *connected smooth formal group* (or a *formal Lie group*) if it is a connected smooth formal scheme with multiplication by p inducing a finite flat morphism $\widehat{G} \to \widehat{G}$; so, in particular, $\widehat{G}$ is an fppf p-divisible abelian sheaf over B. Let $LIE_{/B}$ (resp., $CBT_{/B}$) be the full subcategory of connected smooth formal groups over B in $COF(CL_{/B}, AB)$ (resp., of connected Barsotti–Tate groups over B). The multiplicative group $\widehat{\mathbb{G}}_m$ is a prototypical example of objects in $LIE_{/B}$.

For a free $\mathbb{Z}_p$-module L of rank n, similarly to Example 4.58 (4), we can think of $\widehat{\mathbb{G}}_m \otimes_{\mathbb{Z}_p} L$, which sends R to $(1+\mathfrak{m}_R) \otimes_{\mathbb{Z}_p} L \cong (1+\mathfrak{m}_R)^n = \widehat{\mathbb{G}}_m^n(R)$. Then $L \cong \mathrm{Hom}_{LIE_{/B}}(\widehat{\mathbb{G}}_m, \widehat{\mathbb{G}}_m \otimes_{\mathbb{Z}_p} L) =: X_*(\widehat{\mathbb{G}}_m \otimes L)$ canonically. We put $X^*(\widehat{\mathbb{G}}_m \otimes L) = \mathrm{Hom}_{LIE_{/B}}(\widehat{\mathbb{G}}_m \otimes_{\mathbb{Z}_p} L, \widehat{\mathbb{G}}_m)$. They are $\mathbb{Z}_p$-dual each other canonically in the same manner as in Example 4.58 (4). The group of the form $\widehat{\mathbb{G}}_m \otimes_{\mathbb{Z}_p} L$ is called a *formal (split) torus*. If a formal subscheme Z of a formal torus is a formal subgroup given by $\widehat{\mathbb{G}}_m \otimes_{\mathbb{Z}_p} L'$ for a direct summand L' of L, it is called a *formal subtorus*. Note that multiplication by a prime p of $\widehat{\mathbb{G}}_a$ over a field of characteristic p is not finite; so $\widehat{\mathbb{G}}_{a/B}$ is not an object in $LIE_{/B}$, since p is not invertible in B.

A morphism $f : G \to G'$ in $CBT_{/B}$ is a compatible collection of group morphisms $f_n : G[p^n] \to G'[p^n]$ of finite flat group schemes; so the following diagram is commutative:

$$\begin{array}{ccc} G[p^n] & \xrightarrow{\hookrightarrow} & G[p^{n+1}] \\ {\scriptstyle f_n}\downarrow & & \downarrow{\scriptstyle f_{n+1}} \\ G[p^n] & \xrightarrow{\hookrightarrow} & G[p^{n+1}]. \end{array}$$

Since $\widehat{G}$ in $LIE_{/B}$ is smooth, we have $\widehat{G} = \mathrm{Spf}_B(B[[X_1, \ldots, X_r]])$, where the identity element $\mathbf{0}$ is given by $X_i \mapsto 0$ for all i. The multiplication $\widehat{G} \times \widehat{G} \to$

$\widehat{G}$ induces the comultiplication $B[[X]] \to B[[X]]\widehat{\otimes}_B B[[X]] = B[[X,Y]]$, where $X_i \otimes 1$ is X_i in $B[[X,Y]]$ and $1 \otimes X_i = Y_i$ in $B[[X,Y]]$. This comultiplication is determined by the image of X_i, which we write as $\Phi_i(X,Y)$. We simply write $\Phi(X,Y) = (\Phi_i(X,Y))_i$. Then group law requires the following relation:

1. $X = \Phi(X,0) = \Phi(0,X)$ (**0** gives the identity);
2. $\Phi(X, \Phi(Y,Z)) = \Phi(\Phi(X,Y),Z)$ (associativity);
3. $\Phi(X,Y) = \Phi(Y,X)$ (commutativity).

Since $\widehat{G} \cong \widehat{\mathbb{G}}_a^r$ as formal schemes, we have a set-theoretic identity $\widehat{G}(R) \cong \mathfrak{m}_R^r$. Thus, $\widehat{G}(R)$ has order $|\mathfrak{m}_R|^r = p^N$ for $0 < N \in \mathbb{Z}$ if $R \in ALT_{/B}$. In other words, we have $\widehat{G}(R) = \widehat{G}[p^N](R)$. Thus, $\widehat{G}(R) = \widehat{G}[p^\infty](R) := \bigcup_n \widehat{G}[p^n](R)$ for $R \in ALT_{/B}$. Since $p : \widehat{G} \to \widehat{G}$ is finite flat, $\widehat{G}[p^n]$ is a finite flat group scheme; so $\widehat{G}[p^\infty]$ is a Barsotti–Tate group over B. Thus, we get the following:

Lemma 4.82 *Suppose that B is a p-profinite local ring. Then the functor $\widehat{G} \mapsto \widehat{G}[p^\infty]$ gives an equivalence of category between $LIE_{/B}$ and $CBT_{/B}$.*

For $\widehat{G}_{0/k} \in LIE_{/k}$ $(k = B/\mathfrak{m}_B)$, consider the deformation functor

$$D^{LIE}_{\widehat{G}_0}(R) = \{(\widehat{G}_{/R}, \iota) | \widehat{G}_{/R} \in LIE_{/R}, \iota : \widehat{G} \times_R k \cong \widehat{G}_{0/k}\}/\cong .$$

By Lemma 4.81, μ_{p^n} is rigid (i.e., we have only constant deformations over Artinian B-algebras); so μ_{p^∞} is also rigid, and by the equivalence between $LIE_{/B}$ and $CBT_{/B}$, the rigidity of μ_{p^∞} (Lemma 4.81) tells us that

Corollary 4.83 *The deformation functor $D^{LIE}_{\widehat{\mathbb{G}}_m}$ is represented by B.*

Lemma 4.84 *Let G be a commutative formal Lie group over a complete local ring B with residue field inside $\mathbb{F} = \overline{\mathbb{F}}_p$. Write the group operation additively as $(x,y) \mapsto x+y$. Then $N_G : G \to G$ given by $x \mapsto Nx$ induces multiplication by N on the tangent space $\mathcal{T}_{G/B}$ of the identity* **0**.

Proof. Assume first that $S = \mathrm{Spec}(k)$ for a field k; so $G = \mathrm{Spf}_B(\mathcal{R})$ for $\mathcal{R} := B[[X]]$ with $X = (X_1, \ldots, X_r)$. Let $\mathcal{T} = \mathcal{T}_{G/k} := \mathrm{Hom}((X)/(X)^2, k) = \mathrm{Hom}_{\mathcal{R}}(\Omega_{G/k}(\mathbf{0}), k)$ [the tangent space at the identity $\mathbf{0}$; so $\Omega_{G/k}(\mathbf{0}) = \Omega_{G/k} \otimes_{\mathcal{R}} \mathcal{R}/(X)$]. Then the addition "+": $G \times G \to G$ induces a k-linear binary operation $\mathcal{T} \oplus \mathcal{T} \to \mathcal{T}$, which we write for the moment as a"+"b. This operation is associative, commutative (by the associativity/commutativity of the group structure on G), and k-linear. We claim that on $\mathcal{T}$, "+" and the natural addition on $\mathcal{T}$ coincide up to k-linear automorphism of $\mathcal{T}$. Since $G[p] = \mathrm{Ker}(p_G)$ and G share the same tangent space (as $G[p]$ is a closed connected subgroup of G finite flat over k), we may replace G by $G[p]$. If $r = 1$, then $\mathcal{T} \cong k$, and there is a unique k-vector space structure on k up to isomorphism; so the claim holds. If $r > 1$, by extending scalars to $\overline{k} = \mathbb{F}$, we may assume to have an exact sequence $0 \to G_1 \to G \to G_2 \to 0$ with $0 < \dim G_i < r$ as the category of height-1 commutative group schemes over an algebraically closed field k of characteristic p is equivalent to the category of abelian Lie algebras over k (cf. [ABV] Sects. 11 and 13). Then, by induction on $\dim G$, we get the desired result. In particular, N_G acts on $\mathcal{T}$ by multiplication by N. In general, this is true fiber by fiber [or just by the same argument in the case of $S = \mathrm{Spec}(k)$], and it is true over a general base S.

Let W be a discrete valuation ring finite flat over $\mathbb{Z}_p$ inside $\overline{\mathbb{Q}}_p$, and take $B = W$. Identify $\Lambda = W[[T]]$ with $\widehat{W[t, t^{-1}]} = \varprojlim_n W[t]/(t^{p^n} - 1)$ by $t = 1+T$. Then the formal completion $\widehat{\mathbb{G}}_m$ of the multiplicative group $\mathbb{G}_m$ along the identity $1 \in \mathbb{G}_m(\mathbb{F})$ is given by $\mathrm{Spf}(\Lambda) = \varinjlim_n \mathrm{Spf}(W[t]/(t^{p^n} - 1)) = \varinjlim_n \mu_{p^n}$ (over $ART_{/W}$). Write $\widehat{\mathbb{G}}_m \times \widehat{\mathbb{G}}_m = \mathrm{Spf}(\widehat{W[t, t^{-1}, t', t'^{-1}]})$ for the variable t of the left factor and t' of the right factor, and let $z \in \mathbb{Z}_p^\times$ act on $\widehat{\mathbb{G}}_m \times \widehat{\mathbb{G}}_m$ by $(t, t') \mapsto (t^z, t'^z)$ for $z \in \mathbb{Z}_p^\times$, where $t^z = (1 + T)^z = \sum_{n=0}^{\infty} \binom{z}{n} T^n$. We now prove Lemma 3.30 formulated in a slightly different way. Our formulation here is more geometric than Lemma 3.30, and we leave to the readers the verification of equivalence between Lemma 3.30 and the following result.

Lemma 4.85 *Let $\Phi(t, t') = t'^d + a_1(t)t'^{d-1} + \cdots + a_d(t) \in \Lambda[t']$ with $\Phi(1, 1) = 0$ be an irreducible polynomial over $Q = Frac(\Lambda)$. Suppose that the integral closure of $\mathbb{Z}_p$ in $\mathbb{J} := \widehat{W[t, t^{-1}, t', t'^{-1}]}/(\Phi(t, t'))$ is equal to W. If the integral formal closed subscheme $Z = \mathrm{Spf}(\mathbb{J})$ of $\widehat{\mathbb{G}}_m \times \widehat{\mathbb{G}}_{m/W}$ of codimension 1 is stable under the action of an open subgroup U of $\mathbb{Z}_p^\times$, we have $\Phi(t, t') = t'^d - t^b$ for $b \in \mathbb{Z}_p$ with $d\mathbb{Z}_p + b\mathbb{Z}_p = \mathbb{Z}_p$ for a p-power d.*

Proof. Write $\mathbb{J} := \Lambda[t']/(\Phi(t, t'))$. Regard $\mathbb{J}$ as embedded in an algebraic closure $\overline{Q}$ of Q. Since $\overline{\mathbb{Q}}_p \cap \mathbb{J} = W$ in $\overline{Q}$, $\mathbb{J} \otimes_W \overline{\mathbb{Q}}_p$ is an integral domain by [TCF] §3.5. Thus, $\mathrm{Spec}(\mathbb{J})$ [and $\mathrm{Spec}(\mathbb{J} \otimes_W \overline{\mathbb{Q}}_p)$] is geometrically irreducible with dominant projection to the right factor $\widehat{\mathbb{G}}_m = \mathrm{Spf}(\Lambda)$. By assumption, $Z \subset \widehat{\mathbb{G}}_m \times \widehat{\mathbb{G}}_m$ is stable under the diagonal action $(t, t') \mapsto (t^s, t'^s)$ for $s \in U$. We may assume that $U = 1 + p^r\mathbb{Z}_p$ for $r > 0$. Since Z is flat of relative dimension 1 over W, replacing W by a finite extension if necessary, we find in Z a W-point $(t_0, t_0') \in \widehat{\mathbb{G}}_m^2(W)$ of infinite order. Thus, we have an infinite set $\Xi = \{(t_0^s, t'^s_0) | s \in U\} = \left\{(t_0 t_0^{p^r u}, t'_0 t'^{p^r u}_0) \middle| u \in \mathbb{Z}_p\right\}$ inside Z. By translation, $\tau : (t, t') \mapsto (t t_0^{-1}, t' t'^{-1}_0)$, $\tau(Z)$ contains $\Xi_0 = \left\{(t_0^{p^r u}, t'^{p^r u}_0) \middle| u \in \mathbb{Z}_p\right\}$.

Since Z is finite flat over $\mathrm{Spf}(\Lambda)$ with dominant projection to the right factor $\widehat{\mathbb{G}}_m$, by flatness, $\mathbb{J}$ is a free module over Λ (with basis $1, t', \ldots, t'^{d-1}$), and $\bigcap_{P \in \Xi_0} (P \cap \Lambda)\mathbb{J} = 0$ [regarding $P \in \Xi_0$ as a prime ideal $P \in \mathrm{Spec}(\mathbb{J})$]. So $\tau(Z)$ is the Zariski closure of the infinite subgroup Ξ_0 of $\widehat{\mathbb{G}}_m^2$. Then $\tau(Z)$ must be a formal subgroup of $\widehat{\mathbb{G}}_m^2$ of codimension 1, and Z is a coset $(t_0, t_0')\tau(Z)$ in $\widehat{\mathbb{G}}_m \times \widehat{\mathbb{G}}_m$. Since Z contains the identity $\mathbf{1}$ of $\widehat{\mathbb{G}}_m \times \widehat{\mathbb{G}}_m$ as $\Phi(1, 1) = 0$, we must have $\tau(Z) = Z$. Since the left projection $\pi : Z \to \mathrm{Spf}(\Lambda) = \widehat{\mathbb{G}}_m$ is finite flat of degree d, the kernel $\mathrm{Ker}(\pi)$ is a finite flat group scheme over W, which is therefore a finite flat subgroup of $\widehat{\mathbb{G}}_m^2[p^n] = \mu_{p^n} \times \mu_{p^n}$ for sufficiently large n. Since the Cartier dual of $\mu_{p^n} \times \mu_{p^n}$ is the constant group $(\mathbb{Z}/p^n\mathbb{Z}) \times (\mathbb{Z}/p^n\mathbb{Z})$ (see Exercise 4.64), $\mathrm{Ker}(\pi)$ is isomorphic to $\mu_{p^m} \times \mu_{p^{m'}}$ (for some $m \le n$ and $m' \le n$). Thus, Z is an extension of $\widehat{\mathbb{G}}_m$ by $\mu_{p^m} \times \mu_{p^{m'}}$.

By the geometric irreducibility of $Z = \mathrm{Spec}(\mathbb{J})$, multiplication by p induces a dominant morphism on $Z \to Z$, and hence $Z[p^\infty]$ is p-divisible. Since $\mathrm{Ker}(\pi)$ is finite flat, $Z[p^\infty]$ is a Barsotti–Tate group fitting into an exact sequence $0 \to \mathrm{Ker}(\pi) \to Z[p^\infty] \to \mu_{p^\infty} \to 0$; so we may assume that $m' = 0$. By Cartier duality (see Sect. 4.3.5) applied to this sequence, the dual of $Z[p^\infty]$ is étale, and hence $Z[p^\infty]$ must be isomorphic to μ_{p^∞}. Therefore, $Z \cong \varinjlim_n \mu_{p^n}$ is a formal multiplicative group, and Z is a formal subtorus of $\widehat{\mathbb{G}}_m^2$. Writing $X^*(\widehat{\mathbb{G}}_m \times \widehat{\mathbb{G}}_m) \cong \mathbb{Z}_p^2$ for the formal character group of $\widehat{\mathbb{G}}_m \times \widehat{\mathbb{G}}_m$, we have that Z is the kernel of $\chi \in X^*(\widehat{\mathbb{G}}_m \times \widehat{\mathbb{G}}_m)$. Writing $\chi(t,t') = t^a t'^{-b'}$, we have $Z = \{(t,t')|t^{b'} = t'^a\}$ for $a, b' \in \mathbb{Z}_p$. We may assume that $a = up^n$ for $u \in \mathbb{Z}_p^\times$ by the original form of Φ. Then $Z = \{(t,t')|t^b = t'^{d'}\}$ for $d' = p^n$ and $b = b'u^{-1}$. Thus, we have $d = d'd''$ for an integer d'' prime to p. Since we can extract the d''th root inside $1 + \mathfrak{m}_\Lambda$ for the maximal ideal $\mathfrak{m}_\Lambda$ of Λ (as $1 + \mathfrak{m}_\Lambda$ is p-profinite), the irreducibility of $\Phi(t,t')$ in $Q[t']$ implies $d = d'$, and $\Phi(t,t') = (t'^d - t^b)u(t,t')$ for a unit $u(t,t') \in W\widehat{[t,t^{-1},t',t'^{-1}]}$. Since $\Phi(t,t')$ is irreducible of degree $d = d'$ in $Q[t']$, $u(t,t')$ must be a unit in Λ. Since $\Phi(t,t')$ and $t'^d - t^b$ are monic with respect to t', we conclude $u(t,t') = 1$. This finishes the proof.

Remark 4.86 In the above proof, the two conditions $\Phi(1,1) = 0$ and geometric irreducibility are both indispensable. If we do not suppose $\Phi(1,1) = 0$, for a one-dimensional formal subtorus $G \subset \widehat{\mathbb{G}}_m \times \widehat{\mathbb{G}}_m$, a coset $(\zeta,\zeta')G$ for $(1,1) \neq (\zeta,\zeta') \in \mu_{p^m}(W) \times \mu_{p^m}(W)$ is stable under $U = 1 + p^m\mathbb{Z}_p$ but is not a formal subgroup. As a special instance of this example, we have the case when $\Phi(t,t')$ is an irreducible factor $\Psi(t')$ of $t'^{p^n} - 1$ [and in this case, $Z = \widehat{\mathbb{G}}_m \times \zeta$ for a root ζ of $\Psi(t') = 0$ in W; so, actually, $\Psi(t,t') = t' - \zeta$ by geometric irreducibility]. Similarly, if we do not suppose the geometric irreducibility of $\mathrm{Spec}(\mathbb{J})$, $Z \subset \widehat{\mathbb{G}}_m \times \widehat{\mathbb{G}}_m$ could be of the form $(\mu_{p^m} \times \mu_{p^r}) \cdot G$ (with $m + r > 0$), which is not a formal subtorus but a geometrically reducible formal subgroup. In the above example of $\Phi(t,t') = \Psi(t')$, the geometric irreducibility of $\mathrm{Spec}(\mathbb{J})$ forces $\Phi(t') = t' - 1$ (and hence $d = 1$ and $b = 0$ under the notation of the above lemma).

4.4.4 Infinitesimal Splitting Implies Local Splitting

By Corollary 4.37, if A is an excellent local ring, $\widehat{A} = \varprojlim_n A/\mathfrak{m}^n$ is reduced and faithfully flat over A.

Lemma 4.87 *Let A be an excellent reduced local ring with maximal ideal $\mathfrak{m}$. Suppose that $pA = 0$ or $A \subset A \otimes_{\mathbb{Z}} \mathbb{Q}$ (characteristic 0). Then $\overbrace{\widehat{A} \otimes_A \cdots \otimes_A \widehat{A}}^{n}$ is reduced.*

Proof. We first deal with $\widehat{A} \otimes_A \widehat{A}$. Let $\widehat{Q}$ (resp., Q) be the total ring of fractions of $\widehat{A}$ (resp., A). Since A and $\widehat{A}$ are reduced, we have $\widehat{A} \otimes_A \widehat{A} \hookrightarrow \widehat{Q} \otimes_A \widehat{Q}$. Since $\widehat{Q}$ is flat over $\widehat{A}$ and $\widehat{A}$ is faithfully flat over A, for any nonzero divisor $a \in A$, $\widehat{Q}$ is a-torsion-free. Thus, $\widehat{Q} \otimes_A \widehat{Q}$ is a-torsion-free. Then $a(\frac{1}{a}x \otimes_A y) = x \otimes y = a(x \otimes_A \frac{1}{a}y)$ implies $\frac{1}{a}x \otimes_A y = x \otimes_A \frac{1}{a}y$; so $\widehat{Q} \otimes_A \widehat{Q} = \widehat{Q} \otimes_Q \widehat{Q}$. Thus, we need to show that $\widehat{Q}$ is separable over Q. In the characteristic-0 case, this is true for any ring. In the same way as above,

$$\widehat{Q} \otimes_A \widehat{Q} \otimes \cdots \otimes_A \widehat{Q} = \widehat{Q} \otimes_Q \widehat{Q} \otimes \cdots \otimes_Q \widehat{Q},$$

and the same argument works.

Now, we assume that $pA = 0$. Since the inclusion $A \hookrightarrow \widehat{A}$ is regular [see (e2) in Sect. 4.1.11 and [EGA] IV.7.8.3 (v)], the natural map

$$A = \overbrace{A \otimes_A \cdots \otimes_A A}^{n} \to \overbrace{\widehat{A} \otimes_A \cdots \otimes_A \widehat{A}}^{n}$$

is regular. Indeed, for any prime ideal $\mathfrak{p} \subset A$ and its residue field $k(\mathfrak{p})$,

$$\overbrace{(\widehat{A} \otimes_A \cdots \otimes_A \widehat{A})}^{n} \otimes_A k' \cong \overbrace{(\widehat{A} \otimes_A k') \otimes_{k(\mathfrak{p})} \cdots \otimes_{k(\mathfrak{p})} (\widehat{A} \otimes_A k')}^{n}$$

is regular for any finite extension $k'/k(\mathfrak{p})$ if this is true for $n = 1$. In particular, $\widehat{A} \otimes_A \cdots \otimes_A \widehat{A}$ is reduced. $\square$

Let $S = \mathrm{Spec}(\mathcal{O}_S)$ be an integral excellent affine scheme of finite type flat over $\mathbb{Z}_{(p)}$ (i.e., $\mathcal{O}_S$ is an excellent integral domain of finite type flat over $\mathbb{Z}_{(p)}$). Let $Y \to S$ be a Barsotti–Tate group over S. If there exists a finite field extension of $\mathbb{Q}_p$ with integer ring O having a ring embedding $O \hookrightarrow \mathrm{End}(Y_{/S})$ sending the identity to the identity map on Y, we call Y a Barsotti–Tate O-module. The group Y (resp., a finite flat group scheme) is called *ordinary* if it is an extension of an étale Barsotti–Tate group (resp., an étale group scheme) $Y_s^{\mathrm{et}} \to \widehat{S}_s$ by a toric Barsotti–Tate group $Y_s^{mult} \to \widehat{S}_s$ (resp., a multiplicative finite flat group scheme) over all closed points $s \in S$, where $\widehat{S}_s$ is the formal completion of S along s, thus, étale locally over $\widehat{S}_s$, $Y_{/\widehat{S}_s}^{mult} \cong \mu_{p^\infty}^{m(s)}$ and $Y_{/\widehat{S}_s}^{\mathrm{et}} \cong (\mathbb{Q}_p/\mathbb{Z}_p)^{e(s)}$. We call an ordinary Barsotti–Tate O-module $Y_{/S}$ a *$GL(2)$-type*, if we have $\mathrm{rank}_{\mathbb{Z}_p} O = e(s) = m(s)$ for all closed points s of S [as the Galois representation on its Tate module has values in $GL_2(O)$]. The following result is from [Ch1] Proposition 8.4:

Proposition 4.88 *Let $S = \mathrm{Spec}(\mathcal{O}_S)$ be an integral excellent affine scheme of finite type flat over $\mathbb{F}$ for a subfield $\mathbb{F}$ of $\overline{\mathbb{F}}_p$. Let $Y_{/S}$ be an ordinary Barsotti–Tate O-module of GL(2)-type. Suppose that we have a closed point s_0 of S such that the extension*

$$0 \to (Y \times_S \widehat{S})^{mult} \to Y \times_S \widehat{S} \to (Y \times_S \widehat{S})^{\mathrm{et}} \to 0$$

splits into $Y \times_S \widehat{S} = (Y \times_S \widehat{S})^{mult} \times \underline{Y}^{\mathrm{et}}_{/\widehat{S}}$ over the formal completion $\widehat{S}$ of S at s_0. Then there exists an open neighborhood U of s_0 in S such that we have a decomposition $Y = G_1 \times G_2$ over U of Barsotti–Tate O-modules that induces a splitting of $0 \to Y^{mult}_{u/\widehat{U}_u} \to Y \times_U \widehat{U}_u \to Y^{\mathrm{et}}_{u/\widehat{U}_u} \to 0$ for all closed points $u \in U$.

The splitting in the proposition is unique because $\mathcal{O}_S$ is reduced. We may assume G_1 gives Y_u^{mult} as the connected component is uniquely determined in characteristic p. Though we do not treat the characteristic-0 case, a difficulty is that even if G_js exist, G_1 may not be universally connected all over U (so we cannot make descent as in the following proof, which is based on the uniqueness of the connected component).

Proof. We follow Chai's proof in [Ch1]. Let $S_0 = \mathrm{Spec}(\mathcal{O}_{S,s_0})$ for the stalk $\mathcal{O}_0 := \mathcal{O}_{S,s_0}$ at s_0, and regard $\widehat{S}$ as an S_0-scheme. We consider the self-product $\widehat{S} \times_{S_0} \widehat{S}$ and $\widehat{S} \times_{S_0} \widehat{S} \times_{S_0} \widehat{S}$ and projections $\pi_i : \widehat{S} \times_{S_0} \widehat{S} \to \overset{i}{\widehat{S}}$ $(i = 1, 2)$ and $\pi_{ij} : \widehat{S} \times_{S_0} \widehat{S} \times_{S_0} \widehat{S} \to \overset{i}{\widehat{S}} \times_{S_0} \overset{j}{\widehat{S}}$ $(1 \leq i \leq j \leq 3)$. Let $G_{2/\widehat{S}} \subset \widehat{Y}$ be the splitting image of $\widehat{Y}^{\mathrm{et}}$ over $\widehat{S}$ as a closed subscheme of $\widehat{Y}$; so, for $G_1 = Y^{mult}_{/\widehat{S}}$, we have a product decomposition $Y_{/\widehat{S}} = G_{1/\widehat{S}} \times G_{2/\widehat{S}}$. Since $\mathcal{O}_{S,s}$ is excellent, by Lemma 4.87, $S' := \widehat{S} \times_{S_0} \widehat{S}$ and $S'' := \widehat{S} \times_{S_0} \widehat{S} \times_{S_0} \widehat{S}$ are reduced and satisfy the assumption of Lemma 4.71. The uniqueness of étale-connected splitting is valid as we are in characteristic p. In characteristic p, the connected component $(\pi_j^* Y)^{mult}$ is defined globally over S' and is equal to $\pi_j^*(Y^{mult})$; so the ordering of splitting is given a priori, and no problem arises with the order of the factors. Since G_1 and G_2 are simple Barsotti–Tate O-modules, we can have only one possibility over each connected component of S': $\pi_1^* G_1 = \pi_2^* G_1$. In the same way, we get $\pi_1^* G_2 = \pi_2^* G_2$. Similarly, we have $\pi_{ij}^* \pi_i^* G_k = \pi_{ij}^* \pi_j^* G_k$ over S'' for $k = 1, 2$, and these identities give rise to a decent datum (for closed group subschemes of Y; e.g., [GME] Sect. 1.11). By fpqc descent (see [CBT] Remark I.2.4.2), $G_{1/\widehat{S}}$ and $G_{2/\widehat{S}}$ descend to Barsotti–Tate subgroups G_1 and G_2 of $Y \times_S S_0$, respectively, over S_0.

The splitting $Y = G_1 \times G_{2/S_0}$ implies $Y[p^n]_{/S_0} = G_1[p^n] \times G_2[p^n]_{/S_0}$. Since these factors are finite flat over S_0, they are defined by finitely many equations with coefficients in $\mathcal{O}_{S,s_0} = \varinjlim_{U \ni s_0} \mathcal{O}_U$; so they are defined over $\mathcal{O}_{U_n}$ for an open subset U_n. Since $G_2[p^n]$ is étale over U_n (and $G_1[p^n]$ are multiplicative over U_n), the open subscheme U_n does not depend on n. Thus, we can take $U = U_n$, and we find the desired open subscheme $U \subset S$ with $S_0 \subset U$ over which $Y = G_1 \times G_2$.

5

Geometry of Variety

In this short chapter, we summarize geometric properties of a variety and a tower of varieties we need in the book. One reason for adding this chapter is to make the book logically complete, and another is to give the foundation of the theory of towers of varieties in the language of proschemes, since the Shimura variety is a tower of varieties fundamental to the number-theoretic study of automorphic forms. If the reader is familiar with the subject, he or she can take a brief look at the content of this chapter and go directly to Chap. 6.

5.1 Variety over a Field

In this section, the base ring B is assumed to be a field k (and often it is supposed to be a perfect field). We only deal with varieties over a field, in particular, their geometric properties.

5.1.1 Rational Map

Let V and V' be varieties over a field k. Morphisms $f : U \to V'$ and $f' : U' \to V'$ for open dense subvarieties $U, U' \subset V$ are said to be equivalent if $f|_{U \cap U'} = f'|_{U \cap U'}$. This is an equivalence relation because $U \cap U'$ is still open dense in V. An equivalence class of such a morphism is called a *rational map* from V to V'. Note that a rational map may be defined only on an open dense subvariety of V; hence, we write it as $f : V \rightsquigarrow V'$. A rational map $f : V \rightsquigarrow V'$ is defined at $x \in V$ if there exists a representative $U \to V'$ for U containing x.

Theorem 5.1 *A rational map $f : V \rightsquigarrow V'$ of finite type from a normal variety V over a perfect field k to a proper variety $V'_{/k}$ is defined on an open subvariety U of V whose complement $V - U$ has codimension at least* 2.

Here is a standard proof of this fact.

H. Hida, *Elliptic Curves and Arithmetic Invariants*, Springer Monographs in Mathematics, DOI 10.1007/978-1-4614-6657-4_5,

Proof. We may assume that V is irreducible, restricting the map to irreducible components. Assume first that V is a curve. Since V is normal, it is smooth over a perfect field k, as already remarked in Sect. 4.1.8. Thus, we are given a smooth curve V and a morphism $f : U \to V'$ from an open subvariety U of V. We claim that f extends to V. Consider the following diagram:

$$\begin{array}{ccc} U & \xrightarrow{\subset} & V \\ \| \uparrow & & \uparrow \text{projection} \\ U & \xrightarrow{x \mapsto (x, f(x))} & V \times V' \\ \| \downarrow & & \downarrow \text{projection} \\ U & \xrightarrow[f]{} & V'. \end{array}$$

Let U' be the image of the middle arrow in $V \times V'$ and let Z be the Zariski closure of U'. The properness of V' tells us that $V \times V' \to V$ is a closed map (by the definition in Sect. 4.1.6); hence, the image of Z in V is closed, and it contains U, and so Z maps onto V. Since Z is a curve and V is normal, the morphism $Z \to V$ is an isomorphisms (as a smooth curve is determined by its function field, as remarked at the end of Sect. 4.1.8). Therefore, f is defined everywhere.

Now we deal with the general case. Take a representative of f: $U \to V'$. The schematic complement $V - U$ is a closed subscheme X. Since $\operatorname{codim} X$ is at least 1, we may assume that X has codimension 1 (otherwise, the theorem follows). Then X contains an irreducible component V_1, which gives a prime divisor of V. Since V has a dense affine open subvariety, S_A with $S_A \cap V_1$ is still an irreducible prime divisor in S_A, that is, $S_A \cap V_1$ is defined by a prime ideal $P \subset A$. Since V is normal, A is a normal domain, and hence A_P is a discrete valuation ring. Since $S_A \cap U$ is a dense open subvariety of V, f induces the morphism of the spectrum of the field of fraction of A into V'. By a valuative criterion of properness (Proposition 4.25), we then have $\operatorname{Spec}(A_P) \xrightarrow{f} V'$ well defined. This implies that at the generic point of V_1, f is defined, a contradiction. Thus, f extends to an open subvariety U with $\operatorname{codim}(V - U) \geq 2$.

If we have rational maps $f : V \rightsquigarrow V'$ and $g : V' \rightsquigarrow V$ with $f \circ g$ equivalent to id_V and $g \circ f$ equivalent to $\mathrm{id}_{V'}$, we call f and g *birational*, and V is *birational* to V'.

Proposition 5.2 *Let $f : V' \to V$ be a morphism of irreducible varieties making V birational to V'. Let V^{nfl} be the Zariski closure in V of all maximal points $x \in V$ such that there exists a maximal point $y \in V'$ with $\mathcal{O}_{V',y}$ not flat over $\mathcal{O}_{V,x}$. If V is proper, V^{nfl} has at least codimension 2.*

The closed scheme V^{nfl} is called the *nonflat locus* in V.

Proof. Over an open dense subscheme $U \subset V$, we have an open immersion $g : U \to V'$ giving the inverse of f by birationality. Thus, $f : g(U) \cong U$. Thus, the nonflat locus V^{nfl} is a proper closed subset. Since $V' \to V$ is separated, $f^{-1}(U) = g(U)$. Thus, by the above theorem, if we take the maximal open subvariety U on which $g = f^{-1}$ is defined, $V - U$ has at least codimension 2 and contains V^{nfl}.

Proposition 5.3 *Let $\phi : X \to Y$ be a proper dominant morphism of irreducible varieties over a field k. Let Y^{nfl} be the Zariski closure of all maximal points $y \in Y$ such that $(\phi_* \mathcal{O}_X)_y$ is not flat over $\mathcal{O}_{Y,y}$. If Y is normal, then Y^{nfl} is a closed subscheme of at least codimension 2.*

The closed scheme Y^{nfl} is called the *nonflat locus* in Y.

Proof. Since f is proper and dominant, f is a surjection of topological spaces. At a generic point, $k(X)_{/k(Y)}$ is a field extension; in particular, $k(X)$ is flat over $k(Y)$. Thus, $Y - Y^{\mathrm{nfl}}$ is nonempty. By Proposition 4.5 (1), on an open dense subscheme $U \subset X$, $\phi_U : U \to Y$ is flat. By Proposition 4.5 (2), ϕ_U is an open separated map, and Y^{nfl} is a proper closed subset of Y. If Y^{nfl} has an irreducible subscheme of codimension 1 (that is a prime divisor), the localization at the generic point of the prime divisor is the spectrum of a discrete valuation ring V_1 (as it is the normal noetherian domain of dimension 1). Since f is dominant, $k(X)$ has a discrete valuation ring V flat over V_1 with $V_1 = V \cap k(Y)$. Then we have a commutative diagram:

$$\begin{array}{ccc} \mathrm{Spec}(k(X)) & \longrightarrow & X \\ \downarrow & & \downarrow \\ \mathrm{Spec}(V) & \longrightarrow & Y. \end{array}$$

Then, by the valuative criterion (Proposition 4.25), the morphism $\mathrm{Spec}(k(X)) \hookrightarrow X$ extends to $\mathrm{Spec}(V) \hookrightarrow Y$; hence, the codimension-1 point $\mathrm{Spec}(V_1/\mathfrak{m}_{V_1})$, which is the generic point of the prime divisor, is in the flat locus $Y - Y^{\mathrm{nfl}}$, a contradiction (as we have taken the prime divisor inside Y^{nfl}). Thus, Y^{nfl} must have codimension of at least 2.

Proposition 5.4 *Let $f : V' \to V$ be a dominant morphism of irreducible varieties. Then*

1. *The variety V' is flat over a dense open subscheme O of V,*
2. *Let O be as above. For a point $x \in f^{-1}(O)(K)$ for a field K, the dimension of the fiber $V' \times_V f(x)$ at $f(x)$ is constant given by $\dim V' - \dim V$.*

Proof. By taking open affine coverings of V and V' (compatible with f), we may assume $V' = S_{A'}$ and $V = S_A$ for integral noetherian k-domains A' and A. Let U be the subset of $\mathrm{Spec}(A')$ made up of points x such that A'_x is flat over $A_{f(x)}$. Since f is dominant, the fraction field K' of A' is an extension field of the field of fractions K of A. Thus, U is a nonempty open subscheme by Proposition 4.5 (1) (applying it to $M = A'$). Then $f|_U : U \to S_A$ is an open map by Proposition 4.5 (2); so we just define $O = f(U)$, which is an open subscheme of S_A. This proves (1).

Cover O by affine open subschemes S_{A_i}, and cover $f^{-1}(S_{A_i})$ by affine open subschemes $S_{A'_{ij}}$. Then A'_{ij} is flat over A_i. Then the constancy of dimension of fibers of $S_{A'_{ij}} \to S_{A_i}$ follows from ring theory [see [CRT] Theorem 15.1 (ii)]. Since these covering open subschemes overlap at the generic point, the desired result follows.

5.1.2 Zariski's Main Theorem

We begin by preparing some lemmas.

Lemma 5.5 *Suppose that $f : X \to Y$ is a proper dominant morphism of irreducible varieties. Then $f_*(\mathcal{O}_X)$ is a coherent $\mathcal{O}_Y$-algebra.*

Since a line bundle $\mathcal{L}$ over X is locally isomorphic to $\mathcal{O}_X$, this lemma tells us that $f_*(\mathcal{L})$ is a coherent $\mathcal{O}_Y$-module.

Proof. Write $k(X)$ and $k(Y)$ for the function fields of X and Y. Since f is dominant, we may regard $k(Y)$ as a subfield of $k(X)$. Since f is of finite type (as a proper morphism is of finite type), $k(X)$ is a finitely generated field over $k(Y)$. Since coherency is a local property, we may assume that $Y = \mathrm{Spec}(A)$ for an integral noetherian domain A with quotient field $k(Y)$. We have a canonical open immersion $\mathrm{Spec}(k(X)) \hookrightarrow X$. Take a valuation ring V of $k(X)$ containing A. Then we have a commutative diagram:

$$\begin{array}{ccc} \mathrm{Spec}(k(X)) & \longrightarrow & X \\ \downarrow & & \downarrow f \\ \mathrm{Spec}(V) & \longrightarrow & Y. \end{array}$$

Then, by the valuative criterion of properness, we have a morphism $\mathrm{Spec}(V) \to X$ making the above diagram commutative after inserting the diagonal morphism $\mathrm{Spec}(V) \to X$. Thus, $f_*(\mathcal{O}_X)(Y) \subset V$. Note that $\bigcap_{V \supset A} V$ is the integral closure of A in $k(X)$ (cf. [CRT] Theorem 10.4). Thus, $f_*(\mathcal{O}_X)(Y)$ is a subring of the integral closure $\tilde{A}$ of A in a field finitely generated over the field of fractions of A. Thus, $\tilde{A}$ is the integral closure of A in the integral closure of $k(Y)$ in $k(X)$, which is a finite algebraic extension of $k(Y)$. Then, by [TCF] Theorem 3.9.5, $\tilde{A}$ is an A-module of finite type. Since A is noetherian, $f_*(\mathcal{O}_X)(Y)$ is an A-module M of finite type, and so $f_*(\mathcal{O}_X) = \tilde{M}$ is an $\mathcal{O}_Y$-module of finite type; in other words, $f_*(\mathcal{O}_X)$ is a coherent $\mathcal{O}_Y$-module.

Lemma 5.6 *Let $f : X \to Y_{/k}$ be a proper morphism of irreducible varieties. Then for a maximal point $y \in Y$, define $X_n = X \times_Y \mathrm{Spec}(\mathcal{O}_{Y,y}/\mathfrak{m}_y^n)$. Then we have*

$$\varprojlim_n (f_*\mathcal{O}_X) \otimes_{\mathcal{O}_X} \mathcal{O}_{Y,y}/\mathfrak{m}_y^n \cong \varprojlim_n f_*(\mathcal{O}_{X_n}).$$

Proof. We have a natural map $\mathcal{O}_X \to \mathcal{O}_{X_n}$ that induces a morphism $f_*(\mathcal{O}_X) \to f_*(\mathcal{O}_{X_n})$ factoring through $(f_*\mathcal{O}_X) \otimes_{\mathcal{O}_X} \mathcal{O}_{Y,y}/\mathfrak{m}_y^n$. Thus, we get the morphism in question after passing to the projective limit. We may assume that $Y = \mathrm{Spec}(A)$. Then $f_*\mathcal{O}_X = \tilde{A}'$ for $A' = f_*(\mathcal{O}_X)(Y)$, which is an A-module of finite type by the above lemma. The right-hand side is $f_*(\mathcal{O}_{\widehat{X}})$ for the formal completion $\widehat{X}$ along the fiber at y. Thus, $f_*(\mathcal{O}_{\widehat{X}})(Y) = A' \otimes_A \widehat{A}$ for $\widehat{A} = \varprojlim_n A/\mathfrak{m}_y^n = \varprojlim_n \mathcal{O}_{Y,y}/(\mathfrak{m}_y\mathcal{O}_{Y,y})^n$, which is isomorphic to $\widehat{A}' = \varprojlim_n A'/\mathfrak{m}_y^n$ by [CRT] Theorem 8.7, as A is noetherian and A' is an A-module of finite type. Note that $\widehat{A}'$ is the left-hand side. This finishes the proof.

Theorem 5.7 (Zariski's connectedness theorem) *Let $f : X \to Y$ be a proper morphism of irreducible varieties over a field k, and assume that $f_*(\mathcal{O}_X) = \mathcal{O}_Y$. Then $f^{-1}(y)$ is connected for every maximal point $y \in Y$.*

If $y \in Y(K)$ is a nonmaximal point for an extension field K/k, by a base change to K and applying the theorem to $X_K := X \times_k K \to Y_K := Y \times_k K$, $f^{-1}(Y)$ is connected for any point of Y not necessarily maximal.

Proof. Suppose, on the contrary, that $f^{-1}(y) = X' \sqcup X''$, where X' and X'' are disjoint closed subvarieties. Let $X_n = X \times_Y \mathrm{Spec}(\mathcal{O}_{Y,y}/\mathfrak{m}_y^n)$ as above and let $\widehat{X}$ be the formal completion of X along $f^{-1}(y)$. Then, for each n, we would have $f_*(X_n) = f_*(X'_n) \oplus f_*(X''_n)$. By Lemma 5.6, we have

$$\mathcal{O}_{\widehat{X}} = \varprojlim_n f_*(\mathcal{O}_{X_n}) = (\varprojlim_n f_*(X'_n)) \oplus (\varprojlim_n f_*(X'_n))$$

as a sum of two nontrivial rings. Since $f_*(\mathcal{O}_X) = \mathcal{O}_Y$, again by Lemma 5.6, $\mathcal{O}_{\widehat{X}}(\widehat{X}) = \widehat{\mathcal{O}}_Y = \varprojlim_n \mathcal{O}_{Y,y}/\mathfrak{m}_y^n$, which may not be an integral domain but may be a local ring (by Hensel's lemma; see [BCM] III.2.13). A local ring cannot be a direct sum of two rings, a contradiction. Thus, $f^{-1}(y)$ has to be connected.

Theorem 5.8 (Zariski's main theorem) *Let $f : X \to Y$ be a birational proper morphism of irreducible varieties, and assume that Y is normal. Then for every point $y \in Y$, $f^{-1}(y)$ is connected.*

Proof. By Theorem 5.7, we need to prove $f_*\mathcal{O}_X = \mathcal{O}_Y$. We may then assume that $Y = \mathrm{Spec}(A)$ is affine. By birationality, $k(X) = k(Y)$. By the proof of Lemma 5.5, $A' := f_*\mathcal{O}_X(Y)$ is an A-subalgebra of $k(X) = k(Y)$ integral over A. Since Y is normal, A is integrally closed, and we get $A' = A$. Thus, we have verified $f_*\mathcal{O}_X = \tilde{A}' = \tilde{A} = \mathcal{O}_Y$.

5.1.3 Stein Factorization

Start with a proper morphism $\phi : X \to Y$ of varieties over k. Then by Lemma 5.5, $\phi_*(\mathcal{O}_X)$ is a coherent $\mathcal{O}_Y$-algebra. Define $Y^{\mathrm{st}} = Spec_Y(\phi_*(\mathcal{O}_X))$ as an Y-scheme. Write the morphism as $\phi^f : Y^{\mathrm{st}} \to Y$. Since $\mathcal{O}_X$ is a $\phi_*(\mathcal{O}_X)$-algebra, the structure morphism $\iota_{\mathcal{O}_X} \in \mathrm{Hom}_{\mathcal{O}_Y\text{-alg}}(\phi_*(\mathcal{O}_X), \mathcal{O}_X) = \mathrm{Hom}_{SCH_{/Y}}(Spec_X(\mathcal{O}_X), Y^{\mathrm{st}})$ induces a morphism $\phi^c : X \to Y^{\mathrm{st}}$ with $\phi = \phi^f \circ \phi^c$.

Theorem 5.9 *Let $\phi : X \to Y$ be a proper morphism of varieties over a field k. Then we can naturally factor $X \xrightarrow{\phi^c} Y^{\mathrm{st}} \xrightarrow{\phi^f} Y$ for a finite morphism $\phi^f : Y^{\mathrm{st}} \to Y$ and a proper morphism $\phi^c : X \to Y^{\mathrm{st}}$ such that $\phi^c_*(\mathcal{O}_X) = \mathcal{O}_{Y^{\mathrm{st}}}$ and ϕ^c has connected fibers everywhere.*

Proof. By Lemma 5.5, $\phi^f_*(\mathcal{O}_{Y^{\mathrm{st}}}) = \phi_*\mathcal{O}_X$ is a coherent $\mathcal{O}_Y$-algebra, and so ϕ^f is finite. Since $\phi^c_*\mathcal{O}_X = \phi_*\mathcal{O}_X = \mathcal{O}_{Y^{\mathrm{st}}}$, by Zariski's connectedness theorem (and by the remark following the theorem), ϕ^c has a connected fiber. □

Any finite morphism $f : X \to Y$ is proper by Corollary 4.26 (and is even projective). Then the above theorem proves again that a finite morphism is an affine morphism (Lemma 4.24), as $X = Y^{\mathrm{st}} = Spec_Y(f_*\mathcal{O}_X)$.

Corollary 5.10 *Any proper quasi-finite morphism of varieties is finite.*

Proof. If f is quasi-finite and proper, every fiber of $X \to Y^{\mathrm{st}}$ has to be one point. Thus, by separatedness, $X = Y^{\mathrm{st}}$, and so f is finite.

Corollary 5.11 *Let $\phi : X \to Y$ be a proper dominant morphism of irreducible varieties such that for at least one point $y_0 \in Y$, $\phi^{-1}(y_0) \to y_0$ is a finite morphism. Define the nonfinite locus $Y^{\mathrm{nfi}} \subset Y$ by the Zariski closure of all points $y \in Y$ such that $\phi^{-1}(y)$ is not finite over y. Then $Y^{\mathrm{nfi}} \subset Y^{\mathrm{nfl}}$ and has codimension at least 2.*

Proof. Since ϕ is proper and dominant, ϕ is surjective [so $\phi^{-1}(y_0)$ is nonempty]. Also, ϕ is flat over an open dense subscheme $O \subset Y$. We may assume that O is maximal under this property, since a union of open subschemes is still open. Thus, we may assume that $Y^{\mathrm{nfl}} = Y - O$. Take the Stein factorization $X \to Y^{\mathrm{st}} \to Y$. Replacing Y by Y^{st}, we may assume that ϕ has connected fibers. Since at y_0, it has one point fiber, we have $[k(X) : k(Y)] < \infty$. Thus, over O, fibers of ϕ have constant dimension 0. Since $\phi : f^{-1}(O) \to O$ is proper and quasi-finite, it is finite flat by the above result. In other words, we get $Y^{\mathrm{nfi}} \subset Y - O = Y^{\mathrm{nfl}}$, as desired.

Remark 5.12 Consider the category $CV_{/k}$ of smooth geometrically irreducible proper curves (i.e., irreducible proper varieties of dimension 1) over a filed k. If $V_{/k}$ is a geometrically irreducible proper curve, for its normalization $\tilde{V}$, $\mathcal{O}_{\tilde{V},x}$ is a normal integral domain of dimension 1; hence, it must be a valuation ring. Therefore, $\tilde{V}$ is smooth with $k(V) = k(\tilde{V})$. If V and V' are two smooth irreducible curves with $k(V) \cong k(V')$, then V is birational to V' over k. By Zariski's main theorem, the birational map $f : V \to V'$ is well defined outside a codimension-2 closed subset, which is empty. By the above theorem, f has to be an isomorphism. Thus, $V \mapsto k(V)$ gives an equivalence of categories between $CV_{/k}$ and the category of field extensions of k (of finite type) of transcendental degree 1 in which k is algebraically closed.

5.1.4 Provariety

We study Zariski topology on a projective limit of varieties and schemes (provariety/proschemes); in particular, we explain how to detect closed/open subvarieties out of their projection to varieties appearing in the limit. Since the Shimura variety is a projective limit of varieties of finite type, we need the results later. A main reference is [EGA] IV.8.

Let I be a connected partially ordered index set. Hence, we have a partial ordering denoted by "$\prec$" on I, and if $i \prec j$ and $i \prec j'$, there exists $j'' \in I$ such that $j \prec j''$ and $j' \prec j''$ (connectedness). We first discuss projective limits of affine morphisms. Consider a variety S and the coherent sheaf of reduced $\mathcal{O}_S$-algebras $\{\mathcal{A}_i\}_{i\in I}$ with $\mathcal{O}_S$-algebra homomorphism $\underline{\iota}_{ij} : \mathcal{A}_i \to \mathcal{A}_j$ if $i \prec j$ such that $\underline{\iota}_{j'j} \circ \underline{\iota}_{ij} = \underline{\iota}_{ij'}$. We write $\mathcal{A}_\infty = \varinjlim_i \mathcal{A}_i$, which is a quasi-coherent sheaf of $\mathcal{O}_S$-algebras. We assume that $\underline{\iota}_{ij}$ is finite flat over $\mathcal{A}_i$ for all $i \prec j$. Since $\mathcal{A}_i$ is reduced, $S_i = Spec_S(\mathcal{A}_i)$ finite flat over S is a variety. Since the category of $\mathcal{O}_S$-algebras and the category of S-affine schemes are anti-equivalent, $\{S_i, \pi_{ij} := \iota_{ij} : S_j \to S_i\}_{i\prec j}$ is a projective system of S-schemes. For any $\mathcal{O}_S$-algebra $\mathcal{B}$, by definition, we have

$$\mathrm{Hom}_{\mathcal{O}_S\text{-alg}}(\mathcal{A}_\infty, \mathcal{B}) = \varprojlim_i \mathrm{Hom}_{\mathcal{O}_S\text{-alg}}(\mathcal{A}_i, \mathcal{B}).$$

Thus, as a functor in $\mathcal{C}_S := COF(ALG_{/\mathcal{O}_S}, SETS)$, $\varprojlim_i S_i$ is well defined and given by $S_{\mathcal{A}_\infty}$. Clearly, the functor $S_{\mathcal{A}_\infty}$ is local and coincides with the functor associated with the scheme $S_\infty = Spec_S(\mathcal{A}_\infty)$. Thus, the projective limit S_∞ exists in the category of schemes. Write $\pi_i : S_\infty \to S_i$ for the projection to S_i. Define the projective limit topology on S_∞ to be the weakest topology such that all π_i is continuous. If $U_j \subset S_j$ is open, $\pi_{ij}(U_j) \subset S_i$ is open for $i \prec j$, as flat π_{ij} is an open map. Thus, $U \subset S_\infty$ is open under the projective limit topology if $U = \pi_j^{-1}(U_j)$ for an open subscheme U_j for sufficiently large j. Since $\pi_{ij} : S_j \to S_i$ is finite flat, by the going-up/-down theorems ([CRT] Sect. 9), it induces a surjective open map on their topological space.

Lemma 5.13 *Let $\{S_i = Spec_S(\mathcal{A}_i), \pi_{ij}\}_{i\in I}$ be a projective system of affine finite flat morphisms of S-varieties. Then $S_\infty = Spec_S(\mathcal{A}_\infty)$ for $\varinjlim_i \mathcal{A}_i$ gives a projective limit of $\{S_i, \pi_{ij}\}_{i\in I}$ as a scheme, and the Zariski topology on S_∞ is the projective limit topology of the Zariski topology on S_i.*

This fact is valid actually without much hypothesis as long as $\mathcal{A}_i$ are quasi-coherent $\mathcal{O}_S$-algebras (see [EGA] IV.8.2.3 for a general treatment).

Proof. We have already shown that S_∞ as a functor in $\mathcal{C}_S$ is equal to the scheme $S_{\mathcal{A}_\infty}$; hence, the limit exists in $SCH_{/S}$. We prove therefore the assertion on topological space. First, assume that S is affine; thus, $S_i = S_{A_i}$ for all i for a k-algebra A_i. Then $A_\infty = \varinjlim_i A_i$. By definition, $D_i(f) = \mathrm{Spec}(A_i[\frac{1}{f}])$ for f running in A_i gives a base of open subsets of S_i (for $i \in I \cup \{\infty\}$). Since $f \in A_\infty$ is in the image of A_i for sufficiently large i, we find $D_\infty(f) = \pi_i^{-1}(D_i(f))$. This finishes the proof when S is affine. In general, we cover S by finitely many affine open subschemes $S = \bigcup_{\alpha \in J} S_{A_\alpha}$. Then the inverse image of S_{A_α} in S_i is affine as $S_i \to S$ is finite flat. We therefore write this inverse image as $S_{A_{i,\alpha}} \subset S_i$. Since $\{S_{A_{i,\alpha}}\}_\alpha$ is a finite open covering of S_i with fixed finite set J, applying the result already proven for the affine scheme to $\{S_{A_{i,\alpha}}\}_i$, we conclude the result in the general case.

Corollary 5.14 *Let the notation and assumption be as in the above lemma. If closed subschemes $Z_i \subset S_i$ $(i \in I)$ form a projective system by $\pi_{ij}|_{Z_j}$, the projective limit $Z_\infty = \varprojlim_i Z_i$ is a closed subscheme of S_∞.*

Proof. Since Z_i is a subfunctor of S_i, the projective limit $Z_\infty(R) = \varprojlim_i Z_i(R)$ gives a subfunctor of S_∞. Since Z_i is local and passing to the projective limit is left exact, Z_∞ is a local functor. If $S_{A_\infty} \subset S_\infty$ is an affine open subscheme with a nontrivial intersection with Z_∞, its image in S_i is open (as π_i is flat and hence an open map), which is an affine open subscheme S_{A_i} of S_i, and $A_\infty = \varinjlim_i A_i$. Then $S_{A_i} \cap Z_i$ is nontrivial and gives an affine open subscheme $S_{A'_i}$ of Z_i. Then $S_{A'_\infty} = \varprojlim_i S_{A'_i} = S_{A_\infty} \cap Z_\infty$ as a functor; hence, it is an open affine subscheme of Z_∞. Thus, Z_∞ has an affine open covering induced by the affine open covering of S_∞. Thus, Z_∞ is a subscheme of S_∞. Writing $U_i = S_i - Z_i$, we have $S_\infty - Z_\infty = \bigcup_i \pi_i^{-1}(U_i)$, which is an open subscheme of S_∞. Thus, Z_∞ is a closed subscheme of S_∞.

Let $\{S_i, \pi_{ij}\}_{i \prec j \in I}$ be a projective system as above of finite flat affine morphisms of S-varieties. Let $Z_i \subset S_i$ be a closed subvariety forming a projective system $\{Z_i, \pi_{ij}^Z := \pi_{ij}|_{Z_j}\}$ of proper and dominant morphisms of S-varieties. Since π_{ij}^Z is proper and dominant, its image is closed Zariski-dense; thus, it induces a surjection of associated topological spaces. Let $Z_\infty = \varprojlim_i Z_i$, which is a subfunctor of S_∞ and is a closed subscheme of S_∞ by the above corollary. Since a proper morphism is a closed map, $V \subset Z_\infty$ is a closed subscheme if $V = \pi_j^{-1}(V_j)$ for a closed subscheme V_j of Z_j for a sufficiently large j.

6

Elliptic and Modular Curves over Rings

In Chap. 1, we studied nonvanishing properties modulo p of Dirichlet L-values, and in Sect. 3.5, we outlined the proof of the vanishing of the μ-invariant of p-adic Hecke L-functions. To fill details of the proof, obviously, we need a good amount of the theory of elliptic curves over general rings. In later chapters, replacing $\mathbb{G}_m$ by elliptic curves and modular curves, we study nonvanishing questions for more general Hecke L-values.

In this chapter, we review the (integral) theory of elliptic curves and modular curves over general rings to the extent we need later. Chapter 2 of the book [GME] contains a more complete exposition of the theory.

6.1 Basics of Elliptic Curves over a Scheme

When studying curves C over a field k, classically, mathematicians analyzed its set $C(K)$ of K-points for a field extension K/k. To study a curve C over a general base ring B (rather than a field k), obviously we need to study $C(R)$ directly for an arbitrary B-algebra R. A main point of scheme theory is to consider C a covariant functor $R \mapsto C(R)$ from the category $ALG_{/B}$ of B-algebras into the category $SETS$ of sets. We have many interesting functors $F : ALG_{/B} \to SETS$. For example, we may associate with R the set $[E_{/R}]$ of all isomorphism classes of elliptic curves defined over R [or Spec(R)]. This is a covariant functor since a B-algebra homomorphism $R \xrightarrow{\rho} R'$ induces the base change $E_{/R} \mapsto (E \otimes_R R')$, which is an elliptic curve defined over R'. We can extend this functor to a contravariant functor from the category $SCH_{/B}$ of schemes over B to $SETS$ in an obvious manner. The functor is contravariant because $R \mapsto \mathrm{Spec}(R)$ reverses the direction of morphisms. If the functor $R \mapsto [E_{/R}]$ is isomorphic to a functor $R \mapsto \mathfrak{M}(R)$ for a B-scheme $\mathfrak{M}$, we could use algebraic geometry to study the functor (and at the same time we may use the functoriality to study the geometry of the moduli scheme $\mathfrak{M}$). Finding $\mathfrak{M}$ is a classification problem of elliptic curves. Of course, we may add some extra structures ϕ to E to classify pairs (E, ϕ), and if successful, the moduli $\mathfrak{M}'$ of

H. Hida, *Elliptic Curves and Arithmetic Invariants*, Springer Monographs in Mathematics, DOI 10.1007/978-1-4614-6657-4_6,

the functor $R \mapsto [(E, \phi)_{/R}]$ constitutes a scheme over $\mathfrak{M}$ by the morphism $(E, \phi) \mapsto E$. We study this type of problem in this and the following sections. In this section, we summarize necessary facts on elliptic curves defined over a scheme.

6.1.1 Definition of Elliptic Curves

Here is a geometric definition of an elliptic curve over a general scheme S. For a given scheme S, a proper smooth curve $f : E \to S$ is called an *elliptic curve* if it satisfies the following three conditions:

(E1) E has a section $\mathbf{0} = \mathbf{0}_E \in E(S)$ (thus, $f \circ \mathbf{0} = 1_S$).
(E2) $\dim_S E = 1$, and E is geometrically irreducible fiber by fiber (this means that each geometric fiber of E over an algebraically closed field is smooth, irreducible, and of dimension 1).
(E3) For the $\mathcal{O}_E$-module $\Omega_{E/S}$ of 1-differentials (e.g., Sect. 4.1.7), $f_*\Omega_{E/S}$ (equivalently, $R^1 f_*\mathcal{O}_E$; see [GME] Sect. 2.1.2) is locally free of rank 1.

There is no harm in assuming that S is connected, as we do from now on. For any S-scheme $\phi : T \to S$, the fibered product $E_T = E \times_S T$ is again an elliptic curve with the zero section $\mathbf{0}_T = \mathbf{0}_E \times 1_T$. For two elliptic curves E and E' over S, an S-morphism $h : E \to E'$ is always supposed to take $\mathbf{0}_E$ to $\mathbf{0}_{E'}$. The totality of elliptic curves forms a *fiber category* ELL over SCH. In other words, each elliptic curve $E_{/S}$ is defined relative to its base scheme S, giving rise to a fiber functor $ELL \ni E_{/S} \mapsto S \in SCH$. Thus, $\mathrm{Hom}_{ELL}(E_{/S}, E'_{/S'})$ is made up of two commutative diagrams:

$$\begin{array}{ccc} E & \longrightarrow & E' \\ \downarrow & & \downarrow \\ S & \longrightarrow & S' \end{array} \quad \text{and} \quad \begin{array}{ccc} E & \longrightarrow & E' \\ {\scriptstyle \mathbf{0}_E}\uparrow & & \uparrow{\scriptstyle \mathbf{0}_{E'}} \\ S & \longrightarrow & S'. \end{array}$$

The covariant functor $ELL \ni E_{/S} \mapsto S \in SCH$ is called the fiber functor of $ELL_{/SCH}$.

We have seen in Sect. 2.2.2, by the Riemann–Roch theorem, that an elliptic curve E over an algebraically closed field k can be embedded into the projective space $\mathbf{P}^2_{/k}$. In this chapter, more generally, we will see that $E_{/S}$ can be embedded into the projective space $\mathbf{P}^2_{/S}$ over S of dimension 2 as a curve of degree 3. If $S = \mathrm{Spec}(k)$ for an algebraically closed field k, the intersection of $E \subset \mathbf{P}^2_{/k}$ with a linear hyperplane has three points $\{P, Q, R\}$ (counting with multiplicity). As we have seen in Sect. 2.2.1, defining $P + Q + R = \mathbf{0}_E$, we can put a group structure on $E(k)$. Indeed, even $E_{/S}$ over general scheme S has a unique structure of a group scheme, making $\mathbf{0}_E$ the identity.

Lemma 6.1 *An abelian scheme $f : A \to S$ of relative dimension 1 is an elliptic curve.*

In Sect. 6.1.3, we will see not only that an elliptic curve $E_{/S}$ has the natural structure of a group scheme with identity $\mathbf{0}_E$, but also that any morphism $f : E \to E'$ over S between elliptic curves sending $\mathbf{0}_E$ to $\mathbf{0}_{E'}$ is a group homomorphism. Thus, the category of elliptic curves is identical to the category of abelian schemes of dimension 1. We prove here that an abelian variety of dimension 1 over S is an elliptic curve.

Proof. By definition, an abelian scheme of relative dimension 1 is a proper curve with a specific point given by the identity $\mathbf{0}_A$. Thus, we need to prove $f_*\Omega_{A/S}$ is locally S-free of rank 1. Since A/S is smooth proper, $f_*\Omega_{A/S}$ is a locally free $\mathcal{O}_S$-module of finite type (by Lemma 5.5). To compute $\mathrm{rank}_{\mathcal{O}_S} f_*\Omega_{A/S}$, we may assume that $S = \mathrm{Spec}(k)$ for an algebraically closed field [by pulling back via $\mathrm{Spec}(k) \to S$].

First, suppose that k has characteristic p. Since $A[p^\infty] = \bigcup_n A[p^n]$ for $A[p^n] = \mathrm{Ker}(x \mapsto p^n x)$ is a Barsotti–Tate group, its connected component $A^\circ[p^\infty]$ is prorepresented by the smooth, connected, formal group $\widehat{A} = A|_{CL_k}$ (that is, $\widehat{A}$ is the formal completion of A along the origin $\mathbf{0}_A$; see Lemma 4.82). Thus, $\widehat{A} = \mathrm{Spec}(k[[T]])$ for a parameter T at the identity $\mathbf{0}_A$, and it has a unique invariant differential ω on $\widehat{A}$ (induced by the cotangent vector at $\mathbf{0}_A$) up to constant multiple. Since $A(k)$ acts freely and transitively on $A(k)$, we have $\Omega_{A/k}(x) \cong \Omega_{A/k}(\mathbf{0}_A) \cong k$ at each point $x \in A(k)$ by translation $a \mapsto a + x$. Then the invariant differential ω extends uniquely to an invariant differential over A, which generates $H^0(A, \Omega_{A/k})$; thus, $H^0(A, \Omega_{A/k}) \cong k$ (see [LAG] 9.2, [RAG] 7.7 or [ABV] Sect. 11). Therefore, $f_*\Omega_{A/S}$ is $\mathcal{O}_S$-locally free of rank 1, and hence $A_{/S}$ is an elliptic curve.

Suppose now that k has characteristic 0 under the circumstance of the lemma. Since A is a curve, by the Riemann–Roch theorem, A has an ample line bundle; hence, A is projective defined by finitely many equations (and its group structure is also defined by finitely many equations). Thus, the group scheme A is actually defined over a subfield of k over $\mathbb{Q}$ of finite type. Thus, we may assume that $k \hookrightarrow \mathbb{C}$, and by base change, we may assume that $k = \mathbb{C}$. Then, by the Archimedean uniformization described in Sect. 2.3.3, $A(\mathbb{C}) \cong \mathbb{C}/L$ for a lattice $L \cong \pi_1^{top}(A(\mathbb{C}), \mathbf{0}_A)$ and A is defined by a Weierstrass equation. □

Here is another basic lemma:

Lemma 6.2 *Any nonconstant S-morphism $f : E \to E'$ of elliptic curves over S is finite flat. In particular,* Ker(f) *is a locally free group scheme of finite rank if f is nonconstant.*

Proof. Taking a geometric point $\mathrm{Spec}(k) \to S$ and writing $E_k = E \times_S \mathrm{Spec}(k)$ and $E'_k = E' \times_S \mathrm{Spec}(k)$, we need to prove that the induced morphism $f_k : E_k \to E'_k$ is flat. Once this is done for any geometric points of S, by the flatness criterion fiber by fiber (cf. [EGA] IV.11.3.11), f is proven to be flat. Replacing S by $\mathrm{Spec}(k)$ and E, E' by E_k, E'_k, we may assume $S = \mathrm{Spec}(k)$ for an algebraically closed field. Since E_k is of dimension 1 connected irreducible (by smoothness; Theorem 5.7), $f_k : E_k \to E'_k$ is a dominant morphism inducing an inclusion of function fields $k(E') \hookrightarrow k(E)$. Thus, $f_k : E_k \to E'_k$ is flat at a generic point [as $k(E)$ is a vector space over $k(E')$, which is $k(E')$-flat]. Since E'_k is smooth over k, its local rings are regular (they are, at maximal points, discrete valuation rings); hence, E'_k is normal. Thus, the nonflat locus of f_k is a proper closed subset of codimension at least 2 by Proposition 5.3, which is an empty set (as $\dim E_k = 1$). Hence, f_k is flat for all geometric points $\mathrm{Spec}(k) \to S$, and thus $f : E \to E'$ is flat. Thus, $\dim f^{-1}(x)$ is independent of $x \in E'$ and equal to 0 by Proposition 5.4 (2). Since $f^{-1}(x)$ is a noetherian scheme (as E is a noetherian scheme), the topological space of $f^{-1}(x)$ is a finite set. Thus, f is quasi-finite and proper, and so f is finite by Corollary 5.10 if $S = \mathrm{Spec}(k)$ for a field k. To prove this generally, we may assume that $S = \mathrm{Spec}(A)$. For every geometric point $x' = \mathrm{Spec}(k) \in E'$ over $x \in S$, write $E_x \to E'_x$ for the fiber. Then $f^{-1}(x')$ is nontrivial by the properness of E over E', so take $y \in f^{-1}(x')$, which is over x. Then $\mathcal{O}_{E,y} \otimes_A k = \mathcal{O}_{E_y,y}$ is an $\mathcal{O}_{E_x,x'}$-module of finite type by the result over k. Since $\mathcal{O}_{E_x,x'} = \mathcal{O}_{E',x'} \otimes_A k$, by Nakayama's lemma (Lemma 10.8), $\mathcal{O}_{E,y}$ is an $\mathcal{O}_{E',x'}$-module of finite type. Thus, f is a finite morphism.

Since $\mathrm{Ker}(f) = E \times_{E',\mathbf{0}_{E'}} S$ is the base change of f to S, which is finite flat (and hence locally free of finite rank). □

If $f : E \to E'_{/S}$ is a nonconstant S-morphism of elliptic curves, by the above lemma, $f_*\mathcal{O}_E$ is a locally free $\mathcal{O}_{E'}$-module of finite rank. Thus, if S is connected, $\mathrm{rank}_{\mathcal{O}_{E'}} f_*\mathcal{O}_E$ is a positive integer, which we write as $\deg(\phi)$ (the *degree* of ϕ). If S is nonconnected, we define the degree for each connected component in $\pi_0(S)$, and so we get $\deg(\phi) \in \mathbb{Z}^{\pi_0(S)}$. By Lemma 4.66, multiplication by $N_E : x \mapsto Nx$ for $0 < N \in \mathbb{Z}$ is a nonconstant endomorphism of an elliptic curve $E_{/S}$. It is well known that $\deg(N_E) = N^2$, which we will see using moduli theory of elliptic curves later.

6.1.2 Cartier Divisors

A closed subscheme $D = \mathit{Spec}_E(\mathcal{O}_E/I(D)) \subset E_{/S}$ defined by an invertible sheaf of ideals $I(D) \subset \mathcal{O}_E$ is called an effective Cartier divisor (relative to S) on E if $f_*\mathcal{O}_D = f_*(\mathcal{O}_E/I(D))$ is $\mathcal{O}_S$-flat (thus locally free over S). We define $\mathcal{L}(D) = I(D)^{-1}$ and put $\deg(D) = \deg(\mathcal{L}(D)) := \mathrm{rank}_S f_*(\mathcal{O}_D)$. The $\mathbf{0}$ section gives rise to a divisor $[\mathbf{0}]$ of degree 1 given by $\mathcal{O}_{[\mathbf{0}]} \overset{\mathbf{0}}{\cong} \mathcal{O}_S$. Consider $I(m[\mathbf{0}]) = I([\mathbf{0}])^m$ and $\mathcal{L}(m[\mathbf{0}]) = I(m[\mathbf{0}])^{-1}$ for $m \in \mathbb{Z}$. The line bundle $\mathcal{L}(m[\mathbf{0}])$ can be regarded as the sheaf of meromorphic functions on E smooth outside $\mathbf{0}$ and with possible singularity at $\mathbf{0}$ having a pole of order $\le m$.

If $S = \mathrm{Spec}(k)$ for an algebraically closed field k, k-rational effective divisors can be identified with positive linear combinations of points on $E(k)$. We have $\deg(\sum_P m_P[P]) = \sum_P m_P$. We can thus think of the group $\mathrm{Div}(E/k)$ of all formal linear combinations (including negative coefficients) of points on E. Then $\deg : \mathrm{Div}(E/k) \to \mathbb{Z}$ is a well-defined homomorphism given by the above formula. In particular, for any divisor $D \in \mathrm{Div}(E/k)$, we have $\mathcal{L}(D) = \mathcal{L}(D_+) \otimes \mathcal{L}(D_-)^{-1}$, writing $D = D_+ - D_-$ for effective divisors D_+ and D_-, and we can verify $\deg(\mathcal{L}(D)) = \deg(D)$.

6.1.3 Picard Schemes

For any scheme X, we define $\mathrm{Pic}(X)$ as the set of all isomorphism classes of invertible sheaves on X. The association $X \mapsto \mathrm{Pic}(X)$ is a contravariant functor by the pullback of invertible sheaves, and $\mathrm{Pic}(X)$ is actually a group by the tensor product. In short, $\mathrm{Pic} : SCH \to AB$ is a contravariant group functor.

Let $E \xrightarrow{f} S$ be an elliptic curve. We define, for each S-scheme $\phi : T \to S$,

$$\mathrm{Pic}_{E/S}(T) = \mathrm{Pic}(E_T)/f_T^* \,\mathrm{Pic}(T) \quad \text{for} \quad E_T = E \times_S T \to T.$$

Since $\mathbf{0}_E$ is a section of f, $\mathbf{0}_E^* : \mathrm{Pic}(E) \to \mathrm{Pic}(S)$ is a section of $f^* : \mathrm{Pic}(S) \to \mathrm{Pic}(E)$. Thus, we have another expression:

$$\mathrm{Pic}_{E/S}(T) = \mathrm{Ker}(\mathbf{0}_{E_T}^* : \mathrm{Pic}(E_T) \to \mathrm{Pic}(T)).$$

Since invertible sheaves are determined by its restriction to open covering, the functor Pic is local [in the sense of (4.1.7)], and hence its subfunctor $\mathrm{Pic}_{E/S}$ is local, giving a local group functor from $SCH_{/S}$ into AB.

We can extend the degree map to $\deg : \mathrm{Pic}_{E/S}(T) \to \mathbb{Z}^{\pi_0(T)}$ for the set of connected components $\pi_0(T)$. Indeed, for any algebraically closed field k and a geometric point $s : \mathrm{Spec}(k) \hookrightarrow T$, the fiber $E(s) = E \times_{S,\phi\circ s} s = E_T \times_T s$ is an elliptic curve over the field k and $\deg(\mathcal{L}) = \deg(\mathcal{L}(s))$ for the pullback $\mathcal{L}(s)$ at s, which is well defined independently of the choice of s in a connected component. We embed $\mathbb{Z}$ in $\mathbb{Z}^{\pi_0(T)}$ diagonally, and we define

$$\mathrm{Pic}^r_{E/S}(T) = \left\{\mathcal{L} \in \mathrm{Pic}_{E/S}(T) \middle| \deg(\mathcal{L}) = r\right\}.$$

Here is a generalization of Abel's theorem that is proven as Theorem 2.32 when $S = \mathrm{Spec}(k)$ for an algebraically closed field k:

Theorem 6.3 *We have as functors*

$$\mathrm{Pic}^r_{E/S}(T) \cong E(T) = \mathrm{Hom}_S(T, E) \tag{6.1.1}$$

by $\mathcal{L}([P]) \otimes \mathcal{L}([\mathbf{0}])^{r-1} \leftrightarrow P$.

We repeat here a proof given in [AME] Sect. 2.1 and [GME] Sect. 2.2.2.

Proof. We write E_T for $E \times_S T$. Since the degree is kept by pullback, Pic^r is a contravariant functor. Note that $\mathrm{Pic}^r \cong \mathrm{Pic}^0$ via $\mathcal{L} \mapsto \mathcal{L} \otimes I(r[\mathbf{0}])$.

We want to show

$$\underline{E} \cong \mathrm{Pic}^1 \cong \mathrm{Pic}^0 \quad \text{via} \quad P \mapsto I(P)^{-1} \mapsto I(P)^{-1} \otimes I(\mathbf{0}). \tag{6.1.2}$$

Since Pic^0 is a group functor with the identity $\mathcal{O}_E$ under the multiplication $\mathcal{L}\cdot\mathcal{L}' = \mathcal{L}\otimes\mathcal{L}'$, if (6.1.2) is true, the theorem is proven except for the uniqueness of the group structure. Before stating the theorem we showed that

$$\mathrm{Pic}^1 \text{ is local on } S \text{ in the sense of (4.1.7).} \tag{6.1.3}$$

Since Pic^1 is local, we may assume that $S = \mathrm{Spec}(A)$ is affine. Since $\mathcal{L} \in \mathrm{Pic}^1(S)$ is fiber by fiber of degree 1, $f_*\mathcal{L}$ is locally free and $\mathrm{rank}_{\mathcal{O}_S}(f_*\mathcal{L}) = 1$. Since S is affine, any quasi-coherent nontrivial sheaf $\mathcal{F}$ comes out of a nonzero A-module M so that the stalk $\mathcal{F}_P$ at a prime P is given by M_P (localization); in particular, $H^0(S,\mathcal{F}) = M \neq 0$ (see Remark 4.20 and [GME] Sect. 1.2.3). Thus, $f_*\mathcal{L}$ has a section ℓ. By further shrinking S, we may assume that $f_*\mathcal{L} \cong \mathcal{O}_S$ and that ℓ is the generator of the sheaf $f_*\mathcal{L}$. Then we have an exact sequence: $0 \to \mathcal{O}_E \xrightarrow{\ell} \mathcal{L} \to (\mathcal{L}/\mathcal{O}_E) \to 0$, whose cohomology long exact sequence gives another exact sequence (see (4.1.9)):

$$0 \to f_*\mathcal{O}_E \to f_*\mathcal{L} \to f_*(\mathcal{L}/\mathcal{O}_E) \to R^1 f_*\mathcal{O}_E \overset{(E3)}{\cong} \mathcal{O}_S \to R^1 f_*\mathcal{L} = 0.$$

The vanishing of $R^1 f_*\mathcal{L}$ follows from the Riemann–Roch theorem fiber by fiber (or the relative Riemann–Roch theorem in [GME] Sect. 2.1.4). Since fiber by fiber, this sequence is reduced to $0 \to k(s) \cong k(s) \to k(s) \cong k(s) \to 0$, we conclude that $f_*(\mathcal{L}/\mathcal{O}_E) \cong R^1 f_*\mathcal{O}_E \cong \mathcal{O}_S$. This shows that $\mathcal{L}$ comes from an effective relative Cartier divisor relativeto S; in other words, $\mathcal{L}^{-1} \subset \mathcal{O}_E$ is an ideal $I(D)$ of an effective Cartier divisor D of degree 1 and $\mathcal{L} = I(D)^{-1}$. Then we know that $\mathcal{L}/\mathcal{O}_E \cong \mathcal{O}_S$, because $\mathrm{Supp}(\mathcal{L}/\mathcal{O}_E)$ is finite over S and $f_*(\mathcal{L}/\mathcal{O}_E) \cong \mathcal{O}_S$. Hence, there is a section $P : S \hookrightarrow E$ such that $\mathcal{L} = I(P)^{-1}$ (i.e., $D = [P]$), and the natural map $\iota : \underline{E}(S) \to \mathrm{Pic}^1(S)$ given by $\iota(P) = I(P)^{-1}$ is surjective. From $(\mathcal{L},\ell)$, we can recover the section P as the zero locus of ℓ with $\mathcal{O}_P \cong \mathcal{O}_S$. Thus, P is uniquely determined by $I(P)^{-1}$, and hence ι is injective. □

By this theorem, $\mathrm{Pic}_{E/S} = \bigsqcup_r \mathrm{Pic}^r_{E/S}$ is a group scheme. We also confirm that an elliptic curve is an abelian scheme (proving the converse of Lemma 6.1).

If $\phi : C_{/S} \to C'_{/S}$ is a nonconstant S-morphism of two smooth geometrically connected curves, $|\phi^{-1}(s')| = \dim_{k(s')} \mathcal{O}_{\phi^{-1}(s')}$ (counting with multiplicity) is constant for geometric points s' of C'; that is,

$$\phi_*\mathcal{O}_C \text{ is locally free of finite rank.} \tag{LF}$$

The proof of this fact is the same as in the case of elliptic curves (Lemma 6.2; see also [GME] Sect. 4.1.3). Since a flat module of finite type over connected affine scheme $\mathrm{Spec}(A)$ is locally free of constant rank (e.g., [CRT] Theorem 7.12), by the connectedness of C' (and covering C' by connected affine open subschemes), $\mathrm{rank}\, \phi_*\mathcal{O}_C$ is constant everywhere. We write this number as $\deg(\phi)$. Thus, $\bigwedge^{\deg(\phi)} \phi_* \mathcal{O}_C$ is an invertible sheaf on C'. This construction gives a covariant functoriality to the originally contravariant functor $\mathrm{Pic}_{C/S}$ (called the *Albanese functoriality*). If $\phi : E \to E'$ is an S-morphism of elliptic curves, by our convention, ϕ takes $\mathbf{0}_E$ to $\mathbf{0}_{E'}$ and, hence, at the side of the Picard scheme, it is just $\mathcal{L} \mapsto \bigwedge^{\deg(\phi)} \phi_* \mathcal{L}$. Thus, obviously, ϕ is a homomorphism of group functors. In particular, a morphism of elliptic curves (taking zero to zero) is a homomorphism of group schemes (so the group structure on E is unique and only dependent on $\mathbf{0}_E$).

Exercise 6.4 *Let the notation and the assumption be as above. Give a detailed proof of the commutativity of the following diagram:*

$$\begin{array}{ccc} E & \xrightarrow{\phi} & E' \\ \wr\downarrow \text{Abel} & & \wr\downarrow \text{Abel} \\ \mathrm{Pic}^0(E) & \xrightarrow[\mathcal{L} \mapsto \bigwedge^{\deg(\phi)} \phi_* \mathcal{L}]{} & \mathrm{Pic}^0(E'). \end{array}$$

Let $f : E \to E'$ be an S-morphism of two elliptic curves over S. Suppose $f \neq 0$. By Lemma 6.2, we find that f is locally free of finite rank, and on each connected component of S, the degree of f is constant (as E over each connected component of S is geometrically connected). We have a morphism of functors

$$f^* : \mathrm{Pic}^0_{E'/S} \to \mathrm{Pic}^0_{E/S}$$

by the pullback under f. This induces, by Abel's theorem and Yoneda's lemma (Lemma 4.17), a homomorphism of S-elliptic curves $f^* : E' \to E$. Obviously, we see $(f \circ g)^* = g^* \circ f^*$ and $(f + g)^* = f^* + g^*$; that is, $f \mapsto f^*$ is an involution.

Theorem 6.5 *Suppose that* $\deg(f) = N \in \mathbb{Z}$. *Then* $f^* \circ f = N_E$, $f \circ f^* = N_{E'}$, $(f^*)^* = f$, $N_E^* = N_E$, *and* $\deg(f^*) = N$.

Proof. It is sufficient to prove the result assuming that $S = \mathrm{Spec}(k)$ for an algebraically closed field k. To see this, we note that we may assume that S is connected. For example, $f^* \circ f - N_E \neq 0$ at one fiber implies $\deg(f^* \circ f - N_E)$ is a nonzero constant, and so $f^* \circ f - N_E \neq 0$ everywhere. Thus, $f^* \circ f = N_E$ at one fiber implies $f^* \circ f = N_E$ everywhere.

In this proof, we assume the identity $\deg(N_E) = N^2$ (which we will see later using a certain universal elliptic curve; see Corollary 6.16). We have $\underline{E}(T) \cong \mathrm{Pic}^0(E/T)$ by the correspondence $P \leftrightarrow I(P)^{-1} \otimes I(0)$. The identity $f^* \circ f = N_E$ is therefore equivalent to

$$(I(P)^{-1} \otimes I(0))^{\otimes N} \cong f^*(I(f(P))^{-1} \otimes I(0)),$$

where "$\otimes N$" indicates an N-fold tensor product. Since $D = \mathrm{Ker}(f)$ is a relative effective Cartier divisor of degree N, $D = [Q_1] + \cdots + [Q_N]$ for N-points

Q_i counting with multiplicity (because k is an algebraically closed field). We have the exact sequence defining $I(f(P))$ [cf. [GME] (2.3)]:

$$0 \to I(f(P)) \to \mathcal{O}_{E'} \to k(f(P)) \to 0.$$

Since f is flat, by Exercise 4.22, f^* is left exact, and

$$0 \to f^*I(f(P)) \to \mathcal{O}_E(= f^*\mathcal{O}_{E'}) \to \mathcal{O}_{f^{-1}(f(P))}(= f^*k(f(P))) \to 0$$

is exact. Thus, we have $f^*I(f(P)) = I(f^{-1}(f(P))$, and $x \mapsto x + P$ induces an isomorphism $T_P : \mathrm{Ker}(f) \cong f^{-1}(f(P))$. Therefore,

$$I(f^{-1}(f(P)) = T^*_{-P} I(\mathrm{Ker}(f))$$

and in particular, $f^*I(0) = I(\mathrm{Ker}(f))$. We need to show

$$I(P)^{-N} \otimes I(0)^N \cong I(P + f^{-1}(0))^{-1} \otimes I(f^{-1}(0)) = \bigotimes_{i=1}^{N} (I(P + Q_i)^{-1} \otimes I(Q_i)).$$

By Abel's theorem, we have

$$I(P + Q_i)^{-1} \otimes I(0) \cong I(P)^{-1} \otimes I(Q_i)^{-1} \otimes I(0)^2$$
$$\iff I(P + Q_i)^{-1} \otimes I(Q_i) \cong I(P)^{-1} \otimes I(0),$$

which shows the claim. Now we have $f^* \circ f = N_E$. Thus, $f^* \neq 0$, and hence f^* is locally free by Lemma 6.2. In particular, $\deg(f^*)$ is well defined. Then

$$N^2 = \deg(N_E) = \deg(f^* \circ f) = \deg(f^*) \deg(f) = N \deg(f^*) \Rightarrow \deg(f^*) = N.$$

Since $S = \mathit{Spec}(k)$ with k algebraically closed, f is surjective [because $f(E)$ is closed and of dimension 1]. Thus, $f \circ f^*(f(P)) = f(N \cdot P) = Nf(P)$ implies $f \circ f^* = N$ by the surjectivity. Since $\deg(f^*) = N$, we have $(f^*)^* \circ f^* = N_{E'} = f \circ f^*$. Thus, $((f^*)^* - f)f^* = 0$. Since f^* is surjective, $(f^*)^* = f$. Finally,

$$N_E \circ N_E^* = [\deg(N_E)] = N_E^2 = N_E \circ N_E \Rightarrow (N_E^* - N_E) \circ N_E = 0,$$

which shows the last assertion: $N_E^* = N_E$. □

Theorem 6.6 (Hasse) *Let $f : E \to E$ be an S-endomorphism of an elliptic curve. Then we have*

(i) $\mathrm{Tr}(f) = f^* + f$ *is an integer in* $\mathrm{End}_S(E)$.
(ii) $f^2 - \mathrm{Tr}(f)f + \deg(f) = 0$ *in* $\mathrm{End}_S(E)$.
(iii) *The discriminant of the quadratic polynomial* $X^2 - \mathrm{Tr}(f)X + \deg(f)$ *is negative if f is not in* $\mathbb{Z} \cdot \mathrm{id}_E$.

Proof. Since $\mathrm{End}_S(E)$ is a ring under the composition product $(f, g) \mapsto f \circ g$, we hereafter simply write fg for $f \circ g$. Then

(i) $\mathbb{Z} \ni \deg(\mathrm{id}_E + f) = (\mathrm{id}_E + f)(\mathrm{id}_E + f^*) = \mathrm{id}_E + (f + f^*) + \deg(f) \Rightarrow \mathrm{Tr}(f) \in \mathbb{Z}$.
(ii) $f^2 - \mathrm{Tr}(f)f + \deg(f) = f^2 - (f + f^*)f + f^*f = 0$.
(iii) Put $P(X,Y) = X^2 - \mathrm{Tr}(f)XY + \deg(f)Y^2$. We need to prove the quadratic form $P(X,Y)$ is positive definite over $\mathbb{Q}$. Since $P(\frac{n}{m}, \frac{n'}{m}) = m^{-2}P(n,n')$, we only need to show $P(m,n) > 0$ for every pair of integers $(m,n) \neq 0$. Since $f \notin \mathbb{Z} \cdot \mathrm{id}_E$, $nf + m \neq 0$ for all pairs $(m,n) \neq 0$ by (ii). Thus, $P(m,n) = (nf+m)(nf+m)^* = \deg(nf+m) > 0$ for all $(m,n) \neq (0,0)$.

6.1.4 Invariant Differentials Are Nowhere Vanishing

By (E3), for a dense affine open subset $\mathrm{Spec}(A)$ of S, we have $H^0(E, \Omega_{E/A}) = A\omega$ for a 1-differential ω. For each point $P \in E(S)$, $T_P : x \mapsto x + P$ gives an automorphism of the curve E (sending $\mathbf{0}_E$ to P). Thus, we can bring any given cotangent vector at $\mathbf{0}$ isomorphically to a cotangent vector at P, and as seen in the proof of Lemma 6.1, each cotangent vector at $\mathbf{0}$ extends to a global section of $\Omega_{E/S}$. Thus, $T_P^*\omega = \omega$. Or differently, if we write $\lambda(P)$ for the constant given by $T_P^*\omega = \lambda(P)\omega$, $\lambda : E(S) \mapsto \mathbb{G}_a(S)$ clearly gives rise to a morphism of functors $E \to \mathbb{G}_a$ defined over S; thus, it is a morphism of schemes by Lemma 4.17. Since $\mathbb{G}_a$ is affine and E is proper and geometrically connected over S, the image of λ is a single point. Thus, $\lambda = 1$ as $\lambda(\mathbf{0}_E) = 1$.

The proof of Lemma 4.84 tells us the following fact, for which we give a shorter proof.

Lemma 6.7 *Let E be an elliptic curve over a general base S. Write the group operation additively as $(x,y) \mapsto x + y$. Then $N_E : E \to E$ given by $x \mapsto Nx$ induces multiplication by N on the tangent space of the identity and on the invariant differentials $\Omega_{E/S}$.*

Proof. We only need to prove this for $\Omega_{E/S}$ as the other case follows from duality. First, suppose that E has an invariant differential ω generating $\Omega_{E/S}$. Consider the diagonal embedding $\Delta : E \hookrightarrow E^N := \overbrace{E \times_S E \times_S \cdots \times_S E}^{N}$ given by $\Delta(x) = (x, x, \ldots, x)$ and the sum $s : E^N \to E$ given by $(x_1, \ldots, x_n) \mapsto x_1 + x_2 + \cdots + x_n$. Then $\Delta^* s^* \omega = N\omega$ as ω is invariant under addition. However, $s \circ \Delta = N_E$. This shows the result for $\Omega_{E/S}$. In general, we cover S by an open affine subschemes U such that $E_U := E \times_S U$ has an invariant differential. Then N_E^* is multiplication by N over $\Omega_{E_U/U}$. Since $\omega_{E/S}$ is a sheaf, this implies that $N_E^* = N$ over the entire $\Omega_{E/S}$.

By the above lemma, multiplication by $N > 0$ is a nonconstant morphism $N_E : E \to E$ if N is not identically 0 over S. We will see later in Corollary 6.16 that multiplication by N is nonconstant always (though a sketch of a different proof of this was already given in Lemma 4.66).

Lemma 6.8 *Let $E_{/S}$ be an elliptic curve. If $0 < N \in \mathbb{Z}$ is invertible on S, $E[N] = \mathrm{Ker}(N_E)$ is étale finite over S.*

Proof. Since multiplication by N on $\Omega_{E/S}$ is invertible, we have

$$\Omega_{E/E,N_E} = \mathrm{Coker}(N : N_E^*\Omega_{E/S} \to \Omega_{E/S}) = \Omega_{E/S}/N\Omega_{E/S} = 0.$$

Then, by Lemma 4.32, $N_E : E \to E$ is étale finite.

6.1.5 Classification Functors

An important fact from functorial algebraic geometry is that we can associate, with each S-scheme X, a contravariant functor $\underline{X} : SCH_{/S} \to SETS$ such that $\underline{X}(T) = \mathrm{Hom}_S(T, X)$. This association is fully faithful; in other words, recalling that $SCH_{/S}$ is a full subcategory of the category $\mathcal{C}_S$ of contravariant functors from S-schemes to $SETS$, we have

$$\mathrm{Hom}_S(X, Y) \cong \mathrm{Hom}_{\mathcal{C}_S}(\underline{X}, \underline{Y})$$

via $(X \xrightarrow{\phi} Y) \mapsto (\phi(T) : \underline{X}(T) \to \underline{Y}(T))$ given by $\phi(T)(T \xrightarrow{f} X) = \phi \circ f$. This is intuitively clear because an algebraic variety is just a function associating to each ring R its R-integral points $X(R) = \underline{X}(\mathrm{Spec}(R))$. The verification of this is the same as the proof of Lemma 4.17 and is left to the reader as an exercise [the inverse is given by $\mathrm{Hom}_{CTF}(\underline{X}, \underline{Y}) \ni F \mapsto F(X)(1_X)$, where $F(X) : \underline{X}(X) \to \underline{Y}(X) = \mathrm{Hom}_S(X, Y)$]. We often write X for $\underline{X}$ hereafter.

As we will see later in Corollary 6.16, $\deg N_E = N^2$, and $E[N](k) \cong (\mathbb{Z}/N\mathbb{Z})^2$ for all geometric points $\mathrm{Spec}(k) \hookrightarrow S$ as long as N is invertible in k (see Lemma 4.70). We consider the following functor:

$$\wp'_{\Gamma_1(N)}(R) = \left[(E, P, \omega)_{/R}\right] \tag{6.1.4}$$

from the category ALG of $\mathbb{Z}$-algebras into $SETS$, where ω is a nowhere vanishing invariant differential, P is a point of order exactly N, that is, $m \mapsto mP$ induces a closed immersion $(\mathbb{Z}/N\mathbb{Z})_{/R} \hookrightarrow E$ of group schemes defined over R, and $[\ \ ] = \{\ \ \}/\cong$ is the set of all isomorphism classes of the objects inside the brackets. Here $\mathbb{Z}/N\mathbb{Z}$ as a group functor associates with T the group $(\mathbb{Z}/N\mathbb{Z})^{\pi_0(T)}$ as in Sect. 4.3.5. Therefore, $\mathcal{O}_{(\mathbb{Z}/N\mathbb{Z})_{/\mathbb{Z}}} = \bigoplus_{\mathbb{Z}/N\mathbb{Z}} \mathbb{Z}$ [and $\mathcal{O}_{(\mathbb{Z}/N\mathbb{Z})_{/R}} = \bigoplus_{\mathbb{Z}/N\mathbb{Z}} R$], and the structure sheaf of $\mathbb{Z}/N\mathbb{Z}$ is R-free of finite rank N.

Recall another prototypical example of locally free group schemes: Starting with the multiplicative group $\mathbb{G}_m$ [as a group functor $\mathbb{G}_m(R) = R^\times$ and as an affine scheme $\mathbb{G}_m = \mathrm{Spec}(\mathbb{Z}[t, t^{-1}])$], we consider the kernel μ_N of N [as a group functor $\mu_N(R) = \{\zeta \in R | \zeta^N = 1\}$ and as a scheme $\mu_N = \mathrm{Spec}(\mathbb{Z}[t]/(t^N - 1)) = \mathrm{Spec}(\mathbb{Z}[(\mathbb{Z}/N\mathbb{Z})])$]. Then μ_N is a locally free group scheme of rank N. If $N > 1$, it is nonisomorphic to $\mathbb{Z}/N\mathbb{Z}$, since for any prime p, $\mu_p(\mathbb{F}_p) = \{1\}$, but $(\mathbb{Z}/p\mathbb{Z})(\mathbb{F}_p) = \mathbb{Z}/p\mathbb{Z}$. We consider a version of the functor $\wp'_{\Gamma_1(N)}$ defined as follows:

$$\wp_{\Gamma_1(N)}(R) = \left[(E, \phi_N : \mu_N \hookrightarrow E[N], \omega)_{/R}\right], \tag{6.1.5}$$

where ϕ_N is a closed immersion of group schemes. In particular, $\wp = \wp_{\Gamma_1(1)}$ classifies just pairs $(E, \omega)_{/R}$.

6.1.6 Cartier Duality

The two functors $\wp_{\Gamma_1(N)}$ and $\wp'_{\Gamma_1(N)}$ are isomorphic by the duality of locally free group schemes of finite rank (see Sect. 4.3.5): Here is a brief recall of the theory. If G is a locally free group scheme of rank N over S, write $G_{/T} = G \times_S T$ for an S-scheme T. Then there exists a group scheme $G^*_{/S}$ such that $G^*(T) = \mathrm{Hom}_T(G_{/T}, \mathbb{G}_{m/T}) = \mathrm{Hom}_T(G_{/T}, \mu_{N/T})$, where $\mathbb{G}_{m/S} = \mathbb{G}_m \times S$ and $\mu_{N/S} = \mu_N \times S$ over $\mathrm{Spec}(\mathbb{Z})$. We have $(G^*)^* \cong G$ in an obvious manner, and $(\mathbb{Z}/N\mathbb{Z})^* = \mu_N$ by $\zeta(m) = \zeta^m$ for $\zeta \in \mu_N(R)$ and $m \in (\mathbb{Z}/N\mathbb{Z})(R)$.

Let $E \xrightarrow{f} S$ be an elliptic curve and $f_T : E_{/T} \to T$ be the base change to an S-scheme $T \to S$ with zero section $\mathbf{0}_T : T \hookrightarrow E_{/T}$. Recall the splitting for the identity section $\mathbf{0} : S \to E$:

$$\mathrm{Pic}(E_{/T}) = f_T^* \mathrm{Pic}(T) \oplus \mathrm{Ker}(\mathbf{0}_T^*) \quad \text{and} \quad \mathrm{Ker}(\mathbf{0}_T^*) = \mathrm{Pic}_{E/S}(T),$$

regarding $\mathrm{Pic}_{E/S}(T)$ as a set of isomorphism classes of invertible sheaves whose restriction to $\mathbf{0}$ is trivial; that is, $\mathbf{0}^*\mathcal{L}$ is isomorphic to $\mathcal{O}_S$.

Let $P \in \mathrm{Ker}(\pi)$ for a nonconstant S-morphism $\pi : E \to E'$. Then

$$\pi^* : E' \cong \mathrm{Pic}^0_{E'/S} \to \mathrm{Pic}^0_{E/S} \cong E$$

is an S-homomorphism.

Lemma 6.9 *If π is nonconstant,* $\mathrm{Ker}(\pi)$ *is locally free of finite rank, and we have* $\mathrm{Ker}(\pi^*) = \mathrm{Ker}(\pi)^*$ *for the dual* $\mathrm{Ker}(\pi)^*$ *of* $\mathrm{Ker}(\pi)$. *In particular,* $E[N]$ *is a Cartier self-dual.*

Proof. Since π is nonconstant, it is finite flat (Lemma 6.2), and so $0 < \deg(\pi) \in \mathbb{Z}$. Since $\pi^* \circ \pi = \deg(\pi)$ (Theorem 6.6), π^* is nonconstant, and π^* is finite flat. Take $\mathcal{L} \in \mathrm{Ker}(\pi^*)$. If $U = \mathrm{Spec}(A) \subset E'$ is an affine open subset, $\mathcal{L}|_U = \tilde{L}$ for a locally free A-module L. Shrinking U if necessary, we may assume that $L \cong \phi_U^{-1} A \subset Q(A)$ for the total quotient ring $Q(A)$ of A. Thus, we may write $\mathcal{L}|_U = \phi_U^{-1}\mathcal{O}_U$ for a small open subscheme $U \subset E'$. Take an open covering $E' = \bigcup_i U_i$ such that $\mathcal{L}|_{U_i} = \phi_i^{-1}\mathcal{O}_{U_i}$ for a generator ϕ_i^{-1}. Since $\mathbf{0}^*\mathcal{L} = \mathcal{O}_S$, we may normalize ϕ_i so that $\phi_i \circ \mathbf{0}_{E'} = \phi_j \circ \mathbf{0}_{E'}$ for all $i \neq j$ on $\mathbf{0}_{E'}^{-1}(U_i \cap U_j)$. Let $\varphi_i = \phi_i \circ \pi$; we have $\pi^*\mathcal{L}|_{V_i} = \varphi_i^{-1}\mathcal{O}_{V_i}$ for $V_i = \pi^{-1}(U_i)$. Let $P \in \mathrm{Ker}(\pi)$; then

$$\varphi_i \circ P = \phi_i \circ \pi \circ P = \phi_i \circ \mathbf{0}_{E'} = \phi_j \circ \mathbf{0}_{E'} = \varphi_j \circ P.$$

This implies that the $\varphi_i \circ P$s glue to give a global section $\varphi \circ P \in \Gamma(S, \mathcal{O}_S^\times) = \mathbb{G}_m(S)$, getting a homomorphism $\mathrm{Ker}(\pi^*) \to \mathrm{Ker}(\pi)^*$. Thus, we get a pairing

$$e_\pi : \mathrm{Ker}(\pi) \times \mathrm{Ker}(\pi^*) \to \mathbb{G}_m.$$

Since E is a $\mathrm{Ker}(\pi)$-torsor over E', we have $E \times_{E'} E \cong \mathrm{Ker}(\pi) \times_S E'$. Thus, for any homomorphism $\zeta : \mathrm{Ker}(\pi) \to \mathbb{G}_m$, we find $\phi : \mathrm{Ker}(\pi) \times_S E' \to \mathbf{P}^1$ such that $\phi(y+t) = \zeta(t)\phi(y)$ for $t \in \mathrm{Ker}(\pi)$. This function ϕ gives rise to a divisor D on $E'_E = E' \times_S E$. By definition, $\pi_E^* \mathcal{L}(D) = \mathcal{O}_{E \times_S E}$ for $\pi_E = \pi \times 1 : E \times_S E \to E' \times_S E$, and $e_\pi(x, \mathcal{L}(D)) = \zeta(x)$. Thus, over E, $e_{\pi/E} : \mathrm{Ker}(\pi)_{/E} \times \mathrm{Ker}(\pi^*)_{/E} \to \mathbb{G}_m$ is a perfect pairing. Since $E \to S$ is faithfully flat, we find that the original e_π is perfect. Applying this to $\pi = N_E$, since $N_E^* = N_E$ (Theorem 6.5), we get $E[N]^* = E[N]$. □

Write this pairing as $\langle\ ,\ \rangle : E[N] \times_S E[N] \to \mu_{N/S}$, which is sometimes called the *Weil pairing*. Then we get $\langle \phi(P), Q\rangle = \langle P, \phi^*(Q)\rangle$; so $\phi \mapsto \phi^*$ is an involution with $\phi^* \circ \phi = \deg(\phi) \geq 0$ (a positive involution; see Theorem 6.6).

For a given level-N structure $\phi_N : \mathbb{Z}/N\mathbb{Z} \hookrightarrow E[N]$; by duality, we get $\pi_N : E[N] \twoheadrightarrow \mu_N$, which has a section ϕ'_N well determined modulo $C = \phi_N(\mathbb{Z}/N\mathbb{Z})$. Thus, $(E/C, \phi'_N : \mu_N \hookrightarrow (E/C)[N], \omega')$ is well defined as an element of $\wp_{\Gamma_1(N)}(R)$, where ω' coincides with ω at the identity (because the projection $E \to E/C$ is a local isomorphism, that is, an étale morphism; see [GME] Proposition 1.8.4). The inverse map, $\wp \to \wp'$, is given similarly by

$$(E', \phi'_N : \mu_N \hookrightarrow E'[N], \omega') \mapsto (E'' = E'/\mathrm{Im}(\phi'_N), \phi_N : \mathbb{Z}/N\mathbb{Z} \hookrightarrow E''[N], \omega'').$$

Since $(E/C)/\phi'_N(\mu_N) = E/E[N] \cong E$, we have $\wp' \cong \wp \cong \wp'$.

Corollary 6.10 *For each $N \geq 1$, we have a canonical isomorphism of functors: $\wp_{\Gamma_1(N)} \cong \wp'_{\Gamma_1(N)}$ defined over $ALG_{/\mathbb{Z}[\frac{1}{N}]}$.*

Let p be a prime and $E_{/S}$ be an elliptic curve over a connected base S. Since S is connected, $\mathcal{O}_S$ cannot be decomposed into a direct sum of two nontrivial sheaves of rings (cf. [GME] Sect. 1.2.4). Since $E[p^n]_{/S}$ is a locally free group scheme of rank p^{2n}, we have $E[p^n] = Spec_S(\mathcal{A})$ for a locally free sheaf of $\mathcal{O}_S$-bialgebras $\mathcal{A}$. By Proposition 4.68, over the henselization $\tilde{S}_x$ of S at each closed point x, we have an exact sequence of abelian fppf sheaves:

$$0 \to E[p^n]^\circ_{/\tilde{S}_x} \to E[p^n]_{/\tilde{S}_x} \to E[p^n]^{\mathrm{et}}_{/\tilde{S}_x} \to 0,$$

where $E[p^n]^\circ$ is the connected component containing the identity section.

Suppose that p is locally nilpotent on S [so, for any stalk $\mathcal{O}_{S,x}$ at a closed point x, $p^{m(x)} = 0$ in $\mathcal{O}_{S,x}$ for some positive integer $m(x)$ possibly depending on x]. Because of the local nilpotence of p, the connected component is well defined globally (without making a base change to the henselization); so this exact sequence extends to an exact sequence of fppf abelian sheaves over S:

$$0 \to E[p^n]^\circ_{/S} \to E[p^n]_{/S} \to E[p^n]^{\mathrm{et}}_{/S} \to 0,$$

where $E[p^n]^\circ_{/S}$ is defined in the same way as above, replacing $\tilde{S}_x$ by S. Here $E[p^n]^{\mathrm{et}}_{/S}$ is for the moment just the quotient abelian sheaf of $E[p^n]_{/S}$ by $E[p^n]^\circ_{/S}$. The abelian fppf sheaf $E[p^n]^\circ$ may not be flat but affine over the base S; so we write $E[p^n]^\circ = \mathit{Spec}_S(\mathcal{A}^\circ)$ and $E[p^n] = \mathit{Spec}_S(\mathcal{A})$. Thus, the nonflat locus $E[p^n]^{\circ,nfl}$ is over a proper closed subscheme $S^{nfl} \subset S$. If S^{nfl} is empty, $E[p^n]^\circ$ is finite flat over S and the quotient $E[p^n]^{\mathrm{et}}$ is étale finite over S. Suppose that $E[p^n]^\circ$ is flat over S. By the group structure, over an étale extension S'/S so that $P \in E^{\mathrm{et}}(S')$ with $P \neq \mathbf{0}_{E[p^n]^{\mathrm{et}}}$, regarding $P \in E[p^n](S')$, the translation by P brings $\mathcal{A}^\circ$ isomorphic to the maximal indecomposable $\mathcal{O}_S$-algebra direct summand of $\mathcal{A}$ through which $\underline{P} : \mathcal{A} \to \mathcal{O}_{S'}$ factors. Thus, over a finite étale extension S', $E[p^n]^{\mathrm{et}}_{/S'}$ is a disjoint union of schemes isomorphic to S'. Thus, $E[p^n]^{\mathrm{et}}_{/S'}$ is étale, and by descent, $E[p^n]^{\mathrm{et}}_{/S}$ is étale over S. The étale group scheme $E[p^n]^{\mathrm{et}}$ is called the *maximal étale quotient* of $E[p^n]$.

If p is invertible over S, $E[p^n] = E[p^n]^{\mathrm{et}}$ as p^n_E is étale. Suppose that p is not invertible over S. Since μ_{p^n} is connected, by self-duality $E[p^n] \times E[p^n] \to \mu_{p^n}$, $E[p^n]^\circ$ is nontrivial. Thus, $\operatorname{rank} E[p]^{\mathrm{et}} \leq p$. Writing $\operatorname{rank} E[p]^{\mathrm{et}} = p^{r(E)}$, the number $r(E)$ is called the *p-rank* of E. We call E *supersingular* at p if $r(E) = 0$, and otherwise, we call E *ordinary* at p (or p-ordinary). By the self-duality, we have $(E[p^n]^{\mathrm{et}})^* \hookrightarrow E[p^n]^\circ$. Thus, $\operatorname{rank} E[p] = p^{2r(E)+s(E)}$ for $s(E)$ given by $p^{s(E)} = \operatorname{rank} E[p]^\circ / \operatorname{rank}(E[p]^{\mathrm{et}})^*$. Summarizing this, we have the following:

Lemma 6.11 *Suppose that p is locally nilpotent over S. Then over an open dense subscheme of S, $E[p^n]^\circ$ is finite flat. If $E[p^n]^\circ$ is flat over S, locally under étale topology, the étale quotient $E[p]^{et}$ is isomorphic to $(\mathbb{Z}/p\mathbb{Z})^{r(E)}$ for a constant $r(E)$ called the p-rank of E, and $E[p]^\circ$ has the maximal multiplicative part $(E[p]^{et})^*$ isomorphic to $\mu_p^{r(E)}$. In particular, we have $0 \leq r(E) \leq 1$.*

Exercise 6.12 *Where did we use the assumption of the local nilpotency of p in the proof of the above lemma?*

6.2 Moduli of Elliptic Curves

As already described, we may regard a scheme X over a ring B as a contravariant functor $\underline{X} : SCH_{/B} \to SETS$ by $\underline{X}(S) = \mathrm{Hom}_{SCH_{/B}}(S, X)$. A scheme $M_{/B}$ is a coarse modulus of a contravariant functor $\mathcal{F} : SCH_{/B} \to SETS$ if we have a morphism $f : \mathcal{F} \to \underline{M}$ of functors such that $f(k) : \mathcal{F}(\mathrm{Spec}(k)) \cong \underline{M}(\mathrm{Spec}(k))$ for any algebraically closed field k and for any other morphism $g : \mathcal{F} \to \underline{M}'$ for a B-scheme M', we have a morphism $M \xrightarrow{\pi} M'$ with $g = \pi \circ f$ [see [GME] Sect. 2.3.2 (CM1–3) for more details on coarse moduli schemes]. If the above morphism f is an isomorphism of functors, we say that $\mathcal{F}$ is *representable* by $M_{/B}$, and $M_{/B}$ is a *fine moduli* scheme. A fine or coarse modulus of a functor is uniquely determined up to isomorphisms of B-schemes (by Yoneda's lemma). We now study the scheme over $\mathbb{Z}[\frac{1}{N}]$ representing $\wp_{\Gamma_1(N)}$.

6.2.1 Moduli of Level 1 over $\mathbb{Z}[\frac{1}{6}]$

Hereafter, we assume until Sect. 6.3 (for simplicity) that 6 is invertible in any algebra we consider unless otherwise mentioned (see [GME] Sects. 2.6–9 for the theory valid over $\mathbb{Z}$). Let $(E,\omega)_{/R}$ be a pair of an elliptic curve and a nowhere vanishing differential. We choose a parameter T at $\mathbf{0}$ so that

$$\omega = (1 + \text{higher terms of } T)dT.$$

By the integral version of the Riemann–Roch theorem (see [GME] Sect. 2.1.4), we verify (as in [GME] Sect. 2.2.5) the following result:

Lemma 6.13 *For an elliptic curve $f : E \to S$,* $\mathrm{rank}\, f_*\mathcal{L}(m[\mathbf{0}]) = m$ *if* $m > 0$.

Here is a sketch of a proof.

Proof. We first assume $S = \mathrm{Spec}(k)$ for an algebraically closed field. Since $f : E \to \mathrm{Spec}(k)$ is proper with $f^*\mathcal{O}_E = k$, $\mathcal{L}(-D) = 0$ if $D = \sum_P m_P P \neq 0$ with $m_P \geq 0$. Then by the Riemann–Roch theorem (see Theorem 2.26 in the text or [ALG] IV.1.3),

$$\dim_k H^0(E, \mathcal{L}(D)) = \deg(D) - g + 1 + \dim_k H^0(\mathcal{L}(-D) \otimes \Omega_{E/k})$$

for $g = \mathrm{rank}_k\, H^0(E, \Omega_{E/k}) = 1$. Since $\Omega_{E/k} \cong \mathcal{O}_E$, we have

$$\dim_k H^0(E, \mathcal{L}(m[\mathbf{0}])) = m + \dim_k H^0(E, \mathcal{L}(-m[\mathbf{0}])) = m$$

if $m > 0$.

Now we treat general S. Write $f : E \to S$ for structure morphism and $I([\mathbf{0}])$ for the $\mathcal{O}_E$-ideal defining the image $[\mathbf{0}] \cong S$ as the subscheme of E. Then $\mathcal{L}(m[\mathbf{0}]) = I([\mathbf{0}])^{-m}$. We first consider $I([\mathbf{0}])^{-m}$, which is of degree $m \geq g = 1$. Since $[\mathbf{0}]$ is a relative Cartier divisor, $m[\mathbf{0}]$ is a relative Cartier divisor. We have an exact sequence:

$$0 \to \mathcal{O}_E \longrightarrow I([\mathbf{0}])^{-m} \longrightarrow I([\mathbf{0}])^{-m}/\mathcal{O}_E \to 0,$$

which yields a long exact sequence:

$$0 \to f_*\mathcal{O}_E \to f_*I([\mathbf{0}])^{-m} \to f_*(I([\mathbf{0}])^{-m}/\mathcal{O}_E) \to R^1f_*\mathcal{O}_E \to R^1f_*I([\mathbf{0}])^{-m},$$

where R^1f_* is the right-derived functor on the Zariski site (see Sect. 4.1.10). Since $\deg(I([\mathbf{0}])^{-m}) = m \geq 2g-1 = 1$, fiber by fiber (e.g., Corollary 2.27), $R^1f_*(I([\mathbf{0}])^{-m}) \otimes k(s) = 0$ $(s \in S)$ because $\deg(I([\mathbf{0}])^m \otimes \Omega_{E/S}) < 0$. This shows that $R^1f_*(I([\mathbf{0}])^{-m}) = 0$ and

$$\begin{aligned} 0 \to f_*\mathcal{O}_E \to f_*I([\mathbf{0}])^{-m} \to f_*(I([\mathbf{0}])^{-m}/\mathcal{O}_E) \\ \to R^1f_*\mathcal{O}_E(\cong \mathrm{Hom}_{\mathcal{O}_S}(f_*\Omega_{E/S}, \mathcal{O}_S)) \to 0 \end{aligned}$$

is exact. Since $f_*(I([\mathbf{0}])^{-m}/\mathcal{O}_E)$ is locally $\mathcal{O}_S$-free of finite rank and $R^1f_*\mathcal{O}_E$ is, by the Grothendieck–Serre duality (e.g., [GME] Sect. 2.1.2), isomorphic to the rank 1 locally $\mathcal{O}_S$-free sheaf given by $\mathrm{Hom}_{\mathcal{O}_S}(f_*\Omega_{E/S}, \mathcal{O}_S)$, $f_*I([\mathbf{0}])^{-m}$ is locally free of rank m if $m \geq 1$ over $\mathcal{O}_S$. □

Replacing S by its sufficiently small affine open subscheme $\mathrm{Spec}(R)$, we have $H^0(E, \mathcal{L}(2[\mathbf{0}]) = R1 + Rx$ and $H^0(E, \mathcal{L}(3[\mathbf{0}]) = R1 + Rx + Ry$, and then $H^0(E, \mathcal{L}(m[\mathbf{0}])) = H^0(E, \mathcal{L}([\mathbf{0}])^{\otimes m})$ is locally R-free of rank m for $m > 3$. In particular, by Corollary 2.19, we have two morphisms $x, y : E \to \mathbf{P}^1$ such that

1. x has a pole of order 2 at $\mathbf{0}$ with the leading term T^{-2} in its Laurent expansion in T (removing the constant term by translation);
2. y has a pole of order 3 with leading term $-T^{-3}$.

Out of these functions x and y, we can create a basis of $H^0(E, \mathcal{L}(m[\mathbf{0}]))$, showing $H^0(E, \mathcal{L}(m[\mathbf{0}]))$ is R-free of rank m:

- $H^0(E, \mathcal{L}(2[\mathbf{0}])) = R + Rx$, $H^0(E, \mathcal{L}(3[\mathbf{0}])) = R + Rx + Ry$. This implies that x has a pole of order 2 at $\mathbf{0}$ and y has order 3 at $\mathbf{0}$. They are regular outside $\mathbf{0}$.
- From these functions $1, x, y$, we create functions with a pole of order n at $\mathbf{0}$ as follows:

$$\begin{aligned} n &\leq 4 : 1, x, y, x^2 \ (\dim = 4) \\ n &\leq 5 : 1, x, y, x^2, xy \ (\dim = 5) \\ n &\leq 6 : 1, x, y, x^2, xy, x^3, y^2 \ (\dim = 6). \end{aligned}$$

Comparing with the leading term of T^{-6}, one sees that the seven sections

$$1, x, y, x^2, xy, x^3, y^2$$

in the space $H^0(E, \mathcal{L}(6[\mathbf{0}]))$ have to be linearly dependent and satisfy the following relation:

$$y^2 + a_1 xy + a_3 y = x^3 + a_2 x^2 + a_4 x + a_6.$$

We can kill in a unique way the terms involving xy and y by a variable change $y \mapsto y + ax + b$. Indeed, by the variable change $y \mapsto y - \frac{a_1}{2}x - \frac{a_3}{2}$, we get a simplified equation:

$$y^2 = x^3 + b_2 x^2 + b_4 x + b_6.$$

Again, a variable change $x \mapsto x - \frac{b_2}{3}$ simplifies the equation to

$$y^2 = x^3 + c_2 x + c_3.$$

Since $\mathcal{L}(3[\mathbf{0}])$ is very ample $[\deg(\mathcal{L}(3[\mathbf{0}])) = 3 \geq 2g+1]$, this equation is not an equation of a quotient curve of E. Finally, making a variable change $2y \mapsto y$ (so now the T-expansion of y begins with $-2T^{-3}$), we get a unique equation from $(E, \omega)_{/R}$:

$$y^2 = 4x^3 - g_2(E, \omega)x - g_3(E, \omega) \ \text{ for } g_2(E, \omega), g_3(E, \omega) \in R.$$

In other words, $E \subset \mathbf{P}^2_{/R}$ is given by

$$\mathrm{Proj}(R[X, Y, Z]/(ZY^2 - 4X^3 + g_2(E, \omega)XZ^2 + g_3(E, \omega)Z^3)).$$

Exercise 6.14 *Prove that the equation* $ZY^2 - 4X^3 + g_2XZ^2 + g_3Z^3 = 0$ *in* $R[X, Y, Z]$ *gives a smooth curve of genus* 1 *having the origin* $\mathbf{0} := (0, 1, 0)$ *in* $\mathbf{P}^2_{/R}$ *if and only if* $\Delta := g_2^3 - 27g_3^2 \in R^\times$ (see Example 2.5 and Exercise 2.38).

We recover the differential ω by $\frac{dx}{y}$.

We assumed that $S = \mathrm{Spec}(R)$ so that $f_*\mathcal{L}(2[\mathbf{0}])$ and $f_*\mathcal{L}(3[\mathbf{0}])$ are $\mathcal{O}_S$-free of rank 2 and 3 and normalized the sections x and y to satisfy the Weierstrass equation. We found that the two sections x and y as above are uniquely determined by ω. For a general S not necessarily affine, we cover S by affine open subschemes $\mathrm{Spec}(R_i)$ with the above property. The Weierstrass equations we get over R_i and R_j must be equal on $\mathrm{Spec}(R_i) \cap \mathrm{Spec}(R_j)$ in S, as they are determined by the same $\omega|_{\mathrm{Spec}(R_i)\cap\mathrm{Spec}(R_j)}$. Thus, $g_i(E, \omega)$ $(i = 2, 3)$ are well-defined global sections over S, and $E_{/S}$ is defined by the relative projective scheme (see Sect. 4.1.9)

$$\mathcal{P}roj_S(\mathcal{O}_S[X, Y, Z]/(ZY^2 - 4X^3 + g_2(E, \omega)XZ^2 + g_3(E, \omega)Z^3))$$

having the corresponding nowhere vanishing differential $\frac{dX}{Y}$.

Summing up these arguments, we conclude the following fact:

Theorem 6.15 *Define a polynomial ring* $\mathcal{R} = \mathcal{R}_{\Gamma_1(1)} = \mathbb{Z}[\frac{1}{6}, g_2, g_3, \frac{1}{\Delta}]$ $(\Delta = g_2^3 - 27g_3^2$ *inverted) with variables* g_2 *and* g_3. *Then we have a canonical isomorphism*

$$\wp_{\Gamma_1(1)}(R) \cong \mathrm{Hom}_{\mathbb{Z}[\frac{1}{6}]-alg}(\mathcal{R}, R) = \mathcal{M}_1(R),$$

where $\mathcal{M}_1 = \mathrm{Spec}(\mathcal{R})$ *for* $\mathcal{R} = \mathbb{Z}[\frac{1}{6}, g_2, g_3, \frac{1}{\Delta}]$. *We have the universal elliptic curve and the universal differential* $\boldsymbol{\omega}$ *given by*

$$(\mathbf{E}, \boldsymbol{\omega})_{/\mathcal{M}_1} = \left(\mathrm{Proj}(\mathcal{R}[X, Y, Z]/(ZY^2 - 4X^3 + g_2XZ^2 + g_3Z^3)), \frac{dX}{Y}\right).$$

For each pair $(E, \omega)_{/S}$, *we have a unique morphism* $\varphi : S \to \mathcal{M}_1$ *over* $\mathbb{Z}[\frac{1}{6}]$) *with* $\varphi^*g_j = g_j(E, \omega) \in H^0(S, \mathcal{O}_S)$ $(\Leftrightarrow \underline{\varphi}(g_j) = g_j(E, \omega))$; *in other words,*

$$(E, \omega)_{/S} \cong \varphi^*(\mathbf{E}, \boldsymbol{\omega}) = (\mathbf{E}, \boldsymbol{\omega}) \times_{\mathcal{M}_1} S.$$

In short, the classification functor $\wp_{\Gamma_1(1)}$ is represented by the fine moduli scheme $\mathcal{M}_1 = \mathrm{Spec}(\mathcal{R})$.

If we change ω by $\lambda\omega$ for $\lambda \in H^0(S, \mathcal{O}_S^\times) = \mathbb{G}_m(\mathcal{O}_S)$, the parameter T will be changed to λT and hence (x, y) is replaced by $(\lambda^{-2}x, \lambda^{-3}y)$. Thus, $(E, \lambda\omega)_{/S}$ will be defined by

$$(\lambda^{-3}y)^2 = 4(\lambda^{-2}x)^3 - g_2(E, \lambda\omega)(\lambda^{-2}x) - g_3(E, \lambda\omega).$$

This equation has to be equivalent to the original equation by the uniqueness of the Weierstrass equation, and we have

$$g_j(E, \lambda\omega) = \lambda^{-2j} g_j(E, \omega).$$

Again by the uniqueness of the Weierstrass equation, we find that

$$\mathrm{Aut}((E, \omega)_{/S}) = \{\mathrm{id}_E\}$$

as long as 6 is invertible in S.

Since the moduli scheme$\mathcal{M}_1$ is a fine modulus, it is determined by the restriction of $\wp = \wp_{\Gamma_1(1)}$ to $ALG_{/\mathbb{Z}[\frac{1}{6}]}$.

Corollary 6.16 *The multiplication N_E by N on an elliptic curve E over a scheme S is a finite flat endomorphism of E with* $\deg(N_E) = N^2$. *If* $S =$ $\mathrm{Spec}(k)$ *for an algebraically closed field k,* $E[N](k) \cong (\mathbb{Z}/N\mathbb{Z})^2$ *as long as N is prime to the characteristic of k.*

Proof. We give a proof assuming 6 is invertible over S. Shrinking S if necessary, we assume that E carries a nowhere vanishing differential ω. Then we have a unique $\varphi : S \to \mathcal{M}_1$ with $(E, \omega) \cong \varphi^*(\mathbf{E}, \boldsymbol{\omega})$. Thus, we only need to prove this for the universal curve $(\mathbf{E}, \boldsymbol{\omega})$. Thus, we can pick a point $x \in \mathcal{M}_1(\mathbb{C})$ for the complex number field $\mathbb{C}$. Write $(E_x, \omega_x)_{/\mathbb{C}} = x^*(\mathbf{E}, \boldsymbol{\omega})$. By Theorem 2.51, a genus-1 curve with given origin is analytically isomorphic to $\mathbb{C}/L$ for a lattice $L \subset \mathbb{C}$. Thus, $E(\mathbb{C}) = \mathbb{C}/L$ and the fiber of $N_{E_x} : E_x \to E_x$ is isomorphic to its fiber at $\mathbf{0}$, that is, $N^{-1}L/L$, whose cardinality is N^2. Thus, N_{E_x} is nonconstant, hence finite flat (Lemma 6.2), and has degree N^2. If $N_{\mathbf{E}} : \mathbf{E} \to \mathbf{E}$ is constant, N_{E_x} has to be constant; so $N_{\mathbf{E}} : \mathbf{E} \to \mathbf{E}$ is nonconstant and is finite flat. Since $\mathcal{M}_1$ is irreducible, the degree $N_{\mathbf{E}}$ is constant over $\mathcal{M}_1$, and hence $\deg(N_{\mathbf{E}}) = N^2$, which implies $\deg(N_E) = N^2$.

If 6 is not invertible on S, we can take a different type moduli problem and argue in the same way. For example, we can take the moduli functor $\mathcal{E}_{\Gamma(pq)}$ classifying $(E, \phi_{pq} : (\mathbb{Z}/pq\mathbb{Z})^2 \cong E[pq])$ for primes $p, q > 3$ prime to N. This functor is represented by a smooth affine curve $Y(pq)$ over $\mathbb{Z}[\frac{1}{pq}]$ (see Sect. 6.2.10 in the text and also [GME] Theorem 2.6.8), and we can go in the same way, taking the universal elliptic curve over $Y(pq)$.

There are purely algebro-geometric proofs analyzing the degree of N_E^* on $\mathrm{Pic}^0_{E/S}$ (see, for example, [ABV] Sect. 16).

Since $N_E : E \to E$ is finite flat, $E(k)$ is a divisible group; so $E[l^\infty]$ is l-divisible for a prime l. If l is different from the characteristic of k, l_E is étale by Lemma 6.8, and so $E[l](k) \cong (\mathbb{Z}/l\mathbb{Z})^r$ as abelian groups. Since $\deg(l_E) = l^2$, we find $E[l](k) \cong (\mathbb{Z}/l\mathbb{Z})^2$. By l-divisibility, we then find $E[l^\infty](k) = (\mathbb{Q}_l/\mathbb{Z}_l)^2$; so $E[l^m](k) \cong (\mathbb{Z}/l^m\mathbb{Z})^2$. If $N = \prod_l l^{e(l)}$ is the prime decomposition of N, we find $E[N](k) = \bigoplus_l E[l^{e(l)}](k) \cong (\mathbb{Z}/N\mathbb{Z})^2$.

Corollary 6.17 *Let R be a local $\mathbb{Z}[\frac{1}{6}]$-algebra and I be an ideal of R. If $(E_0, \omega_0)_{/(R/I)}$ is a pair in $\wp_{\Gamma_1(1)}(R/I)$, then there exists a pair $(E, \omega)_{/R} \in \wp_{\Gamma_1(1)}(R)$ such that $i_0 : E_0 \cong E \times_{\mathrm{Spec}(R)} \mathrm{Spec}(R/I)$ and $\omega_0 = i_0^*\omega$.*

Proof. By the above theorem,

$$E_0 = \mathrm{Proj}((R/I)[X,Y,Z]/(Y^2Z - 4X^3 + g_2(E_0,\omega_0)XZ^2 + g_3(E_0,\omega_0)Z^3))$$

with $\Delta(E_0,\omega_0) \in (R/I)^\times$. Pick $g_j(E,\omega) \in R$ so that $g_j(E,\omega) \cong g_j(E_0,\omega_0)$ mod I. Define E by

$$E = \mathrm{Proj}(R[X,Y,Z]/(Y^2Z - 4X^3 + g_2(E,\omega)XZ^2 + g_3(E,\omega)Z^3))$$

and put $\omega = \frac{dX}{Y}$. Since $\Delta(E,\omega) \equiv \Delta(E_0,\omega_0) \mod I$, we have $\Delta(E,\omega) \in R^\times$ and $(E,\omega) \in \wp_{\Gamma_1(1)}(R)$.

6.2.2 Compatible System of Tate Modules

Let $(E,\omega)_{/F}$ be an elliptic curve with nowhere vanishing differential ω defined over a number field $F \subset \overline{\mathbb{Q}}$ with integer ring $O = O_F$. Then (E,ω) gives rise to a point of $P \in \mathcal{M}_1(F)$. Pick a prime l, and we define the l-adic *Tate module* $T_lE = \varprojlim_n E[l^n](\overline{\mathbb{Q}}) \cong \mathbb{Z}_l{}^2$. Since $E[l^n](\overline{\mathbb{Q}}) \subset E(\overline{\mathbb{Q}}) \subset \mathbf{P}^2(\overline{\mathbb{Q}})$, the Galois group $\mathrm{Gal}(\overline{\mathbb{Q}}/F)$ acts on $T_lE \cong \mathbb{Z}_l{}^2$, getting a representation $\rho_{E,l} : \mathrm{Gal}(\overline{\mathbb{Q}}/F) \to GL_2(\mathbb{Z}_l)$. Since $P \in \mathcal{M}_1(F)$ is an algebra morphism $P : \mathbb{Z}[\frac{1}{6}, g_2, g_3, \frac{1}{\Delta}] \to F$, its image lands in $O[\frac{1}{N}]$ for some integer N. Thus, (E,ω) is well defined over $S = \mathrm{Spec}(O[\frac{1}{N}])$. Taking a closed point $\mathfrak{p} \nmid l$ of S, we can think of $E_\mathfrak{p} = E \times_S \mathbb{F}_\mathfrak{p}$ for the residue field $\mathbb{F}_\mathfrak{p}$ of $\mathfrak{p}$. Note that $E[l^n]_{/S}$ is a finite flat group scheme. Thus, taking the integral closure $\overline{\mathbb{Z}}$ of $\mathbb{Z}$ in $\overline{\mathbb{Q}}$ gives $E[l^n](\overline{\mathbb{Z}}[\frac{1}{N}]) \cong (\mathbb{Z}/l^n\mathbb{Z})^2$. Taking a prime $\mathfrak{P}$ over $\mathfrak{p}$ in $\overline{\mathbb{Z}}[\frac{1}{N}]$, we see that the residue field $\mathbb{F}_\mathfrak{P}$ of $\mathfrak{P}$ is an algebraic closure of $\mathbb{F}_p$. We get the reduction map

$$\pi_\mathfrak{P} : E[l^n](\overline{\mathbb{Q}}) = E[l^n]_{/\overline{\mathbb{Z}}[\frac{1}{N}]} \times_{\overline{\mathbb{Z}}[\frac{1}{N}]} \overline{\mathbb{Q}} \overset{(*)}{\cong} E[l^n]_{/\overline{\mathbb{Z}}[\frac{1}{N}]} \times_{\overline{\mathbb{Z}}[\frac{1}{N}]} \mathbb{F}_\mathfrak{P} = E[l^n](\overline{\mathbb{F}}_\mathfrak{p}).$$

The middle isomorphism $(*)$ is just an isomorphism of abelian groups. The decomposition subgroup $D_\mathfrak{P} = \{\sigma \in \mathrm{Gal}(\overline{\mathbb{Q}}/F) | \sigma(\mathfrak{P}) = \mathfrak{P}\}$ commutes with $\pi_\mathfrak{P}$; so its inertia group $I_\mathfrak{P}$ acts trivially on $E[l^n]$, and hence on T_lE. Thus, $\rho_{E,l}$ is unramified outside $N \cdot l$. Let Φ be the Frobenius map of $\mathrm{Gal}(\mathbb{F}_\mathfrak{P}/\mathbb{F}_\mathfrak{p})$. Then Φ induces $(x{:}y{:}z) \mapsto (x^q{:}y^q{:}z^q)$ on $\mathbf{P}^3_{/\mathbb{F}_\mathfrak{p}}$ for $q = N(\mathfrak{p}) = |\mathbb{F}_\mathfrak{p}|$, which is a morphism of schemes. Since $E_\mathfrak{p} \subset \mathbf{P}^3$ is stable under Φ, we get the relative Frobenius morphism $\phi : E_\mathfrak{p} \to E_\mathfrak{p}$ induced by Φ (see Sect. 6.2.11 for a coordinate-free definition of the relative Frobenius map). Since Φ is the qth power morphism, it induces a qth power map of the stalk $\mathcal{O}_{E_\mathfrak{p},\mathbf{0}}$ over $\mathbb{F}_\mathfrak{p}$; so $\deg(\phi) = q = [\mathcal{O}_{E_\mathfrak{p},\mathbf{0}} : \Phi(\mathcal{O}_{E_\mathfrak{p},\mathbf{0}})]$, and ϕ satisfies

$$P_\mathfrak{p}(X) = X^2 - (\phi + \phi^*)X + q = 0$$

in $\mathrm{End}(E_\mathfrak{p})$ (by Theorem 6.6). Here $\mathrm{Tr}(\phi) = \phi + \phi^* \in \mathbb{Z}$. The Frobenius element $Frob_\mathfrak{p}$ relative to $\mathfrak{P}$ in $D_\mathfrak{P}$ acts via the relative Frobenius map $\phi : E \to E$ on T_lE, and $Frob_\mathfrak{P}$ satisfies $P_\mathfrak{p}(X) = 0$, which is independent of l. Thus, $\{T_lE\}_l$ gives rise to a compatible system of Galois representations.

Proposition 6.18 *Let E and E' be elliptic curves defined over a field k. The module $\mathrm{Hom}_k(E, E')$ is torsion-free of finite rank ≤ 4 over $\mathbb{Z}$.*

Proof. Take $0 \ne f \in \mathrm{Hom}_k(E, E')$, and let $L = \mathrm{Hom}_k(E, E')$ and $V = \mathrm{Hom}_k(E, E') \otimes_{\mathbb{Z}} \mathbb{Q}$. Then $f^*f = \deg(f) \ne 0$ by Lemma 6.2, and we find that $\mathrm{Hom}_k(E, E')$ is torsion-free; hence, $L \subset V$. If $\phi = \phi^*$ for $\phi \in \mathrm{End}_k(E)$, then $2\phi = \mathrm{Tr}(\phi) \in \mathbb{Z}$; so $*$-invariant endomorphism has to be in $\frac{1}{2}\mathbb{Z} \cdot \mathrm{id}_E$. For linearly independent $f_1, \dots, f_r \in V$, set $f = x_1 f_1 + \cdots + x_r f_r$ for integers x_j. Then

$$Q(x_1, \dots, x_r) := (x_1, \dots, x_r) \mapsto \\ \deg(f) = f^*f = (x_1 f_1^* + \cdots + x_r f_r^*)(x_1 f_1 + \cdots + x_r f_r)$$

is a homogeneous polynomial of x_j of degree 2 whose coefficients are $*$-invariant endomorphisms of E. Thus, $Q(x_1, \dots, x_r) \in \mathbb{Q}[x_1, \dots, x_r]$. Pick any finite-rank submodule $M \subset \mathrm{Hom}_k(E, E')$. Consider the $\mathbb{Q}$-span $\mathbb{Q} \cdot M \subset V$. For a basis $f_1, \dots, f_r$ of M, $M \ni f \mapsto \deg(f)$ is a positive definite quadratic form in x_i. Thus, $U = \{|Q(f)| < 1 | f \in \mathbb{Q} \cdot M\}$ is a neighborhood of 0 under the Euclidean topology on $\mathbb{Q} \cdot M$, and $U \cap M = 0$. Thus, M is a lattice in $\mathbb{Q} \cdot M$, and $L_M = L \cap (\mathbb{Q} \cdot M)$ is another lattice of the same rank as M.

Pick a prime l different from the characteristic of k. Look at the l-adic Tate modules $T_l E = \varprojlim_n E[l^n](\overline{k})$ and $T_l E'$ for an algebraic closure $\overline{k}$ of k. Then $T_l E \cong \mathbb{Z}_l^2$, and we have a natural linear map $T_l : L \to \mathrm{Hom}_{\mathbb{Z}_l}(T_l E, T_l E') \cong M_2(\mathbb{Z}_l)$. Since E has dimension 1, $E[l^\infty](\overline{k}) \cong (\mathbb{Q}_l/\mathbb{Z}_l)^2$ is Zariski dense in $E_{/\overline{k}}$; so the above map is injective, inducing a linear map $T_l \otimes 1 : L \otimes_{\mathbb{Z}} \mathbb{Z}_l \to M_2(\mathbb{Z}_l)$. If $T_l \otimes 1$ is not injective, we find a $\mathbb{Z}$-linearly independent $f_1, \dots, f_r \in \mathrm{Hom}_k(E, E')$ such that $M = \sum_i \mathbb{Z} f_i = (\mathbb{Q} \cdot M) \cap L$ and $\sum_i \alpha_i T_l(f_i) = 0$ for $\alpha_i \in \mathbb{Z}_l$. By dividing the greatest common divisor of α_is, we may assume that one of α_i, say α_1, is an l-adic unit; so f is not in $l \cdot M$. Pick integers n_i so that $n_i \equiv \alpha_i \mod l$. Then $f := \sum_i n_i f_i$ with $f \equiv \sum_i \alpha_i f_i = 0 \mod l$ kills $E[l](\overline{k}) = T_l E / l \cdot T_l E$. Thus, f factors through $E \xrightarrow{l_E} E$; that is, we have a commutative diagram

$$\begin{array}{ccc} E & \xrightarrow{l_E} & E \\ {\scriptstyle f}\downarrow & & \downarrow{\scriptstyle g} \\ E' & = & E'. \end{array}$$

Thus, $g \in \mathbb{Q} \cdot M \cap L = M$ and $f = l \cdot g$, a contradiction (as $f \notin l \cdot M$ by our choice), and so $T_l \otimes 1$ is an injection. This implies $L \otimes_{\mathbb{Z}} \mathbb{Z}_l \hookrightarrow M_2(\mathbb{Z}_l)$, and hence, $\mathrm{rank}\, L \le \mathrm{rank}\, M_2(\mathbb{Z}_l) = 4$.

Exercise 6.19 *If E is defined over a finite field of $q = p^f$ elements and if $D = \mathrm{End}(E_{/\mathbb{F}_q}) \otimes_{\mathbb{Z}} \mathbb{Q}$ has rank 4, prove that $D_l := D \otimes_{\mathbb{Q}} \mathbb{Q}_l = M_2(\mathbb{Q}_l)$ for all primes $l \ne p$ and $D \otimes_{\mathbb{Q}} \mathbb{R}$ is the Hamilton quaternion algebra. From this, conclude that D_p is a division algebra.*

6.2.3 Moduli of $\wp_{\Gamma_1(N)}$

Let $\mathbf{E}_{/\mathcal{M}_1}$ be the universal elliptic curve. Since $\mathbf{E}[l] \to \mathcal{M}_1$ for a prime l is a finite morphism (Lemma 6.2), it is affine. Write $\mathbf{E}[l] = \mathrm{Spec}(A)$. Then $I(\mathbf{0})^2$ is principal generated by ϕ (the image of x^{-1} in A), and we remove $\mathbf{0}$ by adding $\frac{1}{\phi}$ to A, we get an affine scheme $\mathbf{E}[l] - \{\mathbf{0}\} = \mathrm{Spec}(A[\frac{1}{\phi}])$.

Consider $(E, P, \omega)_{/R}$ for a point $P \in E[l](R)$ of order l for a prime l. We have a unique $\varphi \in \mathcal{M}_1(R)$ such that

$$\varphi_E : (E, \omega)_{/R} \cong \varphi^*(\mathbf{E}, \boldsymbol{\omega}) = (\mathbf{E}, \boldsymbol{\omega}) \times_{\mathcal{M}_1} \mathrm{Spec}(R).$$

We thus have a commutative diagram:

$$\begin{array}{ccccc} & & \mathbf{E} & \to & \mathcal{M}_1 \\ & & {\scriptstyle \varphi_E}\uparrow & & \uparrow{\scriptstyle \varphi} \\ \mathrm{Spec}(R) & \xrightarrow{P} & E & \to & \mathrm{Spec}(R) \end{array}.$$

Then P induces a unique morphism $\varphi_P = \varphi_E \circ P : \mathrm{Spec}(R) \to (\mathbf{E}[l] - \{\mathbf{0}\})$. This shows that, over $\mathbb{Z}[\frac{1}{6l}]$,

$$\wp_{\Gamma_1(l)}(R) \cong \wp'_{\Gamma_1(l)}(R) \cong (\mathbf{E}[l] - \{\mathbf{0}\})\,(R).$$

Recall $\mathbf{E}[l] = \mathrm{Spec}(A)$. Then $\underline{\mathbf{0}} : A \to \mathcal{R}$ is a section of the structure morphism $\underline{\pi} : \mathcal{R} \to A$; so $A = \mathcal{R} \oplus I(\mathbf{0})$ as $\mathcal{R}$-modules. Since $l : \mathbf{E} \to \mathbf{E}$ is étale (Lemma 6.8), the above decomposition has to be a ring direct sum. Thus, $\mathbf{E}[l] - \{\mathbf{0}\}$ is a closed subscheme of $\mathbf{E}[l]$ and hence is proper over $\mathbb{Z}[\frac{1}{6l}]$. Thus, $\mathbf{E}[l] - \{\mathbf{0}\}$ is proper quasi-finite étale and hence is finite étale over $\mathcal{M}_{1/\mathbb{Z}[\frac{1}{6l}]}$.

Similarly, over $\mathbb{Z}[\frac{1}{6N}]$,

$$\wp_{\Gamma_1(N)}(R) \cong \wp'_{\Gamma_1(N)}(R) \cong \left(\mathbf{E}[N] - \bigcup_{N>d|N} \mathbf{E}[d]\right)(R).$$

We put $\mathcal{M}_{\Gamma_1(N)} = \mathbf{E}[N] - \bigcup_{N>d|N} \mathbf{E}[d]$. Similarly as above, $\mathcal{M}_{\Gamma_1(N)}$ is finite étale over $\mathcal{M}_{1/\mathbb{Z}[\frac{1}{6N}]}$. Thus, we have proven

Theorem 6.20 *There is an affine scheme $\mathcal{M}_{\Gamma_1(N)} = \mathrm{Spec}(\mathcal{R}_{\Gamma_1(N)})$ defined over $\mathbb{Z}[\frac{1}{6N}]$ such that*

$$\wp_{\Gamma_1(N)}(R) \cong \wp'_{\Gamma_1(N)}(R) \cong \mathrm{Hom}_{\mathbb{Z}[\frac{1}{6N}]-alg}(\mathcal{R}_{\Gamma_1(N)}, R) = \mathcal{M}_{\Gamma_1(N)}(R)$$

for all $\mathbb{Z}[\frac{1}{6N}]$-algebras R. The scheme $\mathcal{M}_{\Gamma_1(N)}/\mathcal{M}_1$ is an étale finite covering over $\mathcal{M}_{1/\mathbb{Z}[\frac{1}{6N}]}$ of degree $\varphi(N) = |(\mathbb{Z}/N\mathbb{Z})^\times|$ for the Euler function φ.

The fact that the covering is étale finite follows from the same fact for $\mathbf{E}[N]$ since $\mathbf{E}[N](k) \cong (\mathbb{Z}/N\mathbb{Z})^2$ for all algebraically closed fields with characteristic not dividing N. Since $\mathcal{M}_1$ is affine, any finite covering of $\mathcal{M}_1$ is affine. Fix a base $\mathbb{Z}[\frac{1}{6N}]$-algebra B. Then the statement of the theorem is plainly valid for the functors $\wp_{\Gamma_1(N)}$ and $\wp'_{\Gamma_1(N)}$ restricted to $ALG_{/B}$ (by base change to B from $\mathbb{Z}[\frac{1}{6N}]$).

If we find an S-scheme $\mathfrak{M}$ for a given contravariant functor $F : SCH_{/S} \to SETS$ such that we have an isomorphism of functors $\underline{\mathfrak{M}} \cong F$ for $\underline{\mathfrak{M}}(S') = \mathrm{Hom}_S(S', \mathfrak{M})$, we say that F is *representable* (or represented) by an S-scheme $\mathfrak{M}$. The scheme $\mathfrak{M}$ is called the *moduli scheme* of the functor F. Then the statement of the theorem is equivalent to the representability of the functors $\wp_{\Gamma_1(N)}$ and $\wp'_{\Gamma_1(N)}$ by an affine $\mathbb{Z}[\frac{1}{6N}]$-scheme $\mathcal{M}_{\Gamma_1(N)}$.

Even if F is not representable by a scheme, we often find a rough approximation of the moduli called a *coarse moduli* scheme so that $F(\mathrm{Spec}(k)) \cong \mathfrak{M}(k)$ for the spectrum of any algebraically closed field k that is an S-scheme. Here we do not give the scheme-theoretic characterization of coarse moduli schemes, only referring to [GME] Sect. 2.3.2, though we have some coarse moduli schemes in this book.

6.2.4 Definition of Modular Forms

Take a base-$\mathbb{Z}[\frac{1}{6N}]$-algebra B. Write $\wp_{\Gamma_1(N)/B}$ for the restriction of $\wp_{\Gamma_1(N)}$ to $ALG_{/B} \subset ALG_{/\mathbb{Z}[\frac{1}{6N}]}$. The group scheme $\mathbb{G}_m$ acts on the functor $\wp_{\Gamma_1(N)/B}$ in the following way:

$$(E, \phi_N, \omega)_{/R} \mapsto (E, \phi_N, \lambda\omega)_{/R} \quad \text{for } \lambda \in \mathbb{G}_m(R).$$

This induces (by the functoriality described in Sect. 6.1.5) an action of $\mathbb{G}_m$ on $\mathcal{M}_{\Gamma_1(N)}$ and hence on $\mathcal{R}_{\Gamma_1(N)}$.

The action of $\mathbb{G}_{m/B}$ on $\mathcal{M}_{\Gamma_1(N)}$ makes $\mathcal{R}_{\Gamma_1(N)}$ a functorial $\mathbb{G}_m$-module (because it was defined by a functorial action; see Example 4.62 in Sect. 4.3.3). Thus, we can split

$$\mathcal{R}_{\Gamma_1(N)/B} = \bigoplus_{\kappa \in \mathbb{Z}} \mathcal{R}_\kappa(\Gamma_1(N); B),$$

where on $f \in \mathcal{R}_\kappa(\Gamma_1(N); B)$, $\mathbb{G}_m$ acts by the character $-\kappa$. Thus, $\mathcal{R}_{\Gamma_1(N)/B}$ is a graded algebra.

Since $f \in \mathcal{R}_\kappa(\Gamma_1(N); B)$ is a morphism of functors:

$$\mathcal{M}_{\Gamma_1(N)}(R) = \wp_{\Gamma_1(N)}(R) \xrightarrow{f} \mathbb{G}_a(R) = R,$$

we may regard f as a function of $(E, \phi_N, \omega)_{/R}$ (for any B-algebra R) with $f((E, \phi_N, \omega)_{/R}) \in R$ satisfying the following:

(G0) $f((E, \phi_N, \lambda\omega)_{/R}) = \lambda^{-\kappa} f((E, \phi_N, \omega)_{/R})$ for $\lambda \in R^\times = \mathbb{G}_m(R)$.

(G1) If $(E, \phi_N, \omega)_{/R} \cong (E', \phi'_N, \omega')_{/R}$, then we have
$$f((E, \phi_N, \omega)_{/R}) = f((E', \phi'_N, \omega')_{/R}).$$

(G2) If $\rho : R \to R'$ is a morphism of B-algebras, then we have
$$f((E, \phi_N, \omega)_{/R} \times_R R') = \rho(f((E, \phi_N, \omega)_{/R}))).$$

6.2.5 Hecke Operators

Start with a functorial rule $f \in \mathcal{R}_\kappa(\Gamma_1(N); B)$ satisfying (G0–2). For a prime l and a test object $(E, \phi_N : \mu_N \hookrightarrow E, \omega)_{/R}$, let $\mathcal{C}_l = \mathcal{C}_l(E)$ be the set of locally free subgroups (defined over an fppf extension R' of R) of E of rank l disjoint from $\phi_N(\mu_l)$. The disjointness means that $C \cap \phi_N(\mu_l) := C \times_E \phi_N(\mu_l) = [\mathbf{0}_E]$ for $C \in \mathcal{C}_l$. Then the quotient E/C is well defined over R' first as an abelian fppf sheaf, which is represented by an elliptic curve as remarked in Sect. 4.3.7. We therefore regard E/C as an elliptic curve over R' with quotient morphism $\pi = \pi_C : E \to E/C$. We have a commutative diagram:

$$\begin{array}{ccc} \mu_N & = & \mu_N \\ {\scriptstyle \phi_N}\downarrow\cap & & \cap\downarrow{\scriptstyle \pi\circ\phi_N} \\ E & \xrightarrow[\pi]{} & E/C, \end{array}$$

since $\mathrm{Ker}(\pi) \cap \phi_N(\mu_l) = E \times_{E/C, \mathbf{0}_{E/C}} \mathrm{Spec}(R) \times_{E,\phi_N} \mu_l = C \times_{E,\phi_N} \mu_l = [\mathbf{0}_E]$. Thus, the inclusion $\pi_*\phi_N = \pi \circ \phi_N : \mu_N \hookrightarrow E/C$ gives rise to a level-$\Gamma_1(N)$ structure. Since $C \cap \phi(\mu_l) = [\mathbf{0}_E]$, C is an étale subgroup, $\pi : E \to E/C$ is therefore an étale covering. Thus, $\pi^* : \Omega_{(E/C)/R} \cong \Omega_{E/R}$, and so we can define $\pi_*\omega := (\pi^*)^{-1}\omega$.

Definition 6.21 *We define a new rule* $f|T(l) : \wp_{\Gamma_1(N)/B} \to \mathbb{G}_{a/B}$ *by*

$$f|T(l)(E, \phi_N, \omega) = \frac{1}{l} \sum_{C \in \mathcal{C}_l} f(E/C, \pi_{C,*}\phi_N, \pi_{C,*}\omega),$$

assuming that l *is invertible in* B.

We will later show by the q-expansion principle that this is well defined even when l is not invertible in B. More generally, if n is invertible in B and is prime to N, we define $\mathcal{C}_n$ to be the set of rank-n locally free subgroups of E. Then the projection $\pi_C : E \to E/C$ is an isomorphism on $E[N]$ and is étale; thus, $\pi_C \circ \phi_N = \pi_{C,*}\phi_N$ is a well-defined level-$\Gamma_1(N)$ structure and $\pi_{C,*}\omega$ is a nowhere vanishing differential on E/C. Thus, we define

$$f|T(n)(E, \phi_N, \omega) = \frac{1}{n} \sum_{C \in \mathcal{C}_n} f(E/C, \pi_{C,*}\phi_N, \pi_{C,*}\omega),$$

which is a B-linear operator called the *Hecke operator* acting on $\mathcal{R}_\kappa(\Gamma_1(N); B)$. Often, for a prime factor l of N, the operator $T(l)$ is written as $U(l)$ because the definition of $\mathcal{C}_l$ excludes the image of μ_l. For any positive integer n, taking a decomposition $n = n' \prod_{l|N} l^{e(l)}$ for n' prime to N, we put $T(n) = T(n') \prod_{l|N} U(l)^{e(l)}$. We write $\langle l \rangle = (T(l)^2 - T(l^2))$ for primes l outside N, and put $\langle n \rangle = \prod_{l \nmid N} \langle l \rangle^{e(l)}$ if $n = \prod_{l \nmid N} l^{e(l)}$. We call $\langle n \rangle$ the *diamond operator*.

Exercise 6.22 *Prove the following formulas* (cf. [IAT] Theorem 3.24)*:*

(1) $T(l^m)T(l^n) = \sum_{i=0}^{m} \langle l^i \rangle T(l^{m+n-2i})$ *if* $m \leq n$ *and* $l \nmid N$*;*
(2) $T(m)T(n) = \sum_{0<d|(m,n),(d,N)=1} \langle d \rangle T(\frac{mn}{d^2})$.

We compute $T(l)^2$ for $l \nmid N$. For each $C \in \mathcal{C}_l(E)$ and $C' \in \mathcal{C}_l(E/C)$, $\pi_C^{-1}(C')$ is cyclic of order l^2 except for $C^\perp = E[l]/C$. Then we have a commutative diagram:

$$\begin{array}{ccc} E & \xrightarrow{\pi_C} & E/C \\ {\scriptstyle l_E}\downarrow & & \downarrow{\scriptstyle \pi_{C^\perp}} \\ E & \xrightarrow[\iota_l]{\sim} & (E/C)/C^\perp. \end{array}$$

By this diagram, $\pi_{C^\perp,*}\pi_{C,*}\phi_N = \phi_N \circ l_E$ and $\pi_{C^\perp,*}\pi_{C,*}\omega = l^{-1} \cdot \omega$, as the action of l_E on invariant differential is the multiplication by l (see Lemma 6.7). If $C' \neq C^\perp$, then $\pi_C^{-1}(C') \in \mathcal{C}_{l^2}(E)$ is a cyclic subgroup of order l^2. By a simple computation, identifying $E[l^2]$ with $(\mathbb{Z}/l^2\mathbb{Z})^2$, $\mathcal{C}_{l^2}$ has $l^2 + l$ cyclic subgroups and one subgroup killed by l, and more easily, we can show that $\mathcal{C}_l$ has $l + 1 = |\mathbf{P}^1(\mathbb{F}_l)|$ elements. We then find

$$f|T(l)^2 = \frac{1}{l^2} \sum_{C \in \mathcal{C}_l(E)} \sum_{C' \in \mathcal{C}_l(E/C)} f((E/C)/C', \pi_{C',*}\pi C, *\phi_N, \pi_{C',*}\pi C, *\omega)$$

$$= \frac{1}{l^2} \left(\sum_{C'' \in \mathcal{C}_{l^2}} f(E/C'', \pi_{C'',*}\phi_N, \pi_{C'',*}\omega) + l \cdot f(E, l_E \circ \phi_N, l^{-1} \cdot \omega) \right)$$

$$= f|T(l^2)(E, \phi_N, \omega) + l^{k-1} f(E, \phi_N \circ l, \omega). \quad (6.2.1)$$

This shows the following:

Corollary 6.23 *We have* $f|\langle l \rangle(E, \phi_N, \omega) = l^{k-1} f(E, \phi_N \circ l, \omega)$ *for a prime* $l \nmid N$*.*

Exercise 6.24 *Prove that* $|\mathcal{C}_{l^2}| = 1 + l + l^2$ *for a prime* $l \nmid N$*.*

6.2.6 Moduli of Elliptic Curves with Level-$\Gamma_1(N)$ Structure

If a graded ring $\mathcal{A} = \bigoplus_j \mathcal{A}_j$ has a unit u of degree 1, we see that $\mathcal{A} = \mathcal{A}_0 \otimes_\mathbb{Z} \mathbb{Z}[u, u^{-1}]$ and $\mathrm{Spec}(\mathcal{A}) = \mathrm{Spec}(\mathcal{A}_0) \times \mathbb{G}_m$ by definition; so $\mathrm{Proj}(\mathcal{A}) = \mathrm{Spec}(\mathcal{A})/\mathbb{G}_m = \mathrm{Spec}(\mathcal{A}_0)$. If $\mathcal{A}$ has a unit of degree $n > 0$, then we still have

$$\mathrm{Proj}(\mathcal{A}) = \mathrm{Proj}(\mathcal{A}^{(n)}) = \mathrm{Spec}(\mathcal{A}_0)$$

for $\mathcal{A}^{(n)} = \bigoplus_j \mathcal{A}_{nj}$ (see Sect. 4.1.9). Since $\Delta^{-1} \in \mathcal{R} \subset \mathcal{R}_{\Gamma_1(N)}$, the graded ring $\mathcal{R}_{\Gamma_1(N)}$ has a unit of degree 12, and hence, we have

$$\mathbb{G}_m \backslash \mathcal{M}_{\Gamma_1(N)/B} = \mathrm{Proj}(\mathcal{R}_{\Gamma_1(N)/B}) \cong \mathrm{Spec}(\mathcal{R}_0(\Gamma_1(N); B)) =: Y_1(N)_{/B}.$$

Let B be a $\mathbb{Z}[\frac{1}{N}]$-algebra. We consider the functor defined over $ALG_{/B}$ given by

$$\mathcal{E}_{\Gamma_1(N)/B}(R) = \left[(E, \phi_N : \mu_N \hookrightarrow E[N])_{/R}\right].$$

By definition, $\mathcal{E}_{\Gamma_1(N)/B} = (\wp_{\Gamma_1(N)/B})/\mathbb{G}_m$ almost. Indeed, for a pair $(E,P)_{/R}$, we can find an fppf extension R'/R such that $H^0(E\otimes_R R', \Omega_{E\otimes_R R'/R'}) = R'\omega$. Then, for the universal triple $(\mathbf{E},\mathbf{P},\boldsymbol{\omega})_{\mathcal{R}_{\Gamma_1(N)}}$, we have a unique morphism $\varphi \in \mathcal{M}_{\Gamma_1(N)}(R')$ such that $(E\otimes_R R', P, \omega)_{/R'} \cong \varphi^*(\mathbf{E},\mathbf{P},\boldsymbol{\omega})$. The corresponding algebra homomorphism $\underline{\varphi} : \mathcal{R}_{\Gamma_1(N)} \to R'$ restricted to $\mathcal{R}_0 := \mathcal{R}_0(\Gamma_1(N);B)$ does not depend on the choice of ω, as other choices are just $\lambda\omega$, and because of this independence of the choice of ω, $\underline{\varphi}_0 := \underline{\varphi}|_{\mathcal{R}_0}$ has values in R [strictly speaking, by fppf descent under the condition that $\mathrm{Aut}_R(E,P) = \{\mathrm{id}_E\}$ for all $E_{/R}$ with a $\Gamma_1(N)$-level structure; see [GME] Remark 2.6.1]. Hence, we get a unique morphism $\varphi_0 : \mathrm{Spec}(\mathcal{R}) \to \mathrm{Spec}(\mathcal{R}_0(\Gamma_1(N);B))$ associated with $(E,P)_{/R}$ as long as $\mathrm{Aut}_R(E,P) = \{\mathrm{id}_E\}$. When R is an algebraically closed field k, there is no need to take an fppf extension, as a nowhere vanishing differential ω exists over k. Thus, we get φ_0 associated with $(E,P)_{/k}$ without any assumption. As is well known, we always have $\mathrm{Aut}_R(E,P) = \{\mathrm{id}_E\}$ if $N \geq 4$ (e.g., [GME] Sect. 2.6.4). In particular, the universal curve $\mathbf{E}_{/\mathcal{M}_1}$ descends to the universal curve over $\mathrm{Spec}(\mathcal{R}_0(\Gamma_1(N);B))$ if $N \geq 4$. Writing down its equation over $\mathcal{R}_0$ is not too difficult if N is small, and for a multiple MN of such small N, the universal curve of level MN is just the pullback of that of level N (e.g., [GME] Sect. 2.2.6). Since $\mathrm{Proj}(\mathcal{R}_{\Gamma_1(N)}) \cong \mathrm{Spec}(\mathcal{R}_0(\Gamma_1(N);B))$ gives the quotient by $\mathbb{G}_m$ of $\mathrm{Spec}(\mathcal{R}_{\Gamma_1(N)})$ (cf. [GME] Theorem 1.8.2), we conclude

Theorem 6.25 (Shimura, Igusa) *We have an open affine curve*

$$Y_1(N) = \mathrm{Spec}(\mathcal{R}_0(\Gamma_1(N);\mathbb{Z}[\frac{1}{6N}])) \cong \mathrm{Proj}(\mathcal{R}_{\Gamma_1(N)}) = \mathcal{M}_{\Gamma_1(N)}/\mathbb{G}_m$$

defined over $\mathbb{Z}[\frac{1}{6N}]$, *which is locally free of finite rank over* $M_1 = \mathrm{Proj}(\mathcal{R}) = \mathbf{P}^1(J) - \{\infty\}$. *For all geometric points* $\mathrm{Spec}(k) \hookrightarrow \mathrm{Spec}(\mathbb{Z}[\frac{1}{6N}])$, *we have* $\mathcal{E}_{\Gamma_1(N)}(k) = \left[(E,\phi_N)_{/k}\right] = Y_1(N)(k)$. *The above assertion holds for any* $\mathbb{Z}[\frac{1}{6N}]$*-algebras R in place of algebraically closed fields k if $N \geq 4$.*

Here a "geometric point" means that k is an algebraically closed field. This theorem over $\mathbb{Z}[\frac{1}{M}]$ for an unspecified M with $N|M$ was proved by Shimura in the late 1950s (see [CPS] I [57a]), and the result over $\mathbb{Z}[\frac{1}{N}]$ was proved by Igusa in [I] soon after Shimura's work. Again, the result is valid over any $\mathbb{Z}[\frac{1}{N}]$-algebra B by base change.

Returning to the classical setting, we define an arithmetic subgroup $\Gamma_1(N)$ of $SL_2(\mathbb{Q})$ by

$$\Gamma_1(N) = \left\{\left(\begin{smallmatrix} a & b \\ c & d\end{smallmatrix}\right) \in SL_2(\mathbb{Z}) \middle| c \equiv 0 \mod N,\ a \equiv d \equiv 1 \mod N\right\}.$$

Since $SL_2(\mathbb{Z})$ acts on the *upper half-complex plane* $\mathfrak{H}$ discretely by $z \mapsto \frac{az+b}{cz+d}$, we can make a quotient Riemann surface $\Gamma_1(N)\backslash\mathfrak{H}$ (see [IAT] Chap. 1).

By the association $z \mapsto (E_z(\mathbb{C}) = \mathbb{C}/(\mathbb{Z}z + \mathbb{Z}), P_z = \frac{1}{N} \in E_z(\mathbb{C}))$, it is well known that $\Gamma_1(N)\backslash\mathfrak{H}$ classifies all elliptic curves with a point P of order N over $\mathbb{C}$ (cf. [IAT] Chap. 4 and [GME] Sect. 2.4); so we conclude that

$$Y_1(N)(\mathbb{C}) = \Gamma_1(N)\backslash\mathfrak{H}.$$

Thus, $Y_1(N)(\mathbb{C})$ is an open Riemann surface.

There is a $\Gamma_0(N)$-version of the above theorem for

$$\Gamma_0(N) = \left\{ \left(\begin{smallmatrix} a & b \\ c & d \end{smallmatrix}\right) \in SL_2(\mathbb{Z}) \middle| c \equiv 0 \mod N \right\}.$$

Note that $\Gamma_0(N) \rhd \Gamma_1(N)$ and $\Gamma_0(N)/\Gamma_1(N) \cong (\mathbb{Z}/N\mathbb{Z})^\times$ by sending $\sigma_d := \left(\begin{smallmatrix} a & b \\ c & d \end{smallmatrix}\right)$ to the class $(d \mod N)$ in $(\mathbb{Z}/N\mathbb{Z})^\times$. Thus, $d \in (\mathbb{Z}/N\mathbb{Z})^\times$ acts on the Riemann surface $Y_1(N)(\mathbb{C})$ via $z \mapsto \sigma_d(z)$ $(z \in \mathfrak{H})$, and we can make the quotient $Y_0(N)(\mathbb{C})$. This has a moduli-theoretic interpretation. The group $(\mathbb{Z}/N\mathbb{Z})^\times$ can be identified with the automorphism group of μ_N for $d \in (\mathbb{Z}/N\mathbb{Z})^\times$ sending $\zeta \in \mu_N$ to ζ^d. Thus, $d \in (\mathbb{Z}/N\mathbb{Z})^\times$ acts on $\wp_{\Gamma_1(N)}$, sending each pair $(E, \phi_N : \mu_N \hookrightarrow E)$ to $(E, \phi_N \circ d)$, where $\phi_N \circ d$ is the composite of ϕ_N with the action of d on μ_N. This induces an action of $(\mathbb{Z}/N\mathbb{Z})^\times$ on $Y_1(N)_{/B}$ and on $\mathcal{R}_{\Gamma_1(N)/B}$. Let $\mathcal{R}_{\Gamma_0(N)/B} = H^0((\mathbb{Z}/N\mathbb{Z})^\times, \mathcal{R}_{\Gamma_1(N)/B})$. Writing the action of $d \in \mathbb{Z}$ (prime to N) on $\mathcal{R}_{\Gamma_1(N)/B}$ by the operator $\langle d \rangle : \mathcal{R}_{\Gamma_1(N)/B} \to \mathcal{R}_{\Gamma_1(N)/B}$, we have verified in Corollary 6.23 that

$$f|\langle d \rangle(E, \phi_N, \omega) = d^{k-1} f(E, \phi_N \circ d, \omega).$$

So the action preserves the graded component $\mathcal{R}_k(\Gamma_1(N); B)$, and therefore $\mathcal{R}_{\Gamma_0(N)/B}$ is a graded algebra: $\mathcal{R}_{\Gamma_0(N)/B} = \bigoplus_k \mathcal{R}_k(\Gamma_0(N); B)$. We put $\mathcal{M}_{\Gamma_0(N)/B} = \mathrm{Spec}(\mathcal{R}_{\Gamma_0(N)/B})$ for any $\mathbb{Z}[\frac{1}{6N}]$-algebra B. By definition, $\mathcal{M}_{\Gamma_0(N)/B}$ represents the following functor from $ALG_{/B}$ to $SETS$:

$$\wp_{\Gamma_0(N)/B}(R) = \left[(E, C_N, \omega)_{/R} \middle| (E, \omega) \in \wp_{\Gamma_1(1)}(R), C_N \cong \mu_N \right],$$

where C_N is a finite flat subgroup over R locally isomorphic to μ_N under étale topology. The representability of $\wp_{\Gamma_0(N)}$ can be proven in the same manner as in the case of $\wp_{\Gamma_1(N)}$. However, if we remove the datum ω of differential and look into

$$\mathcal{E}_{\Gamma_0(N)/B}(R) = \left[(E, C_N)_{/R} \middle| C_N \cong \mu_N \right],$$

we have a problem because $\mathrm{Aut}_R(E, C_N)$ contains $\{\pm 1\}$; so similarly to the case of $\mathcal{E} = \mathcal{E}_{\Gamma_1(1)}$, this functor is not representable, but, in any case, we define $Y_0(N)_{/B} = \mathrm{Spec}(\mathcal{R}_0(\Gamma_0(N); B)) = \mathcal{M}_{\Gamma_0(N)/B}/\mathbb{G}_m$. Then we have the following:

Corollary 6.26 *The curve $Y_0(N)_{/\mathbb{Z}[\frac{1}{6N}]}$ gives a coarse moduli scheme for $\mathcal{E}_{\Gamma_0(N)}$, in particular, $Y_0(N)(k) = \mathcal{E}_{\Gamma_0(N)}(k)$ for each algebraically closed field k over $\mathbb{Z}[\frac{1}{N}]$.*

See [GME] Sect. 2.9.4 for some more details of this fact.

Exercise 6.27 *Verify that $Y_0(N)(\mathbb{C}) = \mathcal{E}_{\Gamma_0(N)}(\mathbb{C}) = \Gamma_0(N)\backslash\mathfrak{H}$.*

6.2.7 Compactification and Modular Line Bundles

Let $\Gamma = \Gamma_1(N)$ or $\Gamma_0(N)$. For any $\mathbb{Z}[\frac{1}{6}]$-algebra B, we put

$$G(B) = G_{\Gamma_1(1)}(B) := B[g_2, g_3] = \mathbb{Z}\left[\frac{1}{6}, g_2, g_3\right] \otimes_{\mathbb{Z}[\frac{1}{6}]} B.$$

Let $G_\Gamma(\mathbb{Z}[\frac{1}{6N}])$ be the integral closure of the ring $G(\mathbb{Z})$ in the graded ring $\mathcal{R}_{\Gamma/\mathbb{Z}[\frac{1}{6N}]}$. To see that $G_\Gamma(\mathbb{Z}[\frac{1}{6N}])$ is a graded ring, we write $\tilde{r}$ for the nontrivial homogeneous projection of the highest degree of $r \in \mathcal{R}_{\Gamma/\mathbb{Z}[\frac{1}{6N}]}$. If $r \in \mathcal{R}_{\Gamma/\mathbb{Z}[\frac{1}{6N}]}$ is integral over $G(\mathbb{Z})$, r satisfies an equation $P(X) = X^n + a_1X^{n-1} + \cdots + a_n = 0$ with $a_j \in G(\mathbb{Z})$. Then $\tilde{r}$ satisfies $\tilde{P}(X) = X^n + \tilde{a}_1X^{n-1} + \cdots + \tilde{a}_n = 0$, and $\tilde{r}$ is integral over $G(\mathbb{Z})$. Then by induction of the degree of $\tilde{r}$, we see that $G_\Gamma(\mathbb{Z}[\frac{1}{6N}])$ is graded (cf. [BCM] V.1.8).

We define $G_k(\Gamma; B)$ by, for any $\mathbb{Z}[\frac{1}{6N}]$-algebra B,

$$G_\Gamma(B) = G_\Gamma\left(\mathbb{Z}\left[\frac{1}{6N}\right]\right) \otimes B = \bigoplus_{k=0}^{\infty} G_k(\Gamma; B).$$

We then define $X_?(N)_{/B} = \mathrm{Proj}(G_{\Gamma_?(N)}(B))$ for $? = 0, 1$. By definition, $X_?(N)$ is the normalization of $\mathrm{Proj}(G) = \mathrm{Proj}(G^{(12)}) = \mathbf{P}^1(J)$ $[J = \frac{(12g_2)^3}{\Delta}]$ for $G^{(12)} = \bigoplus_{k=0}^{\infty} G_{12k}$ in $Y_?(N)$. Thus, it is a normal projective curve. Since a normal noetherian integral local domain of dimension 1 is a discrete valuation ring ([CRT] Theorem 11.2), any local ring of $X_?(N)$ is regular, and hence $X_?(N)$ is a smooth projective curve over $\mathbb{Z}[\frac{1}{6N}]$ (cf. Sect. 5.1.2).

Since its generic fiber $X_?(N)(\mathbb{C}) = \Gamma_?(N)\backslash(\mathfrak{H} \cup \mathbf{P}^1(\mathbb{Q}))$ is connected, by Zariski's connectedness theorem (see Theorem 5.7), we get the following:

Lemma 6.28 *The curve $X_?(N)_{/\mathbb{F}_p} = X_?(N) \times_{\mathbb{Z}[\frac{1}{6N}]} \mathbb{F}_p$ $(? = 0, 1)$ is a smooth projective irreducible curve over a finite field $\mathbb{F}_p$ for any prime $3 < p \nmid N$.*

Definition 6.29 *Since $X_?(N)_{/B} = \mathrm{Proj}(G_{\Gamma_?(N)}(B))$, $G_k(\Gamma_?(N); B)$ gives rise to a unique invertible sheaf $\underline{\omega}^k = \underline{\omega}^{\otimes k}$ such that*

$$H^0(X_?(N)_{/B}, \underline{\omega}^k_{/B}) = G_k(\Gamma_?(N); B).$$

Thus, $\underline{\omega}$ is an ample line bundle over $X_?(N)$ that gives rise to the projective embedding of $X_?(N)$ via homogeneous coordinates given by modular forms; hence, we could have written $\underline{\omega}^k$ as $\mathcal{O}_{X_?(N)}(k)$.

Here is a version of $G_k(\Gamma_0(N); B)$ with "Neben" character:

Definition 6.30 *For each character $\psi : (\mathbb{Z}/N\mathbb{Z})^\times \to B^\times$, we define*

$$G_k(\Gamma_0(N), \psi; B) = \left\{f \in G_k(\Gamma_1(N); B) \middle| f|\langle d\rangle = d^{k-1}\psi(d)f \ \forall d \in (\mathbb{Z}/N\mathbb{Z})^\times\right\}.$$

6.2.8 q-Expansion

As classically known, J^{-1} has a q-expansion starting with q, that is, $J^{-1} \in q\mathbb{Z}[[q]]$ [see [IAT] (4.6.1)]. Thus, the completion of the local ring of $\mathbf{P}^1(J)$ at the cusp ∞ is isomorphic to $\mathbb{Z}[\frac{1}{6}][[q]]$. Moreover, we have the Tate curve introduced in Sect. 2.4.2:

$$\mathrm{Tate}(q) = \mathrm{Proj}(\mathbb{Z}[[q]][\frac{1}{6}][X,Y,Z]/(ZY^2 - 4X^3 + g_2(q)XZ^2 + g_3(q)Z^3)),$$

which by definition extends the universal curve over $\mathbf{P}^1(J) - \{\infty\}$ to $\mathbf{P}^1(J)$ locally at the cusp ∞.

Since $\mathrm{Tate}(q)(B[[q]]) \supset (B[[q]]^\times)/q^{\mathbb{Z}}$ by Theorem 2.54, we may regard $\mathrm{Tate}(q)$ as the algebraization of the formal quotient $\widehat{\mathbb{G}}_{m/\mathbb{Z}[[q]]}/q^{\mathbb{Z}}$ of the formal multiplicative group $\widehat{\mathbb{G}}_m$; hence, it has a canonical level structure

$$\phi_N^{can} : \mu_N \hookrightarrow \widehat{\mathbb{G}}_m \twoheadrightarrow \mathrm{Tate}(q).$$

The Tate couple $(\mathrm{Tate}(q), \phi_N^{can})_{/\mathbb{Z}[[q]]}$ is an elliptic curve over $\mathbb{Z}[[q]][q^{-1}]$; thus, by the universality of $Y_?(N)$ $(? = 0, 1)$, we have a morphism

$$\iota_\infty : \mathrm{Spec}(\mathbb{Z}[\frac{1}{N}][[q]][q^{-1}]) \to Y_?(N).$$

Since we may regard the Tate curve as a universal formal deformation of a stable curve of genus 1 (with the level structure ϕ_N^{can}) centered at the $\mathbb{Z}[\frac{1}{N}]$-point represented by an ideal (q) of $\mathbb{Z}[\frac{1}{N}][[q]]$ ([GME] Sects. 2.5.2–3), the morphism ι_∞ is an infinitesimal isomorphism centered at the cusp ∞ [by the universality of the $Y_?(N)$ and the universality of the Tate curve].

Since $X_?(N)$ is the normalization of $\mathbf{P}^1(J)$ in $Y_?(N)$, we conclude that the formal completion along the cusp ∞ on $X_?(N)$ is canonically identified with $B[[q]]$ by ι_∞. Replacing the level structure ϕ_N^{can} with $\phi_N^{can} \circ \alpha$ for $\alpha \in SL_2(\mathbb{Z}/N\mathbb{Z})$, basically by the same argument, the local ring at the cusp $\alpha(\infty)$ of $X_?(N)_{/B}$ is given by $B[\mu_d][[q^{1/d}]]$ for a suitable divisor $d|N$. We need to extend the scalars to $B[\mu_d][[q^{1/d}]]$ because the Tate curve $(\mathrm{Tate}(q), \phi_N^{can} \circ \alpha)$ is only defined over $B[\mu_d][[q^{1/d}]]$ for a suitable divisor $d|N$ dependent on the choice of α. This point is slightly more technical, and we refer the reader to a more detailed account, which can be found in [AME] Chap. 10 and [GME] Sect. 3.1.1. Thus, $X_?(N)$ is smooth at the cusps, and moreover, $f \in G_k(\Gamma_?(N); B)$ is a function of (E, ϕ_N, ω) satisfying (G0–2) and

(G3) $f(\mathrm{Tate}(q), \phi_N, \omega) \in B[\zeta_N][[q^{1/N}]]$ for any choice of ϕ_N and ω.

Since $\Gamma_?(N)\backslash(\mathfrak{H} \cup \mathbf{P}^1(\mathbb{Q}))$ is a smooth compact Riemann surface and is the normalization of $\mathbf{P}^1(J)$ in $Y_?(N)(\mathbb{C})$ ([IAT] Chap. 1), we conclude

$$X_?(N)(\mathbb{C}) = \Gamma_?(N)\backslash(\mathfrak{H} \cup \mathbf{P}^1(\mathbb{Q})).$$

The space $G_k(\Gamma_?(N);\mathbb{C})$ is the classical space of modular forms on $\Gamma_?(N)$ of weight k. Since $\mathrm{Tate}(q)(B[[q]]) \supset (B[[q]]^\times)/q^{\mathbb{Z}}$, we may consider $\mathrm{Tate}(q)$ to be a "quotient" of $\mathbb{G}_{m/\mathbb{Z}[[q]]}/q^{\mathbb{Z}}$; thus, it has a canonical level structure $\phi_N^{can} : \mu_N \hookrightarrow \mathbb{G}_m \twoheadrightarrow \mathrm{Tate}(q)$ and a canonical differential ω_{can} induced by $\frac{dt}{t}$ identifying $\mathbb{G}_m = \mathrm{Spec}(\mathbb{Z}[t,t^{-1}])$. In particular,

$$f(q) = f(\mathrm{Tate}(q), \phi_N^{can}, \omega_{can}) = \sum_{n=0}^{\infty} a(n,f)q^n \tag{6.2.2}$$

coincides with the Fourier expansion of f at infinity if $f \in G_k(\Gamma_?(N);\mathbb{C})$.

To see this fact, we note that, by construction, the q-expansion of $g_j(q)$ coincides the Fourier expansion of the modular form $g_j(E,\omega)$. Since $G_{\Gamma_1(1)}(B)$ is generated by g_j over B, the identity for the general modular form f of level 1 follows from the definition. Since $G_{\Gamma_?(N)}$ is the integral closure of $G_{\Gamma_1(1)}$, the identity is still valid for any level N. Since $\widehat{X_?(N)}_{/B} \cong \mathrm{Spf}(B[[q]])$ for the formal completion $\widehat{X_?(N)}$ of $X_?(N)$ at the infinity, the irreducibility of $X_?(N)$ over any field k over $\mathbb{Z}[\frac{1}{6N}]$ tells us that $f(q)$ is determined by its q-expansion:

Lemma 6.31 *If $f \in G_k(\Gamma_?(N);B)$ $(? = 0,1)$ for a $\mathbb{Z}[\frac{1}{6N}]$-algebra B, $f(q) = 0 \Leftrightarrow$ we have $f((E,\phi_N,\omega)_{/R}) = 0$ for all elliptic curve with level-$\Gamma_?(N)$ structure ϕ_N defined over any B-algebra R.*

Recall a complete discrete valuation ring $W \subset \mathbb{C}_p$ over $\mathbb{Z}_p$ with residue field $\mathbb{F}$ (which can be an algebraic closure of $\mathbb{F}_p$) and $\mathcal{W} = W \cap \overline{\mathbb{Q}} = i_p^{-1}(W)$ for the embedding $i_p : \overline{\mathbb{Q}} \hookrightarrow \mathbb{C}_p$ we fixed.

Corollary 6.32 *Let $x \in X_1(N)(A)$ for $A = W$ and $\mathcal{W}$ for $N \geq 4$ carrying a test object $(E,\phi_N)_{/K}$ for the field of fractions K of A. If the J-invariant $J(E)$ belongs to A, E extends to an elliptic curve over A and $x \in Y_1(N)(A)$.*

Proof. Note that J^{-1} has W-integral q-expansion starting with q; hence, J^{-1} gives the parameter at ∞-cusp of $\mathbf{P}^1(J)$. First assume that $A = W$. Since $J(E) \in W$, x is not in $\widehat{X_1(N)}_s(W)$ for the formal completion of $X_1(N)$ along any cusp $s \in X_1(N)(\overline{\mathbb{F}}_p)$ above $\infty \in \mathbf{P}^1(J)$. Thus, $x \in Y_1(N)(W)$, which is a fine modulus of the functor $\mathcal{E}_{\Gamma_1(N)/W}$. So the point x carries a test object $(\mathcal{E},\tilde{\phi}_N)$ with $\mathcal{E} \times_W K = E$. In other words, E extends to an elliptic curve over W.

Since $Y_1(N)(W) \cap X_1(\overline{\mathbb{Q}}) = X_1(\mathcal{W})$ if $A = \mathcal{W}$ as $X_1(N)$ is a projective smooth curve over $\mathcal{W}$, we can find the test object $(\mathcal{E},\tilde{\phi}_N)_{/\mathcal{W}}$ over x that extends E to $\mathcal{W}$. $\square$

In the above proof, the representability of $\mathcal{E}_{\Gamma_1(N)}$ by $Y_1(N)$ is indispensable, and so the corollary is not necessarily valid for $X_0(N)$.

6.2.9 Hecke Operators on q-Expansion

We look into locally free subgroup of rank l in Tate(q) for a prime l. Let us fix a primitive lth root of unity $\zeta \in \mu_l \subset \widehat{\mathbb{G}}_m/q^{\mathbb{Z}}$. We may assume that $\zeta = i_p(i_\infty^{-1}(\exp(\frac{2\pi i}{l})))$. Write $\langle x\rangle$ for the subgroup generated by x in Tate(q). For a prime l, the set $\mathcal{C}_l$ rank-l subgroups in Definition 6.21 for Tate(q) is then given by

$$\mathcal{C}_l = \begin{cases} \{\langle \zeta^u q^{1/l}\rangle | u \in \mathbb{Z}/l\mathbb{Z}\} & \text{if } l|N, \\ \{\langle \zeta^u q^{1/l}\rangle | u \in \mathbb{Z}/l\mathbb{Z}\} \sqcup \{\langle\zeta\rangle\} & \text{if } l \nmid N. \end{cases}$$

Write $C_{\zeta^u} = \langle \zeta^u q^{1/l}\rangle$ and $C' = \langle\zeta\rangle = \mu_l$. Write $\pi_{\zeta^u} : \mathrm{Tate}(q) \to \mathrm{Tate}(q)/C_{\zeta^u}$ and $\pi' : \mathrm{Tate}(q) \to \mathrm{Tate}(q)/C'$ for the projections. This tells us the following fact:

Lemma 6.33 *We have*

$$\mathrm{Tate}(q)/C_{\zeta^u} = \mathrm{Tate}(\zeta^u q^{1/l})_{/\mathbb{Z}[\zeta^u, q^{1/l}]} \quad \textit{and} \quad \mathrm{Tate}(q)/C' = \mathrm{Tate}(q^l)_{/\mathbb{Z}[[q]]}$$

and

$$\pi_{\zeta^u,*}\omega_{can} = \omega_{can} \quad \textit{and} \quad \pi'_*\omega_{can} = l^{-1}\cdot\omega_{can}.$$

Proof. If we write $\widehat{\mathbb{G}}_m = \mathrm{Spf}(\widehat{\mathbb{Z}[w, w^{-1}]})$, the l isogeny $w \mapsto w^l$ on $\mathbb{G}_m$ induces an exact sequence $1 \to C' \to \mathrm{Tate}(q) \xrightarrow{w\mapsto w^l} \mathrm{Tate}(q^l) \to 1$. This shows the second identity. Clearly,

$$\mathrm{Tate}(q)(K)/C_{\zeta^u} = (K^\times/q^{\mathbb{Z}})/\langle\zeta^u q^{1/l}\rangle = K^\times/(\zeta^u q^{1/l})^{\mathbb{Z}}$$

for any p-adically complete field $K \subset \mathbb{C}_p$ containing μ_l. This shows that $\mathrm{Tate}(q)/C_{\zeta^u} = \mathrm{Tate}(\zeta^u q^{1/l})$ over $\mathbb{Z}[\zeta^u, q^{1/l}]$ because the left-hand and right-hand sides match at Zariski-dense fibers over the scheme $\mathrm{Spec}(\mathbb{Z}((q)))$ of relative dimension 1 over $\mathbb{Z}$.

As for the differential, we have $\omega_{can} = \frac{dw}{w}$. Thus, $w \mapsto w^l$ ends up giving $(\pi')^*\omega_{can} = \frac{dw^l}{w^l} = l\cdot\omega_{can}$ and $\pi'_*\phi_N = \phi_N^{can}\circ l$. This implies $\pi'_*\omega_{can} = l^{-1}\omega_{can}$. On the other hand, π_{ζ_u} does not affect the variable w; hence, $\pi_{\zeta^u,*}\omega_{can} = \omega_{can}$ and $\pi_{\zeta^u,*}\phi_N^{can} = \phi_N^{can}$. $\square$

By definition, we get

$$\begin{aligned} f(\mathrm{Tate}(q)/C_{\zeta^u}, \pi_{\zeta^u,*}\phi_N^{can}, \pi_{\zeta^u,*}\omega_{can}) &= f(\mathrm{Tate}(\zeta^u q^{1/l})) = f(\zeta^u q^{1/l}), \\ f(\mathrm{Tate}(q)/C', \pi'_*\phi_N, \pi'_*\omega_{can}) &= f(\mathrm{Tate}(q^l), \phi_N \circ l, l^{-1}\cdot\omega) = l\cdot f|\langle l\rangle(q^l). \end{aligned} \tag{6.2.3}$$

Then, by a direct computation via the definition of $T(l)$, we get the following:

Corollary 6.34 *We have*

$$a(n, f|T(l)) = a(nl, f) + a(\frac{n}{l}, f|\langle l\rangle) \text{ if } l \nmid N, \text{ and } a(n, f|U(l)) = a(nl, f),$$

where we put $a(\xi, f) = 0$ *if* $0 \leq \xi \notin \mathbb{Z}$.

Exercise 6.35 *Prove*

$$a(m, f|T(n)) = \sum_{0<d|(m,n),(d,N)=1} a(\frac{mn}{d^2}, f|\langle d\rangle).$$

6.2.10 Moduli of Principal Level Structure

Fix a base $\mathbb{Z}[\frac{1}{6N}, \zeta_N]$-algebra B, where $\zeta_N = \exp(2\pi i \frac{1}{N})$. We consider the following classification problem of level $\Gamma(N)$:

$$\wp_{\Gamma(N),\zeta_N/B}(R) = \left[(E, \omega, \phi_N : (\mathbb{Z}/N\mathbb{Z})^2 \cong E[N]) \middle| \langle\phi_N(1,0), \phi_N(0,1)\rangle = \zeta_N\right],$$

for all B-algebras R. Writing f for the map $\langle\cdot,\cdot\rangle : \mathbf{E}[N] \times_{\mathcal{M}_1} \mathbf{E}[N] \to \mu_N$ as a morphism of group schemes over B, we clearly see that $\wp_{\Gamma(N),\zeta_N/B}$ is represented over B by the closed affine subscheme $\mathcal{M}_{\Gamma(N),\zeta_N/B} = f^{-1}(\zeta_N)$ of $\mathbf{E}[N] \times_{\mathcal{M}_1} \mathbf{E}[N]$. Writing $\mathcal{M}_{\Gamma(N),\zeta_N/B} = \mathrm{Spec}(\mathcal{R}_{\Gamma(N),\zeta_N/B})$ for a graded algebra $\mathcal{R}_{\Gamma(N),\zeta_N/B} = \bigoplus_k \mathcal{R}_k(\Gamma(N); B)$, we define an affine curve

$$Y(N)_{\zeta_N/B} = \mathrm{Spec}(\mathcal{R}_0(\Gamma(N); B)) = \mathcal{M}_{\Gamma(N),\zeta_N/B}/\mathbb{G}_m,$$

which represents the following functor from $ALG_{/B}$ into $SETS$:

$$\mathcal{E}_{\Gamma(N),\zeta_N/B}(R) = \left[(E, \phi_N : (\mathbb{Z}/N\mathbb{Z})^2 \cong E[N])_{/R} \middle| \langle\phi_N(1,0), \phi_N(0,1)\rangle = \zeta_N\right]$$

if $N \geq 3$ (see [GME] Theorem 2.6.8). Here $\mathcal{R}_0(\Gamma(N); B)$ is the degree-0 component of the graded algebra $\mathcal{R}_{\Gamma(N),\zeta_N/B}$.

Taking $B = \mathbb{Z}[\frac{1}{6N}, \zeta_N]$ and the integral closure $G_{\Gamma(N),\zeta_N}$ of the graded algebra $G(\mathbb{Z}[\frac{1}{6N}, \zeta_N])$ in $\mathcal{R}_{\Gamma(N),\zeta_N}$, we define the compactification $X(N)_{\zeta_N} = \mathrm{Proj}(G_{\Gamma(N),\zeta_N})$ over $\mathbb{Z}[\frac{1}{6N}, \zeta_N]$. For general B, we put

$$X(N)_{\zeta_N/B} = X(N)_{\zeta_N} \times_{\mathbb{Z}[\frac{1}{N}, \zeta_N]} B.$$

Again, we have a unique modular ample line bundle $\underline{\omega}^k$ such that

$$H^0(X(N)_{\zeta_N/B}, \underline{\omega}^k_{/B}) = G_k(\Gamma(N); B)$$

for each $\mathbb{Z}[\frac{1}{6N}, \zeta_N]$-algebra B, and we have a well-defined q-expansion at ∞ of each section in $H^0(X(N)_{\zeta_N/B}, \underline{\omega}^k_{/B})$. Here $G_k(\Gamma(N); B)$ is the degree-k component of $G_{\Gamma(N),\zeta_N/B} = G_{\Gamma(N),\zeta_N} \otimes_{\mathbb{Z}[\frac{1}{N}, \zeta_N]} B$.

We can let a constant group $\alpha \in SL_2(\mathbb{Z}/N\mathbb{Z})$ act on $Y(N)_{\zeta_N}$ [and hence on $X(N)_{\zeta_N}$] by $(E, \phi) \mapsto (E, \phi \circ \alpha)$. Since $\langle\phi \circ \alpha(1,0), \phi \circ \alpha(0,1)\rangle = \zeta_N^{\det(\alpha)}$, the same action of $\alpha \in GL_2(\mathbb{Z}/N\mathbb{Z})$ induces an automorphism of $Y(N) = Y(N)_{\zeta_N}$ [and $X(N) = X(N)_{\zeta_N}$] regarded as schemes over $\mathbb{Z}[\frac{1}{6N}]$ (not over $\mathbb{Z}[\frac{1}{6N}, \zeta_N]$), which coincides with the Galois action $\zeta_N \mapsto \zeta_N^{\det(\alpha)}$ on $\mathbb{Z}[\zeta_N]$. Over $\mathbb{Z}[\frac{1}{6N}, \zeta_N]((q^{1/N}))$, Tate($q$) has a canonical level $\Gamma(N)$-structure $\phi_N^{can} : (\mathbb{Z}/N\mathbb{Z})^2 \cong \mathrm{Tate}(q)[N]$ defined over $\mathbb{Z}[\mu_N][[q^{1/N}]$ given by $(a, b) \mapsto \zeta_N^a q^{b/N}$.

Returning to general $B_{/\mathbb{Z}[\frac{1}{6N}, \zeta_N]}$, define the B-integral space $G_k(\Gamma(N); B)$ of geometric modular forms on $\Gamma(N)$, replacing the level-$\Gamma_1(N)$ structure in (G0–3) with the level-$\Gamma(N)$ structure. Fix $\alpha \in SL_2(\mathbb{Z}/N\mathbb{Z})$. Then we can think of $\phi_N^{can} \circ \alpha$. We define the q-expansion of $f \in G_k(\Gamma(N); B)$ at the cusp $\alpha(\infty)$ for $\alpha \in SL_2(\mathbb{Z}/N\mathbb{Z})$ by

$$f_\alpha(q) = f((\mathrm{Tate}(q), \phi_N^{can} \circ \alpha, \omega_{can})_{/\mathbb{Z}[\frac{1}{N}, \zeta_N]((q^{1/N}))}) \in B((q)).$$

We define $S_k(\Gamma(N); B)$ be the B-submodule of $f \in G_k(\Gamma(N); B)$ made up of f satisfying the following condition in place of (G3):

(s3) $f_\alpha((\mathrm{Tate}(q), \phi_N, \omega)) \in q^{1/N} B[[q^{1/N}]]$ for any choice of ϕ_N and ω.

Then again we have the following:

Lemma 6.36 *Fix $\alpha \in SL_2(\mathbb{Z}/N\mathbb{Z})$ and put*

$$f_\alpha(q) = f((\mathrm{Tate}(q), \phi_N^{can} \circ \alpha, \omega_{can})_{/\mathbb{Z}[\frac{1}{N}, \zeta_N]((q^{1/N}))}).$$

If $f \in G_k(\Gamma(N); B)$ for a $\mathbb{Z}[\frac{1}{6N}, \zeta_N]$-algebra B, $f_\alpha(q) = 0 \Leftrightarrow$ we have $f((E, \phi_N, \omega)_{/R}) = 0$ for all elliptic curves with level-$\Gamma(N)$ structure ϕ_N defined over any B-algebra R.

In particular, for a prime p outside N, supposing $\mathcal{W}$ is a $\mathbb{Z}[\zeta_N]$-algebra, this implies that $f \equiv 0 \mod p\mathcal{W}[[q_N]]$ ($q_N = q^{1/N}$) for $f \in G_k(\Gamma(N); \mathcal{W}) = H^0(X(N)_{\zeta_N/\mathcal{W}}, \underline{\omega}^k)$ if and only if the image of $\overline{f}$ in $G_k(\Gamma(N); \mathbb{F})$ vanishes.

Since $X(N)_{\zeta_N/\mathcal{W}} = \mathrm{Proj}(G_{\Gamma(N)}(\mathcal{W}))$ for the graded algebra $G_{\Gamma(N)}(\mathcal{W}) = \bigoplus_{k \geq 0} G_k(\Gamma(N); \mathcal{W})$, the function field $\mathcal{K}(X(N)_{\zeta_N})$ ($\mathcal{K} = Frac(\mathcal{W})$) is made up of $\frac{f}{g}$ for some $f, g \in G_k(\Gamma(N); \mathcal{W})$ and integer $k \geq 0$. This implies $\mathcal{K}(X(N)_{\zeta_N})$ is embedded into the Laurent series ring $\mathcal{K}((q_N))$; hence, $\frac{f}{g} = \sum_{n \gg -\infty} c(n, \frac{f}{g}) q_N^n$ with $c(n, \frac{f}{g}) \in \mathcal{K}$. Writing ord_p for the standard additive p-adic valuation of $\mathcal{K}$ with valuation ring $\mathcal{W}$ and $\mathrm{ord}_p(p) = 1$, $\overline{\frac{f}{g}}$ is a well-defined element of the mod-p function field $\mathbb{F}(X(N)_{\zeta_N/\mathbb{F}})$ if and only if $v(\frac{f}{g}) := \mathrm{Inf}_n \, \mathrm{ord}_p(c(n, \frac{f}{g})) \geq 0$.

Note that the modular curve $X(N)_{\zeta_N}$ covers $X_1(N)$. We write $\pi : X(N)_{\zeta_N} \to X_1(N)$ for the morphism given by $(E, \phi_N, \omega) \mapsto (E, \phi_N|_{\mu_N}, \omega)$ identifying the first factor $(\mathbb{Z}/N\mathbb{Z})$ of $(\mathbb{Z}/N\mathbb{Z})^2$ with μ_N by $a \leftrightarrow \zeta_N^a$. For any $\mathbb{Z}[\frac{1}{6N}]$-algebra B, writing $B[\zeta_N] = B[t]/(\Phi(t))$ ($\zeta_N \leftrightarrow t$) for the cyclotomic polynomial $\Phi(t) \in \mathbb{Z}[t]$ of primitive Nth roots, we have a pullback map

$$\begin{aligned}\mathrm{Res}_{X(N)_{\zeta_N}/X_1(N)} = \pi^* : H^0(X_1(N), \underline{\omega}^k_{/B}) = G_k(\Gamma_1(N); B)\\ \to G_k(\Gamma(N); B[\zeta_N]) = H^0(X(N)_{\zeta_N}, \underline{\omega}^k_{/B[\zeta_N]}).\end{aligned}$$

We then define the space of *cusp forms* by

$$S_k(\Gamma_1(N); B) = \mathrm{Res}^{-1}_{X(N)_{\zeta_N}/X_1(N)}(S_k(\Gamma(N); B[\zeta_N])). \tag{6.2.4}$$

We can similarly define $S_k(\Gamma_0(N), \chi; B)$ for a character $\chi : (\mathbb{Z}/N\mathbb{Z})^\times \to B^\times$.

If B is a local ring in which p is not invertible, we extend the above definition of cusp forms to level Np^r for N prime to p, just ignoring the p-part. Writing $(E, \phi_p, \phi_N, \omega)$ for level-p structure ϕ_p and level-N structure separately, we define $f_\alpha(q) = f(\mathrm{Tate}(q), \phi_p^{can} : \mu_{p^r} \hookrightarrow \mathrm{Tate}(q), \phi_N \circ \alpha, \omega_{can})$ for each $\alpha \in SL_2(\mathbb{Z}/N\mathbb{Z})$. Then

$$\begin{aligned}S_k(\Gamma_1(Np^r); B) = \{f \in G_k(\Gamma_1(N); B) | f_\alpha(q) \in q^{1/N} B[\zeta_N][[q^{1/N}]]\}\\ S_k(\Gamma_0(Np^r), \chi; B) = G_k(\Gamma_0(N), \chi; B) \cap S_k(\Gamma_1(Np^r); B).\end{aligned} \tag{6.2.5}$$

Here $\chi : (\mathbb{Z}/Np^r\mathbb{Z})^\times \to B^\times$ is a character.

6.2.11 Hasse Invariant

Let R be a ring of characteristic p and let (E, ω) be an elliptic curve over $S = \mathrm{Spec}(R)$. On each affine open subset $U = \mathrm{Spec}(\Gamma(U, \mathcal{O}_E))$ in E, the Frobenius endomorphism $x \mapsto x^p$ induces a morphism $F_{abs} : U \to U$. These glue each other to the *absolute Frobenius endomorphism* $F_{abs} : E_{/R} \to E_{/R}$. Note here that F_{abs} acts nontrivially on the coefficient ring R. We can define the relative Frobenius map $F : E \to E^{(p)} = E \times_{S, F_{abs}} S$ by $F_{abs} \times_S f$ for the structure morphism $f : E \to S$. This *relative Frobenius map* F is the classical map taking homogeneous coordinates on E to their p-powers. Taking the dual $V := F^*$, we have the relative Verschiebung map $V : E^{(p)} \to E$.

Let $\mathcal{T}_{E/S}$ be the relative tangent bundle. By definition, its sections are derivations; in particular, the space of global sections $H^0(E, \mathcal{T}_{E/S})$ is the R-dual of $H^0(E, \Omega_{E/S})$, and $H^0(E, \mathcal{T}_{E/S})$ is spanned by the dual base $\eta = \eta(\omega)$ of ω. One can identify $H^0(E, \mathcal{T}_{E/S})$ with the module of $\mathcal{O}_S$-derivations $Der_{\mathcal{O}_S}(\mathcal{O}_{E,\mathbf{0}}, \mathcal{O}_S)$ (cf. [GME] Sect. 1.5.1). For each derivation D of $\mathcal{O}_{E,\mathbf{0}}$, by the Leibniz formula, we have

$$D^p(xy) = \sum_{j=0}^{p} \binom{p}{j} D^{p-j}x D^j y = x D^p y + y D^p x.$$

Thus, D^p is again a derivation. The association $D \mapsto D^p$ induces an F_{abs}-linear endomorphism F^* of $\mathcal{T}_{E/S}$. Then we define the *Hasse invariant* $H(E, \omega) \in R$ by $F^*\eta = H(E, \omega)\eta$. Since $\eta(\lambda\omega) = \lambda^{-1}\eta(\omega)$, we see

$$\begin{aligned}H(E, \lambda\omega)\eta(\lambda\omega) = F^*\eta(\lambda\omega) = F^*(\lambda^{-1}\eta(\omega)) = \lambda^{-p} F^*\eta(\omega)\\ = \lambda^{-p} H(E, \omega)\eta(\omega) = \lambda^{-p} H(E, \omega)(\lambda \cdot \eta(\lambda\omega)) = \lambda^{1-p} H(E, \omega)\eta(\lambda\omega).\end{aligned}$$

Thus, we get

$$H(E, \lambda\omega) = \lambda^{1-p} H(E, \omega).$$

Then H is a modular form of weight $p-1$ defined over $\mathbb{F}_p$:

$$H(E, \omega) \in G_{p-1}(\Gamma_1(1); \mathbb{F}_p).$$

We compute $H(\mathrm{Tate}(q), \omega_{can})$ for $\omega_{can} = \frac{dt}{t}$, writing the coordinate of $\mathbb{G}_m$ as t [i.e., $\mathbb{G}_m = \mathrm{Spec}(\mathbb{Z}[t, t^{-1}])$]. The dual of $\frac{dt}{t}$ is given by $D = t\frac{d}{dt}$. The action of F keeps D intact, because $D(t) = t$ [so $D^p(t) = t$]. On the tangent space, F acts as identity, and hence $H(\mathrm{Tate}(q), \omega_{can}) = 1$.

Proposition 6.37 *The Hasse invariant is a level-*1 *modular form of weight* $p-1$ *over* $\mathbb{F}_p$. *We have* $H(\mathrm{Tate}(q), \omega_{can}) = 1 \in \mathbb{F}_p[[q]]$, *and for an elliptic curve* (E, ω) *over a field of characteristic* p, $H(E, \omega) = 0$ *if and only if* E *is not* p*-ordinary.*

Proof. We only need to prove the last assertion. Suppose that $(E, \omega)_{/k}$ is an ordinary elliptic curve over a field k of characteristic p. Replacing k by its finite extension, we have $E[p] \cong \mu_p \times \mathbb{Z}/p\mathbb{Z}$. The differential ω induces a differential $\lambda\frac{dw}{w}$ ($\lambda \in k^\times$) on

$$\mu_p = \mathrm{Spec}(k[w]/(w^p - 1)).$$

Since $\mu_p \subset \mathbb{G}_m$, and they share the tangent space at 1, we see from the above argument $H(E, \omega) = \lambda^{1-p} \neq 0$.

Suppose that $(E, \omega)_{/k}$ is supersingular. Let C be the connected component of $E[p]$. If $C \neq E[p]$, then we have locally under the étale topology a constant group scheme $(\mathbb{Z}/p\mathbb{Z})^r$ as a quotient of the locally free group scheme $E[p]$ for $0 < r = r(E) \leq 2$. Since $E[p]$ is isomorphic to its Cartier dual (see Lemma 6.9) and the dual of $\mathbb{Z}/p\mathbb{Z}$ is the connected group scheme μ_p, r has to be 1 and E has to be ordinary, which is a contradiction against supersingularity. Thus, $C = E[p]$ and therefore, $E[p] = \mathrm{Spec}(R)$ for a local ring R of characteristic p. Since E is one-dimensional, the local ring $\mathcal{O}_{E,\mathbf{0}}$ at the origin $\mathbf{0}$ of E is a valuation ring with residue field k. If $C' = \mathrm{Spec}(R') \subset E$ is a connected (locally free) subgroup scheme of E of rank m, $R' = \mathcal{O}_{E,\mathbf{0}}/\mathfrak{m}^m$ for the maximal ideal $\mathfrak{m}$ of $\mathcal{O}_{E,\mathbf{0}}$ because C' is connected. Thus, there is at most one connected subgroup of E for a given rank. This shows that $\mathrm{Ker}(F^2) = C$ for the Frobenius map $F : E \to E^{(p)}$, because $\deg(F^2) = p^2 = \mathrm{rank}(C)$. Thus, $F^2 = p$ up to an automorphism of E, and hence $F_{abs}^*\eta = 0$. That is, $H(E, \omega) = 0$, as desired.

Remark 6.38 The zero locus of the nonzero section H of $\underline{\omega}^{p-1}$ is an effective divisor; hence, for any algebraically closed field k of characteristic p for $p \nmid N$, the points in $X_1(N)(k)$ carrying supersingular elliptic curves are finitely many and algebraic over $\mathbb{F}_p$.

Corollary 6.39 *Let* (E, ω) *be an elliptic curve and an invariant differential defined over a finite field* $\mathbb{F}_{p^n}$ *of characteristic* p. *Let* $F : E \to E$ *be the*

Frobenius endomorphism associated with $x \mapsto x^{p^n}$ *and* $V = F^*$. *We put* $a(p) = F + V \in \mathbb{Z}$. *Then*

$$H(E,\omega) = 0 \iff a(p) \equiv 0 \mod p.$$

Proof. If $a(p) \equiv 0 \mod p$, then $\mathrm{Ker}(F)[p] = \mathrm{Ker}(V)[p]$ and hence, $\mathrm{Ker}(p^n) = \mathrm{Ker}(FV) = \mathrm{Ker}(F^2)$. Thus, $\mathrm{Ker}(p)$ is connected; hence, E is supersingular. If $a(p) \not\equiv 0 \mod p$, then $\mathrm{Ker}(F)[p] \neq \mathrm{Ker}(V)[p]$, and thus E has two subgroup schemes of rank p. Since E can have only one connected subgroup of rank p ($\dim E = 1$), E has to be p-ordinary. In this case, $\mathrm{Ker}(V)$ is étale.

Corollary 6.40 *Let the notation be as in the above corollary. The elliptic curve* E *is ordinary defined over a finite field* $\mathbb{F}_{p^n}$ *of characteristic* p *if and only if the endomorphism algebra* $\mathrm{End}_{\mathbb{F}}(E)\otimes_{\mathbb{Z}}\mathbb{Z}_{(p)}$ *is isomorphic to the integral closure* $\mathfrak{O}_{(p)}$ *of* $\mathbb{Z}_{(p)}$ *in an imaginary quadratic field* M *in which the prime* p *splits into* $\mathfrak{p}\overline{\mathfrak{p}}$. *We can normalize the splitting of* (p) *so that* $E[\mathfrak{p}^n] = \mathrm{Ker}(F)$ *and* $E[\overline{\mathfrak{p}}^n] = \mathrm{Ker}(V)$. *The ring* $\mathfrak{O}_{(p)}$ *is generated over* $\mathbb{Z}_{(p)}$ *by the Frobenius map* F, *and* $\mathfrak{O}_{(p)} = \mathrm{End}_{\mathbb{F}}(\overline{E}) \otimes_{\mathbb{Z}} \mathbb{Z}_{(p)}$ *for an algebraic closure* $\mathbb{F}$ *of* $\mathbb{F}_{p^n}$ *and* $\overline{E} := E \otimes_{\mathbb{F}_{p^n}} \mathbb{F}$.

Proof. Suppose that E is ordinary. By the above corollary, F satisfies

$$X^2 - a(p)X + p^n = 0.$$

Since the discriminant of the quadratic equation is negative by Theorem 6.6 (iii), $M := \mathbb{Q}[F] \subset \mathrm{End}_{\mathbb{F}_{p^n}}(E) \otimes_{\mathbb{Z}} \mathbb{Q}$ is an imaginary quadratic field. Since $a(p) \not\equiv 0 \mod p$, F generates $\mathfrak{p}^n$ for a factor $\mathfrak{p}|p$ and $\overline{\mathfrak{p}} \neq \mathfrak{p}$. Thus, $M = \mathbb{Q}[F^m]$ for any $0 < m \in \mathbb{Z}$. Write $D = \mathrm{End}_{\mathbb{F}_{p^n}}(E) \otimes_{\mathbb{Z}} \mathbb{Q}$ and $\overline{D} = \mathrm{End}_{\mathbb{F}}(\overline{E}) \otimes_{\mathbb{Z}} \mathbb{Q}$. By Proposition 6.18, $D \subset \overline{D} \subset M_2(\mathbb{Q}_l)$ for any prime $l \neq p$ and D is a division algebra at most dimension 4 over $\mathbb{Q}$. Pick $\phi \in \mathrm{End}_{\mathbb{F}}(\overline{E})$. Then ϕ is defined over a finite extension $\mathbb{F}'/\mathbb{F}_{p^n}$ of relative degree d. Then, by the definition of F, for the Frobenius map σ (relative to $\mathbb{F}'$) of $\mathrm{Gal}(\mathbb{F}/\mathbb{F}')$, $F^d\phi F^{-d} = \phi^\sigma = \phi$; so ϕ commutes with $\mathbb{Q}[F^d] = \mathbb{Q}[F] \subset D$. Since D has at most dimension 4 over $\mathbb{Q}$, $\mathbb{Q}[F]$ is the commutant of $\overline{D}$. Thus, ϕ is defined over $\mathbb{F}_{p^n}$, and $\mathrm{End}_{\mathbb{F}_{p^n}}(E) = \mathrm{End}_{\mathbb{F}}(\overline{E})$. This implies $D = \overline{D} = \mathbb{Q}[F]$. Let $\mathfrak{o} = \mathrm{End}_{\mathbb{F}_{p^n}}(E)$ as a subring of $M := \mathbb{Q}[F]$. Write $\mathfrak{o}_p = \varprojlim_n \mathfrak{o}/p^n\mathfrak{o}$. Since

$$X^2 - a(p)X + p^n \equiv (X - \alpha)X \mod p$$

for a unit $\alpha \in (\mathbb{Z}/p\mathbb{Z})^\times$, by Hensel's lemma (cf. [CRT] Theorem 8.3), we have $\mathfrak{o}_p \cong \mathbb{Z}_p \oplus \mathbb{Z}_p$; thus, p splits in $\mathfrak{o}$, and hence, $\mathfrak{o}_{(p)}$ is equal to $\mathfrak{O}_{(p)}$ (see the exercise below).

Since $\mathfrak{O}_p = \mathbb{Z}_p \oplus \mathbb{Z}_p$ acts in the Barsotti–Tate group $E[p^\infty]$, we have $E[p^\infty] = \mathcal{C} \oplus \mathcal{E}$ for $\mathcal{C} = (1,0)E[p^\infty]$ and $\mathcal{E} = (0,1)E[p^\infty]$. Since the first factor $\mathbb{Z}_p$ is identified with $\mathfrak{O}_{\mathfrak{p}}$ for a prime factor $\mathfrak{p}$ of p in $\mathfrak{O}_{(p)}$, we have $\mathcal{C} = E[\mathfrak{p}^\infty]$ and $\mathcal{E} = E[\overline{\mathfrak{p}}^\infty]$. Since E is ordinary, $E[p^\infty]$ is an ordinary Barsotti–Tate group (see Sect. 4.3.9). We have the connected étale exact sequence

$$0 \to E[p^\infty]^\circ \to E[p^\infty] \to E[p^\infty]^{\mathrm{et}} \to 0,$$

which is stable under the action of $\mathfrak{O}_p$ and $\overline{E}[p^\infty]^\circ \cong \mu_{p^\infty/\overline{\mathbb{F}}}$. Thus, one of $E[\mathfrak{p}^\infty]$ and $E[\overline{\mathfrak{p}}^\infty]$ is the connected component; so we normalize the factorization $(p) = \mathfrak{p}\overline{\mathfrak{p}}$ so that $E[\mathfrak{p}^\infty]$ is the connected component. Then $E[\mathfrak{p}^\infty] = E[F^\infty] = \bigcup_n \mathrm{Ker}(F^n)$ as F is an inseparable isogeny. Thus, $\mathrm{Ker}(F) = E[\mathfrak{p}]$, and this determines the factor $\mathfrak{p}$ of p.

Conversely, if $\mathrm{End}_{\mathbb{F}}(E) \otimes_{\mathbb{Z}} \mathbb{Z}_{(p)}$ is the integral closure $\mathfrak{O}_{(p)}$ of $\mathbb{Z}_{(p)}$ in an imaginary quadratic field where p splits, we have $E[p^\infty] = E[\mathfrak{p}^\infty] \oplus E[\overline{\mathfrak{p}}^\infty]$. If $E[p^\infty]$ is connected, the one-dimensional formal group $\widehat{E}$ has to split into the direct product of $\widehat{E}_{\mathfrak{p}}$ and $\widehat{E}_{\overline{\mathfrak{p}}}$, corresponding to the Barsotti–Tate groups $E[\mathfrak{p}^\infty]$ and $E[\overline{\mathfrak{p}}^\infty]$ by Lemma 4.82. This is impossible as each of $\widehat{E}_{\mathfrak{p}}$ and $\widehat{E}_{\overline{\mathfrak{p}}}$ has dimension at least 1; so one of $E[\mathfrak{p}^\infty]$ and $E[\overline{\mathfrak{p}}^\infty]$ must be étale. Thus, $E[p^\infty]$ is ordinary, and hence E is ordinary. This finishes the proof.

Exercise 6.41 *For a subring $\mathfrak{o}$ in the integer ring $\mathfrak{O}$ of a quadratic field M generating M over $\mathbb{Q}$, prove that $\mathfrak{o}_p = \varprojlim_n \mathfrak{o}/p^n\mathfrak{o} \cong \mathbb{Z}_p \oplus \mathbb{Z}_p$ implies that the localization $\mathfrak{o}_{(p)}$ at (p) is the integral closure of $\mathbb{Z}_{(p)}$ in M.*

6.2.12 Igusa Curves

Let $W = \mathbb{Z}_p$ and $W_m = W/p^mW$. Fix N with $p \nmid N$. As $\underline{\omega}$ is ample, we find a lift of a power H^j of the Hasse invariant H in $G_{j(p-1)}(1; \mathbb{Z}_p)$, which is an Eisenstein series $\mathcal{E}$. The standard Eisenstein series E_k of weight k on $SL_2(\mathbb{Z})$ has the following q-expansion (see (1.3.1)):

$$E_k(q) = \frac{\zeta(1-k)}{2} + \sum_{n=1}^{\infty} \left(\sum_{0<d|n} d^{k-1} \right) q^n.$$

Define $\mathcal{E}(q) = \frac{2}{\zeta(2-p)} E_{p-1}$. If $p \geq 5$, by the Von Staut theorem, the q-expansion $\mathcal{E}(q)$ of $\mathcal{E}$ is congruent to 1 modulo $p\mathbb{Z}_{(p)}$; so $\mathcal{E} \mod p$ coincides with H by Proposition 6.37 (and the q-expansion principle). If $p = 3$, we do not have level-1 lift of the Hasse invariant, but taking the weight-4 Eisenstein series E_4 and defining $\mathcal{E} = \frac{4}{\zeta(3)} E_4$, $\mathcal{E}$ lifts H^2 (and this $\mathcal{E}$ does the job). For simplicity, we assume in this subsection that $p \geq 5$, though the argument works for $p = 3$.

Let $M = X_1(N)_{/W}$ and $(\mathbf{E}, \phi_N)_{/M}$ be the genus 1 semistable curve (completed by appropriate Tate curves at the cusps; see [GME] Sect. 4.1). Let

$$M_m = X_1(N)_{/W_m} = X_1(N) \times_W W_m.$$

Define $S_m \subset M_m$ by the maximal open subscheme of M_m on which $\mathcal{E}$ is invertible, and symbolically write $S_m = M_m\left[\frac{1}{\mathcal{E}}\right]$. Similarly, we write $S_0 = M\left[\frac{1}{\mathcal{E}}\right]$ for the maximal open subscheme of M on which $\mathcal{E}$ is invertible. If $\mathcal{E} \equiv \mathcal{E}' \equiv H \mod p$, the congruence tells us that $\mathcal{E}' = \mathcal{E} \cdot (1 + n)$ for n

nilpotent; so $\mathcal{E}$ and $\mathcal{E}'$ are invertible at the same time. Thus, the scheme S_m does not depend on the choice of the lift $\mathcal{E}$ as long as p is nilpotent in the base ring. We write S_∞ for the formal completion $\varprojlim_m S_m$ of S_0 along S_1.

Since we have defined $X_1(N)$ by $\mathrm{Proj}(G_{\Gamma_1(N)})$, the invertible sheaf $\underline{\omega}^k$ ($k > 0$) associated with the kth graded piece is ample. To see for which k, $\underline{\omega}^k$ becomes very ample, we recall that an invertible sheaf of degree $\geq 2g+1$ over a curve of genus g is very ample by the Riemann–Roch theorem (see [GME] Proposition 2.1.4). Computing the genus of $X_1(N)$ (e.g., [GME] Theorem 3.1.2), the invertible sheaf $\underline{\omega}^k_{/R}$ corresponding to $G_k(\Gamma_1(N); R)$ is very ample if $k \geq 2$ and $N \geq 4$ (or $k > 2$). Thus, S_m is affine, and $S_m = \mathrm{Spec}(V_{m,0})$ for a W_m-flat algebra $V_{m,0}$ (the degree-0 component of $G_{\Gamma_1(N)}(W_m)[\frac{1}{\mathcal{E}}]$).

Let $pALG_{/W_m}$ be the category of p-adically complete W_m-algebras R with p-adically continuous morphisms. Here the words "p-adically complete" mean that $R = \varprojlim_n R/p^nR$. When we write W_∞, it means $W = \varprojlim_m W_m$. We define the following two functors, $\mathcal{E}^{ord}_\alpha, \mathcal{E}'^{ord}_\alpha : pALG_{/W_m} \to SETS$, by

$$\mathcal{E}'^{ord}_\alpha(R) = \left[(E, P, \phi_N)_{/R}\right] \quad \text{and} \quad \mathcal{E}^{ord}_\alpha(R) = \left[(E, \mu_{p^\alpha} \hookrightarrow E[p^\alpha], \phi_N)_{/R}\right],$$

where P is a point of order p^α. Similarly to the proof of $\wp'_{\Gamma_1(N)} \cong \wp_{\Gamma_1(N)}$ (Corollary 6.10), we can verify that the two functors are isomorphic. For a given test triple $(E, P, \phi_N)_{/R}$, we have a unique commutative diagram (up to isomorphisms)

$$\begin{array}{ccc} E & \xrightarrow{\iota} & \mathbf{E} \\ \downarrow & & \downarrow \\ \mathrm{Spec}(R) & \longrightarrow & Y_1(N) \end{array}$$

and $\iota(P) \in \left(\mathbf{E}[p^\alpha]^{\mathrm{et}} - \mathbf{E}[p^{\alpha-1}]^{\mathrm{et}}\right)(R)$. The association $(E, P, \phi_N)_{/R} \mapsto \iota(P)$ gives rise to the following isomorphism:

$$\mathcal{E}^{ord}(R) \cong \mathcal{E}'^{ord}(R) \cong \left(\mathbf{E}[p^\alpha]^{\mathrm{et}} - \mathbf{E}[p^{\alpha-1}]^{\mathrm{et}}\right)_{/S_m}(R),$$

if m is finite. Here $\mathbf{E}[p^\alpha]^{\mathrm{et}}$ is the maximal étale quotient of $\mathbf{E}[p^\alpha]$; thus, writing $\mathbf{E}[p^\alpha]$ as a relative affine spectrum $Spec_{S_m}(R)$ (cf. [GME] Sect. 1.5.4) of a sheaf of $\mathcal{O}_{S_m}$-algebras R, we have $\mathbf{E}[p^\alpha]^{\mathrm{et}} = Spec_{S_m}(R^{\mathrm{et}})$, where R^{et} is the maximal subalgebra of R étale over $\mathcal{O}_{S_m}$ [see Sect. 4.3.8 in the text and [T1] (1.4)]. We write $I_{m,\alpha/S_m} = \left(\mathbf{E}[p^\alpha]^{\mathrm{et}} - \mathbf{E}[p^{\alpha-1}]^{\mathrm{et}}\right)_{/S_m}$, which is an étale Galois covering with the Galois group $(\mathbb{Z}/p^\alpha\mathbb{Z})^\times$. Then the scheme $I_{m,\alpha}$ represents the functors $\mathcal{E}'^{ord}_\alpha$ and $\mathcal{E}^{ord}_\alpha$ over W_m. The curve $I_{m,\alpha/W_m}$ is called the *Igusa curve* of level Np^α. We also define a procurve $I_{m,\infty/W_m} = \varprojlim_\alpha I_{m,\alpha}$, which represents the functor $\mathcal{E}^{ord}_\infty$ over W_m.

Here is a theorem of Igusa (e.g., [GME] Sect. 2.9.3):

Theorem 6.42 *The curve $I_{1,\alpha/S_1}$ ($\alpha = 1, 2, \ldots, \infty$) is irreducible.*

We prove this in the following chapter in Sect. 7.2 after reinterpreting the tower of modular curves $\{Y(N)\}_N$ as an example of a Shimura variety $\varprojlim_N Y(N)$.

Since we have a universal elliptic curve $\mathbf{E}$, we can identify the functor $T \mapsto I_{1,\alpha}(T) = \mathcal{E}_\alpha^{ord}(T)$ with

$$\mathrm{Isom}_{S_1}(\mu_{p^\alpha/S_1}, \mathbf{E}[p^\alpha]^\circ_{/S_1})(T) := \{\phi_p : \mu_{p^\alpha/T} \cong \mathbf{E}[p^\alpha]^\circ_{/T} := \mathbf{E}[p^\alpha]^\circ \times_{S_1} T\},$$

where the right-hand side is the collection of isomorphisms $\mu_{p^\alpha/T} \cong \mathbf{E}[p^\alpha]^\circ_{/T}$ over T of finite flat group schemes for each S_1-scheme T. We will prove the irreducibility as a scheme of this "Isom" functor in Sect. 7.2.

6.3 Deformation of Ordinary Elliptic Curves

Let W be a complete discrete valuation ring with residue field $\mathbb{F} = \overline{\mathbb{F}}_p$ (we fix an algebraic closure of the finite field $\mathbb{F}_p$). We regard $W \subset \mathbb{C}_p$ and put $\mathcal{W} = i_p^{-1}(W)$. If $W = W(\mathbb{F})$ (the ring of Witt vectors with coefficients in $\mathbb{F}$; i.e., completion of the integer ring of the maximal unramified extension of $\mathbb{Q}_p$ inside $\mathbb{C}_p$), then $\mathcal{W}$ is a strict henselization of $\mathbb{Z}_{(p)}$. See [BCM] IX.1 for Witt vectors and [NMD] Sect. 2.3 for Henselian rings. In this section, we relate the deformation theory of abelian schemes over local W_m-algebras for $W_m = W/p^mW$ to that of Barsotti–Tate groups. We describe deformation theory of a given Barsotti–Tate group and a given elliptic curve following Katz's exposition principally [K1]. There is a more global version of the theory due to Chai [Ch1] and another due to Moonen [Mo].

6.3.1 A Theorem of Drinfeld

Let B be a local W-algebra complete under $\mathfrak{m}_B$-adic topology with $B/\mathfrak{m}_B = \mathbb{F}$ for the maximal ideal $\mathfrak{m}_B$ of B. Recall the category $CL_{/B}$ of complete local B-algebras sharing the same residue field with B. Let $G : CL_{/B} \to AB$ be an abelian fppf-sheaf. Recall the notation $\mathcal{S}(B_{fppf})$, which is the category of abelian fppf sheaves over B. Let I be an ideal of B such that $I^{\nu+1} = 0$ and $NI = 0$ for an integer N equal to a power of p. Define two functors $G_I, \widehat{G} : CL_{/B} \to AB$ by

$$G_I(R) = \mathrm{Ker}(G(R) \to G(R/IR)) \quad \text{and} \quad \widehat{G}(R) = \mathrm{Ker}(G(R) \to G(R/\mathfrak{m}_R)),$$

where $\mathfrak{m}_R$ is the maximal ideal of R. When $G(R) = \mathrm{Hom}_{CL_{/B}}(\mathcal{R}, R) (= \widehat{G}(R))$ for $\mathcal{R} = B[[T_1, \ldots, T_n]]$ [that is, $G_{/B} = \mathrm{Spf}(\mathcal{R})_{/B}$] and the identity element $\mathbf{0}$ is corresponding to the ideal $(T_1, \ldots, T_n)$, we call G a formal (Lie) group. If G is formal, then the map $\mathrm{Hom}_{CL_{/B}}(\mathcal{R}, R) \ni \phi \mapsto (\phi(T_1), \ldots, \phi(T_n))$ identifies $G_I(R)$ with the set $IR \times IR \times \cdots \times IR$ (n times) endowed with a formal group law.

Suppose that $G_{/B}$ is a formal Lie group. For any integer m, the endomorphism m_G sending x to mx of G induces a continuous algebra endomorphism

$m_G^* : \mathcal{R} \to \mathcal{R}$. By Lemma 4.84, it induces multiplication by m on the tangent space $\mathcal{T}_{G/B}$. Thus,

$$N_G(T_i) \equiv NT_i \mod (T_1, \dots, T_n)^2,$$

and $N_G(G_I(R)) = G_{I^2}(R)$ because $NI = 0$. Similarly, we have inductively $N_G(G_{I^a}(R)) = G_{I^{a+1}}(R)$. Thus, $N_G^{\nu}(G_I) = \{\mathbf{0}\}$. We get the next result.

Lemma 6.43 *We have $G_I \subset G[N^\nu]$ if G is a formal Lie group, where $G[m] = \mathrm{Ker}(m_G : G \to G)$ is the kernel of m_G.*

Theorem 6.44 (Drinfeld) *Let G and H be abelian $fppf$ sheaves over $ALG_{/B}$ and I be as above. Let G_0 and H_0 be the restriction of G and H to $CL_{/(B/I)}$. Suppose*

(i) *G is a p-divisible fppf sheaf;*
(ii) *H is formally smooth [so $H(R) \to H(R/J)$ is surjective for any nilpotent ideal J];*
(iii) *$\widehat{H}$ is a formal Lie group.*

Then

(1) *The modules $\mathrm{Hom}_{\mathcal{S}(B_{fppf})}(G, H)$ and $\mathrm{Hom}_{\mathcal{S}((B/I)_{fppf})}(G_0, H_0)$ are p- torsion-free, where the symbol "$\mathrm{Hom}_{\mathcal{S}(X_{fppf})}$" stands for the homomorphisms of abelian $fppf$ sheaves over $CL_{/X}$.*
(2) *The natural map*

$$\text{“reduction} \mod I\text{”} : \mathrm{Hom}_{\mathcal{S}(B_{fppf})}(G, H) \to \mathrm{Hom}_{\mathcal{S}((B/I)_{fppf})}(G_0, H_0)$$

is injective.
(3) *For any $f_0 \in \mathrm{Hom}_{\mathcal{S}((B/I)_{fppf})}(G_0, H_0)$, there exists a unique homomorphism $\Phi \in \mathrm{Hom}_{\mathcal{S}(B_{fppf})}(G, H)$ such that $\Phi \mod I = N^\nu f_0$. We write as in [K1] "$N^\nu f$" for Φ even if f exists only in $\mathrm{Hom}_{\mathcal{S}(B_{fppf})}(G, H) \otimes_{\mathbb{Z}} \mathbb{Q}$.*
(4) *In order for $f \in \mathrm{Hom}_{\mathcal{S}(B_{fppf})}(G, H)$, it is necessary and sufficient that "$N^\nu f$" kills $G[N^\nu]$.*

Proof. The first assertion follows from p-divisibility, because if $pf(x) = 0$ for all x, taking y with $py = x$, we find $f(x) = pf(y) = 0$ and hence $f = 0$.

We have an exact sequence: $0 \to H_I \to H \to H_0 \to 0$; so we have another exact sequence:

$$0 \to \mathrm{Hom}(G, H_I) \to \mathrm{Hom}(G, H) \xrightarrow{\mod I} \mathrm{Hom}(G, H_0) = \mathrm{Hom}(G_0, H_0),$$

which tells us the injectivity since H_I is killed by N^ν and $\mathrm{Hom}(G, H)$ is p-torsion-free.

To show (3), take $f_0 \in \mathrm{Hom}(G_0, H_0)$. By the surjectivity of $H(R) \to H_0(R/I)$, we can lift $f_0(x \mod I)$ to $y \in H(R)$. The class $y \mod \mathrm{Ker}(H \to H_0)$ is uniquely determined. Since $\mathrm{Ker}(H \to H_0)$ is killed by N^ν, for any

$x \in G(R)$, $N^\nu y$ is uniquely determined; so $x \mapsto N^\nu y$ induces a morphism of functors: "$N^\nu f$": $G(R) \to H(R)$. Then (3) follows from Lemma 4.17.

Assertion (4) is then obvious from the p-divisibility of G. The uniqueness of f follows from the p-torsion-freeness of $\mathrm{Hom}(G, H)$.

6.3.2 A Theorem of Serre–Tate

Let $\mathcal{A}_{/B}$ be the category of abelian schemes defined over B. We consider a category $DEF(B, B/I)$ of triples (A_0, D, ϵ), where A_0 is an abelian scheme over B/I, D is a p-divisible Barsotti–Tate group over B, and $\epsilon : D_0 \cong A_0[p^\infty]$, where $D_0 = D \otimes_{\mathrm{Spec}(B)} \mathrm{Spec}(B/I)$. A morphism

$$\phi : (A_0, D, \epsilon) \to (A'_0, D', \epsilon')$$

of $DEF(B, B/I)$ is a pair of morphisms $\phi_{BT} : D \to D'$ in $\mathrm{Hom}_{BT_{/B}}(D, D')$ and $\phi_{\mathcal{A}} : A_0 \to A'_0$ in $\mathrm{Hom}_{GSCH_{/B/I}}(A_0, A'_0)$, making the following diagram commutative:

$$\begin{array}{ccc} A_0[p^\infty] & \xrightarrow{\phi_{\mathcal{A}}} & A'_0[p^\infty] \\ {\scriptstyle \epsilon}\uparrow & & \uparrow{\scriptstyle \epsilon'} \\ D_0 & \xrightarrow[\phi_{BT}]{} & D'_0. \end{array}$$

We have a natural functor $\mathcal{A}_{/B} \to DEF(B, B/I)$ given by

$$A \mapsto (A_0 = (A \mod I) = A \otimes_B B/I, A[p^\infty], \iota)$$

for the canonical isomorphism $\iota : A[p^\infty] \otimes_B B/I \cong A_0[p^\infty]$.

Theorem 6.45 (Serre–Tate) *Let the notation be as above. The above functor, $\mathcal{A}_{/B} \to DEF(B, B/I)$, is a canonical equivalence of categories.*

Our proof is self-contained without quoting any outside reference if A is an elliptic curve. In general, we assume the existence of a lift of A_0 to an abelian scheme over B (a result of Grothendieck; see [CBT] Chap. 5).

Proof. We use the notation introduced in Theorem 6.44; thus, N is a p-power such that $NI = 0$, and $\nu > 0$ is an integer such that $I^{\nu+1} = 0$. As $A, A', A[p^\infty], A'[p^\infty], D, D'$ are p-divisible objects in $\mathcal{S}(B_{fppf})$, we can apply Theorem 6.44, taking G, H to be any of two groups as above.

We first show that the functor is fully faithful. We have a natural map of sets:

$$\mathrm{Hom}_{\mathcal{A}}(A, A') \to \mathrm{Hom}_{DEF}((A_0, A[p^\infty], \mathrm{id}_{A_0}), (A'_0, A'[p^\infty], \mathrm{id}_{A'_0}))$$

sending a morphism $\phi : A \to A'$ of group schemes to $(\phi|_{A_0}, \phi|_{A[p^\infty]})$, where A_0 and A'_0 are the fibers of A and A' over $\mathrm{Spec}(B/I)$, respectively. Applying Theorem 6.44 (2) to $G = A$ and $H = A'$, this map is injective, as an abelian variety is a p-divisible abelian fppf sheaf. This proof of injectivity only uses $f_0 : A_0 \to A_0'$ induced by f.

We now show surjectivity using $f|_{A[p^\infty]}$. Applying Theorem 6.44 (3) again to $G = A$ and $H = A'$, for any given $f_0 \in \mathrm{Hom}_{\mathcal{A}_{/B/I}}(A_0, A_0')$, we have a lift g:="$N^\nu f$" in $\mathrm{Hom}_{\mathcal{A}}(A, A')$ of $N^\nu f_0$. By Theorem 6.44 (4), we need to show that g kills $A[N^\nu]$. By the unicity of the lift, $N^\nu f = g$ over $A[p^\infty]$. Since N is a p-power, $A[N^\nu] \subset A[p^\infty]$ is killed by $N^\nu f$; hence, $A[N^\nu] \subset \mathrm{Ker}(g)$, as desired. Thus, $f = g/N^\nu$, and the map is surjective.

We now show that $(A_0, D, \epsilon) \in DEF(B, B/I)$ comes from an abelian scheme over B. When E_0 is an elliptic curve, by Corollary 6.17, there exists an elliptic curve $E_{/B}$ such that $E \times_S S_0 = E_0$ for $S = \mathrm{Spec}(B)$ and $S_0 = \mathrm{Spec}(B/I)$. In general, it is known that we can lift A_0 to an abelian scheme over B. This follows from the deformation theory of Grothendieck ([GIT] Sect. 6.3 and [CBT] Chap. V). This lifting is particularly easier if A_0 is ordinary and B/I is $\mathbb{F}$. Indeed, by a theorem of Tate's, A_0 has complex multiplication. By the theory of abelian varieties with complex multiplication, A_0 can be lifted to a unique abelian scheme over B with (the same) complex multiplication (the canonical lift), because isomorphism classes of such abelian varieties of the CM type correspond bijectively to lattices (up to scalar multiplication) in a CM field. In any case, we do not need to specifically lift A_0 to a CM abelian variety (just the existence of a lift is sufficient).

Write $X_{/B}$ for a lift of A_0; hence, $X \times_S S_0 \cong A_0$. Then we have an isomorphism $\alpha_0 : X_0[p^\infty] \to A_0[p^\infty]$. Then we have a unique lifting $g : X[p^\infty] \to D$ of $N^\nu \alpha_0$ [by Theorem 6.44 (4) applied to $G = X_0[p^\infty]$ and $H = D$]. Clearly, g is an isogeny, whose (quasi-)inverse is the lift of $N^\nu(\alpha_0)^{-1}$. Thus, $\mathrm{Ker}(g)$ is a finite flat group subscheme of X. The geometric quotient of X by a finite flat group subscheme exists as remarked in Sect. 4.3.7 (see also [ABV] Sect. 12 and [GME] Sect. 1.8). The quotient has to be a proper geometrically connected group scheme and hence is an abelian scheme over B. Thus, dividing X by the finite flat group scheme $\mathrm{Ker}(g)$, we get the desired $A_{/B} \in \mathcal{A}_{/B}$.

6.3.3 Group Structure on Deformation Space

Recall that $\mathbb{F}$ is the fixed algebraic closure of $\mathbb{F}_p$. Pick a complete local ring B with residue field $\mathbb{F}$. Recall the category $CL_{/B}$ of complete local B-algebras with residue field $\mathbb{F}$. Then $ART_{/B}$ is the full subcategory of $CL_{/B}$ made up of Artinian local B-algebras with residue field $\mathbb{F}$.

Let $(E_0, \omega_0)_{/\mathbb{F}}$ be a pair of an elliptic curve E over $\mathbb{F}$ and a generator ω_0 of $H^0(E_0, \Omega_{E_0/\mathbb{F}})$ over $\mathbb{F}$. Write $\pi_0 : E_0 \to \mathrm{Spec}(\mathbb{F})$ for the structure morphism with the zero section $\mathbf{0}$. Let $\widehat{E}_0$ be the formal completion of E_0 along the origin $\mathbf{0}$. Suppose that $(E_0, \omega_0)_{/\mathbb{F}}$ is an ordinary elliptic curve. By Lemma 4.72, we have $E_0[p^n] \cong \mu_{p^n} \times (\mathbb{Z}/p^n\mathbb{Z})$. This shows $\widehat{E}_0 = \mu_{p^\infty}$ as functors from $ART_{/\mathbb{F}}$ into AB (Sect. 4.4.3); so $\widehat{\eta} : \widehat{E}_0 \cong \widehat{\mathbb{G}}_m$ if we fix an isomorphism $\eta_p^\circ : \mu_{p^\infty} \cong E[p^\infty]^\circ$.

Write $TE[p^\infty]^{et} = \varprojlim_n E[p^n](\mathbb{F})$ for the Tate module of the maximal étale quotient of $E[p^\infty]$, which is a free $\mathbb{Z}_p$-module of rank 1 (if we ignore the Galois action). We consider the following deformation functor $\widehat{\mathcal{E}} : CL_{/B} \to SETS$:

$$\widehat{\mathcal{E}}_{E_0}(R) = \left[(E_{/R}, \iota_E) \middle| \, E_{/R} \text{ is an abelian scheme and } \iota_E : E \times_{S_R} S_{\mathbb{F}} \cong E_0\right]. \tag{6.3.1}$$

As before "[ ]" indicates the set "{ }/ ≅" of isomorphism classes of the objects inside the straight brackets, and $f : (E, \iota_E)_{/S} \cong (E', \iota_{E'})_{/S}$ if $f : E \to E'$ is an isomorphism of abelian schemes with the following commutative diagram:

$$\begin{array}{ccc} E \otimes_B \mathbb{F} & \xrightarrow{f_0} & E' \otimes_B \mathbb{F} \\ {\scriptstyle \iota_E}\downarrow\wr & & {\scriptstyle \iota_{E'}}\downarrow\wr \\ E_0 & = & E_0. \end{array}$$

Since the category of abelian schemes over S of relative dimension 1 is equivalent to the category of elliptic curves over S (Lemma 6.1), we have the following definition equivalent to (6.3.1):

$$\widehat{\mathcal{E}}_{E_0}(R) = \left[(E_{/R}, \iota_E) \middle| \, E_{/R} \text{ is an elliptic curve and } \iota_E : E \times_{S_R} S_{\mathbb{F}} \cong E_0\right]. \tag{6.3.2}$$

Lemma 6.46 *Assume that E_0 is ordinary. Take $(E, \iota_E)_{/B}$ in $\widehat{\mathcal{E}}_{E_0}(R)$ for an Artinian local B-algebra R with residue field $\mathbb{F}$. Let $\widehat{E}$ be the formal completion of E along the origin of E. Then there exists a canonical $\mathbb{Z}_p$-linear homomorphism $\varphi_{E/R} : T_pE[p^\infty](\mathbb{F}) \to \widehat{E}(R)$.*

The construction of $\varphi_{E/R}$ is given in the following proof.

Proof. Since R is an Artinian local ring with maximal ideal $\mathfrak{m}_R$, we find a positive integer ν so that $\mathfrak{m}_R^{\nu+1} = 0$, and hence $p^{\nu+1}\mathfrak{m}_R = 0$. Since $G = \widehat{E}$ is a formal Lie group over R, by Lemma 6.43, p^ν kills $G_{\mathfrak{m}_R} = \widehat{E}$. Since E is ordinary, the connected component $E[p^\infty]^\circ$ is a deformation of $E_0[p^\infty]^\circ \cong \mu_{p^\infty}$ over $\mathbb{F}$, and by Corollary 4.83, $E[p^\infty]^\circ \cong \mu_{p^\infty}$ over R. Thus, the Barsotti–Tate group $E[p^\infty]$ is ordinary, and we have the multiplicative/étale extension of Barsotti–Tate groups over $ART_{/B}$

$$0 \to E[p^\infty]^{mult} \to E[p^\infty] \to E[p^\infty]^{et} \to 0.$$

Since $E[p^\infty]^{mult}$ over $ART_{/R}$ is prorepresented by $\widehat{E}$ in $CL_{/R}$, we may rewrite the above exact sequence as an extension of fppf abelian sheaves:

$$0 \to \widehat{E} \to E[p^\infty] \to E[p^\infty]^{et} \to 0. \tag{6.3.3}$$

Taking the prime-to-p part into our account, we have the following exact sequence of abelian groups:

$$0 \to \widehat{E}(R) \to E(R) \to E(\mathbb{F}) \to 0.$$

Pick $x \in E(\mathbb{F})$, and lift it to $\tilde{x} \in E(R)$. If $\tilde{x}'$ is another lift, we have $\tilde{x} - \tilde{x}' \in \widehat{E}(R)$; hence, $p^\nu(\tilde{x} - \tilde{x}') = 0$. In other words, $p^\nu\tilde{x} \in \widehat{E}(R)$ is uniquely determined by x, getting a homomorphism "p^ν": $E(\mathbb{F})[p^\nu] = E[p^\nu](\mathbb{F}) = E[p^\nu]^{\mathrm{et}}(\mathbb{F}) \to \widehat{E}$. The above argument only uses the fact that $\mathfrak{m}_R^{\nu+1} = 0$; thus, it works for any integer $n \geq \nu$ in place of ν, getting a commutative diagram for any $n \geq \nu$:

$$\begin{array}{ccc} E[p^{n+1}]^{\mathrm{et}}(\mathbb{F}) & \xrightarrow{\text{“}p^{n+1}\text{”}} & \widehat{E}(R) \\ {\scriptstyle p_E}\downarrow & & \downarrow \| \\ E[p^n]^{\mathrm{et}}(\mathbb{F}) & \xrightarrow{\text{“}p^n\text{”}} & \widehat{E}(R). \end{array}$$

Passing to the limit, we get $\varphi_{E/R} : T_pE[p^n]^{\mathrm{et}}(\mathbb{F}) \to \widehat{E}(R)$, sending $x = \varprojlim_n x_n \in T_pE[p^\infty]^{\mathrm{et}}(\mathbb{F})$ to $\lim_{n\to\infty}$ "p^n"(x_n). Here $x_n \in E[p^n](\mathbb{F})$ with $p_E(x_{n+1}) = x_n$. □

Theorem 6.47 (Serre–Tate) *Let $T_pE_0 = \varprojlim_n E_0[p^n](\mathbb{F})$ for a given ordinary elliptic curve $E_{0/\mathbb{F}}$, and consider it a constant group scheme over R (for any given Artinian local B-algebra R with residue field $\mathbb{F}$). The functor $\widehat{\mathcal{E}}_{E_0}$ is representable by the formal torus*

$$R \mapsto \mathrm{Hom}_{\mathbb{Z}_p}(T_pE_0 \otimes_{\mathbb{Z}_p} T_pE_0, \widehat{\mathbb{G}}_m(R)) \cong \widehat{\mathbb{G}}_m(R),$$

and each deformation $(E_{/R}, i_E) \in \widehat{\mathcal{E}}_{E_0}(R)$ gives rise to the Serre–Tate coordinate $t_{E/R} = t_{E/R,E_0} : T_pE_0 \times T_pE_0 \to \widehat{\mathbb{G}}_m(R)$. The last isomorphism is induced by a chosen isomorphism $\eta^\circ : \mu_{p^\infty} \cong E_0[p^\infty]^{\mathrm{mult}}$ and its dual inverse $\eta^{\mathrm{et}} = (\eta^\circ)^* : \mathbb{Q}_p/\mathbb{Z}_p \cong E[p^\infty](\mathbb{F})$ [so $T_pE_0 \cong \mathbb{Z}_p$ by η^{et}].

Here we simply wrote $\mathrm{Hom}_{\mathbb{Z}_p}$ for the abelian group of $\mathbb{Z}_p$-linear maps; thus, strictly speaking, it is $\mathrm{Hom}_{MOD_{/\mathbb{Z}_p}}$.

Proof. This follows from Theorem 6.45 and the above lemma. An elliptic curve $E_{/R} \in \widehat{\mathcal{E}}_{E_0}(R)$ deforming $E_{0/\mathbb{F}}$ is determined by the Serre–Tate theorem by the extension (6.3.3), which gives rise to an extension class in $\mathrm{Ext}_{\mathcal{S}(R_{fppf})}(E[p^\infty]^{\mathrm{et}}, \widehat{E})$. By the Cartier self-duality of $E[N]$ (see Lemma 6.9),

$$\begin{aligned} \widehat{E}[p^n] = E[p^n]^{\mathrm{mult}} &= \mathrm{Hom}_{GSCH_{/R}}(E[p^n]^{\mathrm{et}}, \mu_{p^n}) \\ &\cong \mathrm{Hom}_{GSCH_{/R}}((T_pE[p^\infty]^{\mathrm{et}}) \otimes_{\mathbb{Z}} p^{-n}\mathbb{Z}/\mathbb{Z}, \mu_{p^n}) \\ &\cong \mathrm{Hom}_{GSCH_{/R}}(T_pE_0 \otimes_{\mathbb{Z}} p^{-n}\mathbb{Z}/\mathbb{Z}, \mu_{p^n}), \end{aligned}$$

where we consider $T_pE_0 \otimes_{\mathbb{Z}} p^{-n}\mathbb{Z}/\mathbb{Z}$ a constant group scheme over R whose fiber is given by $T_pE_0 \otimes_{\mathbb{Z}} p^{-n}\mathbb{Z}/\mathbb{Z}$. Passing to the limit, we get

$$\widehat{E}(R) = \mathrm{Hom}_{GSCH_{/R}}(T_pE_{0/R}, \widehat{\mathbb{G}}_{m/R}) = \mathrm{Hom}_{\mathbb{Z}_p}(T_pE_0, \widehat{\mathbb{G}}_m(R)). \quad (6.3.4)$$

The right-hand side is independent of the choice of $E_{/R} \in \widehat{\mathcal{E}}_{E_0}(R)$. Thus, the extension group $\mathrm{Ext}_{BT}(E[p^\infty]^{et}, E[p^\infty]^{mult})$ is isomorphic to

$$\mathrm{Ext}_{\mathcal{S}(R_{fppf})}\left(T_pE_0 \otimes \mathbb{Q}_p/\mathbb{Z}_p, \mathrm{Hom}_{\mathbb{Z}_p}(T_pE_0, \widehat{\mathbb{G}}_m)\right).$$

By Lemma 6.46, we can associate $(\widehat{E} \hookrightarrow E[p^\infty] \twoheadrightarrow T_pE_0 \otimes_{\mathbb{Z}} \mathbb{Q}_p/\mathbb{Z}_p)$ in $\mathrm{Ext}_{\mathcal{S}(R_{fppf})}(T_pE_0 \otimes \mathbb{Q}_p/\mathbb{Z}_p, \mathrm{Hom}_{\mathbb{Z}_p}(T_pE_0, \widehat{\mathbb{G}}_m))$ with a homomorphism $\varphi_{E/R}$: $T_pE_0 \to \widehat{E} = \mathrm{Hom}_{\mathbb{Z}_p}(T_pE_0, \widehat{\mathbb{G}}_m)$. In this way, we get a homomorphism

$$\tau : \mathrm{Ext}_{\mathcal{S}(R_{fppf})}\left(T_pE_0 \otimes \mathbb{Q}_p/\mathbb{Z}_p, \mathrm{Hom}_{\mathbb{Z}_p}(T_pE_0, \widehat{\mathbb{G}}_m)\right) \to \mathrm{Hom}_{\mathbb{Z}_p}\left(T_pE_0, \mathrm{Hom}_{\mathbb{Z}_p}(T_pE_0, \widehat{\mathbb{G}}_m)\right) \quad (6.3.5)$$

sending $E_{/R}$ to $\varphi_{E/R}$.

Starting with $\varphi \in \mathrm{Hom}_{\mathbb{Z}_p}\left(T_pE_0, \mathrm{Hom}_{\mathbb{Z}_p}(T_pE_0, \widehat{\mathbb{G}}_m)\right)$, we can push out the module extension

$$0 \to T_pE_0 \xrightarrow{i} (T_pE_0) \otimes_{\mathbb{Z}_p} \mathbb{Q}_p \to (T_pE_0) \otimes_{\mathbb{Z}_p} \mathbb{Q}_p/\mathbb{Z}_p \to 0$$

to the following extension in $\mathcal{S}(R_{fppf})$:

$$0 \to \mathrm{Hom}_{\mathbb{Z}_p}(T_pE_0, \widehat{\mathbb{G}}_m) \to E_\varphi[p^\infty] \to (T_pE_0) \otimes_{\mathbb{Z}_p} \mathbb{Q}_p/\mathbb{Z}_p \to 0. \quad (*)$$

Here $\mathrm{Hom}_{\mathbb{Z}_p}(T_pE_0, \widehat{\mathbb{G}}_m)$ as an abelian fppf sheaf sends an object $S_{/R}$ in $CL_{/R}$ to the module homomorphism group $\mathrm{Hom}_{\mathbb{Z}_p}(T_pE_0, \widehat{\mathbb{G}}_m(S))$, and

$$E_\varphi[p^\infty] = \mathrm{Coker}\left(i - \varphi : T_pE_0 \to ((T_pE_0) \otimes_{\mathbb{Z}_p} \mathbb{Q}_p) \times \mathrm{Hom}_{\mathbb{Z}_p}(T_pE_0, \widehat{\mathbb{G}}_m)\right), \quad (6.3.6)$$

as the cokernel exists in the abelian category $\mathcal{S}(R_{fppf})$ and by the exact sequence $(*)$, it is a Barsotti–Tate group. By construction, $\tau(E_\varphi[p^\infty]) = \varphi$, and we recover the original extension of Barsotti–Tate modules. Thus, τ is bijective. Thus, we have

$$\widehat{\mathcal{E}}(R) \overset{\text{Theorem 6.45}}{=} \mathrm{Ext}_{\mathcal{S}(R_{fppf})}\left(T_pE_0 \otimes \mathbb{Q}_p/\mathbb{Z}_p, \mathrm{Hom}_{\mathbb{Z}_p}(T_pE_0, \widehat{\mathbb{G}}_m)\right) \overset{\tau}{\cong} \mathrm{Hom}_{\mathbb{Z}_p}\left(T_pE_0, \mathrm{Hom}_{\mathbb{Z}_p}(T_pE_0, \widehat{\mathbb{G}}_m)\right).$$

However, we also have

$$\mathrm{Hom}_{\mathbb{Z}_p}\left(T_pE_0, \mathrm{Hom}_{\mathbb{Z}_p}(T_pE_0, \widehat{\mathbb{G}}_m)\right) \cong \mathrm{Hom}_{\mathbb{Z}_p}\left(T_pE_0 \otimes_{\mathbb{Z}_p} T_pE_0, \widehat{\mathbb{G}}_m\right),$$

sending $\varphi_{E/R} : T_pE_0 \to \mathrm{Hom}_{\mathbb{Z}_p}(T_pE_0, \widehat{\mathbb{G}}_m)$ to $t_{E/R} : T_pE_0 \otimes_{\mathbb{Z}_p} T_pE_0 \to \widehat{\mathbb{G}}_m$ given by $t_{E/R}(a \otimes b) = \varphi_{E/R}(a)(b)$. The identification $T_pE_0 \cong \mathbb{Z}_p$ comes from $(\eta^\circ)^* : E_0[p^\infty](\mathbb{F}) = E_0[p^\infty]^{et}(\mathbb{F}) \cong \mathbb{Q}_p/\mathbb{Z}_p$. Then this isomorphism induces

$$\mathrm{Hom}_{\mathbb{Z}_p}\left(T_pE_0 \otimes_{\mathbb{Z}_p} T_pE_0, \widehat{\mathbb{G}}_m\right) \cong \mathrm{Hom}_{\mathbb{Z}_p}(\mathbb{Z}_p, \widehat{\mathbb{G}}_m) \cong \widehat{\mathbb{G}}_m.$$

This finishes the proof.

We are going to give an explicit description of $t_{E/R}$. We recall some facts from the Cartier self-duality of $E[p^n]_{/R}$ (see Lemma 6.9). Let $\phi : E \to S$ and $\phi' : E' \to S$ be the structure morphisms for $S = \mathrm{Spec}(R)$ for an Artinian B-algebra R. Let $f : E \to E'_{/R}$ be an isogeny over R; hence, $\mathrm{Ker}(f)$ is a finite flat group scheme over R. Pick $x \in \mathrm{Ker}(f)$, and let $\mathcal{L} \in \mathrm{Ker}(f^*) \subset \mathrm{Pic}(E'_{/R})$ be the line bundle on E' with $\mathbf{0}_{E'}^*\mathcal{L} = \mathcal{O}_S$. Thus, $f^*\mathcal{L} = \mathcal{O}_S$ because $\mathcal{L} \in \mathrm{Ker}(f^*)$. Cover E' by open affine subschemes U_i so that $\mathcal{L}|_{U_i} = \phi'^{-1}_i\mathcal{O}_{U_i}$. Since $\mathbf{0}_{E'}^*\mathcal{L} = \mathcal{O}_S$, we may assume that $(\phi'_i/\phi'_j) \circ \mathbf{0}_{E'} = 1$. Since $f : E \to E'$ is finite, it is affine. Write $V_i = f^{-1}(U_i) = \mathrm{Spec}(\mathcal{O}_{U_i})$. Then $f^*\mathcal{L}|_{V_i} = \phi_i^{-1}\mathcal{O}_{V_i}$ with $\phi_i = \phi'_i \circ f$, and we have, regarding $x : S \to \mathrm{Ker}(f)$,

$$\frac{\phi_i \circ x}{\phi_j \circ x} = \frac{\phi'_i \circ f \circ x}{\phi'_j \circ f \circ x} = \frac{\phi_i \circ \mathbf{0}_{E'}}{\phi_j \circ \mathbf{0}_{E'}} = 1.$$

Thus, $\phi_i \circ x$ glue into a morphism $[x, \mathcal{L}] : S \to \mathbb{G}_m$, and we get the perfect pairing in Lemma 6.9

$$e_f : \mathrm{Ker}(f) \times \mathrm{Ker}(f^*) \to \mathbb{G}_m$$

inducing $\mathrm{Ker}(f)^* \cong \mathrm{Ker}(f^*)$.

We apply the above argument to $f = p_E^n : E \to E$, write the pairing as e_n, and record the following points already shown from the construction in the above proof:

Lemma 6.48 *We have*

(P1) $e_n(\alpha(x), y) = e_n(x, \alpha^*(y))$ *for* $\alpha \in \mathrm{End}(E_{/R})$.

(P2) *Fix the identification* $E_0[p^n]^{\mathrm{mult}} = \mu_{p^n} \subset E_0[p^n]$. *Then* e_n *induces an isomorphism of group schemes:* $E[p^n]^{\mathrm{mult}}_{/R} \cong \mathrm{Hom}_{GSCH_{/R}}(E_0[p^n]^{\mathrm{et}}, \mu_{p^n/R})$, *where we regard* $E_0[p^n]$ *as a constant group scheme over* R *with fiber* $E_0[p^n]$.

(P3) *Taking the limit of the above isomorphisms with respect to* n, *we find*

$$\widehat{E} \cong \mathrm{Hom}(T_pE[p^\infty]^{\mathrm{et}}, \widehat{\mathbb{G}}_m) \cong \mathrm{Hom}(T_pE_0, \widehat{\mathbb{G}}_m)$$

as formal groups, where we again regard T_pE_0 *as a constant group scheme over* R *with fiber* $T_pE_0 = \varprojlim_n E_0[p^n](\mathbb{F})$. *In particular,* $\widehat{E} \cong \widehat{\mathbb{G}}_m$.

We are now ready to describe the Serre–Tate coordinate $t_{E/R}$. In large part, this is a summary of the proof of Theorem 6.47. Since $R \in CL_{/B}$ is a projective limit of local B-algebras R_n with a nilpotent maximal ideal, we have $\widehat{\mathbb{G}}_m(R) = \varprojlim_n \widehat{\mathbb{G}}_m(R_n)$; hence, $t_{E/R} = \varprojlim_n t_{E_n/R_n}$ for $E_n = E \otimes_R R_n$.

Therefore, we may assume that R is a local Artinian B-algebra with a nilpotent maximal ideal $\mathfrak{m}_R$. Then, by Lemma 6.43, $\widehat{E}(R)$ is killed by p^ν for sufficiently large ν. Taking a lift $\tilde{x} \in E(R)$ of $x \in E(\mathbb{F}) = E_0(\mathbb{F})$ (such that $\tilde{x} \bmod \mathfrak{m}_R = x$), $\tilde{x}$ is determined modulo $\mathrm{Ker}(E(R) \to E(\mathbb{F})) = \widehat{E}(R)$, which is a subgroup of $E[p^n]$ if $n \geq \nu$. By the smoothness of $E_{/R}$, a lift $\tilde{x} \in E(R)$ of $x \in E(\mathbb{F}) = E_0(\mathbb{F})$ always exists. Thus, $p^n\tilde{x} \in E(R)$ is uniquely determined by $x \in E_0(\mathbb{F})$. If $x \in E[p^n](\mathbb{F}) = E_0[p^n](\mathbb{F})$, $p^n\tilde{x} =$"p^n"$x \in \widehat{E}(R)$ by definition, getting a homomorphism "p^n":$E_0[p^n](\mathbb{F}) \to \widehat{E}(R)$. We have an obvious commutative diagram (if $n \geq \nu$):

$$\begin{array}{ccccc} E_0[p^{n+1}](\mathbb{F}) & \xrightarrow{\sim} & E_0[p^{n+1}](\mathbb{F}) & \xrightarrow{\text{“}p^{n+1}\text{”}} & \widehat{E}(R) \\ {\scriptstyle p_{E_0}}\downarrow & & {\scriptstyle p_{E_0}}\downarrow & & \downarrow\| \\ E_0[p^n](\mathbb{F}) & \xrightarrow{\sim} & E_0[p^n](\mathbb{F}) & \xrightarrow{\text{“}p^{n}\text{”}} & \widehat{E}(R), \end{array}$$

which gives rise to a morphism $T_pE_0 \to \widehat{E}(R)$. Thus, the structure of the Barsotti–Tate group $E[p^\infty]_{/R}$ is uniquely determined by the extension class of the exact sequence of fppf sheaves:

$$0 \to \widehat{E}[p^\infty]_{/R} \to E[p^\infty]_{/R} \xrightarrow{\pi} E[p^\infty]^{\mathrm{et}}_{/R} \to 0. \tag{6.3.7}$$

Take $x = \varprojlim_n x_n \in T_pE_0 = T_pE[p^\infty]^{\mathrm{et}}$ with $x_n \in E_0[p^n](\mathbb{F})$. Lift x_n to $v_n \in E(R)$ so that $\pi(v_n) = x_n$. Then, for "p^n": $E_0[p^n](\mathbb{F}) \to \widehat{E}(R)$,

$$t_n(x) = \text{“}p^n\text{”}v_n \in \widehat{E}(R).$$

The value $t_n(x_n)$ becomes stationary if $n \geq n_0$, and taking the limit of $t_n(x_n)$ as $n \to \infty$, we get $t(x) \in \widehat{E}(R) \cong \mathrm{Hom}_{\mathbb{Z}_p}(T_pE_0, \widehat{\mathbb{G}}_m(R))$. Then we define $t_{E/R}(x,y) = t(x)(y)$. Taking the dual $(\eta_0^\circ)^* : E_0[p^\infty](\mathbb{F}) \cong \mathbb{Q}_p/\mathbb{Z}_p$ of the level structure $\eta_0^\circ : \mu_{p^\infty} \cong E_0[p^\infty]^{\mathrm{mult}}$, we identify T_pE_0 with $\mathbb{Z}_p$. Then we redefine $t_{E/R} \in \widehat{\mathbb{G}}_m(R)$ by $t_{E/R}(1)(1)$ for $1 \in \mathbb{Z}_p$.

Theorem 6.49 (Serre–Tate) *Fix a level structure $\eta_0^\circ : \mu_{p^\infty} \cong E_0[p^\infty]^{\mathrm{mult}}$. Then we have*

(1) *The functor $\widehat{\mathcal{E}}_{E_0}$ is represented by the formal scheme*

$$\mathrm{Hom}_{\mathbb{Z}_p}(T_pE_0 \otimes_{\mathbb{Z}_p} T_pE_0, \widehat{\mathbb{G}}_m) \cong \widehat{\mathbb{G}}_m$$

by $t_{E/R}(x,y) \mapsto t_{E/R} = t_{E/R,E_0} = t_{E/R,E_0}(1,1)$.

(2) *$t_{E/R}(x,y) = t_{E/R}(y,x)$ under Abel's identification: $E = \mathrm{Pic}^0_{E/R} = E^*$.*

(3) *Let $f_0 : E_{0/\mathbb{F}} \to E'_{0/\mathbb{F}}$ be a morphism of two ordinary elliptic curves with the dual map $f_0^* : E'_0 \to E_0$. Then f_0 is induced by a homomorphism $f : E_{/R} \to E'_{/R}$ for $E \in \widehat{\mathcal{E}}_{E_0}(R)$ and $E' \in \widehat{\mathcal{E}}_{E'_0}(R)$ if and only if $t_{E/R}(x, f_0^*(y)) = t_{E'/R}(f_0(x), y)$.*

Proof. The first assertion is an explicit version of Theorem 6.47 as described before the statement of the theorem.

We now prove (3). By Theorem 6.45, we have a lift $f : E \to E'$ of f_0 if and only if we have a lift $f|_{E[p^\infty]} : E[p^\infty] \to E'[p^\infty]$. By our push-out construction (6.3.6), having a lift $f|_{E[p^\infty]}$ is equivalent to the existence of a morphism ? of Barsotti–Tate groups making the following diagram commutative:

$$\begin{array}{ccccc} \mathrm{Hom}_{\mathbb{Z}_p}(T_pE_0, \widehat{\mathbb{G}}_m) & \xrightarrow{\hookrightarrow} & E[p^\infty] & \xrightarrow{\twoheadrightarrow} & T_pE_0 \otimes_{\mathbb{Z}_p} (\mathbb{Q}_p/\mathbb{Z}_p) \\ {\scriptstyle \circ f_0^*}\downarrow & & \downarrow {\scriptstyle ?} & & \downarrow {\scriptstyle f_0 \otimes 1} \\ \mathrm{Hom}_{\mathbb{Z}_p}(T_pE_0', \widehat{\mathbb{G}}_m) & \xrightarrow{\hookrightarrow} & E'[p^\infty] & \xrightarrow{\twoheadrightarrow} & T_pE_0' \otimes_{\mathbb{Z}_p} (\mathbb{Q}_p/\mathbb{Z}_p). \end{array}$$

There are two ways of constructing ?: We may take the vertical arrow $?_L : E[p^\infty] \to E'[p^\infty]$ made by the push-out of the left square, and we may take $?_R : E[p^\infty] \to E'[p^\infty]$ made by the pullback (a fiber product) of the right square. The morphism $f|_{E[p^\infty]}$ (making the above diagram commutative) exists if and only if $?_L = ?_R (= f|_{E[p^\infty]})$. By the construction of the Serre–Tate coordinate, the push-out $?_L$ is determined by an element $t_{E/R}(\cdot, f^*(\cdot))$ of

$$\begin{array}{ccc} \mathrm{Ext}(T_pE(\mathbb{F}) \otimes \frac{\mathbb{Q}_p}{\mathbb{Z}_p}, \mathrm{Hom}(T_pE'(\mathbb{F}), \widehat{\mathbb{G}}_m)) & = & \mathrm{Ext}(T_pE_0 \otimes \frac{\mathbb{Q}_p}{\mathbb{Z}_p}, \mathrm{Hom}(T_pE_0', \widehat{\mathbb{G}}_m)) \\ \wr\downarrow & & \wr\downarrow \\ \mathrm{Hom}(T_pE(\mathbb{F}) \otimes T_pE'(\mathbb{F}), \widehat{\mathbb{G}}_m) & = & \mathrm{Hom}(T_pE_0 \otimes T_pE_0', \widehat{\mathbb{G}}_m). \end{array}$$

Here $\mathrm{Ext} = \mathrm{Ext}_{\mathcal{S}(R_{\mathrm{fppf}})}$ and $\mathrm{Hom} = \mathrm{Hom}_{\mathbb{Z}_p}$. In the same way, the pullback $?_R$ is determined by an element $t_{E'/R}(f(\cdot), \cdot)$ of the same extension group. Thus, having $f_{E[p^\infty]}$ is equivalent to $t_{E/R}(\cdot, f^*(\cdot)) = t_{E'/R}(f(\cdot), \cdot)$.

Assertion (2) follows from the self-duality $E \cong E^* = \mathrm{Pic}^0_{E/S}$ (Albanese covariant functoriality) and the canonical identity $E \cong (E^*)^*$.

6.3.4 Computation of Deformation Coordinates

Let $S = \mathbb{G}_{m/\mathbb{Z}}$ and $\widehat{S} = \widehat{\mathbb{G}}_{m/W}$. Thus, $\widehat{S} = \mathrm{Spf}(\widehat{W[t,t^{-1}]})$ for $\widehat{W[t,t^{-1}]} = \varprojlim_n W[t,t^{-1}]/(t^n - 1)$, which is isomorphic to the formal power series ring $W[[T]]$ for $T = t - 1$. Fix a level structure $\eta_0^\circ : \mu_{p^\infty} \cong E_0[p^\infty]^{\mathrm{mult}}$. By Cartier self-duality, we get the identification $\eta_0^{\mathrm{et}} : \mathbb{Q}_p/\mathbb{Z}_p \cong E_0[p^\infty]^{\mathrm{et}}$. We consider the universal deformation $\mathbb{E}_{/\widehat{S}}$ of $E_{0/\mathbb{F}}$. Since $\mathbb{E}[p^n]^{\mathrm{et}}_{/\widehat{S}}$ is étale and $\widehat{S} = \mathrm{Spf}(\widehat{W[t,t^{-1}]})$, $\mathbb{E}[p^n]^{\mathrm{et}} = \mathrm{Spec}(A_n)$ for an étale algebra A_n over $\widehat{W[t,t^{-1}]}$. Since $\mathbb{E}[p^n]^{\mathrm{et}} \cong p^{-n}\mathbb{Z}/\mathbb{Z}$ by η_0^{et}, $A_n \otimes_{W[[T]]} \mathbb{F} = \mathbb{F}^{p^n} = \prod_{p^{-n}\mathbb{Z}/\mathbb{Z}} \mathbb{F}$ as $\mathbb{F}$-algebras. Then, by Hensel's lemma (see [BCM] III.4.6), $A_n \cong \prod_{p^{-n}\mathbb{Z}/\mathbb{Z}} W[[T]]$, so that we have a commutative diagram:

$$\begin{array}{ccc} A_n & \xrightarrow{\sim} & \prod_{p^{-n}\mathbb{Z}/\mathbb{Z}} W[[T]] \\ {\scriptstyle \mathrm{mod}\ p}\downarrow & & \downarrow {\scriptstyle \mathrm{mod}\ (p,T)} \\ \mathbb{F}^{p^n} & \xrightarrow[\eta_0^{\mathrm{et}}]{\sim} & \prod_{p^{-n}\mathbb{Z}/\mathbb{Z}} \mathbb{F}. \end{array}$$

Thus, η_0^{et} automatically induces $\eta^{\mathrm{et}} : \mathbb{Q}_p/\mathbb{Z}_p \cong \mathbb{E}[p^\infty]^{\mathrm{et}}$. Then, by duality, we get $\eta^\circ : \mu_{p^\infty} \cong \mathbb{E}[p^\infty]^\circ$. This shows the following result:

Lemma 6.50 *The new deformation functor*

$$\widehat{\mathcal{E}}_{Ig}(R) = \Big[(E, \iota_E, \eta_E^{\mathrm{et}}, \eta_E^\circ)_{/R} \big| (E_{/R}, \iota_E) \in \widehat{\mathcal{E}}_{E_0}(R)\Big]$$

is again represented by $\widehat{S}$ *and* $\widehat{\mathcal{E}}_{Ig} \cong \widehat{\mathcal{E}}_{E_0}$ *by the association*

$$(E, \iota_E, \eta_E^{\mathrm{et}}, \eta_E^\circ)_{/R} \mapsto (E, \iota_E)_{/R}.$$

Here $\eta_E^{\mathrm{et}} : \mathbb{Q}_p/\mathbb{Z}_p \cong E[p^\infty]^{\mathrm{et}}$ *(resp.,* $\eta_E^\circ : \mu_{p^\infty} \cong E[p^\infty]^\circ$*) is the level structure inducing* η_0^{et} *(resp.,* η_0°*) on the special fiber* E_0.

Thus, $\mathbb{E}$ has the level structure $\boldsymbol{\eta}^{\mathrm{et}} : \mathbb{Q}_p/\mathbb{Z}_p \cong \mathbb{E}[p^\infty]^{\mathrm{et}}$ and $\boldsymbol{\eta}^\circ : \mu_{p^\infty} \cong \mathbb{E}[p^\infty]^\circ$ induced by η_0^{et} and η_0°. Let $C \subset \mathbb{E}[p^n]$ be a lift of $\boldsymbol{\eta}^{\mathrm{et}}(p^{-n}\mathbb{Z}/\mathbb{Z})$. Since $0 \to \mu_{p^n/\widehat{S}} \to \mathbb{E}[p^n]_{/\widehat{S}} \to p^{-n}\mathbb{Z}/\mathbb{Z} \to 0$ is nonsplit, C is not defined over $\widehat{S}$ but is defined over an fppf extension of $\widehat{S}$. The following lemma, determining the minimal fppf extension carrying C, is taken from [Br1] Sect. 7, though our proof is slightly different from the one given there.

Proposition 6.51 *The above group scheme* C *cyclic of order* p^n *is étale over* $\mathcal{O}_{\widehat{S}}[\mu_{p^n}, t^{1/p^n}] = \widehat{W[t,t^{-1}]}[\mu_{p^n}, t^{1/p^n}]$. *Once we embed* W *in* $\overline{\mathbb{Q}}_p$, *there are* p^n *different choices of étale group scheme* C_ζ *cyclic of order* p^n *indexed by* $\zeta \in \mu_{p^n}(\overline{\mathbb{Q}}_p)$, *which is finite flat over* $\widehat{W[t,t^{-1}]}[\zeta \cdot t^{1/p^n}]$ *but cannot descend to a proper* $\widehat{W[t,t^{-1}]}$*-subalgebra of* $\widehat{W[t,t^{-1}]}[\zeta \cdot t^{1/p^n}]$.

In the following proof, we make explicit a canonical way of associating $\mu_{p^n} \ni \zeta \mapsto C_\zeta$.

Proof. Since C is étale cyclic of order p^n, its image covers $\mathbb{E}[p^n]^{\mathrm{et}}$; hence, we can lift $\boldsymbol{\eta}^{\mathrm{et}}(\frac{1}{p^n})$ to a generator γ of C. Note that taking a lift x_m of $\boldsymbol{\eta}^{\mathrm{et}}(\frac{1}{p^m})$ with $x_n = \gamma$, we have $t := \lim_{m\to\infty} \text{“}p^m\text{”}(x_m)$. In other words, γ gives rise to $\gamma = (\text{“}p^n\text{”})^{-1}(\lim_{n\to\infty} \text{“}p^m\text{”}(x_m))$, which is a p^nth root of t. Thus, $\gamma = t^{1/p^n}$ up to $\zeta \in \mu_{p^n}$, and hence, C is étale over $\widehat{W[t,t^{-1}]}[\zeta, t^{1/p^n}]$.

Another way of seeing this is to look into the following commutative diagram:

$$\begin{array}{ccccc} \mu_{p^\infty} & \overset{\hookrightarrow}{\longrightarrow} & \mathbb{E}[p^\infty] & \overset{\twoheadrightarrow}{\longrightarrow} & \mathbb{Q}_p/\mathbb{Z}_p \\ {\scriptstyle t\mapsto t^{p^n}}\big\downarrow & & {\scriptstyle F^n}\big\downarrow & & \big\downarrow\| \\ \mu_{p^\infty} & \underset{\hookrightarrow}{\longrightarrow} & \mathbb{E}[p^\infty]/\mathbb{E}[p^n]^\circ & \overset{\twoheadrightarrow}{\longrightarrow} & \mathbb{Q}_p/\mathbb{Z}_p. \end{array} \tag{6.3.8}$$

From construction, we get $t_{\mathbb{E}[p^\infty]/\mathbb{E}[p^n]^\circ_{/\widehat{\mathbb{G}}_m}} = t^{p^n}_{\mathbb{E}/\widehat{\mathbb{G}}_m}$. Strictly speaking, the coordinate $t_{\mathbb{E}[p^\infty]/\mathbb{E}[p^n]^\circ_{/\widehat{\mathbb{G}}_m}}$ is the Serre–Tate coordinate $t_{\mathbb{E}[p^\infty]/\mathbb{E}[p^n]^\circ_{/\widehat{\mathbb{G}}_m}, E_0^{(p^n)}}$ of the deformation space of $E_0^{(p^n)} = E_0/E_0[p^n]^\circ$ relative to the p^∞-level structure at the lower exact sequence of the above diagram. We have another commutative diagram, making a quotient of $\mathbb{E}[p^\infty]/\mathbb{E}[p^n]^\circ$ by the image of C in $\mathbb{E}[p^\infty]/\mathbb{E}[p^n]^\circ$ (which is isomorphic to the original $\mathbb{E}$):

$$\begin{array}{ccccc} \mu_{p^\infty} & \hookrightarrow & \mathbb{E}[p^\infty]/\mathbb{E}[p^n]^\circ & \twoheadrightarrow & \mathbb{Q}_p/\mathbb{Z}_p \\ \| \downarrow & & V^n \downarrow & & \downarrow x\mapsto p^n x \\ \mu_{p^\infty} & \hookrightarrow & \mathbb{E}[p^\infty] & \twoheadrightarrow & \mathbb{Q}_p/\mathbb{Z}_p. \end{array} \tag{6.3.9}$$

Since the endomorphism $[p] : x \mapsto px$ on $\mathbb{E}$ factors $[p] = F'V = V'F$ for $F' : \mathbb{E}^{(1/p)} \to \mathbb{E}$, $V : \mathbb{E} \to \mathbb{E}^{(1/p)}$, $F : \mathbb{E} \to \mathbb{E}^{(p)}$, and $V' : \mathbb{E}^{(p)} \to \mathbb{E}$, $C = \mathrm{Ker}(V^n)$ can only be defined over $\widehat{W[t,t^{-1}]}[t^{1/p^n}]$ for a choice of p^nth root t^{1/p^n} of t; hence, it is of the form $\zeta \cdot t^{1/p^n}$ once we make a choice t^{1/p^n}. By abusing the notation, we hereafter write V' and F' also as V and F if confusion is unlikely. Let $C_1 \subset \mathbb{E}[p^n]$ be the cyclic subgroup of order p^n étale over $\widehat{W[t,t^{-1}]}[\mu_{p^n}, t^{1/p^n}]$ such that its specialization at $t = 1$ is the unique étale factor of $\mathbb{E}_0[p^n]$ of the canonical lift $\mathbb{E}_0$ (having complex multiplication). We hereafter write t^{1/p^n} for the canonical choice of the p^nth root of t giving the Serre–Tate coordinate of $\mathbb{E}/C_1$.

We want to make ζ explicit. Note that E_0 has complex multiplication by an order R of the imaginary quadratic field M in which p splits: $pR = \mathfrak{p}\overline{\mathfrak{p}}$, $E_0[\mathfrak{p}] = \mathrm{Ker}(F)$, $E_0[\overline{\mathfrak{p}}] = \mathrm{Ker}(V)$ for the Frobenius map $F : E_0 \to E_0^{(p)}$ with dual $V : E_0 \to E_0^{(1/p)}$, and $E_0[p^\infty] = E_0[\mathfrak{p}^\infty] \oplus E_0[\overline{\mathfrak{p}}^\infty]$ (see Corollary 6.40). Thus, $E_0[\mathfrak{p}^\infty]$ is connected. Take the lift $\mathbb{E}_0$ of E_0 corresponding to the origin $t_{\mathbb{E}_0/W} = 1$. By our construction, $\mathbb{E}_0$ has complex multiplication by R. Then the exact sequence $\mathbb{E}_0[p^\infty]^\circ \hookrightarrow \mathbb{E}_0[p^\infty] \twoheadrightarrow \mathbb{E}[p^\infty]^{\mathrm{et}}$ splits over W: $\mathbb{E}_0[p^\infty] = \mathbb{E}_0[p^\infty]^\circ \oplus \mathbb{E}[p^\infty]^{\mathrm{et}}$ as $t_{\mathbb{E}_0/W} = 1$. Thus, we must have $\mathbb{E}[p^\infty]^\circ = \mathbb{E}_0[\mathfrak{p}^\infty] \cong \mu_{p^\infty}$, and we have a canonical projection $\pi^\circ : \mathbb{E}_0[p^\infty](\overline{\mathbb{Q}}_p) \twoheadrightarrow \mu_{p^\infty}(\overline{\mathbb{Q}}_p)$ (induced by $\boldsymbol{\eta}^\circ$). Let $C_0 = C \times_{\mathbb{E}} \mathbb{E}_0 \subset \mathbb{E}_0$. We then have $\pi^\circ(C_0)(\overline{\mathbb{Q}}_p) \subset \mu_{p^\infty}(\overline{\mathbb{Q}}_p)$, and this subgroup $\pi^\circ(C_0)(\overline{\mathbb{Q}}_p)$ is generated by $\zeta = \pi^\circ(\gamma)$. If C' is another étale cyclic subgroup with $\pi^\circ(C_0) = \pi^\circ(C'_0)$ for $C'_0 = C' \times_{\mathbb{E}} \mathbb{E}_0 \subset \mathbb{E}_0$, then C and C'_0 are étale coverings of $\widehat{W[t,t^{-1}]}[\zeta \cdot t^{1/p^n}]$ whose fibers at $t = 1$ coincide. This implies that $C = C'$ under the universal étale covering studied in Sect. 4.2.4. Write this group C as C_ζ, since the isomorphism class of C as an étale covering of $\widehat{\mathbb{G}}_m$ is determined by ζ. Identify $\widehat{W[[t,t^{-1}]}$ with $W[[T]]$ for $T = t-1$ and embed it into $\overline{\mathbb{Q}}_p[[T]]$. We then let $\mathrm{Gal}(\overline{\mathbb{Q}}_p/\mathbb{Q}_p)$ act on $\overline{\mathbb{Q}}_p[[T]]$ through coefficients: $(\sum_n a_n T^n)^\sigma = \sum_n a_n^\sigma T^n$ for $\sigma \in \mathrm{Gal}(\overline{\mathbb{Q}}_p/\mathbb{Q}_p)$. Then $\sigma(C_\zeta) = C_{\zeta^\sigma}$ and hence $W[\zeta]$ is the minimal ring over which C_ζ is defined.

Here is another look at Corollary 6.40:

Corollary 6.52 *Suppose that E_0 is defined over $\mathbb{F}_q$ for $q = p^f$. Let $\mathbb{E}_0$ be the elliptic curve over W with $t_{\mathbb{E}_0/W} = 1$. Then $\mathbb{E}_{0/W}$ has endomorphism ϕ lifting the qth-power Frobenius endomorphism F^f of $E_{0/\mathbb{F}_q}$, and hence $\mathbb{E}_0$ has* *complex multiplication by an imaginary quadratic field isomorphic to* $\mathbb{Q}[F^f] \subset \mathrm{End}_W(\mathbb{E}_0) \otimes_{\mathbb{Z}} \mathbb{Q}$.

Proof. By the above proof, $t_{\mathbb{E}_0/W} = 1$ implies the connected–étale sequence of $\mathbb{E}_0[p^\infty]$ splits; so $\mathbb{E}_0[p^\infty] = \mathbb{E}_0[p^\infty]^\circ \oplus \mathbb{E}_0[p^\infty]^{\mathrm{et}}$. Since $\mathbb{E}_0[p^\infty]^{\mathrm{et}}$ is étale over W, as a Galois module, we have $\mathbb{E}_0[p^\infty]^{\mathrm{et}} \cong E_0[p^\infty](\mathbb{F})$ on which F^f acts; hence, F^f lifts to $\phi^{\mathrm{et}} \in \mathrm{End}_W(\mathbb{E}_0[p^\infty]^{\mathrm{et}})$. By the Cartier self-duality of $\mathbb{E}_0[p^\infty]$, F^f lifts to $\phi^\circ \in \mathrm{End}_W(\mathbb{E}_0[p^\infty]^\circ)$; so we have the lift $\phi^{\mathrm{div}} = (\phi^\circ \oplus \phi^{\mathrm{et}}) \in \mathrm{End}_W(\mathbb{E}_0[p^\infty]) = \mathrm{End}_{DEF}((E_0, \mathbb{E}_0[p^\infty], \mathrm{id}_{E_0}))$ of F^f. Then by the equivalence of category in Theorem 6.45,

$$\mathrm{End}_{DEF}((E_0, \mathbb{E}_0[p^\infty], \mathrm{id}_{E_0})) = \mathrm{End}_W(\mathbb{E}_0),$$

which implies $\phi \in \mathrm{End}_W(\mathbb{E}_0)$ corresponding to ϕ^{div} lifts $F^f \in \mathrm{End}_{\mathbb{F}}(E_0)$. By Theorem 6.6, $\mathbb{Q}[\phi] \cong \mathbb{Q}[F^f]$ is an imaginary quadratic field, and also by $F^f + V^f \not\equiv 0 \mod p$, p splits in $\mathbb{Q}[F^f]$. This finishes the proof.

The elliptic curve $\mathbb{E}_{0/W}$ as in the above corollary is called the *canonical lift* of $E_{0/\mathbb{F}}$.

6.4 Elliptic Curves with Complex Multiplication

In the previous section, we started with an elliptic curve defined over a finite field $\mathbb{F}_q$ and lifted it to $W(\mathbb{F})$. By the existence of Frobenius endomorphism, any ordinary elliptic curve over a finite field has complex multiplication by an imaginary quadratic field M. In this section, as was traditionally done by Deuring and more generally by Shimura–Taniyama–Weil [ACM], we start with a CM elliptic curve over $\mathbb{C}$ and describe its canonical association to orders of M. We fix a prime p and an identification $\mathbb{C}_p \cong \mathbb{C}$ as fields compatible with our embeddings $\mathbb{C}_p \overset{i_p}{\hookleftarrow} \overline{\mathbb{Q}} \overset{i_\infty}{\hookrightarrow} \mathbb{C}$.

6.4.1 CM Elliptic Curves

Fix an imaginary quadratic field $M \subset \overline{\mathbb{Q}}$ in which the fixed prime p splits into $(p) = \mathfrak{p}\overline{\mathfrak{p}}$, where $i_p : \overline{\mathbb{Q}} \hookrightarrow \mathbb{C}_p$ indices the prime $\mathfrak{p}$; in other words, writing $\mathfrak{O}$ for the integer ring of M, $\mathfrak{p} = \left\{x \in \mathfrak{O} \middle| |x|_p < 1\right\}$.

For each lattice $\mathfrak{a} \subset M$ whose p-adic completion $\mathfrak{a}_p$ is identical to $\mathfrak{O}_p = \mathfrak{O} \otimes_{\mathbb{Z}} \mathbb{Z}_p$, we consider the complex torus $E(\mathfrak{a})(\mathbb{C}) = \mathbb{C}/\mathfrak{a}$, where we wrote $\mathfrak{a}$ for the lattice $\{i_\infty(a) \in \mathbb{C} | a \in \mathfrak{a}\}$ in $\mathbb{C}$. By a theorem in [ACM] 12.4, this complex torus is algebraizable to an elliptic curve $E(\mathfrak{a})$ having CM by M over a number field.

Let $\mathbb{F}$ be an algebraic closure of the finite field $\mathbb{F}_p$ of p-elements. We write W for the p-adically closed discrete valuation ring inside $\mathbb{C}_p$ unramified over $\mathbb{Z}_p$ with residue field $\mathbb{F}$. Thus, W is isomorphic to the ring $W(\mathbb{F})$ of Witt vectors with coefficients in $\mathbb{F}$. Let $\mathcal{W} = i_p^{-1}(W)$, which is a strict henselization of $\mathbb{Z}_{(p)} = \mathbb{Q} \cap \mathbb{Z}_p$. In general, we write W for a finite extension of $W(\mathbb{F})$ in $\mathbb{C}_p$, which is a complete discrete valuation ring. We suppose that p is unramified in $M/\mathbb{Q}$. Then the main theorem of complex multiplication ([IAT] Sects. 5.3–4 for elliptic curves and more generally [ACM] 18.6 for abelian varieties) combined with the criterion in [ST] of a good reduction over W tells us that $E(\mathfrak{a})$ is actually defined over the field of fractions $\mathcal{K}$ of $\mathcal{W}$ and extends to an abelian scheme over $\mathcal{W}$ [still written as $E(\mathfrak{a})_{/\mathcal{W}}$]. In the case of elliptic curves, it is easy to see a good reduction as follows. Since the isomorphism class over $\mathbb{C}$ of an elliptic curve with complex multiplication is determined by its $\mathbb{C}$-points, $E(\mathfrak{a})(\mathbb{C}) = \mathbb{C}/\mathfrak{a}$. Indeed, by the Weierstrass equation we described in Sect. 2.3.4, the Riemann surface $\mathbb{C}/\mathfrak{a}$ can be embedded into $\mathbf{P}^2(\mathbb{C})$, giving a unique elliptic curve $E(\mathfrak{a})_{/\mathbb{C}}$ up to isomorphisms. Multiplication by $\alpha \in M$ induces $E(\mathfrak{a})_{/\mathbb{C}} \cong E(\alpha\mathfrak{a})_{/\mathbb{C}}$. Thus, the ideal classes of the order

$$\mathfrak{O}(\mathfrak{a}) = \{\alpha \in M | \alpha\mathfrak{a} \subset \mathfrak{a}\}$$

are in bijection to the isomorphism classes of CM elliptic curves $E(\mathfrak{a})_{/\mathbb{C}}$ with $\mathrm{End}_{\mathbb{C}}(E(\mathfrak{a})) = \mathfrak{O}(\mathfrak{a})$. The ideals and its classes are at most countably many. For $\sigma \in \mathrm{Aut}(\mathbb{C})$, we have $\mathrm{End}_{\mathbb{C}}(E(\mathfrak{a})^{\sigma}) = \mathfrak{O}(\mathfrak{a})$. Thus, the J-invariant $J(E(\mathfrak{a}))$ can have at most countably many conjugates under field automorphisms of $\mathbb{C}$; hence, $J(E(\mathfrak{a}))$ is in a finite extension of $\mathbb{Q}$, and as we have seen in Corollary 2.41, $E(\mathfrak{a})$ is defined in a finite extension K of $\mathbb{Q}$. In particular, the ideal classes are finitely many (see [IAT] Exercise 4.12). Since the p-adic Tate module $V_p = T_pE(\mathfrak{a}) \otimes_{\mathbb{Z}} \mathbb{Q}$ has rank 1 over $M_p = M \otimes_{\mathbb{Q}} \mathbb{Q}_p$, the action of $\mathrm{Gal}(\overline{\mathbb{Q}}/K)$ on $T_pE(\mathfrak{a})$ produces a character $\varphi : \mathrm{Gal}(\overline{\mathbb{Q}}/K) \to M_p^{\times}$. Thus, the field of rationality of $E(\mathfrak{a})[p^{\infty}]$ over K is an abelian extension of K. Write $\widehat{K}$ for the p-adic closure of K in $\mathbb{C}_p$. If $E(\mathfrak{a})$ is a specialization of the Tate curve over any finite extension of $O = W \cap \widehat{K}$, its Tate period q is equal to $J^{-1}(E(\mathfrak{a}))$ up to units in O, and the field of rationality of $E(\mathfrak{a})[p^{\infty}]$ is a nonabelian infinite extension of $\widehat{K}$ by Corollary 2.56. This is impossible as it is an abelian extension of K. In other words, we have $|J^{-1}(E(\mathfrak{a}))|_p \geq 1$, and so $J(E(\mathfrak{a})) \in O$. Then $E(\mathfrak{a})$ extends to an elliptic curve over $\mathcal{W} \cap K$ by Corollary 6.32. Moreover, since p is arbitrary, $J(E(\mathfrak{a}))$ is an algebraic integer (this fact can be proven directly by an elementary way; see [IAT] Sect. 4.6). Thus, we get

Proposition 6.53 *An elliptic curve with complex multiplication extends to an elliptic curve with complex multiplication over the integer ring of a finite extension of* $\mathbb{Q}$.

In particular, all endomorphisms of $E(\mathfrak{a})_{/\mathcal{W}}$ are defined over $\mathcal{W}$. We write $\theta : M \hookrightarrow \mathrm{End}(E(\mathfrak{a})) \otimes_{\mathbb{Z}} \mathbb{Q}$ for the embedding of M taking $\alpha \in M$ to the complex multiplication by α on $E(\mathfrak{a})(\mathbb{C}) = \mathbb{C}/\mathfrak{a}$.

6.4.2 Order of an Imaginary Quadratic Field

Let $\mathfrak{O}(\mathfrak{a}) = \{\alpha \in \mathfrak{O} | \alpha\mathfrak{a} \subset \mathfrak{a}\}$. Then $\mathfrak{O}(\mathfrak{a})$ is an *order* of M, that is, a subring of the integer ring $\mathfrak{O}$ that spans M. The order $\mathfrak{O}(\mathfrak{a})$ is determined by its *conductor* ideal $\mathfrak{c} \subset \mathbb{Z}$, which is the annihilator of the $\mathbb{Z}$-module $\mathfrak{O}/\mathfrak{O}(\mathfrak{a})$. In other words, $\mathfrak{O}(\mathfrak{a}) = \mathfrak{O}_{\mathfrak{c}} := \mathbb{Z} + \mathfrak{c}\mathfrak{O}$.

Exercise 6.54 *If $\mathfrak{c}$ is the annihilator of $\mathfrak{O}/\mathfrak{O}(\mathfrak{a})$ in $\mathbb{Z}$, prove that $\mathfrak{O}(\mathfrak{a}) = \mathbb{Z} + \mathfrak{c}\mathfrak{O}$ and that any order is of the form $\mathbb{Z} + \mathfrak{c}\mathfrak{O}$ for an ideal $\mathfrak{c}$ of $\mathbb{Z}$.*

Fix a prime ideal $\mathfrak{l} = (l) \subset \mathbb{Z}$ and set $\mathfrak{O}_e = \mathbb{Z} + \mathfrak{l}^e\mathfrak{O}$ (the order of conductor $\mathfrak{l}^e$). The following three conditions for a fractional $\mathfrak{O}_e$-ideal $\mathfrak{a}$ are equivalent (cf. [IAT] Proposition 4.11 and (5.4.2) and [CRT] Theorem 11.3):

(I1) $\mathfrak{a}$ is $\mathfrak{O}_e$-projective;
(I2) $\mathfrak{a}$ is locally principal;
(I3) $\mathfrak{a}$ is a *proper $\mathfrak{O}_e$-ideal* [that is, $\mathfrak{O}_e = \mathfrak{O}(\mathfrak{a})$].

Here $\mathfrak{a}$ is *locally principal* means that for each prime ideal $\mathfrak{q}$ of $\mathfrak{O}_e$, the localization $\mathfrak{a}_{(\mathfrak{q})}$ is a principal ideal of the local ring $\mathfrak{O}_{e,\mathfrak{q}}$. An R-module M for a noetherian ring R is *R-projective* if R is isomorphic to a direct summand of an R-free module of finite rank. An $\mathfrak{O}_e$-ideal is $\mathfrak{O}_e$-projective if it is $\mathfrak{O}_e$-projective as an $\mathfrak{O}_e$-module. Thus, $Cl_e := \mathrm{Pic}(\mathfrak{O}_e)$ is the group of $\mathfrak{O}_e$-projective fractional ideals modulo principal ideals. The group Cl_e is finite and is called the ring class group modulo $\mathfrak{l}^e$.

6.4.3 Algebraic Differential on CM Elliptic Curves

Since $f_*\Omega_{E(\mathfrak{a})/\mathcal{W}} = \mathbf{0}^*_{E(\mathfrak{a})}\Omega_{E(\mathfrak{a})/\mathcal{W}}$ is locally free of rank 1 for the stricture morphism $f : E(\mathfrak{a}) \to \mathrm{Spec}(\mathcal{W})$, it is free of rank 1 as $\mathcal{W}$ is a discrete valuation ring. We choose and fix a differential $\omega = \omega(\mathfrak{O})$ on $E(\mathfrak{O})_{/\mathcal{W}}$ so that

$$H^0(E(\mathfrak{O}), \Omega_{E(\mathfrak{O})/\mathcal{W}}) = \mathcal{W}\omega. \qquad (\Omega 1)$$

As long as $E(\mathfrak{O} \cap \mathfrak{a})$ is an étale covering of both $E(\mathfrak{a})$ and $E(\mathfrak{O})$, $\omega(\mathfrak{O})$ induces a differential $\omega(\mathfrak{a})$ first by pullback to $E(\mathfrak{O} \cap \mathfrak{a})$ and then by pullback inverse from $E(\mathfrak{O} \cap \mathfrak{a})$ to $E(\mathfrak{a})$. Writing $\mathfrak{a}_p$ for the p-adic closure of $\mathfrak{a}$ in M_p, if $\mathfrak{a}_p = \mathfrak{O}_p$, this étaleness of the covering $E(\mathfrak{O}) \xleftarrow{\pi_L} E(\mathfrak{O} \cap \mathfrak{a}) \xrightarrow{\pi_R} E(\mathfrak{a})$ holds, as $\mathrm{Ker}(\pi_L)$ and $\mathrm{Ker}(\pi_R)$ have order prime to p and hence étale. So $\omega(\mathfrak{a})$ is well defined with $H^0(E(\mathfrak{a}), \Omega_{E(\mathfrak{a})/\mathcal{W}}) = \mathcal{W}\omega(\mathfrak{a})$.

Exercise 6.55 *Prove the following facts:*

1. *An isogeny $E \xrightarrow{\pi} E'$ of elliptic curves over $\mathcal{W}$ is étale if* $\deg(\pi)$ *is prime to p;*
2. *$E(\mathfrak{a} \cap \mathfrak{O})_{/\mathcal{W}} \to E(\mathfrak{a})_{/\mathcal{W}}$ is étale if $\mathfrak{a}_p = \mathfrak{O}_p$.*

There is one more case where $\omega(\mathfrak{a})$ is well defined. Take the étale subgroup C_ζ for a primitive p^nth root $\zeta \in \mu_{p^n}(\mathbb{C}_p)$ of unity as in Proposition 6.51 of the universal deformation $\mathbb{E}_{/W}$ of $E_0 = E(\mathfrak{O}) \otimes_{\mathcal{W}} \mathbb{F}$. Note that the canonical lift $\mathbb{E}_0$ of E_0 is exactly $E(\mathfrak{O}) \times_{\mathcal{W}} W$ by Corollary 6.52. Let $\overline{C}_\zeta = C_\zeta \cap E(\mathfrak{O}) = C_\zeta \times_{\mathbb{E}} E(\mathfrak{O})$, regarding $E(\mathfrak{O})$ as the fiber of $\mathbb{E}$ at $t = 1$ for the canonical coordinate t of the Serre–Tate deformation space. Then, for a suitable proper ideal $\mathfrak{a}_\zeta$ of $\mathfrak{O}_n$ for $l = p$ which we describe later, we have $E(\mathfrak{a}_\zeta) = E(\mathfrak{O})/\overline{C}_\zeta$ and $\pi : E(\mathfrak{O}) \to E(\mathfrak{a}_\zeta)$ is an étale covering defined over $W[\zeta]$ by Proposition 6.51, where p^n is the rank of the locally free group scheme $\overline{C}_\zeta$ over $W[\zeta]$. Thus, the pullback inverse $(\pi^*)^{-1} : \Omega_{E(\mathfrak{O})/\mathcal{W}[\zeta]} \to \Omega_{E(\mathfrak{a}_\zeta)/\mathcal{W}[\zeta]}$ is a surjective isomorphism, and for $\omega(\mathfrak{a}_\zeta) = (\pi^*)^{-1}\omega(\mathfrak{O})$, we thus have

$$H^0(E(\mathfrak{a}_\zeta), \Omega_{E(\mathfrak{a}_\zeta)/\mathcal{W}[\zeta]}) = \mathcal{W}[\zeta]\omega(\mathfrak{a}_\zeta) \ \text{ for } \zeta \in \mu_{p^n}(\mathbb{C}_p). \qquad (\Omega 2)$$

The differential $\omega(\mathfrak{a})$ is sometimes called the *Néron differential on* $E(\mathfrak{a})$.

6.4.4 CM Level Structure

Since $E(\mathfrak{O}_e)_{/\mathcal{W}}$ has a subgroup, $C(\mathfrak{O}_e) = \mathfrak{O}/(\mathbb{Z} + \mathfrak{l}^e\mathfrak{O}) \subset E(\mathfrak{O}_e)$ isomorphic étale locally to $\mathbb{Z}/\mathfrak{l}^e$. This subgroup $C(\mathfrak{O}_e)$ is sent by multiplication by a to $C(\mathfrak{a}) \subset E(\mathfrak{a})_{/\mathcal{W}}$, giving rise to a $\Gamma_0(\mathfrak{l}^e)$-level structure $C(\mathfrak{a})$ on $E(\mathfrak{a})$.

For $\mathfrak{O}_e$-ideals $\mathfrak{m}$ of $\mathfrak{b}$, we write $\mathfrak{b}_\mathfrak{m}$ for the $\mathfrak{m}$-adic completion $\mathfrak{b}_\mathfrak{m} = \varprojlim_n \mathfrak{b}/\mathfrak{m}^n\mathfrak{b}$. Abusing language, for an ideal $\mathfrak{f} \subset \mathbb{Z}$, we put $\mathfrak{b}_\mathfrak{f} = \varprojlim_n \mathfrak{b}/\mathfrak{f}^n\mathfrak{b}$. Further abusing notation, we write $\mathfrak{a}_\mathfrak{F}$ for an $\mathfrak{O}$-ideal $\mathfrak{F}$, the $\mathfrak{F}$-adic closure of $\mathfrak{a}$ in $M_\mathfrak{F} = \mathfrak{O}_\mathfrak{F} \otimes_\mathfrak{O} M$. Let $\mathfrak{a}$ be locally principal fractional ideal of $\mathfrak{O}_e$. We choose a local generator $a_\mathfrak{F}$ of $\mathfrak{a}_\mathfrak{F}$. Multiplication by $a_\mathfrak{F}$ induces an isomorphism $\mathfrak{O}_{e,\mathfrak{F}} \cong \mathfrak{a}_\mathfrak{F}$.

For our later use, we choose ideals $\mathfrak{F}$ and $\mathfrak{F}_c$ of $\mathfrak{O}$ so that $\mathfrak{F} \subset \mathfrak{F}_c^c$ and $\mathfrak{F}+\mathfrak{F}_c = \mathfrak{O}$. The product $\mathfrak{C} = \mathfrak{F}\mathfrak{F}_c$ will be the conductor of the Hecke character we study. We let $\mathfrak{f} = \mathfrak{F} \cap \mathbb{Z}$ and $\mathfrak{f}' = \mathfrak{F}_c \cap \mathbb{Z}$, and so $\mathfrak{f} \subset \mathfrak{f}'$. We define a level-$\mathfrak{f}^2$ structure on $E(\mathfrak{a})$: Supposing that $\mathfrak{a}$ is prime to $\mathfrak{f}\mathfrak{O}_e$, we have $\mathfrak{a}_\mathfrak{f} \cong \mathfrak{O}_\mathfrak{f} = \mathfrak{O}_\mathfrak{F} \times \mathfrak{O}_{\mathfrak{F}^c}$, which induces a canonical identification:

$$i_\mathfrak{f}(\mathfrak{a}) : \mathfrak{f}^{-1}/\mathbb{Z} = \mathfrak{f}^{-1}/\mathbb{Z} \cong \mathfrak{F}^{-1}/\mathfrak{O} \xrightarrow[x \mapsto a_\mathfrak{F} x]{\sim} \mathfrak{F}^{-1}\mathfrak{a}_\mathfrak{F}/\mathfrak{a}_\mathfrak{F} \subset E(\mathfrak{a})[\mathfrak{f}]. \qquad (6.4.1)$$

This level structure induces $i'_\mathfrak{f}(\mathfrak{a}) : E(\mathfrak{a})[\mathfrak{f}] \twoheadrightarrow \mathfrak{f}^{-1}/\mathbb{Z}$ by the Cartier self-duality of $E(\mathfrak{a})[\mathfrak{f}]$. In this way, we get many quintuples:

$$(E(\mathfrak{a}), i_\mathfrak{f}(\mathfrak{a}), i'_\mathfrak{f}(\mathfrak{a}), C(\mathfrak{a})[\mathfrak{l}], \omega(\mathfrak{a})) \in \mathcal{M}(\Gamma_0(\mathfrak{f}^2) \cap \Gamma_1(\mathfrak{f}) \cap \Gamma_0(\mathfrak{l}))(\mathcal{W}) \qquad (6.4.2)$$

as long as $\mathfrak{l}^e$ is prime to p, where $C(\mathfrak{a})[\mathfrak{l}] = \{x \in C(\mathfrak{a}) | \mathfrak{l}x = 0\}$. We often write $\phi_\mathfrak{f}$ for the pair $(i_\mathfrak{f}, i'_\mathfrak{f})$ of level-$\mathfrak{f}$ structures as above. A precise definition of the moduli scheme of Γ_0-type: $\mathcal{M}(\mathfrak{c}\mathfrak{l}^{-e}(\mathfrak{a}\mathfrak{a}^c)^{-1}; \mathfrak{f}^2, \Gamma_0(\mathfrak{l}))$ classifying such quintuples will be given in Sect. 8.1.3. The point $x(\mathfrak{a}) = (E(\mathfrak{a}), i_\mathfrak{f}(\mathfrak{a}), i'_\mathfrak{f}(\mathfrak{a}))$ of the moduli scheme $Y(\Gamma_0(\mathfrak{f}^2) \cap \Gamma_1(\mathfrak{f})) = \mathcal{M}(\Gamma_0(\mathfrak{f}^2) \cap \Gamma_1(\mathfrak{f}))/\mathbb{G}_m$ forgetting the differential $\omega(\mathfrak{a})$ is called a *CM point* $x(\mathfrak{a})$ associated with $E(\mathfrak{a})$.

6.4.5 Adelic CM Level Structure

In Corollaries 6.40 and 6.52, we started by choosing a p^∞-level structure $\eta^\circ : \mu_{p^\infty} \hookrightarrow E_0[p^\infty]$ of characteristic-0 CM elliptic curve. If we start with a CM elliptic curve over $\mathbb{C}$, we have a canonical p^∞-level structure. First, we suppose $\mathfrak{a}_p = \mathfrak{O}_p$ [and suppose that $(p) = \mathfrak{p}\overline{\mathfrak{p}}$ in $\mathfrak{O}$ as before]. Then $E(\mathfrak{a}) = \mathbb{C}/i_\infty(\mathfrak{a})$ by complex uniformization with respect to the Weierstrass equation for the differential $\omega_\infty(\mathfrak{a}) = \frac{dx}{y} = du$. By $\mathfrak{a}_p = \mathfrak{O}_p = \mathfrak{O}_{\overline{\mathfrak{p}}} \oplus \mathfrak{O}_{\mathfrak{p}}$ (with canonical isomorphism $\mathfrak{O}_{\mathfrak{p}} = \mathbb{Z}_p = \mathfrak{O}_{\overline{\mathfrak{p}}}$ induced by the inclusion $\mathbb{Z} \subset \mathfrak{O}$), we have

$$E(\mathfrak{a})(\mathbb{C}) = M_{\overline{\mathfrak{p}}}/\mathfrak{O}_{\overline{\mathfrak{p}}} \oplus M_{\mathfrak{p}}/\mathfrak{O}_{\mathfrak{p}}.$$

Here the ordering of writing—that is, $M_{\mathfrak{p}}/\mathfrak{O}_{\mathfrak{p}}$ comes at the right side of $M_{\overline{\mathfrak{p}}}/\mathfrak{O}_{\overline{\mathfrak{p}}}$—is important. In complex uniformization $E(\mathfrak{a})(\mathbb{C}) = \mathbb{C}/(\mathbb{Z}z + \mathbb{Z}) \overset{\exp}{\cong} \mathbb{C}^\times/q^{\mathbb{Z}}$ for a basis $(z, 1)$ of $\mathfrak{a}$ and $q = \exp(2\pi i z)$ by the exponential map $u \mapsto \mathbf{e}(u) := \exp(2\pi u)$, $\left(\bigcup_n \frac{1}{p^n}\mathbb{Z}\right) + \mathbb{Z}z/\mathbb{Z}z$ is sent to the canonical level structure $\mu_{p^\infty} \hookrightarrow E(\mathfrak{a})(\mathbb{C})$. This corresponds to $M_{\mathfrak{p}}/\mathfrak{O}_{\mathfrak{p}} \hookrightarrow E(\mathfrak{a})[p^\infty]$, and 1 in the basis $(z, 1)$ is put on the right end traditionally.

Identify $M_{\mathfrak{p}}/\mathfrak{O}_{\mathfrak{p}}$ with $\mathbb{Q}_p/\mathbb{Z}_p$ and then $\mathbb{Q}_p/\mathbb{Z}_p$ with $\bigcup_n p^{-n}\mathbb{Z}/\mathbb{Z} \xrightarrow{u \mapsto \mathbf{e}(u)} \mu_{p^\infty}(\mathbb{C})$. We get $\eta^\circ_\infty(\mathfrak{a}) : \mu_{p^\infty} \cong E(\mathfrak{a})[\mathfrak{p}^\infty]$ canonically given. By the Cartier dual inverse, we get $\eta^{\text{et}}_\infty(\mathfrak{a}) : \mathbb{Q}_p/\mathbb{Z}_p \cong E(\mathfrak{a})[\overline{\mathfrak{p}}^\infty]$. Since $\mathcal{W}$ is strictly Henselian, $\eta^{\text{et}}_\infty(\mathfrak{a})$ descends to the étale level structure $\eta^{\text{et}}(\mathfrak{a}) = \eta^{\text{et}}_p(\mathfrak{a})$ over $\mathcal{W}$, again by the Cartier dual inverse, we get the descent $\eta^\circ(\mathfrak{a}) = \eta^\circ_p(\mathfrak{a})$ to $\mathcal{W}$ of the connected level structure $\eta^\circ_\infty(\mathfrak{a})$. Since $E(\mathfrak{a})_{/W} = E(\mathfrak{a}) \times_{\mathcal{W}} W$ is the canonical lift of $E(\mathfrak{a}) \times_{\mathcal{W}} \mathbb{F}$, we recover a canonically chosen two-level structure

$$\eta^{ord}_p(\mathfrak{a}) = \eta^{\text{et}}_p(\mathfrak{a}) \times \eta^\circ_p(\mathfrak{a}) : \mathbb{Q}_p/\mathbb{Z}_p \times \mu_{p^\infty/\mathcal{W}} \cong E(\mathfrak{a})[p^\infty]_{/\mathcal{W}} \text{ if } \mathfrak{a}_p = \mathfrak{O}_p. \quad (6.4.3)$$

Outside p, for a prime $q \neq p$, $E(\mathfrak{a})[q^\infty]$ is an étale q-divisible group [as $q_{E(\mathfrak{a})} : x \mapsto q \cdot x$ is an étale map]. Since $\mathcal{W}$ is strictly Henselian, the identification $E(\mathfrak{a})[q^\infty](\mathbb{C}) = M_q/\mathfrak{a}_q$ over $\mathbb{C}$ extends uniquely to give $\eta^{\text{div}} : M^{(p\infty)}_{\mathbb{A}}/\widehat{\mathfrak{a}}^{(p)} \cong \bigcup_{N, p\nmid N} E(\mathfrak{a})[N]$, where $\widehat{\mathfrak{a}}^{(p)} = \varprojlim_{N, p\nmid N} \mathfrak{a}/N\mathfrak{a}$ and $M^{(p\infty)}_{\mathbb{A}} = \widehat{\mathfrak{a}}^{(p)} \otimes_{\mathbb{Z}} \mathbb{Q}$ (the prime-to-p finite adeles of M). By taking the Pontryagin dual, we have

$$\eta^{(p)} = \eta^{(p)}(\mathfrak{a}) : \widehat{\mathfrak{a}}^{(p)} \cong T^{(p)}E(\mathfrak{a}) = \varprojlim_{N, p\nmid N} E(\mathfrak{a})[N]. \quad (6.4.4)$$

Here $T^{(p)}E(\mathfrak{a})$ is the prime-to-p Tate module of $E(\mathfrak{a})$.

Replacing $E(\mathfrak{a})$ by isomorphic $E(\alpha\mathfrak{a})$ for $\alpha \in M^\times$, we may assume that $\mathfrak{a} \supset \mathfrak{O}$ with $\mathfrak{a}_p = \mathfrak{O}_p$. Then $C = \mathfrak{a}/\mathfrak{O} \subset E(\mathfrak{O})$, and $E(\mathfrak{a})(\mathbb{C}) = E(\mathfrak{O})/C$ over $\mathbb{C}$. Since $\mathfrak{a}_p = \mathfrak{O}_p$, C has order prime to p; hence, $\eta^{\text{et}}(\mathfrak{O})$ brings C into $E(\mathfrak{O})$. We now take the subgroup $\overline{C}_\zeta = C_\zeta \times_{\mathbb{E}} \mathbb{E}_0$ of $\mathbb{E}_0 = E(\mathfrak{O}) \times_{\mathcal{W}} W$ in the computation of C_ζ above Corollary 6.52. Then $\pi^\circ(\gamma) = \zeta$ in $\mathbb{E}_0/\overline{C}_\zeta$ for $\gamma = \eta^{\text{et}}(\frac{1}{p^n})$. This implies $\overline{C}_\zeta$ is a cyclic subgroup generated by $\gamma\zeta^{-1}$ over $\mathbb{C}$.

Exercise 6.56 *Explain why* $\pi^\circ(\gamma) = \zeta$ *implies that* $\overline{C}_\zeta$ *is a cyclic subgroup generated by* $\gamma\zeta^{-1}$.

Definition 6.57 *Taking* $\mathfrak{a}_\zeta \supset \mathfrak{O}$ *such that* $\overline{C}_\zeta \oplus C \cong \mathfrak{a}_\zeta/\mathfrak{O}$, *we may regard* $\overline{C}_\zeta \oplus C$ *as a finite flat group subscheme in* $E(\mathfrak{O})$ *defined over* $\mathcal{W}$. *Then we have* $E(\mathfrak{a}_\zeta) = E(\mathfrak{O})/(\overline{C}_\zeta \oplus C)$, *and the bottom lines of the commutative diagrams* (6.3.8) *and* (6.3.9) *induce the canonical level structure* $\eta^\circ(\mathfrak{a}_\zeta) : \mu_{p^\infty} \hookrightarrow E(\mathfrak{a}_\zeta)[p^\infty]$ *defined over* $\mathcal{W}[\mu_{p^n}]$. *Then, by the Cartier duality, we get the identification* $\eta^{et}(\mathfrak{a}_\zeta) : \mathbb{Q}_p/\mathbb{Z}_p \cong E(\mathfrak{a}_\zeta)[p^\infty]^{et}$.

In short, we get a *test object* $(E(\mathfrak{a}_\zeta), \eta^\circ(\mathfrak{a}_\zeta), \eta^{et}(\mathfrak{a}_\zeta), \eta^{(p)}(\mathfrak{a}_\zeta))_{/\mathcal{W}[\zeta]}$.

6.4.6 Zeta Function of CM Elliptic Curves

By Proposition 6.53, the elliptic curve $E(\mathfrak{a})$ is defined over a number field K and extends to an elliptic curve over the integer ring $\mathfrak{O}_K$. Write $\mathfrak{O}(\mathfrak{a})$ for the order of $\mathfrak{a}$. Since $\mathfrak{a}_p = \mathfrak{O}_p$, we have $\mathfrak{O}(\mathfrak{a})_p = \mathfrak{O}_p$. Thus, we may regard $\mathfrak{p}$ as a proper $\mathfrak{O}(\mathfrak{a})$-ideal [i.e., we just write $\mathfrak{p}$ for $\mathfrak{p} \cap \mathfrak{O}(\mathfrak{a})$]. By the proof of Corollary 6.52, for the Frobenius automorphism $\phi : W \to W$, $\mathbb{E}_0^\phi = \mathbb{E}_0/\mathbb{E}_0[p]^\circ$. Thus, we get $E(\mathfrak{a})^\phi = E(\mathfrak{p}^{-1}\mathfrak{a})$ as $E(\mathfrak{a})[\mathfrak{p}] = E(\mathfrak{a})[p]^\circ$. Consider $T_l(E(\mathfrak{a})) = \varprojlim_n E(\mathfrak{a})[l^n] \cong \mathfrak{a}_l = T_l(E(\mathfrak{p}^{-1}\mathfrak{a}))$ (for a prime $l \neq p$) as $\mathfrak{a}_l = (\mathfrak{p}^{-1}\mathfrak{a})_l$. Thus, the action of ϕ induces a $\mathbb{Z}_l$-linear automorphism

$$\phi_l : \mathfrak{a}_l \overset{\eta^{(p)}(\mathfrak{a})}{=} T_l(E(\mathfrak{a})) \to T_l(E(\mathfrak{p}^{-1}\mathfrak{a})) \overset{\eta^{(p)}(\mathfrak{p}^{-1}\mathfrak{a})}{=} \mathfrak{a}_l.$$

Consider $(\phi_l)_l$ an element in $(\widehat{\mathfrak{O}}^{(p)})^\times$.

Define $\rho^{(p)} : (M_{\mathbb{A}^{(p\infty)}})^\times \to G(\mathbb{A}^{(p\infty)})$ by $\alpha\eta^{(p)}(\mathfrak{a}) = \eta^{(p)}(\mathfrak{a}) \circ \rho^{(p)}(\alpha)$. Since the action of ϕ is multiplication by the unit $(\phi_l)_l$ in $\widehat{\mathfrak{O}}^{(p)} \cap M$, if the level group Γ contains $\rho^{(p)}((\phi_l)_l)$, we can conclude

$$x(\mathfrak{a})^\phi = x(\mathfrak{p}^{-1}\mathfrak{a}) \text{ as a point of the modular curve } Y_\Gamma \text{ of prime to } p \text{ level } \Gamma. \tag{6.4.5}$$

This is a special case of Shimura's reciprocity law we state later as Proposition 7.24 (see [IAT] Chap. 6 and [EDM] Sect. A5).

As described in [IAT] Sect. 7.8 (and [EDM] Sect. 14), the Hasse–Weil zeta function $L(s, E(\mathfrak{a})) := L_1(s, E(\mathfrak{a}))$ (defined in (3.1.5)) can be identified with a Hecke L-function $L(s,\psi)L(s,\overline{\psi})$ of its field of definition K for an everywhere unramified Hecke ideal character ψ : {fractional ideal of K} $\to M^\times$. Here are some details of this fact: Since complex multiplication by $\alpha \in M$ on $E(\mathfrak{a})$ induces multiplication by α on $\Omega_{E(\mathfrak{a})/K}$ as is clear from the complex uniformization $E(\mathfrak{a})(\mathbb{C}) = \mathbb{C}/\mathfrak{a}$, K has to contain M. Write $\Phi = \phi^f$ for the Frobenius element in $\mathrm{Gal}(\overline{\mathbb{Q}}/K)$ for a prime $\mathfrak{P}|p$ in $\mathfrak{O}_K$. Thus, the action of Φ on $T_l(E(\mathfrak{a})) = \varprojlim_n E(\mathfrak{a})[l^n] \cong \mathfrak{a}_l$ (for a prime $l \neq p$) gives rise to a M-linear endomorphism of $T(E(\mathfrak{a}))$. Since Φ satisfies $X^2 - (F+V)X + p^f = 0$ for the Frobenius map F and its dual V (relative to Φ), we find that on $T_l(E(\mathfrak{a}))$, considering it an $\mathfrak{O}(\mathfrak{a})_l$-module by complex multiplication, the action of Φ is the multiplication by $F \in M$. Thus, $\psi(\mathfrak{P}) = F \in M^\times$ for the prime $\mathfrak{P}$ of K. This tells us that ψ has values in $M^\times$, and by (6.4.5),

$$x(N_{K/M}(\mathfrak{P})^{-1}\mathfrak{a}) = x(\mathfrak{a})^{\Phi} \quad \text{on } Y_\Gamma.$$

Here, again, we regard $N_{K/M}(\mathfrak{P})$ as a proper prime ideal of the order $\mathfrak{O}(\mathfrak{a}) = \{\alpha \in M | \alpha\mathfrak{a} \subset \mathfrak{a}\}$ [in other words, for any prime ideal $\mathfrak{L}$ of $\mathfrak{O}$ with $\mathfrak{O}(\mathfrak{a})_{\mathfrak{L}} = \mathfrak{O}_{\mathfrak{L}}$, we consider $\mathfrak{L}$ to be a proper prime ideal of $\mathfrak{O}(\mathfrak{a})$ by taking the intersection $\mathfrak{L} \cap \mathfrak{O}(\mathfrak{a})$]. In particular, $N_{K/M}(\mathfrak{P})$ is a principal $\mathfrak{O}(\mathfrak{a})$-ideal generated by $\psi(\mathfrak{P})$.

Exercise 6.58 *Give a detailed proof of the fact that $N_{K/M}(\mathfrak{P})$ is a principal $\mathfrak{O}(\mathfrak{a})$-ideal generated by $\psi(\mathfrak{P})$.*

7

Modular Curves as Shimura Variety

Modular varieties classifying abelian varieties with a level structure, given endomorphism subring and polarization, by making the level structure finer and finer, give rise to a tower of varieties. This tower has a strikingly big automorphism group, which can be described by an adele group of an algebraic group. Such a tower with a big adelic automorphism group is called a Shimura variety (of PEL type; see [PAF]) because Shimura was the first to realize that the symmetry given by well-described automorphisms is essential in studying arithmetic problems and automorphic forms defined over the variety (see [Sh4] and his earlier papers quoted there, including [Sh2]). We interpret here the modular curves and their tower as a prototypical example of Shimura varieties. Since each elliptic curve has a canonical polarization coming from Abel's map $E_{/S} \cong \mathrm{Pic}^0(E_{/S})$, in this book, we forget about polarization.

7.1 Shimura Curve

A Shimura variety is a fundamental object made out of a pair of a reductive group $G_{/\mathbb{Q}}$ and its hermitian symmetric domain $G(\mathbb{R})/C(\mathbb{R})Z(\mathbb{R})$ for a maximal compact subgroup $C(\mathbb{R})$ and the center $Z(\mathbb{R})$. Such a pair produces a tower of varieties defined over a canonical field of definition characterized by an explicit description of special points under Galois action (local reciprocity law). The tower has an unusually large automorphism group (global reciprocity law) coming from symmetry of the action of the adele group $G(\mathbb{A})$. The interaction of local and global reciprocity laws is a part of the axioms formulated by Deligne (cf. [D5] and [PAF] Sect. 7.2.1) after Shimura's invention of geometric reciprocity laws to characterize such pairs. We study here the simplest example when $G = GL(2)_{/\mathbb{Q}}$, which is deeply related to the classical theory of elliptic modular forms (this was originally treated in [IAT] Chaps. 6 and 7, but our treatment is slightly more ring theoretic). We put more emphasis on p-integral models (for the prime p we fixed) of the Shimura curve

H. Hida, *Elliptic Curves and Arithmetic Invariants*, Springer Monographs in Mathematics, DOI 10.1007/978-1-4614-6657-4_7,

of prime-to-p level and the Igusa tower over it, as we seek applications to integrality of L-values and p-integral modular forms defined over the tower.

7.1.1 Elliptic Modular Function Fields

We recall the elliptic modular function field in [IAT] Chap. 6 (see also [EDM] Sects. 14 and A5). We consider a field $\mathcal{K}$ given by $\bigcup_{p\nmid N}\mathbb{Q}(\mu_N)$ inside $\overline{\mathbb{Q}}$; hence, $\mathcal{K}\overset{i_p}{\hookrightarrow}\overline{\mathbb{Q}}_p$. Take a p-adic place $\mathfrak{P}$ of $\mathcal{K}$ given by i_p and write $\mathcal{W}\subset\mathcal{K}$ for the *discrete* valuation ring of $\mathfrak{P}$. We thus have a continuous embedding $i_p:\mathcal{W}\hookrightarrow\overline{\mathbb{Q}}_p$, and for the maximal ideal $\mathfrak{m}$ of $\mathcal{W}$, $\mathbb{F}=\mathcal{W}/\mathfrak{m}$ is an algebraic closure of $\mathbb{F}_p$. Put $G_1=SL(2)_{/\mathbb{Q}}$ as an algebraic group and $\mathcal{G}_I:=GL_1(\mathbb{Z}_p)\times SL_2(\mathbb{A}^{(p\infty)})$. We embed diagonally $\mathbb{Z}_{(p)}$-points of the standard diagonal torus $T\subset G_1$ (of the upper triangular Borel subgroup $P=\left\{\left(\begin{smallmatrix}a&*\\0&a^{-1}\end{smallmatrix}\right)\middle|a\in GL(1)\right\}$ of $SL(2)$) into $\mathcal{G}_I$ so that $\left(\begin{smallmatrix}a&0\\0&a^{-1}\end{smallmatrix}\right)$ is sent to $a\in GL_1(\mathbb{Z}_p)$ at p and $\left(\begin{smallmatrix}a&0\\0&a^{-1}\end{smallmatrix}\right)\in SL_2(\mathbb{Q}_\ell)$ at all primes $\ell\nmid p$.

Definition 7.1 *Decompose* $P=T\ltimes U$ *for* $U(A)=\{\left(\begin{smallmatrix}1&u\\0&1\end{smallmatrix}\right)|u\in A\}$ *and* $T(A)=\left\{\left(\begin{smallmatrix}a&0\\0&a^{-1}\end{smallmatrix}\right)\middle|a\in A^\times\right\}$.

We consider the modular curve $Y(N)_{/\mathbb{Z}[\frac{1}{N}]}$ for an integer N prime to p that classifies pairs $(E,\phi_N)_{/A}$, where E is an elliptic over A and $\phi_N:(\mathbb{Z}/N\mathbb{Z})^2\cong E[N]=\mathrm{Ker}(N:E\to E)$ is an isomorphism of finite flat group schemes over R. The level structure ϕ_N specifies a primitive root of unity $\zeta_N\in\mu_N$ via the Weil pairing $\zeta_N:=\langle\phi_N(1,0),\phi_N(0,1)\rangle$. Over $\mathbb{C}$, we identify ζ_N with the canonical generator $\zeta_N=\exp\left(\frac{2\pi\sqrt{-1}}{N}\right)$ of $\mu_N(\mathbb{C})$. Thus, $Y(N)$ has a scheme structure over $\mathbb{Z}[\mu_N,\frac{1}{N}]$, but we may consider it defined over $\mathbb{Z}[\frac{1}{N}]$, composing the morphism $\mathrm{Spec}(\mathbb{Z}[\mu_N,\frac{1}{N}])\to\mathrm{Spec}(\mathbb{Z}[\frac{1}{N}])$. We write $Y(N)_{\zeta_N}$ when we consider $Y(N)$ to be defined over $\mathbb{Z}[\zeta_N,\frac{1}{N}]$. Then it classifies $(E,\phi_N)_{/A}$, with $\langle\phi_N(1,0),\phi(0,1)\rangle=\zeta_N$ for a specific primitive Nth root $\zeta_N\in A$. If we think of it as a scheme over $\mathbb{Z}[\frac{1}{N}]$, we just write it as $Y(N)$. For the fixed prime p, let

$$\widehat{\Gamma}_0(p^m)_p=\left\{\left(\begin{smallmatrix}a&b\\c&d\end{smallmatrix}\right)\in G(\mathbb{Z}_p)\middle|c\in p^m\mathbb{Z}_p\right\},$$
$$\widehat{\Gamma}_1(p^m)_p=\left\{\left(\begin{smallmatrix}a&b\\c&d\end{smallmatrix}\right)\in\widehat{\Gamma}_0(p^m)_p\middle|d-1\in p^m\mathbb{Z}_p\right\}.$$

Definition 7.2 *We consider the level structure* ϕ_p *of type* $\Gamma=\widehat{\Gamma}_?(p^m)_p\subset G(\mathbb{Z}_p)$ *(*$?=0,1$*) given as follows:* ϕ_p *is a subgroup étale locally isomorphic to* μ_{p^m} *if* $\Gamma=\widehat{\Gamma}_0(p^m)_p$*, and* $\phi_p:\mu_{p^m}\hookrightarrow E[p^m]$ *(a closed immersion of finite flat group schemes) if* $\Gamma=\widehat{\Gamma}_1(p^m)_p$.

We studied in Sect. 6.2 the moduli scheme $Y(N,\Gamma)_{/B}$ over a base ring $B_{/\mathbb{Z}_{(p)}}$, which classifies triples $(E,\phi_N,\phi_p)_{/R}$ for a level-Γ structure ϕ_p (at p) over B-algebras R. The ring B there is a discrete valuation ring over $\mathbb{Z}_{(p)}$ (or actually any algebra over smaller $\mathbb{Z}[\frac{1}{N}, \mu_N]$). This type of level structure has already been studied in earlier sections (cf. Sect. 2.2), and over $\mathbb{Z}_{(p)}[\mu_{p^n}]$ (or over a complete discrete valuation ring $W \supset \mathbb{Z}_{(p)}[\mu_{p^n}]$), we know the representability (sometimes as a coarse modulus) of the functor

$$R \mapsto \mathcal{E}_{\Gamma(N),\Gamma}(R) := \left[(E, \phi_N, \phi_p)_{/R}\right]$$

by a modular curve $Y(N, \Gamma)$. In Sect. 6.2.12, this functor was denoted by $\mathcal{E}^{ord}_\alpha$ if $\Gamma = \Gamma_1(p^\alpha)_p$ (to emphasize the exclusion of supersingular elliptic curves of characteristic p). We consider the p-integral Shimura curve (of prime-to-p level)

$$Sh^{(p)}_{/\mathbb{Z}_{(p)}} = \varprojlim_{p \nmid N} Y(N)_{/\mathbb{Z}_{(p)}}$$

[regarding these schemes as $\mathbb{Z}_{(p)}$-schemes].

If we regard the open/affine curves $Y(N)$ [resp., $Y(N, \Gamma)$] as a scheme over $\mathrm{Spec}(\mathbb{Q}[\mu_N])$, for N prime to p, $Y(N, \Gamma)_{/\mathbb{Q}[\mu_N]}$ is geometrically irreducible. Inverting all integers, that is, over $\mathbb{Q}$, we denote

$$Sh_{\Gamma/\mathbb{Q}} = \varprojlim_{p \nmid N} Y(N, \Gamma)_{/\mathbb{Q}}, \quad Sh_{/\mathbb{Q}} = \varprojlim_{N} Y(N)_{/\mathbb{Q}} \text{ (including } N \text{ divisible by } p)$$

(regarding these schemes as $\mathbb{Q}$-schemes). We will later see that $Sh_{\Gamma/\mathbb{Q}} = Sh/\Gamma$ as a geometric quotient as in [GME] Sects. 1.8.3 and 3.3. Write $\mu_N^\times \subset \mu_N$ for the set of all primitive Nth roots of unity, and let $\widehat{\mathbb{Z}}(1)^\times = \varprojlim_N \mu_N^\times$. For each $\zeta = (\zeta_N)_N \in \widehat{\mathbb{Z}}(1)^\times$, let

$$Y(N, \Gamma)_{\zeta/B} = Y(N, \Gamma)_{\zeta_N} \times_{\mathbb{Z}_{(p)}[\mu_N]} B, \; Y(N)_{\zeta/B} = Y(N)_{\zeta_N} \times_{\mathbb{Z}_{(p)}[\mu_N]} B.$$

In this chapter, the base ring B is often $\mathcal{W}$, $\mathbb{F}$, or $\mathcal{K}$. The proschemes with Zariski topology as discussed in Sect. 5.1.4

$$Y_{\Gamma,\zeta/\mathcal{K}} = \varprojlim_{p \nmid N} Y(N, \Gamma)_{\zeta_N/\mathcal{K}} \quad \text{and} \quad Y^{(p)}_{\zeta/\mathcal{W}} = \varprojlim_{p \nmid N} Y(N)_{\zeta/\mathcal{W}} \tag{7.1.1}$$

give geometrically irreducible components of $Sh_{\Gamma/\mathbb{Q}} \times_{\mathbb{Q}} \mathcal{K}$ (indexed by ζ). Once we identify $\widehat{\mathbb{Z}}(1) = \varprojlim_N \mu_N(\mathbb{C})$ with $\widehat{\mathbb{Z}}$, for $\zeta = (\zeta_N = \exp(2\pi i/N))_N$ corresponding to $1 \in \widehat{\mathbb{Z}}$,, the component $Y^{(p)}_{\zeta/\mathcal{W}}$ gives the *neutral component* of $Sh^{(p)}_{/\mathbb{Z}_{(p)}} \times_{\mathbb{Z}_{(p)}} \mathcal{W}$ [i.e., the components containing the image of the identity of $G(\mathbb{A}^{(\infty)})$ in the expression of Lemma 7.8]. We simply write the neutral component as $Y^{(p)}_{/\mathcal{W}}$. If convenient, we write $Sh_{G(\mathbb{Z}_p)/\mathbb{Z}_{(p)}}$ for $Sh^{(p)}_{/\mathbb{Z}_{(p)}}$ (abusing the notation). Deligne–Kottwitz's interpretation is functorial, relating $Sh_{\Gamma/B}$ to the covariant functor

$$ALG_{/B} \ni R \mapsto \frac{\{(E, \eta^{(p)} : (\mathbb{A}^{(p\infty)})^2 \cong V^{(p)}(E), \phi_p)_{/R}\}}{\text{prime-to-}p\text{ isogenies}} \in SETS.$$

In place of isomorphisms, an equivalence relation on the category $\{(E, \eta^{(p)}, \phi_p)_{/R}\}$ is given by isogenies of degree prime to p; hence, $(E, \eta^{(p)}, \phi_p)_{/R}$ is equivalent to $(E', \eta'^{(p)}, \phi'_p)_{/R}$ if we have an invertible isogeny $\varphi \in \mathrm{Hom}_{ELL}(E, E') \otimes_{\mathbb{Z}} \mathbb{Z}_{(p)}$ such that $\eta'^{(p)} = \varphi \circ \eta^{(p)}$ and $\phi'_p = \varphi \circ \phi_p$. To understand Shimura's symmetry by the adele group $G(\mathbb{A})$ in a geometric way, it is imperative to replace "isomorphisms" by "isogenies." In short, we have

Theorem 7.3

$$Sh_\Gamma(R) \cong \frac{\{(E, \eta^{(p)} : (\mathbb{A}^{(p\infty)})^2 \cong V^{(p)}(E), \phi_p)_{/R}\}}{\textit{prime-to-p isogenies}}, \tag{7.1.2}$$

where R runs over $\mathbb{Z}_{(p)}$-algebras if $\Gamma = G(\mathbb{Z}_p)$ and B-algebras ($B = \mathbb{F}$ or $\mathbb{Q}$) if $\Gamma = \widehat{\Gamma}_?(p^m)_p$ with $m > 0$ ($? = 0, 1$), $V^{(p)}(E) = \mathbb{A}^{(p\infty)} \otimes_{\mathbb{Z}} \varprojlim_{p\nmid N} E[N]$.

We will prove this theorem after some preparation in Sect. 7.1.4.

We write $X^{(p)}_{/B}$ (resp., $X_{\Gamma/B}$) for $B = \mathcal{K}, \mathcal{W}$ for the normalization of $\mathbf{P}^1(J)$ in $Y^{(p)}_{/B}$ (resp., $Y_{\Gamma/B}$), which gives a smooth compactification of $Y^{(p)}_{/\mathcal{K}}$ and $Y_{\Gamma/\mathcal{K}}$, as seen in Sect. 6.2.7. Write $\mathfrak{F}_\Gamma$ for the function field $\mathcal{K}(X_\Gamma)$ and $\mathfrak{F}^{(p)}$ for $\mathcal{K}(X^{(p)})$ (the *arithmetic modular function fields*). Since $f \in \mathfrak{F}_\Gamma$ is a ratio $\frac{h}{g}$ of modular forms h, g of the same weight, as in (6.2.2), it has Laurent q-expansion $f = \sum_\xi c(\xi, f) q^\xi$ in the Laurent series ring $\mathcal{K}((q)) = \mathcal{K}[[q]][q^{-1}]$.

By the above theorem, $(u, g) \in \mathcal{G}_I$ [$u \in GL_1(\mathbb{Z}_p)$ and $g \in SL_2(\mathbb{A}^{(p\infty)})$] acts on Sh_Γ by

$$(E, \eta^{(p)}, \phi_p) \mapsto (E, \eta^{(p)} \circ g, \phi_p \circ u).$$

The action of produces an embedding

$$\tau : \mathcal{G}_I/\{\pm 1\} \hookrightarrow \mathrm{Aut}(\mathfrak{F}_{\widehat{\Gamma}_1(p^\infty)_p}/\mathcal{K}) = \mathrm{Aut}(Y_{\widehat{\Gamma}_1(p^\infty)_p/\mathcal{K}}).$$

The action of $\tau(z, g)$ on $\mathfrak{F}_\Gamma$ is on the left and has the following property (shown by Shimura; e.g., [IAT] Theorem 6.23 or [PAF] Theorem 4.14). Simply write $a := (a, \left(\begin{smallmatrix} a & 0 \\ 0 & a^{-1} \end{smallmatrix}\right)) \in GL_1(\mathbb{Z}_{(p)}) \times SL_2(\mathbb{Z}_{(p)}) \subset \mathcal{G}_I$ [the matrix $\left(\begin{smallmatrix} a & 0 \\ 0 & a^{-1} \end{smallmatrix}\right)$ in $T(\mathbb{Z}_{(p)})$ being diagonally embedded in $SL_2(\mathbb{A}^{(\infty)})$]. Then, for $f \in \mathcal{F}_\Gamma$, we have

$$\tau(a)(f)(z) = f(a^{-2}z); \tag{7.1.3}$$

hence, we have $\tau(\alpha)(f)(z) = f(\alpha^{-1}(z))$ for $\alpha = \left(\begin{smallmatrix} a & 0 \\ 0 & a^{-1} \end{smallmatrix}\right)$. This formula is valid for general $\alpha \in GL_2(\mathbb{Z}_{(p)})$ if $f \in \mathfrak{F}^{(p)}$ (so our "τ" is the inverse of Shimura's "τ" in [IAT] Chap. 6 and "$[\cdot]$" in [EDM] Sect. 14.2).

Definition 7.4 *We define a valuation $v_\Gamma : \mathfrak{F}_\Gamma \to \mathbb{Z} \cup \{\infty\}$ by*

$$v_\Gamma(f) = \inf_\xi \mathrm{ord}_p(c(\xi, f)) \quad \textit{for} \quad f = \sum_\xi c(\xi, f) q^\xi \in \mathfrak{F}_\Gamma.$$

We write v_m for v_Γ if $\Gamma = \widehat{\Gamma}_1(p^m)_p$.

The valuation v_Γ is well defined because f can be written as $\frac{h}{g}$ for $h, g \in G_k(\Gamma(N) \cap \Gamma, \overline{\mathbb{Q}})$ for sufficiently large k (ampleness of $\underline{\omega}$) and as shown in [IAT] Chap. 8, $G_k(\Gamma(N) \cap \Gamma, \overline{\mathbb{Q}})$ is generated by modular forms with integer q-expansion coefficients; thus, $\mathrm{ord}_p(c(\xi, f))$ is bounded below over ξ. Thus, the valuation $v_0 : \mathfrak{F}^{(p)} \to \mathbb{Z} \cup \{\infty\}$ has a standard **unramified** extension $v_\Gamma : \mathfrak{F}_\Gamma \twoheadrightarrow \mathbb{Z} \cup \{\infty\}$. Here are some easy facts:

Lemma 7.5 *Let the notation be as above.*

1. *If $a \in GL_1(\mathbb{Z}_{(p)}) \cong T(\mathbb{Z}_{(p)})$, then*

$$c(\xi, \tau(a)(f)) = c(a^2\xi, f),$$

and the diagonally embedded $T(\mathbb{Z}_{(p)}) \subset \mathcal{G}_I$ preserves the valuation v_Γ.

2. *The vertical divisor $X^{(p)}_{/\mathbb{F}} := X^{(p)}_{/\mathcal{W}} \otimes_{\mathcal{W}} \mathbb{F}$ of $X^{(p)}_{/\mathcal{W}}$ is a prime divisor (geometrically irreducible) and gives rise to a* **unique** *valuation of $\mathfrak{F}^{(p)}$, whose explicit form is given by the valuation v_0.*

Proof. The first assertion follows directly from (7.1.3). By the existence of a smooth compactification $X^{(p)}_{/\mathcal{W}}$ of $Y^{(p)}$ over $\mathcal{W}$, Zariski's connectedness theorem (Theorem 5.7) tells us that $Y^{(p)}_{/\mathbb{F}} = Y^{(p)} \times_{\mathcal{W}} \mathbb{F}$ is irreducible. Thus, the vertical Weil prime divisor $X^{(p)}_{/\mathbb{F}}$ on the smooth arithmetic surface $X^{(p)}_{/\mathcal{W}}$ gives rise to a unique valuation. To see this, writing $Y^{(p)}_{/\mathcal{W}} = \mathrm{Spec}(\mathcal{R})$ (because it is affine), we have $\mathcal{R} = \bigcup_{p \nmid N} \mathcal{R}_N$ for the noetherian normal domain $\mathcal{R}_N := \mathcal{R}_0(\Gamma(N), \mathcal{W})$ of Krull dimension 2 with $Y(N) = \mathrm{Spec}(\mathcal{R}_0(\Gamma(N), \mathcal{W}))$. So the principal ideal $(p) = p\mathcal{R}$ is the prime ideal corresponding to $Y^{(p)}_{/\mathbb{F}} = \mathrm{Spec}(\mathcal{R}/(p))$. Since $Y(N)$ is finite (integral) over $\mathbf{P}^1(J) - \{\infty\}$ over the discrete valuation ring $\mathcal{W}$, the ring $\mathcal{R}_N$ has Krull dimension 2. The tower $Y^{(p)}_{/\mathcal{W}}$ is an étale cover over $Y(N)_{\zeta_N/\mathcal{W}}$ for $N \geq 3$ prime to p, as $Y(N)_{\zeta_N}$ is a fine modulus of $\mathcal{E}_{\Gamma(N),\zeta_N}$. Thus, the localization $\mathcal{R}_{N,(p)}$ of $\mathcal{R}_N$ at (p) is a normal noetherian local domain of dimension 1, and it is a discrete valuation ring (e.g., [CRT] Theorem 11.2). Again by the étaleness of $Y^{(p)}/Y(N)$, $\mathcal{R}_{(p)} = \bigcup_N \mathcal{R}_{N,(p)}$ is a discrete valuation ring. By the irreducibility of $X^{(p)}_{/\mathbb{F}}$, a $\mathcal{W}$-integral modular form of level away from p vanishes on the divisor $X^{(p)}_{/\mathbb{F}}$ if and only if its q-expansion vanishes modulo p by Lemma 6.36 and the remark following the lemma. Thus, the valuation v_0 is the one corresponding to the valuation ring $\mathcal{R}_{(p)}$ and to the vertical prime divisor $X^{(p)}_{/\mathbb{F}} \subset X^{(p)}_{/\mathcal{W}}$.

7.1.2 Complex Points of Shimura Curve

We study complex points of the Shimura curve. As already remarked, over $\mathbb{C}$, we identify ζ_N with the canonical generator $\zeta_N = \exp\left(\frac{2\pi\sqrt{-1}}{N}\right)$ of $\mu_N(\mathbb{C})$.

Write $G_{/\mathbb{Z}}$ for the affine algebraic group $GL(2) = \mathrm{Spec}(\mathbb{Z}[a,b,c,d,\frac{1}{ad-bc}])$ defined over $\mathrm{Spec}(\mathbb{Z})$ and $Z_{/\mathbb{Z}}$ for its center [see Example 4.58 (5)].

Write the homogeneous coordinate of $\mathbf{P}^1(\mathbb{C})$ as column vectors $\left[\begin{smallmatrix} x \\ y \end{smallmatrix}\right]$. Then $g = \left(\begin{smallmatrix} a & b \\ c & d \end{smallmatrix}\right) \in G(\mathbb{C})$ acts on $\mathbf{P}^1(\mathbb{C})$ by the usual matrix multiplication:

$$\left[\begin{smallmatrix} x \\ y \end{smallmatrix}\right] \mapsto g\left[\begin{smallmatrix} x \\ y \end{smallmatrix}\right] = \left[\begin{smallmatrix} ax+by \\ cx+dy \end{smallmatrix}\right].$$

Since the center $Z(\mathbb{C})$ acts trivially, this action factors through $PGL_2(\mathbb{C}) = G(\mathbb{C})/Z(\mathbb{C})$. Since $G(\mathbb{R})$ leaves $\mathbf{P}^1(\mathbb{R})$ stable, $G(\mathbb{R})$ acts on $\mathbf{P}^1(\mathbb{C}) \setminus \mathbf{P}^1(\mathbb{R}) \cong \mathbb{C} - \mathbb{R}$, where we identify $\mathbf{P}^1(\mathbb{C}) \setminus \mathbf{P}^1(\mathbb{R})$ with $\mathbb{C} - \mathbb{R}$, sending the homogeneous coordinate $\left[\begin{smallmatrix} x \\ y \end{smallmatrix}\right]$ to the inhomogeneous one $z = \frac{x}{y} \in \mathbb{C}$.

Let $\mathfrak{H} = \{z \in \mathbb{C} | i(\overline{z} - z) > 0\}$ (the upper half-complex plane); hence, $\mathbb{C} - \mathbb{R} = \mathfrak{H} \sqcup \overline{\mathfrak{H}}$ for the lower half-plane $\overline{\mathfrak{H}}$ (the complex conjugate of $\mathfrak{H}$). Thus, $G(\mathbb{R})$ acts on $\mathbf{P}^1(\mathbb{C}) \setminus \mathbf{P}^1(\mathbb{R}) \cong \mathfrak{H} \sqcup \overline{\mathfrak{H}}$. Tracking down the above isomorphisms, for $g = \left(\begin{smallmatrix} a & b \\ c & d \end{smallmatrix}\right) \in G(\mathbb{R})$ and $z \in \mathbb{C} - \mathbb{R}$, we have $g(z) = \frac{az+b}{cz+d}$. Put

$$O(\mathbb{R}) = \{x \in G(\mathbb{R}) | x \cdot {}^t x = 1\} \quad \text{and} \quad SO_2(\mathbb{R}) = \{x \in O(\mathbb{R}) | \det(x) = 1\}.$$

Here $x \mapsto {}^t x$ indicate the "transpose" of matrices x. These are closed subgroups of $G(\mathbb{R})$ under Euclidean topology. We start with an easy exercise:

Exercise 7.6 *Prove*

1. $G(\mathbb{R})^+ := \{g \in G(\mathbb{R}) | g(\mathfrak{H}) \subset \mathfrak{H}\}$ *is equal to* $\{g \in G(\mathbb{R}) | \det(g) > 0\}$.
2. *If* $g(z) = z$ *for all* $z \in \mathfrak{H}$, *then* $g \in Z(\mathbb{R})$.
3. $SO_2(\mathbb{R}) = \left\{r(\theta) := \left(\begin{smallmatrix} \cos\theta & \sin\theta \\ -\sin\theta & \cos\theta \end{smallmatrix}\right) \middle| \theta \in [0, 2\pi)\right\}$.

Lemma 7.7 *The map:* $g \mapsto g(i)$ $(i = \sqrt{-1} \in \mathfrak{H})$ *induces an isomorphism of real analytic manifolds*

$$G(\mathbb{R})/Z(\mathbb{R})O(\mathbb{R}) \cong G(\mathbb{R})^+/Z(\mathbb{R})SO_2(\mathbb{R}) \cong SL_2(\mathbb{R})/SO_2(\mathbb{R}) \cong \mathfrak{H}.$$

In particular, $Z(\mathbb{R})SO_2(\mathbb{R}) = \{g \in G(\mathbb{R})^+ | g(i) = i\}$.

Proof. By the exercise above, $SO_2(\mathbb{R}) = \{r(\theta) | 0 \le \theta \le 2\pi\}$, and $G(\mathbb{R})^+$ acts on $\mathfrak{H}$ by $z \mapsto g(z)$. By a simple computation, we confirm $r(\theta)(i) = i$ for $\theta \in \mathbb{R}$. If $g = \left(\begin{smallmatrix} a & b \\ c & d \end{smallmatrix}\right)$ with $\det(x) = 1$ satisfies $g(i) = i$ $(i = \sqrt{-1})$, we have $ai + b = (ci + d)i = -c + di$; hence, $a = d$ and $b = -c$, resulting in $\det(g) = a^2 + b^2 = 1$, and so we can find $\theta \in [0, 2\pi)$ such that $g = r(\theta)$. Thus, $SO_2(\mathbb{R})$ is the stabilizer of $i \in \mathfrak{H}$ in $SL_2(\mathbb{R})$. We thus have $SL_2(\mathbb{R})/SO_2(\mathbb{R}) \hookrightarrow \mathfrak{H}$. The map is surjective because $g_z(i) = z = x + iy$ for $g_z = \sqrt{y}^{-1}\left(\begin{smallmatrix} y & x \\ 0 & 1 \end{smallmatrix}\right)$. By this fact, for $g \in SL_2(\mathbb{R})$, the expression $g = g_z r(\theta)$ uniquely determines $z \in \mathfrak{H}$ and $\theta \in [0, 2\pi)$; thus, we can take $z = x + iy \in \mathfrak{H}$ and θ as a coordinate system (x, y, θ) of $SL_2(\mathbb{R})$, and by this choice, the set-theoretic morphism $g \mapsto g(i)$ is an isomorphism $SL_2(\mathbb{R})/SO_2(\mathbb{R}) \cong \mathfrak{H}$ of real analytic manifolds. Since $G(\mathbb{R})^+ = Z(\mathbb{R})SL_2(\mathbb{R})$ and $G(\mathbb{R}) = G(\mathbb{R})^+O(\mathbb{R})$, we get

$$G(\mathbb{R})/Z(\mathbb{R})O(\mathbb{R}) \cong G(\mathbb{R})^+/Z(\mathbb{R})SO_2(\mathbb{R}) \cong SL_2(\mathbb{R})/SO_2(\mathbb{R})$$

as $Z(\mathbb{R})$ acts trivially on $\mathfrak{H}$.

Here is an adelic version of the above lemma.

Lemma 7.8 *For the adele ring* $\mathbb{A}$, *the procurve* Y *has a canonical expression* $Y(\mathbb{C}) = G(\mathbb{Q})\backslash G(\mathbb{A})/Z(\mathbb{Q})Z(\mathbb{R})SO_2(\mathbb{R}) = G(\mathbb{Q})\backslash((\mathfrak{H}\sqcup\overline{\mathfrak{H}})\times G(\mathbb{A}^{(\infty)}))$. *Writing the class in* $Y(\mathbb{C})$ *of* $(z,g) \in (\mathfrak{H}\sqcup\overline{\mathfrak{H}}) \times G(\mathbb{A}^{(\infty)})$ *as* $[z,g]$, *we have*

$$[z, \gamma^{(\infty)}g] = [\gamma_\infty^{-1}(z), g] \quad \textit{for } \gamma \in G(\mathbb{Q}).$$

Proof. Recall the topology of $GL_n(\mathbb{A})$. We give a profinite (compact) topology on $GL_n(\widehat{\mathbb{Z}}) = \varprojlim_N GL_n(\mathbb{Z}/N\mathbb{Z})$. Thus, $U \subset GL_n(\widehat{\mathbb{Z}})$ is an open neighborhood of 1 if and only if $U \supset U_n(N) = (1 + NM_n(\widehat{\mathbb{Z}})) \cap GL_n(\widehat{\mathbb{Z}})$. We simply write $\widehat{\Gamma}(N)$ for $U_2(N) \subset G(\mathbb{A})$. In $\mathbb{A}^\times$, $\mathbb{Q}^\times \cap \widehat{\mathbb{Z}}^\times = \mathbb{Z}^\times = \{\pm 1\}$. Thus, $\mathbb{Q}^\times$ is closed in $\mathbb{A}^\times$ (this fact is not true for $F_\mathbb{A}^\times \supset F^\times$ for a number field F with infinite unit group $O_F^\times$, for example, a real quadratic field). Thus, $Z(\mathbb{Q}) \subset Z(\mathbb{A}) \subset G(\mathbb{A})$ is closed, and hence $Z(\mathbb{Q}) = \bigcap_{U\subset G(\widehat{\mathbb{Z}})} U \cdot Z(\mathbb{Q})$ for open compact subgroups $U \subset G(\widehat{\mathbb{Z}})$. Choose a complete representative set $\{c\}$ inside $\widehat{\mathbb{Z}}^\times$ for

$$Cl(N) := \mathbb{Q}^\times\backslash\mathbb{A}^\times/\mathbb{R}_+^\times U_1(N) = \mathbb{A}^\times/\mathbb{Q}^\times\mathbb{R}_+^\times U_1(N). \tag{7.1.4}$$

Note that $Cl(N) \cong (\mathbb{Z}/N\mathbb{Z})^\times$ by sending the integer $0 < j < N$ prime to N to its image in $\mathbb{Z}_N^\times \subset \widehat{\mathbb{Z}}^\times$ (see Exercise 7.9). By the strong approximation theorem, $SL_2(\mathbb{Q})$ is dense in $SL_2(\mathbb{A}^{(\infty)}) = SL_2(\mathbb{A})/SL_2(\mathbb{R})$ (see Exercise 7.9 again), $\det : G(\mathbb{A}) \to \mathbb{A}^\times$ induces $G(\mathbb{Q})\backslash G(\mathbb{A})/\widehat{\Gamma}(N)G(\mathbb{R})^+ \cong Cl(N)$. Indeed, from density, we see

$$SL_2(\mathbb{Q})U(N) = SL_2(\mathbb{A}^{(\infty)}) = \mathrm{Ker}(\det : G(\mathbb{A}^{(\infty)} \to \mathbb{G}_m(\mathbb{A}^{(\infty)}))).$$

Thus, we have

$$G(\mathbb{A}) = \bigsqcup_{c\in Cl(N)} G(\mathbb{Q})t_c\widehat{\Gamma}(N)G(\mathbb{R})^+$$

for $t_c = \left(\begin{smallmatrix} c & 0 \\ 0 & 1 \end{smallmatrix}\right)$. The projection $G(\mathbb{R})^+ \ni \left(\begin{smallmatrix} a & b \\ c & d \end{smallmatrix}\right) \mapsto \frac{a\sqrt{-1}+b}{c\sqrt{-1}+d} \in \mathfrak{H}$ induces

$$G(\mathbb{Q})\backslash G(\mathbb{Q})t_c\widehat{\Gamma}(N)G(\mathbb{R})^+ \cong (G(\mathbb{Q}) \cap t_c^{-1}\widehat{\Gamma}(N)G(\mathbb{R})^+t_c)\backslash\mathfrak{H} \cong \Gamma(N)\backslash\mathfrak{H},$$

because t_c normalizes $\widehat{\Gamma}(N)$; so

$$G(\mathbb{Q}) \cap t_c^{-1}\widehat{\Gamma}(N)G(\mathbb{R})^+t_c = \left\{\gamma \in SL_2(\mathbb{Z}) \middle| \gamma \equiv 1 \mod N\right\}.$$

Since $Y(N)_{\zeta_N^c/\mathbb{Q}[\mu_N]}(\mathbb{C}) = \Gamma(N)\backslash\mathfrak{H}$ and $Y(N)_{/\mathbb{Q}}(\mathbb{C}) = \bigsqcup_c Y(N)_{\zeta_N^c/\mathbb{Q}[\mu_N]}(\mathbb{C})$, we have $Y(N)_{/\mathbb{Q}}(\mathbb{C}) = G(\mathbb{Q})\backslash G(\mathbb{A})/\widehat{\Gamma}(N)Z(\mathbb{Q})$. Passing to the limit, we get

$$\begin{aligned} Y(\mathbb{C}) &= \varprojlim_N Y(N)(\mathbb{C}) \\ &= G(\mathbb{Q})\backslash((\mathfrak{H}\sqcup\overline{\mathfrak{H}}) \times G(\mathbb{A}^{(\infty)}))/\bigcap_N(Z(\mathbb{Q})\widehat{\Gamma}(N)) \qquad (7.1.5) \\ &= G(\mathbb{Q})\backslash((\mathfrak{H}\sqcup\overline{\mathfrak{H}}) \times G(\mathbb{A}^{(\infty)}))/Z(\mathbb{Q}), \end{aligned}$$

since $\bigcap_N Z(\mathbb{Q})\widehat{\Gamma}(N) = Z(\mathbb{Q})$ inside $G(\mathbb{A}^{(\infty)})$.

For $\gamma \in G(\mathbb{Q}) \subset G(\mathbb{A})$, by definition, we have $[\gamma_\infty(z'), \gamma^{(\infty)}g] = [z', g]$. Making a variable change, $z = \gamma_\infty(z')$, we get $[z, \gamma^{(\infty)}g] = [\gamma_\infty^{-1}(z), g]$ as in the lemma.

Exercise 7.9 *Prove the following facts:*

1. $Z(\mathbb{Q}) = \bigcap_{U \subset G(\widehat{\mathbb{Z}})} U \cdot Z(\mathbb{Q})$ *for open compact subgroups* $U \subset G(\widehat{\mathbb{Z}})$.
2. $(\mathbb{Z}/N\mathbb{Z})^\times \cong Cl(N) := \mathbb{A}^\times/\mathbb{Q}^\times\widehat{\Gamma}_1(N)\mathbb{R}_+^\times$.
3. $SL_2(\mathbb{Z}_p)$ *is generated by* $U_p = \{\alpha(u) | u \in \mathbb{Z}_p\}$ *and its transpose* ${}^tU_p = \{{}^t\alpha(u) | u \in \mathbb{Z}_p\}$, *where* $\alpha(u)$ *is the upper unipotent matrix with upper right corner entry* u.
4. $SL_2(\mathbb{Z})$ *is dense in* $SL_2(\widehat{\mathbb{Z}})$ *[use (3)].*
5. $SL_2(\mathbb{Q})$ *is dense in* $SL_2(\mathbb{A}^{(\infty)})$ *(strong approximation theorem).*

It is standard to identify $\mathbb{C} - \mathbb{R} = \mathfrak{H} \sqcup \overline{\mathfrak{H}}$ with the set of complex structure on $V(\mathbb{R}) = V \otimes_\mathbb{Q} \mathbb{R}$. For each $z \in \mathbb{C} - \mathbb{R}$, the basis $(z, 1)$ over $\mathbb{R}$ on $\mathbb{C}$ gives rise to the identification $V(\mathbb{R}) = \mathbb{R}^2 \cong \mathbb{C}$ (i.e., a complex structure) by $\left(\begin{smallmatrix} a \\ b \end{smallmatrix}\right) \mapsto -a + bz = (a, b)J_1\left(\begin{smallmatrix} z \\ 1 \end{smallmatrix}\right)$ for $J_1 = \left(\begin{smallmatrix} 0 & -1 \\ 1 & 0 \end{smallmatrix}\right)$. If we identify $V(\mathbb{R}) = \mathbb{C}$ as two-dimensional real vector spaces as above, multiplication by $u \in \mathbb{C}^\times$ on $V(\mathbb{R})$ gives rise to an embedding $h_z : \mathbb{C}^\times \hookrightarrow G(\mathbb{R})$ [having values actually in $G(\mathbb{R})^+$]. Recall the group homomorphism $\rho_z : \mathbb{C}^\times \hookrightarrow G(\mathbb{R})$ given in Sect. 1.3.4 by $\rho_z(u)\left[\begin{smallmatrix} z \\ 1 \end{smallmatrix}\right] = \left[\begin{smallmatrix} zu \\ u \end{smallmatrix}\right]$. Noting

$$\left[\begin{smallmatrix} \rho_z(u)(z) \\ 1 \end{smallmatrix}\right] j(\rho_z(u), z) = \rho_z(u)\left[\begin{smallmatrix} z \\ 1 \end{smallmatrix}\right] = \left[\begin{smallmatrix} zu \\ u \end{smallmatrix}\right], \tag{$*$}$$

we have $h_z(u) = J_1\rho_z(u)J_1^{-1} = {}^t\rho_z^\iota(u)$, where $j(\left(\begin{smallmatrix} a & b \\ c & d \end{smallmatrix}\right), z) = cz + d$ and $g^\iota = \det(g)g^{-1}$. The above identity $(*)$ implies $\rho_z(u)(z) = z$ and $j(\rho_z(u), z) = u$, and we see easily from Lemma 7.7 that $\rho_z(\mathbb{C}^\times)$ is the stabilizer of z, and hence $\mathrm{Im}(h_z)$ is the stabilizer of $-z^{-1} = J_1(z)$. Since we have

$$\mathfrak{X} = \{h : \mathbb{C}^\times \hookrightarrow G(\mathbb{R})\} \cong \{\mathrm{Im}(h)\} \cong \mathbb{C} - \mathbb{R}$$

by associating z with the stabilizer in $G(\mathbb{R})$ of z, we find

$$ {}^tgh_z{}^tg^{-1} = h_{g^{-1}(z)} \quad \text{for the linear fractional transformation } z \mapsto g(z), \tag{7.1.6}$$

as $J_1{}^tgJ_1^{-1} = g^\iota = \det(g)g^{-1}$, where h is supposed to be induced from a morphism of group schemes from $\mathrm{Res}_{\mathbb{C}/\mathbb{R}}\mathbb{G}_m$ into $GL(2)_{/\mathbb{R}}$. In this way, we identify $\mathfrak{H} \sqcup \overline{\mathfrak{H}}$ with $\mathfrak{X}$. The isomorphism $\mathfrak{H} \sqcup \overline{\mathfrak{H}} \cong \mathfrak{X}$ sending z to h_z is equivariant under the action of $G(\mathbb{R})$, making the following diagram commutative:

$$\begin{array}{ccc} V(\mathbb{R}) & \xrightarrow{v \mapsto gv} & V(\mathbb{R}) \\ {\scriptstyle \rho^\iota_{g^{-1}(z)}(u) = {}^th_{g^{-1}(z)}(u)} \downarrow & & \downarrow {\scriptstyle \rho^\iota_z(u) = {}^th_z(u)} \\ V(\mathbb{R}) & \xrightarrow[v \mapsto gv]{} & V(\mathbb{R}). \end{array} \tag{7.1.7}$$

Since any $z = x + iy \in \mathfrak{H}$ $(x, y \in \mathbb{R})$ can be written as $z = g_z(i)$ for $g_z = \sqrt{y}^{-1}\left(\begin{smallmatrix} y & x \\ 0 & 1 \end{smallmatrix}\right)$, we conclude from the above exercise that z is the unique fixed point of ${}^t h_z(\mathbb{C}^\times)$ in $\mathfrak{H}$.

7.1.3 Elliptic Curves Up to Isogenies

Recall $\mathbb{A}^{(\Sigma\infty)} = \{x \in \mathbb{A} | x_l = x_\infty = 0 \text{ for } l \in \Sigma\}$ for a set of rational primes Σ. We fix such a finite set Σ. Let $V = \mathbb{Q}^2$ be a column vector space, and put $V(\mathbb{A}^{(\Sigma\infty)}) = V_{\mathbb{A}^{(\Sigma\infty)}} := V \otimes_{\mathbb{Q}} \mathbb{A}^{(\Sigma\infty)}$. We fix an alternating pairing $\Lambda_0 : V \wedge V \cong \mathbb{Q}$ given by $\Lambda_0((a,b),(a',b')) = a'b - ab'$. Then $V(\mathbb{A}^{(\Sigma\infty)})$ is an $\mathbb{A}^{(\Sigma\infty)}$-free module of rank 2. Let $\mathbb{Z}_{(\Sigma)} = \mathbb{Q} \cap \prod_{l \in \Sigma} \mathbb{Z}_l$. We consider the fibered category $ELL^{(\Sigma)}$ over $SCH_{/\mathbb{Z}_{(\Sigma)}}$ defined by the following data:

(Objects) elliptic curves;
(Morphisms) $\mathrm{Hom}^{(\Sigma)}(E_{/S}, E'_{/S}) = \mathrm{Hom}_{ELL}(E, E') \otimes_{\mathbb{Z}} \mathbb{Z}_{(\Sigma)}$.

A linear morphism $\phi : E \to E'_{/S}$ gives rise to an isomorphism in $ELL^{(\Sigma)}$ if $\deg(\phi) \neq 0$ and is prime to Σ. Such a morphism is called a *prime-to-Σ isogeny*, and if $\Sigma = \emptyset$, we just call a prime-to-$\emptyset$ isogeny an *isogeny.*

For an object $E_{/S}$ of $ELL^{(\Sigma)}$, we take a maximal point $s = \mathrm{Spec}(k(s)) \in S$ and a geometric point $\overline{s} = \mathrm{Spec}(k(\overline{s})) \in S(k(\overline{s}))$ over s, and we consider the *Tate module* $T^{(\Sigma)}(E) = T^{(\Sigma)}_{\overline{s}}(E) = \varprojlim_N E[N](k(\overline{s}))$ *prime-to-Σ*, where N runs over all positive integers prime to Σ. We then define $V^{(\Sigma)}(E) = V^{(\Sigma)}_{\overline{s}}(E) = T^{(\Sigma)}(E) \otimes_{\mathbb{Z}} \mathbb{A}^{(\Sigma\infty)}$. The module $V^{(\Sigma)}(E)$ is an $\mathbb{A}^{(\Sigma\infty)}$-free module of rank 2 and has an $\widehat{\mathbb{Z}}^{(\Sigma)}$-lattice $T^{(\Sigma)}(E)$, where we have written $\widehat{\mathbb{Z}}^{(\Sigma)} = \prod_{l \notin \Sigma} \mathbb{Z}_l$. Since N is invertible in $\mathbb{Z}_{(\Sigma)}$, $E[N]$ is étale over S; so, through the projection $\pi_1(S, \overline{s}) \twoheadrightarrow \mathrm{Gal}(k(\overline{s})/k(s))$ (Corollary 4.55), the algebraic fundamental group $\pi_1(S, \overline{s})$ naturally acts on $E[N](k(\overline{s}))$, $T^{(\Sigma)}_{\overline{s}}(E)$, and $V^{(\Sigma)}_{\overline{s}}(E)$.

A *full-level structure* prime to Σ on A is an isomorphism $\eta^{(\Sigma)} : V(\mathbb{A}^{(\Sigma\infty)}) \cong V^{(\Sigma)}_{\overline{s}}(E) = V_{\overline{s}}(E) \otimes_{\mathbb{A}} \mathbb{A}^{(\Sigma\infty)}$ of $\mathbb{A}^{(\Sigma\infty)}$-modules, picking a geometric point $\overline{s}$ in each connected component of S. Here Σ is a set of rational primes containing all primes not invertible over the base S (often $\Sigma = \emptyset$ or is made of a single prime p). We write $\mathbb{Z}_\Sigma = \prod_{l \in \Sigma} \mathbb{Z}_l$. When $\Sigma = \emptyset$, we write η (resp., $ELL^{\mathbb{Q}}$) instead of $\eta^{(\emptyset)}$ [resp., $ELL^{(\emptyset)}$]. When $\Sigma = \{p\}$, as before, we write $\eta^{(p)}$ for $\eta^{(\Sigma)}$ and $ELL^{(p)}$ for $ELL^{(\Sigma)}$.

Since we can restrict things to each connected component, we may assume that S is connected. Take the fundamental functor $F_{\overline{s}}$ as in Remark 4.52, and fix a fundamental pro-object $P = \{P_i\}_{i \in I}$ in item 3 of Sect. 4.2.3. Let $\varphi = \varprojlim_i \varphi_i$ be a universal element in $\varprojlim_i F_{\overline{s}}(P_i)$. Fix an elliptic curve $E_{/S}$. Consider the small étale site S_{et} over S (see Sect. 4.1.10). Take the Galois category $\mathcal{C} \subset S_{et}$ of all connected coverings of S in S_{et} as in Remark 4.52. The sheaf of sets $LV^{(\Sigma)}$ whose values at a connected covering $S'_{/S}$ in $\mathcal{C}$ with the unique $\phi_i : P_i \to S'$ (see Sect. 4.2.3 item 3) are given by the set $H^0(\pi_1(S',s'),\mathrm{Isom}(V(\mathbb{A}^{(\Sigma\infty)}),V^{(\Sigma)}_{s'}(E_{/S'})))$ [of $\pi_1(S',s')$-fixed points in $\mathrm{Isom}(V(\mathbb{A}^{(\Sigma\infty)}),V^{(\Sigma)}_{s'}(E_{/S'}))$], where s' is a unique geometric point of S' such that $s' = F_{\overline{s}}(\phi_i)(\varphi_i)$. Here $\mathrm{Isom}(V(\mathbb{A}^{(\Sigma\infty)}),V^{(\Sigma)}_{s'}(E_{/S'}))$ is the set of $\mathbb{A}^{(\Sigma\infty)}$-linear isomorphisms of the $V(\mathbb{A}^{(\Sigma\infty)})$ onto $V^{(\Sigma)}_{s'}(E_{/S'})$. This is the *sheaf of level structures*. Pick $\eta^{(\Sigma)} \in \mathrm{Isom}(V(\mathbb{A}^{(\Sigma\infty)}),V^{(\Sigma)}_{s'}(E_{/S'}))$. Then $\gamma \in \pi_1(S',s')$ [resp., $g \in G(\mathbb{A}^{(\Sigma\infty)})$] acts on $\eta^{(\Sigma)}$ through its action on $V^{(\Sigma)}_{s'}(E_{/S'}))$ [resp., the natural left action of $G(\mathbb{A}^{(\Sigma\infty)})$ on $V(\mathbb{A}^{(\Sigma\infty)})$].

Lemma 7.10 *The right action of $G(\mathbb{A}^{(\Sigma\infty)})$ and the action of $\pi_1(S',s')$ commute, and the definition of the sheaf is independent of the choice of s'.*

Proof. Clearly, $S' \mapsto \mathrm{Isom}(V(\mathbb{A}^{(\Sigma\infty)}),V^{(\Sigma)}_{s'}(E_{/S'}))$ is a local functor (hence a sheaf of sets). Then $g \in G(\mathbb{A}^{(\Sigma\infty)})$ acts through $V(\mathbb{A}^{(\Sigma\infty)})$ as $\eta^{(\Sigma)} \mapsto \eta^{(\Sigma)} \circ g$. On the other hand, $E[N]_{/S'}$ for N invertible is an étale group scheme over S'; so the fundamental group $\pi_1(S',s')$ acts on its fiber $E[N]_{s'}$ at s' and on its projective limit $V^{(\Sigma)}_{s'}(E_{/S'})$. Thus, the action of the fundamental group is through the target $V^{(\Sigma)}_{s'}(E_{/S'})$; so they commute.

We consider the level structures $\eta : V_{\mathbb{A}^{(\infty)}} \cong V_{\overline{s}}(A)$ and $\eta^{(p)} : V_{\mathbb{A}^{(p\infty)}} \cong V^{(p)}_{\overline{s}}(A)$. An element $g \in G(\mathbb{A})$ acts on η [and on $\eta^{(p)}$] by $\eta \mapsto \eta \circ g$. In particular, $\sigma \circ \eta$ with $\sigma \in \pi_1(S,\overline{s})$ can be written as $\eta \circ k_{\overline{s}}(\sigma)$ for a unique $k_{\overline{s}}(\sigma) \in G(\mathbb{A})$. For a closed subgroup $\Gamma \subset G(\mathbb{A}^{(\infty)})$, the level-$\Gamma$ structure $\overline{\eta} = \overline{\eta}_\Gamma := \eta \circ \Gamma$ is defined over S if $\sigma \circ \eta = \eta \circ k_{\overline{s}}(\sigma)$, with $k_{\overline{s}}(\sigma) \in \Gamma$ for all $\sigma \in \pi_1(S,\overline{s})$ for each chosen $\overline{s}$. Changing $\overline{s}$ to another point $\overline{t}$ in the same connected component gives isomorphisms (called paths) $\iota_{\overline{s},\overline{t}} : \pi_1(S,\overline{s}) \cong \pi_1(S,\overline{t})$ and $\iota^{\mathcal{T}}_{\overline{s},\overline{t}} : \mathcal{T}_{\overline{s}}(A) \cong \mathcal{T}_{\overline{t}}(A)$, which satisfy the compatibility condition $\iota^{\mathcal{T}}_{\overline{s},\overline{t}}(\sigma(x)) = \iota_{\overline{s},\overline{t}}(\sigma)(\iota^{\mathcal{T}}_{\overline{s},\overline{t}}(x))$ (see Sect. 4.2 in the text and [ECH] Sect. I.5); so if $\overline{\eta}$ is defined over S, we have

$$\iota^{\mathcal{T}}_{\overline{s},\overline{t}}(\eta \circ k_{\overline{s}}(\sigma)v) = \iota^{\mathcal{T}}_{\overline{s},\overline{t}}(\sigma \circ \eta(v)) = \iota_{\overline{s},\overline{t}}(\sigma) \circ (\iota^{\mathcal{T}}_{\overline{s},\overline{t}}(\eta(v))).$$

Thus, giving $\overline{\eta}$ defined over S at $\overline{s}$ is equivalent to giving $\iota^{\mathcal{T}}_{\overline{s},\overline{t}} \circ \overline{\eta}$ again defined over S at $\overline{t}$, and so we may forget about the choice $\overline{s}$. □

For a closed subgroup $\Gamma \subset G(\mathbb{A}^{(\infty)})$ with $G(\mathbb{Z}_\Sigma) \times \Gamma^{(\Sigma)}$, we consider the quotient sheaf $LV^{(\Sigma)}/\Gamma$, which is the sheafication of the presheaf

$$S'_{/S} \mapsto \{\Gamma^{(\Sigma)}\text{-orbits fixed by } \pi_1(S',s') \text{ of } \eta^{(\Sigma)}\}$$

for the right action $\eta^{(\Sigma)} \mapsto \eta^{(\Sigma)} \circ u$ [$u \in \Gamma^{(\Sigma)}$]. Here $\eta^{(\Sigma)}$ runs over $\mathrm{Isom}(V(\mathbb{A}^{(\Sigma\infty)}),V^{(\Sigma)}_{s'}(E_{/S'}))$, and if S' is not connected, the right-hand side is the disjoint union of the sets (each defined for each connected component). We define a level-Γ *structure* $\overline{\eta}^{(\Sigma)}$ over S to be a section $\overline{\eta}^{(\Sigma)} \in (LV^{(\Sigma)}/\Gamma)(S)$. When Γ is sufficiently small (i.e., a so-called *neat* subgroup, for example,

$\Gamma = \widehat{\Gamma}(N)$ with $N \geq 3$ or $\Gamma = \widehat{\Gamma}_1(N)$ with $N \geq 4$), by Theorem 6.6, we have $\mathrm{Aut}_{GSCH}(E, \overline{\eta}) = \{\mathrm{id}_E\}$ for all possible pairs $(E, \overline{\eta})_{/S}$ over all $S_{/\mathbb{Q}}$. Then we do not need to take sheafication; so $\overline{\eta}$ is just a Γ-orbit of $\eta^{(\Sigma)}$. In [PAF] (4.25) in 4.2.1, $\overline{\eta}^{(\Sigma)}$ is defined to be the Γ-orbit $\eta^{(\Sigma)} \circ \Gamma$ as a global section of $LV^{(\Sigma)}/\Gamma$ over S. Note here that a literal Γ-orbit for nonneat Γ may not have necessary local functoriality, and the correct definition of the level-Γ structure for nonneat Γ has to be this sheaf-theoretic orbit. All the assertions in [PAF] Chap. 4 are valid, taking sheaf-theoretic orbits even for nonneat Γ. For many instances, we assume Γ to be open compact.

We consider the following functor from $SCH_{/\mathbb{Q}}$ into $SETS$:

$$\mathcal{E}^{\mathbb{Q}}_{\Gamma}(S) = \left[(E, \overline{\eta}_{\Gamma})_{/S}\right],$$

where $\overline{\eta}_{\Gamma}$ is a level-Γ structure as defined above for $\Sigma = \emptyset$, and $[\ \ ] = \{\ \ \}/\cong$ indicates the set of isomorphism classes in $ELL^{\mathbb{Q}}$ of the objects defined over S in the brackets. A $\mathbb{Q}$-linear morphism $\phi \in \mathrm{Hom}^{\mathbb{Q}}_F(E, E')$ is an isomorphism between pairs $(E, \overline{\eta})_{/S}$ and $(E', \overline{\eta}')_{/S}$ if $\phi \circ \overline{\eta} = \overline{\eta}'$. Pairs $(E, \overline{\eta})_{/S}$ giving rise to an element $\mathcal{E}^{\mathbb{Q}}_{\Gamma}(S)$ are often called a *test object of level* Γ.

We restate Theorem 7.3 in a slightly different fashion.

Theorem 7.11 *The Shimura curve* $Sh_{/\mathbb{Q}}$ *represents the functor* $\mathcal{E}^{\mathbb{Q}}_1$ *over* $\mathbb{Q}$ *for the trivial subgroup* 1 *made of the identity element of* $G(\mathbb{A}^{(\infty)})$.

This was stated in a basically equivalent but possibly slightly different way and proved in [Sh3], [IAT], and [D3] 4.16–21. Through the action of $G(\mathbb{A}^{(\infty)})$ on ${\mathbb{A}^{(\infty)}}^2$, $g \in G(\mathbb{A}^{(\infty)})$ acts on the level structure by $\eta \mapsto \eta \circ g$ and hence on the variety Sh from the right. If Γ is open and sufficiently small [so that $\mathrm{Aut}((E, \overline{\eta})_{/S}) = \{1\}$ for all test objects $(E, \overline{\eta})_{/S}$], by definition, we have, as a geometric étale quotient (cf. [GME] Proposition 1.8.4), $Sh_{\Gamma} = (Sh/\Gamma)_{/\mathbb{Q}}$ [whose complex points are given by $G(\mathbb{Q})\backslash(\mathfrak{X} \times G(\mathbb{A}^{(\infty)}))/\Gamma$] represents $\mathcal{E}^{\mathbb{Q}}_{\Gamma}$ over $\mathbb{Q}$. This point will be made clear in the proof.

We now give an outline of the proof of the representability, reducing it to the representability of a functor classifying elliptic curves up to isomorphisms, not up to isogenies.

Proof. Since $\mathcal{E}^{\mathbb{Q}}_1(S) = \varprojlim_{\Gamma} \mathcal{E}^{\mathbb{Q}}_{\Gamma}(S) = \varprojlim_{N} \mathcal{E}^{\mathbb{Q}}_{\widehat{\Gamma}(N)}(S)$ for Γ running over open compact subgroups with the obvious projection $(E, \overline{\eta}_{\Gamma'}) \mapsto (E, \overline{\eta}_{\Gamma})$ if $\Gamma' \subset \Gamma$, we study $\mathcal{E}^{\mathbb{Q}}_{\Gamma}$ for (sufficiently small) open compact subgroups Γ.

Let $G_1 = SL(2) \subset G$ and $G^{ad} = G/Z = PGL(2)$. First, note that the images of $g\Gamma g^{-1} \cap G_1(\mathbb{Q})$ and $g\Gamma g^{-1} \cap G(\mathbb{Q})^+$ in $G^{ad}(\mathbb{Q})$ are equal, where $G(\mathbb{R})^+ = \{x \in G(\mathbb{R}) | \det(x) > 0\}$ and $G(\mathbb{Q})^+ = G(\mathbb{Q}) \cap G(\mathbb{R})^+$. Indeed, if $\gamma = g\Gamma g^{-1} \cap G(\mathbb{Q})^+$, $\det(\gamma) \in \mathbb{Q}^{\times}_+ \cap \det(\Gamma) \subset \mathbb{Q}^{\times}_+ \cap \widehat{\mathbb{Z}}^{\times} = \{1\}$; hence, $\gamma \in SL_2(\mathbb{Z})$. So, by Lemma 7.8, $Sh_{\Gamma}(\mathbb{C})$ can be embedded into $Sh_{\Gamma_1}(\mathbb{C})$ for $\Gamma_1 = G_1(\mathbb{A}^{(\infty)}) \cap \Gamma$. This also follows from the fact that the moduli problem with respect to Γ_1 is neat without having any nontrivial automorphisms. Let

$L \subset V$ be a lattice. We may assume that $L = \mathfrak{a}^* \oplus \mathfrak{b}$ for a pair $(\mathfrak{a}, \mathfrak{b})$ of two fractional ideals in $\mathbb{Q}$, where $\mathfrak{a}^* = \mathfrak{a}^{-1}$ [but we regard $\mathfrak{a}^*$ as the dual of $\mathfrak{a}$ by the natural pairing $(x, y) \mapsto xy$ on $\mathbb{Q} \times \mathbb{Q}$, hence the notation $\mathfrak{a}^*$ rather than $\mathfrak{a}^{-1}$]. Since all ideals of $\mathbb{Z}$ are principal, we hereafter assume that $\mathfrak{a} = \mathfrak{b} = \mathbb{Z}$, but we keep this notation $(\mathfrak{a}, \mathfrak{b})$ to distinguish the left/right factor. For each point $h_z \in \mathfrak{X}$, we have a unique point $z \in (\mathbb{C} - \mathbb{R})$ fixed by $h_z(\mathbb{C}^\times)$ (in this way, we have identified $\mathfrak{H}$ with the connected component $\mathfrak{X}^+$ of $\mathfrak{X}$ containing h_i; see Sect. 7.1.2). By changing the identification $V = \mathbb{Q}^2$, we may assume that $z \in \mathfrak{H}$. The action of $h_z(\mathbb{C}^\times)$ on $V_{\mathbb{R}} = V \otimes_{\mathbb{Q}} \mathbb{R}$ gives a structure of a complex vector space of dimension 1 on $V_{\mathbb{R}}$; that is, $V_{\mathbb{R}} = \mathbb{C}$ via $(a, b) \mapsto -a + bz = (a, b)J_1 \cdot {}^t(z, 1)$ for $J_1 = \left(\begin{smallmatrix} 0 & -1 \\ 1 & 0 \end{smallmatrix}\right)$. Then $L \subset V$ gives rise to the lattice $L_z = \mathfrak{a}^* + \mathfrak{b}z$ (the image of L in $\mathbb{C} = V_{\mathbb{R}}$). Set $\widehat{L} = L \otimes_{\mathbb{Z}} \widehat{\mathbb{Z}} \subset V_{\mathbb{A}}^{(\infty)} = V \otimes_{\mathbb{Q}} \mathbb{A}^{(\infty)}$, and define an elliptic curve $E_{z/\mathbb{C}}$ with $E_z(\mathbb{C}) = \mathbb{C}/L_z$. Then we have $T(E_z) = \widehat{L}$, which induces $\eta_z : V_{\mathbb{A}^{(\infty)}} \cong V(E_z)$ and gives rise to a level-N structure $\phi_N : N^{-1}L/L \cong E_z[N]$ for any $N > 0$.

Define $Cl(\Gamma) = \mathbb{A}^{(\infty)\times}/\det(\Gamma)\mathbb{Q}_+^\times \cong \mathbb{A}^\times/\mathbb{Q}^\times \det(\Gamma)\mathbb{R}_+^\times$, which is the quotient of $Cl(N)$ if $\Gamma \supset \widehat{\Gamma}(N)$ and hence is a finite group if Γ is open. Let $Cl(N) = (\mathbb{Z}/N\mathbb{Z})^\times$, which is naturally a quotient of $\widehat{\mathbb{Z}}^\times$. Let $\widehat{\mathbb{Z}}(1) = \varprojlim_N \mu_N(\overline{\mathbb{Q}})$, and identify it with $\widehat{\mathbb{Z}} = \varprojlim_N \mathbb{Z}/N\mathbb{Z}$ by $\mathbb{Z}/N\mathbb{Z} \ni m \leftrightarrow \zeta_N^m$ for $\zeta_N := \exp\left(\frac{2\pi i}{N}\right)$ for all positive integers N. We set $Cl(N) = \mu_N$ and define μ_Γ by the following commutative diagram:

$$\begin{array}{ccc} \widehat{\mathbb{Z}}^\times & \twoheadrightarrow & Cl_\Gamma \\ \| \downarrow & & \downarrow \| \\ \widehat{\mathbb{Z}}(1)^\times & \twoheadrightarrow & \mu_\Gamma, \end{array}$$

and write ζ_Γ for the "standard generator" of the group μ_Γ. On μ_Γ and on $\widehat{\mathbb{Z}}(1)$, the absolute Galois group acts by the natural (nontrivial) Galois action, and $(\mathbb{Z}/N\mathbb{Z}) \times \mu_N \ni (m, \zeta) \mapsto \zeta^m \in \mu_N$ gives rise to a natural action of $Cl(\Gamma)$ on μ_Γ (we write this action multiplicatively as $\zeta \mapsto \zeta^z$). We fix a complete representative set $\{c \in \mathbb{A}^{(\infty)\times}\}$ for $Cl(\Gamma)$ and put $\mathfrak{c} = c\widehat{\mathbb{Z}} \cap \mathbb{Q}$. Though this is not necessary, we often assume that $c \in \widehat{\mathbb{Z}}^\times$; and under this choice, $\mathfrak{c} = \mathbb{Z}$, and we may assume that $\mathfrak{a} = \mathfrak{b} = \mathbb{Z}$ and $\Gamma \subset G(\widehat{\mathbb{Z}})$. Thus, we take $L = \mathbb{Z} \oplus \mathbb{Z} \subset V$.

For each isogeny class of $(E, \overline{\eta})_{/S} \in \mathcal{E}_\Gamma^{\mathbb{Q}}(S)$, we functorially find a unique test object $(E', \overline{\eta}')_{/S}$ representing a unique class in $Cl(\Gamma)$ such that $\overline{\eta}'(\widehat{L}) = T(E)$. Take a complex point $s : \mathrm{Spec}(\mathbb{C}) \to S$ with fiber $E_s = E \times_S s$ and consider $\pi_1(E_s, \mathbf{0}_s)$ for $\mathbf{0}_s = \mathbf{0}_{E_s}$. Note that $E_s(\mathbb{C}) = \mathbb{C}/\pi$ for a lattice $\pi \subset \mathbb{C}$. Since $\mathbb{C}$ is simply connected, $\pi_1^{top}(E_s(\mathbb{C}), \mathbf{0}_s) = \pi$; hence,

$$\pi_1(E_s, \mathbf{0}_s) = \widehat{\pi} = \pi \otimes_{\mathbb{Z}} \widehat{\mathbb{Z}} = \varprojlim_N N^{-1}\pi/\pi = \varprojlim_N E[N](k(s)) = T(E_s). \quad (7.1.8)$$

Thus, for any $\widehat{\mathbb{Z}}$-lattice M of $V(E) = T(E_s) \otimes_{\mathbb{Z}} \mathbb{Q}$, we have a unique elliptic curve $E_{M/S}$ and an isogeny $f : E_M \to E$ such that $\pi(f) \otimes \mathbb{Q} : V(E_M) \to$

$V(E)$, sending $T(E_M) = \pi_1(E_M, \mathbf{0})$ to M. Apply this to $M = \eta(\widehat{L})$ and write E_M as E'. We have the natural exact sequence $\pi_1(E_s, \mathbf{0}_s) \hookrightarrow \pi_1(E, \mathbf{0}) \twoheadrightarrow \pi_1(S, s)$ as in Theorem 4.54; thus, we have the conjugate action of $\pi_1(S, s)$ on $\pi_1(E_s, \mathbf{0}_s)$, which is compatible with $\pi(f)$ as E and $\eta'(\widehat{L})$ are invariant under the action. Thus, E' descends to S, and we get the desired object $(E', \overline{\eta}')_{/S}$. The uniqueness of this test objects follows from the triviality of the automorphism group $\mathrm{Aut}_S(E', \overline{\eta}')$. See [PAF] pages 135–136 for a slightly different argument of the construction of $(E', \overline{\eta}')_{/S}$.

We now have a unique test object $(E', \overline{\eta}')_{/S}$ with $\overline{\eta}'(\widehat{L}) = T(E')$. If two such choices are isogenous, the isogeny between them has to be an isomorphism. Thus, we get an isomorphism of functors: $\mathcal{E}^{\mathbb{Q}}_{\Gamma}(S) \cong \mathcal{E}'_{\Gamma}(S) := \bigsqcup_{c \in Cl(\Gamma)} \mathcal{E}'_{\Gamma,c}(S)$, where

$$\mathcal{E}'_{\Gamma,c}(S) = \left\{ (E', \overline{\eta}')_{/S} \middle|\, \overline{\eta}'(\widehat{L}) = T(E'),\ \langle \eta'(1,0), \eta'(0,1) \rangle_{E'} = \zeta_{\Gamma}^{c} \in \mu_{\Gamma} \right\} / \cong .$$

Here $\langle \cdot, \cdot \rangle_{E'} : T(E') \times T(E') \to \widehat{\mathbb{Z}}(1) (\twoheadrightarrow \mu_{\Gamma})$ is the projective limit of the self-duality pairing $E'[N] \times E'[N] \to \mu_N$, and $\cong$ is the isomorphism (not an isogeny). As we now see, following the argument we presented earlier, this functor $\mathcal{E}'_{\Gamma,c}$ is represented by a scheme $Y(c, \Gamma)$ over a specific abelian extension k_{Γ} of $\mathbb{Q}$ dependent on Γ (see below for a description of k_{Γ} for some specific Γ's). See [PAF] Sect. 4.2 for this process for more general families of abelian varieties (of higher dimension).

Recall the principal congruence subgroup of level N

$$\widehat{\Gamma}(N) = \mathrm{Ker}(GL(\widehat{L}) \twoheadrightarrow GL(L/NL)) = \mathrm{Ker}(GL_2(\widehat{\mathbb{Z}}) \twoheadrightarrow GL_2(\mathbb{Z}/N\mathbb{Z}))$$

of $G(\mathbb{A}^{(\infty)})$ for an integer $N > 0$. Having a level-$\widehat{\Gamma}(N)$ structure $\overline{\eta}'$ is equivalent to having a level-$\Gamma(N)$ structure $\phi_N := \overline{\eta}' \mod \widehat{\Gamma}(N)$; in other words, giving $\overline{\eta}'$ is equivalent to giving an isomorphism of locally free group schemes

$$\phi_N : (\mathbb{Z}/N\mathbb{Z})^2 \cong N^{-1}L/L \overset{\overline{\eta}'}{\cong} E'[N].$$

Thus, $\mathcal{E}'_{\widehat{\Gamma}(N),c}$ is the standard modulifunctor classifying the level structure for the principal congruence subgroup $\widehat{\Gamma}(N)$:

$$\mathcal{E}'_{\widehat{\Gamma}(N),c}(S) = \mathcal{E}_{\Gamma(N),\zeta_N^c}(S) = \left\{ (E, \phi)_{/S} \middle| \begin{matrix} \phi : (\mathbb{Z}/N\mathbb{Z})^2 \cong E[N] \\ \text{and } \langle \phi(1,0), \phi(0,1) \rangle = \zeta_N^c \end{matrix} \right\} / \cong .$$

As we have verified in Sect. 6.2.10, this functor is represented by a geometrically irreducible affine curve $Y(N)_{\zeta_N^c/\mathbb{Q}[\mu_N]}$. Define $c(\phi) \in (\mathbb{Z}/N\mathbb{Z})^{\times} = Cl(N)$ by $\langle \phi(1,0), \phi(0,1) \rangle = \zeta_N^{c(\phi)}$. Extending this definition, for a level-Γ test object $(E, \overline{\eta})$, taking N so that $\Gamma \supset \widehat{\Gamma}(N)$, we define $c(\eta)$ to be the image of $(c(\eta) \mod \widehat{\Gamma}(N))$ in $Cl(\Gamma)$.

Over $k_{\Gamma(N)} = \mathbb{Q}[\mu_N]$, the component $Y(N)_{\zeta_N^c}$ of $Sh_{\widehat{\Gamma}(N)}$ represents the functor $\mathcal{E}'_{\Gamma(N),c}$. This irreducible component in turn corresponds to the component $G(\mathbb{Q})\backslash\left(\mathfrak{X} \times G(\mathbb{Q})t_c\widehat{\Gamma}(N)\right)/\widehat{\Gamma}(N) \subset Sh_{\widehat{\Gamma}(N)}(\mathbb{C})$ for $t_c = \left(\begin{smallmatrix} c & 0 \\ 0 & 1 \end{smallmatrix}\right)$. The choice $\zeta = \zeta_N$ gives rise to the identification $\mathbb{Q}[\mu_N] = \mathbb{Q}[X]/(\Phi_N(X))$ with $\mathbb{Q}[\zeta_N]$ for the cyclotomic polynomial $\Phi_N(X)$, and an automorphism $\sigma \in \mathrm{Gal}(\mathbb{Q}[\zeta]/\mathbb{Q})$ changes the identification by $\zeta \mapsto \zeta^\sigma$, whose action is induced by $\phi_N \mapsto \phi_N \circ \left(\begin{smallmatrix} c & 0 \\ 0 & 1 \end{smallmatrix}\right)$ for a unit $c \in \widehat{\mathbb{Z}}^\times$ such that $\zeta^\sigma = \zeta^c$. In other words, the action of $\left(\begin{smallmatrix} c & 0 \\ 0 & 1 \end{smallmatrix}\right) \in G(\mathbb{A}^{(\infty)})$ on $Sh_{/\mathbb{Q}}$ brings $Y(N)_{\zeta_N/\mathbb{Q}[\zeta]} \subset Sh_{\widehat{\Gamma}(N)}$ to its σ-conjugate $Y(N)^\sigma_{\zeta_N} = Y(N)_{\zeta_N^\sigma/\mathbb{Q}[\zeta]}$ in $Sh_{\widehat{\Gamma}(N)/\mathbb{Q}}$.

Summing up all these, we have

$$\mathcal{E}'_{\widehat{\Gamma}(N)} = \bigsqcup_{c \in Cl(N)} \mathcal{E}'_{\widehat{\Gamma}(N),c} \quad \text{over } SCH_{/\mathbb{Q}[\mu_N]}, \tag{7.1.9}$$

which implies

$$Sh_{\widehat{\Gamma}(N)/\mathbb{Q}[\mu_N]} = \bigsqcup_{c \in Cl(N)} Y(N)_{\zeta_N^c} \quad \text{over } \mathbb{Q}[\mu_N]. \tag{7.1.10}$$

Since $\det(t_c) = c$, this corresponds to the decomposition

$$G(\mathbb{A}^{(\infty)}) = \bigsqcup_{c \in Cl(N)} G(\mathbb{Q})t_c\widehat{\Gamma}(N).$$

Since the Galois group permutes $Y(N)_\zeta$, the right-hand side of (7.1.10) descends to the base $\mathbb{Q}$, and we obtain the model $Sh_{\widehat{\Gamma}(N)}$ over $\mathbb{Q}$, because $Y(N)$ is a quasi-projective normal curve (if $N \geq 3$). Since a normal curve over a field is smooth (cf. Corollary 2.11), $Y(N)$ (with $N \geq 3$) is smooth over $\mathbb{Q}$. Since $\mathcal{E}_1^{\mathbb{Q}}(S) = \varprojlim_N \mathcal{E}^{\mathbb{Q}}_{\widehat{\Gamma}(N)}(S)$, the procurve $Sh_{/\mathbb{Q}} = \varprojlim_N Y(N)_{/\mathbb{Q}}$ represents $\mathcal{E}_1^{\mathbb{Q}}$, and also $Sh_{/\mathbb{Q}}$ is smooth. This finishes the proof of the theorem. We will give a more direct proof of smoothness below in the proof of Theorem 7.13.

Remark 7.12 The set $\wp(\mathbb{C})$ of isomorphism classes of elliptic curves over $\mathbb{C}$ with differential is in one-to-one onto correspondence to the set of lattices Lat in $\mathbb{C}$ as seen in Sect. 2.3.3. The set $\mathcal{E}_1^{\mathbb{Q}}(\mathbb{C})$ is a union of $G(\mathbb{A}^{(\infty)})$-orbits each isomorphic to

$$\widehat{Lat} = \{\widehat{L} = \widehat{\mathbb{Z}}w_1 + \widehat{\mathbb{Z}}w_2 | (w_1, w_2) \text{ is a basis of } (\mathbb{A}^{(\infty)})^2 \text{ over } \mathbb{A}^{(\infty)}\}.$$

To describe this fact in a more down-to-earth manner, pick $(E, \eta)_{/\mathbb{C}} \in \mathcal{E}_1^{\mathbb{Q}}(\mathbb{C})$ and write $E(\mathbb{C}) = \mathbb{C}/L_w$ for $L_w = \mathbb{Z}w_{1,\infty} + \mathbb{Z}w_{2,\infty}$ with a basis $w = (w_{1,\infty}, w_{2,\infty})$ of $\mathbb{C}$ over $\mathbb{R}$. Let $\widehat{L} = \varprojlim_N L/NL$, which has a basis $w = (w_1, w_2)$ over $\widehat{\mathbb{Z}}$ induced by $(w_{1,\infty}, w_{2,\infty})$. Since $E[N] \cong L/NL$ canonically, we have $T(E) = \widehat{L} = \varprojlim_N L/NL$; thus, we get $\eta_w : V_{\mathbb{A}^{(\infty)}} \ni (a, b) \mapsto aw_1 + bw_2 \in T(E)$. Then we have a bijection

$$\{(E, \eta_w \circ g) | g \in G(\mathbb{A}^{(\infty)})\} \leftrightarrow \widehat{Lat}$$

given by $\eta_w \circ g \mapsto (\eta_w \circ g)^{-1}(T(E)) = g^{-1}\widehat{L}$. This explains well why the action $\eta \mapsto \eta \circ g$ is interpreted as the action of g^{-1} on $\widehat{Lat}$.

7.1.4 Integral Shimura Curve

We fix a rational prime p. To construct a canonical scheme $Sh^{(p)}_{/\mathbb{Z}_{(p)}}$ with $Sh^{(p)} \times_{\mathrm{Spec}(\mathbb{Z}_{(p)})} \mathbb{Q} \cong Sh_{GL_2(\mathbb{Z}_p)}$, we use the functor $\mathcal{E}^{(p)}_{\Gamma}$, taking $\Sigma = \{p\}$. This functor was studied by Kottwitz in [Ko] for general Shimura varieties of PEL type (though for modular curves, it is implicit in the work of Shimura and Igusa). A closed subgroup $\Gamma \subset G(\mathbb{A}^{(\infty)})$ is *maximal at* p if $\Gamma = G(\mathbb{Z}_p) \times \Gamma^{(p)}$. Here we have written $\Gamma^{(p)} = \{x \in \Gamma | x_p = 1\}$. In this section, we suppose that Γ is maximal at p.

Consider the following fibered category $ELL^{(p)}$ over $SCH_{/\mathbb{Z}_{(p)}}$ given by

(Objects) elliptic curves;
(Morphisms) $\mathrm{Hom}_{ELL^{(p)}}(E, E') = \mathrm{Hom}_{ELL}(E, E') \otimes_{\mathbb{Z}} \mathbb{Z}_{(p)}$.

This means that to classify test objects, we now allow only isogenies with degree prime to p (called "prime-to-p isogenies").

Recall the lattice $L = \mathbb{Z}^2 \subset V = \mathbb{Q}^2$; hence, $\Lambda_0(L \wedge L) = \mathbb{Z}$. Consider test objects $(E, \overline{\eta}^{(p)})_{/S}$. The level-$\Gamma$ structure $\overline{\eta}^{(p)}$ is the Γ-orbit in $LV^{(p)}/\Gamma$ (in the sheaf-theoretic sense as before) of an $\mathbb{A}^{(p\infty)}$-linear isomorphism

$$\eta^{(p)} : V(\mathbb{A}^{(p\infty)}) = V \otimes_{\mathbb{Q}} \mathbb{A}^{(p\infty)} \cong V^{(p)}(E) = T(E) \otimes_{\mathbb{Z}} \mathbb{A}^{(p\infty)},$$

which is supposed to satisfy the following requirement:

$$T^{(p)}(E) \wedge T^{(p)}(E) \xrightarrow{\langle \cdot, \cdot \rangle} \mathbb{A}^{(p\infty)}$$

is proportional to $\Lambda_0 : V \wedge V \cong \mathbb{Q}$ up to a scalar $c(\eta^{(p)})$ in $\mathbb{A}^{(p\infty)\times}$; in short, we have

$$\langle \eta^{(p)}((1,0)), \eta^{(p)}((0,1)) \rangle = c(\eta^{(p)})\Lambda_0((1,0),(0,1)) = c(\eta^{(p)}).$$

Here $\langle \cdot, \cdot \rangle : T^{(p)}E \times T^{(p)}E \to \widehat{\mathbb{Z}}^{(p)}(1)$ is the limit bilinear form $\varprojlim_{N, p \nmid N} \langle \cdot, \cdot \rangle_N$ of the self-duality pairing $\langle \cdot, \cdot \rangle_N$ on $E[N]$. Identifying $\widehat{\mathbb{Z}}$ with $\widehat{\mathbb{Z}}(1)$, taking $1 \in \widehat{\mathbb{Z}}$ to $\varprojlim_N \zeta_N$ for $\zeta_N = \exp(\frac{2\pi i}{N})$, we consider $\langle \cdot, \cdot \rangle$ to have values in $\widehat{\mathbb{Z}}^{(p)} \subset \mathbb{A}^{(p\infty)}$. Then we consider the following functor from $\mathbb{Z}_{(p)}$-schemes into $SETS$.

$$\mathcal{E}^{(p)}_{\Gamma}(S) = \left[(E, \overline{\eta}^{(p)})_{/S}\right]. \tag{7.1.11}$$

Here $[\cdot]$ indicates isomorphism classes in $ELL^{(p)}$ (so, in the classical language, "prime-to-p isogeny classes").

Theorem 7.13 *The p-integral Shimura curve $Sh^{(p)}_{/\mathbb{Z}_{(p)}}$ represents the functor $\mathcal{E}^{(p)}_{G(\mathbb{Z}_p)}$ over $\mathbb{Z}_{(p)}$ for the closed subgroup $G(\mathbb{Z}_p)$ of $G(\mathbb{A}^{(\infty)})$. The procurve $Sh^{(p)}_{/\mathbb{Z}_{(p)}}$ is smooth over $\mathbb{Z}_{(p)}$ and satisfies the following canonical identities:*

$$Sh^{(p)}_{/\mathbb{Z}_{(p)}} = \varprojlim_{N,p\nmid N} Y(N)_{/\mathbb{Z}_{(p)}}, \quad Sh^{(p)}_{/\mathbb{Z}_{(p)}} \times_{\mathrm{Spec}(\mathbb{Z}_{(p)})} \mathbb{Q} \cong Sh_{G(\mathbb{Z}_p)} = \varprojlim_{N,p\nmid N} Y(N)_{/\mathbb{Q}}.$$

Proof. Let $\mathbb{Z}^\times_{(p)+} = \mathbb{Z}_p^\times \cap \mathbb{Q}_+^\times$. As long as Γ is maximal at p, we can identify $Cl(\Gamma) = \mathbb{A}^{(\infty)\times}/\mathbb{Q}_+^\times \det(\Gamma)$ with $\mathbb{A}^{(p\infty)\times}/\mathbb{Z}^\times_{(p)+}\det(\Gamma^{(p)})$. Thus, we may choose the representative set $\{c\}$ in $(\widehat{\mathbb{Z}}^{(p)})^\times$ (and hence we may assume the self-duality of L at p). By the same process of bringing $\mathcal{E}^{\mathbb{Q}}_\Gamma$ isomorphically to $\mathcal{E}'_{\Gamma/\mathbb{Q}}$ [for an open, compact subgroup $\Gamma \subset G(\widehat{\mathbb{Z}})$ maximal at p], the functor is equivalent to $\mathcal{E}'_{\Gamma/\mathbb{Z}_{(p)}}$ defined over $SCH_{/\mathbb{Z}_{(p)}}$; thus, it is representable over $\mathbb{Z}_{(p)}$, giving a canonical model $Sh^{(p)}_{\Gamma/\mathbb{Z}_{(p)}}$ over $\mathbb{Z}_{(p)}$. The functor $\mathcal{E}'_{\Gamma/\mathbb{Z}_{(p)}}$ is a disjoint union of functors $\mathcal{E}'_{\Gamma,c}$ indexed by $c \in Cl(\Gamma)$, where

$$\mathcal{E}'_{\Gamma,c}(S) = \left[(E, \overline{\eta}^{(p)})_{/S} \middle| \overline{\eta}^{(p)}(\widehat{L}) = T^{(p)}(E_s) \text{ and } c(\eta^{(p)}) \equiv c \mod \det(\Gamma)\right]. \tag{7.1.12}$$

Since $\overline{\eta}^{(p)} \in (LV^{(p)}/\Gamma)(S)$, at the geometric point $s \in S$, we have a small étale neighborhood $U \ni s$ such that $\overline{\eta}^{(p)}$ is the image of a full prime-to-p-level structure $\eta^{(p)}$ over U, and we define $\overline{\eta}^{(p)}(\widehat{L}_{\mathfrak{c}}) := \eta^{(p)}(\widehat{L}_{\mathfrak{c}})$ [which is independent of the choice of $\eta^{(p)}$ because $\Gamma \cdot \widehat{L} = \widehat{L}$].

A subtle point is to relate $Sh^{(p)}_{/\mathbb{Z}_{(p)}}$ to $Sh_{/\mathbb{Q}}$. Since the above process of identifying functors $\mathcal{E}^{\mathbb{Q}}_{\Gamma(N)}$ and $\mathcal{E}^{(p)}_{\Gamma(N)}$ ends up with the functor $\mathcal{E}'_{\Gamma(N)}$ independent of p [which is represented by $Y(N)_{/\mathbb{Z}_{(p)}}$] as long as N is prime to p, we conclude that $Sh^{(p)} = \varprojlim_{N,p\nmid N} Y(N)_{/\mathbb{Z}_{(p)}}$ and $Sh^{(p)} \otimes_{\mathbb{Z}_{(p)}} \mathbb{Q} \cong (Sh/G(\mathbb{Z}_p))$. Since $\mathbb{Q}$ has a strict class number 1, we have $\widehat{L}' \cong \widehat{L}$ for any other lattice $L' \subset V$. Therefore, we were able to fix the standard one L.

In the Hilbert modular cases, if the base field has nontrivial ideal classes, it is essential to allow all isomorphism classes of standard lattices L to define a functor like $\mathcal{E}'_{\Gamma(N)}$ classifying abelian varieties of $GL(2)$ type over an order of the base totally real field. This problem is more acute at p because over $\mathbb{Z}_{(p)}$, $T_p(E)$ does not determine L_p. Over $\mathbb{F}_p$ or $\overline{\mathbb{F}}_p$, the T_pE has rank lower than $2 = \operatorname{rank} L$. The self-duality at p of L has to be imposed to overcome this point (see the argument just above Remark 7.4 in Sect. 7.1.5 of [PAF]). Writing G_1 for the derived group (i.e., the commutator subgroup) of G, we also implicitly used the density of $G_1(\mathbb{Q}) = SL_2(\mathbb{Q})$ in $G_1(\mathbb{A}^{(\infty)})$ (the strong approximation theorem: Exercise 7.9) in order to know that geometrically irreducible components of $Sh^{(p)}_\Gamma$ are indexed by the class group $Cl(\Gamma)$: $\pi_0(Sh^{(p)}_{\Gamma/\overline{\mathbb{Q}}}) \cong Cl(\Gamma)$.

We now give a proof of the smoothness of $Sh^{(p)}_{/\mathbb{Z}_{(p)}}$, assuming $p \geq 5$ (and after that, we explain how to do it when $p = 2, 3$). To see the flatness of $X := Sh_{\widehat{\Gamma}(N)}$ over $\mathbb{Z}_{(p)}$, we look at the functor $\wp_{\Gamma(N)}$, which is represented by $\mathcal{M}_{\Gamma(N)}$. Since $\mathcal{M}_{\Gamma(N)}$ is a closed subscheme of $\mathbf{E}[N] \times_{\mathcal{M}_1} \mathbf{E}[N]$ étale over $\mathcal{M}_1$ by the construction in Sect. 6.2.10 and $\mathcal{M}_1$ is smooth over $\mathbb{Z}[\frac{1}{6}]$, $\mathcal{M}_{\Gamma(N)}$ is smooth over $\mathbb{Z}_{(p)}$. As seen before, $\mathbb{G}_m$ acts on $\mathcal{M}_{\Gamma(N)}$, and $Y(N)_{\zeta_N} = \mathrm{Proj}(\mathcal{R}_{\Gamma(N)}) = \mathcal{M}_{\Gamma(N)}/\mathbb{G}_m$; so $Y(N)_{\zeta_N} = \mathrm{Spec}(A)$ for the degree-0 component A of $\mathcal{R}_{\Gamma(N)}$. The action of $\mathbb{G}_m$ is free if $N \geq 3$ as $\mathrm{Aut}((E, \phi_N)_{/R}) = \{1\}$ for all test objects $(E, \phi_N)_{/R}$ of level $\Gamma(N)$ (see [GME] Sect. 2.6.4). Thus, $\mathcal{M}_{\Gamma(N)}$ is a principal homogeneous space over $Y(N)_{\zeta_N}$ if $N \geq 3$. If $\widehat{\Gamma}(N) \subset \widehat{\Gamma}(24)$, $t = \sqrt[12]{\Delta}$ is actually a unit of degree 1 in $\mathcal{R}_{\Gamma(N)}$, and hence $\mathcal{R}_{\Gamma(N)} = A[t, t^{-1}] = A \otimes_{\mathbb{Z}} \mathbb{Z}[t, t^{-1}]$, which shows that $\mathcal{M}_{\Gamma(N)} = \mathbb{G}_m \times Y(N)_{\zeta_N}$. This shows that $Y(N)_{\zeta_N}$ is smooth over $\mathbb{Z}[\frac{1}{6N}, \mu_N]$, since $\mathcal{M}_{\Gamma(N)}/\mathcal{M}_{\Gamma(1)/\mathbb{Z}[\mu_N, \frac{1}{6N}]}$ is étale and $\mathcal{M}_{\Gamma(1)/\mathbb{Z}[\mu_N, \frac{1}{6N}]}$ is smooth. Since $\mathbb{Z}[\frac{1}{6N}, \mu_N]_{/\mathbb{Z}[\frac{1}{6N}]}$ is étale, $Y(N) = Sh_{\widehat{\Gamma}(N)} = \mathrm{Res}_{\mathbb{Z}[\frac{1}{6N}, \mu_N]/\mathbb{Z}[\frac{1}{6N}]} Y(N)_{\zeta_N}$ is smooth over $\mathbb{Z}[\frac{1}{6N}]$ and hence smooth over $\mathbb{Z}_{(p)}$. Then $Sh^{(p)} = \varprojlim_{N, p \nmid N, 24 | N} Sh_{\widehat{\Gamma}(N)/\mathbb{Z}_{(p)}}$ is smooth over $\mathbb{Z}_{(p)}$.

Now we give a sketch of the proof when $p = 2, 3$. By an explicit construction of the moduli $Y(lq)$ of $\mathcal{E}_{\Gamma(lq)}$ already used in the proof of Corollary 6.16 for sufficiently large two distinct primes l, q different from p, we know that $Y := Y(lq)$ is smooth over $\mathbb{Z}_{(p)}$. Then we can check in the same manner of the proof of representability of $\wp_{\Gamma(N)}$ in Sect. 6.2.10 that $Y(lqN)$ is a closed subscheme in $\mathbf{E}[lqN] \times_Y \mathbf{E}[lqN]$ over $\mathbb{Z}[\frac{1}{lqN}]$ for the universal elliptic curve $\mathbf{E}$ over Y. Then the above argument works again, and we get smoothness of $Sh_{\widehat{\Gamma}(lqN)}$ over $\mathbb{Z}_{(p)}$ if $p \nmid lqN$. Thus, the limit $Sh^{(p)} = \varprojlim_{N, p \nmid N} Sh_{\widehat{\Gamma}(lqN)/\mathbb{Z}_{(p)}}$ is smooth over $\mathbb{Z}_{(p)}$.

There is a different argument. We remark that for a local Artinian $\mathbb{Z}_{(p)}$-algebra R with finite residue field, any test object (E_0, ω_0) over R/I (for an R-ideal I) can be lifted to a test object (E, ω) over R (by Corollary 6.17). If E_0 has a $\Gamma(N)$-level structure ϕ_0 for N prime to p, ϕ_0 lifts uniquely to $\phi : (\mathbb{Z}/N\mathbb{Z})^2 \cong E[N]$ by the étaleness of $E[N]_{/R}$. Further, if $N \geq 3$, $\mathrm{Aut}((E, \phi_N)_{/R})$ is trivial; hence, the lift of ϕ_N is unique.

Pick a maximal point x_0 of $X := Sh_{\widehat{\Gamma}(N)}$. Consider the stalk $\mathcal{O}_{X,x_0}$ with maximal ideal $\mathfrak{m}_0$, which is a local ring over $\mathbb{Z}_{(p)}$ with finite residue field $k(x_0)$. By the definition of smoothness (taking flatness for granted), we need to show that for any square-zero ideal I of an Artinian local ring R over $\mathbb{Z}_{(p)}$, any $\mathbb{Z}_{(p)}$-algebra homomorphism $\underline{\varphi}_0 : \mathcal{O}_{X,x_0} \to R/I$ lifts to $\underline{\varphi} : \mathcal{O}_{X,x_0} \to R$. The map $\underline{\varphi}_0$ induces $\varphi_0 : \mathrm{Spec}(R/I) \to \mathrm{Spec}(\mathcal{O}_{X,x_0}) \subset Sh_{\widehat{\Gamma}(N)}$. Take $(E_0, \phi_0) = \varphi_0^*(\mathbf{E}, \boldsymbol{\phi})$ for the universal test object $(\mathbf{E}, \boldsymbol{\phi})_{/Sh_{\widehat{\Gamma}(N)}}$. Then we can lift (E_0, ϕ_0) to $(\mathbf{E}, \boldsymbol{\phi})$ over R, which induces $\varphi : \mathrm{Spec}(R) \to Sh_{\widehat{\Gamma}(N)}$ by the universality such that $\varphi^*(\mathbf{E}, \boldsymbol{\phi}) \cong (E, \phi)_{/R}$. Since $(E, \phi)_{/R}$ is the lift of $(E_0, \phi_0)_{/(R/I)}$, $\phi \mod I = \phi_0$, and hence φ factors through $\mathrm{Spec}(\mathcal{O}_{X,x_0})$, getting the corresponding $\underline{\varphi} : \mathcal{O}_{X,x_0} \to R$, which lifts $\underline{\varphi}_0$. Thus, by definition, $Sh_{\widehat{\Gamma}(N)}$ and its limit $Sh^{(p)}$

are smooth over $\mathbb{Z}_p$ and hence over $\mathbb{Z}_{(p)}$ (as smoothness only depends on the completion).

Example 7.14 Let us return to the setting of Sect. 6.4; thus, we start with an imaginary quadratic field M in which the prime p splits into $(p) = \mathfrak{p}\overline{\mathfrak{p}}$, with $\mathfrak{p}$ corresponding to i_p. For each lattice $\mathfrak{a} \subset M$, we made a tuple in (6.4.4) $x(\mathfrak{a})_{/\mathcal{W}} = (E(\mathfrak{a}), \eta^{(p)} : \widehat{\mathfrak{a}}^{(p)} \cong T^{(p)}E(\mathfrak{a}))$ of a CM elliptic curve $E(\mathfrak{a})$ defined over $\mathcal{W}$ and level structures. Fixing a basis of $\widehat{\mathfrak{a}}^{(p)}$ over $\widehat{\mathbb{Z}}^{(p)}$ and hence identifying $\widehat{\mathfrak{a}}^{(p)} \cong (\widehat{\mathbb{Z}}^{(p)})^2$, we get a $\mathcal{W}$-point $(E(\mathfrak{a}), \eta^{(p)})_{/\mathcal{W}}$ of $Sh^{(p)}(\mathcal{W})$ for $\eta^{(p)}$ in (6.4.4). This is an example of a CM point. After specifying the identification $\widehat{\mathfrak{a}}^{(p)} \cong (\widehat{\mathbb{Z}}^{(p)})^2$, we often write this point as $x(\mathfrak{a}) \in Sh^{(p)}(\mathcal{W})$. If we add the information of a p-adic level structure $\eta^\circ : \mu_{p^\infty} \hookrightarrow E(\mathfrak{a})[p^\infty]$ and $\eta^{et} : \mathbb{Q}_p/\mathbb{Z}_p \cong E(\mathfrak{a})[p^\infty]^{et}$ in Definition 6.57, we get a point, again written as $x(\mathfrak{a})$, of the Igusa tower I over $Sh^{(p)}$ we describe in the following Sect. 7.2.

7.1.5 Finite-Level Structure

Slightly more generally, we take a discrete valuation ring $\mathcal{W}$ flat over $\mathbb{Z}_{(p)}$ as a base ring B. Hence, hereafter $\mathcal{W}$ is not necessarily equal to the discrete valuation ring $i_p^{-1}(W \cap i_p(\bigcup_{p\nmid N} \mathbb{Q}[\mu_N]))$. We consider the base change $Sh^{(p)}_{\Gamma/\mathcal{W}} = Sh^{(p)}_{/\mathbb{Z}_{(p)}} \times_{\mathrm{Spec}(\mathbb{Z}_{(p)})} \mathrm{Spec}(\mathcal{W})$. Consider the following open compact subgroups for an integer $N \geq 1$:

$$\begin{aligned}
\widehat{\Gamma}_0(N) &= \left\{ \left(\begin{smallmatrix} a & b \\ c & d \end{smallmatrix}\right) \in G(\widehat{\mathbb{Z}}) \middle| c \in N\widehat{\mathbb{Z}} \right\} \\
\widehat{\Gamma}_1(N) &= \left\{ \left(\begin{smallmatrix} a & b \\ c & d \end{smallmatrix}\right) \in \widehat{\Gamma}_0(N) \middle| d-1 \in N\widehat{\mathbb{Z}} \right\} \\
\widehat{\Gamma}_1^1(N) &= \left\{ \left(\begin{smallmatrix} a & b \\ c & d \end{smallmatrix}\right) \in \widehat{\Gamma}_1(N) \middle| a-1 \in N\widehat{\mathbb{Z}} \right\} \\
\widehat{\Gamma}(N) &= \left\{ \left(\begin{smallmatrix} a & b \\ c & d \end{smallmatrix}\right) \in \widehat{\Gamma}_1^1(N) \middle| b \in N\widehat{\mathbb{Z}} \right\}.
\end{aligned} \tag{7.1.13}$$

We write Γ for $\widehat{\Gamma}_j(N)$ $(j = 0, 1)$ or $\widehat{\Gamma}_1^1(N)$. Then we consider the functor $\mathcal{E}_\Gamma^{(p)}$. We assume that $\mathcal{W}$ contains a primitive Nth root. We suppose that N is prime to p [hence, $\Gamma_j(N)$ is maximal at p]. Note that $Cl(\Gamma) = \{1\}$ for $\Gamma = \widehat{\Gamma}_j(N)$. Then the disjoint union $\bigsqcup_{c\in Cl(\Gamma)}$ as in the previous section is over a singleton for such Γ, and we find $\mathcal{E}_\Gamma^{(p)} \cong \mathcal{E}'_{\Gamma,1}$ if $\Gamma = \widehat{\Gamma}_j(N)$ $(j = 0, 1)$, and $\mathcal{E}_\Gamma^{(p)} \cong \bigsqcup_{c\in(\mathbb{Z}/N\mathbb{Z})^\times} \mathcal{E}'_{\Gamma,c}$ for $\Gamma = \widehat{\Gamma}_1^1(N)$. Thus, $Y_j(N) = Y(\widehat{\Gamma}_j(N))$ is geometrically irreducible for $j = 0, 1$. The functor $\mathcal{E}'_\Gamma$ assigns to each $\mathcal{W}$-algebra R the set of all isomorphism classes of test objects $(E, \phi_N)_{/R}$ defined over R, where ϕ_N is the following data:

- an exact sequence $\mu_N \hookrightarrow E[N] \twoheadrightarrow \mathbb{Z}/N$ over R if $\Gamma = \widehat{\Gamma}_1^1(N)$;
- an embedding $\phi_N : \mu_{N/R} \hookrightarrow E[N]_{/R}$ if $\Gamma = \widehat{\Gamma}_1(N)$;

- a locally free group subscheme in $E[N]$ isomorphic to μ_N locally over $ET_{/R}$ if $\Gamma = \widehat{\Gamma}_0(N)$.

As before, identify $\mathbb{Z}/N\mathbb{Z}$ with $\mu_{N/\mathcal{W}}$ by $1 \mapsto \zeta_N$, and regard the self-duality pairing $\langle\cdot,\cdot\rangle : E[N] \times E[N] \to \mu_N$ as an isomorphism of group schemes $\langle\cdot,\cdot\rangle : E[N] \wedge E[N] \cong \mathbb{Z}/N\mathbb{Z}(1)$. Pick a level-$\widehat{\Gamma}_1^1(N)$ structure $\phi_N : \mu_N \hookrightarrow E[N] \twoheadrightarrow \mathbb{Z}/N$. Then the image $\phi_N(\zeta_N)$ of $\zeta_N \in \mu_N$ in $E[N]$ and any choice of $\phi_N(1) \in E[N]$ projecting down to $1 \in \mathbb{Z}/N\mathbb{Z}$ give rise to a unique $\langle \phi_N(\zeta_N), \phi_N(1)\rangle \in \mathbb{Z}/N\mathbb{Z}^\times(1)$ as $\langle \mu_N, \mu_N\rangle = 0$.

Exercise 7.15 *Prove that*

$$\langle \phi_N(\zeta_N), x\rangle = \langle \phi_N(\zeta_N), x'\rangle$$

for any $x, x' \in E[N]$ projecting down to $1 \in \mathbb{Z}/N\mathbb{Z}$ for a given exact sequence $\phi_N : \mu_N \hookrightarrow E[N] \twoheadrightarrow \mathbb{Z}/N$ (of finite flat group schemes).

Then we have the following expression of $\mathcal{E}'_{\widehat{\Gamma}_1^1(N),c}$:

$$\mathcal{E}'_{\widehat{\Gamma}_1^1(N),c}(R) = \left[(E, \phi_N : \mu_N \hookrightarrow E[N] \twoheadrightarrow \mathbb{Z}/N\mathbb{Z})_{/R} \middle| \langle \phi_N(\zeta_N), \phi_N(1)\rangle = \zeta_N^c\right]. \tag{7.1.14}$$

Exercise 7.16 *Verify the expression* (7.1.14) *if $N \geq 4$.*

If $N \geq 4$, $\mathrm{Aut}((E,\phi_N)_{/R}) = \{1\}$ for $(E,\phi_N) \in \mathcal{E}'_{\widehat{\Gamma}_1^1(N),c}(R)$ [so if $\eta^{(p)}(\widehat{L}) = T^{(p)}(E)$, $\overline{\eta}^{(p)}$ is the really the orbit $\eta^{(p)} \circ \widehat{\Gamma}_1^1(N)$ before sheafication of $LV/\widehat{\Gamma}_1^1(N)$]. This point is essential to answer the exercise above.

By this triviality of the automorphism of level-$\widehat{\Gamma}_1^1(N)$ test objects, $\mathcal{E}'_{\widehat{\Gamma}_1^1(N),c}$ is representable by $Y(\Gamma_1^1(N))_{\zeta_N^c/\mathcal{W}}$ (if $p \nmid N \geq 4$). In any case, if $p \nmid N$, we have a coarse moduli scheme $Y(\Gamma)_{/\mathcal{W}}$ for $\mathcal{E}'_{\Gamma,c}$ with respect to the above three choices of Γ, since we have $Y(\Gamma) = Y(N)/\Gamma$, where $\gamma \in \Gamma$ acts on level-$\widehat{\Gamma}(N)$ test objects $(E, \phi_N : (\mathbb{Z}/N\mathbb{Z})^2 \cong E[N])$ by $\phi_N \mapsto \phi_N \circ \overline{\gamma}$ for the image $\overline{\gamma}$ of γ in $G(\mathbb{Z}/N\mathbb{Z})$ (see [GME] Sect. 2.6).

Let $T_G \subset G$ be the diagonal torus given by

$$T_G(R) = \left\{\left(\begin{smallmatrix} a & 0 \\ 0 & d\end{smallmatrix}\right) \in G(R) \middle| a, d \in \mathbb{G}_m(R)\right\}.$$

By definition, $T_G \cong \mathbb{G}_m^2$ by sending $\left(\begin{smallmatrix} a & 0 \\ 0 & d\end{smallmatrix}\right)$ to $(a,d) \in \mathbb{G}_m^2$. Often we write (a,d) for the diagonal matrix $\left(\begin{smallmatrix} a & 0 \\ 0 & d\end{smallmatrix}\right)$. On $(E, \phi_N : \mu_N \xrightarrow{t} E[N] \xrightarrow{b} \mathbb{Z}/N)_{/R} \in \mathcal{E}'_{\widehat{\Gamma}_1^1(N)}(R)$, let $(a,d) \in T_G(\widehat{\mathbb{Z}})$ act by $(a,d)(E,\phi_N) = (E, \phi'_N)$, where $\phi'_N : \mu_N \xrightarrow{t \circ a} E[N] \xrightarrow{d \circ b} \mathbb{Z}/N$ with $t \circ a(\zeta) = t(\zeta^a)$ and $d \circ b(x) = d(b(x))$. This action sends $\mathcal{E}'_{\widehat{\Gamma}_1^1(N),c}$ isomorphically into $\mathcal{E}'_{\widehat{\Gamma}_1^1(N),adc}$, and hence $T_G(\widehat{\mathbb{Z}})$ transitively permutes geometrically irreducible components of $Sh^{(p)}_{\widehat{\Gamma}_1^1(N)/\mathcal{W}}$.

7.1.6 Adelic Action on Shimura Curves

We can let $g \in G(\mathbb{A}^{(\infty)})$ act on $\mathcal{E}_1^{\mathbb{Q}}(R)$ by

$$(E, \eta)_{/R} \mapsto (E, \eta \circ g)_{/R},$$

which gives a right action of $G(\mathbb{A}^{(\infty)})$ on the functor. This action obviously extends the action of $u \in \Gamma \subset G(\widehat{\mathbb{Z}})$ (defined earlier) to $G(\mathbb{A}^{(\infty)})$. By Lemma 4.17, this induces a right action of $G(\mathbb{A}^{(\infty)})$ on $Sh_{/\mathbb{Q}}$. If $g \in G(\mathbb{A}^{(p\infty)})$, the same action as above, $(E, \eta^{(p)}) \mapsto (E, \eta^{(p)} \circ g)$, is well defined on $\mathcal{E}^{(p)}_{/\mathcal{W}}$ and therefore extends to a p-integral action on $Sh^{(p)}_{/\mathcal{W}}$.

For a general open compact subgroup $\Gamma \subset G(\mathbb{A}^{(\infty)})$ maximal at p [not necessarily in $G(\widehat{\mathbb{Z}})$], we define $Y(\Gamma) = Sh^{(p)}_{\Gamma/\mathbb{Z}_{(p)}}$ by the (geometric) quotient $Sh^{(p)}/\Gamma$, as Γ acts on $Sh^{(p)}$ through a constant finite group (see [GME] 1.8.3 for geometric quotients). Over $\mathbb{Q}$, for any open compact subgroup Γ, we have $Sh_{\Gamma/\mathbb{Q}} = Sh/\Gamma$. Let $\Sigma = \emptyset$ or $\Sigma = \{p\}$.

Lemma 7.17 *If Γ is sufficiently small with trivial $\mathrm{Aut}((E, \overline{\eta}^{(\Sigma)}_\Gamma))$ for all test objects $(E, \overline{\eta}^{(\Sigma)}_\Gamma)$, $Sh^{(\Sigma)}$ is étale over $Sh^{(\Sigma)}_\Gamma$, and each geometric fiber of $Sh^{(\Sigma)}$ over $Sh^{(\Sigma)}_\Gamma$ is isomorphic to the group Γ.*

We give a proof assuming $\frac{1}{6} \in \mathcal{W}$ [general cases can be treated using $Y(lq)$ for two primes l and q as we have done many times now].

Proof. Since the proof is the same for any choice of Σ, we assume $\Sigma = \emptyset$. Since we have a commutative diagram

$$\begin{array}{ccc} Sh & \xrightarrow[g]{\sim} & Sh \\ \downarrow & & \downarrow \\ Sh_\Gamma & \xrightarrow[g]{\sim} & Sh_{g^{-1}\Gamma g} \end{array}$$

by the action of $g \in G(\mathbb{A}^{(\infty)})$, étaleness of the left vertical arrow is equivalent to étaleness of the right vertical arrow. Thus, we may assume $g^{-1}\Gamma g \subset \widehat{\Gamma}(24)$ if Γ is sufficiently small [as maximal compact subgroups of $G(\mathbb{A}^{(\infty)})$ are all conjugate to $G(\widehat{\mathbb{Z}})$]. Thus, we may suppose $\Gamma = \widehat{\Gamma}(24)$. Since $\mathcal{M}_{\Gamma(24N)}$ is a closed subscheme of $\mathbf{E}[24N] \times_{\mathcal{M}_1} \mathbf{E}[24N]$ (for the universal elliptic curve $\mathbf{E}_{/\mathcal{M}_1}$), which is étale over $\mathcal{M}_{\Gamma(24)} \subset \mathbf{E}[24] \times_{\mathcal{M}_1} \mathbf{E}[24]$. Write $Y(24N)_{\zeta_{24N}} = \mathrm{Spec}(A_N)$. Then A_N is the degree-0 component of $\mathcal{R}_{\Gamma(24N)}$. As we have already observed, $t = \sqrt[12]{\Delta}$ is a unit of degree 1 in $\mathcal{R}_{\Gamma(24)}$. Thus, $\mathcal{M}_{\Gamma(24N)} = \mathrm{Spec}(A_N[t, t^{-1}]) = \mathbb{G}_m \times Y(24N)_{\zeta_{24N}}$, and $Sh_{\widehat{\Gamma}(24N)} = Y(24N)$ is étale over $Y(24) = Sh_{\widehat{\Gamma}(24)}$ as $Y(24N) = \bigsqcup_{c \in (\mathbb{Z}/24N\mathbb{Z})^\times} Y(24N)_{\zeta^c_{24N}}$. Passing to the projective limit, Sh is étale (infinite) over $Y(24)$. Since $Sh_{\widehat{\Gamma}(24)} = Sh/\widehat{\Gamma}(24)$ and

$\widehat{\Gamma}(24)$ acts freely on Sh [because $\mathrm{Aut}((E,\phi_{24})_{/R}) = \{1\}$], we find that any geometric fiber is one-to-one onto $\widehat{\Gamma}(24)$, as desired.

We can further make a quotient by $Z(\mathbb{A}^{(\Sigma\infty)})\Gamma$ for the center $Z \subset G$. The action of $z \in Z(\mathbb{Z}_{(\Sigma)})$ on (E,η) is induced by the prime-to-Σ isogeny $z : E \to E$, and $Z(\mathbb{Z}_{(\Sigma)})$ acts trivially on Sh. Since $Z(\mathbb{A}^{(\Sigma\infty)}) = Z(\widehat{\mathbb{Z}}^{(\Sigma)})Z(\mathbb{Z}_{(\Sigma)})$,

$$Sh^{(\Sigma)}_{Z(\mathbb{A}^{(\Sigma\infty)})\Gamma} = Sh^{(\Sigma)}_{Z(\widehat{\mathbb{Z}}^{(\Sigma)})\Gamma}.$$

If $\Sigma = \{p\}$, Γ is maximal at p; hence, $Z(\widehat{\mathbb{Z}}^{(p)})\Gamma = Z(\widehat{\mathbb{Z}})\Gamma$. Thus, assuming $\Sigma \subset \{p\}$, we have

$$Sh^{(\Sigma)}_{Z(\mathbb{A}^{(p\infty)})\Gamma} = Sh^{(\Sigma)}_{Z(\widehat{\mathbb{Z}})\Gamma}.$$

The action of $(a,d) \in T_G(\widehat{\mathbb{Z}})$ on $\mathcal{E}'_\Gamma$ sends $\mathcal{E}'_{\Gamma,c}$ to $\mathcal{E}'_{\Gamma,adc}$, as we have observed. Thus, $z \in Z(\widehat{\mathbb{Z}})$ acts by sending the index c to z^2c [and hence, $Y(\Gamma_1^1(N))_{\zeta_N^c} \cong Y(\Gamma_1^1(N))_{\zeta_N^{cz^2}}$]. This shows

$$Sh^{(p)}_{Z(\widehat{\mathbb{Z}})\widehat{\Gamma}_1^1(N)/\mathcal{W}} \cong \bigsqcup_{c \in Cl(N)/(Cl(N))^2} Y(\Gamma_1^1(N))_{\zeta_N^c/\mathcal{W}} \quad \text{and}$$

$$Sh_{\widehat{\Gamma}_0(N)/\mathcal{W}} = Sh_{Z(\widehat{\mathbb{Z}})\widehat{\Gamma}_1(N)/\mathcal{W}} = Y_0(N)_{/\mathcal{W}}. \qquad (7.1.15)$$

7.1.7 Isogeny Action

Recall the right action of $g \in G(\mathbb{A}^{(\infty)})$ on $Sh_{/\mathbb{Q}}$ given by $(E,\eta) \mapsto (E,\eta\circ g)$. Consider $\mathcal{G} := G(\mathbb{A}^{(\infty)})/Z(\mathbb{Q})$ (see [IAT] Theorem 6.23). The following fact has been shown in [IAT] Theorem 6.23 (see also [PAF] Theorem 4.14):

Theorem 7.18 *The action of the group $G(\mathbb{A}^{(\infty)})$ on Sh leaves any geometrically irreducible component $Y \subset Sh$ stable, and we have $\mathcal{G} = G(\mathbb{A}^{(\infty)})/Z(\mathbb{Q}) = \mathrm{Aut}_{SCH}(Y_{/\mathbb{Q}})$. Each geometrically irreducible component Y is geometrically defined over the maximal cyclotomic extension $\mathbb{Q}^{ab}/\mathbb{Q}$. The right action of $g \in \mathcal{G}$ on $[x,g']$ is given by*

$$[x,g'] \mapsto [x,g'g].$$

The image $\mathcal{G}_+$ in $\mathcal{G}$ of $\{g \in G(\mathbb{A}^{(\infty)}) | 0 < \det(g) \in \mathbb{Q}^\times\}$ [containing $SL_2(\mathbb{A}^{(\infty)})$] falls in $\mathrm{Aut}_{SCH}(Y_{/\mathbb{Q}^{ab}})$.

Here is an outline of the proof. For each elliptic curve $E_{/\mathbb{C}}$, we have $E(\mathbb{C}) \cong \mathbb{C}/L_z$ with $L_z = \mathbb{Z}z + \mathbb{Z}$ for $z \in \mathfrak{H}$. Note that $TE = \varprojlim_N E[N](\mathbb{C}) = \varprojlim_N N^{-1}L_z/L_z = L_z \otimes_{\mathbb{Z}} \widehat{\mathbb{Z}}$. Identifying the $\mathbb{Q}$-span $\mathbb{Q}\cdot L_z$ with $V = \mathbb{Q}^2$ by $(a,b) \mapsto az+b$, we get from this identification $\eta_z : V(\mathbb{A}^{\infty)}) \cong V(E)$. The point $[z,1]$ carries $(E,\eta_z)_{/\mathbb{C}}$ by the moduli interpretation; hence, $g[z,1] = [z,g]$

corresponds to $(E, \eta_z \circ g)_{/\mathbb{C}}$. Thus, the action $(E, \eta) \mapsto (E, \eta \circ g)$ induces a homomorphism $\tau : G(\mathbb{A}^{(\infty)}) \to \mathrm{Aut}_{SCH}(Sh)$ realized by $\tau(g) : [z, 1] \mapsto [z, g]$. Since Y contains a point of the form $[z, g]$ by (7.1.5), $\tau(g)$ preserves $Y \subset Sh$, and we get a continuous group homomorphism $\tau : G(\mathbb{A}^{(\infty)}) \to \mathrm{Aut}_{SCH}(Y_{/\mathbb{Q}})$.

Since $Z(\mathbb{Q}) \cap \widehat{\Gamma}(N) = \{1\}$, $Z(\mathbb{Q}) \subset G(\mathbb{A}^{(\infty)})$ is a closed subgroup. The isogeny induced by $\alpha \in Z(\mathbb{Q}) = \mathbb{Q}^\times$ is an endomorphism of $E_{/S}$ as $\mathrm{End}_S(E) \otimes_{\mathbb{Z}} \mathbb{Q}$ contains $\mathbb{Q} \cdot \mathrm{id}_E$; so it acts trivially on $\mathcal{E}^{\mathbb{Q}}$ and hence trivially on Sh. Since the value of the duality pairing on the N-torsion points of the universal elliptic curve covers all roots of unity, Y has a scheme structure over $\mathrm{Spec}(\mathbb{Q}^{ab})$. It is proven in [IAT] Chap. 6 that the algebraic closure of $\mathbb{Q}$ in the function field of Y is equal to $\mathbb{Q}^{ab}$; so Y is geometrically irreducible over $\mathbb{Q}^{ab}$. The map τ is compatible with the Artin reciprocity map $[\cdot, \mathbb{Q}] : \mathbb{A}^\times/\mathbb{Q}^\times\mathbb{R}_+^\times \cong \mathrm{Gal}(\mathbb{Q}^{ab}/\mathbb{Q})$ sending $p \in \mathbb{Q}_p^\times$ to the geometric Frobenius element at p. In short, the following diagram is commutative (see [IAT] Theorem 6.23):

$$\begin{array}{ccc} Y & \xrightarrow{\tau(g)} & Y \\ \downarrow & & \downarrow \\ \mathrm{Spec}(\mathbb{Q}^{ab}) & \xrightarrow[{[\det(g),\mathbb{Q}]}]{} & \mathrm{Spec}(\mathbb{Q}^{ab}). \end{array}$$

Thus, $\tau(g)$ with $0 < \det(g) \in \mathbb{Q}^\times$ preserves each component.

By the following exercise, we conclude $\mathcal{G} \hookrightarrow \mathrm{Aut}_{SCH}(Y)$. A difficult point is to show the surjectivity of τ. For the function field $\mathfrak{F} = \mathfrak{F}_Y$ of Y, the morphism $g \mapsto \tau(g^{-1})$ is surjective onto $\mathrm{Aut}(\mathfrak{F})$ by [IAT] Theorem 6.23. This is sufficient as the category of algebraic function field of transcendental degree 1 over a field k is equivalent to the category of smooth proper, geometrically irreducible curves over k (see Remark 5.12). A proof covering higher-dimensional Shimura varieties is in [Sh4], and a scheme-theoretic proof is in [PAF]. Since in this book, we only need injectivity of τ, we do not touch surjectivity.

Exercise 7.19 *Using the fact* $Sh(\mathbb{C}) = G(\mathbb{Q})\backslash \left(\mathfrak{X} \times G(\mathbb{A}^{(\infty)})\right)/Z(\mathbb{Q})$ *(see (7.1.5)), prove that* $\tau : G(\mathbb{A}^{(\infty)}) \to \mathrm{Aut}_{SCH}(Y)$ *has kernel* $Z(\mathbb{Q})$.

Taking the integral model $Sh^{(p)}_{/\mathbb{Z}_{(p)}}$, we can think of $\mathrm{Aut}_{SCH}(Sh^{(p)}_{/\mathbb{Z}_{(p)}})$. In the same way as above, we get $G(\mathbb{A}^{(p\infty)})/Z(\mathbb{Z}_{(p)}) = \mathrm{Aut}_{SCH}(Sh^{(p)}_{/\mathbb{Z}_{(p)}})$ (see [PAF] Theorem 4.14). Since $Sh^{(p)}_{/\mathcal{W}}$ is smooth, we can think of $Sh^{(p)}_{/\mathbb{F}} = Sh^{(p)} \times_{\mathbb{Z}_{(p)}} \mathbb{F}$ for $\mathbb{F} = \overline{\mathbb{F}}_p$ and its ordinary locus $Sh^{\mathrm{ord}}_{/\mathbb{F}} = Sh^{(p)}[\frac{1}{H}]_{/\mathbb{F}}$ for the Hasse invariant H. Each mod-p elliptic curve over $x \in Sh^{\mathrm{ord}}(\mathbb{F})$ has a canonical lift giving rise to $\tilde{x} \in Sh^{(p)}(\mathcal{W})$ (assuming that $\mathcal{W}$ has residue field $\mathbb{F}$) and by definition $\tau(g)(\tilde{x}) = \widetilde{\tau(g)(x)}$ for $g \in G(\mathbb{A}^{(p\infty)})$. Note that $\{\tilde{x} \in Sh^{(p)}(\mathcal{W}) | x \in Sh^{\mathrm{ord}}(\mathbb{F})\}$ is Zariski dense in $Sh^{(p)}_{/\mathcal{W}}$. Thus, $\{g \in G(\mathbb{A}^{(p\infty)}) | \det(g) \in \mathbb{Z}_{(p)}\}/Z(\mathbb{Z}_{(p)}) \hookrightarrow \mathrm{Aut}_{SCH}(Sh^{(p)}_{/\mathbb{F}})$, and by smoothness, the set $\pi_0(Sh^{(p)}_{/\mathbb{C}})$ of connected components is in bijection to the corresponding mod p set $\pi_0(Sh^{(p)}_{/\mathbb{F}})$ (Zariski's connectedness theorem: Theorem 5.7). Therefore, we get

$$\{g \in G(\mathbb{A}^{(p\infty)}) | \det(g) \in \mathbb{Z}_{(p)}\}/Z(\mathbb{Z}_{(p)}) \hookrightarrow \mathrm{Aut}_{SCH}(Y^{(p)}_{/\mathbb{F}})$$

for each irreducible component $Y^{(p)}$ of $Sh^{(p)}_{/\mathbb{F}}$. We have proved the injectivity of

Theorem 7.20 *We have*

$$\{g \in G(\mathbb{A}^{(p\infty)}) | \det(g) \in \mathbb{Z}_{(p)}\}/Z(\mathbb{Z}_{(p)}) \cong \mathrm{Aut}_{SCH}(Y^{(p)}_{/\mathbb{F}}).$$

For the surjectivity, see [H06b].

Exercise 7.21 *Why does the Zariski density of* $\{\tilde{x} \in Sh^{(p)}(\mathcal{W}) | x \in Sh^{\mathrm{ord}}(\mathbb{F})\}$ *in* $Sh^{(p)}_{/\mathcal{W}}$ *imply the injectivity of the above theorem?*

7.1.8 Reciprocity Law at CM Points

Since $Sh(\mathbb{C}) = G(\mathbb{Q})\backslash\left(\mathfrak{X} \times G(\mathbb{A}^{(\infty)})\right)/Z(\mathbb{Q})$, we write $[z, g] \in Sh(\mathbb{C})$ for the image of $(z, g) \in \mathfrak{X} \times G(\mathbb{A}^{(\infty)})$. A point $[z, g]$ is called a *CM point* (a complex multiplication point) if h_z restricted to $\mathbb{G}_m \times 1 \subset \mathbb{S}_{/\mathbb{C}} = \mathbb{G}_m \times \mathbb{G}_m$ has image in a maximal torus $T_z \subset G := GL(2)$ defined over $\mathbb{Q}$. Since $h_z = J_1 \rho_z J_1^{-1}$, to have ρ_z defined over $\mathbb{Q}$ means that $(1, z)$ has to be a basis of an imaginary quadratic field $M_z = \mathbb{Q}(z)$ over $\mathbb{Q}$. Then we have $T_z \cong \mathrm{Res}_{M_z/\mathbb{Q}}\mathbb{G}_m$ for an imaginary quadratic field M_z over $\mathbb{Q}$. If confusion is unlikely, we write M for M_z. As before, we write $\mathfrak{O} = \mathfrak{O}_z$ for the integer ring of M. The $\mathbb{Z}$-span $\mathfrak{a} = \mathbb{Z} + \mathbb{Z}z$ is a lattice in M; so we have CM elliptic curve $E(\mathfrak{a})$ and the adelic level structure

$$\eta^{(p)}_z : (\widehat{\mathbb{Z}}^{(p)})^2 \cong (\widehat{\mathbb{Z}}^{(p)})^2 J_1 \left[\begin{smallmatrix} z \\ 1 \end{smallmatrix}\right] = \widehat{\mathfrak{a}}^{(p)} \xrightarrow{\sim} T^{(p)}E(\mathfrak{a})$$

as in Sect. 6.4.5. Then $(E(\mathfrak{a}), \eta^{(p)}_z)$ gives the point $[z, 1] \in Sh^{(p)}(\mathcal{W})$. We can extend $\eta^{(p)}_z$ to

$$\eta_z : \widehat{\mathbb{Z}}^2 \cong \widehat{\mathbb{Z}}^2 J_1 \left[\begin{smallmatrix} z \\ 1 \end{smallmatrix}\right] = \widehat{\mathfrak{a}} \xrightarrow{\sim} TE(\mathfrak{a}) = \varprojlim_N E(\mathfrak{a})[N](\mathbb{C}),$$

and $(E(\mathfrak{a}), \eta_z)$ is sitting over $[z, 1] \in Sh(\mathbb{C})$. If we choose another basis w_1, w_2 of $\widehat{\mathfrak{a}}$ over $\widehat{\mathbb{Z}}$, we have $g \in G(\widehat{\mathbb{Z}})$ such that $J_1 \left[\begin{smallmatrix} z \\ 1 \end{smallmatrix}\right] = g \left[\begin{smallmatrix} w_1 \\ w_2 \end{smallmatrix}\right]$; then $(E(\mathfrak{a}), \eta_z \circ g)$ is sitting over $[z, g] \in Sh$. If $\mathfrak{a}_p = \mathfrak{O}_p$, we will choose $g \in G(\widehat{\mathbb{Z}})$, whose p-component $g_p \in G(\mathbb{Z}_p)$ is compatible with $(\eta^\circ_p, \eta^{\mathrm{et}}_p)$; specifically, identifying $\mathfrak{a}_p = \mathfrak{O}_\mathfrak{p} \times \mathfrak{O}_{\overline{\mathfrak{p}}}$, we take g_p given by $g_p \left[\begin{smallmatrix} z \\ 1 \end{smallmatrix}\right] = \left[\begin{smallmatrix} w_1 \\ w_2 \end{smallmatrix}\right]$ for $w_1 = (1, 0) \in \mathfrak{O}_\mathfrak{p} \times \mathfrak{O}_{\overline{\mathfrak{p}}}$ and $w_2 = (0, 1) \in \mathfrak{O}_\mathfrak{p} \times \mathfrak{O}_{\overline{\mathfrak{p}}}$. If $\mathfrak{a}_p \neq \mathfrak{O}_p$, we take $\mathfrak{a}_\zeta$ as in Sect. 6.4.5 for $\zeta = \exp(2\pi i u/p^n)$ for $u \in (\mathbb{Z}/p^n\mathbb{Z})^\times$; so $\mathfrak{a}_\zeta$ spanned by

$$\begin{bmatrix} \frac{-u+z}{p^n} \\ 1 \end{bmatrix} = \begin{pmatrix} p^{-n} & -p^{-n}u \\ 0 & 1 \end{pmatrix} \begin{bmatrix} z \\ 1 \end{bmatrix}$$

over $\mathbb{Z}$. By (7.1.7), this operation corresponds to $v \mapsto \left(\begin{smallmatrix} p^n & u \\ 0 & 1 \end{smallmatrix}\right) v = J_1 {}^t g'^{-1} J_1^{-1} v$ ($v \in V(\mathbb{R})$) for $g' = \begin{pmatrix} p^{-n} & -p^{-n}u \\ 0 & 1 \end{pmatrix}$. Thus, if we choose $g_\zeta = \gamma_u g$ for $\gamma_u = \left(\begin{smallmatrix} p^n & u \\ 0 & 1 \end{smallmatrix}\right) \in G(\mathbb{Q})$ [$\zeta = \exp(2\pi i u/p^n)$], noting $\gamma_u^{-1}(z) = \frac{z-u}{p^n}$, we get, from Lemma 7.8,

$$[z, g_\zeta] = [z, \gamma_u^{(\infty)} g] = [\gamma_u^{-1}(z), g],$$

which implies

$$(E(\mathfrak{a}), \eta_z \circ g_\zeta) \approx (E(\mathfrak{a}_\zeta), \eta_{\frac{z-u}{p^n}} \circ g)$$

sits on $[z, g_\zeta]$, where "$\approx$" indicates that there is an isogeny $\alpha_\zeta : E(\mathfrak{a}_\zeta) \to E(\mathfrak{a})$ such that $\alpha_\zeta \circ \eta_{\frac{z-u}{p^n}} = \eta_z \circ \gamma_u$. Thus, we get the following result:

Lemma 7.22 *Let the notation be as above. If $\mathfrak{a}_p = \mathfrak{O}_p$, $(E(\mathfrak{a}), \eta_z \circ g)$ is the fiber at $[z, g] \in Sh(\mathbb{C})$ of the universal elliptic curve over Sh. More generally, for $\mathfrak{a}_\zeta$ with $\zeta = \exp(2\pi i u/p^n)$, the test object $(E(\mathfrak{a}), \eta_z \circ g_\zeta) \approx (E(\mathfrak{a}_\zeta), \eta_{\frac{z-u}{p^n}} \circ g)$ is the fiber at $[z, g_\zeta] \in Sh(\mathbb{C})$ of the universal elliptic curve over Sh.*

Exercise 7.23 *Compute an explicit form of complex number $a \in \mathbb{C}^\times$, which, by the multiplication $u \mapsto a \cdot u$, induces the isogeny $\alpha_\zeta : E(\mathfrak{a}_\zeta)(\mathbb{C}) = \mathbb{C}/\mathfrak{a}_\zeta \to \mathbb{C}/\mathfrak{a} = E(\mathfrak{a})(\mathbb{C})$. Why is this number a uniquely determined by α_ζ?*

We let $G(\mathbb{Q})$ act on the column vector space $V = \mathbb{Q}^2$ through the matrix multiplication. The action of T_z on V makes V a vector space over M of dimension 1 and gives rise to an embedding $r_z : T_z \hookrightarrow G$. Then the subspace $V_z = V \otimes_{\mathbb{Q}} \mathbb{C}$ on which h_z acts by its restriction $\mu_z = h_z|_{\mathbb{G}_m \times 1}$ is preserved by multiplication by M, yielding an embedding $\mathrm{id}_z : M \hookrightarrow \mathbb{C}$ of M. Since $M/\mathbb{Q}$ is imaginary quadratic, $\mu_z \times \overline{\mu}_z = h_z$, we find that $\{\mathrm{id}_z, c = c \circ \mathrm{id}_z\}$ for complex conjugation c is the full set I_z of complex embeddings of M_z into $\mathbb{C}$. Taking an elliptic curve A sitting on $z \in \mathfrak{X}$, we find that A has complex multiplication by M_z. For each $b \in T_z(\mathbb{A}^{(\infty)}) = M^\times_{\mathbb{A}^{(\infty)}}$, we have the Artin reciprocity image $[b, M] \in \mathrm{Gal}(M^{ab}/M)$, where M^{ab} is the maximal abelian extension of M. Since $T_z(\mathbb{R})$ is the centralizer of z, $[z, \gamma g] = [\gamma^{-1}(z), g]$ for $\gamma \in T_z(\mathbb{Q})$, and hence $[z, g] \mapsto [z, r_z(b) g]$ depends only on $[b, M]$ by class field theory. Now we are ready to state Shimura's reciprocity law for the CM point $[z, 1]$ (see [IAT] Theorem 6.31 and [EDM] Theorem A5.4):

Proposition 7.24 *Let $[z, 1]$ be a CM point in $Sh_{/\mathbb{Q}}$. Then the point $[z, 1]$ is M^{ab}-rational, and for any $b \in T_z(\mathbb{A}^{(\infty)})$ and $\gamma \in T_z(\mathbb{Q})$, we have*

$$[b^{-1}, M_z]([z, 1]) = [z, 1] \cdot r_z(b) = [z, r_z(b)].$$

In [IAT] Theorem 6.31, for a modular function $f \in \mathfrak{F}$ finite at $[z, 1]$, it is proven that $f([z, 1])^{[b, M_z]} = f^{\tau(r_z(b))}([z, 1]) = f([z, 1] \cdot r_z(b))$, which is equivalent to the above proposition as functions determine the point. Also, note that

Shimura's identification $\mathcal{G} = \mathrm{Aut}(\mathfrak{F})$ is given by $g \mapsto \tau(g^{-1})$, as his action on $\mathfrak{F}$ is from the right while ours is from the left by our normalization [we let $g \in G(\mathbb{A}^{(\infty)})$ act on Sh from the right: $(E, \eta) \mapsto (E, \eta \circ g)$; thus, its pullback action on $\mathfrak{F}$ is a left action].

More generally, we take a point $[z, g] = [z, 1] \cdot g$. By definition, $[z, g]$ is also a CM point (if we start with a CM point $[z, 1]$). Then we write $(E_x, T_x, M_x, r_x, \eta_x)$ for $(E_z, T_z, M_z, g^{-1} r_z g, \eta_z \circ g)$. Since the action of $g \in G(\mathbb{A}^{(\infty)})$ is defined over $\mathbb{Q}$, it commutes with the Galois action; so, writing $\sigma = [b^{-1}, M_x]$, we have

$$\sigma([z, g]) = \sigma([z, 1] \cdot g) = \sigma([z, 1]) \cdot g = [z, r_z(b) g] = [z, g] \cdot r_x(b).$$

Thus, we get the following interpretation:

Corollary 7.25 *Let $x = [z, g]$ be a CM point in $Sh_{/\mathbb{Q}}$. Then the point $[z, g]$ is M^{ab}-rational, and for any $b \in T_x(\mathbb{A}^{(\infty)})$ and $\gamma \in T_x(\mathbb{Q})$, we have*

$$[b^{-1}, M_x](x) = x \cdot r_x(b).$$

If $b \in T_x(\mathbb{Q}) = M^\times$, $[b^{-1}, M] = \mathrm{id}$; so we have

$$x \cdot r_x(\alpha) = x \quad \text{for } \alpha \in T_x(\mathbb{Q}) = M^\times. \tag{7.1.16}$$

To see this fact, note that $\mathrm{End}(E_x) \otimes_{\mathbb{Z}} \mathbb{Q} = M$. So $(E_x, \eta_x \circ r_x(\alpha)) \approx (E_x, \alpha \circ \eta_x)$; therefore, as an element in $\mathcal{E}^{\mathbb{Q}}(\mathbb{C})$, (E_x, η_x) is fixed by $r_x(T_x)(\mathbb{Q})$, as stated. Since we only need (7.1.16), we do not give a proof of the reciprocity law above, instead referring to Theorem 6.31 in [IAT].

Corollary 7.26 *Let (E_x, η_x) be the test object sitting over a geometric point $x \in Sh(\overline{\mathbb{Q}})$ projecting down to $x \in Y(N)(K)$ $(N \geq 3)$ for a finite extension $K/\mathbb{Q}$. Suppose E_x has complex multiplication by an imaginary quadratic field M. Suppose that all elements in $\mathrm{End}_{\overline{\mathbb{Q}}}(E_x)$ are defined over K. Then the J-invariant $J(E_x)$ is an algebraic integer in K and E_x extends to an elliptic curve with complex multiplication over the integer ring of K.*

There is a more elementary proof in [IAT] Sect. 4.6 for the integrality of $J(E_x)$. Here we give a moduli-theoretic proof.

Proof. Take any valuation ring V of K in which N is invertible. By Corollary 6.32, if $J(E_x) \in V$, E_x extends to an elliptic curve over V. Since the integer ring O_K is the intersection of all valuation rings of K, we conclude that $J(E_x)$ is an algebraic integer, and E_x extends to an elliptic curve over O_K as we can change N.

Thus, we need to prove $J(E_x) \in V$. Otherwise, for the residual characteristic l of V, we have $|q(E_x)|_l < 1$ since $J = q^{-1}(1 + f(q))$ for $f(q) \in q\mathbb{Z}[[q]]$, as is well known (see Sect. 2.4.1). Then E over the completion $\widehat{V}$ of V is a Tate curve with Tate period $q(E_x)$ (see Theorem 2.54). By the construction of the moduli $Y(nN)$ inside $\mathbf{E}[nN] \times \mathbf{E}[nN]$ in Sect. 6.2.10, the field of rationality

$K(x_n)$ of the image x_n of x in $Y(nN)$ for any $n \geq 1$ is a subfield of the field of rationality of $E_x[nN]$ over K (with bounded index). By Shimura's reciprocity law above, $K(x_n)/K$ is abelian, which is impossible for infinitely many choices of n by Corollary 2.56. Thus, $J(E_x) \in V$. This finishes the proof.

7.1.9 Degeneracy Operators

Let $\mathfrak{l} = (l)$ for a prime $l > 0$ be a prime ideal in $\mathbb{Z}$. Then, for a level-N prime to $\mathfrak{l}$ with $\Gamma \supset \widehat{\Gamma}(N)$, if $\mathfrak{l}A = A$ (for a $\mathbb{Z}[\frac{1}{6N}]$-algebra A), we define a morphism of functors

$$[\mathfrak{l}] : \wp_{\Gamma \cap \Gamma_0(\mathfrak{l})}(A) = \left[(E, \phi_N, C, \omega)_{/A}\right] \to \wp_\Gamma(A) = \left[(E, \phi_N, \omega)_{/A}\right]$$

by sending $(E, \phi_N, C, \omega)_{/A}$ to $(E/C, \pi \circ \phi_N, (\pi^*)^{-1}\omega)_{/A}$ for the projection $\pi : E \to E/C$, where C is a locally free subgroup of E étale-locally isomorphic to μ_l. Here the phrase "étale-locally" means that there exists a faithfully flat étale extension S of $\mathrm{Spec}(A)$ such that $C \times_A S \cong \mu_{l/S}$. Obviously, $[\mathfrak{l}]$ is a natural transformation; so, by pullback, we get the *degeneracy operator* $[\mathfrak{l}] : G_k(\Gamma; A) \to G_k(\Gamma \cap \Gamma_0(\mathfrak{l}); A)$ given by $f|[\mathfrak{l}](E, \phi_N, C, \omega) = f(E/C, \pi \circ \phi_N, (\pi^*)^{-1}\omega)$. Replacing ϕ_N by a level-p^∞ structure $\phi_p : \mu_{p^\infty} \hookrightarrow E$, we get the corresponding map $[\mathfrak{l}] : V_\Gamma(R) \to V_{\Gamma \cap \Gamma_0(\mathfrak{l})}$. Choose the positive generator l of $\mathfrak{l}$. Over complex numbers, we have

$$\mathrm{Tate}(q) = \mathbb{C}^\times/q^{\mathbb{Z}} \xleftarrow{\exp} \mathbb{C}/(2\pi i)(\mathbb{Z} + \mathbb{Z}z)$$

via $w = \exp(u)$, which has a canonical $\Gamma_0(\mathfrak{l})$-structure $\mu_l \subset \mathbb{C}^\times$. Then

$$\mathrm{Tate}(q)/\mu_l = \mathbb{C}/(2\pi i)(\frac{1}{l}\mathbb{Z} + \mathbb{Z}z) \overset{u \mapsto lu}{\cong} \mathbb{C}/(2\pi i)(\mathbb{Z} + \mathbb{Z}lz) = \mathrm{Tate}(q^l).$$

In the middle identity via $u \mapsto lu$, the differential du is sent to $\frac{1}{l}du$ (as its action on the tangent is multiplication by l). Thus, ω_{can} is sent to $\frac{1}{l}\omega_{can}$ on $\mathrm{Tate}(q^l)$. Similarly, $\phi_{N,can}$ is sent to $\phi_{N,can} \circ l$. So we have

$$f|[\mathfrak{l}](q) = f(\mathrm{Tate}(q^l), \phi_{N,can} \circ l, \frac{1}{l}\omega_{can}) = l^k \cdot \psi(l) f(q^l)$$

if the Nebentypus of f is ψ. This also applies to p-adic modular forms, as they are the p-adic limit of classical modular forms (limit principle in Sect. 1.3.3).

In terms of Shimura varieties, the operator $[\mathfrak{l}] : Sh \to Sh$ corresponds to $(E, \eta) \mapsto (E, \eta \circ \begin{pmatrix} 1 & 0 \\ 0 & l_{\mathfrak{l}}^{-1} \end{pmatrix})$, where $l_{\mathfrak{l}} = (1, \ldots, 1, \overset{\mathfrak{l}}{l}, 1, \ldots, 1, \ldots) \in \mathbb{A}^\times$. The reason for this expression is twofold: First, $\eta \mapsto \eta \circ g$ corresponds to $\widehat{L} \mapsto g^{-1}\widehat{L}$ for the lattice $\widehat{L} = \eta^{-1}(TE)$; second, on classical modular forms, the action is given by $z \mapsto \left(\begin{smallmatrix} l & 0 \\ 0 & 1 \end{smallmatrix}\right)(z) = lz$, and by the effect of J_1 as in Sect. 7.1.8, we have $J_1 \left(\begin{smallmatrix} l & 0 \\ 0 & 1 \end{smallmatrix}\right) J_1^{-1}$ is equivalent to $\left(\begin{smallmatrix} 1 & 0 \\ 0 & l \end{smallmatrix}\right)$. Since this action only affects the $\mathfrak{l}$-factor $T_{\mathfrak{l}}E \subset TE$ (or, equivalently, the $\mathfrak{l}$-divisible group $E[\mathfrak{l}^\infty] \subset E$), we can have only a nontrivial component at the place $\mathfrak{l}$.

7.2 Igusa Tower

We defined the *Igusa tower* $I = I_{1,\infty} \to \cdots \to I_{1,n} \to \cdots \to I_{1,1} \to S_{1/\mathbb{F}}$ in Sect. 6.2.12 for the ordinary locus $S_1 = X(N)[\frac{1}{H}] \subset X(N)_{/\mathbb{F}}$. In this section, we give a proof of the irreducibility of $I_{1,m}$ for each $m = 1, 2, \dots, \infty$ (Theorem 6.42). A key point is that the automorphism group of $X^{(p)}_{/\overline{\mathbb{F}}_p}$ is big, containing $\mathcal{G}_1 = SL_2(\mathbb{A}^{(p\infty)})$, and the second key point is that $\mathrm{Aut}(I/S_1)$ contains $GL_1(\mathbb{Z}_p)$ and the two actions on $I_{/\overline{\mathbb{F}}_p}$ commute. There are at least two ways of showing irreducibility via the symmetry action on the Shimura variety: (i) the use of the automorphism group of the function field of the Shimura variety of characteristic 0 (cf. [PAF] Sects. 6.4.3 and 8.4.4), which uses characteristic-0 results to prove the characteristic-p assertion; and (ii) a purely characteristic-p proof following a line close to (i) (see [H09a]). In this book, we follow the first method. There are some other arguments purely in characteristic p (for example, Igusa's original argument in [I] relying on ramification at a supersingular point) to prove the same result (covering different families of reductive groups giving the Shimura variety) as sketched in [Ch4] for the Siegel modular variety. In this section, we write $\mathbb{F} = \overline{\mathbb{F}}_p$ (an algebraic closure of $\mathbb{F}_p$) and take $W = W(\mathbb{F})$ (the ring of Witt vectors with coefficients in $\mathbb{F}$ in [BCM] IX.1, or equivalently, the completed p-adic integer ring of the maximal unramified extension of $\mathbb{Q}_p$ in $\mathbb{C}_p$). Hereafter we assume $\mathcal{W} = i_p^{-1}(W)$, which is a strict henselization of $\mathbb{Z}_{(p)}$ (see [NMD] Sect. 2.2 for Henselian rings).

7.2.1 Axiomatic Approach to Irreducibility

Here is an axiomatic approach to prove the irreducibility of an étale covering $\pi : I \to S$ of a smooth irreducible variety S over $\mathbb{F} = \overline{\mathbb{F}}_p$. Write $\pi_0(I)$ for the set of connected components of I. We start with the following two axioms:

(A1) A group $\mathcal{G}_I = \mathcal{T} \times \mathcal{G}_1$ acts on I and S compatibly so that $\mathcal{T}$ acts on $I_{/S}$, $\mathcal{G}_1$ acts on S, and $\mathcal{G}_1$ acts trivially on $\pi_0(I)$.
(A2) $\mathcal{T}$ acts on each fiber of I/S **transitively**; so $\mathcal{T}$ acts transitively on $\pi_0(I)$.

Under (A1–2), we study the stabilizer subgroup T_x in $\mathcal{G}_I$ of a point x in a connected component I° of I and try to prove the following conclusion:

(G) $\{T_x\}_x$ for a collection of points x and $\mathcal{G}_1$ generate a dense subgroup of $\mathcal{G}_I$.

Once we reach the conclusion (G), by the transitivity (A2), we obtain $I^\circ = I$, getting the irreducibility of I.

A Shimura variety is said to be of PEL type (see [PAF] Sect. 7.1) if it is a moduli variety of abelian schemes (of fixed dimension d) with a given type of polarization P, having a given algebra B (with positive involution) as a subalgebra of its endomorphism algebra, and a level structure η. The elliptic modular Shimura curve classifies elliptic curves $E_{/S}$ (abelian schemes of dimension 1; Lemma 6.1, Theorems 2.32 and 6.3) and level structure $\eta : \mathbb{A}^2 \cong V(E)$. In this simplest case, the fixed endomorphism algebra is $\mathbb{Q} \subset \mathrm{End}(E_{/S}) \otimes_{\mathbb{Z}} \mathbb{Q}$, and the polarization is well hidden as each elliptic curve has a canonical polarization whose polar divisor is $[\mathbf{0}_E]$. Indeed, the Néron–Severi group $NS(E_{/S}) := \mathrm{Pic}(E_{/S})/\mathrm{Pic}^0(E_{/S}) \xrightarrow[\deg]{\sim} \mathbb{Z}$ has a canonical generator $[\mathbf{0}_E]$. For a higher-dimensional abelian scheme $A_{/S}$, $NS(A_{/S})$ has a higher rank, and a single A could have different polarizations.

In the setting of the Shimura variety Sh of PEL type (of level away from a given finite set Σ of places), assuming we have a smooth integral compactification of Sh over a p-adic discrete valuation ring $\mathcal{W}$ (see [ACS] 6.4.1), we can verify the axioms (A1–2) for the following reasons: Compatibility with the action in (A1) and the transitivity in (A2) follow from the definition. In the case of a Shimura variety of PEL type, S is usually the ordinary locus (the open locus on which a generalized Hasse invariant is invertible) of the modulo p Shimura variety $Sh_{/\mathbb{F}}$ of level away from a given finite set Σ of places containing p and ∞. Then for the adele ring $\mathbb{A}^{(\Sigma)}$ away from Σ, $\mathcal{G}_1$ is often the adele group $G_1(\mathbb{A}^{(\Sigma)})$ for a semisimple group $G_{1/\mathbb{Q}}$ (G_1 is the derived group of the starting reductive group in Shimura's data), and $\mathcal{T}$ is the $\mathbb{Z}_p$-points $T(\mathbb{Z}_p)$ of the reductive part T of a parabolic subgroup P of G_1 [as this notation suggests, we apply our axiomatic approach to (G_1, P, T) in Definition 7.1 to prove irreducibility of the elliptic Igusa tower]. Here P modulo its unipotent radical is reductive by definition, and the quotient can be realized as a subgroup of P, called the Levi subgroup (or the reductive part; see [LAG] Sect. 30). If we choose Σ so that $G_1(\mathbb{Q}_l)$ is generated by unipotent elements for all $l \notin \Sigma$, $\mathcal{G}_1$ has no nontrivial finite quotient group (as unipotent groups over characteristic-0 field are divisible). For any finite subcovering I'/S of I, $\mathcal{G}_1$ acts on the finite set $\pi_0(I')$ through a finite quotient of $\mathcal{G}_1$; so the action is trivial, getting (A1).

In the above discussion of how to verify (A1), a key ingredient is that $\mathcal{G}_1$ is large enough not to have finite (nontrivial) quotient. For the general Shimura variety, this is deduced from the existence of smooth toroidal compactification (if the Shimura variety is not projective; see [CSM], [DAV], and [ACS]) and a characteristic-0 determination of the automorphism group of the Shimura variety (principally due to Shimura; cf. [Sh4] II, [IAT] Theorem 6.23, and [PAF] Theorems 4.14 and 7.7). The compactification in our Shimura curve case is as described in Sect. 6.2.7. Alternatively, one can prove that $\mathcal{G}_1$ is large by showing that the l-adic monodromy representation for primes $l \neq p$ has a large open image in $G_1(\mathbb{A}^{(\Sigma)})$ (here an l-adic monodromy representation is given by the action of the fundamental group of the Shimura variety on the l-adic Tate module of the fiber at the base point). Indeed, C.-L. Chai [Ch2] (in the symplectic case) has deduced the open image result via group theory from the semisimplicity theorem of Grothendieck–Deligne of the l-adic representation. Chai's method would work for $l \notin \Sigma$ (for an appropriate Σ) in our setting.

Let $I^\circ_{/\mathbb{F}}$ be an irreducible component of $I_{/\mathbb{F}}$. We want to prove $I^\circ = \pi^{-1}(S) = I$ (irreducibility). Then $\mathrm{Gal}(I^\circ/S)$ is the stabilizer of $I^\circ \in \pi_0(I)$, and hence $\mathrm{Gal}(I^\circ/S) \subset \mathcal{T}$. If $\mathcal{T} = \mathrm{Gal}(I^\circ/S)$, we get $I^\circ = \pi^{-1}(S)$. Let D be the stabilizer of $I^\circ \in \pi_0(I)$ in $\mathcal{G}_I$. Pick a point $x \in I$ (which can be a generic point), and look at the stabilizer $T_x \subset \mathcal{G}_I$ of x. Since $g_x(x) \in I^\circ$ ($g_x \in \mathcal{T}$) by the transitivity of the action, we have $g_x T_x g_x^{-1} \subset D$. Then we show that $\mathcal{T} = \mathcal{G}_I/\mathcal{G}_1$ is generated topologically by $\{g_x T_x g_x^{-1} | x \in I\}$, which implies $\mathcal{T} = \mathrm{Gal}(I^\circ/S) = \mathrm{Gal}(I/S)$ from (G); so $I = I^\circ$.

In the setting of the Igusa tower of a Shimura variety, we can have at least three choices for the points $x \in I$:

(∞) A cusp, assuming that the group $G = \mathrm{Res}_{F/\mathbb{Q}} G_0$ for a quasi-split group G_0 over a number field F (acting on a tube domain). This is the proof given in $GSp(2n)$ in [DAV].

(cm) A closed point $x \in I(\mathbb{F})$ is fixed by a maximal torus T_x of G anisotropic at ∞; so $g_x T_x(\mathbb{Z}_{(p)}) g_x^{-1} \subset D$ for $\mathbb{Z}_{(p)} = \mathbb{Q} \cap \mathbb{Z}_p$. A well-chosen finite set of closed points $X := \{x\}$ is enough to generate a dense subgroup of D by $\{g_x T_x(\mathbb{Z}_{(p)}) g_x^{-1}\}_{x \in X}$. This is Ribet's choice in [Ri1] and [DRi] for Hilbert modular varieties and is also taken in [H09a]. If one uses a CM point (the so-called hypersymmetric point) carrying a product of copies of CM elliptic curves, often one such point is sufficient.

(gn) Taking a coordinate system $T = (T_1, \ldots, T_d)$ around $x \in Sh(W)$ with (x mod p) $\in I^\circ(\mathbb{F})$ (so that $\widehat{\mathcal{O}}_{Sh,x} \cong W[[T_1, \ldots, T_d]]$) and take the valuation

$$v_x(\sum_\alpha c(\alpha, f) T^\alpha) = \mathrm{Inf}_\alpha \, \mathrm{ord}_p(c(\alpha, f)).$$

Then the decomposition group D of v_x contains T_x (for all $x \in I^\circ$), and D is the stabilizer of the generic point of I° containing x (this choice is taken in [PAF] 8.4.4). The valuation v_x corresponds to the generic point of I°. The point x is chosen to be a cusp in our case of the elliptic modular Igusa tower (see Sect. 7.2.3). For the Hilbert–Siegel modular variety and $U(n, n)$ Shimura variety, the choice of the infinity cusp works as well (cf. [PAF] 6.4.3).

Actually, there is (at least) one more choice. Igusa [I] completed his tower over modular curves, adding supersingular points, and used such points to prove his irreducibility theorem in the 1950s. Here we describe the method (gn) for our Shimura curve, and the base point x we use is the infinity cusp.

Fix a prime p and an algebraic closure $\mathbb{F}$ of $\mathbb{F}_p$. Throughout this section, proofs of the results claimed are given assuming $p > 3$ (just for simplicity; see [H09a] for the treatment in the case $p = 2, 3$).

7.2.2 Mod-p Connected Components

We define a generic version of the Igusa tower and prepare necessary lemmas to count the number of generic points of the Igusa tower (so, our goal in the next subsection is to show that this number is equal to 1, i.e., irreducibility).

Recall the neutral component $Y^{(p)}$ of $Sh^{(p)}$ and its function field $\mathfrak{F}^{(p)}$ over the fraction field $\mathcal{K}$ of $\mathcal{W}$. Let S be the ordinary locus $Y^{(p)}[\frac{1}{H}]_{/\mathbb{F}}$ for the Hasse invariant H. Then S is irreducible over $\mathbb{F}$, because H is a global section of the ample line bundle $\underline{\omega}^{\otimes(p-1)}$ over the compactification $X^{(p)}_{/\mathbb{F}}$ of $Y^{(p)}_{/\mathbb{F}}$.

We look into the Igusa tower in a local neighborhood of the generic point of the vertical irreducible divisor $X^{(p)}_{/\mathbb{F}} \subset X^{(p)}_{/\mathcal{W}}$. Consider the valuation ring $\mathfrak{V}$ of $\mathfrak{F}^{(p)}$ of the valuation v_0 in Definition 7.4. Thus, the residue field $\mathfrak{V}/\mathfrak{m}_{\mathfrak{V}}$ is the function field $\mathbb{F}(S)$ of S. Let $\mathbf{E}_{/Y^{(p)}}$ be the universal elliptic curve. Then we consider the Cartesian diagram for $\mathbf{E}_{\mathfrak{V}} = \mathbf{E} \times_{Y^{(p)}} \mathrm{Spec}(\mathfrak{V})$:

$$\begin{array}{ccc} \mathbf{E}_{\mathfrak{V}} & \xrightarrow{\hookrightarrow} & \mathbf{E} \\ \downarrow & & \downarrow \\ \mathrm{Spec}(\mathfrak{V}) & \xrightarrow[\hookrightarrow]{} & Y^{(p)}. \end{array}$$

Since any lift of a power of H is inverted in $\mathfrak{V}$, $\mathbf{E}_{\widehat{\mathfrak{V}}} = \mathbf{E}_{\mathfrak{V}} \times_{\mathfrak{V}} \widehat{\mathfrak{V}}$ is an ordinary elliptic curve for the completed valuation ring $\widehat{\mathfrak{V}} = \varprojlim_n \mathfrak{V}/p^n\mathfrak{V}$. Thus, we can think of the functor $I_{\widehat{\mathfrak{V}},m} = \mathrm{Imm}_{\widehat{\mathfrak{V}}}(\mu_{p^m}, \mathbf{E}_{\widehat{\mathfrak{V}}}[p^m])$ from the category of p-adic $\widehat{\mathfrak{V}}$-algebras to sets, which assigns to each p-adic $\widehat{\mathfrak{V}}$-algebra $R = \varprojlim_n R/p^nR$ the set of closed immersions of finite flat groups: $\mu_{p^m/R} \to \mathbf{E}_{\widehat{\mathfrak{V}}}[p^m]_{/R}$ defined over R. Here the category of p-adic $\widehat{\mathfrak{V}}$-algebras is made up of p-adically complete $\widehat{\mathfrak{V}}$-algebras with p-adically continuous $\widehat{\mathfrak{V}}$-algebra homomorphisms.

Lemma 7.27 *We have canonical isomorphisms of formal schemes:*

$$I_{\widehat{\mathfrak{V}},m} = \mathrm{Isom}_{\widehat{\mathfrak{V}}}(\mu_{p^m/\widehat{\mathfrak{V}}}, \mathbf{E}_{\widehat{\mathfrak{V}}}[p^m]^\circ) \overset{(*)}{\cong} \mathrm{Isom}_{\widehat{\mathfrak{V}}}((\mathbb{Z}/p^m\mathbb{Z})_{/\widehat{\mathfrak{V}}}, \mathbf{E}_{\widehat{\mathfrak{V}}}[p^m]^{\mathrm{et}})$$
$$\overset{(**)}{\cong} \mathbf{E}_{\widehat{\mathfrak{V}}}[p^m]^{et} - \mathbf{E}_{\widehat{\mathfrak{V}}}[p^{m-1}]^{\mathrm{et}},$$

where the identity $(*)$ is given by taking the inverse of the Cartier dual map and $(**)$ is given by $\phi \mapsto \phi(1)$ $(1 \in \mathbb{Z}/p^m\mathbb{Z})$ for $\phi \in \mathrm{Isom}_{\widehat{\mathfrak{V}}}((\mathbb{Z}/p^m\mathbb{Z}), \mathbf{E}[p^m]^{\mathrm{et}})$. In particular, $I_{\widehat{\mathfrak{V}},m}/\mathrm{Spf}(\widehat{\mathfrak{V}})$ is an étale finite covering over $\widehat{\mathfrak{V}}$, and $GL_1(\mathbb{Z}_p)$ naturally acts on $I_{\widehat{\mathfrak{V}},m}$ by its action on $(\mathbb{Z}/p^m\mathbb{Z})_{/\widehat{\mathfrak{V}}}$.

Proof. Note that $\widehat{\mathfrak{V}}$ is p-adically complete; so $\widehat{\mathfrak{V}}$ is a Henselian local ring by Hensel's lemma (see [CRT] Theorem 8.3). Thus, the connected component $\mathbf{E}_{\widehat{\mathfrak{V}}}[p^m]^\circ$ of $\mathbf{E}_{\widehat{\mathfrak{V}}}[p^m]$ is well defined by Proposition 4.68. If $\mu_{p^m} \hookrightarrow \mathbf{E}_{\widehat{\mathfrak{V}}}[p^m]$ is a closed immersion sending identity to identity, it factors through $\mathbf{E}_{\widehat{\mathfrak{V}}}[p^m]^\circ$. Comparing the rank, we find that it is an isomorphism. Thus, we get

$$I_{\widehat{\mathfrak{V}},m} = \mathrm{Imm}_{\widehat{\mathfrak{V}}}(\mu_{p^m/\widehat{\mathfrak{V}}}, \mathbf{E}_{\widehat{\mathfrak{V}}}[p^m]) \cong \mathrm{Isom}_{\widehat{\mathfrak{V}}}(\mu_{p^m/\widehat{\mathfrak{V}}}, \mathbf{E}_{\widehat{\mathfrak{V}}}[p^m]^\circ).$$

In any case, $\mathbf{E}_{\widehat{\mathfrak{V}}}[p^m]^\circ$ is isomorphic to $\mu_{p^m/\widehat{\mathfrak{V}}}$ étale locally. Since the Cartier dual of μ_{p^m} is isomorphic to $\mathbb{Z}/p^m\mathbb{Z}$ (Exercise 4.64), by the self-duality of $\mathbf{E}[p^m]$ (see Lemma 6.9), we get the dual-inverse isomorphism

$$\mathrm{Isom}_{\widehat{\mathfrak{V}}}(\mu_{p^m}, \mathbf{E}_{\widehat{\mathfrak{V}}}[p^m]^\circ) \cong \mathrm{Isom}_{\widehat{\mathfrak{V}}}(\mathbb{Z}/p^m\mathbb{Z}, \mathbf{E}_{\widehat{\mathfrak{V}}}[p^m]^{\mathrm{et}}) \subset \mathbb{Z}/p^m\mathbb{Z} \times_{\widehat{\mathfrak{V}}} \mathbf{E}_{\widehat{\mathfrak{V}}}[p^m]^{\mathrm{et}}.$$

Thus, $I_{\widehat{\mathfrak{V}},m}/\mathrm{Spf}(\widehat{\mathfrak{V}})$ is an étale finite covering over $\widehat{\mathfrak{V}}$. □

Since $I_{\widehat{\mathfrak{V}},m}$ is étale finite over $\mathrm{Spf}(\widehat{\mathfrak{V}})$, it is affine (Lemma 4.24), and we may write $I_{\widehat{\mathfrak{V}},m} = \mathrm{Spf}(\widehat{\mathfrak{V}}_m)$. Then $\widehat{\mathfrak{V}}_m$ is a semilocal normal $\widehat{\mathfrak{V}}$-algebra étale finite flat over $\widehat{\mathfrak{V}}$; so it is a product of complete discrete valuation rings whose maximal ideal is, by étaleness, generated by the rational prime p.

Recall $W = \varprojlim_n \mathcal{W}/p^n\mathcal{W}$, and take a modular form $\mathcal{E}$ on $X^{(p)}_{/\mathcal{W}}$ lifting the Hasse invariant H (as in Sect. 6.2.12). Let $\widehat{X}^{(p)}$ [resp., $\widehat{Y}^{(p)}$] be a formal completion of $X^{(p)}[\frac{1}{\mathcal{E}}]_{/W}$ (resp., $Y^{(p)}[\frac{1}{\mathcal{E}}]_{/W}$) along $X^{(p)}[\frac{1}{H}]_{/\mathbb{F}}$ (resp., $S := Y^{(p)}[\frac{1}{H}]_{/\mathbb{F}}$).

Lemma 7.28 *The p-adic formal scheme $\widehat{X}^{(p)}$ [resp., $\widehat{Y}^{(p)}$] does not depend on the choice of the lift $\mathcal{E}$.*

Proof. Since the proof is the same, we deal with the case of $\widehat{X}^{(p)}$. Take another lift $\mathcal{E}'$ of H^a. By replacing $\mathcal{E}$ and $\mathcal{E}'$ by its power, we may assume that both $\mathcal{E}$ and $\mathcal{E}'$ are lifts of H^a. Since $\mathcal{E} \equiv H \mod p$, $\mathcal{E}$ is invertible on the topological space S of $\widehat{X}^{(p)}[\frac{1}{\mathcal{E}}]_{/W}$. Since $X^{(p)}[\frac{1}{\mathcal{E}}]_{/W} = \mathrm{Spec}(R)$ is affine, writing $\widehat{R} = \varprojlim_n R/p^nR$, we have $\mathcal{E}'/\mathcal{E} \in R$, and $\mathcal{E}' \equiv \mathcal{E} \mod p\widehat{R}$ implies $\mathcal{E}'/\mathcal{E} \equiv 1 \mod p$; so $\mathcal{E}'/\mathcal{E}$ is invertible in $\widehat{R}$, whose inverse is given by $1+\sum_{n=1}^{\infty}(\mathcal{E}'/\mathcal{E}-1)^n \in \widehat{R}$. In other words, if we write $\widehat{X}^{(p)}[\frac{1}{\mathcal{E}'}]_{/W} = \mathrm{Spec}(R')$, $R' \subset \widehat{R}$ and $R \subset \widehat{R}'$. Thus, we get $\widehat{R} = \widehat{R}'$, and hence, $\widehat{X}^{(p)}[\frac{1}{\mathcal{E}}]_{/W} = \widehat{X}^{(p)}[\frac{1}{\mathcal{E}'}]_{/W}$. □

Define, for $\Gamma = \widehat{\Gamma}_1(p^m)_p$, a $\mathcal{W}$-scheme $Y_\Gamma = \mathrm{Imm}_{Y^{(p)}}(\mu_{p^m}, \mathbf{E})$ and a p-adic formal scheme $\widehat{Y}_{\Gamma/W} = \mathrm{Imm}_{\widehat{Y}^{(p)}}(\mu_{p^m}, \mathbf{E}) \cong \mathbf{E}[p^m]^{\mathrm{et}} - \mathbf{E}[p^{m-1}]^{\mathrm{et}}$ over $\widehat{Y}^{(p)}$, which is étale finite over $\widehat{Y}^{(p)}$. We obtain $Y_{\Gamma/\mathcal{K}}$ as in (7.1.1) by the base change $Y_{\Gamma/\mathcal{K}} \cong Y_{\Gamma/\mathcal{W}} \times_{\mathcal{W}} \mathcal{K}$. We may regard $\widehat{Y}_{\Gamma/W}$ as the formal completion of $Y_{\Gamma/\mathcal{W}}$ along $Y_{\Gamma/\mathbb{F}} := \mathrm{Isom}_S(\mu_{p^m}, \mathbf{E}[p^m]^\circ)$. By definition, we have an open immersion

$$I_{\widehat{\mathfrak{V}},m} \hookrightarrow \widehat{Y}_{\widehat{\Gamma}_1(p^m)_p/W} \times_{\widehat{Y}^{(p)}_{/\mathcal{W}}} \mathrm{Spf}(\widehat{\mathfrak{V}}),$$

and $\widehat{\mathfrak{V}}_m$ is the product of the completions of valuation rings of $\mathfrak{F}_{\widehat{\Gamma}_1(p^m)_p}$ unramified over $\mathfrak{V}$. Thus, $\mathfrak{V}_m = \widehat{\mathfrak{V}}_m \cap \mathfrak{F}_{\widehat{\Gamma}_1(p^m)_p}$ inside $\mathfrak{F}_{\widehat{\Gamma}_1(p^m)_p} \otimes_{\mathfrak{V}} \widehat{\mathfrak{V}}$ is a semilocal ring $\mathfrak{V}_m$ with $\widehat{\mathfrak{V}}_m = \varprojlim_n \mathfrak{V}_m/p^n\mathfrak{V}_m = \mathfrak{V}_m \otimes_{\mathfrak{V}} \widehat{\mathfrak{V}}$.

Similarly, we define, for N prime to p,

$$Y(N,\Gamma)_{/\mathbb{F}} = \mathrm{Imm}_{Y(N)[\frac{1}{H}]}(\mu_{p^m}, \mathbf{E}[p^m]) = \mathrm{Isom}_{Y(N)[\frac{1}{H}]}(\mu_{p^m}, \mathbf{E}[p^m]^\circ).$$

By definition, the moduli scheme $Y(N,\Gamma)_{/\mathbb{F}}$ is a scheme over $Y(N)[\frac{1}{H}]$ (forgetting the level-p structure). We put $I_{\mathfrak{V},m} = \mathrm{Spec}(\mathfrak{V}_m)$ and $Y_{\Gamma/\mathbb{F}} = \varprojlim_{p\nmid N} Y(N,\Gamma)_{/\mathbb{F}}$. Then

$$Y_{\widehat{\Gamma}_1(p^m)_p/\mathbb{F}} = \mathrm{Isom}_S(\mu_{p^m}, \mathbf{E}[p^m]^\circ) =: I_m$$

gives rise to the Igusa tower $I \twoheadrightarrow \cdots \twoheadrightarrow I_m \twoheadrightarrow \cdots \twoheadrightarrow I_1 \twoheadrightarrow S$ over S. The set of generic points $\{\eta_{I_m^\circ} \in I^\circ_{m/\mathbb{F}} | I^\circ_{m/\mathbb{F}} \in \pi_0(I_{m/\mathbb{F}})\}$ is in bijection to $\pi_0(I_m)$, and

$$\widehat{\mathfrak{V}}_m \otimes_{\mathbb{Z}_p} \mathbb{F} = \mathfrak{V}_m \otimes_{\mathbb{Z}_{(p)}} \mathbb{F} = \prod_{I_m^\circ \in \pi_0(I_m)} \mathbb{F}(I_m^\circ)\ (\Leftrightarrow I_{\mathfrak{V},m} \otimes_{\mathbb{Z}_{(p)}} \mathbb{F} = \bigsqcup_{I^\circ \in \pi_0(I_m)} \{\eta_{I_m^\circ}\}).$$

By the definition of the action of $(a,g) \in \mathcal{G}_I$:

$$(E, \eta^{(p)}, \phi_p) \mapsto (E, \eta^{(p)} \circ g, \phi_p \circ a),$$

the group $\mathcal{G}_I := GL_1(\mathbb{Z}_p) \times SL_2(\mathbb{A}^{(p\infty)})$ acts on $I_{\widehat{\mathfrak{V}},m}$ and hence on $I_{\mathfrak{V},m}$ ($m = 1, 2, \ldots, \infty$), $\mathrm{Spec}(\mathfrak{V})$ [by Lemma 7.5 (2)], $\mathfrak{F}_\Gamma$, I_m, $Y_{\Gamma/\mathbb{F}}$, and $Y_{\Gamma/\mathcal{K}}$. Thus, we can make the étale quotient $I_{\Gamma_0(p^m)} := I_{\mathfrak{V},m}/GL_1(\mathbb{Z}/p^m\mathbb{Z})$.

Lemma 7.29 *We have $I_{\Gamma_0(p^m)} = \mathrm{Spec}(\mathfrak{V}_{\Gamma_0(p^m)})$, and $\mathfrak{V}_{\Gamma_0(p^m)}$ is a valuation ring finite flat over $\mathfrak{V}$ sharing the same residue field with $\mathfrak{V}$.*

Proof. As seen in Sect. 4.3.8, the connected component of a finite flat group scheme over a Henselian local ring R with residue field $\mathbb{F}$ is uniquely determined. Thus, there is a unique connected subgroup of E (isomorphic to μ_{p^m} étale locally) if $(E, \phi_N)_{/R}$ gives rise to an R-point of $Y(N, \Gamma_0(p^m))_{/W}$. Thus, for any $m > 0$, we have $S_{/\mathbb{F}} = \varprojlim_{p\nmid N} Y(N, \Gamma_0(p^m))_{/\mathbb{F}}$. This shows that the residue field of $\mathfrak{V}_{\Gamma_0(p^m)}$ is the function field of S and that the quotient field of $\mathfrak{V}_{\Gamma_0(p^m)}$ is the field $\mathfrak{F}_{\Gamma_0(p^m)}$. □

Since $\mathfrak{V}_m/\mathfrak{V}$ is étale, we have

$$\widehat{\mathfrak{V}}_{\Gamma_0(p^m)} = \varprojlim_m \mathfrak{V}_{\Gamma_0(p^m)}/p^m\mathfrak{V}_{\Gamma_0(p^m)} = \varprojlim_m \mathfrak{V}/p^m\mathfrak{V} = \widehat{\mathfrak{V}},$$

and $\mathfrak{V}_m$ is étale finite over $\mathfrak{V}_{\Gamma_0(p^m)}$. This shows the following:

Lemma 7.30 *We have the following one-to-one onto correspondences:*

$$\left\{v : \mathfrak{F}_{\widehat{\Gamma}_1(p^m)_p} \to \mathbb{Z} : v|_{\mathfrak{F}_{\Gamma_0(p^m)}} = v_{\Gamma_0(p^m)} \text{ unramified over } v_0\right\}$$
$$\leftrightarrow \mathrm{Max}(\mathfrak{V}_m) \leftrightarrow \pi_0(I_m) \leftrightarrow \{\eta_{I_m^\circ}\},$$

where v is a p-adic valuation of $\mathfrak{F}_{\widehat{\Gamma}_1(p^m)_p}$ unramified (of degree 1) over v_0 and $\mathrm{Max}(\mathfrak{V}_m)$ is the set of maximal ideals of $\mathfrak{V}_m$.

The correspondence is given by

$$v \leftrightarrow \mathfrak{m}_v = \{x \in \mathfrak{V}_m | v(x) > 0\} \leftrightarrow I_m^\circ \quad \text{with } \mathbb{F}(I_m^\circ) = \mathfrak{V}_m/\mathfrak{m}_v.$$

Lemma 7.31 *The action of $\mathcal{G}_1 := SL_2(\mathbb{A}^{(p\infty)})$ fixes $v_m = v_{\widehat{\Gamma}_1(p^m)_p}$ and each element of $\pi_0(I_m)$.*

Proof. Since $\mathfrak{F}_{\widehat{\Gamma}_1(p^m)_p}/\mathfrak{F}_{\widehat{\Gamma}_0(p^m)_p}$ is a finite Galois extension, the set of extensions of $v_{\Gamma_0(p^m)}$ to $\mathfrak{F}_{\widehat{\Gamma}_1(p^m)_p}$ is a finite set, and by the above lemma, it is in bijection with $\pi_0(I_m)$. Thus, the action of $SL_2(\mathbb{A}^{(p\infty)})$ on $\pi_0(I_m)$ gives a finite permutation representation of $SL_2(\mathbb{A}^{(p\infty)})$. Since $SL_2(k)$ of any field k of characteristic 0 does not have any nontrivial finite quotient group (because it is generated by divisible unipotent subgroups by the following lemma), the action of $SL_2(\mathbb{A}^{(p\infty)})$ fixes every irreducible component in $\pi_0(I_m)$.

Exercise 7.32 *Why does unipotent generation imply that $SL_2(k)$ of any field k of characteristic 0 does not have any nontrivial finite quotient group?*

Here is the lemma on the unipotent generation of $SL(2)$:

Lemma 7.33 *Let K be a field. Then $SL_2(K)$ and $PSL_2(K)$ are generated by unipotent elements.*

Proof. Let H be a subgroup generated by all unipotent elements in $SL_2(K)$ and $\overline{H}$ be its image in $PSL_2(K)$. If u is a nontrivial unipotent element, we find an eigenvector $v \neq 0$ of $V = K^2$ fixed by u. Taking a base (w, v) of V over K, we find $u\left(\begin{smallmatrix} w \\ v \end{smallmatrix}\right) = \left(\begin{smallmatrix} aw+bv \\ v \end{smallmatrix}\right) = \left(\begin{smallmatrix} a & b \\ 0 & 1 \end{smallmatrix}\right)\left(\begin{smallmatrix} w \\ v \end{smallmatrix}\right)$. Taking the determinant, we find $a = 1$; so u is a conjugate of $\left(\begin{smallmatrix} 1 & b \\ 0 & 1 \end{smallmatrix}\right)$. This shows that $g\overline{H}g^{-1} = \overline{H}$. Thus, $\overline{H}$ is a normal subgroup of $PSL_2(K)$. Since the upper triangular unipotent group acts by translation on the affine space $\mathbf{A}^1(K) = K$, it acts transitively on $\mathbf{A}^1(K)$. The opposite unipotent subgroup ${}^tU$ acts transitively on $\mathbf{P}^1(K) - \{0\} \cong \mathbf{A}^1(K)$, and H acts transitively on $\mathbf{P}^1(K)$. Thus, for any $g \in PSL_2(K)$, we find $h \in \overline{H}$ such that $h(\infty) = g(\infty)$; so $h^{-1}g \in B$, where B is the subgroup of upper triangular matrices. Thus, if $\overline{T} \subset \overline{H}$ for the diagonal torus $\overline{T}$, we find $\overline{T} \cdot \overline{H} = PSL_2(K)$.

Since $\left(\begin{smallmatrix} 1 & 0 \\ x & 1 \end{smallmatrix}\right) = \left(\begin{smallmatrix} 1 & x^{-1} \\ 0 & 1 \end{smallmatrix}\right)\left(\begin{smallmatrix} 0 & -x^{-1} \\ x & 0 \end{smallmatrix}\right)\left(\begin{smallmatrix} 1 & x^{-1} \\ 0 & 1 \end{smallmatrix}\right)$, we have $j(x) = \left(\begin{smallmatrix} 0 & -x^{-1} \\ x & 0 \end{smallmatrix}\right) \in H$. Since $j(-x)j(1) = \left(\begin{smallmatrix} x^{-1} & 0 \\ 0 & x \end{smallmatrix}\right)$, we find that $T \subset H$. Thus, $SL_2(K)$ and $PSL_2(K)$ are always generated by unipotent elements.

7.2.3 Reciprocity Law and Irreducibility of Igusa Tower

Let $v_\infty = v_{\Gamma_1(p^\infty)}$, and define

$$D = \left\{ x \in (\mathbb{Z}_p^\times \times SL_2(\mathbb{A}^{(p\infty)})) \middle| v_\infty \circ \tau(x) = v_\infty \right\}.$$

Exercise 7.34 *Prove that the subgroup $\mathbb{Z}_{(p)}^\times$ is p-adically dense in $\mathbb{Z}_p^\times$; hence, $T(\mathbb{Z}_{(p)}) \xrightarrow[\iota]{\sim} \mathbb{Z}_{(p)}^\times$ is dense in $T(\mathbb{Z}_p) \cong \mathbb{Z}_p^\times$ and $T(\mathbb{Z}_{(p)}) \cdot SL_2(\mathbb{A}^{(p\infty)})$ is dense in $\mathcal{G}_I = \mathbb{Z}_p^\times \times SL_2(\mathbb{A}^{(p\infty)})$. Here an element $\alpha \in T(\mathbb{Z}_{(p)})$ is sent to $(\iota(\alpha), \alpha^{(p)}) \in \mathbb{Z}_p^\times \times SL_2(\mathbb{A}^{(p\infty)})$ with $\alpha^{(p)} = (\alpha, \dots, \overset{l}{\alpha}, \dots) \in SL_2(\mathbb{A}^{(p\infty)})$ at primes $l \neq p$.*

Since $T(\mathbb{Z}_{(p)})$ and $SL_2(\mathbb{A}^{(p\infty)})$ fix v_∞ (Lemmas 7.5 and 7.31), by the density in Exercise 7.34, we conclude (G) for x given by the generic point associated with v_∞ in Sect. 7.2.1:

Theorem 7.35 *We have $D = \mathcal{G}_I/\{\pm 1\}$, and hence I is irreducible.*

Note that $\mathrm{Gal}(I^\circ/S)$ is the stabilizer of I° in $\mathrm{Aut}(I/S)$. Then, the irreducibility of I follows from $D = \mathcal{G}_I/\{\pm 1\}$, because this implies $\mathrm{Aut}(I/S) = \mathrm{Gal}(I^\circ/S)$, as argued in Sect. 7.2.1.

Let $K^{(p)}$ be an open compact subgroup of $SL_2(\mathbb{A}^{(p\infty)})$ and $K = K^{(p)} \times GL_2(\mathbb{Z}_p)$. Put $Y_K = Y^{(p)}/K^{(p)}$ (which is the level-K modular curve). Let $I_K = I/K^{(p)}$, which is the Igusa tower over Y_K. Since I is irreducible by the above theorem, the quotient $I_K = I/K^{(p)}$ is irreducible (as $I \twoheadrightarrow I_K$ is a dominant morphism). Thus, we have reproved

Corollary 7.36 (Igusa) *The Igusa tower I_K over $Y_{K/\mathbb{F}}$ is irreducible for $K = GL_2(\mathbb{Z}_p) \times K^{(p)}$ for any open compact subgroup $K^{(p)}$ of $SL_2(\mathbb{A}^{(p\infty)})$.*

The following corollary is called the q-expansion principle and is a consequence of the irreducibility of I_α combined with the fact that the local q-parameter gives the uniformizing parameter at a cusp of $Y_1(N)$ over $\mathbb{Z}[\frac{1}{N}]$.

Corollary 7.37 (q-expansion principle) *Let $\mathcal{L}_{/\overline{\mathbb{F}}_p}$ be a line bundle realized in the function field of $I_{\alpha/\overline{\mathbb{F}}_p}$ $(\alpha = 1, 2, \dots, \infty)$. Let s be a cusp and U be any Zariski open neighborhood of $s \in I_{\alpha/\overline{\mathbb{F}}_p}$. If the q-expansion at s of a section $f \in H^0(U, \mathcal{L})$ vanishes, then f vanishes on U.*

7.2.4 p-Adic Elliptic Modular Forms

Pick a subgroup $\Gamma \subset SL_2(\mathbb{Z})$ between $\Gamma(N)$ and $\Gamma_0(N)$, we consider the Igusa tower $I_{m,\alpha}$ over $X_\Gamma = X(\Gamma)$ [the compactification of $Y(\Gamma)$; see Sect. 6.2.12]. Write $\tilde{\mathbf{E}}_{/X_\Gamma}$ for the smooth locus of the universal stable curve [extending the universal elliptic curve over $Y(\Gamma)$ by the Tate curve over the infinitesimal neighborhood of each cusp; see [DRa] for more details]. The scheme $\tilde{\mathbf{E}}$ has a group law over X_Γ extending that of $\mathbf{E}$. To have q-expansion finite at cusps, we use $\tilde{\mathbf{E}}_{/X_\Gamma}$ in place of $\mathbf{E}_{/Y_\Gamma}$. Writing $S_m = X_\Gamma[\frac{1}{E}] \otimes_W W_m$ for $W_m = W/p^m W$ with a lift E of (a power of) the Hasse invariant, $I_{m,\alpha} = \mathrm{Isom}_{S_m}(\mu_{p^\alpha}, \tilde{\mathbf{E}}[p^\alpha]^\circ)$. Note here that $\tilde{\mathbf{E}}[p^\alpha]^\circ \times_{X_\Gamma} S_m$ is finite flat over S_m, because $\mathrm{Tate}(q)[p^\alpha]^\circ = \mu_{p^\alpha}$, though $\mathrm{Tate}(q)[p^m]$ itself is quasi-finite nonfinite over $\mathrm{Spec}(W[[q]])$. Thus, $I_{m,\alpha}$ is still étale finite over S_m. In other words, by an abuse of the notation, writing $\tilde{\mathbf{E}}[p^\alpha]^{\mathrm{et}}$ for the Cartier dual $\mathrm{Hom}_{GSCH}(\tilde{\mathbf{E}}[p^\alpha]^\circ, \mu_{p^\alpha})$, $\tilde{\mathbf{E}}[p^\alpha]^{\mathrm{et}}$ is an étale group over S_m isomorphic to $\mathbb{Z}/p^\alpha\mathbb{Z}$ étale locally extending $\mathbf{E}[p^\alpha]^{\mathrm{et}}_{/Y_\Gamma}$ to X_Γ, and we have $I_{m,\alpha} \cong \mathrm{Isom}_{S_m}(\mathbb{Z}/p^\alpha\mathbb{Z}, \tilde{\mathbf{E}}[p^\alpha]^{\mathrm{et}})$, which is étale finite by definition. Since S_m is affine, $I_{m,\alpha}$ is also affine. We write $I_{m,\alpha} = \mathrm{Spec}(V_{m,\alpha})$. We have a tower of W_m-flat algebras:

$$V_{m,0} \subset V_{m,1} \subset \cdots \subset V_{m,\alpha} \subset \cdots .$$

These algebras are étale over $V_{m,0}$ and $\mathrm{Gal}(V_{m,\alpha}/V_{m,0}) \cong (\mathbb{Z}/p^\alpha\mathbb{Z})^\times$. By construction, we have $I_{m,\infty} = \mathrm{Spec}(V_{m,\infty})$ for $V_{m,\infty} = \bigcup_\alpha V_{m,\alpha}$. Simply write P_α for $\tilde{\mathbf{E}}[p^\alpha]^{\mathrm{et}}$ as a locally free étale group scheme over S_m. Over $V_{m,\alpha}$, we have a canonical isomorphism (the universal level-p structure)

$$I_{can} = \boldsymbol{\phi}_p : \mu_{p^\alpha} \cong \tilde{\mathbf{E}}[p^\alpha]^\circ \quad \text{and} \quad {}^t I_{can}^{-1} : \mathbb{Z}/p^\alpha\mathbb{Z} \cong P_\alpha = \tilde{\mathbf{E}}[p^\alpha]^{\mathrm{et}},$$

where ${}^t I_{can}^{-1}$ is the Cartier dual inverse of I_{can}. We then define $V_{m,\infty} = \bigcup_\alpha V_{m,\alpha}$ and

$$\mathcal{V} = \mathcal{V}_\Gamma = \varinjlim_m V_{m,\infty} \quad \text{and} \quad V = V_\Gamma = V(\Gamma; W) = \varprojlim_m V_{m,\infty}.$$

The space V_Γ is the space of p-adic modular forms on Γ (originally defined in Sect. 1.3.3). Indeed, by universality, we may regard $f \in V_{m,\alpha}$ as a function of $(E, \phi_p : \mu_{p^\alpha} \hookrightarrow E[p^\alpha], \phi_N)$ satisfying the following conditions similar to (G1–3) for p-adically complete W-algebras $R = \varprojlim_n R/p^n R$.

(V1) $(E, \phi_p, \phi_N)_{/R} \cong (E', \phi'_p, \phi'_N)_{/R} \Rightarrow f(E, \phi_p, \phi_N) = f(E', \phi'_p, \phi'_N) \in R$.
(V2) If $\rho : R \to R'$ is a continuous W-algebra homomorphism, then

$$f((E, \phi_p, \phi_N) \times_{R,\rho} R') = \rho(f(E, \phi_p, \phi_N)).$$

(V3) For all level structures ϕ_N of type Γ on $\mathrm{Tate}(q)$, we have

$$f(\mathrm{Tate}(q), z\phi_p^{can}, \phi_N) \in W[[q^{1/N}]] \quad \text{for all } z \in \mathbb{Z}_p^\times.$$

(s) We call $f \in V_\Gamma$ a *p-adic cusp form* if

$$f(\mathrm{Tate}(q), z\phi_p^{can}, \phi_N) \in q^{1/N} W[[q^{1/N}]]$$

for any choice of ϕ_N and all $z \in \mathbb{Z}_p^\times$.

Let $V_{\Gamma/B} = V(\Gamma; B) := V_{\Gamma/W}\widehat{\otimes}_W B$ (p-adic completion of $V_{\Gamma/W} \otimes_W B$) for a W-algebra $B = \varprojlim_m B/p^m B$. Each element $f \in V_{\Gamma/B}$ is a function of $(E, \phi_p, \phi_N)_{/R}$ satisfying conditions (V1–3) for p-adically complete B-algebras R (replacing W in the conditions by B).

Again, we note the consequence of the irreducibility of the Igusa curve.

Corollary 7.38 (q-Expansion Principle) *If the q-expansion at the infinity cusp of a p-adic modular form $f \in V_{\Gamma/B}$ over a p-adically complete W-algebra B vanishes, f itself vanishes; that is, $f(E, \phi_p, \phi_N) = 0$ for all test objects $(E, \phi_p, \phi_N)_{/R}$ defined over p-adically complete B-algebras R.*

The level-p structure ϕ_p of a p-adic test object $(E, \phi_p, \phi_N)_{/R}$ as in (V1–3) is an isomorphism $\phi_p : \mu_{p^\infty} \cong E[p^\infty]$, which induces by Lemma 4.82, $\widehat{\phi}_p : \widehat{\mathbb{G}}_m \cong \widehat{E}$. Here $\widehat{E}$ is the formal completion of E along $\mathbf{0}_E \in E(R/pR)$. Thus, the invariant differential $\frac{dt}{t}$ of $\widehat{\mathbb{G}}_m = \mathrm{Spf}(\widehat{R[t, t^{-1}]})$ induces an invariant differential $\widehat{\omega}_p$ so that $\widehat{\phi}_p^* \widehat{\omega}_p = \frac{dt}{t}$. Since $\omega \in H^0(E, \Omega_{E/R})$ induces an invariant differential over $\widehat{E}$, we have $H^0(E, \Omega_{E/R}) = H^0(\widehat{E}, \Omega_{\widehat{E}/R})$; so we have a unique invariant differential ω_p induced by $\widehat{\omega}_p$ on E such that $H^0(E, \Omega_{E/R}) = R\omega_p$.

Definition 7.39 *For each $f \in G_k(\Gamma_1(Np^\alpha); R)$, we define $\widehat{f}((E, \phi_p, \phi_N)) = f((E, \phi_p|_{\mu_{p^\alpha}}, \phi_N, \omega_p)$.*

This association $(E, \phi_p, \phi_N)_{/R} \mapsto \widehat{f}((E, \phi_p, \phi_N))$ clearly satisfies (V1–3) by (G0–4), and we have a natural R-linear map

$$G_k(\Gamma_1(Np^\infty); R) := \bigcup_\alpha G_k(\Gamma_1(Np^\alpha); R) \to V_{\Gamma_1(N)/R}.$$

Exercise 7.40 *Prove the identity $\widehat{f}(q) = f(q)$ of q-expansion at ∞.*

By the above exercise, $f \mapsto \widehat{f}$ preserve q-expansion; so, by the q-expansion principle (applied to $\mathcal{L} = \underline{\omega}^k$), $f \mapsto \widehat{f}$ is an inclusion. If $f, g \in G_k(\Gamma_1(Np^\infty); R)$ has congruence $f(q) \equiv g(q) \mod p^n R[[q]]$, then again by the q-expansion principle, $(\widehat{f} - \widehat{g})((E, \phi_p, \phi_N)_{R/p^n R})$ vanishes for any p-adic test objects (E, ϕ_p, ϕ_N) over $R/p^n R$; so $\widehat{f} \equiv \widehat{g} \mod p^n V_{\Gamma_1(N)/R}$ by the p-adic q-expansion principle. In other words, the map $f \mapsto \widehat{f}$ is p-adically continuous. Thus, we get

Proposition 7.41 *We have a canonical p-adically continuous inclusion of classical modular forms in $G_k(\Gamma_1(Np^\infty); R)$ into the space of p-adic modular forms $V_{\Gamma_1(N)/R}$, preserving the q-expansion at ∞.*

Remark 7.42 By this proposition, for $G = G_k(\Gamma_1(Np^\infty); R)$, the p-adic completion $\widehat{G}$ is a subspace of $V_{\Gamma_1(N)/R}$. Though we can prove $\widehat{G} = V_{\Gamma_1(N)/R}$ if $k \geq 2$ (cf. [H88b] Corollary 5.2), we do not need this fact. This justifies the construction done in Sect. 1.3.6 of the Katz derivative df for $f \in G_k(\Gamma_1(Np^\infty); R)$ as a p-adic modular form.

7.2.5 Reciprocity Law for Deformation Space

We start with a fixed CM point $x = [z, g]$ and the associated elliptic curve (E_x, η_x) with complex multiplication by M_x. See Sect. 7.1.8 for the notation. We choose $z \in \mathfrak{H} \cap \mathrm{id}_x(M)$ so that $\mathfrak{O} = \mathbb{Z} + \mathbb{Z} \cdot z$. Unless confusion seems likely, we write M for M_x. We suppose that $\mathfrak{O} \hookrightarrow \mathrm{End}(E_x)$ for the integer ring $\mathfrak{O}$ of M. Take $W = W(\overline{\mathbb{F}}_p)$ and consider the reduction E_0 modulo (p) of E_x for the maximal ideal $(p) \subset W$. Suppose that E_0 is ordinary; thus, p splits $(p) = \mathfrak{p}\overline{\mathfrak{p}}$ in $\mathfrak{O}$. We assume $E_0[\mathfrak{p}^\infty] \cong \mu_{p^\infty}$ (see Corollary 6.40). By this choice, $\mathrm{id}_x = \mathrm{id}_z$ defined in Sect. 7.1.8 coincides with $i_p|_M$ for the embedding $i_p : \overline{\mathbb{Q}} \hookrightarrow \mathbb{C}_p$ we fixed at the beginning. We write v for the p-adic place of M associated with id_x. Thus, the ordinarity of E_0 is equivalent to

$$v \neq v \circ c \text{ for the complex conjugation } c \in \mathrm{Gal}(M/\mathbb{Q}). \qquad \text{(ord)}$$

Since $x = [z, g]$, $\eta_x = \eta_z \circ g$, and so the choice of g is tantamount to the choice of the base of $M_{\mathbb{A}^{(\infty)}}$ over $\mathbb{A}^{(\infty)}$, sending the basis $\{-1, z\}$ of $\widehat{L}_z = L_z \otimes_{\mathbb{Z}} \widehat{\mathbb{Z}}$ to $g^{-1}\left(\begin{smallmatrix} -1 \\ z \end{smallmatrix}\right)$. We have chosen -1 in order to incorporate the conjugation by J_1 we made in (7.1.6) [because $\left(\begin{smallmatrix} -1 \\ z \end{smallmatrix}\right) = J_1 \left(\begin{smallmatrix} z \\ 1 \end{smallmatrix}\right)$]. We then have $E_x(\mathbb{C}) = L\backslash(M \otimes_{\mathbb{Q}} \mathbb{R})$ for a fractional ideal $L \subset M$ identifying $M \otimes_{\mathbb{Q}} \mathbb{R}$ with $\mathbb{C}$ through $a \otimes t \mapsto \mathrm{id}_x(a)t$. This induces $\eta^{(p)} = \eta_z^{(p)} \circ g^{(p)} : M \otimes_{\mathbb{Q}} \mathbb{A}^{(p\infty)} \cong V^{(p)}(E_x)$. Since $\mathfrak{O} \subset \mathrm{End}_W(E_x)$, we may assume that $L_p = \mathfrak{O}_p = \mathfrak{O} \otimes_{\mathbb{Z}} \mathbb{Z}_p$ [so $g_p \in G(\mathbb{Z}_p)$ by our choice of z, so that $L_z = \mathfrak{O}$ inside M]. We are dealing with Kottwitz's moduli problem (as in (7.1.11) in Sect. 7.1.3). By reduction mod p, $\eta^{(p)}$ induces a prime-to-p-level structure $\eta_0^{(p)}$ on E_0. Let $(E, \iota_E)_{/R}$ be any deformation of $(E_0, \iota_0)_{/\mathbb{F}}$ ($\mathbb{F} = \overline{\mathbb{F}}_p$) over $\mathrm{Spec}(R)$ for an Artinian W-algebra R. Since $E[N]$ for N prime to p is étale over $\mathrm{Spec}(R)$, the level structure $\eta_0^{(p)}$ at the special fiber extends uniquely to a level structure $\eta_E^{(p)}$ on $E_{/R}$. Thus, the level structure is insensitive to the deformation of the underlying triple (E_0, ι_0). Therefore, for the deformation functor similar to the one in (6.3.1):

$$\widehat{\wp}(R) = \left[(E, \iota_E, \eta_E^{(p)})_{/R} \middle| (E, \iota_E, \eta_E^{(p)}) \mod \mathfrak{m}_R = (E_0, \iota_0, \eta_0^{(p)})\right],$$

the forgetful morphism: $(E, \iota_E, \eta_E^{(p)})_{/R} \mapsto (E, \iota_E)_{/R}$ of $\widehat{\wp}$ into the original deformation functor $\widehat{\mathcal{E}}_{E_0}$ in (6.3.2) induces an isomorphism of functors; so they have canonically isomorphic deformation spaces (though it would be better to view that they are not equal).

We consider the Serre–Tate deformation space $\widehat{S}_x$ representing $\widehat{\wp}$. We take the Shimura–Igusa–Kottwitz model $Sh^{(p)}_{/W}$ over W and consider $x = [z, g]$ as a W-point of a geometrically connected component $Y^{(p)}_{/W}$ of $Sh^{(p)}_{/W}$. Abusing the notation, we let $S = Y^{(p)}[\frac{1}{E}]$ [instead of $X^{(p)}$]; that is, we invert over $Y^{(p)}$ a lift E of a power of the Hasse invariant H. Recall that the formal completion $\widehat{Y}^{(p)}$ of $Y^{(p)}$ along $S_m = S \otimes_W W_m$ is uniquely determined independently of the choice of E and gives the ordinary locus of the universal elliptic curve $\mathbf{E}$ (see Lemma 7.28). By our choice, x gives rise to a closed point of $\widehat{Y}^{(p)}$.

Since the Serre–Tate deformation space $\widehat{\mathbb{G}}_m$ carries the universal deformation $\underline{\mathbb{E}} = (\mathbb{E}, \eta^{(p)})_{/\widehat{S}_x}$, which is an element of $\mathcal{E}^{(p)}_{G(\mathbb{Z}_p)}(\widehat{\mathbb{G}}_m) = \mathcal{E}^{(p)}(\widehat{\mathbb{G}}_m)$, by the universality of the Shimura variety, we have an inclusion

$$\varphi : \widehat{S}_x \hookrightarrow \widehat{Y}^{(p)} \text{ such that } \varphi^*\underline{\mathbf{E}}^{ord} = \underline{\mathbb{E}} \tag{7.2.1}$$

for the universal quadruple $\underline{\mathbf{E}}^{ord}$ over $\widehat{Y}^{(p)}$. Since $\eta^{(p)}$ lacks the information about $E_0[p^\infty]$, the identification of $\widehat{S}_x$ with $\widehat{\mathbb{G}}_m$ has not yet been specified.

We can lift the morphism φ to the Igusa tower over the formal completion $\widehat{Y}^{(p)}$ of $Y^{(p)}[\frac{1}{E}]_{/W}$ along $S_{1/\mathbb{F}}$. Then the identification $\bigcup_j \overline{\mathfrak{p}}^{-j}/\mathfrak{O} \cong E_x[(\overline{\mathfrak{p}})^\infty]$ induces

$$\eta_p^{\mathrm{et}} : \mathbb{Q}_p/\mathbb{Z}_p = M_{\overline{\mathfrak{p}}}/\mathfrak{O}_{\overline{\mathfrak{p}}} \cong E_x[p^\infty]^{\mathrm{et}} = E_x[\overline{\mathfrak{p}}^\infty].$$

By the Cartier dual inverse, η^{et} induces $\eta^\circ : \mu_{p^\infty} \cong E_x[p^\infty]^\circ = E_x[\mathfrak{p}^\infty]$. Let $\widehat{\mathcal{K}}$ be the field of fractions of W. Over the field $\widehat{\mathcal{K}}$, we can extend $\eta^{(p)}$ to $\eta_p^{\mathrm{ord}} = \eta_p^{\mathrm{et}} \times \eta_p^\circ : \mathbb{Z}_p \times \mathbb{Z}_p(1) = \mathfrak{O}_p \cong T_pE_x$. This choice is tantamount to the choice of g_p, which brings the base of L_p to the base given by the idempotents $-1_{\overline{\mathfrak{p}}} := (0,1)$ of $\mathfrak{O}_{\overline{\mathfrak{p}}}$ and $1_{\mathfrak{p}} := (1,0)$ of $\mathfrak{O}_{\mathfrak{p}}$ in $\mathfrak{O}_{\overline{\mathfrak{p}}} \times \mathfrak{O}_{\mathfrak{p}} = \mathbb{Z}_p \times \mathbb{Z}_p(1)$. We write $\eta^{\mathrm{ord}} = \eta_p^{\mathrm{ord}} \times \eta^{(p)}$ [this η^{ord} incorporates the Galois action on $\mathbb{Z}_p \times \mathbb{Z}_p(1)$, and if we want to make explicit about ignoring the action, we write $\eta = \eta_p^{\mathrm{ord}} \times \eta^{(p)}$]. This ordering of $\mathbb{Z}_p$ and $\mathbb{Z}_p(1)$ is important as the canonical inclusion $\mu_{p^n} \hookrightarrow \mathbb{C}^\times/q^{\mathbb{Z}}$ for $q = \exp(2\pi i z)$ is induced by

$$\frac{1}{p^n} \mapsto \frac{1}{p^n} \in (2\pi i)(\mathbb{Z}\frac{1}{p^n} + \mathbb{Z}z)/(2\pi i)(\mathbb{Z} + \mathbb{Z}z),$$

and as a convention, the basis of $(2\pi i)(\mathbb{Z} + \mathbb{Z}z)$ is written as $2\pi i^t(z,1)$ [i.e., the right-hand side 1 corresponding to the level-$\Gamma_1(p^n)$ structure]. In other words, as we have remarked, the action $\eta \mapsto \eta \circ g$ corresponds to $J_1 g^{-1} J_1^{-1}$ on the lattice side for $J_1 = \left(\begin{smallmatrix} 0 & -1 \\ 1 & 0 \end{smallmatrix}\right)$, and by this choice $J_1 \left[\begin{smallmatrix} 1_{\overline{\mathfrak{p}}} \\ -1_{\mathfrak{p}} \end{smallmatrix}\right] = \left[\begin{smallmatrix} 1_{\mathfrak{p}} \\ 1_{\overline{\mathfrak{p}}} \end{smallmatrix}\right]$, we do not need to worry about conjugating by J_1 (so we forget about J_1).

We can think of the deformation of $(E_0, \iota_0, \eta_0^{ord})_{/\overline{\mathbb{F}}_p}$ for $\eta_0^{ord} = \eta^{ord} \mod p$. The p-part of the level-p structure η_0^{ord} provides the canonical identification of the deformation space $\widehat{S}$ with $\widehat{\mathbb{G}}_m$. For any complete local W-algebra C and any deformation $E_{/C}$ of E_0, $E[p^\infty]^{et}$ is étale over $\mathrm{Spec}(C)$; so, again, the deformation is insensitive to the ordinary level structure. Thus, we get a canonical immersion:

$$\varphi_{\eta_0^{ord}} : \widehat{\mathbb{G}}_m \hookrightarrow I \text{ such that } \varphi^*\underline{\mathbf{E}}^{ord} = \underline{\mathbb{E}}. \tag{7.2.2}$$

Here $\underline{\mathbf{E}}^{ord}$ (resp., $\underline{\mathbb{E}}$) denotes the universal ordinary triple over I (resp., the universal triple over $\widehat{\mathbb{G}}_m$).

The elliptic curve $\underline{E}_x = (E_x, i, \eta^{ord})$ is the fiber of $\mathbb{E}$ at a point $t_{E_x/W} \in \widehat{\mathbb{G}}_m(W)$. Here $t_{E_x/W}$ is an $\mathbb{Z}_p$-bilinear form on $TE_0[p^\infty]^{et}$. Since any element $a \in \mathfrak{O} \subset \mathrm{End}(E_{0/\mathbb{F}})$ can be lifted to E_x, by Theorem 6.49 (3), we have

$$t_{E_x/W}(i(\alpha)y, y') = t_{E_x/W}(y, i(\overline{\alpha})y') \ (\alpha \in \mathfrak{O}), \tag{7.2.3}$$

where $\overline{\alpha} = c(\alpha)$ for $1 \neq c \in \mathrm{Gal}(M/\mathbb{Q})$. This forces $t_{E_x/W}$ to be 0, that is, $t_{E_x/W}(y, y')$ is the constant 1 of the group $\widehat{\mathbb{G}}_m$ identically, because $t_{E_x/W}$ is also $\mathbb{Z}_p$-linear and $\mathbb{Z}_p = \mathfrak{O}_\mathfrak{p}$. Indeed, the connected étale exact sequence of $E_x[p^\infty]$ does split by complex multiplication, and hence $t_{E_x/W} = 1$ by definition.

We now compute the effect of the isogeny $\alpha : E_0 \to E_0$ ($E_0 = E_{x/\mathbb{F}}$ and $\alpha \in \mathfrak{O}$) on the deformation space $\widehat{S}$. Pick a deformation $E_{/R}$ of $E_0 = E_{x/\mathbb{F}}$ for an Artinian $R \in CL_{/W}$, and we look into the following diagram with exact rows:

$$\begin{array}{ccccc} \mathrm{Hom}(TE_0[p^n]^{et}, \widehat{\mathbb{G}}_m(R)) & \hookrightarrow & E[p^n](R) & \xrightarrow{\pi} & E_0[p^n]^{et}(R) \\ \alpha\uparrow & & & & \downarrow\alpha^{-c} \\ \mathrm{Hom}(TE_0[p^n]^{et}, \widehat{\mathbb{G}}_m(R)) & \hookrightarrow & E[p^n](R) & \xrightarrow{\pi} & E_0[p^n]^{et}(R). \end{array} \tag{7.2.4}$$

Take $u = \varprojlim_n u_n \in TE_0[p^\infty]^{et}$, and lift it to $v = \varprojlim_n v_n$ for $v_n \in E(R)$ (but $v_n \notin E[p^n]$). Then

$$t(u) = \varprojlim_n t_n(u_n) \in \mathrm{Hom}(TE_0[p^\infty]^{et}, \widehat{\mathbb{G}}_m) \ \text{ for } t_n(u_n) = \text{“}p^n\text{”} v_n.$$

Note that the identification of $\mathrm{Hom}(TE_0[p^\infty]^{et}, \widehat{\mathbb{G}}_m)$ with the formal group $\widehat{E}$ of E is given by the Cartier duality composed with the polarization; so if α is p-integral, α sends t to $t^{\alpha^{1-c}} = \varprojlim_n \alpha(\text{“}p^n\text{”}\alpha^{-c}(v_n))$ (if we allow fraction power of t, this is valid without assuming the integrality of α). Thus, the effect of α on t is given by $t \mapsto t^{\alpha^{1-c}}$. Once the identification of $\widehat{S}$ with $\widehat{\mathbb{G}}_m$ is given (that is, a level-p^∞ structure $\eta_0^{ord} : \mathbb{Q}_p/\mathbb{Z}_p \cong E_0[p^\infty]^{et}$ is chosen), nonzero $\alpha \in \mathfrak{O}$ acts on the coordinate t (of $\widehat{\mathbb{G}}_m$) by $t \mapsto t^{\alpha^{1-c}}$.

Write $\widehat{\mathbb{Z}_p[t, t^{-1}]}$ for the formal completion of $\mathbb{Z}_p[t, t^{-1}]$ at the origin $1 \in S(\mathbb{F})$ for $S = \mathbb{G}_m$. Identify $\widehat{\mathbb{Z}_p[t, t^{-1}]}$ with the ring made up of formal series: $\sum_{\xi \in \mathbb{Z}} a(\xi) t^\xi$ for $a(\xi) \in \mathbb{Z}_p$. Let $T_0 = \mathbb{G}_{m/\mathbb{Q}}$. Since $\widehat{S}_x = \widehat{\mathbb{G}}_m = \mathrm{Spf}(\widehat{\mathbb{Z}_p[t, t^{-1}]})$ for the completion at the origin $1 \in \mathbb{G}_m(\mathbb{F})$, $s \in \mathbb{Z}_p^\times = T_0(\mathbb{Z}_p)$ acts on $\widehat{S}_x$ by $t \mapsto t^s = \sum_{n=1}^\infty \binom{s}{n}(t-1)^n$, which induces an automorphism of the formal group $\widehat{\mathbb{G}}_m$, and all $\mathbb{Z}_p$-linear automorphisms are obtained in this way. On a point $u \in \widehat{\mathbb{G}}_m(W) = 1 + \mathfrak{m}_W$, the action induces $u \mapsto u^s$.

The inclusion $\mathbb{Z} \hookrightarrow \mathfrak{O}$ induces an identification of p-adic rings $\mathfrak{O}_\mathfrak{p}$ with $\mathbb{Z}_p$, which we fix in this paper and use always in the sequel. Note that $\mathfrak{O}_p = \mathfrak{O}_\mathfrak{p} \times \mathfrak{O}_{\overline{\mathfrak{p}}}$. This same inclusion, $\mathbb{Z} \hookrightarrow \mathfrak{O}$, induces an inclusion of $\mathbb{Z}_{(p)}$-tori

$T_0 \hookrightarrow T_x$. Let $\mathcal{T} := T_x/T_0$. By the identification above, the map $\mathfrak{O}_{(p)}^\times \to \mathfrak{O}_\mathfrak{p}^\times$ given by $\alpha \mapsto \alpha^{1-c}$ induces an injective homomorphism

$$\mathcal{T}(\mathbb{Z}_{(p)}) \to \mathbb{Z}_p^\times = T_0(\mathbb{Z}_p). \tag{7.2.5}$$

Thereby, the action of $T(\mathbb{Z}_p)$ on $\widehat{S}$ and that of $\mathcal{T}(\mathbb{Z}_{(p)})$ are compatible. The torus $\mathcal{T}(\mathbb{Z}_{(p)})$ is isomorphic to the image of $T_x(\mathbb{Z}_{(p)})$ in Shimura's automorphism group $\mathrm{Aut}(X^{(p)}) = G(\mathbb{A}^{(p\infty)})/Z(\mathbb{Q})$, and its action on $\widehat{S}_x$ factors through the action of $\mathcal{G}_I$ on I via (7.2.2). The $\mathbb{Z}_p$-module structure of $\widehat{S}$ given by $t \mapsto t^s$ therefore commutes with the isogeny action of T_x on $\widehat{S}$.

By the level structure η_p^{ord} (and its dual), we identify $\widehat{E}_x$ with $\widehat{\mathbb{G}}_m$ and $E_x[p^\infty]^{et}$ with $\mathbb{Q}_p/\mathbb{Z}_p$. In this way, we may identify the torus T with the diagonal torus T of $G_1 = SL(2)$. The action of $t \in T_0(\mathbb{Z}_p) = \mathbb{Z}_p^\times$ on the quotient $E_x[p^\infty]^{et}$ is given by the multiplication by $t \in \mathbb{Z}_p^\times$, and hence the two tori are identified by $T_0(\mathbb{Z}_p) \ni t \mapsto \left(\begin{smallmatrix} t^{-1} & 0 \\ 0 & t \end{smallmatrix}\right) \in T(\mathbb{Z}_p) \subset SL_2(\mathbb{Z}_p)$, taking the lower diagonal entry of T as the coordinate of the quotient.

Change of level structure $\eta^{ord} \mapsto \eta^{ord} \circ a$ for $a \in T_0(\mathbb{Z}_p)$ is given by the action of an element $\left(\begin{smallmatrix} a^{-1} & 0 \\ 0 & a \end{smallmatrix}\right)$ of the diagonal torus in $T(\widehat{\mathbb{Z}}) \subset SL_2(\mathbb{A}^{(\infty)})$, which moves the point $x \in I$ to a different point $y = a(x)$ étale over the image of x in $Sh_K = \mathbf{P}^1(j)$ [for $K = G(\widehat{\mathbb{Z}})$] and brings the canonical coordinate at x to that of the image y. In other words, the action of a by the change of η^{ord} to $\eta^{ord} \circ a$:

$$(E, \iota_E, \eta^{ord}) \mapsto (E, \iota_E, \eta^{ord} \circ a),$$

sends the deformation space $\widehat{S}_x$ centered at $(E_0, \iota_0, \eta_0^{ord})$ on x to the different deformation space $\widehat{S}_y$ centered at $(E_0, \iota_0, \eta_0^{ord} \circ a)$ on $y = a(x)$ (as long as the two triples are not isogenous). The action of $r_x : T_x(\mathbb{Q}) \hookrightarrow G(\mathbb{A}^{(\infty)})$ and the action of $T(\mathbb{Z}_p)$ via the change of level structure are compatible, since the intersection of the images of the two groups in $\mathcal{G}_I$ is trivial [and the p-component $T(\mathbb{Z}_p)$ and the projection of $r_x(T_x(\mathbb{Q}))$ to the p-component $G(\mathbb{Q}_p)$ are both diagonal].

By the definition of r_x given above, we have $\alpha \circ \eta^{ord} = \eta^{ord} \circ r_x(\alpha)$ automatically. If $r_x(\alpha) \in T_x(\mathbb{Z}_{(p)})$, it acts on I as an automorphism, while $r_x(\alpha) \in T_x(\mathbb{Q})$ may expand or shrink $\mathcal{O}_I$ because it would induce a morphism like the Frobenius map on the special fiber. The action of $\tau(r_x(\alpha))$ sends the canonical coordinate t into $t^{\alpha^{1-c}}$ (identifying α with its image in $\mathfrak{O}_\mathfrak{p} = \mathbb{Z}_p$).

Lemma 7.43 *If $h \in \mathcal{G}_I$ fixes x and is an image of $\tilde{h} \in G(\mathbb{A}^{(\infty)})$ with $\tilde{h}_p \in G(\mathbb{Z}_p)$, then it is induced by an endomorphism $\alpha \in \mathrm{End}_\mathbb{Q}(E_x) \otimes_\mathbb{Z} \mathbb{Q} = M$, and h induces $t \mapsto t^{\alpha^{1-c}}$.*

Proof. Since h fixes x, it has to preserve I and $\widehat{S}$ by the irreducibility of I (a theorem of Igusa; see Corollary 7.36). Take $\tilde{h} \in G(\mathbb{A}^{(\infty)})$ with $\tilde{h}_p \in G(\mathbb{Z}_p)$ projecting down to h. Thus, $\tilde{h}_p$ is in the upper triangular Borel subgroup $B(\mathbb{Z}_p)$ by Theorem 7.35. The Borel subgroup B is upper triangular with respect to the coordinate given by η^{ord} (and its dual) under, which we identified as T_0 and T. By the universality of $Sh_{/\mathbb{Q}}$, there exists an isogeny $\alpha : E_x \to E_x$ such that $\eta^{(p)} \circ \tilde{h} = \alpha \circ \eta^{(p)}$ and $\eta^{ord} \circ \tilde{h} = \alpha \circ \eta^{ord}$. Since $\alpha \in \mathrm{End}(E_x) = \mathfrak{O}$, we have $\tilde{h} = r_x(\alpha)$ modulo $Z(\mathbb{Q})$, and therefore, h is the image of $r_x(\alpha)$ in $\mathcal{G}_I$. Then the assertion follows from the above discussion.

Summing up the above discussion, we have the following fact:

Proposition 7.44 *Let $a \in T_0(\mathbb{Z}_p)$ for $T_0 = \mathbb{G}_{m/\mathbb{Q}}$. Then the action*

$$(E, \iota_E, \eta^{ord}) \mapsto (E, \iota_E, \eta^{ord} \circ a)$$

induces an isomorphism: $\widehat{S}_y \cong \widehat{S}_x$ ($y = a(x)$) sending $f = \sum_\xi c(\xi) t'^\xi \in \Gamma(\widehat{S}_y, \mathcal{O}_{\widehat{S}_y})$ to $f \circ a(t) = \sum_\xi c(\xi) t^\xi \in \Gamma(\widehat{S}_x, \mathcal{O}_{\widehat{S}})$, where t (resp., t') is the canonical coordinate of $\widehat{S}_x$ (resp., $\widehat{S}_y$). For an isogeny $\alpha \in \mathrm{End}^{\mathbb{Q}}_F(E_x)$ regarded as an element of $T_x(\mathbb{Q})$ by r_x, we have $t \circ \tau(r_x(\alpha)) = t^{\alpha^{1-c}}$.

7.2.6 Katz Differential Operator

We have defined the two-level structure η_p^{et} and η_p° of the canonical lift E_x of $E_0 := E_x \otimes_W \mathbb{F}$, which automatically (and uniquely) extends to the level structure $\boldsymbol{\eta}_p^{\mathrm{et}} : \mathbb{Q}_p/\mathbb{Z}_p \cong \mathbb{E}[p^\infty]^{\mathrm{et}}$ and $\boldsymbol{\eta}_p^\circ : \mu_{p^\infty} \cong \mathbb{E}[p^\infty]^\circ$ over $\widehat{\mathbb{G}}_{m/W}$. Let $\zeta_n \overset{i_p}{=} \mathbf{e}(\frac{1}{p^n})$ [the canonical generator of $\mu_{p^n}(\mathbb{C}_p)$ we chose in Proposition 6.51]. By Proposition 6.51, after extending scalars to $\mathrm{Spf}(W[\widehat{t^{1/p^n}, t^{-1/p^n}}][\zeta_n])$, the finite flat group scheme $\mathbb{E}[p^n]$ splits as $\mathbb{E}[p^n]^\circ \times C_1$, and $\boldsymbol{\eta}_p \mod p^n$ induces this splitting (though the splitting is up to the truncated Barsotti–Tate group of order p^n). We consider an open compact subgroup $\Gamma \subset G(\widehat{\mathbb{Z}}^{(p)})$ containing

$$\widehat{\Gamma}(N)^{(p)} = \left\{ x \in G(\widehat{\mathbb{Z}}^{(p)}) \middle| x \equiv 1 \mod N \cdot M_2(\widehat{\mathbb{Z}}^{(p)}) \right\}$$

for a positive integer N prime to p. Let I_Γ be the Igusa tower over $Sh_\Gamma^{(p)} = Sh^{(p)}/\Gamma$. Then we have the following result in [Br1] Sect. 7:

Lemma 7.45 *Consider $\widehat{S}$ embedded in I_Γ, and take a positive integer n with $p^n \equiv 1 \mod N$. Let $\zeta = \zeta_n^u$ for $0 \le u < p^n$ ($u \in \mathbb{Z}$). Then the isogeny as in* Proposition 6.51

$$\begin{array}{ccc}
\mathbb{E} \times_{\widehat{\mathbb{G}}_m} \mathrm{Spf}(W[\widehat{t^{1/p^n}, t^{-1/p^n}}][\zeta]) & \xrightarrow{\ \pi\ } & \mathbb{E}/C_{\zeta}{}_{/\mathrm{Spf}(W[\widehat{t^{1/p^n}, t^{-1/p^n}}][\zeta])} \\
\downarrow & & \downarrow \\
\mathrm{Spf}(W[\widehat{t^{1/p^n}, t^{-1/p^n}}][\zeta]) & \xrightarrow[t \mapsto t^{1/p^n}\zeta]{} & \mathrm{Spf}(W[\widehat{t^{1/p^n}, t^{-1/p^n}}][\zeta])
\end{array}$$

brings η^{ord} to $\eta^{ord} \circ g$, where $g_p = \left(\begin{smallmatrix} p^n & u \\ 0 & 1 \end{smallmatrix}\right)$ and $g_l = \left(\begin{smallmatrix} p^n & 0 \\ 0 & 1 \end{smallmatrix}\right)$ for all primes $l \neq p$.

Note here that $\mathbb{E}/C_\zeta$ is defined over $\mathrm{Spf}(\widehat{W[t^{1/p^n}, t^{-1/p^n}]}[\zeta])$ and do not descend to $\widehat{\mathbb{G}}_m = \mathrm{Spf}(\widehat{W[t,t^{-1}]})$ (as C_ζ is only well defined over $\mathrm{Spf}(\widehat{W[t^{1/p^n}, t^{-1/p^n}]}[\zeta])$.

Proof. Our proof is slightly different from the proof given in [Br1]. By Proposition 6.51, the induced map in the bottom line of the commutative diagram in the proposition is given by $t \mapsto \zeta t^{1/p^n}$ for the canonical coordinate t with respect to $\boldsymbol{\eta}$. Thus, we need to determine g. Since the connected-étale exact sequence of $\mathbb{E}[p^n]$ splits over an fppf extension of $\widehat{\mathbb{G}}_m$, it is canonical over $\widehat{W[t^{1/p^n}, t^{-1/p^n}]}$, and the splitting is given by C_1 (i.e., $u = 0$). Thus, up to $\mathbb{E}[p^n]$, we have a well-defined $\boldsymbol{\eta}_p \mod p^n : \mathbb{Z}/p^n\mathbb{Z} \times \mu_{p^n} \cong C_1 \times \mathbb{E}[p^n]^\circ$. More audaciously, if we work over $\mathcal{R} := \bigcup_n \widehat{W[t^{1/p^n}, t^{-1/p^n}]}$, we can split the sequence even at the level of the Barsotti–Tate group $\mathbb{E}[p^\infty]$ canonically. Thus, $g \in G(\mathbb{A}^{(\infty)})$ is well defined over $\mathcal{R}$, and we can compute g at any characteristic-0 fiber of $\mathbb{E}$. Thus, we can detect g by the fiber at $t = 1$ (i.e., at E_x), and we work at the specialization at $t = 1$. Writing simply $\eta = \eta^{\mathrm{ord}}$ and putting $\widehat{L} = \eta^{-1}(TE_x) \subset (\mathbb{A}^{(\infty)})^2$, then we need to have $g^{-1}\widehat{L}/\widehat{L} = C_{\zeta^u}$ for the canonical generator $\zeta \overset{i_p}{=} \mathbf{e}(\frac{1}{p^n})$. Writing $\gamma = \eta_p^{\mathrm{et}}(\frac{1}{p^n})$, as in the proof of Proposition 6.51, C_ζ is generated by $\gamma\zeta^{-1} = \gamma\zeta_n^{-u}$. Note that at p, we have chosen the basis $(1_{\overline{\mathfrak{p}}}, -1_{\mathfrak{p}})$ of $\widehat{L}_p = \mathfrak{O}_p$ in the proof of Proposition 6.51. Writing additively, we have $\eta_p^{-1}(\gamma\zeta^{-1}) = \frac{-u}{p^n}(-1_{\mathfrak{p}}) + \frac{1}{p^n}1_{\overline{\mathfrak{p}}}$. Thus, we may take g_p given by

$$g_p^{-1}\begin{bmatrix} 1_{\overline{\mathfrak{p}}} \\ -1_{\mathfrak{p}} \end{bmatrix} = \begin{bmatrix} \frac{-u}{p^n}(-1_{\mathfrak{p}}) + \frac{1}{p^n}1_{\overline{\mathfrak{p}}} \\ -1_{\mathfrak{p}} \end{bmatrix}.$$

This shows g_p is as given in the lemma. Outside p, $g_p\Gamma = g\Gamma$ because of $p^n \equiv 1 \mod N$; so the effect of g_p and g is the same on I_Γ; hence, the result follows.

Exercise 7.46 *Let the assumption be as in the above lemma. Prove that the quotient map* $\mathbb{E} \to \mathbb{E}/\boldsymbol{\eta}(\mu_{p^n}) = \mathbb{E}/\mathbb{E}[p^n]^\circ$ *is defined over* $\widehat{\mathbb{G}}_m$ *induced by* $\boldsymbol{\eta} \mapsto \boldsymbol{\eta} \circ h$ *for* h *with* $h_l = \left(\begin{smallmatrix} 1 & 0 \\ 0 & p^n \end{smallmatrix}\right)$ *for all primes* l *and that the following diagram is commutative:*

$$\begin{array}{ccc} \mathbb{E} & \xrightarrow{\pi} & \mathbb{E}/\mathbb{E}[p^n]^\circ \\ \downarrow & & \downarrow \\ \mathrm{Spf}(\widehat{W[t,t^{-1}]}) & \xrightarrow[t \mapsto t^{p^n}]{} & \mathrm{Spf}(\widehat{W[t,t^{-1}]}). \end{array}$$

Here is another result from [Br1]:

Lemma 7.47 *Let* $\alpha(\frac{u}{p^m}) = \begin{pmatrix} 1 & \frac{u}{p^m} \\ 0 & 1 \end{pmatrix} \in G(\mathbb{Q}_p)$ *regarded as an element of* $G(\mathbb{A}^{(\infty)})$ *for a positive integer* m *and* $u \in \mathbb{Z}_p$ *[so* $\alpha(\frac{u}{p^m})_l = 1$ *for any prime* $l \neq p$*]. Then the isogeny action of* $\alpha(\frac{u}{p^m})$ *on the Igusa tower* I *over* $Sh^{(p)}$

preserves the universal deformation space of E_x (after extending scalars to $W[\zeta_n^u]$ and induces $t \mapsto t\zeta_n^u$).

Since we computed the isogeny action of g and h incorporating the conjugation by J_1, the isogeny action in the lemma already incorporates the conjugation by J_1 (so this corresponds to the action of $z \mapsto \alpha(u)(z) = z + \frac{u}{p^n}$ on the upper half-plane).

Proof. Since $I = \varprojlim_\Gamma I_\Gamma$, we need to prove the result only for I_Γ for an open compact subgroup $\Gamma \subset G(\mathbb{A}^{(p)})$ as in Lemma 7.45. For any $\frac{u}{p^m}$ with $n \geq m$, we have $\frac{u}{p^m} = \frac{p^{n-m}u}{p^n}$, and we can take n in Lemma 7.45 as large as we want so that $p^n \equiv 1 \mod N$. Let $\zeta = \zeta_n^u$. Inside $E_{0/\mathbb{F}} = E_x \otimes_W \mathbb{F}$, the image $\overline{C}_\zeta$ of $C_\zeta \subset \mathbb{E} \times_{\widehat{\mathbb{G}}_m} W\widehat{[t^{1/p^n}, t^{-1/p^n}]}[\zeta]$ is equal to $\overline{C}_1 = E_0[\overline{\mathfrak{p}}^n]$. We have $E_0 \xrightarrow{g} E_0' := E_0/\overline{C}_\zeta = E_0/E_0[\overline{\mathfrak{p}}^n] \xrightarrow{h} E_0'/E_0'[\mathfrak{p}^n] = E_0/(\overline{C}_1 + E_0[\mathfrak{p}^n]) = E_0$, since $\overline{C}_1 = E_0[\overline{\mathfrak{p}}^n]$. Thus, the action of gh preserves the origin of $\widehat{S} = \widehat{\mathbb{G}}_m$, and the isogeny action of gh preserves the deformation space. As for the coordinates, we have

$$t \xrightarrow[g]{\text{Lemma 7.45}} \zeta_n^u t^{1/p^n} \xrightarrow[h]{\text{Exercise 7.46}} t\zeta_n^u.$$

Note that the action of $gh = \begin{pmatrix} p^n & u \\ 0 & p^n \end{pmatrix} = p^n\alpha(\frac{u}{p^n})$. Since the global $p = (p, p, \ldots, p, \ldots) \in G(\mathbb{A}^{(\infty)})$ diagonally embedded in $G(\mathbb{A})$ acts trivially on Sh as it is induced by self-isogeny $\mathbb{E} \xrightarrow{x \mapsto px} \mathbb{E}$, the action of the scalar p^n on $\widehat{\mathbb{G}}_m$ is the identity. Hence, we get the desired result.

Proposition 7.48 *Let t be the canonical Serre–Tate coordinate of the universal deformation space $\widehat{S} := \widehat{\mathbb{G}}_m$ of (E_x, η^{ord}). Then the Katz differential operator d induces $t\frac{d}{dt}$ on $\widehat{S} := \widehat{\mathbb{G}}_m$.*

This follows from [K1] Theorem 4.3.1, but we give a different proof without using the Gauss–Manin connection on the deformation space.

Proof. We use the notation introduced between (1.3.9) and (1.3.10). Since a p-adic modular form is a p-adic limit of classical modular forms, we only need to verify the result for classical forms. Then for a classical modular form f with q-expansion $f(q) = \sum_{m \in \mathbb{Z}} a(m, f)q^m$ and t-expansion $f(t) = \sum_{m \in \mathbb{Z}} c(m, f)t^m$ (over $\widehat{S}$), we defined in (1.3.10) the operator d as

$$df(q) = \lim_{n \to \infty} f|\phi_n(q) = \sum_n na(n, f)q^n,$$

where we have chosen ϕ_n so that $\lim_{n\to\infty} \phi_n(m) = m$ p-adically. By (1.3.9), we have

$$f|\phi_n = p^{-n} \sum_{u \in \mathbb{Z}/p^n\mathbb{Z}} \phi_n^*(-u) f|\alpha(\frac{u}{p^n}),$$

where $f|\alpha(a)(z) = f(\alpha(a)(z)) = f(z+a)$ for $z \in \mathfrak{H}$. By the q-expansion principle combined with the t-expansion principle, we have $df(t) = \lim_{n\to\infty} f|\phi_n(t)$. By definition, the action of $\alpha(\frac{u}{p^n})$ on $\mathfrak{H}$ is given by $z \mapsto \alpha(\frac{u}{p^n})(z) = z + \frac{u}{p^n}$ for $z \in \mathfrak{H}$ and by Lemma 7.47, we have $\alpha(\frac{u}{p^n})(t) = \zeta_n^u t$ for $t \in \widehat{\mathbb{G}}_m$. Then, by the definition of the Fourier transform, $\phi_n(m) = p^{-n}\sum_{u\in\mathbb{Z}/p^n\mathbb{Z}} \phi_n^*(-u)\zeta_n^{mu}$. In particular, we have

$$\begin{aligned} f|\phi_n(t) &= p^{-n}\sum_m \sum_{u\in\mathbb{Z}/p^n\mathbb{Z}} \phi_n^*(-u)c(m,f)(\zeta_n^u t)^m \\ &= \sum_m p^{-n}\left(\sum_{u\in\mathbb{Z}/p^n\mathbb{Z}} \phi_n^*(-u)\zeta_n^{mu}\right) c(m,f)t^m = \sum_m \phi_n(m)c(m,f)t^m. \end{aligned}$$

Passing to the p-adic limit with respect to n, we get

$$df(t) = \lim_{n\to\infty} f|\phi_n(t) = \sum_m m\cdot c(m,f)t^m = t\frac{df}{dt},$$

as desired.

Remark 7.49 The action of $\rho(\alpha)$ for $\alpha \in T_x(\mathbb{Z}_{(p)})$ induces $t \mapsto t^{\alpha^{1-c}}$ by Proposition 7.44. Then the above lemma supplies us with the final missing item in the proof of Lemma 1.33; so Lemma 1.33 has now been proven.

7.3 Elliptic Modular Forms of Slope 0

We keep the notation in the previous section and recall the one introduced in Sect. 6.2.12. Take $\widehat{\Gamma}(N)^{(p)} \subset \Gamma \subset \widehat{\Gamma}_0(N)^{(p)}$, and write X for the compactified modular curve X_Γ. We have $S_0 = X[\frac{1}{\mathcal{E}}]$, $S_m = S_0 \otimes_W W_m = \mathrm{Spec}(V_{m,0})$, $S_\infty = \varprojlim_m S_m$, and $I_{m,\alpha} = \mathrm{Spec}(V_{m,\alpha})$ for the αth layer of the Igusa tower over S_m. We have a p-divisible module $\mathcal{V}_\Gamma$ on which $\mathrm{Gal}(V_{m,\infty}/V_{m,0}) = \mathbb{Z}_p^\times = \mathbb{G}_m(\mathbb{Z}_p)$ acts continuously. Here $T = \mathbb{G}_m$. We construct a projector e acting on $\mathcal{V}$ out of the Hecke operator $U(p)$ commuting with the action of $\mathbb{Z}_p^\times = \mathrm{Gal}(V_{m,\infty}/V_{m,0})$. The important features of e (we are going to show) are

- $e = \lim_{n\to\infty} U(p)^{n!}$ (we recall the definition of $U(p)$ in Sect. 7.3.3);
- $\mathcal{V}^{ord} = e\mathcal{V}$ has the Pontryagin dual, which is projective over $W[[\mathbb{G}_m(\mathbb{Z}_p)]]$;
- for any $k \geq 3$, there is a canonical isomorphism

  $$\mathcal{V}^{ord}[k] \cong eH^0(S_0, \underline{\omega}^k \otimes \mathbb{T}_p) = eH^0(X, \underline{\omega}^k \otimes \mathbb{T}_p) \quad (\mathbb{T}_p = \mathbb{Q}_p/\mathbb{Z}_p),$$

  where $\mathcal{V}[k] = \{f \in \mathcal{V} | zf = z^k f \text{ for all } z \in \mathbb{Z}_p^\times\}$ and $\mathcal{V}^{ord}[k] = e\mathcal{V}[k]$.

Here is how to prove the convergence of the projector $e = \lim_{n\to\infty} U(p)^{n!}$. For any finite p-torsion module M with an endomorphism u, the power $u^{n!}$

stabilizes as n grows, giving an idempotent e_u in the endomorphism algebra $\mathrm{End}(M)$, since $\mathrm{End}(M)$ is a finite ring. For a p-divisible module M with an endomorphism u, we assume that $M = \varinjlim_{i\in I} M_i$ for finite modules M_i stable under u. Then the endomorphism $e_u = \lim_{n\to\infty} u^{n!}$ is well defined over M_i for all i and hence over M. For a compact p-profinite torsion-free module M with an endomorphism u, we take the Pontryagin dual pair (M^*, u^*), which is p-divisible (cf. [FAN] Chaps. 1 and 3). Thus, e_{u^*} exists on M^*. Then $e_u = \lim_{n\to\infty} u^{n!}$ (which is the dual of e_{u^*}) exists as an endomorphism of M.

7.3.1 Control Theorems of p-Adic Modular Forms

Recall the invertible sheaf $\underline{\omega}^k$ over $X_1(N)$ as in Definition 6.29. Computing the genus of $X_1(N)$, the Riemann–Roch theorem tells us that $\underline{\omega}^k$ is very ample if $k \geq 3$ (see [GME] Proposition 2.1.4 and Theorem 3.1.2).

Let $(\mathbf{E}, \boldsymbol{\phi}_N, \boldsymbol{\omega})$ be the universal elliptic curve over $Y_1(N)$. For each element $(E, \phi_N, \omega) \in \wp_{\Gamma_1(N)}(R)$ defined over R (called a *test object*), we have a unique morphism $\iota : \mathrm{Spec}(R) \to Y_1(N)$ such that $\iota^*(\mathbf{E}, \boldsymbol{\phi}_N, \boldsymbol{\omega}) = (E, \phi_N, \omega)$. For each section $f \in H^0(Y_1(N), \underline{\omega}^k)$, we define $\iota^* f = f(E, \phi_N, \omega)\omega^{\otimes k}$. The function $(E, \phi_N, \omega) \mapsto f(E, \phi_N, \omega)$ satisfies (G0–2). Condition (G3) ensures that f extends to $X_1(N)$. This reconfirms Definition 6.29:

$$H^0(X_1(N)_{/R}, \underline{\omega}^k_{/R}) \cong G_k(\Gamma_1(N); R) \quad \text{(canonically)} \qquad (7.3.1)$$

for all $\mathbb{Z}[\frac{1}{6N}]$-algebras R.

Let $(\mathbf{E}, \boldsymbol{\phi}_p, \boldsymbol{\phi}_N)$ be the universal elliptic curve over S_m. Pick a section $f \in H^0(S_m, \underline{\omega}^k)$. Since μ_{p^∞} carries a canonical differential $\omega_{can} = \frac{dt}{t}$, writing $\mu_{p^\alpha} = \mathrm{Spec}(\mathbb{Z}[t]/(t^{p^\alpha} - 1))$, we may regard f as a function of (E, ϕ_p, ϕ_N) by $f(E, \phi_p, \phi_N) = f(E, \phi_N, \phi_{p,*}\omega_{can})$. For each $(E, \phi_p, \phi_N) \in \mathcal{E}^{ord}_\infty(R)$ for a W_m-algebra R, we have a unique morphism $\iota : \mathrm{Spec}(R) \to I_{m,\infty} = \mathrm{Spec}(V_{m,\infty})$ such that $(E, \phi_p, \phi_N) = \iota^*(\mathbf{E}, \boldsymbol{\phi}_p, \boldsymbol{\phi}_N)$. Then $\iota^* f$ is just a function on $f(E, \phi_p, \phi_N)$ such that $f(E, z^{-1}\phi_p, \phi_N) = z^k f(E, \phi_p, \phi_N)$ for $z \in \mathrm{Gal}(V_{m,\infty}/V_{m,0}) = \mathbb{Z}_p^\times$. We recall that the action of $\mathbb{G}_m(\mathbb{Z}_p) = \mathbb{Z}_p^\times$ is normalized so that $z \in \mathbb{Z}_p^\times$ acts on the dual identification $\phi_p^{\mathrm{et}} : E[p^\infty]^{\mathrm{et}} \cong \mathbb{Q}_p/\mathbb{Z}_p$ of $\phi_p : \mu_{p^\infty} \hookrightarrow E[p^\infty]$ via the multiplication by z; so z acts on ϕ_p by $\phi_p \mapsto \phi_p \circ z^{-1}$. In other words, we have $z \cdot f(E, \phi_p, \phi_N) = f(E, z^{-1}\phi_p, \phi_N) = z^k f(E, \phi_p, \phi_N)$. This shows that

$$V_{m,\infty}[k] = H^0(S_m, \underline{\omega}^k)$$
$$\text{and } \mathcal{V}[k] = H^0(S_{\infty/W}, \underline{\omega}^k \otimes \mathbb{T}_p) = H^0(S_{\infty/W}, \underline{\omega}^k) \otimes \mathbb{T}_p, \quad (7.3.2)$$

where $\mathbb{T}_p = \mathbb{Q}_p/\mathbb{Z}_p$. The last identity follows from the affineness of S_0. When we generalize this point to classical groups bigger than $G = GL(2)_{/\mathbb{Q}}$, the affineness is no longer true; so we need to work out this point (the base-change property) carefully to bring $\mathbb{T}_p$ inside the cohomology group outside the cohomology group. In the elliptic modular case, simply by the affineness of S_0, (7.3.2) shows that $\mathcal{V}[k]$ is p-divisible, and its direct summand $e\mathcal{V}[k]$ is also p-divisible. We show that e contracts $\mathcal{V}[k]$ into $H^0(S_{0/W}, \underline{\omega}^k \otimes \mathbb{T}_p)$.

We consider the following condition:

(F) $\mathrm{corank}_W\, e\mathcal{V}[k] = \mathrm{rank}_W \mathrm{Hom}(e\mathcal{V}[k], \mathbb{T}_p)$ *is finite for an integer* k.

In practice, this condition is often proven by showing

$$eH^0(S_{0/W}, \underline{\omega}^k \otimes \mathbb{T}_p) = eH^0(X_1(N)_{/W}, \underline{\omega}^k \otimes \mathbb{T}_p) = G_k^{ord}(\Gamma_1(N); \mathbb{T}_p), \quad (7.3.3)$$

where

$$G_k^{ord}(\Gamma_1(N); R) = eG_k(\Gamma_1(N); R) \text{ and } G_k(\Gamma_1(N); \mathbb{T}_p) = G_k(\Gamma_1(N); \mathbb{Z}_p) \otimes \mathbb{T}_p.$$

The left-hand side (LHS) of (7.3.3) is p-divisible, since S_0 is affine. The right-hand side (RHS) is of finite corank, since $X_1(N)$ is projective. Thus, $e\mathcal{V}[k]$ is p-divisible of finite corank.

Decompose $\mathbb{Z}_p^\times = \Gamma_T \times \Delta$ for a p-profinite group Γ_T and a prime-to-p finite group Δ. For simplicity, suppose that $p > 2$. Then Δ has order $p-1$ prime to p, and Γ_T is isomorphic to $\mathbb{Z}_p$. For the generator $\gamma = 1+p$ of Γ_T, we have $W[[\Gamma_T]] \cong W[[x]] = \Lambda$ via $\gamma \mapsto 1+x$ [i.e., $\gamma^s \mapsto (1+x)^s = \sum_{j=0}^{\infty} \binom{s}{j} x^j$], and $W[[\mathbb{Z}_p^\times]] = \Lambda[\Delta]$. Let $\mathbf{V}^{ord}$ be the Pontryagin dual module of $e\mathcal{V}$. See [FAN] Chaps. 1 and 3 for generalities on topological groups (including Pontryagin duality and profinite groups), and [LFE] Chap. 8 also contains a short exposition on this subject. The Pontryagin dual for p-torsion modules M (equipped with discrete topology) is given by $M^* = \mathrm{Hom}_{cont}(M, \mathbb{T}_p)$ for $\mathbb{T}_p = \mathbb{Q}_p/\mathbb{Z}_p$ with discrete topology. Writing $M = \varinjlim_{i \in I} M_i$ for finite submodules M_i of order $p^{n(i)}$, we have

$$M^* = \varprojlim_{i,n} \mathrm{Hom}(M_i, p^{-n}\mathbb{Z}/\mathbb{Z}) \cong \varprojlim_{i} \mathrm{Hom}(M_i, p^{-n(i)}\mathbb{Z}/\mathbb{Z}),$$

and the topology on M^* is the profinite topology with respect to the finite discrete groups $\mathrm{Hom}(M_i, p^{-n}\mathbb{Z}/\mathbb{Z})$. Then M^* is a compact profinite group. The duality is perfect; so if $M \hookrightarrow Y \twoheadrightarrow Z$ is an exact sequence of p-torsion modules, $Z^* \hookrightarrow Y^* \twoheadrightarrow M^*$ is an exact sequence of profinite groups. In particular, if M is p-divisible and p-torsion, the multiplication by p is surjective; so, by duality, multiplication by p on M^* is injective, which shows that M^* is torsion-free if M is p-divisible and p-torsion.

Assume that a profinite (noetherian) W-algebra R acts on M continuously. For any closed ideal $\mathfrak{a} \subset R$, Pontryagin's dual of $M[\mathfrak{a}] = \{x \in M | \mathfrak{a}x = 0\}$ is given by $M^*/\mathfrak{a}M^*$, where R acts on M^* continuously by the pullback (dual) action. To see this, we choose a finite set of generators $(a_1, \ldots, a_r)$ of $\mathfrak{a}$ and look into the exact sequence: $0 \to M[\mathfrak{a}] \to M \xrightarrow{x \mapsto (a_1x, \ldots, a_rx)} M^r$. Taking the dual, we find that $M^{*r} \xrightarrow{(x_1, \ldots, x_r) \mapsto \sum_i a_i x_i} M^* \to M^*/\mathfrak{a}M^* \to 0$ is exact, which shows $M[\mathfrak{a}]^* = M^*/\mathfrak{a}M^*$. For any (continuous) W-algebra homomorphism $\chi : R \to W$, applying the above fact to $\mathfrak{a} = \mathrm{Ker}(\chi)$, the Pontryagin dual of the χ-eigen-subspace $M[\chi] = \{x \in M | ax = \chi(a)x \text{ for all } a \in R\}$ is given by the maximal χ-eigen-quotient $M^*/\mathfrak{a}M^*$ for the χ-augmentation ideal $\mathfrak{a} = \mathrm{Ker}(\chi) = (a - \chi(a))_{a \in R} \subset R$.

Let $\widehat{\Delta} = \mathrm{Hom}(\Delta, W^\times)$, and write $\mathcal{V}[\chi]$ for the χ-eigen-subspace of $\chi \in \widehat{\Delta}$:

$$\mathcal{V}[\chi] = \{v \in \mathcal{V} | \delta v = \chi(\delta)v \text{ for all } \delta \in \Delta\}.$$

If $e\mathcal{V}[k]$ is of finite corank for one k, taking $\chi \in \widehat{\Delta}$ with $\chi = k|_\Delta$, we have $\mathcal{V}[k] = \mathcal{V}[\chi][k]$, and then by duality, we have

$$\begin{aligned}
\mathbf{V}^{ord} \otimes_{W[[\mathbb{G}_m(\mathbb{Z}_p)]],k} W &= \mathbf{V}^{ord}[\chi]/(x+1-\gamma^k)\mathbf{V}^{ord}[\chi] \\
&= \mathrm{Hom}_{cont}(\mathcal{V}^{ord}[\chi, (x+1-\gamma^k)], \mathbb{T}_p) \\
&\cong \mathrm{Hom}_{\mathbb{Z}_p}(\varinjlim_m H^0_{ord}(X_1(N)_{/W_m}, \underline{\omega}^k_{/W_m}), \varinjlim_m p^{-m}\mathbb{Z}/\mathbb{Z}) \\
&\cong \varprojlim_m \mathrm{Hom}_{\mathbb{Z}_p}(H^0_{ord}(X_1(N)_{/W_m}, \underline{\omega}^k_{/W_m}), p^{-m}\mathbb{Z}/\mathbb{Z}) \\
&\cong \varprojlim_m \mathrm{Hom}_W(H^0_{ord}(X_1(N)_{/W_m}, \underline{\omega}^k_{/W_m}), W_m) \\
&= \mathrm{Hom}_W(H^0_{ord}(X_1(N), \underline{\omega}^k), W) \cong \mathrm{Hom}_W(G_k^{ord}(\Gamma_1(N); W), W). \quad (7.3.4)
\end{aligned}$$

In the middle equality, we have assumed (7.3.3). Here the subscript or superscript *ord* indicates the image of e.

Write $\widehat{\Delta} = \mathrm{Hom}(\Delta, W^\times)$. Note here that each $\chi \in \widehat{\Delta}$ is induced by infinitely many $0 < k \in X^*(T)$. Decompose $\mathbf{V}^{ord}$ into the sum of χ-eigenspaces,

$$\mathbf{V}^{ord} = \bigoplus_{\chi \in \widehat{\Delta}} \mathbf{V}^{ord}[\chi].$$

This is possible, as we assumed that Δ has order prime to p. If $z \mapsto z^k$ coincides with χ on Δ, then $\mathbf{V}^{ord}[\chi] \otimes_{\Lambda,k} W = \mathbf{V}^{ord} \otimes_{W[[\mathbb{G}_m(\mathbb{Z}_p)]],k} W$. By Nakayama's lemma (Lemma 10.8), we have a surjective homomorphism of Λ-modules:

$$\pi : \Lambda^{s(\chi)} \twoheadrightarrow \mathbf{V}^{ord}[\chi],$$

where $s = s(\chi) = \mathrm{corank}_W e\mathcal{V}[k]$. If (F) holds for one k, it holds for all $\kappa \in X^*(T)$ inducing χ, and π is an isomorphism for the following reason: s is the minimum number of generators of $\mathbf{V}^{ord}[\chi] \otimes_{\Lambda,\kappa} W$ over W. We know that this module is W-free, because its dual $\mathcal{V}[\kappa]$ is p-divisible; so it is free of rank s. The morphism π induces an isomorphism modulo $(1+x) - \gamma^\kappa$ for all κ inducing χ. Then $\mathrm{Ker}(\pi) \subset \bigcap_\kappa \mathrm{Ker}(\pi \mod (1+x-\gamma^\kappa)) = 0$, and we get

Theorem 7.50 *Suppose that* (F) *holds for one* k. *Then* $\mathbf{V}^{ord} = \bigoplus_\chi \mathbf{V}^{ord}[\chi]$ *is* Λ*-free of finite rank, and if* (7.3.3) *holds for* $\kappa \in X^*(T)$ *with* $\kappa|_\Delta = k|_\Delta$, *then*

$$\mathbf{V}^{ord} \otimes_{W[[\mathbb{Z}_p^\times]],\kappa} W \cong \mathrm{Hom}_W(G_\kappa^{ord}(\Gamma_1(N);W),W).$$

We write the corresponding subspace of cusp forms satisfying (s3) in Sect. 6.2.10 and (s) in Sect. 7.2.4 adding superscript "cusp"; so $\mathbf{V}^{ord,cusp} \subset \mathbf{V}^{ord}$ except for $S_k = G_k^{cusp}$. By the same argument, we get the corresponding assertion:

$$\mathbf{V}^{ord,cusp} \otimes_{W[[\mathbb{Z}_p^\times]],\kappa} W \cong \mathrm{Hom}_W(S_\kappa^{ord}(\Gamma_1(N);W),W).$$

We can also replace $\Gamma_1(N)$ by $\Gamma_0(N),\chi$ for a character $\chi : (\mathbb{Z}/N\mathbb{Z})^\times \to W^\times$.

Remark 7.51 Since we get a parallel statement replacing $\Gamma_1(N)$ (resp., the phrase "modular form") by $\Gamma_0(N),\chi$ (resp., "cusp form"), hereafter we state the result only for G_k and $\Gamma_1(N)$. Unless we note specifically, all the results are valid for cusp forms and for $(\Gamma_0(N),\chi)$.

7.3.2 Bounding the p-Ordinary Rank

Since S_0 is affine, we have $H^0(S_1,\underline{\omega}^k) = H^0(S_{0/W},\underline{\omega}^k) \otimes_W W_1$. If we have a sequence $\overline{f}_1,\overline{f}_2,\dots,\overline{f}_j$ of linearly independent sections in $H^0_{ord}(S_1,\underline{\omega}^k)$, we can lift them to $f_i \in H^0(S_{0/W},\underline{\omega}^k)$ so that $\overline{f}_i = (f_i \mod p)$. Since $S_0 = X[\frac{1}{\mathcal{E}}]$ for a lift $\mathcal{E}$ of a power of the Hasse invariant H, we have

$$H^0(S_{0/W},\underline{\omega}^k) = \varinjlim_n H^0(X_{/W},\underline{\omega}^{k+n(p-1)})/\mathcal{E}^n.$$

Thus, $\mathcal{E}^m f_i \in H^0(X_{/W},\underline{\omega}^{k+m(p-1)})$ for all $i = 1,\dots,j$ for sufficiently large m, and they are linearly independent. We now assume

(C) $e(\mathcal{E}f) = \mathcal{E}(ef)$ for all $f \in H^0(S_1,\underline{\omega}^k)$.

By this, $e(\mathcal{E}^m f_i)$ are still linearly independent in $H^0_{ord}(X_{/K},\underline{\omega}^{k+m(p-1)})$; so we have $\dim_K H^0_{ord}(X_{/K},\underline{\omega}^{k+m(p-1)}) \geq j$. If $\mathrm{rank}_{W_1} H^0_{ord}(S_1,\underline{\omega}^k)$ is infinite, the rank of $H^0_{ord}(X_{/K},\underline{\omega}^{k+m(p-1)})$ grows as m grows. Thus, condition (F) for all k follows from

(F′) $\dim_K G_k^{ord}(\Gamma_1(N);\mathbb{Q}_p)$ is bounded (independently of k).

Indeed, the Eichler–Shimura isomorphism combined with a calculation of group cohomology $H^1_{ord}(\Gamma_1(N),L(k;K))$ proves much stronger:

(E) If $k \geq 3$, $\dim_K G_k^{ord}(\Gamma_1(N);K)$ depends only on $k \mod p-1$

for the projector e [H86b] Section 4. Here $L(k;K) = \sum_{j=0}^k KX^jY^{k-j}$ is the K-vector space of homogeneous polynomials $\phi(X,Y)$ in (X,Y) of degree k on which $\gamma \in \Gamma_1(N)$ acts by $\gamma\cdot\phi(X,Y) = \phi((X,Y)\gamma)$. We can ease the condition $k \geq 3$ to $k \geq 2$ if we replace $G_k^{ord}(\Gamma_1(N);K)$ by $G_k^{ord}(\Gamma_1(N)\cap\Gamma_0(p);K)$ in (E) (see [LFE] 7.2).

7.3.3 Vertical Control Theorem

Let $(E, \phi_p, \phi_N)_{/R}$ be a p-adic test object (i.e., $\phi_p : \mu_{p^\infty} \hookrightarrow E[p^\infty]$ and R is a p-adic ring). For the moment, we assume that R is of characteristic 0. Each subgroup C of order p outside the image of ϕ_p is étale; so we can think of the quotient $(E/C, \phi_p, \phi_N)$ defined over a (possibly "radiciel") finite extension R' of R. To make this R' explicit, we may assume that R is a complete local p-adic algebra (by taking a completion). Then, writing t for the Serre–Tate coordinate of $(E, \phi_p, {}^t\phi_p^{-1})$ for the Cartier dual inverse ${}^t\phi_p^{-1}$ of ϕ, we have $R' = R[t^{1/p}]$ by (the version specialized to E of) Proposition 6.51. Thus, R' is R-free of finite rank; so, identifying $R' \cong R^p$ (column vector space) as an R-module, we can write multiplication by $r \in R'$ on R' as a matrix form $\rho(r) \in M_p(R)$; hence, ${}^t(a_1 r, \dots, a_p r) = \rho(r){}^t(a_1, \dots, a_p)$. Then we define $\mathrm{Tr} = \mathrm{Tr}_{R'/R} : R' \to R$ by $\mathrm{Tr}(r) = \mathrm{Tr}(\rho(r))$, where right-hand side is the matrix trace of $\rho(r)$.

Definition 7.52 *We define the $U(p)$ operator acting on a (p-adic or classical) geometric modular form f by*

$$f|U(p)(E, \phi_p, \phi_N) = \frac{1}{p}\mathrm{Tr}(f(E/C, \phi_p, \phi_N)) \overset{(*)}{=} \frac{1}{p}\sum_C f(E/C, \phi_p, \phi_N).$$

Since $C \cap \mathrm{Im}(\phi_p)$ is trivial, ϕ_p induces ϕ_p for E/C. Since $(E/C, \phi_p, \phi_N)$ is defined over R', $(f(E/C, \phi_p, \phi_N)) \in R'$, and $\mathrm{Tr}(f(E/C, \phi_p, \phi_N))$ is the trace as defined above. The last equality $(*)$ is only valid when p is not a zero divisor in R. In this case, the C's are given by C_ζ for $\zeta \in \mu_p(R[\mu_p])$ as in Proposition 6.51 (or more specifically, the version specialized to E), and $\sigma_\zeta(f(E/C_1, \phi_p, \phi_N)) = f(E/C_\zeta, \phi_p, \phi_N)$ for the automorphism σ_ζ of R', sending $t^{1/p}$ to $\zeta \cdot t^{1/p}$; hence, the sum $\sum_C$ is actually a trace. In this way, the sum $\sum_C$ in the above definition has to be understood as a trace from R' to R. From the definition, as long as the prime p is invertible in the base ring R, we can verify conditions (G0–3) [or (V1–3)] for $f|U(p)$. Computing the q-expansion (see (7.4.1)), we find $a(n, f|U(p)) = a(np, f)$. Then, by the q-expansion principle, the operator $U(p)$ is well defined p-integrally over $\mathcal{W}$ even if the definition of the operator involves division by the prime p. Since $T(n) = T(n')U(p^e)$ if $n = n'p^e$ with $p \nmid n'$, the sum over $C \in \mathcal{C}_n$ in the definition of the operator $T(n)$ can be separated into a p-part and a prime-to-p part. Though the sum of the p-part corresponding to $U(p^e)$ is actually the trace as explained above, we hereafter always pretend it is a sum (since the notation becomes simpler by this abuse and it does not do any harm as long as readers have this fact in mind).

Since $\mathcal{E} \equiv 1 \mod p$, we confirm (C). Recall

$$G_{\Gamma_1(N)}(R) = \bigoplus_{k=0}^{\infty} G_k(\Gamma_1(N); R).$$

We have a natural inclusion map: $G_{\Gamma_1(N)}(R) \hookrightarrow V_{\Gamma_1(N)/R}$, since we may regard $f \in G_k(\Gamma_1(Np^\alpha); R)$ as a p-adic modular form by putting $f(E, \phi_p, \phi_N) = f(E, \phi_p|_{\mu_{p^\alpha}}, \phi_N, \omega_p)$ as in Proposition 7.41. One can prove the p-adic density of $D = G_{\Gamma_1(N)}(W)[\frac{1}{p}] \cap V$ in V (see [GME] Corollary 3.2.4). Using this fact, we can show that we can write $\mathcal{V}$ as a union of finite modules stable under $U(p)$; so $e = \lim_{n\to\infty} U(p)^{n!}$ exists. The final result is as follows.

Theorem 7.53 (Vertical control theorem) *We have*

$$\mathbf{V}^{ord} \otimes_{W[[\mathbb{Z}_p^\times]],k} W \cong \begin{cases} \mathrm{Hom}_W(G_k^{ord}(\Gamma_1(N); W), W) & \text{if } k \geq 3, \\ \mathrm{Hom}_W(G_2^{ord}(\Gamma_1(N) \cap \Gamma_0(p); W), W) & \text{if } k = 2. \end{cases}$$

Similarly, if we write $\mathcal{V}^{ord}_{cusp}$ for the subspace of cusp forms in $\mathcal{V}^{ord}$ and write $\mathbf{V}^{ord}_{cusp}$ for its Pontryagin dual (i.e., the cuspidal quotient of $\mathbf{V}^{ord}$), the above result is still valid, replacing $\mathbf{V}^{ord}$ and G_k^{ord} by $\mathbf{V}^{ord}_{cusp}$ and S_k^{ord}.

7.3.4 Families of p-Ordinary Modular Forms

Recall the decomposition $\mathbb{G}_m(\mathbb{Z}_p) = \mathbb{Z}_p^\times = \Gamma_T \times \Delta$ for a finite group Δ and $\Gamma_T \cong \mathbb{Z}_p$. We have chosen a generator $\gamma \in \Gamma_T$ so that $\mathbb{Z}_p \ni s \mapsto \gamma^s \in \Gamma_T$ induces the isomorphism $\mathbb{Z}_p \cong \Gamma_T$ and $\Lambda = W[[\Gamma_T]] \cong W[[x]]$ by $\gamma \leftrightarrow 1 + x$. Here we have the relation $\gamma^{-1}(1 + x) = (1 + T)$ for the variable T in Chap. 3.

Let $a(n) : \mathcal{V}^{ord} \to \mathbb{T}_p$ be the linear map associating f with its coefficient of q^n in the q-expansion; hence, $a(n)$ is in the dual $\mathbf{V}^{ord}$. We now consider

$$G(\chi; \Lambda) = \mathrm{Hom}_\Lambda(\mathbf{V}^{ord}[\chi], \Lambda).$$

With each $\Phi \in G(\chi; \Lambda)$, we associate its q-expansion

$$\Phi(q) = \sum_{n=0}^{\infty} a(n; \Phi) q^n \in \Lambda[[q]] \ \text{ with } a(n, \Phi) = \Phi(a(n)) \in \Lambda.$$

Theorem 7.54 *For each $k \geq 2$, we have*

(1) *$G(\chi; \Lambda)$ is Λ-free of finite rank;*
(2) *$G(\chi; \Lambda) \otimes_{\Lambda,k} W \cong G_k^{ord}(\Gamma_1(N) \cap \Gamma_0(p), \chi\omega^{-k}; W)$, where ω is the Teichmüller character modulo p;*
(3) *the above identification is induced by $\Phi \mapsto \sum_n a(n, \Phi)(\gamma^k - 1) q^n \in W[[q]]$.*

Proof. The Λ-freeness follows from the freeness of $\mathbf{V}^{ord}[\chi]$. We only prove the assertion when k induces χ on Δ; so $\chi\omega^{-k}$ is trivial. We have

$$\begin{aligned} G(\chi; \Lambda) \otimes_{\Lambda,k} W &\cong \mathrm{Hom}_\Lambda(\mathbf{V}^{ord}[\chi], \Lambda) \otimes_{\Lambda,k} W \\ &\cong \mathrm{Hom}_W(\mathbf{V}^{ord}[\chi] \otimes_{\Lambda,k} W, W) \cong \mathrm{Hom}_W(\mathrm{Hom}_W(G_k^{ord}(\Gamma_1(N); W), W)) \\ &= G_k^{ord}(\Gamma_1(N); W) \cong G_k^{ord}(\Gamma_1(N) \cap \Gamma_0(p); W). \end{aligned}$$

Here is the explanation of how to obtain the last isomorphism in the above sequence of isomorphisms. As is well known (see Corollary 6.34 and (7.4.1)), the Hecke operator $T(p)$ acts on $f \in G_k(\Gamma_1(N); W)$ by

$$a(n, f|T(p)) = a(pn, f) + a(\frac{n}{p}, f|\langle p\rangle),$$

where we put $a(\xi, f) = 0$ if $\xi \notin \mathbb{Z}$ and $\langle p\rangle$ is the diamond operator we define after (7.4.1). Thus, in $W[[q]]$, $T(p) \equiv U(p) \mod p$ if $k \geq 2$, and the natural inclusion $G_k(\Gamma_1(N); W) \hookrightarrow G_k(\Gamma_1(N) \cap \Gamma_0(p); W)$ induces an inclusion:

$$G_k^{ord}(\Gamma_1(N); W) \hookrightarrow G_k^{ord}(\Gamma_1(N) \cap \Gamma_0(p); W).$$

By the theory of primitive forms ([MFM] Theorem 4.6.17; see also [H85] Section 3), this inclusion is surjective if $k \geq 3$ (not surjective if $k = 2$, for example, if $N = 1$ and $p = 11$ or $p \geq 17$). We leave the verification of assertion (3) to the reader.

For a given element $\Phi \in G(\chi; \Lambda)$, we get a family of p-ordinary elliptic modular forms $\{\Phi(\gamma^k - 1)\}_{k\geq 2}$ whose q-expansion coefficients depend (p-adic) analytically on the weight $k \in \mathrm{Hom}_{cont}(T(\mathbb{Z}_p), \mathbb{Z}_p^\times)$. Each element $\Phi \in G(\chi; \Lambda)$ is called a p-ordinary *Λ-adic form*, and $\{\Phi(\gamma^k - 1)\}_{k\geq 2}$ is called the *p-adic analytic family* of modular forms associated with a Λ-adic form Φ.

Remark 7.55 As we mentioned in Remark 7.51, there is a version of this for $\Gamma_0(N)$ and cusp forms. For $k \geq 2$, we have

1. $\mathcal{F}(\Gamma_0(N), \chi; \Lambda)$ is Λ-free of finite rank;
2. $\mathcal{F}(\Gamma_0(N), \chi; \Lambda) \otimes_{\Lambda,k} W \cong \mathcal{F}_k^{ord}(\Gamma_0(Np), \chi\omega^{-k}; W)$;
3. the above identification is induced by $\Phi \mapsto \sum_n a(n, \Phi)(\gamma^k - 1)q^n \in W[[q]]$,

where χ is a character modulo Np and $\mathcal{F} = G$ and S (see [GME] Chap. 3).

We put $\mathcal{F}(\Gamma_0(N), \chi; \mathbb{I}) = \mathcal{F}(\Gamma_0(N), \chi; \Lambda) \otimes_\Lambda \mathbb{I}$ for a Λ-algebra $\mathbb{I}$.

7.4 Hecke Algebras

In this section, we introduce Hecke operators' action on the space of p-adic modular forms and $\mathbb{I}$-adic modular forms, as was done early in Sect. 6.2.5 for modular forms. Here $\mathcal{W}$ denotes a discrete valuation ring in $\overline{\mathbb{Q}}$ flat over $\mathbb{Z}_{(p)}$.

7.4.1 Hecke Operators on p-Adic Modular Forms

We call a ring R a *p-adic ring* if $R = \varprojlim_n R/p^nR$. We now assume that the base algebra R is a $\mathcal{W}$-algebra when we consider classical modular forms. When we consider p-adic modular forms, we further suppose that R is a p-adic ring; hence, it is a W-algebra automatically for $W = \varprojlim_n \mathcal{W}/p^n\mathcal{W}$. We use the notation introduced in Sect. 6.2.5. In particular, $\mathcal{C}_n(E)$ for a test object $(E, \phi_N : \mu_N \hookrightarrow E, \phi_p : \mu_{p^\infty} \hookrightarrow E)$ is the set of locally free subgroups of rank n disjoint from $\phi_N(\mu_l)$ and $\phi(\mu_p)$. As before, we make the quotient elliptic curve E/C for $C \in \mathcal{C}_n$ and associated level structure $\pi_{C,*}\phi_N$ and $\pi_{C,*}\phi_p$ for the projection morphism: $\pi_C : E \to E/C$. In particular, we have the following commutative diagram defining $\pi_{C,*}\phi_p$:

$$\begin{array}{ccc} \mu_{p^\infty} & = & \mu_{p^\infty} \\ {\scriptstyle \phi_p}\downarrow & & \downarrow{\scriptstyle \pi_{C,*}\phi_p} \\ E & \xrightarrow[\pi_C]{} & E. \end{array}$$

Definition 7.56 *For $0 < n \in \mathbb{Z}$ prime to p, we define a linear operator $f|T(n) : V_{\Gamma_1(N)/B} \to V_{\Gamma_1(N)/B}$ by*

$$f|T(n)(E, \phi_N, \phi_p) = \frac{1}{n} \sum_{C \in \mathcal{C}_n(E)} f(E/C, \pi_{C,*}\phi_N, \pi_{C,*}\phi_p).$$

For $n = n'p^r$ with $p \nmid n'$, we put $T(n) = U(p)^r T(n')$ for $U(p)$ in Definition 7.52.

It is easy to see that $\widehat{f}|T(n) = \widehat{f|T(n)}$ for $f \in G_k(\Gamma_1(Np^\infty); R)$, and as computed earlier in Sect. 6.2.9, we have

$$a(m, f|T(n)) = \sum_{0<d|(m,n),(d,Np)=1} a(\frac{mn}{d^2}, f|\langle d \rangle). \qquad (7.4.1)$$

Here we recall from Sect. 6.2.5 that $\langle l \rangle = (T(l)^2 - T(l^2))$ for primes l outside N, and as before, we have $\langle n \rangle = \prod_{l \nmid N} \langle l \rangle^{e(l)}$ if $n = \prod_{l \nmid N} l^{e(l)}$. Then, by the q-expansion principle, this is well defined even when n is not invertible in R. In other words, if R has characteristic 0, first we define $T(n)$ on $G_k(\Gamma_1(Np^\infty); R[\frac{1}{n}])$ and verify by the q-expansion principle that $T(n)$ preserves R-integral modular forms. Since $G_k(\Gamma_1(Np^\infty); R)$ is p-adically dense, we get $T(n)$ acting on $V_{\Gamma_1(N)/R}$. Since $V_{\Gamma_1(N)/R/pR} = V_{\Gamma_1(N)/R} \otimes_R R/pR$, this also covers characteristic-p cases.

7.4.2 Hecke Operators on Λ-Adic Modular Forms

By (7.4.1), the operator $T(n)$ is automatically interpolated on $\mathcal{F}(\Gamma_0(N), \chi; \Lambda)$. Indeed, for $\Phi \in \mathcal{F}(\Gamma_0(N), \chi; \Lambda)$, if we put $\Phi|\langle n \rangle = \chi(n)(1+T)^{\log_p(n)/\log_p(\gamma)}\Phi$, we see easily that $\Phi(\gamma^k - 1)|\langle n \rangle = (\Phi|\langle n \rangle)(\gamma^k - 1)$ for integers k (even for p-adic integers k). Then we may define $T(n)$ on $\mathcal{F}(\Gamma_0(N), \chi; \Lambda)$ formally by

$$a(m, \Phi|T(n)) = \sum_{0<d|(m,n),(d,Np)=1} a(\frac{mn}{d^2}, \Phi|\langle d\rangle). \tag{7.4.2}$$

Then it is a matter of computation to see that

$$(\Phi|T(n))(\gamma^k - 1) = \Phi(\gamma^k - 1)|T(n)$$

for all n, and therefore $T(n)$ gives rise to the Λ-linear operator $T(n)$ on $\mathcal{F}(\Gamma_0(N), \chi; \Lambda)$. As seen in [GME] Theorem 3.2.16, we may regard the space $\mathcal{F}(\Gamma_0(N), \chi; \Lambda)$ as a subspace of $V_{\Gamma_1(N)/\Lambda}$, and we can verify the operator $T(n)$ introduced in the previous section preserves $\mathcal{F}_0(\Gamma_0(N), \chi; \Lambda)$, inducing the $T(n)$ formally defined by (7.4.2).

7.4.3 Duality of Hecke Algebra and Hecke Modules

We suppose the following axiom for a character ϕ of $(\mathbb{Z}/N\mathbb{Z})^\times \times \mathbb{Z}_p^\times$ with values in $R^\times$ for a ring R:

(d1) We have an R-free module of finite rank E with commuting R-linear operator $T(n)$ $(n = 1, 2, \dots)$ with $T(1)$ giving the identity operator.
(d2) We have an embedding $E \hookrightarrow qR[[q]]$ for a power series ring $R[[q]]$ given by $E \ni f \mapsto \sum_{n=1}^\infty a(n, f)q^n \in qR[[q]]$.
(d3) $a(m, f|T(n)) = \sum_{0<d|(m,n),(d,Np)=1} \phi(d)a(\frac{mn}{d^2}, f)$ for all positive integers m, n.

Let $h(E)$ be the subring of the R-linear endomorphism algebra $\mathrm{End}_R(E)$ generated by $T(n)$ $(n = 1, 2, \dots)$.

Exercise 7.57 *Using* (d3), *show that* $h(E)$ *is a commutative algebra.*

Definition 7.58 *If* $E = G_k(\Gamma_0(Np^r), \chi; R)$ *[resp.,* $E = S_k(\Gamma_0(Np^r), \chi; R)$*], we write* $H_k(\Gamma_0(Np^r), \chi; R)$ *[resp.,* $h_k(\Gamma_0(Np^r), \chi; R)$*] for* $h(E)$*. If* $E = G(\Gamma_0(N), \chi; \mathbb{I})$*, we write* $\mathbf{H}(\Gamma_0(N), \chi; \mathbb{I})$ *for* $h(E)$*, where* $\chi : (\mathbb{Z}/Np\mathbb{Z})^\times \to W^\times$ *is a character. If* $E = S(\Gamma_0(N), \chi; \mathbb{I}) = S(\Gamma_0(N), \chi; \Lambda) \otimes_\Lambda \mathbb{I}$*, we write* $\mathbf{h}(\Gamma_0(N), \chi; \mathbb{I})$ *for* $h(E)$*.*

We define a pairing $\langle \cdot, \cdot \rangle : E \times h(E) \to R$ by $\langle f, h\rangle = a(1, f|h)$. By (d3), we have $\langle f, T(n)\rangle = a(n, f)$. Then by (d2), $\langle f, T(n)\rangle = 0$ for all n implies $f = 0$. On the other hand, if we assume that $\langle f, h\rangle = 0$ for all $f \in E$, we have

$$\langle f, hT(n)\rangle = a(1, f|hT(n)) \overset{\text{Exercise 7.57}}{=} a(1, f|T(n)h) = \langle f|h, T(n)\rangle.$$

This shows that $f|h = 0$ for all $f \in E$, and by definition $h = 0$, so the pairing is nondegenerate. Thus, we get

Lemma 7.59 *If* R *is a field or a discrete valuation ring and* E *is free of finite rank over* R*, we have* $E \cong \mathrm{Hom}_R(h(E), R)$ *and* $h(E) = \mathrm{Hom}_R(E, R)$ *under the above pairing.*

We state the following easy fact (see [LFE] Chap. 7).

Proposition 7.60 *If* $\mathbb{I}$ *is a* Λ*-algebra and* $E = S(\chi;\mathbb{I})$ *and* $S(\Gamma_0(N),\chi;\mathbb{I})$, $E \cong \mathrm{Hom}_{\mathbb{I}}(h(E),\mathbb{I})$ *and* $h(E) = \mathrm{Hom}_{\mathbb{I}}(E,\mathbb{I})$ *under the above pairing.*

Proof. Since the space over $\mathbb{I}$ is the scalar extension of the space over Λ, we may assume that $\mathbb{I} = \Lambda$. For simplicity, we write $S = S(\Gamma_0(N),\chi;\mathbb{I})$ and $\mathbf{h} = \mathbf{h}(\Gamma_0(N),\chi;\mathbb{I})$. Let $\mathfrak{m}$ be the maximal ideal of Λ with $\mathbb{F} = \Lambda/\mathfrak{m}$. By definition, $\mathbf{h}/\mathfrak{m}\mathbf{h}$ surjects down to $h(S/\mathfrak{m}S)$ as the two algebras are generated by $T(n)$. This shows the morphism $\mathbf{h} \to \mathrm{Hom}_\Lambda(S,\Lambda)$ induced by the pairing gives rise to

$$\mathbf{h}/\mathfrak{m}\mathbf{h} \xrightarrow{i} \mathrm{Hom}_\Lambda(S,\Lambda)\otimes_\Lambda \Lambda/\mathfrak{m} \cong \mathrm{Hom}_{\mathbb{F}}(S/\mathfrak{m}S,\mathbb{F}).$$

The last identity follows as S is Λ-free of finite rank. Since $\mathrm{Hom}_{\mathbb{F}}(S/\mathfrak{m}S,\mathbb{F}) \cong h(S/\mathfrak{m}S)$ by the nondegeneracy over the field $\mathbb{F}$, i factors through $h(S/\mathfrak{m}S)$. Then by Nakayama's lemma (Lemma 10.8), i is surjective. Tensoring the quotient field Q of Λ,

$$i\otimes 1 : \mathbf{h}(\Gamma_0(N),\chi;Q) = \mathbf{h}\otimes_\Lambda Q \to S\otimes_\Lambda Q = S(\Gamma_0(N),\chi;Q),$$

again by the result over now the field Q. Thus, i is an isomorphism. Since $\mathbf{h}$ is the Λ-dual of the Λ-free module S, $\mathbf{h}$ is Λ-free. Then, by applying $\mathrm{Hom}_\Lambda(?,\Lambda)$ to $\mathbf{h} \cong \mathrm{Hom}_\Lambda(S,\Lambda)$, we recover

$$S = \mathrm{Hom}_\Lambda(\mathrm{Hom}_\Lambda(S,\Lambda),\Lambda) \cong \mathrm{Hom}_\Lambda(\mathbf{h},\Lambda),$$

as desired.

By Remark 7.55 and the above proposition, we have now proven the result we used in Sect. 3.1:

Corollary 7.61 *We have a canonical isomorphism*

$$\mathbf{h}/((1+x)-\varepsilon(\gamma)\gamma^k)\mathbf{h} \cong h_k^{ord}(\Gamma_0(Np^{r+1}),\varepsilon\chi\omega^{-k};W)$$

sending $T(n)$ *to* $T(n)$.

Exercise 7.62 *In* Sect. 3.1, *we had a variable* T *different from* x *with* $\Lambda = W[[T]] = W[[x]]$. *Find out the relation between* T *and* x, *and write* χ *here in terms of* ψ *in* Sect. 3.1.

8

Nonvanishing Modulo p of Hecke L-Values

We return to the setting of Sect. 6.4; thus, $M = \mathbb{Q}[\sqrt{-D}] \subset \overline{\mathbb{Q}}$ with discriminant $-D$ is an imaginary quadratic field in which the fixed prime (p) splits into a product of two primes $\mathfrak{p}\overline{\mathfrak{p}}$ with $\mathfrak{p} \neq \overline{\mathfrak{p}}$. As before, assume that $\mathfrak{p} = \{\alpha \in \mathfrak{O} \big| |i_p(\alpha)|_p < 1\}$ and denote by c the nontrivial field automorphism of M (complex conjugation). For an ideal $\mathfrak{C}$ of the integer ring $\mathfrak{O}$ of M, we consider the ideal group $\mathcal{I}(\mathfrak{C})$ of fractional ideals prime to $\mathfrak{C}$. We put

$$\mathcal{P}(\mathfrak{C}) = \{(\alpha) \big| \alpha \in M^\times \text{ and } \alpha \equiv 1 \mod \mathfrak{C} \text{ in } \mathfrak{O}_\mathfrak{C}\}.$$

As is well known, $Cl(\mathfrak{C}) = \mathcal{I}(\mathfrak{C})/\mathcal{P}(\mathfrak{C})$ is a finite group (see [LFE] Sect. 1.2). An ideal Hecke character of M modulo $\mathfrak{C}$ is a character (i.e., a group homomorphism) $\lambda : \mathcal{I}(\mathfrak{C}) \to \overline{\mathbb{Q}}^\times$ such that there exist integers k, j with $\lambda(\alpha) = \alpha^{-k}\overline{\alpha}^{-j} = \alpha^{-k}\alpha^{-jc}$ for all principal ideals (α) in $\mathcal{P}(\mathfrak{C})$. Here $M \ni \alpha \mapsto \alpha \in \mathbb{C}$ is induced by i_∞ and $\overline{\alpha} = \alpha^c$ is the complex conjugation of $\alpha \in M$ induced by i_∞.

The linear combination $\infty(\lambda) := k \cdot i_\infty + j \cdot c \in \mathbb{Z}[\mathrm{Gal}(M/\mathbb{Q})]$ is called the infinity type of λ [note here the sign change from the exponent of $\lambda(\alpha) = \alpha^{-k-jc}$; the reason will be explained later]. Since λ has values in M in the subgroup $\mathcal{P}(\mathfrak{C}) \subset \mathcal{I}(\mathfrak{C})$ of finite index, it has values in a finite extension of M inside $\overline{\mathbb{Q}}$ all over $\mathcal{I}(\mathfrak{C})$. Also, a Hecke character has finite order if and only if $k = j = 0$. A character λ of $\mathcal{I}(\mathfrak{C})$ is called *anticyclotomic* if $\lambda(\mathfrak{A}^c) = \lambda(\mathfrak{A})^{-1}$ for all $\mathfrak{A} \in \mathcal{I}(\mathfrak{C})$, while a character satisfying $\lambda(\mathfrak{A}^c) = \lambda(\mathfrak{A})$ is called *cyclotomic.* If $\mathfrak{C}|\mathfrak{C}'$, we have a natural inclusion $\mathcal{I}(\mathfrak{C}') \to \mathcal{I}(\mathfrak{C})$; hence, we can pull back a Hecke character modulo $\mathfrak{C}$ to a Hecke character modulo $\mathfrak{C}'$. If $\mathfrak{C} \subsetneq \mathfrak{C}'$, the character obtained by pullback is called *imprimitive.* A character is called *primitive* if it is not imprimitive. If λ is primitive modulo $\mathfrak{C}$, the ideal $\mathfrak{C}$ is called the *conductor* of λ. The norm $N(\mathfrak{A}) = \frac{[\mathfrak{O}:\mathfrak{O}\cap\mathfrak{A}]}{[\mathfrak{A}:\mathfrak{O}\cap\mathfrak{A}]}$ is an example of a Hecke character with $\infty(\lambda) = -i_\infty - c$. If a primitive λ has conductor $\mathfrak{C}$, we extend $\lambda : \mathcal{I}(\mathfrak{C}) \to \overline{\mathbb{Q}}^\times$ to $\mathcal{I}(\mathfrak{O}) \to \overline{\mathbb{Q}}$ by putting $\lambda(\mathfrak{A}) = 0$ if $\mathfrak{A} \notin \mathcal{I}(\mathfrak{C})$. Then the (primitive) L-function of λ is defined by an absolutely and locally uniformly convergent Dirichlet series/Euler product on a right half-complex plane:

H. Hida, *Elliptic Curves and Arithmetic Invariants*, Springer Monographs in Mathematics, DOI 10.1007/978-1-4614-6657-4_8,

$$L(s,\lambda)=\sum_{\mathfrak{A}}\lambda(\mathfrak{A})N(\mathfrak{A})^{-s}=\prod_{\mathfrak{L}}\left(1-\frac{\lambda(\mathfrak{L})}{N(\mathfrak{L})^s}\right)^{-1},$$

where $\mathfrak{A}$ runs over all $\mathfrak{O}$-ideals and $\mathfrak{L}$ runs over all prime ideals of $\mathfrak{O}$. It is a result of Hecke that this L-function can be continued to a meromorphic function on the whole complex plane with holomorphy everywhere except for the case of a power of the norm character $\lambda = N^m$ (see [LFE] Sect. 2.7 and Chap. 8). For a prime ideal $0 \neq \mathfrak{q}$ of $\mathbb{Z}$, we put $L^{(\mathfrak{q})}(s,\lambda)=\left\{\prod_{\mathfrak{Q}|\mathfrak{q}}\left(1-\frac{\lambda(\mathfrak{Q})}{N(\mathfrak{Q})^s}\right)\right\}L(s,\lambda)$ (the imprimitive L-function whose Euler $\mathfrak{q}$-factor is removed).

In this chapter, we fix a prime $\mathfrak{l}=(l)\nmid(p)$ of $\mathbb{Z}$. Generalizing the result of Sect. 1.2, we prove nonvanishing modulo p of Hecke L-values $\frac{L^{(\mathfrak{l})}(0,\lambda\chi)}{\Omega}$ for the canonical period Ω for "almost all" anticyclotomic characters χ of finite order with $\mathfrak{l}$-power conductor (under some mild assumptions; Theorem 8.31). This topic is a generalization of Washington's result [Ws] (and Sinnott [Sn2]) from Dirichlet L-values to Hecke L-values, and a more general case of Hecke characters of a CM field has been dealt with in [H04b], [H07a], and [Hs1] by the same technique, resorting to a Hilbert modular Shimura variety in place of a Shimura curve here. There will be much room for generalization of the techniques (we list here only one of them: [Br2]). The phrase "almost all" means "except for finitely many," that is, nonvanishing modulo p outside a (unspecified) finite set.

8.1 Rationality of Hecke L-Values

Here we give more details of the rationality result from [Sh6] as well as Katz's p-integrality results from [K2], covering more general cases than in Sect. 1.3. As before, invariant differential operators play a prominent role.

8.1.1 Differential Operators and Rationality

Recall the Archimedean invariant differential operators we studied in Sect. 1.3.5:

$$\delta_k=\frac{1}{2\pi\sqrt{-1}}\left(\frac{\partial}{\partial z}+\frac{k}{2y\sqrt{-1}}\right)\quad\text{and}\quad\delta_k^r=\delta_{k+2r-2}\cdots\delta_k,\tag{8.1.1}$$

where $0\leq r\in\mathbb{Z}$. As we learned in Sects. 1.3.5 and 1.3.6, an important point is that the differential operator preserves the rationality property at CM points of (arithmetic) modular forms, although it does not preserve holomorphy. We recall the rationality property from Theorems 1.27 and 1.34. For a proper ideal $\mathfrak{a}$ of an order of M, the complex uniformization $\iota: E(\mathfrak{a})(\mathbb{C})\cong\mathbb{C}/i_\infty(\mathfrak{a})$ induces a canonical base $\omega_\infty=\iota^*du$ of $\Omega_{E(\mathfrak{a})/\mathbb{C}}$ over $\mathbb{C}=\mathfrak{O}\otimes_{\mathbb{Z}}\mathbb{R}$, where u is the variable in $\mathbb{C}$. Recall the Néron differential $\omega(\mathfrak{a})$ in Sect. 6.4.3, and define the Archimedean Néron *CM period* $0\neq\Omega_\infty\in\mathbb{C}=\mathfrak{O}\otimes_{\mathbb{Z}}\mathbb{R}$ by

$$\omega(\mathfrak{O}) = \omega(\mathfrak{a}) = \Omega_\infty \omega_\infty. \tag{8.1.2}$$

Here the first identity follows from the fact that $\omega(\mathfrak{a})$ is induced by $\omega(\mathfrak{O})$ on $E(\mathfrak{O})$. We suppose that $\mathfrak{a}$ is prime to p. Recall Shimura's rationality result for $f \in G_k(N;\mathcal{W})$ we reproduced in Theorem 1.27:

$$\frac{(\delta_k^r f)(x(\mathfrak{a}), \omega_\infty)}{\Omega_\infty^{k+2r}} = (\delta_k^r f)(x(\mathfrak{a}), \omega(\mathfrak{a})) \in \overline{\mathbb{Q}}. \tag{Sh}$$

Katz interpreted the differential operator in terms of the Gauss–Manin connection of the universal elliptic curve over $Sh^{(p)}$ and gave a purely algebro-geometric definition of the p-adic invariant differential operator (see [K2] Chap. II and [HT1] Sect. 1). We have given in Sect. 1.3.6 a more elementary construction of his operator. As before, we write his operator corresponding to δ_*^k as $d^k : V_{\Gamma_1(N)/B} \to V_{\Gamma_1(N)/B}$ for a p-adic base $\mathbb{Z}_p$-algebra B. The level-p structure $\eta_p^\circ(\mathfrak{a}) : \mu_{p^\infty} \cong M_\mathfrak{p}/\mathfrak{a}_\mathfrak{p} \hookrightarrow E(\mathfrak{a})[p^\infty]$ ($\mathfrak{p}$ corresponding to $i_p|_M : M \hookrightarrow \mathbb{C}_p$) induces an isomorphism $\widehat{\eta}_p^\circ(\mathfrak{a}) : \widehat{\mathbb{G}}_m \cong \widehat{E}(\mathfrak{a})$ for the p-adic formal group $\widehat{E}(\mathfrak{a})_{/W}$ obtained by formal completion of E along the origin $\mathbf{0}_E \in E(\mathbb{F})$. Then $\omega(\mathfrak{O}) = \omega(\mathfrak{a}) = \Omega_p\omega_p$ ($\Omega_p \in W^\times$) for $\omega_p = \widehat{\eta}_p^\circ(\mathfrak{a})_* \frac{dt}{t}$. By Theorem 1.34, we have, for $f \in G_k(N;\mathcal{W})$,

$$\frac{(d^r f)(x(\mathfrak{a}), \eta_p^\circ(\mathfrak{a}))}{\Omega_p^{k+2r}} = (d^r f)(x(\mathfrak{a}), \omega(\mathfrak{a})) = (\delta_k^r f)(x(\mathfrak{a}), \omega(\mathfrak{a})) \in \mathcal{W}. \tag{K}$$

By the definition in Sect. 1.3.6, the effect of the invariant differential operator d^r on q-expansion of a modular form is given by

$$d^r \sum_n a(n) q^n = \sum_n a(n) n^r q^n. \tag{8.1.3}$$

8.1.2 Arithmetic Hecke Characters

Consider the torus $T_M = \mathrm{Res}_{M/\mathbb{Q}}\mathbb{G}_m$ [see Example 4.58(11) for the definition of $\mathrm{Res}_{M/\mathbb{Q}}G$]. Each integral linear combination $k \cdot i_\infty + j \cdot c \in \mathbb{Z}[\mathrm{Gal}(M/\mathbb{Q})]$ is regarded as a character of T_M by $x \mapsto i_\infty(x)^k c(x)^j$. We fix an arithmetic *Hecke character* λ of infinity type $k \cdot i_\infty + \kappa(i_\infty - c)$ for the integers k and κ. Originally, λ is defined as a character $\lambda : \mathcal{I}(\mathfrak{C}) \to \overline{\mathbb{Q}}^\times$ with $\lambda(\alpha) = \alpha^{-k-\kappa(1-c)}$ for $(\alpha) \in \mathcal{P}(\mathfrak{C})$. We can extend λ to idele characters of $M_\mathbb{A}^\times$ in two ways.

For each prime $\mathfrak{L}$ of M, we choose its local generator $\varpi_\mathfrak{L} \in M$ so that $\mathfrak{L}\mathfrak{O}_{(\mathfrak{L})} = (\varpi_\mathfrak{L})$, where

$$\mathfrak{O}_{(\mathfrak{L})} = \left\{ \frac{b}{a} \middle| a, b \in \mathfrak{O},\ a\mathfrak{O} + \mathfrak{L} = \mathfrak{O} \right\} \text{ (the localization at } \mathfrak{L}).$$

We write $\mathfrak{O}_\mathfrak{L}$ for the completion $\mathfrak{O}_\mathfrak{L} = \varprojlim_n \mathfrak{O}/\mathfrak{L}^n$ and put $M_\mathfrak{L} = \mathfrak{O}_\mathfrak{L} \otimes_\mathbb{Z} \mathbb{Q} = \mathfrak{O}_\mathfrak{L} \otimes_\mathfrak{O} M$.

Exercise 8.1 *Prove that* $\mathfrak{O}_{\mathfrak{L}} \cap M = \mathfrak{O}_{(\mathfrak{L})}$.

Recall that the adele ring $M_{\mathbb{A}}$ of M is an M-subalgebra in the product $\prod_{\mathfrak{L}} M_{\mathfrak{L}}$ made up of $x^{(\infty)} = (x_{\mathfrak{L}})_{\mathfrak{L}}$ with $x_{\mathfrak{L}} \in \mathfrak{O}_{\mathfrak{L}}$ except for finitely many prime ideals $\mathfrak{L}$. Decompose $M_{\mathbb{A}} = M_{\mathbb{A}}^{(\infty)} \times \mathbb{C}$. Embedding $M \subset M_{\mathfrak{L}}$ naturally and in $\mathbb{C}$ by i_∞, we regard $M_{\mathbb{A}}$ as an M-algebra by the diagonal embedding

$$M \ni \xi \mapsto (\dots, \xi, \dots, \overset{\mathfrak{L}}{\xi}, \dots, \xi, \dots) \in M_{\mathbb{A}}.$$

Thus, $T_M(\mathbb{A}) = M_{\mathbb{A}}^{\times}$ and $T_M(\mathbb{Q}) = M^{\times}$. The infinity component of $x \in M_{\mathbb{A}}$ is written as x_∞. See [AQF] Chap. III for a concise description of adeles. Put

$$M_{\mathbb{A}}^{(\mathfrak{C}\infty)} = \{x \in M_{\mathbb{A}}^{(\infty)} | x_{\mathfrak{L}} = 1 \ \text{ if } \mathfrak{L} \supset \mathfrak{C}\},$$

and $U(\mathfrak{C}) = \{x \in \widehat{\mathfrak{O}}^{\times} | x \equiv 1 \mod \widehat{\mathfrak{C}}\}$ for an ideal $0 \neq \mathfrak{C} \subset \mathfrak{O}$. Write $U(\mathfrak{C})^{(\mathfrak{C})} = U(\mathfrak{C}) \cap (M_{\mathbb{A}}^{(\mathfrak{C}\infty)})^{\times}$ and $U(\mathfrak{C}) = U(\mathfrak{C})_{\mathfrak{C}} \times U(\mathfrak{C})^{(\mathfrak{C})}$.

Exercise 8.2 *Prove that we have an isomorphism* $\mathcal{I}(\mathfrak{C}) \cong (M_{\mathbb{A}}^{(\mathfrak{C}\infty)})^{\times}/U(\mathfrak{C})^{(\mathfrak{C})}$ *of groups sending each prime ideal* $\mathfrak{L}$ *to the element in* $(M_{\mathbb{A}}^{(\mathfrak{C}\infty)})^{\times}$, *still written as* $\varpi_{\mathfrak{L}}$, *whose* $\mathfrak{L}$*-component is equal to* $\varpi_{\mathfrak{L}}$ *and all other components are* 1.

By the exercise, identifying $(M_{\mathbb{A}}^{(\mathfrak{C}\infty)})^{\times}/U(\mathfrak{C})^{(\mathfrak{C})}$ with $\mathcal{I}(\mathfrak{C})$, we may regard λ as a character of $(M_{\mathbb{A}}^{(\mathfrak{C}\infty)})^{\times}/U(\mathfrak{C})^{(\mathfrak{C})}$. We can then extend λ to a character $M^{\times}\backslash M_{\mathbb{A}}^{\times} \to \mathbb{C}_v^{\times}$ in the following way for a place v. Note that $M^{\times}$ projected in $M_{\mathfrak{C}} = \prod_{\mathfrak{L}|\mathfrak{C}} M_{\mathfrak{L}}$ is dense; hence, $U(\mathfrak{C})_{\mathfrak{C}} M^{\times} = M_{\mathfrak{C}}$ [resulting in $M_{\mathbb{A}}^{\times} = M^{\times}(M_{\mathbb{A}}^{(v\mathfrak{C})})^{\times} U(\mathfrak{C})$].

Definition 8.3 *For a place* v *of* $\mathbb{Q}$, *put* $\lambda(\alpha x u) = u_v^{k \cdot i_v + \kappa(i_v - c i_v)} \lambda(x) \in \mathbb{C}_v^{\times}$ *for* $\alpha \in M^{\times}$, $u \in U(\mathfrak{C}) M_{\infty}^{\times}$, *and* $x \in (M_{\mathbb{A}}^{(v\mathfrak{C})})^{\times}$, *where* $\mathbb{C}_v = \mathbb{C}$ *or* $\mathbb{C}_p$ *depending on whether* $v = \infty$ *or the prime* p. *If we choose* $v = p$, *we write the character obtained as* $\widehat{\lambda} : M^{\times}\backslash M_{\mathbb{A}}^{\times} \to \overline{\mathbb{Q}}_p^{\times}$, *and if* $v = \infty$, *we just simply write it as* λ.

Because of this definition, $k \cdot i_v + \kappa(i_v - c)$ [abbreviated as $k + \kappa(1-c)$] appears as the exponent at v, so it is called the ∞-*type* of λ and the p-*type* of $\widehat{\lambda}$. The p-adic character $\widehat{\lambda}$ is called the p-*adic avatar* of λ (and it was originally invented by Weil). Note here that $\lambda(\alpha) = \alpha^{-k-\kappa(1-c)}$ [$\alpha \in M^{\times} \cap U(\mathfrak{C})_{\mathfrak{C}}$ in $M_{\mathfrak{C}}$] for the original ideal character λ. The subgroup $T_M(\mathbb{Q}) \subset T_M(\mathbb{A})$ is a discrete subgroup (see [AQF] Theorem 15.7).

Exercise 8.4 *Prove the well-definedness of* $\lambda : T_M(\mathbb{Q})\backslash T_M(\mathbb{A}) \to \mathbb{C}^{\times}$ *and* $\widehat{\lambda} : T_M(\mathbb{Q})\backslash T_M(\mathbb{A})/T_M(\mathbb{R}) \to \mathbb{C}_p^{\times}$ *as a continuous character.*

Exercise 8.5 *Let* $Cl(\mathfrak{C}p^{\infty}) = \varprojlim_n Cl(\mathfrak{C}p^n)$ *as a profinite compact group and* $U(\mathfrak{C}p^{\infty}) = \bigcap_n U(\mathfrak{C}p^n)$. *Then prove that (i)* $T_M(\mathbb{Q})\backslash T_M(\mathbb{A}^{(\infty)})/U(\mathfrak{C}p^n) \cong Cl(\mathfrak{C}p^n)$ *as a group if* $n < \infty$ *and as a compact group if* $n = \infty$ *and (ii)* $\widehat{\lambda}$ *factors through* $Cl(\mathfrak{C}p^{\infty})$.

Let λ be an idele character of $M_{\mathbb{A}}^{\times}$ of infinity type $k+\kappa(1-c)$. Fixing two primes p (we work in the mod p and the p-adic setting) and l (we consider l-power conductor characters), we assume the following three conditions:

(ct) $k > 1$ *and* $\kappa \geq 0$.
(pl) *The conductor* $\mathfrak{C}$ *of* λ *is prime to* p *and* $(l) = \mathfrak{l}$.
(sp) *The ideal* $\mathfrak{C}$ *is a product of primes split over* $\mathbb{Q}$.

The first condition implies criticality at $s = 0$ for $L(s,\lambda)$. The two other conditions are for simplicity and can be removed [see [Hs1] how to proceed without assuming (sp)].

8.1.3 Optimal Eisenstein Series

We are going to define an Eisenstein series whose special values at CM points optimally interpolate the values $L(0,\lambda\chi^{-1})$ for a given Hecke character λ twisted by anticyclotomic characters χ of finite order.

We split the conductor $\mathfrak{C}$ in the following way: $\mathfrak{C} = \mathfrak{F}\mathfrak{F}_c$ with $\mathfrak{F}+\mathfrak{F}_c = \mathfrak{O}$ and $\mathfrak{F} \subset \mathfrak{F}_c^c$. This is possible by our assumption (sp). We then define $\mathfrak{f} = \mathfrak{F}\cap\mathbb{Z}$ and $\mathfrak{f}' = \mathfrak{F}_c \cap \mathbb{Z}$. Then $\mathfrak{f} \subset \mathfrak{f}'$. Here $X = \mathbb{Z}/\mathfrak{f} \cong \mathfrak{O}/\mathfrak{F}$ and $Y = \mathbb{Z}/\mathfrak{f}' \cong \mathfrak{O}/\mathfrak{F}_c$. Let $\psi : X \times Y \to \mathbb{C}$ be a function such that $\psi(\varepsilon^{-1}x, \varepsilon y) = \varepsilon^k \psi(x,y)$ for all $\varepsilon \in \mathbb{Z}^{\times} = \{\pm 1\}$ with the integer k as above. We put $X^* = \mathfrak{f}^{-1}/\mathbb{Z}$; so X^* is naturally the Pontryagin dual module of X under the pairing $(x^*,x) = \mathbf{e}(x^*x)$ ($x^* \in X^*$ and $x \in X$), where $\mathbf{e}(x) = \exp(2\pi i x)$ for $x \in \mathbb{C}$. Let $N(\mathfrak{f}) = |\mathbb{Z}/\mathfrak{f}| = [\mathbb{Z} : \mathfrak{f}]$, which is the positive generator of the ideal $\mathfrak{f}$. We define the *partial Fourier transform* $P\psi : X^* \times Y \to \mathbb{C}$ of ψ by

$$P\psi(x,y) = N(\mathfrak{f})^{-1} \sum_{a \in X} \psi(a,y)\mathbf{e}_{\mathbb{Q}}(ax), \tag{8.1.4}$$

where $\mathbf{e}_{\mathbb{Q}}$ is the restriction of the standard additive character $\mathbf{e}_{\mathbb{Q}} : \mathbb{A}/\mathbb{Q} \to \mathbb{C}^{\times}$ [with $\mathbf{e}_{\mathbb{Q}}(x_\infty) = \exp(2\pi i x_\infty)$ for $x_\infty \in \mathbb{R}$] to the local component $F_{\mathfrak{f}}$ at $\mathfrak{f}$.

Exercise 8.6 *Prove that there is a unique additive character* $\mathbf{e}_{\mathbb{Q}} : \mathbb{A}/\mathbb{Q} \to \mathbb{C}^{\times}$ *with* $\mathbf{e}_{\mathbb{Q}}(x_\infty) = \exp(2\pi i x_\infty)$ *for* $x_\infty \in \mathbb{R}$ (e.g., [LFE] Sect. 8.3).

We now relate the function ψ with a function on $\wp_{\Gamma}(\mathbb{C})$ (a function of elliptic curve with level-Γ structure). Here Γ is given by

$$\widehat{\Gamma}_1^1(\mathfrak{f}^2) = \left\{ \left(\begin{smallmatrix} a & b \\ c & d \end{smallmatrix}\right) \in G(\widehat{\mathbb{Z}}) \middle| a \equiv d \equiv 1 \mod \widehat{\mathfrak{f}}^2, c \equiv 0 \mod \mathfrak{f}^2 \right\}. \tag{8.1.5}$$

Since $\wp_{\Gamma_1(1)}(\mathbb{C})$ is identified with the complex manifold $\mathcal{B}$ made of lattices $L \subset \mathbb{C}$ in Sect. 2.3.4, a function ψ as above can be interpreted as a function of lattices L with level-Γ structure $\phi_{\mathfrak{f}} = (i_{\mathfrak{f}}, i'_{\mathfrak{f}}) = (i,i')$ defined in Sect. 6.4.4. Here $i : X^* \hookrightarrow \mathfrak{f}^{-1}L/L$ is the level-$\mathfrak{f}$ structure. We define an $\mathbb{Z}_{\mathfrak{f}}$-submodule $PV(L) \subset L \otimes_{\mathbb{Z}} \mathbb{Q}_{\mathfrak{f}}$ specified by the following conditions:

$$PV(L) \supset L \otimes_{\mathbb{Z}} \mathbb{Z}_{\mathfrak{f}},\ PV(L)/L_{\mathfrak{f}} = \mathrm{Im}(i)\ (L_{\mathfrak{f}} = L \otimes_{\mathbb{Z}} \mathbb{Z}_{\mathfrak{f}}). \tag{PV}$$

By definition, we may regard

$$i^{-1} : PV(L) \twoheadrightarrow PV(L)/\left(L \otimes_{\mathbb{Z}} \mathbb{Z}_{\mathfrak{f}}\right) \cong \mathfrak{f}^{-1}/\mathbb{Z}.$$

We write $L = (2\pi i)(\mathbb{Z}w_1 + \mathbb{Z}w_2)$ with $z = w_1/w_2 \in \mathfrak{H}$ and equip it with a *self-dual alternating pairing* given by

$$\langle (2\pi i)(aw_1 + bw_2), (2\pi i)(cw_1 + dw_2)\rangle_L = bc - ad. \tag{8.1.6}$$

This pairing depends only on L, not on the choice of z.

Exercise 8.7 *Why does the pairing depend only on L and not on the choice of z?*

This pairing induces a perfect duality pairing $\langle \cdot, \cdot \rangle_{\mathfrak{f}}$ on $\mathfrak{f}^{-1}L/L \times L/\mathfrak{f}L$ given by $\langle \overline{x}, \overline{y}\rangle_{\mathfrak{f}} = \langle x, y\rangle \mod \mathfrak{f} \in \mathbb{Z}/\mathfrak{f}$, where $x \in \mathfrak{f}^{-1}L$ with $x \mod L = \overline{x}$ and $y \in L$ with $y \mod \mathfrak{f}L = \overline{y}$. By the duality induced by $\langle \cdot, \cdot \rangle_{\mathfrak{f}}$, the dual map of i gives rise to $i' : PV(L) \twoheadrightarrow \mathbb{Z}/\mathfrak{f}$. Taking a lift $\tilde{i} : \mathfrak{f}^{-2}/\mathbb{Z} \hookrightarrow PV(L)/\mathfrak{f}L_{\mathfrak{f}}$ with $\tilde{i} \mod L_{\mathfrak{f}} = i$, we have an exact sequence:

$$0 \to \mathfrak{f}^{-2}/\mathbb{Z} \xrightarrow{\tilde{i}} PV(L)/\mathfrak{f}L_{\mathfrak{f}} \xrightarrow{i'} \mathbb{Z}/\mathfrak{f} \to 0.$$

This sequence is kept under $\alpha \in \mathrm{Aut}(L)$ with unipotent reduction modulo $\mathfrak{f}^2$, and the pair (i, i') gives a level-$\Gamma_1^1(\mathfrak{f}^2)$ structure in Sect. 6.4.4.

By the commutativity of the following diagram:

$$\begin{array}{ccccc} \mathfrak{f}^{-2}/\mathbb{Z} & \xrightarrow{\tilde{i}} & PV(L)/\mathfrak{f}L_{\mathfrak{f}} & \xrightarrow{i'} & \mathbb{Z}/\mathfrak{f} \\ \downarrow \wr & & f \downarrow \cap & & f \downarrow \cap \\ \mathfrak{f}^{-2}/\mathbb{Z} & \longrightarrow & E[\mathfrak{f}^2] & \longrightarrow & \mathbb{Z}/\mathfrak{f}^2, \end{array} \tag{8.1.7}$$

giving $\phi_{\mathfrak{f}} = (i, i')$ is equivalent to having the bottom sequence of maps in the above diagram. This explains why the pair (i, i') gives rise to a level-$\Gamma_1^1(\mathfrak{f}^2)$ structure; strictly speaking, the exact level group is given by

$$\widehat{\Gamma}^1_{1,0}(\mathfrak{f}^2) = \left\{ \left(\begin{smallmatrix} a & b \\ c & d \end{smallmatrix}\right) \in G(\widehat{\mathbb{Z}}) \middle| a \equiv d \equiv 1 \mod \widehat{\mathfrak{f}},\ c \equiv 0 \mod \widehat{\mathfrak{f}^2} \right\}. \tag{8.1.8}$$

We regard $P\psi$ as a function of $L \otimes_{\mathbb{Z}} \mathbb{Q}$ supported on $(\mathfrak{f}^{-2}L) \cap PV(L)$ by

$$P\psi(w) = \begin{cases} P\psi(i^{-1}(w), i'(w)) & \text{if } (w \mod L) \in \mathrm{Im}(i), \\ 0 & \text{otherwise.} \end{cases} \tag{8.1.9}$$

For any positive integer $k > 0$, we can now define the Eisenstein series E_k. Writing $\underline{L} = (L, \phi = \phi_{\mathfrak{f}})$ for simplicity, we define the value $E_k(\underline{L}; \psi)$ by

$$E_k(\underline{L};\psi) = (-1)^k \Gamma(k+s) \sum_{w\in\mathfrak{f}^{-1}L/\mathbb{Z}^\times}{}' \left.\frac{P\psi(w)}{w^k|w|^{2s}}\right|_{s=0}. \tag{8.1.10}$$

Here "$\sum'$" indicates that we are excluding the term corresponding to $w=0$ from the summation. As shown by Hecke, this type of series is convergent when the real part of s is sufficiently large and can be continued to a meromorphic function well defined at $s=0$ [as long as either $k\geq 3$ or $\psi(a,0)=0$ for all a; see Sect. 9.1.1 in the text and [LFE] Sect. 2.5 for analytic continuation].

Lemma 8.8 *If either $k\geq 3$ or $\psi(a,0)=0$ for all a, the function $E_k(\psi)$ gives an element in $G_k(\Gamma_1(\mathfrak{f}^2);\mathbb{C})$, whose q-expansion at the cusp ∞ is given by*

$$E_k(\psi)(q) = 2^{-1}L(1-k,\psi) + \sum_{0<n\in\mathbb{Z}} \sum_{\substack{(a,b)\in(\mathbb{Z}\times\mathbb{Z})/\mathbb{Z}^\times \\ ab=n}} \psi(a,b)\frac{a}{|a|}a^{k-1}q^n, \tag{8.1.11}$$

where $L(s,\psi)$ is the partial L-function given by the Dirichlet series

$$\sum_{n\in\mathbb{Z}/\{\pm1\}^\times}{}' \psi(n,0)\left(\frac{N(n)}{|N(n)|}\right)^k |N(n)|^{-s}.$$

When $k=2$, $\mathfrak{f}=1$, and $\psi=\mathbf{1}$ is the constant equal to 1, $E_k(\mathbf{1})$ is non-holomorphic and its Fourier expansion contains an extra term $\frac{1}{8\pi\,\mathrm{Im}(z)}$ in addition to the above q-expansion specialized at $k=2$. Moreover, we have $E_2(\mathbf{1}) = -\Delta^{-1}\delta_{12}\Delta/24$ for the discriminant function Δ.

If $\psi(x,y)=\psi_X(x)\psi_Y(y)$ for two functions $\psi_X : X\to\mathbb{C}$ and $\psi_Y : Y\to\mathbb{C}$ with ψ_Y factoring through $\mathbb{Z}/\mathfrak{f}'$, we can check easily that $E_k(\psi)\in G_k(\Gamma_1(\mathfrak{f}\mathfrak{f}');\mathbb{C})$. We will compute in Sect. 9.1.1 the Fourier/q-expansion of the Eisenstein series, applying the Poisson summation formulas (which actually proves the q-expansion formulas of the cotangent function and its derivatives quoted after (1.3.1)). Computation in a more general Hilbert modular case is given in [K2] Chap. III and [HT1] Sect. 2. The last formula about $E_2(\mathbf{1})$ follows from taking the logarithmic derivative of the product form of Δ in (1.3.4) [see [EDM] (8.18)].

8.1.4 Hecke Eigenvalues

We take a Hecke character λ as in Sect. 8.1.2. Write $\lambda_{\mathfrak{C}}$ for the composite $M_{\mathfrak{C}}^\times = \prod_{\mathfrak{L}|\mathfrak{C}} M_{\mathfrak{L}}^\times \hookrightarrow M_{\mathbb{A}}^\times \xrightarrow{\lambda} \mathbb{C}^\times$. Then the restriction $\lambda_{\mathfrak{C}}^{-1} : \mathfrak{O}_{\mathfrak{F}}^\times \times \mathfrak{O}_{\mathfrak{F}_c}^\times \to \mathcal{W}^\times$ induces a locally constant function $\psi : (\mathbb{Z}/\mathfrak{f})\times(\mathbb{Z}/\mathfrak{f}')\to\mathcal{W}$ supported on $(\mathbb{Z}/\mathfrak{f})^\times\times(\mathbb{Z}/\mathfrak{f}')^\times$, because $\lambda_{\mathfrak{C}}$ factors through $(\mathfrak{O}/\mathfrak{C})^\times$, which is canonically isomorphic to $(\mathbb{Z}/\mathfrak{f})^\times\times(\mathbb{Z}/\mathfrak{f}')^\times$. Since λ is trivial on $M^\times$, ψ satisfies

$$\psi(\varepsilon x,\varepsilon y) = \varepsilon^{k\cdot i_\infty+\kappa(1-c)}\psi(x,y) = \varepsilon^k\psi(x,y)$$

for any unit $\varepsilon\in\mathbb{Z}^\times=\{\pm1\}$.

We consider the local uniformizer $\varpi_{\mathfrak{q}} = q \in \mathbb{Z}_{\mathfrak{q}}$ to be an idele. For each fractional $\mathbb{Z}$-ideal $\mathfrak{A}$ in $\mathbb{Q}$, decomposing $\mathfrak{A} = \prod_{\mathfrak{q}} \mathfrak{q}^{e(\mathfrak{q})}$ for primes $\mathfrak{q}$ (allowing possibly negative exponents), we define $\varpi^{e(\mathfrak{A})} = \prod_{\mathfrak{q}} \varpi_{\mathfrak{q}}^{e(\mathfrak{q})} \in \mathbb{A}^{\times}$. We then define another *partial Fourier transform* $\psi^{\circ} : X \times Y \to \mathcal{W}$ by

$$\psi^{\circ}(a,b) = \sum_{u \in \mathbb{Z}/\mathfrak{f}} \psi(u,b)\mathbf{e}_{\mathbb{Q}}(-ua\varpi^{-e(\mathfrak{f})}). \tag{8.1.12}$$

Since ψ is induced by $\lambda_{\mathfrak{C}}^{-1}$, we find by definition that

$$\psi^{\circ}(a,b) = G(\lambda_{\mathfrak{F}}^{-1})\lambda_{\mathfrak{F}}(a)\lambda_{\mathfrak{F}_c}^{-1}(b) \tag{8.1.13}$$

for the *Gauss sum* $G(\lambda_{\mathfrak{F}}^{-1}) = \sum_{u \in \mathbb{Z}/\mathfrak{f}} \lambda_{\mathfrak{F}}^{-1}(u)\mathbf{e}_{\mathbb{Q}}(-u\varpi^{-e(\mathfrak{f})})$. Here we agree to put $\lambda_{\mathfrak{F}}(a)\lambda_{\mathfrak{F}_c}^{-1}(b) = 0$ if $(a,b) \notin (\mathbb{Z}/\mathfrak{f})^{\times} \times (\mathbb{Z}/\mathfrak{f}')^{\times}$.

By the Fourier inversion formula (see [LFE] Sect. 4.1(5a)), we have

$$P\psi^{\circ}(x,y) = \psi(\varpi^{e(\mathfrak{f})}x, y). \tag{8.1.14}$$

Exercise 8.9 *Prove* (8.1.14).

From this and the definition of $E_k(\underline{L}) = E_k(\underline{L};\psi^{\circ})$, we find

$$E_k(E, i \circ x, i' \circ y, a\omega) = a^{-k}\lambda_{\mathfrak{F}}(x)\lambda_{\mathfrak{F}_c}^{-1}(y)E_k(E,i,i',\omega) \tag{8.1.15}$$

for $x \in (\mathbb{Z}/\mathfrak{f})^{\times} = (\mathfrak{O}/\mathfrak{F})^{\times}$ and $y \in (\mathbb{Z}/\mathfrak{f}')^{\times} = (\mathfrak{O}/\mathfrak{F}_c)^{\times}$. Because of this, $E_k(\psi^{\circ})$ actually belongs to $G_k(\Gamma_0(\mathfrak{f}\mathfrak{f}'), \epsilon_{\lambda}; \mathbb{C})$ for $\epsilon_{\lambda} = \lambda_{\mathfrak{F}_c}|_{(\mathbb{Z}/\mathfrak{f}')^{\times}}\lambda_{\mathfrak{F}}|_{(\mathbb{Z}/\mathfrak{f})^{\times}}$, identifying $\mathbb{Z}_{\mathfrak{f}} = \mathfrak{O}_{\mathfrak{F}}$ and $\mathbb{Z}_{\mathfrak{f}'} = \mathfrak{O}_{\mathfrak{F}_c}$.

By (8.1.13), the q-expansion of $E(\psi^{\circ})$ is a constant multiple of

$$a_0 + \sum_{n=1}^{\infty}\left(\sum_{(a,b)\in\mathbb{Z}, ab=n} \lambda_{\mathfrak{F}}(a)\lambda_{\mathfrak{F}^c}^{-1}(b)\frac{a^k}{|a|}\right)q^n \quad \text{for a constant } a_0.$$

Then, from the effect of $T(l)$ and $U(l)$ on q-expansion (as in Exercise 6.22), we verify the following lemma easily.

Lemma 8.10 *Let $\mathfrak{l}$ be a prime ideal of $\mathbb{Z}$ outside $\mathfrak{f}$ generated by $0 < l \in \mathbb{Z}$. Then we have*

$$E_k(\psi^{\circ})|T(\mathfrak{l}) = (\lambda_{\mathfrak{F}_c}^{-1}(l) + \lambda_{\mathfrak{F}}(l)l^{k-1})E_k(\psi^{\circ}), \tag{8.1.16}$$

We introduce a new operation $(\mathfrak{l})$ closethe operator $\langle l^{-1}\rangle$. On the elliptic curve side, for a prime ideal $\mathfrak{l}$ outside $\mathfrak{f}\mathfrak{f}'$, $(\mathfrak{l})(E,i,i',\omega) = (E \otimes_{\mathbb{Z}} \mathfrak{l}, j, j', \omega')$, where as an fppf abelian sheaf, $E \otimes_{\mathbb{Z}} \mathfrak{l}$ is the sheafication of $R \mapsto E \otimes_{\mathbb{Z}} \mathfrak{l}(R) = E(R) \otimes_{\mathbb{Z}} \mathfrak{l}$. Since $\mathfrak{l}_{\mathfrak{f}\mathfrak{f}'} = \mathbb{Z}_{\mathfrak{f}\mathfrak{f}'}$, $E[\mathfrak{f}\mathfrak{f}'] \otimes_{\mathbb{Z}} \mathfrak{l}$ is identical to $E[\mathfrak{f}\mathfrak{f}']$. The tensor E

has the exact sequence $0 \to \mathfrak{l} \to \mathbb{Z} \to \mathbb{Z}/\mathfrak{l} \to 0$ (of constant abelian fppf sheaves). Since E is a divisible fppf abelian sheaf, we have $E \otimes_{\mathbb{Z}} \mathbb{Z}/\mathfrak{l} = 0$, and the sequence

$$0 \to \mathrm{Tor}_1(E, \mathbb{Z}/\mathfrak{l}) \to E \otimes_{\mathbb{Z}} \mathfrak{l} \to E \to 0$$

is exact. Since $\mathbb{Z}$ is a principal ideal domain, we have $\mathrm{Tor}_1(E, \mathbb{Z}/\mathfrak{l}) \cong E[\mathfrak{l}] = E[l]$; hence, we have a commutative diagram for the generator $0 < l \in \mathbb{Z}$ of $\mathfrak{l}$:

$$\begin{array}{ccccc} \mathrm{Tor}_1(E, \mathbb{Z}/\mathfrak{l}) & \xrightarrow{\hookrightarrow} & E \otimes_{\mathbb{Z}} \mathfrak{l} & \xrightarrow{\twoheadrightarrow} & E \\ \wr\downarrow & & \wr\downarrow & & \| \downarrow \\ E[l] & \xrightarrow[\hookrightarrow]{} & E & \xrightarrow[l]{} & E. \end{array}$$

Thus multiplication by l induces $E \cong E \otimes_{\mathbb{Z}} \mathfrak{l}$, which acts on ω by $\omega \mapsto l\omega =: \omega'$ and on (i, i') by $(i, i') \mapsto (i \circ l^{-1}, i' \circ l) =: (j, j')$. On the lattice side, $(\mathbb{C}/L) \otimes_{\mathbb{Z}} \mathfrak{l} = \mathbb{C}/\mathfrak{l}L$; so it is given by multiplication $L \mapsto lL$. For the modular form f, we define $f|(\mathfrak{l})(E, \phi, \omega) = f((\mathfrak{l})(E, \phi, \omega))$.

Lemma 8.11 *Let $\mathfrak{l}$ be a prime outside $\mathfrak{f}$. Suppose that $\mathfrak{l} = (l)$ for a positive $l \in \mathbb{Q}$. Let $\mathbb{E}'_k(\psi) = E_k(\psi^\circ) - l^{-1}E_k(\psi^\circ)|[\mathfrak{l}]$ and $\mathbb{E}_k(\psi) = E_k(\psi^\circ) - E_k(\psi^\circ)|(\mathfrak{l})|[\mathfrak{l}]$. Then we have*

(1) $\mathbb{E}'_k(\psi)|U(l) = \lambda^{-1}_{\mathfrak{F}_c}(l)\mathbb{E}'_k(\psi)$.
(2) $\mathbb{E}_k(\psi)|U(l) = \lambda_{\mathfrak{F}}(l)l^{k-1}\mathbb{E}_k(\psi)$.
(3) Even if ψ is the identity character $\mathbf{1}$, $\mathfrak{f} = 1$, and $k = 2$, $\mathbb{E}'_2(\mathbf{1})$ is a holomorphic modular form.

Proof. The effect of Hecke operators on q-expansion is computed in Sect. 6.2.9 assuming the level structure is "one-sided," in other words, assuming we have a level structure of the form $i : \mu_{\mathfrak{f}} \hookrightarrow E[\mathfrak{f}]$. This corresponds to taking the adelic open subgroup

$$\widehat{\Gamma}_0(\mathfrak{f}) = \left\{ \left(\begin{smallmatrix} a & b \\ c & d \end{smallmatrix}\right) \in G(\widehat{\mathbb{Z}}) \middle| c \equiv 0 \mod \mathfrak{f}\widehat{\mathbb{Z}} \right\}$$

with the operation $i \mapsto i \circ l$ corresponding to the action of $\left(\begin{smallmatrix} 1 & 0 \\ 0 & l \end{smallmatrix}\right) \in G(\mathbb{Z}_{\mathfrak{f}})$. On the other hand, our level-$\Gamma_0(\mathfrak{f}\mathfrak{f}')$ structure is "two-sided" as $\mu_{\mathfrak{f}} \xrightarrow{i} E[\mathfrak{f}\mathfrak{f}'] \xrightarrow{i'} \mathbb{Z}/\mathfrak{f}'\mathbb{Z}$, and the action of $\left(\begin{smallmatrix} l & 0 \\ 0 & l \end{smallmatrix}\right) \in G(\mathbb{Z}_{\mathfrak{f}\mathfrak{f}'})$ corresponds to

$$(i, i') \mapsto (i \circ l, i' \circ l^{-1}) =: (j, j')$$

(as i' is the dual of i under the Cartier self-duality of $E[\mathfrak{f}\mathfrak{f}']$ in Sect. 6.1.6). Note that $f|[\mathfrak{l}](q) = l^k\lambda_{\mathfrak{F}}(l)f$ for $f \in G_k(\Gamma_0(\mathfrak{f}\mathfrak{f}'), \epsilon_\lambda)$ because as seen in Sect. 7.1.9, the action of $[\mathfrak{l}]$ corresponds to the action on the level structure: $\eta \mapsto \eta \circ \left(\begin{smallmatrix} 1 & 0 \\ 0 & l_{\mathfrak{l}}^{-1} \end{smallmatrix}\right)$ with $\left(\begin{smallmatrix} 1 & 0 \\ 0 & l_{\mathfrak{l}}^{-1} \end{smallmatrix}\right) \equiv \left(\begin{smallmatrix} l^{(\mathfrak{l})} & 0 \\ 0 & 1 \end{smallmatrix}\right) \mod Z(\mathbb{Q})$. If one wants to compute the action of $U(l)$ and $T(l)$ via q-expansion, we need to compute the corresponding q-expansion for this two-sided Neben character. Or we can directly compute the effect using the geometric definition of the Hecke operators as in Sect. 6.2.5. This way of using the geometric definition (Definition 6.21) is adopted in [H07a] Sect. 1.3.

Combining the effect of the operator $(\mathfrak{l})$ explained before the lemma with the computation of $[\mathfrak{l}]$ in Sect. 7.1.9, for $f = E_k(\psi^\circ)$, we get

$$f|(\mathfrak{l})|[\mathfrak{l}](q) = l^k \lambda_{\mathfrak{F}}(l) f|(\mathfrak{l})(q^l) = l^k \lambda_{\mathfrak{F}}(l)(l^{-k}\lambda_{\mathfrak{F}}(l)^{-1}\lambda_{\mathfrak{F}_c}^{-1}(l)) f(q^l) = \lambda_{\mathfrak{F}_c}^{-1}(l) f(q^l).$$

If $f|T(l) = (\alpha+\beta)f$ and $\alpha\beta = \epsilon_\lambda(l)l^{k-1}$, by putting $f_\alpha(q) = f(q) - \beta f(q^l)$, we have $f_\alpha|U(l) = \alpha f$ because of $a(n, f|U(l)) = a(nl, f)$ (see Corollary 6.34). Thus, $\mathbb{E}_k(\psi)|U(l) = l^{k-1}\lambda_{\mathfrak{F}}(l)\mathbb{E}_k(\psi)$. The computation for $\mathbb{E}_k'(\psi)$ is similar.

Note that $E_2(\mathbf{1})(z) = \frac{c}{\mathrm{Im}(z)} + f(z)$ for a holomorphic function $f(z)$. Since $\mathbb{E}_2'(\mathbf{1})(z) = E_2(\mathbf{1})(z) - lE_2(lz)$, the term $\frac{c}{\mathrm{Im}(z)}$ cancels out, and we get the holomorphy.

Remark 8.12 By (7.4.1) and $a(n, df) = n \cdot a(n, f)$, the Hecke operators $T(l)$ and $U(l)$ satisfy $T(l) \circ d = l \cdot d \circ T(l)$ and $U(l) \circ d = l \cdot d \circ U(l)$ for the Katz differential operator d. Set $\mathbb{E}(\lambda) = d^\kappa \mathbb{E}_k(\psi)$ (if neither $k = 2$ nor $\psi = \mathbf{1}$) and $\mathbb{E}'(\lambda) = d^\kappa \mathbb{E}_k'(\psi)$. Put $\mathbb{E}(\lambda) = -d^\kappa(\Delta^{-1} d\Delta - \Delta^{-1} d\Delta|(\mathfrak{l})|[\mathfrak{l}])/24$ if $k = 2$ and $\psi = \mathbf{1}$. Then we have, under the notation of Lemma 8.11,

$$\mathbb{E}'(\lambda)|U(l) = \lambda_{\mathfrak{F}_c}^{-1}(l) l^\kappa \mathbb{E}'(\lambda), \quad \mathbb{E}(\lambda)|U(l) = \lambda_{\mathfrak{F}}(l) l^{k-1+\kappa}\mathbb{E}(\lambda). \tag{8.1.17}$$

We now look into the degeneracy operator $[\mathfrak{q}]$ for a prime $\mathfrak{q} = (q)$ (with $0 < q \in \mathbb{Z}$) outside the level $\mathfrak{f}$. This operator brings a level $\Gamma_0(\mathfrak{q})$-test object (E, C, ϕ) with level $\mathfrak{f}$ structure i outside $\mathfrak{q}$ to $(E/C, \phi)$, where the level-$\mathfrak{f}$ structure i is intact under the quotient map: $E \to E/C$. On the lattice side, taking the lattice L_C with $L_C/L = C$, it is defined as follows:

$$f|[\mathfrak{q}](L, C, \phi) = f(L_C, \phi). \tag{8.1.18}$$

The above operator is useful to relate $U(\mathfrak{q})$ and $T(\mathfrak{q})$.

Exercise 8.13 *For $f \in G_k(\Gamma_1(N), \psi; A)$, show that $f|[\mathfrak{q}]$ is an element in $G_k(\Gamma_1(N) \cap \Gamma_0(\mathfrak{q}), \psi; A)$ by proving* (G1–3) *in* Sect. 6.2.4.

By definition,

$$f|U(\mathfrak{q})(L, C, \phi) = q^{-1} \sum_{L', L' \neq L_C} f(L', C', \phi)$$

for $C' = L_C + L'/L' = \mathfrak{q}^{-1}L/L'$. Thus, we have

$$U(\mathfrak{q}) = T(\mathfrak{q}) - q^{-1}[\mathfrak{q}]. \tag{8.1.19}$$

A similar computation yields

$$(f|[\mathfrak{q}])|U(\mathfrak{q})(L,C,\phi) = q^{-1} \sum_{L',L' \neq L_C} f|[\mathfrak{q}](L',C',\phi)$$
$$= q^{-1} \sum_{L'} f(\mathfrak{q}^{-1}L,\phi) = f|\langle \mathfrak{q}^{-1} \rangle(\underline{L}).$$

In short, we have

$$[\mathfrak{q}] \circ U(\mathfrak{q}) = (\mathfrak{q})^{-1}. \tag{8.1.20}$$

Exercise 8.14 *Make explicit the relation between the two operators $\langle l \rangle$ and (l) for a prime $l > 0$ outside the level.*

8.1.5 L-Functions of an Order

Pick a prime l in $\mathbb{Z}$ and write $\mathfrak{l} = l\mathfrak{O}$. Recall from Sect. 6.4.2 the order $\mathfrak{O}_n = \mathbb{Z} + \mathfrak{l}^n\mathfrak{O}$ of conductor $\mathfrak{l}^n$. We determine the type of Hecke L-function obtained by values of the Eisenstein series at CM points. Results equivalent to the ones presented here are more detailed in [LAP] V.3.2 (which were obtained independently by H. Yoshida). Since $\mathfrak{l}\mathfrak{O}_n \subset \mathfrak{O}_{n+1}$, $\mathfrak{a} \mapsto \mathfrak{l}\mathfrak{a}$ gives a map from the set of $\mathfrak{O}_n$-ideals into the set of nonproper $\mathfrak{O}_{n+1}$-ideals.

Proposition 8.15 *Let I_n be the group of all proper fractional $\mathfrak{O}_n$-ideals. Associating with each $\mathfrak{O}_{n+1}$-ideal $\mathfrak{a}$ the $\mathfrak{O}_n$-ideal $\mathfrak{O}_n\mathfrak{a}$, we get the following homomorphism of groups $\pi_n : I_{n+1} \to I_n$. The homomorphism π_n is surjective, and the kernel of π_n is isomorphic to $\mathfrak{O}_{n,\mathfrak{l}}^\times/\mathfrak{O}_{n+1,\mathfrak{l}}^\times$. After taking the quotient by principal ideals, we still have the following exact sequence:*

$$1 \to \mathfrak{O}_{n,\mathfrak{l}}^\times/\mathfrak{O}_{n+1,\mathfrak{l}}^\times\mathfrak{O}_n^\times \to Cl_{n+1} \to Cl_n \to 1,$$

where $Cl_n = I_n/\{\text{principal } \mathfrak{O}_n\text{-ideals}\}$.

If $n \geq 1$ if l is odd and $n > 1$ if $l = 2$, we have $\mathfrak{O}_n^\times = \{\pm 1\}$, and hence if $\mathfrak{O}_n^\times = \{\pm 1\}$, we have $\mathrm{Ker}(Cl_{n+1} \to Cl_n) \cong \mathbb{Z}/l\mathbb{Z}$ (the cyclic group of prime order l). Later in Sect. 8.2.1, we will make this isomorphism explicit.

Proof. Since a proper ideal is locally principal, we find an exact sequence:

$$1 \to M_\infty^\times\widehat{\mathfrak{O}}_n^\times \to M_{\mathbb{A}}^\times \to I_n \to 1.$$

Here we put $\widehat{\mathfrak{O}}_n = \mathfrak{O}_n \otimes_{\mathbb{Z}} \widehat{\mathbb{Z}}$, and to each idele x we associate the ideal $x\mathfrak{O}_n \in I_n$, where $x\mathfrak{O}_n = x\widehat{\mathfrak{O}}_n \cap M$, which is generated locally by $x_\mathfrak{q}$ at a prime ideal $\mathfrak{q}$. Then applying the snake lemma to the following diagram with exact rows:

$$\begin{array}{ccccccc}
1 \longrightarrow & M_\infty^\times\widehat{\mathfrak{O}}_{n+1}^\times & \longrightarrow & M_{\mathbb{A}}^\times & \longrightarrow & I_{n+1} & \longrightarrow 1 \\
& \downarrow & & \| & & \downarrow \pi_n & \\
1 \longrightarrow & M_\infty^\times\widehat{\mathfrak{O}}_n^\times & \longrightarrow & M_{\mathbb{A}}^\times & \longrightarrow & I_n & \longrightarrow 1,
\end{array}$$

we get the expression of the kernel and the surjectivity of π_n.

Again applying the snake lemma to the following diagram with exact rows:

$$\begin{array}{ccccccccc} 1 & \longrightarrow & M_\infty^\times \widehat{\mathfrak{O}}_{n+1}^\times M^\times & \longrightarrow & M_{\mathbb{A}}^\times & \longrightarrow & Cl_{n+1} & \longrightarrow & 1 \\ & & \downarrow & & \| & & \downarrow \pi_n & & \\ 1 & \longrightarrow & M_\infty^\times \widehat{\mathfrak{O}}_n^\times M^\times & \longrightarrow & M_{\mathbb{A}}^\times & \longrightarrow & Cl_n & \longrightarrow & 1, \end{array}$$

we get the last assertion, noting that $M^\times \cap M_\infty^\times \widehat{\mathfrak{O}}_n^\times = \mathfrak{O}_n^\times$. □

Writing a prime element of $\mathfrak{L}$ and $\mathfrak{l}$ as ϖ and l, respectively, we find that

$$\mathfrak{O}_{n+1,\mathfrak{l}}/\mathfrak{O}_{n+1,\mathfrak{l}}^\times = \{1\} \sqcup l\mathfrak{O}_{n,\mathfrak{l}}/\mathfrak{O}_{n+1,\mathfrak{l}}^\times. \tag{8.1.21}$$

When $\mathfrak{l}$ splits into a product of two primes $\mathfrak{L}\overline{\mathfrak{L}}$ in M, $\mathfrak{L}^i\overline{\mathfrak{L}}^j \subset \mathfrak{O}_n$ if and only if $i \geq n$ and $j \geq n$, and hence, we have

$$\mathfrak{O}_{n,\mathfrak{l}}/\mathfrak{O}_{n,\mathfrak{l}}^\times \cong \left(\sqcup_{j=0}^{n-1} (\varpi\overline{\varpi})^j \mathfrak{O}_{n-j,\mathfrak{l}}^\times/\mathfrak{O}_{n,\mathfrak{l}}^\times\right) \bigsqcup \left(\sqcup_{i,j=0}^\infty \varpi^{n+i}\overline{\varpi}^{n+j}\mathfrak{O}_{\mathfrak{l}}^\times/\mathfrak{O}_{n,\mathfrak{l}}^\times\right). \tag{8.1.22}$$

This follows from the fact that the multiplication by $\varpi^j\overline{\varpi}^j$ for $0 \leq j \leq n$ brings $\mathfrak{O}_{n-j,\mathfrak{l}} = \mathbb{Z}_l + \varpi^{n-j}\overline{\varpi}^{n-j}\mathfrak{O}_{\mathfrak{l}}$ into $\mathfrak{O}_{n,\mathfrak{l}} = \mathbb{Z}_l + \varpi^n\overline{\varpi}^n\mathfrak{O}_{\mathfrak{l}}$ and that $\mathfrak{O}_{\mathfrak{l}} = \bigsqcup_{i,j\geq 0} \varpi^i\overline{\varpi}^j\mathfrak{O}_{\mathfrak{l}}$. When $\mathfrak{l}$ ramifies in $M/\mathbb{Q}$, similarly, we get

$$\mathfrak{O}_{n,\mathfrak{l}}/\mathfrak{O}_{n,\mathfrak{l}}^\times \cong \left(\sqcup_{j=0}^{n-1} \varpi^{2j} \mathfrak{O}_{n-j,\mathfrak{l}}^\times/\mathfrak{O}_{n,\mathfrak{l}}^\times\right) \bigsqcup \left(\sqcup_{j=0}^\infty \varpi^j\mathfrak{O}_{\mathfrak{l}}^\times/\mathfrak{O}_{n,\mathfrak{l}}^\times\right). \tag{8.1.23}$$

When $\mathfrak{l}$ is inert in $M/\mathbb{Q}$, we have

$$\mathfrak{O}_{n,\mathfrak{l}}/\mathfrak{O}_{n,\mathfrak{l}}^\times \cong \left(\sqcup_{j=0}^{n-1} \varpi^{j} \mathfrak{O}_{n-j,\mathfrak{l}}^\times/\mathfrak{O}_{n,\mathfrak{l}}^\times\right) \bigsqcup \left(\sqcup_{j=0}^\infty \varpi^j\mathfrak{O}_{\mathfrak{l}}^\times/\mathfrak{O}_{n,\mathfrak{l}}^\times\right). \tag{8.1.24}$$

Let χ be a character of the group of fractional proper ideals of $\mathfrak{O}_n$. Identifying $I_n = M_{\mathbb{A}}^\times/M_\infty^\times\widehat{\mathfrak{O}}_n^\times$, χ gives rise to a unique idele character of $M_{\mathbb{A}}^\times$. In particular, this applies to the norm character $\mathfrak{a} \mapsto N(\mathfrak{a}) = [\mathfrak{O}_n : \mathfrak{a}] = [\mathfrak{O} : \mathfrak{O}\mathfrak{a}]$. We then define an L-function by

$$L^n(s,\chi) = \sum_{\mathfrak{a}\subset\mathfrak{O}_n} \chi(\mathfrak{a})N(\mathfrak{a})^{-s}, \tag{8.1.25}$$

where $\mathfrak{a}$ runs over all proper $\mathfrak{O}_n$-ideals (or equivalently, $\mathfrak{a}$ runs over $\widehat{\mathfrak{O}}_n$-integral ideles in $I_n = M_{\mathbb{A}}^\times/M_\infty^\times\widehat{\mathfrak{O}}_n^\times$). Even if χ is a primitive character of Cl_f with $f > 0$, we could have $\chi(\mathfrak{a}) \neq 0$ even if $\mathfrak{a} + l\mathfrak{O}_n \subsetneq \mathfrak{O}_n$ (for $n \geq f$) in this definition, as the proper $\mathfrak{O}_n$-ideal $\mathfrak{a}$ is equivalent to another proper ideal $\mathfrak{b}$ prime to $l\mathfrak{O}_n$. Thus, the l-Euler factor $L_{\mathfrak{l}}^f(s,\chi)$ of $L^f(s,\chi)$ can be nontrivial.

Exercise 8.16 *Prove that the L-function $L^n(s,\chi)$ has an Euler product whose Euler factor is given by $(1-\chi(\mathfrak{Q})N(\mathfrak{Q})^{-s})^{-1}$ for each prime $\mathfrak{Q} \nmid \mathfrak{l}$.*

We write $L(s,\chi)$ for $L^0(s,\chi)$, which is the classical Hecke L-function. The L-function $L^n(s,\chi)$ depends on n, because the set of proper $\mathfrak{O}_n$-ideals depends on n. We are going to compute the dependence on n. Since each proper ideal $\mathfrak{b} \subset \mathfrak{O}_n$ equivalent to $\mathfrak{a}^{-1}$ can be written as $\mathfrak{b} = \alpha\mathfrak{a}^{-1}$ with $\alpha \in \mathfrak{a}$, we have

$$\sum_{\alpha\in\mathfrak{a}} \chi((\alpha))N(\alpha)^{-s} = \chi(\mathfrak{a})N(\mathfrak{a})^{-s} \sum_{\mathfrak{b}\sim\mathfrak{a}^{-1}} \chi(\mathfrak{b})N(\mathfrak{b})^{-s}.$$

We write $L^n_{\mathfrak{a}^{-1}}(s,\chi) = \sum_{\mathfrak{b}\sim\mathfrak{a}^{-1}} \chi(\mathfrak{b})N(\mathfrak{b})^{-s}$. Then we get

$$L^n(s,\chi) = \sum_{\mathfrak{a}\in Cl_n} L^n_{\mathfrak{a}^{-1}}(s,\chi) = \sum_{\mathfrak{a}\in Cl_n} \chi(\mathfrak{a})^{-1}N(\mathfrak{a})^s \sum_{\alpha\in\mathfrak{a}} \chi((\alpha))N(\alpha)^{-s}.$$

For a primitive character χ of Cl_f, we compute the Euler product of $L^{f+k}(s,\chi)$ ($k > 0$). By Exercise 8.16, we only need to compute the $\mathfrak{l}$-factor

$$L^{f+k}_{\mathfrak{l}}(s,\chi) = \sum_{\mathfrak{a}\in I_{f+k,\mathfrak{l}}} \chi(\mathfrak{a})N(\mathfrak{a})^{-s},$$

where $I_{f+k,\mathfrak{l}}$ is the set of all proper $\mathfrak{O}_{f+k}$-ideals $\mathfrak{a}$ with $N(\mathfrak{a}) = l^m$ for some $m \geq 0$. Note that $I_{f+k,\mathfrak{l}} \cong (\mathfrak{O}_{f+k,\mathfrak{l}} \cap M_{\mathfrak{l}}^\times)/\mathfrak{O}^\times_{f+k,\mathfrak{l}}$ by $\mathfrak{a} \mapsto \alpha\mathfrak{O}^\times_{f+k,\mathfrak{l}}$ for $\alpha \in \mathfrak{O}_{f+k,\mathfrak{l}}$ with $\mathfrak{a}_{\mathfrak{l}} = (\alpha)$.

We first deal with the case where $\mathfrak{l}$ splits in $M/\mathbb{Q}$. As $N(\mathfrak{L}) = l$, by (8.1.22), writing $\chi_{\mathfrak{l}} = \chi|_{M_{\mathfrak{l}}^\times}$, we have $\chi_{\mathfrak{l}}(\varpi\overline{\varpi}) = \chi(l\mathfrak{O}_{f+k})$ and

$$\begin{aligned} L^{f+k}_{\mathfrak{l}}(s,\chi) = &\sum_{j=0}^{f+k-1} \chi_{\mathfrak{l}}(\varpi\overline{\varpi})^j l^{-2sj} \sum_{u\in U_{f+k-j}} \chi_{\mathfrak{l}}(u) \\ &+ \chi_{\mathfrak{l}}(\varpi\overline{\varpi})^f l^{-2(f+k)s} \sum_{i,j} \chi_{\mathfrak{l}}(\varpi^i\overline{\varpi}^j) l^{-(i+j)s} \sum_{u\in U_{f+k}} \chi_{\mathfrak{l}}(u), \end{aligned}$$

where $U_j = \mathfrak{O}^\times_{f+k-j,\mathfrak{l}}/\mathfrak{O}^\times_{f+k,\mathfrak{l}}$. We have $\sum_{u\in U_{f+k-j}} \chi_{\mathfrak{l}}(u) = 0$ unless $j = 0, 1, \ldots, k$, and we get, for $f > 0$, if $\left(\frac{M/\mathbb{Q}}{l}\right) = 1$

$$L^{f+k}_{\mathfrak{l}}(s,\chi) = \sum_{j=0}^{k} \chi_{\mathfrak{l}}(\varpi\overline{\varpi})^j l^{j-2sj} \quad \text{with } \chi_{\mathfrak{l}}(\varpi\overline{\varpi}) = \chi(\mathfrak{l}). \tag{8.1.26}$$

Similarly, for ramified and inert primes, we have, for $f > 0$,

$$\begin{aligned} L^{f+k}_{\mathfrak{l}}(s,\chi) &= \sum_{j=0}^{k} \chi_{\mathfrak{l}}(\varpi)^{2j} l^{j-2sj} \text{ with } \chi_{\mathfrak{l}}(\varpi)^2 = \chi(\mathfrak{l}) \text{ if } \left(\frac{M/\mathbb{Q}}{l}\right) = 0, \\ L^{f+k}_{\mathfrak{l}}(s,\chi) &= \sum_{j=0}^{k} \chi_{\mathfrak{l}}(\varpi)^j l^{j-2sj} \text{ with } \chi_{\mathfrak{l}}(\varpi) = \chi(\mathfrak{l}) \text{ if } \left(\frac{M/\mathbb{Q}}{l}\right) = -1. \end{aligned} \tag{8.1.27}$$

When χ is of conductor 1 (i.e., $f = 0$), we have $\chi_{\mathfrak{l}}(l) = \chi(\mathfrak{l})$ and

$$L_{\mathfrak{l}}^{k}(s,\chi) - \left(l - \left(\frac{M/\mathbb{Q}}{l}\right)\right)\chi(\mathfrak{l})^{k} l^{k-1-2ks} L_{\mathfrak{l}}^{0}(s,\chi) = \sum_{j=0}^{k-1} \chi(\mathfrak{l})^{j} l^{j-2sj} \quad (k > 0). \quad (8.1.28)$$

In summary, we have

$$L_{\mathfrak{l}}^{k+1}(s,\chi) - \chi(\mathfrak{l}) l^{1-2s} L_{\mathfrak{l}}^{k}(s,\chi) = 1 \quad \text{for } k > 0 \text{ and } k \geq f > 0, \quad (8.1.29)$$

$$L_{\mathfrak{l}}^{1}(s,\chi) - \left(l - \left(\frac{M/\mathbb{Q}}{l}\right)\right)\chi(\mathfrak{l}) l^{-2s} L_{\mathfrak{l}}^{0}(s,\chi) = 1 \quad \text{for } f = 0. \quad (8.1.30)$$

8.1.6 Anticyclotomic Hecke L-Values

Pick the Néron differential $\omega(\mathfrak{O})$ on $E(\mathfrak{O})$ in (Ω1) of Sect. 6.4.3, and recall the Néron period $\Omega_\infty \in \mathbb{C}^\times$ given by $\omega(\mathfrak{O}) = \Omega_\infty \omega_\infty$ in (8.1.2). Since Cl_∞ is an almost pro-l group, all finite order characters of Cl_∞ have values in $\mathcal{W}[\mu_{p^m}]$ if every element of the finite p-Sylow subgroup of Cl_∞ is killed by p^m. Replacing W by $W[\mu_{p^m}] = W(\overline{\mathbb{F}}_p)[\mu_{p^m}]$, we hereafter write $\mathcal{W}$ for $i_p^{-1}(W(\overline{\mathbb{F}}_p)[\mu_{p^m}])$. Put $L^{(\mathfrak{l})}(s,\chi) = L^f(s,\chi) L_{\mathfrak{l}}^{f}(s,\chi)^{-1}$ (removing the Euler factor at $\mathfrak{l}$). We will prove, assuming $\mathfrak{C} = 1$, the following theorem at the end of this chapter after a long preparation (a proof in the general case where $\mathfrak{C} \neq 1$ can be found in [H07a]):

Theorem 8.17 *Let p be an odd prime splitting in $M/\mathbb{Q}$. Let λ be a Hecke character of M of conductor $\mathfrak{C}$ and of infinity type $k + \kappa(1-c)$. Assume $p \geq 5$. Suppose* (ct), (sp), *and* (pl) *in* Sect. 8.1.2. *Then for the period Ω_∞ as above, we have $\frac{\pi^\kappa \Gamma(k+\kappa) L^{(\mathfrak{l})}(0,\chi^{-1}\lambda)}{\Omega_\infty^{k+2\kappa}} \in \mathcal{W}$ for all finite order characters $\chi : Cl_\infty \to \mathcal{W}^\times$, where $\Gamma(s)$ is Euler's Gamma function. Moreover, except for finitely many characters χ of Cl_∞, we have*

$$\frac{\pi^\kappa \Gamma(k+\kappa) L^{(\mathfrak{l})}(0,\chi^{-1}\lambda)}{\Omega_\infty^{k+2\kappa}} \not\equiv 0 \mod \mathfrak{m}_{\mathcal{W}}.$$

8.1.7 Values of Eisenstein Series at CM Points

We relate values of an Eisenstein series at CM points to L-values. For a while, we deal with general cases of $\mathfrak{C}$ not necessarily $\mathfrak{C} = 1$ (but assuming $\mathfrak{l}$ is outside $\mathfrak{C}p$). We take a proper $\mathfrak{O}_{n+1}$-ideal $\mathfrak{a}$ prime to $\mathfrak{C}p$ for a variable integer $n \geq 0$, and identify $\mathfrak{a}$ with the lattice $i_\infty(\mathfrak{a})$ in $\mathbb{C}$. Recall the test object $x(\mathfrak{a})_{/\mathcal{W}} = (E(\mathfrak{a}), i_{\mathfrak{f}}(\mathfrak{a}), i'_{\mathfrak{f}}(\mathfrak{a}))$ with $E(\mathfrak{a})(\mathbb{C}) = \mathbb{C}/\mathfrak{a}$ in (6.4.2) for $\mathfrak{f}$ and $\mathfrak{f}'$ specified in Sect. 8.1.3. Since $\mathfrak{a}\mathfrak{O}_n \supset \mathfrak{a}$ by $\mathfrak{O}_n = \mathbb{Z} + \mathfrak{l}^n\mathfrak{O} \supset \mathbb{Z} + \mathfrak{l}^{n+1}\mathfrak{O} = \mathfrak{O}_{n+1}$, $C(\mathfrak{a}) =$

$\mathfrak{a}\mathfrak{O}_n/\mathfrak{a} \subset E(\mathfrak{a})(\mathbb{C})$ gives rise to a canonical level-$\Gamma_0(l)$ structure to $x(\mathfrak{a})$ defined over $\mathcal{W}$ (as $p \neq l$). We hereafter in this section write

$$x(\mathfrak{a})_{/\mathcal{W}} := (E(\mathfrak{a}), i_{\mathfrak{f}}(\mathfrak{a}), i'_{\mathfrak{f}}(\mathfrak{a}), C(\mathfrak{a}))_{/\mathcal{W}}, \tag{8.1.31}$$

which is a test object of level $\Gamma^1_{1,0}(\mathfrak{f}^2) \cap \Gamma_0(l)$. For a p-adic modular form f of the form $d^\kappa g$ with classical $g \in G_k(\Gamma^1_{1,0}(\mathfrak{f}^2) \cap \Gamma_0(l); \mathcal{W})$, we have by Theorem 1.34

$$\frac{d^\kappa f(x(\mathfrak{a}), \omega_p)}{\Omega_p^{k+2\kappa}} = \delta_k^\kappa f(x(\mathfrak{a}), \omega(\mathfrak{a})) = \frac{\delta_k^\kappa f(x(\mathfrak{a}), \omega_\infty)}{\Omega_\infty^{k+2\kappa}}.$$

We write $c_0 = (-1)^k \frac{\pi^\kappa \Gamma(k+\kappa)}{\mathrm{Im}(\delta)^\kappa \sqrt{D} \Omega_\infty^{k+2\kappa}}$ with $2\delta = \sqrt{-D}$. Here $\Gamma(s)$ is the Euler's Gamma function [i.e., $\Gamma(k+\kappa) = (k+\kappa-1)!$]. By definition, we find, for $e = [\mathfrak{O}_{n+1}^\times : \mathbb{Z}^\times]$ (which is equal to 1 if $n > 0$ since a root of unity congruent to ± 1 modulo l^2 is ± 1),

$$\begin{aligned}
&(c_0 e)^{-1} \delta_k^\kappa E_k(x(\mathfrak{a}), \omega(\mathfrak{a})) \\
&= \sum'_{w \in \mathfrak{F}^{-1}\mathfrak{a}/\mathfrak{O}_{n+1}^\times} \frac{\lambda_{\mathfrak{C}}^{-1}(\varpi^{e(\mathfrak{F})} w)}{w^{k+\kappa(1-c)} N_{M/\mathbb{Q}}(w)^s}\Big|_{s=0} \\
&= \lambda_{\mathfrak{C}}^{-1}(\varpi^{e(\mathfrak{F})}) \sum'_{w \in \mathfrak{F}^{-1}\mathfrak{a}/\mathfrak{O}_{n+1}^\times} \frac{\lambda_{\mathfrak{C}}^{-1}(w)\lambda^{-1}(w^{(\infty)})}{N_{M/\mathbb{Q}}(w)^s}\Big|_{s=0} \\
&= \lambda_{\mathfrak{C}}^{-1}(\varpi^{e(\mathfrak{F})}) \sum'_{w \in \mathfrak{F}^{-1}\mathfrak{a}/\mathfrak{O}_{n+1}^\times} \frac{\lambda(w\mathfrak{F})}{N_{M/\mathbb{Q}}(w)^s}\Big|_{s=0} \\
&= \lambda_{\mathfrak{C}}^{-1}(\varpi^{e(\mathfrak{F})})\lambda(\mathfrak{a}) N_{M/\mathbb{Q}}(\mathfrak{F}\mathfrak{a}^{-1})^s \sum'_{w\mathfrak{F}\mathfrak{a}^{-1} \subset \mathfrak{O}_{n+1}} \frac{\lambda(w\mathfrak{F}\mathfrak{a}^{-1})}{N_{M/\mathbb{Q}}(w\mathfrak{F}\mathfrak{a}^{-1})^s}\Big|_{s=0} \\
&= \lambda_{\mathfrak{C}}^{-1}(\varpi^{e(\mathfrak{F})})\lambda(\mathfrak{a}) L^{n+1}_{[\mathfrak{F}\mathfrak{a}^{-1}]}(0, \lambda),
\end{aligned} \tag{8.1.32}$$

where for an ideal class $[\mathfrak{c}] \in Cl_{n+1}$ represented by a proper $\mathfrak{O}_{n+1}$-ideal $\mathfrak{c}$,

$$L^{n+1}_{[\mathfrak{c}]}(s, \lambda) = \sum_{\mathfrak{b} \in [\mathfrak{c}]} \lambda(\mathfrak{b}) N_{M/\mathbb{Q}}(\mathfrak{b})^{-s}$$

is the partial L-function of the class $[\mathfrak{c}]$ whose summation is over $\mathfrak{b}$ running over all $\mathfrak{O}_{n+1}$-proper integral ideals in the class $[\mathfrak{c}]$ prime to $\mathfrak{C}$. In the second and third lines of (8.1.32), we regard λ as an idele character and in the sum of the other lines as an ideal character. For an idele a with $a\widehat{\mathfrak{O}} = \mathfrak{a}\widehat{\mathfrak{O}}$ and $a_{\mathfrak{C}} = 1$, we have $\lambda(a^{(\infty)}) = \lambda(\mathfrak{a})$ and $\widehat{\lambda}(a^{(p)}) = \widehat{\lambda}(\mathfrak{a})$.

We put $\mathbb{E}(\lambda) = d^\kappa \mathbb{E}_k(\psi)$ and $\mathbb{E}'(\lambda) = d^\kappa \mathbb{E}'_k(\psi)$ as in Remark 8.12. We want to evaluate $\mathbb{E}(\lambda)$ and $\mathbb{E}'(\lambda)$ at $x = (x(\mathfrak{a}), \omega(\mathfrak{a})) = (E(\mathfrak{a}), i_{\mathfrak{f}}(\mathfrak{a}), i'_{\mathfrak{f}}(\mathfrak{a}), C(\mathfrak{a}), \omega(\mathfrak{a}))$. Then, by definition, we have

$$\begin{aligned}
\mathbb{E}'(\lambda)(x) &= \delta_k^\kappa E_k(\psi^\circ)(x) - l^{-1}\delta_k^\kappa E_k(\psi^\circ)(x(\mathfrak{a}\mathfrak{O}_n), \omega(\mathfrak{a}\mathfrak{O}_n)) \\
\mathbb{E}(\lambda)(x) &= \delta_k^\kappa E_k(\psi^\circ)(x) - \delta_k^\kappa E_k(\psi^\circ)(x(\mathfrak{l}\mathfrak{a}\mathfrak{O}_n), \omega(\mathfrak{l}\mathfrak{a}\mathfrak{O}_n))
\end{aligned} \tag{8.1.33}$$

because $C(\mathfrak{a}) = \mathfrak{a}\mathfrak{O}_n/\mathfrak{a}$, and hence $[\mathfrak{l}](x(\mathfrak{a})) = x(\mathfrak{a}\mathfrak{O}_n)$ and $(\mathfrak{l})(x(\mathfrak{a})) = x(\mathfrak{l}\mathfrak{a})$.

To simplify the notation, write $\phi([\mathfrak{a}]) = \widehat{\lambda}(\mathfrak{a})^{-1}\phi(x(\mathfrak{a}), \omega(\mathfrak{a}))$. By (8.1.15), for $\phi = \mathbb{E}(\lambda)$ and $\mathbb{E}'(\lambda)$, the value $\phi([\mathfrak{a}])$ only depends on the ideal class $[\mathfrak{a}]$ but not on the individual $\mathfrak{a}$. The formula (8.1.33) combined with (8.1.32) shows

$$\begin{aligned} e^{-1}\lambda_{\mathfrak{C}}(\varpi^{e(\mathfrak{F})})\mathbb{E}'(\lambda)([\mathfrak{a}]) &= c_0\left(L^{n+1}_{[\mathfrak{F}\mathfrak{a}^{-1}]}(0,\lambda) - l^{-1}L^{n}_{[\mathfrak{F}\mathfrak{a}^{-1}\mathfrak{O}_n]}(0,\lambda)\right) \\ e^{-1}\lambda_{\mathfrak{C}}(\varpi^{e(\mathfrak{F})})\mathbb{E}(\lambda)([\mathfrak{a}]) &= c_0\left(L^{n+1}_{[\mathfrak{F}\mathfrak{a}^{-1}]}(0,\lambda) - \lambda(\mathfrak{l})L^{n}_{[\mathfrak{F}\mathfrak{l}^{-1}\mathfrak{a}^{-1}\mathfrak{O}_n]}(0,\lambda)\right), \end{aligned} \tag{8.1.34}$$

where $e = [\mathfrak{O}^{\times}_{n+1} : \mathbb{Z}^{\times}]$ and $c_0 = (-1)^k \frac{\pi^{\kappa}\Gamma(k+\kappa)}{\mathrm{Im}(\delta)^{\kappa}\sqrt{D}\Omega_{\infty}^{k+2\kappa}}$, with $2\delta = \sqrt{-D}$.

Recall $L^n(s,\lambda)$ as in (8.1.25) for $\chi = \lambda$. For each primitive character $\chi : Cl_f \to \overline{\mathbb{Q}}^{\times}$, taking $n = \max(f,1)$, we have

$$\begin{aligned} e^{-1}\lambda_{\mathfrak{C}}(\varpi^{e(\mathfrak{F})}) &\sum_{[\mathfrak{a}]\in Cl_{n+1}} \chi(\mathfrak{a})\mathbb{E}'(\lambda)([\mathfrak{a}]) \\ &= c_0 \cdot \chi(\mathfrak{F})\left(L^{n+1}(0,\lambda\chi^{-1}) - L^n(0,\lambda\chi^{-1})\right) \\ e^{-1}\lambda_{\mathfrak{C}}(\varpi^{e(\mathfrak{F})}) &\sum_{[\mathfrak{a}]\in Cl_{n+1}} \chi(\mathfrak{a})\mathbb{E}(\lambda)([\mathfrak{a}]) \\ &= c_0 \cdot \chi(\mathfrak{F})\left(L^{n+1}(0,\lambda\chi^{-1}) - \lambda\chi^{-1}(l_{\mathfrak{l}})l \cdot L^n(0,\lambda\chi^{-1})\right). \end{aligned} \tag{8.1.35}$$

As computed in Sect. 8.1.5, the Euler $\mathfrak{l}$-factor of $L^k(s,\chi^{-1}\lambda)$ for $k \geq f$ is given by

$$\begin{aligned} &\sum_{j=0}^{k-f} \frac{(\chi^{-1}\lambda(\mathfrak{l})l)^j}{l^{2sj}} \quad \text{if } f > 0, \\ &\sum_{j=0}^{k-1} \frac{(\chi^{-1}\lambda(\mathfrak{l})l)^j}{l^{2sj}} + \frac{\left(l - \left(\frac{M/\mathbb{Q}}{\mathfrak{l}}\right)\right)(\chi^{-1}\lambda(\mathfrak{l})l)^k}{l^{1+2ks}} L^0_{\mathfrak{l}}(s,\chi^{-1}\lambda) \quad \text{if } f = 0, \end{aligned} \tag{8.1.36}$$

where $\left(\frac{M/\mathbb{Q}}{\mathfrak{l}}\right)$ is 1, -1, or 0 depending on whether $\mathfrak{l}$ splits, remains prime, or ramifies in $M/\mathbb{Q}$, respectively, and $L^0_{\mathfrak{l}}(s,\chi^{-1}\lambda)$ is the $\mathfrak{l}$-Euler factor of the primitive L-function $L^0(s,\chi^{-1}\lambda)$. We define a possibly imprimitive L-function

$$L^{(\mathfrak{l})}(s,\chi^{-1}\lambda) = L^0_{\mathfrak{l}}(s,\chi^{-1}\lambda)^{-1}L^0(s,\chi^{-1}\lambda) = L(s,\chi^{-1}\lambda)\prod_{\mathfrak{L}|\mathfrak{l}}\left(1 - \frac{\chi^{-1}\lambda(\mathfrak{L})}{N(\mathfrak{L})}\right)$$

removing the $\mathfrak{l}$-Euler factor.

Though the computation in the case when $f = 0$ is rather involved (as $n = 1 > 0 = f$), combining all these formulas, we find

$$\frac{\lambda\chi^{-1}(\varpi^{e(\mathfrak{F})}l_{\mathfrak{l}})}{el}\sum_{[\mathfrak{a}]\in Cl_{n+1}}\chi(\mathfrak{a})\mathbb{E}'(\lambda)([\mathfrak{a}])=c_0\cdot L(0,\chi^{-1}\lambda)\ \text{ if } f>0, \quad (8.1.37)$$

$$\frac{\lambda\chi^{-1}(\varpi^{e(\mathfrak{F})})}{e}\sum_{[\mathfrak{a}]\in Cl_{n+1}}\chi(\mathfrak{a})\mathbb{E}(\lambda)([\mathfrak{a}])=c_0\cdot L^{(\mathfrak{l})}(0,\chi^{-1}\lambda)\ \text{ if } f\geq 0. \quad (8.1.38)$$

Note that $L(0,\chi^{-1}\lambda)=L^{(\mathfrak{l})}(0,\chi^{-1}\lambda)$ if $f>0$. The right-hand side of (8.1.37) when $f=0$ (hrnvr, $n=1$) is explicitly computed in [H04b] (4.23) and the l-Euler factor of $L(s,\chi\lambda^{-1})$ at $s=1$ shows up in front of $L(0,\chi^{-1}\lambda)$ with a p-adic unit factor. Following [Sh6], we have proven that all these values are algebraic in $\overline{\mathbb{Q}}$ and actually p-integral over $\mathcal{W}$ (most of the cases by Theorem 1.34):

Theorem 8.18 *Let* $c_0=(-1)^k\frac{\pi^\kappa\Gamma(k+\kappa)}{\mathrm{Im}(\delta)^\kappa\sqrt{D}\Omega_\infty^{k+2\kappa}}$ *for integers* $k>1$ *and* $\kappa\geq 0$. *Then the* L*-value of* (8.1.38) *is an algebraic* p*-integer in* $\mathcal{W}$ *if either* $p\geq 5$ *or* $f>0$ *or* $k\neq 2$.

Recall that we have modified $\mathcal{W}$, adding some p-power roots of unity to the strict henselization of $\mathbb{Z}_{(p)}$ given by $i_p^{-1}(W(\overline{\mathbb{F}}_p))$ (so that any finite order character of Cl_∞ has values in $\mathcal{W}^\times$). As we have seen in Sect. 6.3.4, $x(\mathfrak{a})$ is defined over $\mathcal{W}[\mu_{l^n}]$, but $\mathcal{W}\supset\mu_{l^n}(\overline{\mathbb{Q}})$, as it contains a strict henselization of $\mathbb{Z}_{(p)}$; so the result is clear from the formula (8.1.38) and Theorems 1.27 and 1.34 if $d^\kappa\mathbb{E}(\lambda)$ has q-expansion in $\mathcal{W}[[q]]$. Thus, we need to show that $d^\kappa\mathbb{E}(\lambda)$ is $\mathcal{W}$-integral. If $k\neq 2$ or $f>0$, by the explicit q-expansion, $d^\kappa\mathbb{E}(\lambda)$ has $\mathcal{W}$-integral coefficients except for the constant term. If $\kappa>0$ and $k\neq 2$, the constant term vanishes; hence, the integrality follows. Even if $\kappa=0$, if $f>0$, $\mathbb{E}'(\lambda)$ does not have constant term and $\mathcal{W}$-integral; so by (8.1.37), the integrality follows. Thus, we may assume that $k=2$ and $f=0$. Then we have $E_2(\mathbf{1})=-\Delta^{-1}\delta_{12}\Delta/24$ by Lemma 8.8. Recall $\mathbb{E}(\lambda)=-d^\kappa(\Delta^{-1}d\Delta-\Delta^{-1}d\Delta|(\mathfrak{l})|[\mathfrak{l}])/24$. If $p\geq 5$, Δ^{-1} and $\Delta/24$ are in $W[[q]]$, and hence $\mathbb{E}(\lambda)$ is W-integral if $p\geq 5$. By Theorems 1.27 and 1.34 applied to $d^\kappa(\Delta^{-1}d\Delta)/24=-\sum_{m+n=\kappa}\binom{\kappa}{m}d^m\Delta^{-1}d^{n+1}\Delta/24$ and $\delta_2^\kappa(\Delta^{-1}\delta_{12}\Delta/24)=\sum_{m+n=\kappa}\binom{\kappa}{m}\delta_{-12}^m\Delta^{-1}\delta_{12}^{n+1}\Delta/24$ [cf. Exercise 1.26(1)] and their translates by $[\mathfrak{l}]\circ(\mathfrak{l})$, we have $\mathbb{E}(\lambda)([\mathfrak{a}])\in W\cap\overline{\mathbb{Q}}=\mathcal{W}$.

8.2 Nonvanishing Modulo p of L-Values

We construct an $\mathbb{F}$-valued measure ($\mathbb{F}=\overline{\mathbb{F}}_p$) over the anticyclotomic class group $Cl_\infty=\varprojlim_n Cl_n$ modulo $\mathfrak{l}^\infty$ whose integral against a character χ is the Hecke L-value $L(0,\chi^{-1}\lambda)$ (up to a period). The idea is to translate the Hecke relation of the Eisenstein series into a distribution relation on the profinite group Cl_∞. At the end, we relate the nontriviality of the measure to the q-expansion of the Eisenstein series by the density of $\{x(\mathfrak{a})\}_\mathfrak{a}$ for the test objects in (8.1.31). This is the strategy we explained for Dirichlet L-values in Sect. 1.2 and followed in [H04b] and [H07a].

8.2.1 Construction of a Modular Measure

It is slightly more technical in dealing with the nontrivial conductor $\mathfrak{C}$ of the starting character λ. Since the purpose of this book is to present the idea of the author, we simply assume that $\mathfrak{C} = 1$; so $\mathfrak{F} = \mathfrak{F}_c = \mathfrak{f} = \mathfrak{f}' = 1$, referring the general case to the research article [H07a].

Let $B = W$ or $\mathbb{F} = \overline{\mathbb{F}}_p$. Let $f \in V_{\Gamma_0(l)/B}$ be a normalized Hecke eigenform [here normalization means that $f|T(n) = a(n,f)f$, $f|U(l) = a(l,f)f$, and $f(q) = \sum_{n=0}^{\infty} a(n,f)q^n$]. We could have treated $f \in V_{\Gamma_0(N\mathfrak{l})/B}$ adding level-N structure (but for simplicity of notation, we do not explicitly do this here; see again [H07a] for the general cases). A typical example of f can be given as follows: Take a modular form g in $G_k(\Gamma_0(l); B)$ for $B = \mathbb{F}$. Put $f = d^{\kappa} g$ for the differential operator d^{κ} in Sect. 1.3.5. We write $f(x(\mathfrak{a}))$ for the value of at $x(\mathfrak{a})$ in (8.1.31). The Hecke operator $U(l)$ takes the space $V_{\Gamma_0(l)/B}$ into $V_{\Gamma_0(l)/B}$. We regard $U(l)$ as an operator acting on $V_{\Gamma_0(l)/W}$. Suppose that $g|U(l) = a \cdot g$ with $a \in W^{\times}$; hence, $f|U(l) = l^{\kappa} a \cdot f$ for the positive generator l of $\mathfrak{l}$ (see Remark 8.12). The Eisenstein series $\mathbb{E}(\lambda)$ satisfies this condition by Lemma 8.11.

Choosing a basis $w = (w_1, w_2)$ of $\widehat{\mathfrak{O}} = \mathfrak{O} \otimes_{\mathbb{Z}} \widehat{\mathbb{Z}}$, identify $T(E(\mathfrak{O})_{/\overline{\mathbb{Q}}}) = \widehat{\mathfrak{O}}$ with $\widehat{\mathbb{Z}}^2$ by $\widehat{\mathbb{Z}} \ni (a,b) \mapsto aw_1 + bw_2 \in T(E(\mathfrak{O}))$, getting a level structure: $\mathbb{Q}^2 \otimes_{\mathbb{Q}} \mathbb{A}^{(p\infty)} \cong V^{(p)}(E(\mathfrak{O}))$ defined over $\mathcal{W}$. Choose the basis w satisfying the following condition:

(B) $w_{1,\mathfrak{l}} = 1$ and $\mathfrak{O}_{\mathfrak{l}} = \mathbb{Z}_{\mathfrak{l}}[w_{2,\mathfrak{l}}]$.

Let $\mathfrak{a}$ be a proper $\mathfrak{O}_n$-ideal (for $\mathfrak{O}_n = \mathbb{Z} + \mathfrak{l}^n \mathfrak{O}$) prime to $\mathfrak{f}$. Write $l_{\mathfrak{l}} = (1, \ldots, 1, \overset{\mathfrak{l}}{l}, 1, \ldots) \in \mathbb{A}^{\times}$. Then $(w_1, l_{\mathfrak{l}}^n w_2)$ is a basis of $\widehat{\mathfrak{O}}_n$ and gives a level structure $\eta^{(p)}(\mathfrak{O}_n) : \mathbb{Q}^2 \otimes_{\mathbb{Q}} \mathbb{A}^{(p\infty)} \cong V^{(p)}(E(\mathfrak{O}_n))$. We also write $l_{\mathfrak{l}}$ for $l \in \mathbb{Z}_{\mathfrak{l}}$ (if we want to avoid confusion).

We choose a complete representative set $A_n = \{a_1, \ldots, a_H\} \subset M_{\mathbb{A}}^{\times}$ so that $M_{\mathbb{A}}^{\times} = \bigsqcup_{j=1}^{H} M^{\times} a_j \widehat{\mathfrak{O}}_n^{\times} M_{\infty}^{\times}$. More precisely, we first choose a complete representative set A_1 for $Cl_1 = M_{\mathbb{A}^{(\infty)}}^{\times} / \widehat{\mathfrak{O}}_1 M^{\times}$ (inside $M_{\mathbb{A}^{(p\infty)}}^{\times}$) and then inductively on n, we choose A_n ($n > 1$) as follows. Suppose we have chosen $A_{n-1} = \{a_1, \ldots, a_h\}$. For each $a \in A_{n-1}$, writing $\mathfrak{a} = a\widehat{\mathfrak{O}}_{n-1} \cap M$ (the corresponding proper $\mathfrak{O}_{n-1}$-ideal), we choose proper $\mathfrak{O}_n$-ideals $\mathfrak{a}_0, \ldots, \mathfrak{a}_{l-1}$ distinct in Cl_n projecting down to the class of $\mathfrak{a}$ in Cl_{n-1} so that $\mathfrak{a} \subsetneq \mathfrak{a}_j \subsetneq l^{-1}\mathfrak{a}$. Since $\mathrm{Ker}(Cl_n \to Cl_{n-1})$ is isomorphic to $\mathbb{Z}/\mathfrak{l}$, we may assume that the index set $\{0, 1, \ldots, l-1\}$ is actually given by $\mathrm{Ker}(Cl_n \to Cl_{n-1}) \cong \mathbb{Z}/\mathfrak{l}$ (see Proposition 8.15). Then choose $a_j(a) \in M_{\mathbb{A}^{(p\infty)}}^{\times}$ so that $a_j(a)\widehat{\mathfrak{O}}_n = \mathfrak{a}_j$ ($j \in \mathbb{Z}/\mathfrak{l}$). We put $A_n = \bigsqcup_{a \in A_{n-1}} \{a_j(a) | j \in \mathbb{Z}/\mathfrak{l}\}$.

Once A_n is chosen, for each proper $\mathfrak{O}_n$-ideal $\mathfrak{a}$, we write $\mathfrak{a}\widehat{\mathfrak{O}}_n = \alpha a_j \widehat{\mathfrak{O}}_n$ for $\alpha \in M^{\times}$ for some index j. We have a commutative diagram

$$\begin{array}{ccc} T^{(p)}E(\mathfrak{a}) & \xrightarrow{\alpha} & T^{(p)}E(\alpha\mathfrak{a}) \\ {\scriptstyle \eta^{(p)}(\mathfrak{a})}\uparrow & & \uparrow{\scriptstyle \eta^{(p)}(\alpha\mathfrak{a})} \\ \widehat{\mathfrak{a}}^{(p)} & \xrightarrow[\alpha]{} & \alpha\widehat{\mathfrak{a}}^{(p)}. \end{array}$$

Since the top arrow is an identification up to isogenies (by our choice of classification functor), we need to define $\eta^{(p)}(\alpha\mathfrak{a}) = \alpha^{-1}\eta(\mathfrak{a})$. To make level structures compatible with idele multiplication, we are forced to define $\eta^{(p)}(\mathfrak{a}) = \alpha^{-1}a_j^{-1}\eta^{(p)}(\mathfrak{O}_n)$.

There is ambiguity of the choice of α up to units in $\mathfrak{O}_n^\times$, which is up to $\pm 1 \in Z(\mathbb{Q})$ if $n > 1$ or $|D| > 4$. This small ambiguity does not cause any trouble as elements in $Z(\mathbb{Q})$ act trivially on the Shimura curve Sh. Write $x(\mathfrak{a}) = (E(\mathfrak{a}), C(\mathfrak{a}), \omega(\mathfrak{a}))_{/\mathcal{W}}$. This is a test object of level $\Gamma_0(l)$. We pick a subgroup $C \subset E(\mathfrak{O}_n)$ such that $C \cong \mathbb{Z}/\mathfrak{l}^m$ ($m > 0$) but $C \cap C(\mathfrak{O}_n) = \{0\}$. Since $\mathcal{W}$ is strictly Henselian (i.e., $\mathcal{W}/\mathfrak{m}_{\mathcal{W}} = \overline{\mathbb{F}}_p = \mathbb{F}$) and $\mathfrak{l} \nmid p$, $E(\mathfrak{O}_n)[\mathfrak{l}^m]$ is a constant étale group scheme isomorphic to $(\mathbb{Z}/\mathfrak{l})^2$; so making the quotient $E(\mathfrak{O}_n)/C$ is easy (see [GME] Sect. 1.8.3). Then we define $x(\mathfrak{O}_n)/C$ by

$$\left(\frac{E(\mathfrak{O}_n)}{C}, \frac{C + C(\mathfrak{O}_n)}{C}, (\pi^*)^{-1}\omega(\mathfrak{O}_n)\right)$$

for the projection map $\pi : E(\mathfrak{O}_n) \twoheadrightarrow E(\mathfrak{O}_n)/C$. Since C is a subgroup of $E(\mathfrak{O}_n)[l] = l^{-1}\mathfrak{O}_n/\mathfrak{O}_n$, we have a unique lattice $\mathfrak{a} \subset l^{-1}\mathfrak{O}_n$ of index l such that $E(\mathfrak{O}_n)/C = E(\mathfrak{a})$. Since the level structure of $x(\mathfrak{a})$ is the one induced by the quotient process, we write $x(\mathfrak{O}_n)/C$ for $x(\mathfrak{a})$.

Lemma 8.19 *We have*

$$x(\mathfrak{O}_n)/C = x(\mathfrak{a}) \in \mathcal{M}_{\Gamma_0(l)}(\mathcal{W})$$

for a proper $\mathfrak{O}_{n+m}$-ideal $\mathfrak{a} \supset \mathfrak{O}_n$ with $(\mathfrak{a}\mathfrak{a}^c) = \mathfrak{l}^{-2m}$, and for some $u \in \mathbb{Z}_\mathfrak{l}^\times$, we have

$$x(\mathfrak{a}) = x(\mathfrak{O}_n)/C = x(\mathfrak{O}_{m+n})|\begin{pmatrix} 1 & \frac{u}{l_\mathfrak{l}^m} \\ 0 & 1 \end{pmatrix}. \tag{8.2.1}$$

See Sect. 7.1.6 for the action of $g = \begin{pmatrix} 1 & \frac{u}{l_\mathfrak{l}^m} \\ 0 & 1 \end{pmatrix}$ on the point $x(\mathfrak{O}_{m+n})$.

Proof. We only need to prove (8.2.1). The basis of $\mathfrak{O}_{n,\mathfrak{l}}$ is given by $\alpha_n{}^t(1, w_2)$ for $\alpha_n = \begin{pmatrix} 1 & 0 \\ 0 & l_\mathfrak{l}^n \end{pmatrix}$ with a prime element $l_\mathfrak{l}$ of $\mathbb{Z}_\mathfrak{l}$. By Remark 7.12, the action on level structure $\eta \mapsto \eta \circ g$ induces the action $\widehat{L} \mapsto g^{-1}\widehat{L}$ for $\widehat{\mathbb{Z}}$-lattices. Thus, we find that $\alpha_n^{-1}(x(\mathfrak{O})) = x(\mathfrak{O}_n)$ and $\alpha_1^{-1}(x(\mathfrak{O}_{n-1})) = x(\mathfrak{O}_n)$. Since the general case of $m > 1$ follows by iteration of the formula in the case $m = 1$, we suppose $m = 1$. Then the formula becomes, for a suitable $u \in \mathbb{Z}_\mathfrak{l}^\times$,

$$l_\mathfrak{l}^{-1}(x(\mathfrak{a})) = x(l\mathfrak{a}) = x(\mathfrak{O}_{n+1})|\begin{pmatrix} 1 & \frac{u}{l_\mathfrak{l}} \\ 0 & 1 \end{pmatrix} \tag{8.2.2}$$

if $x(\mathfrak{a}) = x(\mathfrak{O}_n)/C$ for C as above. To see this, note that the basis of $l_\mathfrak{l}\mathfrak{a}_\mathfrak{l}$ is given by

$$\begin{pmatrix} 1-l_{\mathfrak{l}}^n u w_2 \\ l_{\mathfrak{l}}^{n+1} w_2 \end{pmatrix} = \begin{pmatrix} 1 & \frac{-u}{l_{\mathfrak{l}}} \\ 0 & 1 \end{pmatrix} \alpha_{n+1} \begin{pmatrix} 1 \\ w_2 \end{pmatrix}.$$

Thus, $\mathfrak{a}_{\mathfrak{l}}/\mathfrak{O}_n$ is generated by $\frac{1-ul_{\mathfrak{l}}^n w_2}{l_{\mathfrak{l}}} \mod \mathfrak{O}_{n,\mathfrak{l}}$, which gives the subgroup C for a suitable choice of u. Since $l^{(l)} \in \widehat{\mathbb{Z}}^\times$, the action of $l_{\mathfrak{l}}$ is equivalent to the action of $l \in Z(\mathbb{Q})$, which is trivial; hence, we can forget about $l_{\mathfrak{l}}$ in (8.2.2).

For each proper $\mathfrak{O}_n$-ideal $\mathfrak{a}$, we have an embedding $\rho_{\mathfrak{a}} : T_M(\mathbb{A}^{(p\infty)}) \to G(\mathbb{A}^{(p\infty)})$ given by $\alpha\eta^{(p)}(\mathfrak{a}) = \eta^{(p)}(\mathfrak{a}) \circ \rho_{\mathfrak{a}}(\alpha)$. Since multiplication by $b \in M_{\mathbb{A}}^\times$ on $\eta^{(p)}(\mathfrak{O}_n)$ commutes with complex multiplication by $\alpha \in M^\times$ and $\eta^{(p)}(\mathfrak{a}_j) = \alpha^{-1}a_j^{-1}\eta^{(p)}(\mathfrak{O}_n)$ for some j, $\rho_{\mathfrak{a}}$ depends only on $\mathfrak{O}_n$, for which $\mathfrak{a}$ is a proper fractional ideal, and its prime-to-l part $\rho_{\mathfrak{a}}^{(l)}$ is equal to $\rho_{\mathfrak{O}}$ (so often, we write simply ρ for $\rho_{\mathfrak{a}}$). Since $\det(\rho(\alpha)) = \alpha\alpha^c \gg 0$, $\alpha \in T_M(\mathbb{Z}_{(p)})$ acts on $Sh^{(p)}$ through $\rho(\alpha) \in G(\mathbb{A})$. Then we have

$$\rho(\alpha)(x(\mathfrak{a})) = (E(\mathfrak{a}), \eta^{(p)}(\mathfrak{a}) \circ \rho(\alpha)) = (E(\alpha\mathfrak{a}), \eta^{(p)}(\alpha\mathfrak{a}))$$

for the prime-to-p isogeny $\alpha \in \mathrm{End}(E(\mathfrak{a})_{/\mathcal{W}}) = \mathfrak{O}_{(p)}$. Thus, $T_M(\mathbb{Z}_{(p)})$ acts on $Sh^{(p)}$, fixing the point $x(\mathfrak{a})$. The pullback $\alpha^*\omega(\mathfrak{a})$ of $\omega(\mathfrak{a})$ via the complex multiplication by $\alpha \in \mathrm{End}(E(\mathfrak{a})_{/\mathcal{W}})$ is equal to $\alpha \cdot \omega(\mathfrak{a}) \in H^0(E(\mathfrak{a}), \Omega_{E(\mathfrak{a})/\mathcal{W}})$, and hence we find

$$g(x(\alpha\mathfrak{a}), \alpha\omega(\mathfrak{a})) = g(\rho(\alpha)(x(\mathfrak{a}), \omega(\mathfrak{a}))) = \alpha^{-k} g(x(\mathfrak{a}), \omega(\mathfrak{a})).$$

From this, we conclude that

$$f(x(\alpha\mathfrak{a}), \alpha\omega(\mathfrak{a})) = f(\rho(\alpha)(x(\mathfrak{a}), \omega(\mathfrak{a}))) = \alpha^{-k-\kappa(1-c)} f(x(\mathfrak{a}), \omega(\mathfrak{a})),$$

because the effect of the differential operator d is identical to the effect of δ at the CM point $x(\mathfrak{a})$ by Theorem 1.34. By our choice of the Hecke character λ, we find

$$\widehat{\lambda}(\alpha\mathfrak{a}) = \alpha^{-k-\kappa(1-c)}\widehat{\lambda}(\mathfrak{a}).$$

If $\mathfrak{a}$ and α are prime to p, the value $\alpha^{-k-\kappa(1-c)}\widehat{\lambda}(\alpha)$ is determined independently of the choice of α for a given ideal $\alpha\mathfrak{a}$, and the value $\widehat{\lambda}(\mathfrak{a})^{-1} f(x(\mathfrak{a}), \omega(\mathfrak{a}))$ is independent of the representative set $A = \{a_j\}$ for Cl_n. Defining

$$f([\mathfrak{a}]) = \widehat{\lambda}(\mathfrak{a})^{-1} f(x(\mathfrak{a}), \omega(\mathfrak{a})) \quad \text{for a proper } \mathfrak{O}_n\text{-ideal } \mathfrak{a} \text{ prime to } p, \tag{8.2.3}$$

we find that $f([\mathfrak{a}])$ depends only on the proper ideal class $[\mathfrak{a}] \in Cl_n$.

We write $x(\mathfrak{a}_u) = x(\mathfrak{a})|\alpha_1^{-1}\left(\begin{smallmatrix} 1 & \frac{u}{l} \\ 0 & 1 \end{smallmatrix}\right)$ $(\alpha_1 = \left(\begin{smallmatrix} 1 & 0 \\ 0 & l_{\mathfrak{l}} \end{smallmatrix}\right))$, where $\mathfrak{l} = (l)$ for the fixed prime $0 < l \in \mathbb{Z}$. Then $\mathfrak{a}_u$ depends only on $u \mod \mathfrak{l}$, and $\{\mathfrak{a}_u\}_{u \mod \mathfrak{l}}$ gives a complete representative set for proper $\mathfrak{O}_{n+1}$-ideal classes, which project down to the ideal class $[\mathfrak{a}] \in Cl_n$. Since $\mathfrak{a}_u\mathfrak{O}_n = l^{-1}\mathfrak{a}$, we find $\widehat{\lambda}(\mathfrak{a}_u) = \widehat{\lambda}(\mathfrak{l})^{-1}\widehat{\lambda}(\mathfrak{a})$. Recalling $f|U(l) = l^\kappa a \cdot f$, we have

$$l^\kappa a \cdot f([\mathfrak{a}]) = \widehat{\lambda}(\mathfrak{a})^{-1} f|U(l)(x(\mathfrak{a})) = \frac{1}{\widehat{\lambda}(\mathfrak{l})l} \sum_{u \mod \mathfrak{l}} f([\mathfrak{a}_u]). \tag{8.2.4}$$

Exercise 8.20 *Prove that a function $\phi : Cl_\infty \to \mathbb{F}$ continuous with respect to the profinite topology on Cl_∞ and the discrete topology on $\mathbb{F}$ factors through Cl_n for a suitable $n > 0$.*

Definition 8.21 *For a continuous function $\phi : Cl_\infty \to \mathbb{F}$, taking $n > 0$ so that ϕ factors through Cl_n, we define a measure φ_f on Cl_∞ with values in $\mathbb{F}$ by*

$$\int_{Cl_\infty} \phi d\varphi_f = b^{-n} \sum_{\mathfrak{a}\in Cl_n} \phi(\mathfrak{a}^{-1}) f([\mathfrak{a}]) \quad (\textit{for } b = l^{\kappa+1} a \widehat{\lambda}(\mathfrak{l})). \tag{8.2.5}$$

We need to have the term $\phi(\mathfrak{a}^{-1})f([\mathfrak{a}])$ instead of $\phi(\mathfrak{a})f([\mathfrak{a}])$ because in our application, $f([\mathfrak{a}])$ for a specific f actually gives the partial L-value for the ideal class $[\mathfrak{a}^{-1}]$, not $[\mathfrak{a}]$.

Exercise 8.22 *Prove that $\int_{Cl_\infty} \phi d\varphi_f$ depends only on ϕ and not on the choice of n, so that ϕ factors through Cl_n* (hint: use (8.2.4)).

8.2.2 Nontriviality of the Modular Measure

A sufficient condition for φ_f not identically 0 modulo p is given in [H04b] Theorems 3.2 and 3.3. To state the result in [H04b], we recall the degeneracy operator $[\mathfrak{q}]$, commuting with $U(l)$, for a prime ideal $\mathfrak{q} \neq \mathfrak{l}$ of $\mathbb{Z}$. For a test object (E, η) of level $\Gamma_0(\mathfrak{l}\mathfrak{q})$, the $\mathfrak{q}$-part $\eta_\mathfrak{q}$ of η is a subgroup $C \cong \mathbb{Z}/\mathfrak{q}$ in E. We made a correspondence $[\mathfrak{q}](E, \eta) = (E', \eta')$ with $E' = E/C$ (see (8.1.18)). If $\mathfrak{q}$ splits into $\mathfrak{Q}\overline{\mathfrak{Q}}$ in $M/\mathbb{Q}$, choosing $\eta_\mathfrak{q}$ induced by $E(\mathfrak{a})[\mathfrak{q}^\infty] \cong M_\mathfrak{Q}/\mathfrak{O}_\mathfrak{Q} \times M_{\overline{\mathfrak{Q}}}/\mathfrak{O}_{\overline{\mathfrak{Q}}} \cong \mathbb{Q}_\mathfrak{q}/\mathbb{Z}_\mathfrak{q} \times \mathbb{Q}_\mathfrak{q}/\mathbb{Z}_\mathfrak{q}$, we always have a level-$\Gamma_0(\mathfrak{q})$ structure $C = E(\mathfrak{a})[\mathfrak{Q}_n]$ for $\mathfrak{Q}_n = \mathfrak{Q} \cap \mathfrak{O}_n$ on $E(\mathfrak{a})$ induced by the choice of the factor $\mathfrak{Q}$. Then we have $[\mathfrak{q}](E(\mathfrak{a}), C) = E(\mathfrak{a}\mathfrak{Q}_n^{-1})$ for a proper $\mathfrak{O}_n$-ideal $\mathfrak{a}$, as $\mathfrak{Q}_n^{-1}\mathfrak{a}/\mathfrak{a} \cong C$ by $\eta_\mathfrak{q}$ [so $E(\mathfrak{a})/C = E(\mathfrak{a})/(\mathfrak{a}\mathfrak{Q}_n^{-1}/\mathfrak{a}) = E(\mathfrak{a}\mathfrak{Q}_n^{-1})$]. When $\mathfrak{q}$ ramifies in $M/\mathbb{Q}$ as $\mathfrak{q} = \mathfrak{Q}^2$, $E(\mathfrak{a})$ has a subgroup $C = E(\mathfrak{a})[\mathfrak{Q}_n]$ isomorphic to $\mathbb{Z}/\mathfrak{q}$; so we can still define $[\mathfrak{q}](E(\mathfrak{a})) = E(\mathfrak{a}\mathfrak{Q}_n^{-1})$. The effect of $[\mathfrak{q}]$ on the q-expansion at the infinity cusp is computed in Sect. 7.1.9 and is given by a unit multiple of the q-expansion of f at the Tate curve $\mathrm{Tate}(q^\varpi)$ for a positive generator ϖ of $\mathfrak{q}$. The operator $[\mathfrak{q}]$ corresponds to the action of $g = \begin{pmatrix} 1 & 0 \\ 0 & \varpi_\mathfrak{q}^{-1} \end{pmatrix} \in GL_2(\mathbb{Q}_\mathfrak{q})$. In Sect. 7.1.9, we also noted that $[\mathfrak{q}]$ induces a linear map well defined on $V(\Gamma_0(l); B)$ into $V(\Gamma_0(\mathfrak{l}\mathfrak{q}); B)$ (for any p-adic W-algebra B).

We fix a decomposition $Cl_\infty = \Gamma^- \times \Delta$ for a finite group Δ and a torsion-free subgroup Γ^- (we may assume that complex conjugation acts on Γ^- by $\gamma^c = \gamma^{-1}$). In this section, for simplicity, we write Γ for Γ^-. Since each fractional $\mathfrak{O}$-ideal $\mathfrak{A}$ prime to $\mathfrak{l}$ defines a class $[\mathfrak{A}]$ in Cl_∞, we can embed the ideal group of fractional ideals prime to $\mathfrak{l}$ into Cl_∞. We write Cl^{alg} for its image. By class field theory, we can identify Cl_n and $\mathrm{Gal}(H_n/M)$ by Artin symbol (see [CFN] IV.6), where H_n is the ring class field corresponding to Cl_n. If n is finite, Chebotarev's density theorem ([CFN] V.6.4) tells us that for any

$\sigma \in \mathrm{Gal}(H_n/M)$, we can find a split prime $\mathfrak{Q}$ outside $\mathfrak{l}$ such that $\sigma = Frob_{\mathfrak{Q}}$; hence, the class in Cl_n corresponding to σ contains $[\mathfrak{Q}]$. In particular, split primes (not just Cl^{alg}) are dense in Cl_∞.

Lemma 8.23 *1. Complex conjugation c acts on $z \in Cl_\infty$ by $z \mapsto z^{-1}$ (and this is the reason why Cl_n are called anticyclotomic class group).*
2. The intersection $\Delta^{alg} = \Delta \cap Cl^{alg}$ is represented by square-free products of prime ideals of M ramified over $\mathbb{Q}$. In other words, Δ^{alg} is isomorphic to the 2-torsion part of the class group of M (i.e., the ambiguous class group of M).
3. The quotient $Cl_\infty/\Gamma\Delta^{alg}$ has a complete representative set in the set of prime ideals split over $\mathbb{Q}$ (prime to $\mathfrak{l}$).
4. Write $[\mathfrak{Q}]_\Gamma$ (resp., $[\mathfrak{Q}]_\Delta$) for the projection of $[\mathfrak{Q}] \in Cl^{alg}$ to Γ (resp., to Δ). If $[\mathfrak{Q}]_\Delta \notin [\mathfrak{Q}']_\Delta\Delta^{alg}$, then $[\mathfrak{Q}]_\Gamma/[\mathfrak{Q}']_\Gamma \notin Cl^{alg}$.

Proof. For any ideal $\mathfrak{Q}$ of $\mathfrak{O}$ prime to $\mathfrak{l}$, $\mathfrak{Q}_n\mathfrak{Q}_n^c$ is principal generated by $N(\mathfrak{Q}) \in \mathbb{Z}$, and hence trivial in Cl_n. Therefore, complex conjugation acts on Cl_∞ by $z \mapsto z^{-1}$. Since we can bring any fractional ideal inside $\mathfrak{O}$ by multiplying a positive integer, we have $Cl^{alg} = \{[\mathfrak{Q}]|\mathfrak{Q} \subset \mathfrak{O}\}$. Thus, Cl^{alg} is generated by prime ideals. If $[\mathfrak{Q}] = 1$ in Cl_∞, we have $\mathfrak{Q}_n = (\alpha_n)$ with $\alpha_n \in \mathfrak{O}_n$ for all n. Since $\bigcap_n \mathfrak{O}_n = \mathbb{Z}$, $\mathfrak{Q}$ is generated by a rational integer. Therefore, $[\mathfrak{Q}]^m = 1$ implies that $\mathfrak{Q}^m$ is generated by a rational integer, but $\mathfrak{Q}$ is not. By making prime factorization, if $\mathfrak{Q}^m = (n)$ for $0 < n \in \mathbb{Z}$, $\mathfrak{Q}$ is a product of $\mathbb{Z}$-ideals and ramified primes, so Δ^{alg} is generated by ramified primes. The ambiguous class group Cl^{amb} is generated by ramified primes, and hence we have a natural surjection $\Delta^{alg} \to Cl^{amb}$. If $[\mathfrak{Q}]$ is trivial, as we have seen, $\mathfrak{Q}$ is either a $\mathbb{Z}$-ideal or a $\mathbb{Z}$-ideal multiplied by $(\sqrt{-D})$, so this map is actually an isomorphism.

Note that $Cl_\infty/\Gamma\Delta^{alg}$ is a finite group. Thus, for sufficiently large but finite n, the natural map $Cl_n \to Cl_\infty/\Gamma\Delta^{alg}$ is surjective. By Chebotarev's density theorem (and class field theory), any class of Cl_n is represented by a prime ideal $\mathfrak{Q}$ (prime to $\mathfrak{l}$). If $\mathfrak{Q}$ is inert, the class is trivial. If $\mathfrak{Q}$ is ramified, the class is in Δ^{alg} as already seen. Thus, for the class $[\mathfrak{Q}] \neq 1$ in $Cl_\infty/\Gamma\Delta^{alg}$, the ideal $\mathfrak{Q}$ has to be split over $\mathbb{Q}$, as otherwise, $[\mathfrak{Q}] \in \Delta^{alg}$. For the trivial class in $Cl_\infty/\Gamma\Delta^{alg}$, making n larger if necessary, we pick a nontrivial class $C \notin \Delta^{alg}$ in Cl_n mapping to $1 \in Cl_\infty/\Gamma\Delta^{alg}$. Then we can choose a split prime $\mathfrak{Q}$ in the class again by Chebotarev density, and we can chose a complete representative set for $Cl_\infty/\Gamma\Delta^{alg}$ among split primes. The last assertion is obvious, as $[\mathfrak{Q}]_\Gamma/[\mathfrak{Q}']_\Gamma \notin Cl^{alg} \Leftrightarrow [\mathfrak{Q}]_\Delta/[\mathfrak{Q}']_\Delta \notin \Delta^{alg}$. □

We choose a complete representative set $\{\mathfrak{R}^{-1}|\mathfrak{r} \in \mathcal{R}\}$ for Δ^{alg} such that the set $\mathcal{R}$ is a subset of the set of all square-free products of primes in $\mathbb{Q}$ ramifying in $M/\mathbb{Q}$, and $\mathfrak{R}$ is a unique ideal in M with $\mathfrak{R}^2 = \mathfrak{r}$. The set $\{\mathfrak{R}|\mathfrak{r} \in \mathcal{R}\}$ is a complete representative set for 2-torsion elements in the class group Cl_0 of $\mathfrak{O}$ (i.e., the ambiguous classes). We fix a character $\nu : \Delta \to \mathbb{F}^\times$, and define

$$f_\nu = \sum_{\mathfrak{r}\in\mathcal{R}} \widehat{\lambda}\nu^{-1}(\mathfrak{R}) f|[\mathfrak{r}]. \tag{8.2.6}$$

Choose a complete representative set $\mathcal{Q}$ for $Cl_\infty/\Gamma\Delta^{alg}$ made of primes of M split over $\mathbb{Q}$ outside $p\mathfrak{l}$. Since Cl^{alg} is dense in Cl_∞, we can choose $\mathfrak{Q}\in\mathcal{Q}$, whose projection to Γ is whateverclose to 1 under the profinite topology (this remark will be useful later). We choose $\eta_n^{(p)}$ out of the basis (w_1, w_2) of $\widehat{\mathfrak{O}}_n$ so that at $\mathfrak{q} = \mathfrak{Q}\cap\mathbb{Q}$, $w_1 = (1,0)\in\mathfrak{O}_{\mathfrak{Q}}\times\mathfrak{O}_{\mathfrak{Q}^c} = \mathfrak{O}_{\mathfrak{q}}$ and $w_2 = (0,1)\in\mathfrak{O}_{\mathfrak{Q}}\times\mathfrak{O}_{\mathfrak{Q}^c} = \mathfrak{O}_{\mathfrak{q}}$. Since all operators $[\mathfrak{q}]$ and $[\mathfrak{r}]$ involved in this definition commute with $U(l)$, $f_\nu|[\mathfrak{q}]$ is still an eigenform of $U(l)$, with the same eigenvalue as f. Thus, in particular, we have a measure φ_{f_ν}. We project it to Γ along ν, which produces a measure φ_f^ν on Γ explicitly given by

$$\int_\Gamma \phi d\varphi_f^\nu = \sum_{\mathfrak{Q}\in\mathcal{Q}} \widehat{\lambda}\nu^{-1}(\mathfrak{Q}) \int_\Gamma \phi|\mathfrak{Q} d\varphi_{f_\nu|[\mathfrak{q}]},$$

where $\phi|\mathfrak{Q}(y) = \phi(y[\mathfrak{Q}]_\Gamma^{-1})$ for the projection $[\mathfrak{Q}]_\Gamma$ in Γ of the class $[\mathfrak{Q}]\in Cl_\infty$.

Lemma 8.24 *If $\chi : Cl_\infty \to \mathbb{F}^\times$ is a character inducing ν on Δ, we have*

$$\int_\Gamma \chi d\varphi_f^\nu = \int_{Cl_\infty} \chi d\varphi_f.$$

Proof. Write $\Gamma_{f,n}$ for the image of Γ in Cl_n. For proper $\mathfrak{O}_n$-ideal $\mathfrak{a}$, by the above definition of these operators,

$$f|[\mathfrak{r}]|[\mathfrak{q}]([\mathfrak{a}]) = \widehat{\lambda}(\mathfrak{a})^{-1} f(x(\mathfrak{Q}^{-1}\mathfrak{R}^{-1}\mathfrak{a}), \omega(\mathfrak{Q}^{-1}\mathfrak{R}^{-1}\mathfrak{a})).$$

For sufficiently large n, χ factors through Cl_n. Since $\chi = \nu$ on Δ, we have

$$\begin{aligned}\int_\Gamma \chi d\varphi_f^\nu &= \sum_{\mathfrak{Q}\in\mathcal{Q}}\sum_{\mathfrak{r}\in\mathcal{R}}\sum_{\mathfrak{a}\in\Gamma_{f,n}} \widehat{\lambda}\chi^{-1}(\mathfrak{Q}\mathfrak{R}\mathfrak{a}) f|[\mathfrak{r}]|[\mathfrak{q}]([\mathfrak{a}])\\ &= \sum_{\mathfrak{a},\mathfrak{Q},\mathfrak{r}} \chi(\mathfrak{Q}\mathfrak{R}\mathfrak{a}) f([\mathfrak{Q}^{-1}\mathfrak{R}^{-1}\mathfrak{a}]) = \int_{Cl_\infty} \chi d\varphi_f,\end{aligned}$$

because $Cl_\infty = \bigsqcup_{\mathfrak{Q},\mathfrak{R}} [\mathfrak{Q}^{-1}\mathfrak{R}^{-1}]\Gamma$.

In the next couple of sections, we prove the following result (given in [H04b] as Theorems 3.2 and 3.3):

Theorem 8.25 *Suppose that $p = \mathfrak{p}\overline{\mathfrak{p}}$ with two primes $\mathfrak{p}\neq\overline{\mathfrak{p}}$ in $M/\mathbb{Q}$ (the factor $\mathfrak{p}$ corresponding to $i_p : M \hookrightarrow \overline{\mathbb{Q}}_p$). Let $f\neq 0$ be an eigenform of $U(l)$ of level $\Gamma_0(l)$ defined over $\mathbb{F}$ with nonzero eigenvalue. Fix a character $\nu : \Delta\to\mathbb{F}^\times$, and define f_ν as in (8.2.6). If f satisfies the following condition:*

(H) *For any given integer $r > 0$ and any congruence class $u\in(\mathbb{Z}/l^r\mathbb{Z})^\times$, there exists $0\le\xi\in u$ such that $a(\xi, f_\nu)\neq 0$,* *then nonvanishing* $\int_{Cl_\infty} \nu\chi d\varphi_f \neq 0$ *holds except for finitely many characters* χ *of* Γ. *Here* $\nu\chi$ *is the character of* $Cl_\infty = \Gamma \times \Delta$ *given by* $\nu\chi(\gamma, \delta) = \nu(\delta)\chi(\gamma)$ *for* $\gamma \in \Gamma$ *and* $\delta \in \Delta$.

8.2.3 Preliminary to the Proof of Theorem 8.25

The association $\mathfrak{a} \mapsto x(\mathfrak{a})$ induces a map $Cl^{(\infty)} = \bigsqcup_n Cl_n \to Y_0(l)$. Over $\mathbb{C}$, the isomorphism class of $E(\mathfrak{a})$ is determined by the ideal class $[\mathfrak{a}]$; so even after projecting down to the J-line $\mathbf{P}^1(J)$, this map is injective (cf. [D2]). Over $\mathbb{F} = \overline{\mathbb{F}}_p$, via canonical lift (see Sect. 6.3.2), the image of $Cl^{(\infty)}$ in $\mathbf{P}^1(J)(\mathbb{F})$ is in bijection with the canonically lifted image in $\mathbf{P}^1(J)(\overline{\mathbb{Q}})$; hence, we regard $Cl^{(\infty)} \subset Y_0(l)(B)$ or $Cl^{(\infty)} \subset Y_1(l)(B)$ for $B = \mathbb{F}$ or $\mathcal{W}$. We now consider $[\mathfrak{a}] \mapsto f([\mathfrak{a}])$ a function of $Cl^{(\infty)} = \bigsqcup_n Cl_n \subset Y_1(l)(\mathbb{F})$.

Suppose that f is defined over a $\mathcal{W}$-algebra A in which l is invertible. By (8.2.4), we have, for an integer $n > m$,

$$\sum_{[\mathfrak{a}]\in Cl_n,\ \mathfrak{a}\mapsto[\mathfrak{l}^{n-m}\mathfrak{A}]=[\mathfrak{A}]\in Cl_m} f([\mathfrak{a}]) = (\widehat{\lambda}(\mathfrak{l})l)^{n-m} f|U(l^{n-m})([\mathfrak{A}]), \qquad (8.2.7)$$

where $[\mathfrak{a}]$ runs over all classes in Cl_n that project down to $[\mathfrak{A}] = [\mathfrak{l}^{m-n}\mathfrak{A}] \in Cl_m$. We suppose that $f|U(l) = (a/\widehat{\lambda}(\mathfrak{l})l)f$ with a unit $a \in A$. For each function $\phi : Cl_\infty \to A$ factoring through Cl_m, we define

$$\int_{Cl_\infty} \phi d\varphi_f = a^{-m} \sum_{\mathfrak{a}\in Cl_m} \phi(\mathfrak{a}^{-1}) f([\mathfrak{a}]). \qquad (8.2.8)$$

Then for $n > m$, we find

$$\begin{aligned} a^{-n} \sum_{\mathfrak{a}\in Cl_n} \phi(\mathfrak{a}^{-1}) f([\mathfrak{a}]) &= a^{-m} \sum_{\mathfrak{A}\in Cl_m} \phi(\mathfrak{A}^{-1}) a^{(m-n)} \sum_{\mathfrak{a}\in Cl_n,\ \mathfrak{a}\mapsto\mathfrak{A}} f([\mathfrak{a}]) \\ &\overset{(8.2.7)}{=} a^{-m} \sum_{\mathfrak{A}\in Cl_m} \phi(\mathfrak{A}^{-1}) a^{(m-n)} f|U(l^{(n-m)})([\mathfrak{A}]) = \int_{Cl_\infty} \phi(x) d\varphi_f(x). \end{aligned}$$

Thus, φ_f gives an A-valued distribution on Cl_∞ well defined independently of the choice of m for which ϕ factors through Cl_m, because $U(l^\alpha) = U(l)^\alpha$.

Classical modular forms are actually defined over a number field; so we assume that f is defined over the localization $\mathcal{V}$ of the integer ring in a number field K containing M over which $E(\mathfrak{a})$ for each class $[\mathfrak{a}] \in Cl_0$ is defined. We write $\mathcal{P}|p$ for the prime ideal of the p-integral closure $\widetilde{\mathcal{V}}$ of $\mathcal{V}$ in $\overline{\mathbb{Q}}$ corresponding to $i_p : \overline{\mathbb{Q}} \hookrightarrow \overline{\mathbb{Q}}_p$. More generally, if $f = d^\kappa g$ for a classical modular form g integral over $\mathcal{V}$, the value $f([\mathfrak{a}])$ is algebraic and $\mathcal{P}$-integral over $\mathcal{V}$ by a result of Shimura's and Katz's (see Theorems 1.27 and 1.34).

Let $f = d^\kappa g$ for $g \in G_k(\Gamma_0(l); \mathcal{V})$. Suppose $f|U(l) = (a/\widehat{\lambda}(\mathfrak{l})l)f$ for a, giving a unit of $\widetilde{\mathcal{V}}/\mathcal{P}$. For the moment, let φ be the measure associated with

f, with values in $B = W$ for a finite extension W of $W(\overline{\mathbb{F}}_p)$ containing $\mathcal{V}$. We have a well-defined measure φ mod $\mathcal{P}$. Let K_f be the field generated by $f([\mathfrak{a}])$ over $K[\mu_{l^\infty}]$. Then K_f/K is an abelian extension unramified outside l, and we have the Frobenius element $\sigma_{\mathfrak{b}} \in \mathrm{Gal}(K_f/K)$ (that is, the image of $\mathfrak{b}$ under the Artin reciprocity map) for each ideal $\mathfrak{b}$ of K prime to l. By Shimura's reciprocity law ((6.4.5) and Proposition 7.24), we find for $\sigma = \sigma_{\mathfrak{b}}$, $x(\mathfrak{a})^\sigma = x(N(\mathfrak{b})^{-1}\mathfrak{a})$ for the norm $N : K \to M$ (this fact is actually explained in Sect. 6.4.6 in more down-to-earth terms). As for $\eta_p^{ord}(\mathfrak{a})$, we find $\sigma \circ \eta_p^{ord}(\mathfrak{a}) = u \cdot \eta_p^{ord}(\mathfrak{a})$ for $u \in \mathfrak{O}_{\mathfrak{p}}^\times$ [see (6.4.3) for $\eta_p^{ord}(\mathfrak{a})$]. Since $\mathfrak{a}_p \cong \mathfrak{O}_p$, we have $E(\mathfrak{O})[p^\infty] \cong E(\mathfrak{a})[p^\infty]$ as a Galois module. As seen in Sect. 6.4.6, we know that $L_1(s, E(\mathfrak{O})_{/K}) = L(s,\psi)L(s,\overline{\psi})$ for a Hecke character $\psi : K_{\mathbb{A}}^\times/K^\times \to \mathbb{C}^\times$. This means that on the p-adic Tate module $T_pE(\mathfrak{O})$, the action of $\sigma_{\mathfrak{b}}$ is given by $\psi(\mathfrak{b}) \in M^\times$ [so the eigenvalues of $\sigma_{\mathfrak{b}}$ on $\mathbb{Q}_p$-vector space $T_pE(\mathfrak{O}) \otimes_{\mathbb{Z}} \mathbb{Q} \cong M_p$ are given by $\psi(\mathfrak{b})$ and $\overline{\psi}(\mathfrak{b})$]. Thus, replacing ψ by $\overline{\psi}$ if necessary, we conclude $u = \psi(\mathfrak{b})$ for the Hecke character ψ of $K_{\mathbb{A}}^\times/K^\times$, giving rise to the zeta function $L_1(s, E(\mathfrak{O}))$ of $E(\mathfrak{O})$. From this, if we extend K further if necessary, we see that $f([\mathfrak{a}])^\sigma = f([N(\mathfrak{b})^{-1}\mathfrak{a}])$ for any ideal $\mathfrak{b}$, since $\psi(\mathfrak{b}) \in M$ generates the ideal $N(\mathfrak{b}) \subset M$ by Exercise 6.58 (see also [IAT] Sect. 7.8 or [EDM] Sect. 14.4) and hence $\psi(\mathfrak{b})^{k+\kappa(1-c)} = \widehat{\lambda}(N(\mathfrak{b}))$. We then have

$$\sigma \cdot \left(\int_{Cl_\infty} \phi(x) d\varphi_f(x) \right) = \int_{Cl_\infty} \sigma \circ \phi(N(\mathfrak{b})x) d\varphi_f(x), \tag{8.2.9}$$

where $N(\mathfrak{b})$ is the norm of $\mathfrak{b}$ over M. If ϕ is a character χ, we have

$$\begin{aligned} \sigma \cdot \left(\int_{Cl_\infty} \chi(x) d\varphi_f(x) \right) &= \int_{Cl_\infty} \sigma \circ \chi(N(\mathfrak{b})x) d\varphi_f(x) \\ &= \sigma \circ \chi(N(\mathfrak{b})) \int_{Cl_\infty} \sigma \circ \chi(x) d\varphi_f(x). \end{aligned} \tag{8.2.10}$$

Writing $\mathbb{F}_{p^r}$ for $\mathcal{V}/\mathcal{P} \cap \mathcal{V}$, any modular form defined over $\mathbb{F}_{p^r}$ is a reduction modulo $\mathcal{P}$ of a classical modular form defined over $\mathcal{V}$ of sufficiently high weight. Thus, the above identity is valid for $\sigma = \Phi^s$ ($s \in \mathbb{Z}$) for the Frobenius element $\Phi \in \mathrm{Gal}(\mathbb{F}/\mathbb{F}_{p^r})$. In this case, $N(\mathfrak{b})$ is a power of a prime ideal $\mathfrak{p}|p$ in M.

We now take the base ring B to be $\mathbb{F} = W/\mathfrak{m}_W = \widetilde{\mathcal{V}}/\mathcal{P}$ and consider the measure φ_f as having values in $\mathbb{F}$. Thus, we simply write $g \in G_k(\Gamma_0(l);\mathbb{F})$ for g mod $\mathcal{P}$ for the original $g \in G_k(\Gamma_0(l);\mathcal{V})$ [and $f = d^\kappa g \in V_{\Gamma_1(l)/\mathbb{F}}$]. Then (8.2.10) shows that if ϕ is a character χ of Cl_∞, for $\sigma \in \mathrm{Gal}(\mathbb{F}/\mathbb{F}_{p^r})$,

$$\int_{Cl_\infty} \chi(x) d\varphi_f(x) = 0 \iff \int_{Cl_\infty} \sigma \circ \chi(x) d\varphi_f(x) = 0. \tag{8.2.11}$$

Recall the decomposition $Cl_\infty = \Gamma \times \Delta$. Let $\mathbb{F}_q$ be the finite subfield of $\mathbb{F}$ generated by all $l|\Delta|$th roots of unity over the field $\mathbb{F}_{p^r}$ of rationality of f and $\widehat{\lambda}$. For any finite extension $\mathbb{F}'/\mathbb{F}_q$, we consider the trace map:

$$\mathrm{Tr}_{\mathbb{F}'/\mathbb{F}_q}(\xi) = \sum_{\sigma \in \mathrm{Gal}(\mathbb{F}'/\mathbb{F}_q)} \sigma(\xi)$$

for $\xi \in \mathbb{F}'$. If $\chi : Cl_n \to \mathbb{F}^\times$ is a character, $d := [\mathrm{Im}(\chi) : \mathrm{Im}(\chi) \cap {\mathbb{F}_q}^\times]$ is not divisible by p (as $|\mathbb{F}_{p^m}^\times| = p^m - 1 \not\equiv 0 \mod p$). Thus, $d \in \mathbb{F}^\times$, and we find

$$\int_{Cl_\infty} \mathrm{Tr}_{\mathbb{F}_q(\chi)/\mathbb{F}_q} \circ \chi(y^{-1}x) d\varphi_f(x) = \frac{d}{a^n} \sum_{\mathfrak{a} \in Cl_n : \chi^{-1}(\mathfrak{a}y^{-1}) \in \mathbb{F}_q} \chi^{-1}(y^{-1}\mathfrak{a}) f([\mathfrak{a}]), \tag{8.2.12}$$

because, by Exercise 1.15, for an lth-power root of unity $\zeta \in \mu_{l^n} - \mu_l$,

$$\mathrm{Tr}_{\mathbb{F}_q(\mu_{l^n})/\mathbb{F}_q}(\zeta^s) = \begin{cases} l^{n-m}\zeta^s & \text{if } \zeta^s \in \mathbb{F}_q \text{ and } \mathbb{F}_q \cap \mu_{l^\infty}(\mathbb{F}) = \mu_{l^m}(\mathbb{F}) \\ 0 & \text{otherwise.} \end{cases}$$

Suppose $\int_{Cl_\infty} \chi(x) d\varphi_f(x) = 0$ for an infinite set $\mathcal{X}$ of characters χ. For sufficiently large m, we always find a character $\chi \in \mathcal{X}$ such that $\mathrm{Ker}(\chi) \subset \Gamma^{l^m}$. Then, writing $\mathrm{Ker}(\chi) = \Gamma^{l^n}$ for $n \geq m$, we have the vanishing from (8.2.11)

$$\int_{Cl_\infty} \sigma \circ \chi d\varphi_f = 0 \text{ for all } \sigma \in \mathrm{Gal}(\mathbb{F}/\mathbb{F}_q).$$

Combining this with (8.2.12), with a variable change $y^{-1}\mathfrak{a} \mapsto \mathfrak{a}$, we find $\sum_{\mathfrak{a} \in \chi^{-1}(\mathbb{F}_q^\times)} \chi^{-1}(\mathfrak{a}) f(y[\mathfrak{a}])) = 0$ for all $y \in \Gamma_n$, where Γ_n is the image of Γ in Cl_n.

8.2.4 Proof of Theorem 8.25

We write $\mathbb{F}_p[f]$ for the minimal field of definition of $f \in V(\mathbb{F})$ [i.e., the field generated by $a(\xi, f) \in \mathbb{F}$ for all $0 \leq \xi \in \mathbb{Z}$]. Similarly, $\mathbb{F}_p[\lambda]$ (resp., $\mathbb{F}_p[\nu]$) is the subfield of $\mathbb{F}$ generated by the values $\widehat{\lambda}([\mathfrak{a}]) \mod \mathcal{P}$ (resp., $\nu([\mathfrak{a}])$) for all $[\mathfrak{a}] \in C^{alg}$. Define $\mathbb{F}_p[f, \lambda, \nu]$ by the composite of these fields and $\mathbb{F}_p[\mu_l]$. Note that $\mathbb{F}_p[f, \lambda, \nu]$ is a finite extension of $\mathbb{F}_p$ as f is mod-p reduction of some classical modular form of some weight ≥ 2. Define $1 \leq r = r(\nu) \in \mathbb{Z}$ by $|\mu_{l^\infty}(\mathbb{F}_p[f, \lambda, \nu])| = l^r$. For general $h \in V(\mathbb{F})$, we write the q-expansion at the infinity cusp as $\sum_{0 \leq \xi} a(\xi, h) q^\xi$.

By definition, the projection $\{[\mathfrak{Q}]_\Gamma\}_{\mathfrak{Q} \in \mathcal{Q}}$ of $[\mathfrak{Q}]$ in Γ are all distinct in Cl_∞/Cl^{alg}. By Lemma 8.24, we need to prove that the integral $\int_\Gamma \chi d\varphi_f^\nu$ vanishes only for finitely many characters χ of Γ. Suppose toward a contradiction that the integral vanishes for characters χ in an infinite set $\mathcal{X}$.

Let $\Gamma(n) = \Gamma^{l^{n-r}}/\Gamma^{l^n}$ for $r = r(\nu)$. By applying (8.2.12) to a character in $\mathcal{X}$ with $\mathrm{Ker}(\chi) = \Gamma^{l^n}$, we find

$$\sum_{\mathfrak{Q} \in \mathcal{Q}} \nu(\mathfrak{Q})^{-1} \sum_{\mathfrak{a} \in y\chi^{-1}(\mu_{l^r})} \chi^{-1}(\mathfrak{a}) f_\nu([\mathfrak{a}\mathfrak{Q}^{-1}][\mathfrak{Q}]_\Gamma) = 0. \tag{8.2.13}$$

Fix $\mathfrak{Q} \in \mathcal{Q}$. By Lemma 8.19, $\{x(\mathfrak{a})|[\mathfrak{a}] \in y\chi^{-1}(\mu_{l^r})\}$ is given by $\alpha(\frac{u}{l_\mathfrak{l}^r})(x(\mathfrak{a}_0))$ for any member $\mathfrak{a}_0 \in y\chi^{-1}(\mu_{l^r})$, where

$$\alpha(t) = \left(\begin{smallmatrix} 1 & t \\ 0 & 1 \end{smallmatrix}\right). \tag{8.2.14}$$

Actually, $\mathfrak{a} \mapsto u \mod \mathfrak{l}^r$ gives a bijection of $y\chi^{-1}(\mu_{l^r})$ onto $O/\mathfrak{l}^r$. We write the element $\mathfrak{a}$ corresponding to u as $\alpha(\frac{u}{l_\mathfrak{l}^r})\mathfrak{a}_0$. This shows, choosing a primitive l^rth root of unity $\zeta = \exp(2\pi i/l^r)$ and $\mathfrak{a}_y \in y\chi^{-1}(\mu_{l^r})$ so that $\chi^{-1}(\alpha(\frac{u}{l_\mathfrak{l}^r})\mathfrak{a}_y) = \zeta^{uv}$ for an integer $0 < v < l^r$ prime to l (independent of y), the inner sum of (8.2.13) is equal to, up to a nonzero factor,

$$\sum_{u \mod \mathfrak{l}^r} \zeta^{uv}(f_\nu|\alpha(\frac{u}{l_\mathfrak{l}^r}))([\mathfrak{a}_y\mathfrak{Q}^{-1}][\mathfrak{Q}]_\Gamma).$$

The choice of v depends on χ. Since $\mathcal{X}$ is infinite, we can choose an infinite subset $\mathcal{X}'$ of $\mathcal{X}$ for which v is independent of the element in $\mathcal{X}'$. Then write n_j for the integers given by $\Gamma^{l^{n_j}} = \mathrm{Ker}(\chi)$ for $\chi \in \mathcal{X}'$ (in increasing order), and define Ξ to be the set of points $x(\mathfrak{a})$ for $\mathfrak{a} \in Cl_{n_j}$ with $[\mathfrak{a}\mathfrak{O}_{n_1}] = [\mathfrak{O}_{n_1}]$ in Cl_{n_1}. Define also $h_\mathfrak{Q} = \sum_{u \mod \mathfrak{l}^r} \zeta^{uv} f_\nu|\alpha(\frac{u}{l_\mathfrak{l}^r})$, because (8.2.13) is now the sum:

$$\sum_{\mathfrak{Q} \in \mathcal{Q}} \nu(\mathfrak{Q})^{-1} h_\mathfrak{Q}|[\mathfrak{q}]([\mathfrak{a}][\mathfrak{Q}]_\Gamma) = 0,$$

where $\mathfrak{q} = \mathfrak{Q} \cap F$. If necessary, as we remarked already, we reselect the representative set $\mathcal{Q}$ so that $[\mathfrak{Q}]_\Gamma \in \mathrm{Ker}(Cl_\infty \to Cl_{n_1})$. This is possible because $\{[\mathfrak{Q}_f] \in \Gamma|\mathfrak{Q} \sim \mathfrak{A}\}$ for all split primes is dense by Chebotarev density, where $\mathfrak{Q} \sim \mathfrak{A}$ means the class of $\mathfrak{Q}$ is equal to the class of $\mathfrak{A}$ in $Cl_\infty/\Gamma\Delta^{alg}$. Take $\mathfrak{Q}, \mathfrak{Q}'$ in $\mathcal{Q}$. Then by Lemma 8.23(4), $[\mathfrak{Q}]_\Gamma/[\mathfrak{Q}']_\Gamma \in Cl^{alg} \Leftrightarrow \mathfrak{Q} = \mathfrak{Q}'$. Thus, we may apply Corollary 8.30 in the following section to the following set of functions: $\{[\mathfrak{a}] \mapsto h_\mathfrak{Q}|[\mathfrak{q}]([\mathfrak{a}][\mathfrak{Q}]_\Gamma)\}$. By the corollary, if $h_\mathfrak{Q}|[\mathfrak{q}] \neq 0$ for one $\mathfrak{Q}$, the above sum is nonzero as a function of $[\mathfrak{a}]$; hence, this implies that $h_\mathfrak{Q}|[\mathfrak{q}] = 0$. By the q-expansion principle, we conclude $h_\mathfrak{Q} = 0$ [as $h|[\mathfrak{q}](q) = h(q^\varpi)$ for the positive generator ϖ of $\mathfrak{q}$].

Writing $f_\nu = \sum_{0 \le \xi \in \mathbb{Z}} a(\xi, f_\nu)q^\xi$, we have $f_\nu|\left(\begin{smallmatrix} 1 & \frac{u}{l^r} \\ 0 & 1 \end{smallmatrix}\right) = \sum_{0 \le \xi \in \mathbb{Z}} a(\xi, f_\nu)\zeta^u q^\xi$ for $\zeta = \exp(\frac{2\pi i}{l^r})$. Then the q-expansion coefficient $a(\xi, h_\mathfrak{Q})$ of $h_\mathfrak{Q}$ is given by $a(\xi, f_\nu)$ if $\xi \equiv -v \mod \mathfrak{l}^r$ and vanishes otherwise. This is a contradiction against assumption (H).

Exercise 8.26 *Give a detailed computation showing that $a(\xi, h_\mathfrak{Q}) = a(\xi, f_\nu)$ if $\xi \equiv -v \mod \mathfrak{l}^r$ and vanishes otherwise.*

8.2.5 Linear Independence

Fix a positive integer $n_1 > 0$. We create a complete representative set R_n for $\mathrm{Ker}(Cl_n \to Cl_{n_1})$ by $\alpha(\frac{u}{l^n})(x(\mathfrak{O}_{n_1}))$ [for $\alpha(t)$ as in (8.2.14)] by choosing suitable integers $0 < u < l^n$. Choose an infinite sequence $\underline{n} := 0 < n_1 <$

$n_2 < \cdots < n_m < \cdots$ of positive integers. Take a geometrically irreducible component $V_{/\mathbb{F}} \subset Sh^{(p)}_{/\mathbb{F}}$ containing $x(\mathfrak{O}_n)$, where $Sh^{(p)}_{/\mathbb{F}} = Sh^{(p)} \times_{\mathcal{W}} \mathbb{F}$. Since V is affine, we can write $V = \mathrm{Spec}(O_V)$ for $O_V = H^0(V, \mathcal{O}_V)$. Sometimes we just write $O = O_V$ if confusion is unlikely. Define

$$\Xi_{\underline{n}} = \Xi = \bigcup_{j=1}^{\infty} \{x(\mathfrak{a}) \in V | \mathfrak{a} \in R_{n_j}\} \subset V.$$

Since $SL_2(\mathbb{A}^{(p\infty)})$ keeps any geometrically irreducible component (see Theorem 7.18), $x(\mathfrak{a})$ as above always resides in one component V.

Let $\mathcal{F} = \mathcal{F}_\Xi$ denote the $\mathbb{F}$-algebra of functions $\phi : \Xi \to \mathbf{P}^1(\mathbb{F}) = \mathbb{F} \sqcup \{\infty\}$ with $|\phi^{-1}(0)| < \infty$ and $|\phi^{-1}(\infty)| < \infty$. The profinite class group $C = C_{n_1} := \mathrm{Ker}(Cl_\infty \to Cl_{n_1})$ acts on $\mathcal{F}$ by translation: $f(x) \mapsto f(xy)$ $(y \in Cl_\infty)$. In particular, $\alpha \in \mathfrak{O}_{(p)}$ with trivial $[(\alpha)] \in Cl_{n_1}$ acts on Ξ and such α is p-adically dense in $T_M(\mathbb{Z}_p)$. For $f \in \mathbb{F}(V)^\times$, for each $x = x(\mathfrak{a}) \in \Xi$, expanding f into a Laurent series $f(t) = \sum_n a_n t^n \in \mathbb{F}[[t]][t^{-1}]$ with leading nonzero term $a_m t^m$ $(m \in \mathbb{Z})$, we may define

$$f(x(\mathfrak{a})) = \begin{cases} \infty & \text{if } m < 0, \\ a_0 & \text{if } m = 0, \\ 0 & \text{if } m > 0. \end{cases}$$

By the Zariski density of Ξ in $V = V^{(p)}_{/\mathbb{F}}$, we can embed into $\mathcal{F}$ the function field $\mathbb{F}(V)$ of V.

Exercise 8.27 *Why is Ξ Zariski dense in V? Why does density imply the injectivity of $\mathbb{F}(V)$ into $\mathcal{F}$?*

Proposition 8.28 *Take a finite set $\Delta = \{\gamma_1, \ldots, \gamma_m\} \subset C_{n_1}$ injecting into Cl_∞/Cl^{alg}. Then the subset $\tilde{\Xi} := \{(x(\delta(\mathfrak{a}))_{\delta \in \Delta} | x(\mathfrak{a}) \in \Xi\}$ is Zariski dense in the product $V^\Delta_{/\mathbb{F}}$ of Δ copies of $V_{/\mathbb{F}}$. This implies that the fields $\gamma_1(\mathbb{F}(V)), \ldots, \gamma_m(\mathbb{F}(V))$ are linearly disjoint over $\mathbb{F}$ in $\mathcal{F}_\Xi$, where $\gamma(\mathbb{F}(V))$ is the image of $\mathbb{F}(V) \subset \mathcal{F}$ under the action of $\gamma \in C_{n_1}$. In other words, we have injectivity of the map $\gamma_1 \otimes \cdots \otimes \gamma_m : \overbrace{O_V \otimes_{\mathbb{F}} O_V \otimes_{\mathbb{F}} \cdots \otimes_{\mathbb{F}} O_V}^{m} \to \mathcal{F}$ sending $f_1 \otimes \cdots \otimes f_m$ to an element in $\mathcal{F}$ given by $x(\mathfrak{a}) \mapsto \prod_j f_j(x(\gamma_j \mathfrak{a}))$.*

Proof. We write $x_m(\mathfrak{a})$ for $(x(\gamma_j \mathfrak{a}))_j \in V^m$. Let X be the Zariski closure of $\tilde{\Xi} := \{x_m(\mathfrak{a}) \in V^m_{/\mathbb{F}} | x(\mathfrak{a}) \in \Xi\}$ in V^m. The last two assertions follow from the density of the above set: $X = V^m$. Thus, we only prove $X = V^m$. We have a homomorphism $\pi : \mathfrak{O}^\times_{(p\mathfrak{l})} \to Cl_\infty$ by sending α to its class $[\alpha]$. Let $T = \pi^{-1}(C_{n_1})$. We have $\alpha \in T \Leftrightarrow (\alpha) \cap \mathfrak{O}_{n_1,\mathfrak{l}}$ is a principal $\mathfrak{O}_{n_1}$-ideal (necessarily generated by one of $\pm\alpha$), where $(\alpha) = \alpha\mathfrak{O}$. Note that $[\alpha] = \varprojlim_n [(\alpha) \cap \mathfrak{O}_{n,\mathfrak{l}}]_n$ for the class $[(\alpha) \cap \mathfrak{O}_{n,\mathfrak{l}}]_n$ in Cl_n. We only need to deal with $Y_1(l)$ as we assumed that the level is just l, but we take care of a slightly more general case of level N whose $\mathfrak{l}$-part is a factor of l^{n_1}. As long as the $\mathfrak{l}$-part of N is a factor of l^{n_1}, writing ρ for $\rho_{\mathfrak{O}}$, we have the equality in $Y_1(N)$:

$$(E(\mathfrak{a}), \alpha\overline{\eta}^{(p)}(\mathfrak{a})) = (E(\mathfrak{a}), \overline{\eta}^{(p)}(\mathfrak{a}) \circ \rho^{(p)}(\alpha)) \\ = (E((\alpha) \cap \mathfrak{O}_n)^{-1}\mathfrak{a}), \overline{\eta}^{(p)}((\alpha) \cap \mathfrak{O}_n)^{-1}\mathfrak{a})) \quad (8.2.15)$$

for proper $\mathfrak{O}_n$-ideals $\mathfrak{a}$. We have $((\alpha) \cap \mathfrak{O}_n)^{-1}$ in the right end, because of Remark 7.12 [i.e., on the lattice, the action of $g \in G(\mathbb{A}^{(\infty)})$ is given by $\widehat{L} \mapsto g^{-1}\widehat{L}$]. A tricky point is that for a principal ideal (α) of $\mathfrak{O}_{n_1}$, $(\alpha) \cap \mathfrak{O}_{n,\mathfrak{l}}$ is no longer the principal $\mathfrak{O}_n$-ideal if $n \gg n_1$. By (8.2.15), the diagonal automorphism $\tau(\rho^{(p)}(\alpha))^{\Delta} \in \mathrm{Aut}(V^{\Delta}_{/\mathbb{F}})$ leaves $\tilde{\Xi}$ stable as multiplication by $\delta \in \Delta$ commutes with multiplication by $[\alpha] \in C_{n_1}$. Thus, the closure X is stable under T. Since X is a union of finitely many irreducible components, the group T permutes them; so, replacing T by a subgroup of finite index, we may assume that T fixes all of the irreducible components. Replacing X by an irreducible component, we may assume that X is irreducible of positive dimension. Since $\dim X > 0$, we have that $\tilde{\Xi}_X = X \cap \tilde{\Xi}$ has infinitely many densely populated distinct points (and stable under T). Thus, we could restart with $\Xi_X = \{x \in \Xi | (\delta(x))_{\delta\in\Delta} \in \tilde{\Xi}_X\}$. Again, moving by the action of Cl_∞, we may assume that X contain $x(\mathfrak{O}_n)$. Note that $\mathfrak{O}_{n,p} = \mathfrak{O}_p$ as $l \neq p$. Thus, we have the canonical coordinate t around $x(\mathfrak{O})$ (by choosing a level-p^∞ structure $\eta_p^\circ \times \eta_p^{\mathrm{et}}$). Then the action of $\tau(\rho^{(p)}(\alpha))$ is given by $t \mapsto t^{\alpha^{1-c}}$ (see Proposition 7.44). In particular, the action is p-adically continuous on the formal completion $\widehat{V}^{\Delta}$ along $(\delta(x(\mathfrak{O}_n))_{\delta\in\Delta} \in V^{\Delta}$. Thus, the p-adic closure $\widehat{T}$ of $\{\tau(\rho^{(p)}(\alpha))^{\Delta} | \alpha \in T\}$ in $\mathrm{Aut}(\widehat{V}^{\Delta})$ leaves stable X. Identifying $\widehat{V}$ with $\widehat{\mathbb{G}}_m$ by t, the action of $\widehat{T}$ contains the diagonal action of an open subgroup U of $\mathrm{Aut}_{GSCH}(\widehat{\mathbb{G}}_m) = \mathbb{Z}_p^\times$. Then if $\dim X < \dim V^{\Delta}$, by Theorem 3.44 and Corollary 11.11 (or by the theorem in Sect. 3.5.3), $X \subset V \times V \times \cdots \times V \times \Delta_{\alpha,\beta}$ after permutation. Thus, we have two distinct elements $\delta, \gamma \in \Delta$ such that $\delta\gamma^{-1} = \tau(\rho^{(p)}(\alpha\beta^{-1}))$ in $\mathrm{Aut}(V_{/\mathbb{F}})$. By Theorem 7.20, this means $\delta\gamma^{-1} = \alpha\beta^{-1}$ modulo $\mathbb{Q}^\times$. Since (n) for $n \in \mathbb{Q}^\times$ is trivial in Cl_∞, we have $\delta\gamma^{-1} = [(\alpha\beta^{-1})] \in Cl^{alg}$, which is a contradiction.

Exercise 8.29 *Explain why density implies the injectivity of* $\gamma_1 \otimes \cdots \otimes \gamma_m$.

The linear independence applied to the global sections of a modular line bundle (regarded as sitting inside the function field) yields the following result:

Corollary 8.30 *Let the notation and the assumption be as in Proposition 8.28. Let* $\underline{\omega}^k$ *be a modular line bundle over the Igusa tower* $Ig_{/\mathbb{F}}$ *over* $V_{/\mathbb{F}}$. *Then for a finite set* $\Delta \subset Cl_\infty$ *injecting into* Cl_∞/Cl^{alg} *and a set* $\{s_\delta \in H^0(I, \underline{\omega}^k)\}_{\delta\in\Delta}$ *of nonconstant global sections* s_δ *of* $\underline{\omega}^k$ *finite at* Ξ, *the functions* $s_\delta \circ \delta$ $(\delta \in \Delta)$ *are linearly independent in* $\mathcal{F}_\Xi$.

Choosing one nonzero section s (different from constant multiples of any of the s_δ's) and replacing s_δ by s_δ/s, which is a modular function, we can bring the situation in the case of modular functions, which is taken care of by the above theorem.

8.2.6 $\mathfrak{l}$-Adic Eisenstein Measure Modulo p

We apply Theorem 8.25 to the Eisenstein series $\mathbb{E}(\lambda)$ in (8.1.17) for the Hecke character λ fixed in Sect. 8.1.2. Choosing a generator π of $\mathfrak{m}_W$, the exact sequence $\underline{\omega}^k_{/W} \xrightarrow{l} \underline{\omega}^k_{/W} \twoheadrightarrow \underline{\omega}^k_{/\mathbb{F}}$ induces a reduction map: $H^0(\mathfrak{M}, \underline{\omega}^k_{/W}) \to H^0(\mathfrak{M}, \underline{\omega}^k_{/\mathbb{F}})$. We write $E_k(\psi^\circ) \mod \mathfrak{m}_W$ for the image of the Eisenstein series $E_k(\psi^\circ)$. Then we put

$$f = (d^\kappa(E_k(\psi^\circ)) \mod \mathfrak{m}_W) \in V_{\Gamma_0(l)/\mathbb{F}}.$$

By definition, the q-expansion of f is the reduction modulo $\mathfrak{m}_W$ of the q-expansion of $\mathbb{E}(\lambda)$ of characteristic 0.

We fix a character $\nu : \Delta \to \mathbb{F}^\times$ as in the previous section and write $\varphi = \varphi_f$ and $\varphi^\nu = \varphi^\nu_f$. By (8.1.38) combined with Lemma 8.24, we have, for a character $\chi : CL_\infty \to \mathbb{F}^\times$ with $\chi|_\Delta = \nu$,

$$\int_\Gamma \chi d\varphi^\nu = \int_{Cl_\infty} \chi d\varphi = C\frac{\pi^\kappa\Gamma(k+\kappa)L^{(\mathfrak{l})}(0,\chi^{-1}\lambda)}{\Omega_\infty^{k+2\kappa}} \mod \mathfrak{m}_W, \qquad (8.2.16)$$

where C is a nonzero constant given by the class modulo $\mathfrak{m}_W$ of

$$\frac{(-1)^k(\mathfrak{O}^\times : \mathbb{Z}^\times)}{\mathrm{Im}(\delta)^\kappa\sqrt{D}}.$$

The nonvanishing of C modulo $\mathfrak{m}_W$ follows from the unramifiedness of p in $M/\mathbb{Q}$. We now prove the following a slightly more specific version of Theorem 8.17, assuming $\mathfrak{C} = 1$:

Theorem 8.31 *Let p be an odd prime splitting in $M/\mathbb{Q}$. Let λ be a Hecke character of M of conductor $\mathfrak{C}$ and of infinity type $k + \kappa(1-c)$. Suppose* (ct), (sp), *and* (pl) *in* Sect. 8.1.2. *Fix a character $\nu : \Delta \to \overline{\mathbb{Q}}^\times$. Then for the Néron period Ω_∞ in (8.1.2), we have $\frac{\pi^\kappa\Gamma_\Sigma(k+\kappa)L^{(\mathfrak{l})}(0,\nu^{-1}\chi^{-1}\lambda)}{\Omega_\infty^{k+2\kappa}} \in \mathcal{W}$ for all nontrivial characters $\chi : Cl_\infty \to \mu_{l^\infty}(\overline{\mathbb{Q}})$ factoring through Γ. Moreover, except for finitely many characters χ in $\mathrm{Hom}_{GP}(\Gamma, \mu_{l^\infty})$, we have*

$$\frac{\pi^\kappa\Gamma(k+\kappa)L^{(\mathfrak{l})}(0,\nu^{-1}\chi^{-1}\lambda)}{\Omega_\infty^{k+2\kappa}} \not\equiv 0 \mod \mathfrak{m}_{\mathcal{W}}.$$

Proof. By Theorem 8.25, we need to verify the condition (H) for $f := \mathbb{E}(\lambda)$. We recall $f_\nu \in V_{\Gamma_0(l)/W}$ in (8.2.6). Thus, we need to prove that for a given congruence class $u \in (\mathbb{Z}/\mathfrak{l}^r)^\times$

$$a(\xi, f_\nu) \not\equiv 0 \mod \mathfrak{m}_W \text{ for at least one } \xi \in u. \tag{8.2.17}$$

Since $a(\xi, d^\kappa h) = \xi^\kappa a(\xi, h)$ (8.1.3), (8.2.17) is achieved if

$$a(\xi, f'_\nu) \not\equiv 0 \mod \mathfrak{m}_W \text{ for at least one } \xi \in u \text{ prime to } p \tag{8.2.18}$$

holds for

$$f' = (E_k(\psi^\circ) - N(\mathfrak{l})E_k(\psi^\circ)|\langle\mathfrak{l}\rangle|[\mathfrak{l}]),$$

because $\mathfrak{l} \nmid p$. Up to nonzero constant, $\psi^\circ(a,b)$ in (8.1.12) is equal to $\phi(a,b) = \lambda_{\mathfrak{F}}(a)\lambda_{\mathfrak{F}_c}^{-1}(b)$ for $(a,b) \in (\mathbb{Z}/\mathfrak{f})^\times$, so by our simplifying assumption that ϕ is a constant function $\mathbf{1}$ with value 1. Thus, we are going to prove

$$a(\xi, f''_\nu) \not\equiv 0 \mod \mathfrak{m}_W \text{ for at least one } \xi \in u \text{ prime to } p, \tag{8.2.19}$$

where $f'' = E_k(\mathbf{1}) - l \cdot E_k(\mathbf{1})|\langle\mathfrak{l}\rangle|[\mathfrak{l}]$. Recall (8.2.6):

$$f''_\nu = \sum_{\mathfrak{r}\in\mathcal{R}} \widehat{\lambda}\nu^{-1}(\mathfrak{R}) f''|[\mathfrak{r}]. \tag{8.2.20}$$

As computed in Sect. 7.1.9, the degeneracy operator $[\mathfrak{r}]$ on q-expansion is basically a variable change $f(q) \mapsto f(q^r)$ up to a constant for the positive generator r of $\mathfrak{r}$. Thus, as long as $\xi \in u$ is prime to lD for the discriminant D of M (note here that $r|D$), $a(\xi, f''_\nu) = a(\xi, E_k(\mathbf{1}))$. Since u is a class in the multiplicative group $(\mathbb{Z}/\mathfrak{l}^r)^\times$, we can choose ξ so that ξ is a prime outside lD (by Dirichlet's theorem asserting infinity of primes in an arithmetic progression). We have an explicit formula $a(\xi, E_k(\mathbf{1})) = 1 + \xi^{k-1}$ (see (1.3.1)). Then it is clear that we can find a prime $\xi \in u$ such that $a(\xi, E_k(\mathbf{1})) \not\equiv 0 \mod p$ (unless $p = 2$). This finishes the proof.

Exercise 8.32 *Give a detailed proof showing* $1+\xi^{k-1} \not\equiv 0 \mod p$ *for infinity of primes* ξ *in a given class* $u \in (\mathbb{Z}/\mathfrak{l}^r)^\times$.

Remark 8.33 If we start with a CM field M in place of an imaginary quadratic field, a similar theorem holds, but there exist well-specified exceptional cases where the canonical integral part of $L(0, \chi^{-1}\lambda)$ vanishes modulo $\mathfrak{m}_W$ for all χ. See details in [H04b] and [H07a] (M1–3).

9

p-Adic Hecke L-Functions and Their μ-Invariants

We first recall the construction of p-adic Hecke L-functions. We follow Katz [K2] and [HT2], which covers general CM fields also. In the case of imaginary quadratic fields, there are different and earlier works making the L-function (see [IEC] and [Y]). There is another work covering possibly non-CM extensions of imaginary quadratic fields (see [CzS]). We state the result in a general form covering an arbitrary prime-to-p conductor, but we give a detailed proof only when the prime-to-p conductor is equal to 1 (so the measure is defined on the ray class group of conductor p^∞). After making the measure, we prove the vanishing of its μ-invariant along the line explained in Sect. 3.5. Again, there are other methods getting the vanishing (see [Gi2], [Scn], and [Fi2]), but at this moment, our path is the only way to prove the vanishing for a general Katz measure of CM fields (see [H10a], [H11b], [Hs2], and [BuH]), although Finis' method in [Fi2] is probably also applicable to CM fields.

9.1 Eisenstein and Katz Measure

We first construct a p-adic measure interpolating Eisenstein series and Hecke L-values (expanding the idea introduced in Sect. 1.3.7 in a slightly different style).

9.1.1 q-Expansion of Eisenstein Series

We generalize the construction in Sect. 8.1.3 done for a finite set $X \times Y$ to a p-adic space $X \times Y$. Let $\Phi : \{\mathbb{Z}_p \times (\mathbb{Z}/\mathfrak{f}')\} \times \{\mathbb{Z}_p \times (\mathbb{Z}/\mathfrak{f}'')\} \to \mathbb{C}$ be a locally constant function such that

$$\Phi(\varepsilon^{-1}x, \varepsilon y) = \varepsilon^k \Phi(x, y) \tag{9.1.1}$$

for all $\varepsilon \in \mathbb{Z}^\times$, where k is a positive integer and $\mathfrak{f}'$ and $\mathfrak{f}''$ are integral ideals prime to p. We put $\mathfrak{f} = \mathfrak{f}' \cap \mathfrak{f}''$. We regard Φ as a function on $X \times Y$ with $X = Y = \mathbb{Z}_p \times (\mathbb{Z}/\mathfrak{f})$ via the natural projection

H. Hida, *Elliptic Curves and Arithmetic Invariants*, Springer Monographs in Mathematics, DOI 10.1007/978-1-4614-6657-4_9,

$$\{\mathbb{Z}_p \times (\mathbb{Z}/\mathfrak{f})\} \times \{\mathbb{Z}_p \times (\mathbb{Z}/\mathfrak{f})\} \twoheadrightarrow \{\mathbb{Z}_p \times (\mathbb{Z}/\mathfrak{f}')\} \times \{\mathbb{Z}_p \times (\mathbb{Z}/\mathfrak{f}'')\}.$$

Recall $N(\mathfrak{f}) = |\mathbb{Z}/\mathfrak{f}| = [\mathbb{Z} : \mathfrak{f}]$, which is the positive generator of the ideal $\mathfrak{f}$. We put $X_\alpha = (\mathbb{Z}/p^\alpha\mathbb{Z}) \times (\mathbb{Z}/\mathfrak{f})$ and define the *partial Fourier transform*

$$P\Phi : \{(\mathbb{Q}_p/\mathbb{Z}_p) \times (\mathfrak{f}^{-1}/\mathbb{Z})\} \times Y = \left\{\bigcup_\alpha p^{-\alpha}\mathfrak{f}^{-1}/\mathbb{Z}\right\} \times Y \to \mathbb{C}$$

of Φ, taking α so that Φ factors through $X_\alpha \times Y$, by

$$P\Phi(x,y) = \begin{cases} p^{-\alpha}N(\mathfrak{f})^{-1}\sum_{a\in X_\alpha}\Phi(a,y)\mathbf{e}_{\mathbb{Q}}(ax) & \text{if } x \in p^{-\alpha}\mathfrak{f}^{-1}/\mathbb{Z}, \\ 0 & \text{if } x \notin p^{-\alpha}\mathfrak{f}^{-1}/\mathbb{Z}, \end{cases} \tag{9.1.2}$$

where $\mathbf{e}_{\mathbb{Q}}$ is the standard additive character of $\mathbb{A}/\mathbb{Q}$ (as in Exercise 8.6) restricted to the local component $\mathbb{Q}_{p\mathfrak{f}}$ at $p\mathfrak{f}$.

Exercise 9.1 *Show that the definition of $P\Phi$ as above does not depend on the choice of α.*

We construct an Eisenstein series $E_k(z;\Phi)$ for a positive integer k and Φ as above as a function of test objects $(L, \phi_{p\mathfrak{f}})$ [for a level structure $\phi_{p\mathfrak{f}} = (i, i'_{\mathfrak{f}})$] we studied in the previous subsection. Here, $i : \mathbb{Q}_p/\mathbb{Z}_p \times (\mathfrak{f}^{-1}/\mathbb{Z}) \hookrightarrow p^{-\infty}L/L \times \mathfrak{f}^{-1}L/L$ is the level-$p^\infty\mathfrak{f}^2$ structure. The $\mathfrak{f}$-part $i_{\mathfrak{f}}$ of i induces, via the duality with respect to a pairing on $\mathfrak{f}^{-1}L/L \times L/\mathfrak{f}L$ induced by $\langle\cdot,\cdot\rangle_L : L \times L \to \mathbb{Z}$ as in (8.1.6), the dual map $i'_{\mathfrak{f}} : L/\mathfrak{f}L \twoheadrightarrow \mathbb{Z}/\mathfrak{f}$, and hence having $i_{\mathfrak{f}}$ is equivalent to having a pair $\phi_{p\mathfrak{f}} = (i, i'_{\mathfrak{f}})$, which is literally of level $\mathfrak{f}^2$ (not just of level $\mathfrak{f}$). We define an $\mathbb{Z}_{\mathfrak{f}p}$-submodule $PV(L) \subset L \otimes_{\mathbb{Z}} \mathbb{Q}_{\mathfrak{f}p}$ specified by the following conditions:

(pv1) $PV(L) \supset L \otimes_{\mathbb{Z}} \mathbb{Z}_{\mathfrak{f}p}$;
(pv2) $PV(L)/(L \otimes_{\mathbb{Z}} \mathbb{Z}_{\mathfrak{f}p}) = \mathrm{Im}(i)$.

By definition, we may consider

$$i^{-1} : PV(L) \twoheadrightarrow PV(L)/(L \otimes_{\mathbb{Z}} \mathbb{Z}_{\mathfrak{f}p}) \cong \mathbb{Q}_p/\mathbb{Z}_p \times \mathfrak{f}^{-1}/\mathbb{Z}.$$

By Pontryagin duality under $\langle x,y\rangle_L$ as in (8.1.6), the dual map i' of i gives rise to

$$i' : PV(L) \twoheadrightarrow \mathbb{Z}_p \times (\mathbb{Z}/\mathfrak{f}).$$

We regard $P\Phi$ as a function on $p^{-\infty}\mathfrak{f}^{-1}L \cap PV(L) = \left(\bigcup_\alpha p^{-\alpha}\mathfrak{f}^{-1}L\right) \cap PV(L)$ by

$$P\Phi(w) = \begin{cases} P\Phi(i^{-1}(w), i'(w)) & \text{if } (w \mod L) \in \mathrm{Im}(i), \\ 0 & \text{otherwise.} \end{cases} \tag{9.1.3}$$

Writing $\underline{L} = (L, \phi_{p\mathfrak{f}})$ for simplicity, we define, similarly to (8.1.10), the value $E_k(\underline{L};\Phi)$ by

$$E_k(\underline{L};\Phi) = (-1)^k \Gamma(k+s) \sum_{w\in p^{-\infty}\mathfrak{f}^{-2}L/\mathbb{Z}^\times}{}' \frac{P\Phi(w)}{w^k|w|^{2s}}\bigg|_{s=0}.$$

We now restate Lemma 8.8:

Lemma 9.2 *Suppose $k \geq 2$ or $\Phi(a,0) = 0$ for all a. We have*

$$E_k(\Phi)(q) = 2^{-1}L(1-k;\Phi) + \sum_{0<n\in\mathbb{Z}} \sum_{\substack{(a,b)\in(\mathbb{Z}\times\mathbb{Z})/\mathbb{Z}^\times \\ ab=n}} \Phi(a,b)\frac{a^k}{|a|}q^n,$$

where $L(s,\Phi)$ is the partial L-function given by the Dirichlet series:

$$\sum_{n\in\mathbb{Z}/\{\pm1\}^\times}{}' \Phi(n,0)\left(\frac{N(n)}{|N(n)|}\right)^k |N(n)|^{-s}.$$

Proof. For simplicity, assume $\mathfrak{f} = 1$ (see [HT2] for the general cases). Let $\underline{L}_z := (L_z, i_z)$ for $L_z = (2\pi i)(\mathbb{Z}+\mathbb{Z}z)$ with $z \in \mathfrak{H}$ and $i_z : p^{-\alpha}\mathbb{Z}/\mathbb{Z} \to \mathbb{Q}\cdot L_z/L_z$ given by $i_z(a \mod \mathbb{Z}) = (2\pi i)a \mod L_z$. Then we have

$$E_k(z;\Phi) := E_k(\underline{L}_z;\Phi) = c(k,s) \sum_{(a,b)\in(p^{-\alpha}\mathbb{Z}\times\mathbb{Z})/\mathbb{Z}^\times}{}' \frac{P\Phi(a,b)}{(a+bz)^k|a+bz|^{2s}}\bigg|_{s=0},$$

where $c(k,s) := \frac{(-1)^k\Gamma(k+s)}{(2\pi i)^k(2\pi)^s}$. Put $c = c(k,0)$. We can separate the sum for $(a,0)$ and (a,b) with $b \neq 0$:

$$\begin{aligned} c^{-1}E_k(z;\Phi) = &\sum_{0<a\in p^{-\alpha}\mathbb{Z}} \frac{P\Phi(a,0)}{a^k|a|^{2s}}\bigg|_{s=0} \\ &+ \sum_{0<b\in\mathbb{Z}} \sum_{a_0\in(p^{-\alpha}\mathbb{Z}/\mathbb{Z})} P\Phi(a_0,b)\sum_{a\in\mathbb{Z}} \frac{1}{(a+a_0+bz)^k|a+a_0+bz|^{2s}}\bigg|_{s=0}. \end{aligned}$$

Put

$$S_k(z;s) = \sum_{a\in\mathbb{Z}} \frac{1}{(a+z)^k|a+z|^{2s}}.$$

Thus, we get, for a, b, a_0 running in the range as above,

$$E_k(z;\Phi) = c\left\{\sum_a \frac{P\Phi(a,0)}{a^k|a|^{2s}}\bigg|_{s=0} + \sum_b\sum_{a_0} P\Phi(a_0,b)S_k(a_0+bz;0)\right\}.$$

By Poisson's summation formula (cf. [EDM] Theorem 2.3 or [LFE] Sect. 8.4), we get

$$S_k(z;s) = \sum_{a\in\mathbb{Z}} C_k(a,z;s)$$

for $C_k(x,z;s) = \int_{-\infty}^{\infty} \frac{\mathbf{e}_{\mathbb{Q}}(-xt)}{(t+z)^k|t+z|^{2s}}dt$. We verify from the definition that

$$C_k(x, a_0 + bz; s) = \mathbf{e}_{\mathbb{Q}}(xa_0)C_k(x, bz; s) \tag{9.1.4}$$

and by computation of the integral (see [EDM] Sect. 2.4), we have

$$C_k(x, bz; 0) = \begin{cases} -2\pi i & \text{if } k = 1 \text{ and } x = 0, \\ \frac{(-1)^k(2\pi i)^k}{\Gamma(k)} \frac{x}{|x|} x^{k-1}\mathbf{e}_{\mathbb{Q}}(xbz) & \text{if } xb > 0, \\ 0 & \text{otherwise.} \end{cases} \tag{9.1.5}$$

By this, we reprove (1.3.2) and (1.3.3), and we confirm

$$c^{-1}E_k(z;\Phi) = \sum_{0<a\in p^{-\alpha}\mathbb{Z}} \left.\frac{P\Phi(a,0)}{a^k|a|^{2s}}\right|_{s=0}$$
$$+ \sum_{0<b\in\mathbb{Z}} \sum_{a\in\mathbb{Z}} \{ \sum_{a_0\in p^{-\alpha}\mathbb{Z}/\mathbb{Z}} P\Phi(a_0,b)\mathbf{e}_{\mathbb{Q}}(-aa_0)\}C_k(a,bz;0).$$

By the Fourier inversion formula, we have

$$\sum_{a_0\in p^{-\alpha}\mathbb{Z}/\mathbb{Z}} P\Phi(a_0,b)\mathbf{e}_{\mathbb{Q}}(-aa_0) = \Phi(a,b).$$

This finishes the proof.

Exercise 9.3 *Prove the Fourier inversion formula used in the above proof.*

9.1.2 Eisenstein Measure

Let $\mathfrak{s}|\mathfrak{f}$ be two nonzero ideals of $\mathbb{Z}$ prime to p. We recall the definition of the Eisenstein measure having values in the space of p-adic modular forms $V_{\Gamma_1^1(\mathfrak{f}\mathfrak{s})/W}$. Recall that $\mathbb{F}$ is an algebraic closure of $\mathbb{F}_p$ and that $W(\mathbb{F})$ is the ring of Witt vectors with coefficients in $\mathbb{F}$. We regard $W(\mathbb{F})$ as a subring of the p-adic completion $\mathbb{C}_p$ of $\overline{\mathbb{Q}}_p$. Let W be a discrete valuation ring finite flat over $W(\mathbb{F})$ inside $\mathbb{C}_p$. We write, for any fractional ideal $\mathfrak{a}$, its prime decomposition as $\prod_{\mathfrak{q}} \mathfrak{q}^{\varepsilon_{\mathfrak{q}}}$; so $\varepsilon_{\mathfrak{q}}$ is an integer with $\varepsilon_{\mathfrak{q}} = 0$ for primes $\mathfrak{q}$ that do not show up in the prime decomposition. We denote $\varepsilon(\mathfrak{a}) = \{\varepsilon_{\mathfrak{q}}\}_{\mathfrak{q}}$ for this set of exponents. We abbreviate the product $\prod_{\mathfrak{q}} \mathfrak{q}^{\varepsilon_{\mathfrak{q}}}$ as $\mathfrak{q}^{\varepsilon(\mathfrak{a})}$, which is equal to $\mathfrak{a}$.

We consider the space $\widetilde{\mathfrak{H}} = (\mathbb{Z}_p \times (\mathbb{Z}/\mathfrak{f})) \times (\mathbb{Z}_p \times (\mathbb{Z}/\mathfrak{s}))$ and write the variable on $\widetilde{\mathfrak{H}}$ as $(x, a; y, b)$ for $x, y \in \mathbb{Z}_p$ and $a \in \mathbb{Z}/\mathfrak{f}$ and $b \in \mathbb{Z}/\mathfrak{s}$. We regard $\widetilde{\mathfrak{H}}$ as a ring; then $\widetilde{\mathfrak{H}}^\times$ is the group of invertible elements in $\widetilde{\mathfrak{H}}$. We embed $\mathbb{Z}^\times$ into $\widetilde{\mathfrak{H}}$ diagonally. We also write $\widetilde{T} = (\mathbb{Z}_p^\times \times (\mathbb{Z}/\mathfrak{f})) \times (\mathbb{Z}_p^\times \times (\mathbb{Z}/\mathfrak{s}))$. We let $\epsilon \in \mathbb{Z}^\times$ act on $\widetilde{\mathfrak{H}}$ by $\epsilon(x, a; y, b) = (\epsilon x, \epsilon a; \epsilon y, \epsilon b)$ and define $T = \widetilde{T}/\mathbb{Z}^\times$ and $T^\times = \widetilde{\mathfrak{H}}^\times/\mathbb{Z}^\times$. These are the profinite compact spaces carrying the Eisenstein measure.

For each continuous function $\Phi(x,a;y,b)$ on $\widetilde{T}$, we consider the following *partial Fourier transform*:

$$\Phi^\circ(x,a;y,b) = \sum_{u\in\mathbb{Z}/\mathfrak{f}} \Phi(x^{-1},u;y,b)\mathbf{e}_{\mathbb{Q}}(-ua\varpi^{-\varepsilon(\mathfrak{f})}), \tag{9.1.6}$$

where we have chosen, for each prime $\mathfrak{q}=(q)$ of $\mathbb{Q}$, the prime element $\varpi_{\mathfrak{q}}=q$ in $\mathbb{Q}_{\mathfrak{q}}^\times \subset \mathbb{A}^\times$ and put $\varpi^e = \prod_{\mathfrak{q}} \varpi_{\mathfrak{q}}^{e_{\mathfrak{q}}}$ in $\mathbb{A}^\times$ for each exponent $e=\{e_{\mathfrak{q}}\in\mathbb{Z}\}_{\mathfrak{q}}$ with $e_{\mathfrak{q}}=0$ almost everywhere (that is, except for finitely many primes). The map $\Phi\mapsto\Phi^\circ$ is a linear operator acting on the space $\mathcal{C}(\widetilde{T},W)$ of all continuous functions on $\widetilde{T}$ with values in W (and invertible by the Fourier inversion formula). If Φ factors through T, then Φ° satisfies the following property:

$$\Phi^\circ(\epsilon x,\epsilon a;\epsilon^{-1}y,\epsilon^{-1}b) = \Phi^\circ(x,a;y,b) \text{ for all } \epsilon\in\mathbb{Z}^\times.$$

This is the property required to define an Eisenstein series (for even weight k) in Sect. 8.1.3. Then there exists a unique measure $\mathbf{E}:\mathcal{C}(T,W)\to V_{\Gamma_1^1(\mathfrak{f}\mathfrak{s})/W}$ with the following two properties:

(1) If Φ has values in $\overline{\mathbb{Q}}$ equipped with discrete topology, then for each positive integer $k>0$,
$$\mathbf{E}(N^{-k}\Phi) = E_k(\Phi^\circ),$$
where $N:\widetilde{T}\to\mathbb{Z}_p^\times$ is given by $N(x,a;y,b)=x$. Note here that $N^{-k}\Phi^\circ$ (for any positive integer k) factors through $T \iff \Phi^\circ$ andsatisfies invariance (9.1.1) under $\mathbb{Z}^\times$ required for the definition of the Eisenstein series.

(2) The q-expansion of $\mathbf{E}(\Phi)$ at the infinity cusp is given by
$$\sum_{0<n\in\mathbb{Z}} q^n \sum_{(a,b)\in(\mathbb{Z}\times\mathbb{Z})/\mathbb{Z}^\times, ab=n} \Phi^\circ(a;b)|a|^{-1}.$$

The verification of the existence and uniqueness of the measure satisfying (1–2) is a consequence of the q-expansion principle (see Corollary 7.38) and the q-expansion of the classical Eisenstein series given in Lemma 9.2. We write $E(\Phi)$ for $E_0(\Phi)$.

9.1.3 Katz Measure

Interpreting p-adic modular forms f as geometric modular forms, we can evaluate them at any test object $(E,\phi_{p\mathfrak{f}})_{/W}$ (as in Sect. 6.4.4) with $\phi_{p\mathfrak{f}} = (\eta^\circ, i_{\mathfrak{f}}, i'_{\mathfrak{s}})$ with $\eta^\circ:\mu_{p^\infty}\hookrightarrow E$ in Definition 6.57 defined over W. This gives rise to a linear form:
$$Ev: V_{\Gamma_1^1(\mathfrak{f}\mathfrak{s})/W}\to W$$
given by $Ev(f)=f(E,\phi_{p\mathfrak{f}})$. Thus, we can think of the evaluation $Ev\circ\mathbf{E}$, which is a bounded measure on $\mathcal{C}(T,W)$ with values in W.

We choose a specific test object. Recall the imaginary quadratic field M with integer ring $\mathfrak{O}$. Let $\mathfrak{A}$ be an $\mathfrak{O}$-ideal prime to p. We take the elliptic curve

$$x(\mathfrak{A}) = (E(\mathfrak{A}), \phi_{p\mathfrak{f}}(\mathfrak{A}), \omega(\mathfrak{A})) \text{ with } \phi_{p\mathfrak{f}}(\mathfrak{A}) = (i_p(\mathfrak{A}), i_\mathfrak{f}, i'_\mathfrak{s})$$

sitting over $x \in Sh^{(p)}_{\Gamma_1^1(\mathfrak{fs})}$. Here $i_p(\mathfrak{A})$ is a slight modification of $\eta^\circ(\mathfrak{A})$ in Definition 6.57 (see below for the modification). Thus, $E = E(\mathfrak{A})$ has complex multiplication by an imaginary quadratic field $M = M_x$. As we have already seen, $x(\mathfrak{A})$ is defined over W. By splitting $(p) = \mathfrak{p}\overline{\mathfrak{p}}$, we see that the special fiber $E(\mathfrak{A})_{/\mathbb{F}}$ at p is an ordinary elliptic curve. Since the residue field $\mathbb{F}$ of W is algebraically closed, $E(\mathfrak{A})[p^\infty] \cong (\mu_{p^\infty} \otimes \mathbb{Z}) \times (\mathbb{Q}_p/\mathbb{Z}_p)$. Thus, $E(\mathfrak{A})_{/\mathbb{F}} := E(\mathfrak{A}) \times_W \mathbb{F}$ has level-p^∞ structure η_0° defined over $\mathbb{F}$. By deformation theory of elliptic curves (see Sect. 6.3), $E(\mathfrak{A})_{/W}$ sits at the origin $1 \in \widehat{\mathbb{G}}_m$ of the deformation space $\widehat{\mathbb{G}}_m$ of $E(\mathfrak{A})_{/\mathbb{F}}$. Thus, we can uniquely lift η_0° to a level-p^∞ structure $\eta^\circ(\mathfrak{A})$; so we may assume that $\phi_{p\mathfrak{f}}(\mathfrak{A})$ contains a level-p^∞ structure defined over W.

We recall the construction of the Katz measure interpolating the L-values of arithmetic Hecke characters of conductor dividing $\mathfrak{C}p^\infty$, where $\mathfrak{C}$ is an integral ideal of M prime to p. We write $\mathfrak{O}$ for the integer ring of M. We decompose $\mathfrak{C} = \mathfrak{F}\mathfrak{F}_c\mathfrak{I}$. Here $\mathfrak{I}$ consists of inert or ramified primes over $\mathbb{Q}$, $\mathfrak{F}\mathfrak{F}_c$ consists of split primes over $\mathbb{Q}$, and

$$\mathfrak{F} + \mathfrak{F}_c = \mathfrak{F} + \mathfrak{F}^c = \mathfrak{F}_c + \mathfrak{F}_c^c = \mathfrak{O} \text{ and } \mathfrak{F}_c \supset \mathfrak{F}^c.$$

We put $\mathfrak{f} = \mathfrak{F}\mathfrak{I} \cap \mathbb{Q}$, $\mathfrak{s} = \mathfrak{F}_c \cap \mathbb{Q}$ and $\mathfrak{i} = \mathfrak{I} \cap \mathbb{Q}$. We have $\mathfrak{O}_p = \mathfrak{O} \otimes_\mathbb{Z} \mathbb{Z}_p = \mathfrak{O}_\mathfrak{p} \times \mathfrak{O}_{\overline{\mathfrak{p}}}$ [so that $E(\mathfrak{O})[p^\infty]^\circ = E(\mathfrak{O})[\mathfrak{p}^\infty]$ over W]. We recall the following four conditions:

1. The lattice $\mathfrak{A}$ is a fractional ideal of M prime to $\mathfrak{C}p$; so we write $E = E(\mathfrak{A})$.
2. Choose $\delta \in M$ so that $\delta^c = -\delta$, and the imaginary part $\mathrm{Im}(i_\infty(\delta))$ is positive and the alternating form $\langle u, v\rangle = (u^c v - uv^c)/2\delta$ induces $\mathfrak{O} \wedge \mathfrak{O} \cong \mathbb{Z}$ (in short, $2\delta = \sqrt{-D}$ for the discriminant $D > 0$ of M).
3. The inclusion $\mathbb{Q} \hookrightarrow M$ induces a canonical isomorphism $\mathbb{Z}_p \cong \mathfrak{O}_\mathfrak{p}$, which in turn induces $\eta^\circ(\mathfrak{A}) : \mu_{p^\infty} \overset{(*)}{\cong} \mathbb{Q}_p/\mathbb{Z}_p \cong M_\mathfrak{p}/\mathfrak{A}\mathfrak{O}_\mathfrak{p} \subset \mathbb{C}/i_\infty(\mathfrak{A}) = E(\mathfrak{A})[\mathfrak{p}^\infty]$. We put $i_p(\mathfrak{A})(x) = \eta^\circ(\mathfrak{A})(2\delta x)$. This $i_p(\mathfrak{A})$ is the p-part of the level structure $x(\mathfrak{A})$. Here $(*)$ is given by $\mathbf{e}(\frac{m}{p^n}) \leftrightarrow \frac{m}{p^n}$.
4. The prime-to-p part $\phi^{(p)} = \phi_\mathfrak{f}$ sends $(x \mod \mathbb{Z})$ to $(x \mod \mathfrak{A})$, which induces $(\mathfrak{f}^2)^{-1}/\mathbb{Z} \hookrightarrow (\mathfrak{F}\mathfrak{i})^{-2}\mathfrak{A}/\mathfrak{A}$.

Let $Cl_M(\mathfrak{i})$ be the ideal class group of M modulo $\mathfrak{i}$ and let $Cl^-(\mathfrak{i})$ be the quotient of $Cl_M(\mathfrak{i})$ by the image of $(\mathbb{Z}/\mathfrak{i})^\times$. Identifying $\mathfrak{O}_\mathfrak{p}$ (resp., $\mathfrak{O}_{\overline{\mathfrak{p}}}$) with the first (resp., last) component of $\mathbb{Z}_p$ of $\widetilde{\mathfrak{O}}$ and embedding $\mathbb{Z}/\mathfrak{f}$ into $\mathfrak{O}/\mathfrak{F}\mathfrak{I}$ resp., identifying $\mathbb{Z}/\mathfrak{s}$ with $\mathfrak{O}/\mathfrak{F}_c$)] through the inclusion $\mathbb{Z} \hookrightarrow \mathfrak{O}$, we bring $T^\times$ into $Z = Cl_M(\mathfrak{C}p^\infty) = \varprojlim_n Cl_M(\mathfrak{C}p^n)$. Then we have the exact sequence

$$T^\times \xrightarrow{\iota} Z \to Cl^-(\mathfrak{i}) \to 1,$$

and the kernel of ι is a finite group. We write $[\mathfrak{A}]$ for the image of the class of an ideal $\mathfrak{A}$ prime to $\mathfrak{C}p$ in Z. For $\alpha \in \mathfrak{O}$, we have $[(\alpha)] = \alpha^{-1}$, where the right-hand side is the image of the inclusion $\mathfrak{O}_{\mathfrak{C}p}^{\times} \to Z$. Choosing a complete representative set $\{\mathfrak{A}\}$ for $Cl^{-}(\mathrm{i})$, we have a decomposition

$$Z = \bigsqcup_{\mathfrak{A}} \mathrm{Im}(\iota)[\mathfrak{A}]^{-1}.$$

For each function $\Phi \in \mathcal{C}(Z, W)$, we define $\Phi_{\mathfrak{A}} \in \mathcal{C}(T, W)$ in the following way: $\Phi_{\mathfrak{A}}(t) = \Phi(t[\mathfrak{A}]^{-1})$ for $t \in T^{\times}$ and extend it by 0 outside $T^{\times}$. Then define

$$\int_Z \Phi d\mu = \sum_{\mathfrak{A}} \int_T \Phi_{\mathfrak{A}} d\mathbf{E}(x(\mathfrak{A})). \tag{9.1.7}$$

The left-hand side is actually independent of the choice of representatives $\mathfrak{A}$ (which can be verified directly, but it will be clear by our evaluation of the measure). We write $E(\Phi)$ for $\mathbf{E}(\Phi)$ for functions $\Phi \in \mathcal{C}(T^{\times}, W)$.

9.1.4 Evaluation of Katz Measure

In [K2] Chap. V and [HT1] Sect. 4, $\int_Z \widehat{\lambda} d\mu$ is computed for the p-adic avatar $\widehat{\lambda}$ of an arithmetic Hecke character λ of conductor that is a factor of $\mathfrak{C}p$. When $\mathfrak{I} = 1$, the computation is basically done in Sect. 8.1.7, though in the p-adic case, as we will see soon, we get an extra modification factor different from the Euler p-factor.

Let us quote the exact formula from [HT1]: We fix a finite idele d of M such that the ideal $d\widehat{\mathfrak{O}} \cap M$ corresponding to d is the different $\mathfrak{d}$ of $M/\mathbb{Q}$. Let $\lambda : M_{\mathbb{A}}^{\times}/M^{\times} \to \mathbb{C}^{\times}$ be a Hecke character such that

$$\lambda(x_\infty) = x_\infty^{k+\kappa(1-c)},$$

where k and κ are integers. Then λ has values in $\overline{\mathbb{Q}}$ on the finite part $M_{\mathbb{A}(\infty)}^{\times}$ of $M_{\mathbb{A}}^{\times}$. Moreover, the map $\widehat{\lambda} : M_{\mathbb{A}(\infty)}^{\times}/M^{\times} \to \overline{\mathbb{Q}}_p^{\times}$ defined by $\widehat{\lambda}(x) = \lambda(x)x_p^{k+\kappa(1-c)}$ is a well-defined continuous character, which is called the *p-adic avatar* of λ. We can associate λ with its dual λ^* given by $\lambda^*(x) = \lambda(x^c)^{-1}|x|_{\mathbb{A}}$. Then the p-adic avatar of λ^* is given by $\widehat{\lambda}^*(x) = \widehat{\lambda}(x^c)^{-1}\mathcal{N}(x)^{-1}$ for the cyclotomic character $\mathcal{N}$. We define a *local Gauss sum* of λ at prime ideals $\mathfrak{Q}$ dividing the conductor of λ by

$$G(d, \lambda_{\mathfrak{Q}}) = \lambda(\varpi_{\mathfrak{Q}}^{-e}) \sum_{u \in (\mathfrak{O}_{\mathfrak{Q}}/\mathfrak{Q}^e)^{\times}} \lambda_{\mathfrak{Q}}(u)\mathbf{e}_M(\varpi_{\mathfrak{Q}}^{-e} d_{\mathfrak{Q}}^{-1} u), \tag{9.1.8}$$

where $\varpi_{\mathfrak{Q}}$ is a prime element of the $\mathfrak{Q}$-adic completion $M_{\mathfrak{Q}}$, $\mathfrak{O}_{\mathfrak{Q}}$ is the $\mathfrak{Q}$-adic integer ring of $M_{\mathfrak{Q}}$, $\lambda_{\mathfrak{Q}}$ is the restriction of λ to $M_{\mathfrak{Q}}^{\times}$, $e = e(\mathfrak{Q})$ is the exponent of $\mathfrak{Q}$ in the conductor of λ, and $\mathbf{e}_M : M_{\mathbb{A}}/M \to \mathbb{C}^{\times}$ is the standard additive character normalized as $\mathbf{e}_M(x_\infty) = \exp(2\pi\sqrt{-1}\mathrm{Tr}(x_\infty))$. Outside the conductor of λ, we simply put $G(d, \lambda_{\mathfrak{Q}}) = 1$. We have defined in (3.5.1) and (8.1.2) the complex and the p-adic period $\Omega_\infty \in \mathbb{C}^\times \cong \mathbb{C}^\times$ and $\Omega_p \in W^\times$ following [K2]. By definition, these numbers are defined uniquely only modulo $\overline{\mathbb{Q}}^\times$, but the ratio "$\Omega_\infty/\Omega_p$" is uniquely determined. As before, we take 2δ or $(2\delta)^c$ as $d_{\mathfrak{Q}}$. Then we define root numbers:

$$W_p(\lambda) = p^{-e(\mathfrak{p})} G(2\delta; \lambda_{\mathfrak{p}}),$$
$$W'(\lambda) = \prod_{\mathfrak{L}|\mathfrak{F}} G((2\delta)^c, \lambda_{\mathfrak{L}}^{-1}) \prod_{\mathfrak{L}|\mathfrak{F}_c} G(2\delta, \lambda_{\mathfrak{L}^c}^{-1}) \prod_{\mathfrak{l}|\mathfrak{I}} G((2\delta)^c, \lambda_{\mathfrak{l}}^{-1}), \quad (9.1.9)$$

where we decomposed $\mathfrak{C} = \mathfrak{F}\mathfrak{F}_c\mathfrak{I}$ so that $\mathfrak{F}\mathfrak{F}_c$ consists of split primes over $\mathbb{Q}$, $\mathfrak{I}$ consists of inert or ramified primes over $\mathbb{Q}$, $\mathfrak{F} + \mathfrak{F}_c = \mathfrak{O}$, and $\mathfrak{F}_c^c \supset \mathfrak{F}$. We computed in [HT1] (following [K2], where the case $\mathfrak{C} = 1$ is treated) that the measure μ in (9.1.7) satisfies the following:

Theorem 9.4 *Let the notation and the assumption be as above. Then we have*

$$\frac{\int_{Z(\mathfrak{C})} \widehat{\lambda} d\mu}{\Omega_p^{k+2\kappa}} = (\mathfrak{O}^\times : \mathbb{Z}^\times) W_p(\lambda) \frac{(-1)^k \pi^\kappa \Gamma(k+\kappa)}{\mathrm{Im}(\delta)^\kappa \Omega_\infty^{k+2\kappa}} \prod_{\mathfrak{L}|\mathfrak{C}} (1 - \lambda(\mathfrak{L}))$$
$$\times \{(1 - \lambda(\mathfrak{p}^c))(1 - \lambda^*(\mathfrak{p}^c))\} L(0, \lambda) \quad (9.1.10)$$

for all Hecke characters λ modulo $\mathfrak{C}p^\infty$ such that (i) its conductor is divisible by all prime factors of $\mathfrak{F}$, (ii) the infinity type of λ is $k + \kappa(1-c)$ for integers k and κ satisfying either $k > 0$ and $\kappa \geq 0$ or $k \leq 1$ and $\kappa \geq 1 - k$.

An element $a \in M$ is considered to be an element of $\mathbb{C}$ via i_∞. The L-function in (9.1.10) is always the primitive one associated with the primitive Hecke character. We also tacitly agree to put $\lambda(\mathfrak{Q}) = 0$ if $\mathfrak{Q}$ divides the conductor of λ.

Remark 9.5 Since Hecke L-functions satisfy functional equation as in [LFE] Chap. 8, we can deduce a p-adic version from the Archimedean functional equation. Denoting the measure μ for $Z(\mathfrak{C}^c)$ by μ_c, we have the following functional equation:

$$\int_{Z(\mathfrak{C})} \widehat{\lambda} d\mu = W'(\lambda) \int_{Z(\mathfrak{C}^c)} \widehat{\lambda}^* d\mu_c$$

as long as the conductor of λ is divisible by all prime factors of $\mathfrak{F}$. We refer the reader to [K2] and [HT1] for the proof of the functional equation.

After stating a preliminary result as an exercise, we begin the proof of the theorem. Let $\psi : \mathfrak{O}_{\mathfrak{p}}^\times \times \mathfrak{O}_{\overline{\mathfrak{p}}}^\times \to \overline{\mathbb{Q}}_p^\times$ be a finite order character of conductor $\mathfrak{p}^{e(\mathfrak{p})}\overline{\mathfrak{p}}^{e(\overline{\mathfrak{p}})}$; hence, $\psi(x, y) = \psi_{\mathfrak{p}}(x)\psi_{\overline{\mathfrak{p}}}(y)$ for $x \in \mathfrak{O}_{\mathfrak{p}}^\times$ and $y \in \mathfrak{O}_{\overline{\mathfrak{p}}}^\times$.

Exercise 9.6 *Let the notation be as above. Prove*

$$P\psi^\circ(x,y)=\begin{cases}G(\psi_{\mathfrak{p}}^{-1})\psi_{\mathfrak{p}}(p^{e(\mathfrak{p})}x)\psi_{\overline{\mathfrak{p}}}(y) & \text{if } e(\mathfrak{p})>0 \text{ and } p^{e(\mathfrak{p})}x\in\mathbb{Z}_p^\times,\\ (\mathbf{1}_{\mathfrak{p}}(x)-\frac{1}{p}\mathbf{1}_{\mathfrak{p}}(\varpi_{\mathfrak{p}}x))\psi_{\overline{\mathfrak{p}}}(y) & \text{if } e(\mathfrak{p})=0,\end{cases}$$

where $\mathbf{1}_{\mathfrak{p}}$ *is the characteristic function of* $\mathfrak{O}_{\mathfrak{p}}=\mathbb{Z}_p$ *in* $M_{\mathfrak{p}}=\mathbb{Q}_p$ *and* $G(\chi)=p^{-e}\sum_{u\in\mathbb{Z}/p^e}\chi(u)\mathbf{e}(\frac{u}{p^e})$ *for a character* χ *of conductor* p^e *of* $\mathbb{Z}_p^\times$.

Proof of Theorem 9.4. By the functional equation mentioned in Remark 9.5, we may assume $\kappa\geq 0$. For simplicity, we assume that $\mathfrak{C}=1$ following [K2]. We refer to [HT1] for a treatment of the general cases. Regard λ as an idele character $M_{\mathbb{A}}^\times/M^\times\to\mathbb{C}^\times$ by $\lambda(\varpi_{\mathfrak{L}})=\lambda(\mathfrak{L})$ with $\varpi_{\mathfrak{L}}=(1,\dots,1,\overset{\mathfrak{L}}{\varpi},1,\dots,1)\in M_{\mathbb{A}}^\times$ for each prime $\mathfrak{L}$ outside p, where ϖ is a prime element $\varpi\in\mathfrak{O}$ with $\mathfrak{L}\mathfrak{O}_{\mathfrak{L}}=(\varpi)$. Since $\lambda(\alpha)=1$ for $\alpha\in M^\times$ prime to p, we have $1=\lambda(\alpha)=\lambda_{\mathfrak{p}}(\alpha)\lambda_{\overline{\mathfrak{p}}}(\alpha)\lambda((\alpha))\alpha^{k+\kappa(1-c)}$. Thus,

$$\lambda((\alpha))=\alpha^{-k-\kappa(1-c)}\lambda_{\mathfrak{p}}^{-1}(\alpha)\lambda_{\overline{\mathfrak{p}}}^{-1}(\alpha). \tag{9.1.11}$$

This implies on the image of ι that $\widehat{\lambda}(x,y)=\lambda_{\mathfrak{p}}^{-1}(x)\lambda_{\overline{\mathfrak{p}}}^{-1}(y)x^{-k}(x^{-1}y)^\kappa$ for $(x,y)\in\mathfrak{O}_{\mathfrak{p}}^\times\times\mathfrak{O}_{\overline{\mathfrak{p}}}^\times$. By definition, the integral over $\mathrm{Im}(\iota)[\mathfrak{A}]$ is the value of the derivative $d^\kappa E_k(\Phi^\circ)$ of the Eisenstein series $E_k(\Phi^\circ)$ at $x(\mathfrak{A})$ for $\Phi(x,y)=\Phi_{\mathfrak{A}}(x,y)$, with $x\in\mathfrak{O}_{\mathfrak{p}}^\times$ and $y\in\mathfrak{O}_{\overline{\mathfrak{p}}}^\times$ given by

$$\Phi_{\mathfrak{A}}(x,y):=\lambda(\mathfrak{A})^{-1}\lambda_{\mathfrak{p}}^{-1}(x)\lambda_{\overline{\mathfrak{p}}}^{-1}(y).$$

By the above exercise, noting that we have twisted $\eta^\circ(\mathfrak{A})$ slightly to define $i_p(\mathfrak{A})(x):=\eta^\circ(\mathfrak{A})(2\delta x)$ (to be in line with the existing literature on p-adic L-functions), we have, for $x\in\mathfrak{p}^{-\max(e(\mathfrak{p}),1)}\mathfrak{A}$,

$$P\Phi_{\mathfrak{A}}^\circ(x,y)=\begin{cases}W_p(\lambda)\lambda_{\mathfrak{p}}^{-1}(x)\lambda_{\overline{\mathfrak{p}}}^{-1}(y) & \text{if } e(\mathfrak{p})>0 \text{ and } x\mathfrak{p}^{e(\mathfrak{p})}\mathfrak{O}_{\mathfrak{p}}=\mathfrak{O}_{\mathfrak{p}},\\ (\mathbf{1}_{\mathfrak{p}}(x)-\frac{1}{p}\mathbf{1}_{\mathfrak{p}}(\varpi_{\mathfrak{p}}x))\lambda_{\overline{\mathfrak{p}}}^{-1}(y) & \text{if } e(\mathfrak{p})=0,\end{cases}$$

where $\mathfrak{p}^{e(\mathfrak{p})}\overline{\mathfrak{p}}^{e(\overline{\mathfrak{p}})}$ is the conductor of λ. Note that multiplication by 2δ sneaks into $G(2\delta,\lambda_{\mathfrak{p}})$ in the process computing $P\Phi_{\mathfrak{A}}(i_p(\mathfrak{A})^{-1}(w),i_p(\mathfrak{A})'(w))$ by our twist of $\eta^\circ(\mathfrak{A})$. First, suppose that $e:=e(\mathfrak{p})>0$. Then, writing E_k for $E_k(\Phi^\circ)$ and $c_0=\frac{(-1)^k\pi^\kappa\Gamma(k+\kappa)}{\mathrm{Im}(\delta)^\kappa\Omega_\infty^{k+2\kappa}}$, we find, for a prime element $\varpi_{\mathfrak{p}}\in M$ with $\varpi_{\mathfrak{p}}\mathfrak{O}_{\mathfrak{p}}=\mathfrak{p}\mathfrak{O}_{\mathfrak{p}}$,

$$
\begin{aligned}
&(c_0(\mathfrak{O}^\times : \mathbb{Z}^\times)W_p(\lambda))^{-1}\delta_k^\kappa E_k(x(\mathfrak{A}), \omega(\mathfrak{A})) \\
&= \lambda(\mathfrak{A})^{-1} \sum_{\alpha \in (\mathfrak{p}^{-e}\mathfrak{A} - \mathfrak{p}^{-e}p\mathfrak{A})/\mathfrak{O}^\times}' \frac{\lambda_{\mathfrak{p}}^{-1}(\alpha)\lambda_{\overline{\mathfrak{p}}}^{-1}(\alpha)}{\alpha^{k+\kappa(1-c)} N_{M/\mathbb{Q}}(\alpha)^s}\Big|_{s=0} \\
&= \lambda(\mathfrak{A})^{-1}\lambda_p(\varpi_{\mathfrak{p}}^e) \sum_{\varpi_{\mathfrak{p}}^e \alpha \in (\mathfrak{B} \cap \mathfrak{O}_p^\times)/\mathfrak{O}^\times}' \frac{\lambda_{\mathfrak{p}}^{-1}(\varpi_{\mathfrak{p}}^e\alpha)\lambda_{\overline{\mathfrak{p}}}^{-1}(\varpi_{\mathfrak{p}}^e\alpha)}{\alpha^{k+\kappa(1-c)} N_{M/\mathbb{Q}}(\alpha)^s}\Big|_{s=0} \\
&\overset{\varpi_{\mathfrak{p}}^e\alpha \mapsto \alpha}{=} \lambda(\mathfrak{A})^{-1}\lambda_p(\varpi_{\mathfrak{p}}^e) \sum_{\alpha \in (\mathfrak{B} \cap \mathfrak{O}_p^\times)/\mathfrak{O}^\times}' \frac{\lambda_{\mathfrak{p}}^{-1}(\alpha)\lambda_{\overline{\mathfrak{p}}}^{-1}(\alpha)}{(\varpi_{\mathfrak{p}}^{-e}\alpha)^{k+\kappa(1-c)} N_{M/\mathbb{Q}}(\varpi_{\mathfrak{p}}^{-e}\alpha)^s}\Big|_{s=0} \\
&= \lambda(\mathfrak{A})^{-1}\lambda_p(\varpi_{\mathfrak{p}}^e)\varpi_{\mathfrak{p}}^{e(k+\kappa(1-c))} \sum_{\alpha \in (\mathfrak{B} \cap \mathfrak{O}_p^\times)/\mathfrak{O}^\times}' \frac{\lambda_{\mathfrak{p}}^{-1}(\alpha)\lambda_{\overline{\mathfrak{p}}}^{-1}(\alpha)}{\alpha^{k+\kappa(1-c)} N_{M/\mathbb{Q}}(\varpi_{\mathfrak{p}}^{-e}\alpha)^s}\Big|_{s=0} \\
&\overset{\text{Exercise 9.7}}{=} \lambda(\mathfrak{B})^{-1} \sum_{\alpha \in (\mathfrak{B} \cap \mathfrak{O}_p^\times/\mathfrak{O}^\times}' \frac{\lambda((\alpha))}{N_{M/\mathbb{Q}}(\alpha)^s}\Big|_{s=0} = L_{\mathfrak{B}^{-1}}(0, \lambda'),
\end{aligned}
\tag{9.1.12}
$$

where $\mathfrak{B} = \mathfrak{A}\mathfrak{p}^{-e}(\varpi_{\mathfrak{p}}^e)$, which is prime to p, and λ' is equal to λ if $e(\overline{\mathfrak{p}}) > 0$ but is an imprimitive character modulo $\overline{\mathfrak{p}}$ induced by λ if $e(\overline{\mathfrak{p}}) = 0$ [so $\lambda'(\overline{\mathfrak{p}}) = 0$ in the imprimitive case]. Summing up over the ideal classes $\mathfrak{B}$, we get

$$
\frac{\int_{Z(\mathfrak{C})} \widehat{\lambda} d\mu}{\Omega_p^{k+2\kappa}} = c_0(\mathfrak{O}^\times : \mathbb{Z}^\times)W_p(\lambda)(1 - \lambda(\mathfrak{p}^c))L(0, \lambda). \tag{9.1.13}
$$

We now assume that $e(\mathfrak{p}) = 0$. Then in the same manner as above, we get

$$
\begin{aligned}
&(c_0(\mathfrak{O}^\times : \mathbb{Z}^\times)W_p(\lambda))^{-1}\delta_k^\kappa E_k(x(\mathfrak{A}), \omega(\mathfrak{A})) \\
&= \lambda(\mathfrak{A})^{-1} \sum_{\alpha \in (\mathfrak{p}^{-1}\mathfrak{A} \cap \mathfrak{O}_{\overline{\mathfrak{p}}}^\times)/\mathfrak{O}^\times}' \frac{(\mathbf{1}_{\mathfrak{p}}(\alpha) - \frac{1}{p}\mathbf{1}_{\mathfrak{p}}(\varpi_{\mathfrak{p}}\alpha))\lambda_{\overline{\mathfrak{p}}}^{-1}(\alpha)}{\alpha^{k+\kappa(1-c)} N_{M/\mathbb{Q}}(\alpha)^s}\Big|_{s=0} \\
&= \text{First sum} - \text{Second sum}.
\end{aligned}
\tag{9.1.14}
$$

The first sum is

$$
\lambda(\mathfrak{A})^{-1} \sum_{\alpha \in (\mathfrak{A} \cap \mathfrak{O}_{\overline{\mathfrak{p}}}^\times))/\mathfrak{O}^\times}' \frac{\lambda_{\overline{\mathfrak{p}}}^{-1}(\alpha)}{\alpha^{k+\kappa(1-c)} N_{M/\mathbb{Q}}(\alpha)^s}\Big|_{s=0} = L_{\mathfrak{A}^{-1}}(0, \lambda').
$$

The second sum is given by

$$
\begin{aligned}
&(p\lambda(\mathfrak{A}))^{-1} \sum_{\alpha\in(\mathfrak{p}^{-1}\mathfrak{A}\cap\mathfrak{O}_{\overline{\mathfrak{p}}}^{\times}))/\mathfrak{O}^{\times}}{}' \lambda_{\mathfrak{p}}(\alpha)\frac{\lambda_{\mathfrak{p}}^{-1}(\alpha)\lambda_{\overline{\mathfrak{p}}}^{-1}(\alpha)}{\alpha^{k+\kappa(1-c)}N_{M/\mathbb{Q}}(\alpha)^s}\Big|_{s=0}\\
&= \frac{\lambda_p(\varpi_{\mathfrak{p}})\varpi_{\mathfrak{p}}^{k+\kappa(1-c)}}{p\lambda(\mathfrak{A})\lambda_{\mathfrak{p}}(\varpi_{\mathfrak{p}})} \sum_{\alpha\in(\mathfrak{p}^{-1}\mathfrak{A}\cap\mathfrak{O}_{\overline{\mathfrak{p}}}^{\times}))/\mathfrak{O}^{\times}}{}' \lambda_{\mathfrak{p}}(\varpi_{\mathfrak{p}}\alpha)\frac{\lambda_{\mathfrak{p}}^{-1}(\varpi_{\mathfrak{p}}\alpha)\lambda_{\overline{\mathfrak{p}}}^{-1}(\varpi_{\mathfrak{p}}\alpha)}{(\varpi_{\mathfrak{p}}\alpha)^{k+\kappa(1-c)}N_{M/\mathbb{Q}}(\alpha)^s}\Big|_{s=0}\\
&= \frac{\lambda_p(\varpi_{\mathfrak{p}})\varpi_{\mathfrak{p}}^{k+\kappa(1-c)}}{p\lambda(\mathfrak{A})\lambda_{\mathfrak{p}}(\varpi_{\mathfrak{p}})} \sum_{\alpha\in(\mathfrak{B}\cap\mathfrak{O}_{\overline{\mathfrak{p}}}^{\times})/\mathfrak{O}^{\times}}{}' \frac{\lambda_{\overline{\mathfrak{p}}}^{-1}(\alpha)}{\alpha^{k+\kappa(1-c)}N_{M/\mathbb{Q}}(\varpi_{\mathfrak{p}}^{-1}\alpha)^s}\Big|_{s=0}\\
&\overset{\text{Exercise 9.7}}{=} \lambda(\mathfrak{B})^{-1}\lambda^*(\mathfrak{p}^c) \sum_{\alpha\in(\mathfrak{B}\cap\mathfrak{O}_{\overline{\mathfrak{p}}}^{\times})/\mathfrak{O}^{\times}}{}' \frac{\lambda((\alpha))}{N_{M/\mathbb{Q}}(\alpha)^s}\Big|_{s=0} = \lambda^*(\mathfrak{p}^c)L_{\mathfrak{B}^{-1}}(0,\lambda'),
\end{aligned}
$$

where $\mathfrak{B} = \mathfrak{A}\mathfrak{p}^{-1}(\varpi_{\mathfrak{p}})$. Summing up over ideal classes the difference between the first and second sums, we get

$$\frac{\int_{Z(\mathfrak{C})}\widehat{\lambda}d\mu}{\Omega_p^{k+2\kappa}} = c_0(\mathfrak{O}^{\times}:\mathbb{Z}^{\times})W_p(\lambda)(1-\lambda(\mathfrak{p}^c))(1-\lambda^*(\mathfrak{p}^c))L(0,\lambda). \tag{9.1.15}$$

This finishes the proof. □

Exercise 9.7 *In the above computation* (9.1.12), *prove that*

$$\lambda(\mathfrak{A})^{-1}\lambda_p(\varpi_{\mathfrak{p}}^e)\varpi_{\mathfrak{p}}^{e(k+\kappa(1-c))} = \lambda(\mathfrak{B})^{-1}.$$

9.2 Computation of the μ-Invariant

For simplicity, we assume that the branch character ψ_0 has prime-to-p conductor 1 and order prime to p; so ψ_0 has values in $W(\mathbb{F})$ (more general cases are dealt with in [H10a], [H11b], and [Hs2]). Take, as a base ring, a sufficiently large discrete valuation ring $W \subset \mathbb{C}_p$ over $W(\mathbb{F})$. We write ψ for an arithmetic Hecke character with $\widehat{\psi}|_{\Delta} = \psi_0$ having prime-to-p conductor 1. We now prove Theorem 3.37, assuming $F = \mathbb{Q}$ and $p > 2$, and the proof concludes in Sect. 9.2.4.

9.2.1 Splitting the Katz Measure

In a general setting, for a finite order character ϵ of the additive group $H \cong \mathbb{Z}_p^r$ with values in $W^{\times}$ and a measure φ on H, we define $\varphi_\epsilon = \epsilon * \varphi$ by $\int_H \phi d\varphi_\epsilon = \int_H \phi\epsilon d\varphi$. It is clear that $\mu(\varphi) = \mu(\varphi_\epsilon)$. However, if we have a nontrivial projection $\pi : H \to H'$ for $H' \cong \mathbb{Z}_p^s$ $(0 < s < r)$ and if ϵ does not factor through π, we often have $\mu(\pi_*\varphi) \neq \mu(\pi_*(\epsilon * \varphi))$, as the convolution product $\varphi \mapsto \epsilon * \varphi$ does not commute with the push-forward along π. Here the direct image $\pi_*(\varphi)$ is a measure on H' defined by $\int_{H'} \phi d\pi_*(\varphi) = \int_H (\phi\circ\pi)d\varphi$. Since $0 \le \mu(\varphi) \le \mu(\pi_*(\epsilon * \varphi))$ for any ϵ from a many-variable version of Theorem 1.36, we get the following result:

Lemma 9.8 *We have* $\mu(\varphi) = 0$ *if* $\liminf_\epsilon \mu(\pi_*(\epsilon * \varphi)) = 0$, *where* ϵ *runs over all finite order characters of* H.

Actually, we can prove the converse of the lemma, but we do not need it.

Exercise 9.9 *Give a detailed proof of the above lemma.*

We apply the splitting techniques in Sect. 8.2.2 to the Katz measure to compute the q-expansion of the corresponding Eisenstein series. We consider the measure $E_{\mathfrak{A}} : \phi \mapsto \int_T \phi d\mathbf{E}(x(\mathfrak{A}))$ for $x(\mathfrak{A}) = E(\mathfrak{A}), i(\mathfrak{A})$ [on the image of $T^\times$ in $Cl_M(p^\infty)$].

Lemma 9.10 *For* $\alpha \in M$ *prime to* p, *we have*

$$\int_{T^\times} \phi(t) d\mathbf{E}(\rho(\alpha)^{-1}(x(\mathfrak{A}))) = \int_{T^\times} \phi(\alpha t) dE_{\mathfrak{A}} = \int_{T^\times} \phi(t) dE_{\alpha\mathfrak{A}}, \tag{9.2.1}$$

where $\alpha(x; y) = (\alpha x, \alpha a; \alpha^c y, \alpha^c b)$ *for* $t = (x; y)$.

Proof. For $\alpha \in M$ prime to p, multiplication by α: $u \mapsto \alpha u$ induces the matrix $\rho(\alpha)$ and an isomorphism: $E(\mathfrak{A})(\mathbb{C}) = \mathbb{C}/\mathfrak{A} \cong \mathbb{C}/\alpha\mathfrak{A} = E(\alpha\mathfrak{A})(\mathbb{C})$. This multiplication by α sends $i(\mathfrak{A})$ [resp., $\omega(\mathfrak{A})$] to $\alpha \circ i(\alpha\mathfrak{A})$, which sends an element $x \in (\mathbb{Q}_p/\mathbb{Z}^*) \times (\mathfrak{f}^2)^*/O^*$ to $\alpha x \mod \alpha\mathfrak{A}$ [resp., $\alpha\omega(\alpha\mathfrak{A})$]. □

For each function ϕ on $\mathrm{Im}(\iota)[\mathfrak{A}]^{-1}$, we define

$$\phi_{\mathfrak{A}}(x) = \phi(x[\mathfrak{A}]^{-1}). \tag{9.2.2}$$

Now we decompose, for an open subgroup H of $T^\times$ containing $\mathrm{Ker}(\iota)$,

$$\iota(T^\times)[\mathfrak{A}]^{-1} = \bigsqcup_{\mathfrak{B}} \iota(H)[\mathfrak{B}]^{-1} \iff T^\times = \bigsqcup_{\mathfrak{B}} H[\mathfrak{B}^{-1}\mathfrak{A}].$$

Thus, we have

$$\int_{T^\times} \phi_{\mathfrak{A}}(t) dE_{\mathfrak{A}}(t) = \sum_{\mathfrak{B}} \int_{T^\times} \mathbf{1}_{H[\mathfrak{B}^{-1}\mathfrak{A}]} \phi_{\mathfrak{A}}(t) dE_{\mathfrak{A}}(t)$$
$$= \sum_{\mathfrak{B}} \int_{T^\times} \mathbf{1}_H(t[\mathfrak{B}\mathfrak{A}^{-1}]) \phi_{\mathfrak{B}}(t[\mathfrak{B}\mathfrak{A}^{-1}]) dE_{\mathfrak{A}}(t) = \sum_{\mathfrak{B}} \int_T \mathbf{1}_H(t) \phi_{\mathfrak{B}}(t) dE_{\mathfrak{B}}(t),$$

where $\mathbf{1}_H$ is the characteristic function of H. Note here that we have $\mathfrak{B} = \alpha\mathfrak{A}$ for $\alpha \in M^\times$.

Decompose $Z = \Gamma_M \times \Delta = \Gamma^+ \times \Gamma^- \times \Delta$ such that

1. $Z = \Gamma_M \times \Delta$ with torsion-free Γ_M and a finite group Δ.
2. Γ_M and Δ is stable under c.
3. $\Gamma^\pm$ is the $\pm$-eigenspace in Γ_M of the action of c (so $\gamma^c = \gamma^{\pm 1}$ if $\gamma \in \Gamma^\pm$).

Exercise 9.11 *Why can we choose Γ_M and Δ stable under c?*

For each $z \in Z$, we define $\pi^-(z) = [z]^- := z^{1-c}$. Recall the torus $T = T_x = \mathrm{Res}_{M/\mathbb{Q}}\mathbb{G}_m \subset G$ fixing the closed point $x \in Sh^{ord}$ over $(E(\mathfrak{O}), i(\mathfrak{O})) \in V$ and its quotient $\mathcal{T}$ as in (7.2.5) [with the injection $\mathcal{T}(\mathbb{Z}_{(p)}) \hookrightarrow T(\mathbb{Z}_p) = \mathbb{Z}_p^\times$ sending $\alpha \in T_x(\mathbb{Z}_{(p)})$ to $\alpha^{1-c} \in \mathbb{Z}_p^\times$]. We have an exact sequence:

$$1 \to \mathfrak{O}_p^\times/\mathfrak{O}^\times \to Z \to Cl_M \to 1,$$

where Cl_M is the class group of M. Since $\mathbb{Z}^\times$ is a subgroup of $\mathfrak{O}^\times$ of finite index and p is unramified in $M/\mathbb{Q}$, $\pi^-(\mathfrak{O}^\times)$ is a finite group of order prime to p. By this fact, we see that

$$\Gamma^- \cap \pi^-\left(\mathfrak{O}_p^\times/\mathfrak{O}^\times\right) \hookrightarrow \mathfrak{O}_p^\times[-1] \cong \mathbb{Z}_p^\times,$$

where $\mathfrak{O}_p^\times[-1] = \{a \in \mathfrak{O}_p^\times | c(a) = a^{-1}\}$. In particular, identifying $\mathfrak{O}_\mathfrak{p}$ with $\mathbb{Z}_p$, for a principal ideal (α) prime to p, $[(\alpha)]^- = (\alpha^{c-1})_\mathfrak{p} \in T(\mathbb{Z}_p) = \mathfrak{O}_\mathfrak{p}^\times = \mathbb{Z}_p^\times$. Here as an element of $T(\mathbb{Z}_p)$, having $(\alpha^{c-1})_\mathfrak{p}$ [instead of $(\alpha^{1-c})_\mathfrak{p}$] is correct because in $Z = M_\mathbb{A}^\times/M^\times(\widehat{\mathfrak{O}}^{(p)})^\times M_\infty^\times$, the diagonal class of $\beta = \alpha^{1-c}$ is trivial, and so the principal ideal (β) corresponds to the ideal class of $\beta^{(p\infty)}$ with $\beta_p\beta^{(p\infty)} = \beta = 1$ in Z [i.e., $\beta_p^{-1} = \beta^{(p\infty)}$]. Therefore, we have, with $\mathcal{T}(\mathbb{Z}_{(p)}) \subset T(\mathbb{Z}_p) = \mathbb{Z}_p^\times$ by (7.2.5),

$$[\mathfrak{A}]^- \in \mathcal{T}(\mathbb{Z}_{(p)}) \iff [\mathfrak{A}] \in \{[(\alpha)] | \alpha \in \mathfrak{O}_{(p)}^\times\} \cdot \left(\Gamma^+ \times \Delta^+\right) \subset Z, \tag{9.2.3}$$

where $\Delta^+ = H^0(\mathrm{Gal}(M/\mathbb{Q}), \Delta)$ and $\mathfrak{O}_{(p)} \subset M$ is the localization (not the completion) of $\mathfrak{O}$ at p.

The translation $\phi(z) \mapsto \phi(z\zeta)$ by $\zeta \in Z$ gives an action of Z on the space of continuous functions $\mathcal{C}(Z, W)$ on Z with values in W. For each character ψ of Z, regard its restriction $\psi|_\Delta$ to Δ as a character of Z pulling $\psi|_\Delta$ by the projection $Z \twoheadrightarrow \Delta$. We write $\mathcal{C}(Z, \mathcal{O})[\psi|_\Delta]$ for the $\psi|_\Delta$-eigenspace. Then the restriction of continuous functions on Z to Γ_M gives rise to an isomorphism $\mathrm{Res} : \mathcal{C}(Z, W)[\psi|_\Delta] \cong \mathcal{C}(\Gamma_M, W)$ sending $\Phi : Z \to W$ to $\Phi|_{\Gamma_M}$. Indeed, any function $\Phi \in \mathcal{C}(Z, W)[\psi|_\Delta]$ is of the form $\Phi(\gamma, \delta) = \phi(\gamma)\psi(\delta)$ with $\phi \in \mathcal{C}(\Gamma_M, W)$ for $\gamma \in \Gamma_M$ and $\delta \in \Delta$. We write Inf_ψ for Res^{-1}.

For a given measure φ on Z, the ψ-projection $\varphi_\psi \in W[[\Gamma_M]]$ and its "$-$" part $\varphi_\psi^- \in W[[\Gamma^-]]$ are defined by

$$\int_{\Gamma_M} \phi d\varphi_\psi = \int_Z \mathrm{Inf}_\psi(\phi \cdot \psi|_{\Gamma_M}) d\varphi \quad \text{and} \quad \varphi_\psi^- = \pi_*^- \varphi_\psi.$$

By definition, we have $\int_{\Gamma_M} \phi d\varphi_{\psi\epsilon} = \int_{\Gamma_M} \phi\epsilon d\varphi_\psi$ if ϵ factors through Γ_M.

9.2.2 Good Representatives

We would like to choose a representative set D for Δ so that the projection $\pi_\Delta : Z \to \Delta$ induces an isomorphism $D \cong \Delta$ if $p \nmid |Cl_M|$. In general, $D \cong Z/\Gamma'$

for the intersection Γ' of Γ_M and the image of $\mathfrak{O}_p^\times$. We would like to choose D so that our computation of q-expansion (of Eisenstein series) becomes easier.

Let $\mathcal{I}(p)$ be the group of fractional ideals of M prime to p, and define

$$\mathcal{I}(p)^+ = \{\mathfrak{A} \in \mathcal{I}(p) \big| \mathfrak{A}^{1-c} = \alpha^{1-c}\mathfrak{O} \text{ for } \alpha \in M^\times\}.$$

Since $\mathfrak{A}$ is prime to p, α^{1-c} is prime to p. Thus, if a prime factor $\mathfrak{p}$ of p divides the principal ideal (α), its conjugate $\mathfrak{p}^c$ divides (α) with the equal multiplicity. Thus, $\alpha = \beta\gamma$ for $\gamma \in F^\times$ with β prime to p. In other words, $(\beta^{1-c}) = (\alpha^{1-c}) = \mathfrak{A}^{1-c}$, and hence we can write

$$\mathcal{I}(p)^+ = \{\mathfrak{A} \in \mathcal{I}(p) \big| \mathfrak{A}^{1-c} = \alpha_{\mathfrak{A}}^{1-c}\mathfrak{O} \text{ for } \alpha_{\mathfrak{A}} \in M^\times \text{ prime to } p\}. \quad (9.2.4)$$

The quotient of $\mathcal{I}(p)^+$ by principal ideals prime to p is a subgroup of the class group Cl_M of M, which we write as Cl_M^+. We see easily that

$$Cl_M^+ \subset H^0(\mathrm{Gal}(M/F), Cl_M).$$

Exercise 9.12 *Prove that* $Cl_M^+ = H^0(\mathrm{Gal}(M/F), Cl_M)$ *and that* Cl_M^+ *is equal to the ambiguous class group* Cl_M^{amb}.

Thus, if, furthermore, the class number of M is odd, $Cl_M^+ = \{1\}$. We take a complete representative set D^- (resp., D^+) for Cl_M/Cl_M^+ [resp., Cl_M^+ in $\mathcal{I}(\mathfrak{C}p)^+$].

When the class number of M is odd, we choose D^- among primes of M split over $\mathbb{Q}$. If the class number is even, we choose $D^+ \cup D^-$ among primes of M split over $\mathbb{Q}$.

We write Γ' for the intersection of Γ_M with the image of $\mathfrak{O}_p^\times$ in the group $Z = Cl_M(p^\infty)$. Then we put $\mathcal{D}$ for a complete representative set in the localization (not the completion) $\mathfrak{O}_{(p)}^\times$ for $\pi(\mathfrak{O}_p^\times)/\Gamma'$ with the projection $\pi : \mathfrak{O}_p^\times \to Z$. Then we have

$$\int_Z \phi d\varphi = \sum_{\mathfrak{A}\in D^+} \sum_{\alpha\in\mathcal{D}} \sum_{\mathfrak{B}\in D^-} \int_{\Gamma'} \phi_{\mathfrak{A}\alpha\mathfrak{B}}(z) dE_{\mathfrak{A}\alpha\mathfrak{B}}.$$

We make use of degeneracy operator. As seen in Sect. 8.2.2, for each split prime $\mathfrak{q} = \mathfrak{Q}\overline{\mathfrak{Q}}$ or a square-free product $\mathfrak{Q}$ of ramified primes with $\mathfrak{Q}^2 = \mathfrak{q}$, we have a unique level-$\Gamma_0(\mathfrak{q})$ structure on $E(\mathfrak{A})$ given by $C = E(\mathfrak{A})[\mathfrak{Q}]$. Thus, for the degeneration operator $[\mathfrak{q}]$,

$$[\mathfrak{q}](E(\mathfrak{A}), i(\mathfrak{A}), \omega(\mathfrak{A}))_{/\mathcal{W}} \cong (E(\mathfrak{A}\mathfrak{Q}^{-1}), i(\mathfrak{A}\mathfrak{Q}^{-1}), \omega(\mathfrak{A}\mathfrak{Q}^{-1}))_{/\mathcal{W}}. \quad (9.2.5)$$

We can always choose $\mathfrak{Q} \in D^+$ so that $\mathfrak{Q}^c + \mathfrak{Q} = \mathfrak{O}$ and $\mathfrak{O}/\mathfrak{Q} \cong \mathbb{Z}/\mathfrak{q}$ for $\mathfrak{q} = \mathfrak{Q} \cap \mathbb{Q}$. This shows, for the characteristic function $\mathbf{1}_{\Gamma'}$ of Γ',

$$\int_{\Gamma'} \phi_{\mathfrak{Q}^{-1}\mathfrak{B}} dE_{\mathfrak{Q}^{-1}\mathfrak{B}} = (\mathbf{E}(\mathbf{1}_{\Gamma'}\phi_{\mathfrak{Q}^{-1}\mathfrak{B}})|[\mathfrak{q}])\,(A(\mathfrak{B}), \lambda(\mathfrak{B}), i(\mathfrak{B})) \quad (9.2.6)$$

for $\mathfrak{Q}$ and $\mathfrak{q}$ as above. As for the effect of $\alpha \in \mathcal{D}$, we may assume $\alpha \in \mathfrak{O} - \mathbb{Z}$ with $\mathfrak{Q} = (\alpha)$ is a prime split over $\mathbb{Q}$. Then we have, for the characteristic function $\mathbf{1}_{\Gamma'}$ of Γ', if $\alpha \notin \mathbb{Z}$,

$$\int_{\Gamma'} \phi_{\alpha^{-1}\mathfrak{B}} dE_{\alpha^{-1}\mathfrak{B}} = (\mathbf{E}(\mathbf{1}_{\Gamma'}\phi_{\alpha^{-1}\mathfrak{B}})|[\alpha\alpha^c])\,(A(\mathfrak{B}), \lambda(\mathfrak{B}), i(\mathfrak{B})). \tag{9.2.7}$$

For any $\alpha \in M^\times$ prime to p and a function ϕ on $\mathrm{Im}(\iota)[\mathfrak{A}]^{-1}$ with values in $\mathbb{C}_p$, we write $\phi|\alpha(z) = \phi([(\alpha)]z) = \phi(\alpha z)$. From the definition of $\phi_{\mathfrak{A}}$, we have immediately that

$$\phi_{\mathfrak{A}}|\alpha(z) = \phi_{\mathfrak{A}}(z\alpha) = \phi(z[(\alpha)][\mathfrak{A}]^{-1}) = \phi(z[\alpha^{-1}\mathfrak{A}]^{-1}) = \phi_{\alpha^{-1}\mathfrak{A}}(z). \tag{9.2.8}$$

9.2.3 Operators on p-Adic Modular Forms

We interpret the ideal action as pullback of the action of an element in $G(\mathbb{A}^{(\infty)})$ on the Shimura variety. Pick an element $g \in G(\mathbb{A}^{(\infty)})$ with positive $\det(g) \in \mathbb{Q}$. Then g induces an automorphism of each geometrically irreducible component of the Shimura variety (see Theorem 7.18), and hence the functorial action of g on test objects. We write

$$g(E, \eta) = (E, \eta_g = \eta \circ g)$$

for the image of a test object (A, η) under the action of g. Here, writing $T(E) = \varprojlim_N E[N]$ for the Tate module, recall the level structure on E given by $\eta : (\mathbb{A}^{(\infty)})^2 \cong T(E) \otimes_{\mathbb{Q}} \mathbb{A}^{(\infty)}$, where $(\mathbb{A}^{(\infty)})^2$ is made up of column vectors on which $G(\mathbb{A}^{(\infty)})$ acts from the left.

On the lattice side, $\eta \mapsto \eta \circ g$ corresponds to the action $\widehat{L} = \eta^{-1}(T(E)) \mapsto g^{-1}\widehat{L} = (\eta \circ g)^{-1}(T(E))$. Suppose $g = z \in Z(\mathbb{Z}_p)$. The p-part of the level structure η_p is given by $i : \mathbb{Q}_p/\mathbb{Z}_p \cong PV(L)/L_p$ and its dual $i' : PV(L) \to \mathbb{Z}_p$ in (pv1–2). This (i, i') is sent to $(i \circ z, z^{-1}i')$ as i' is the dual of i, or, equivalently, (i^{-1}, i') is sent to $(z^{-1}i^{-1}, z^{-1}i')$. The latter expression indicates that this is the action of the central element z. Then we sum up the terms involving $(P\Phi)(z^{-1}i^{-1}(w), z^{-1}i'(w))$ over $PV(L) \cap \mathbb{Q} \cdot L$ to define the Eisenstein series. We have $(P\Phi)(z^{-1}i^{-1}(w), z^{-1}i'(w)) = P(\Phi|z)(i^{-1}(w), i'(w))$ for $\Phi|z(x, y) = \Phi(zx, z^{-1}y)$. Thus, we get the next result.

Lemma 9.13 *If $z \in Z(\mathbb{Z}_p)$, then $E(\Phi^\circ)(E, \eta \circ z) = E(\Phi^\circ|z^{-1})(E, \eta)$. In particular,*

$$E(\Phi^\circ)|z(q) = \sum_{(a,b)\in\mathbb{Z}\times\mathbb{Z}, ab>0} \frac{\Phi(za, z^{-1}b)}{|a|} q^{ab}.$$

Similarly, for a prime ideal $\mathfrak{l}$ of $\mathbb{Z}$ prime to p with positive generator l, we get the following:

Lemma 9.14 *Let* $l_{\mathfrak{l}} = (1, \ldots, 1, \overset{\mathfrak{l}}{l}, 1, \ldots, 1) \in \mathbb{A}^{(\infty)}$, *and put* $g = \begin{pmatrix} 1 & 0 \\ 0 & l_{\mathfrak{l}}^{-1} \end{pmatrix}$. *Then* $E(\Phi^\circ)(E_g, \eta_g) = E(\Phi^\circ)|[\mathfrak{l}](E, \eta)$. *In particular,*

$$E(\Phi^\circ)|[\mathfrak{l}](q) = \sum_{(a,b) \in \mathfrak{l} \times \mathbb{Z}, ab > 0} \frac{\Phi(a,b)}{|a|} q^{ab}.$$

Proof. The action of the degeneracy operator $[\mathfrak{l}]$ is equal to the action of g is shown in Sect. 7.1.9. It corresponds to the action of $h := J_1 g^{-1} J_1^{-1}$ on the lattice side. Thus, we see that $P\Phi|_{hL} = P(\Phi')$ for $\Phi'(x, y) = \Phi(l_{\mathfrak{l}}^{-1} x, y)$. From this, the result follows, as $l_{\mathfrak{l}}^{-1} x \in \widehat{\mathbb{Z}} \Leftrightarrow x \in \mathfrak{l}\widehat{\mathbb{Z}}$ and the multiplication by $l_{\mathfrak{l}}$ does not affect Φ as $\mathfrak{l}$ is prime to the level.

9.2.4 Linear Independence of Eisenstein Series

We look at $\mathbf{E}(N^{-k}\phi)$ for a positive k [for $N(x; y) = x$ as in Sect. 9.1.2]. Take an arithmetic Hecke character (of conductor p) $\lambda = \lambda_k$ with $\lambda((\alpha)) = \alpha^{-k}$ if $\alpha \equiv 1 \mod p$. Recall $T = (\mathbb{Z}_p^\times) \times (\mathbb{Z}_p^\times)/\mathbb{Z}^\times$. Then the variable on T is written as $(x; y)$ with $x, y \in \mathbb{Z}_p$. Write $\pi_x : T \to \mathbb{Z}_p^\times/\mathbb{Z}^\times$ as the projection given by $\pi_x(x; y) = x \mod \mathbb{Z}^\times$. Let $\epsilon_0 : \mathbb{Z}_p^\times/\mathbb{Z}^\times \to \mu_{p^r}(\mathbb{C}_p)$ be a character and $\epsilon : Z \twoheadrightarrow \Gamma_M \to \mu_{p^\infty}(\mathbb{C}_p)$ be a character extending $\epsilon_0 \circ \pi_x$. Enlarge W if necessary so that ϵ has values in $W^\times$. Define a character $\widehat{\psi}_{k,\epsilon} : Z \to W^\times$ by $\widehat{\psi}_{k,\epsilon}(\zeta\gamma) = \widehat{\lambda}_k(\gamma)\epsilon(\gamma)\psi_0(\zeta)$. Then $\widehat{\psi}_{k,\epsilon}$ is the p-adic avatar of an arithmetic Hecke character $\psi_{k,\epsilon}$ of conductor p. We compute the t-expansion of this Eisenstein series at $x = (E(\mathfrak{A}), i(\mathfrak{A}))$ and relate the t-expansion to the Iwasawa power series of φ_ψ^- for $\psi = \psi_{k,\epsilon}$ restricted to $[\mathfrak{A}]\Gamma_M$.

If confusion is not likely, we just write ψ for $\psi_{k,\epsilon}$. Recall that $D \cong Z/\Gamma'$ for the intersection Γ' (in Z) of Γ_M with the image of $\mathfrak{O}_p^\times$. Let $\mathbf{1}_{\Gamma'}$ be the characteristic function of $\Gamma' \subset Z$. We put $\phi = \mathrm{Inf}_{\psi_0} \widehat{\psi}\mathbf{1}_{\Gamma'}$ for a character $\psi_0 : \Delta \to W^\times$. We apply Theorem 3.38 to $\mathbf{E}(\phi)$ to compute the μ-invariant. Recall that we have written $E(\phi^\circ)$ for $E_0(\phi^\circ) = \mathbf{E}(\phi)$ for a suitable choice of $\mathfrak{c}$ in the context. Let $\mathbf{1}^\circ_{\Gamma'}(x; y) = \mathbf{1}_{\Gamma'}(x^{-1}; y) = \mathbf{1}(x; y)$, where $\mathbf{1}$ is the characteristic function of $\{(x; y) | \pi(x; y) \in \Gamma'\}$ for $\pi : \mathfrak{O}_p^\times \to Z$.

Recall $\mathcal{D}$, which is a complete representative set in the localization (not the completion) $\mathfrak{O}_{(p)}^\times$ for $\pi(\mathfrak{O}_p^\times)/\Gamma'$, with the projection $\pi : \mathfrak{O}_p^\times \to Z$. We further split $\mathcal{D} = \bigsqcup_{\alpha \in \mathcal{D}^-} \alpha\mathcal{D}^+$, where $\mathcal{D}^+$ is the subset of $\mathcal{D}$ represented by elements of $\mathbb{Q}^\times$:

$$\mathcal{D}^+ = \left\{\alpha \in \mathcal{D} \middle| \alpha\Gamma' = \beta\Gamma' \text{ with } \beta \in \mathbb{Q}^\times \cap \mathfrak{O}_p^\times\right\}.$$

We choose $\alpha \in \mathcal{D}^-$ so that $(\alpha) = \mathfrak{Q}$ is a prime ideal split in M over $\mathbb{Q}$. Since the class $\mathcal{D}^-$ injects into $\mathfrak{O}_p^\times/\mathbb{Z}_p^\times$, we may assume that $\alpha\alpha^c \equiv \pm 1 \mod p^N$ for sufficiently large N (i.e., $\alpha^c \equiv \pm\alpha^{-1} \mod p^N$). Define $\mathcal{S}$ (resp., $\mathcal{R}$) by a subset $\{\beta_p | \beta \in \mathcal{D}^+\}$ (resp., $\{\alpha_p | \alpha \in \mathcal{D}^-\}$) in the completion $\mathfrak{O}_p^\times$. Note that $\widehat{\psi}(s) = \psi((\beta))^{-1}$ because $s = \beta_p$. Similarly, $\widehat{\psi}(r) = \psi((\alpha))^{-1}$ $(r = \alpha_p \in \mathfrak{O}_p)$.

By Lemma 9.13, we get

$$\sum_{\beta\in\mathcal{D}^+} \psi(\beta)^{-1}E(N^{-k}\epsilon^{-1}\mathbf{1}^\circ_{\Gamma'}|\beta) = E(\Phi^\circ_+) \quad \text{for } \Phi^\circ_+ \text{ defined below,}$$

where "$|\beta$" is the action of the scalar element $\beta \in Z(\mathbb{Q})$, and

$$\Phi^\circ_+(x;y) = \sum_{s\in\mathcal{S}} \widehat{\psi}(sx)\mathbf{1}^\circ(s^{-1}x^{-1};sy), \tag{9.2.9}$$

since $s = \beta_p \in \mathcal{S} \subset \mathfrak{O}_p^\times$ (for $\beta \in \mathcal{D}^+$). We further sum up over $\mathcal{D}^-$:

$$\sum_{\alpha\in\mathcal{D}^-} \psi(\alpha)E(\Phi^\circ_+)|[\alpha\alpha^c]|\rho(\alpha)^{-1} = E(\Phi^\circ), \tag{9.2.10}$$

where we have chosen (α) to be an integral ideal, with $\mathfrak{O}/(\alpha) \cong \mathbb{Z}/((\alpha)\cap\mathbb{Q})$ and $(\alpha)\cap\mathbb{Q} = (\alpha\alpha^c)$, and Φ° is given from (9.2.8) by

$$\Phi^\circ(x;y) \overset{(9.2.8)}{=} \sum_{r\in\mathcal{R}} \widehat{\psi}(r)^{-1}\Phi^\circ_+(rx;r^{-c}y)) \overset{(*)}{=} \sum_{r\in\mathcal{R}} \widehat{\psi}(r)^{-1}\Phi^\circ_+(r(x;y)), \tag{9.2.11}$$

where $(*)$ follows from the fact $r = \alpha_p \equiv \alpha_p^{-c} \equiv r \mod p^N$ for whatever large N we have chosen. In short, we have

$$\sum_{\alpha\in\mathcal{D}^-} \psi(\alpha)\left(\sum_{\beta\in\mathcal{D}^+} \psi(\beta)^{-1}E(N^{-k}\epsilon^{-1}\mathbf{1}^\circ_{\Gamma'}|\beta)\right)|[\alpha\alpha^c]|\rho(\alpha)^{-1} = E(\Phi^\circ). \tag{9.2.12}$$

We have computed $E(\Phi^\circ)$ as a linear combination of transforms of the weight-k Eisenstein series $E(N^{-k}\epsilon^{-1}\mathbf{1}^\circ_{\Gamma'})$. By the definition of φ_ψ, Φ as above is the restriction of $\mathrm{Inf}_\psi\,\widehat{\psi}\mathbf{1}_{\Gamma'}$ to $Z_0 = \mathfrak{O}_p^\times/\mathfrak{O}^\times \subset Z$.

We have chosen $\mathfrak{B} \in D^-$ among prime ideals of M split over $\mathbb{Q}$; so the operator $[\mathfrak{B}\mathfrak{B}^c]$ regarding $\mathfrak{B}\mathfrak{B}^c$ as a prime ideal of $\mathbb{Q}$ also makes sense.

Theorem 9.15 *Suppose $p > 2$ and $\mathfrak{C} = 1$. Let t be the canonical variable of the Serre–Tate deformation space $\widehat{S} = \widehat{\mathbb{G}}_m$ of $(E(\mathfrak{O}), i(\mathfrak{O}))_{/W}$ so that the parameter $T = t - 1$ gives the parameter around the origin $1 \in \widehat{S}$. Write Φ for the restriction of $\mathrm{Inf}_\psi\,\widehat{\psi}\mathbf{1}_{\Gamma'}$ to $Z_0 \subset Z$ for $\psi = \psi_{k,\epsilon}$ for $0 < k \in \mathbb{Z}$ and $\epsilon : Z \to \mu_{p^\infty}(\overline{\mathbb{Q}}_p)$ as above. For each $\mathfrak{B} \in D^-$, let*

$$E_{\mathfrak{B}}(t) = E(\Phi^\circ)(t) \in \mathcal{O}_{\widehat{S}}.$$

Then the t-expansion of

$$\mathcal{E} = \sum_{\mathfrak{B}\in D^-,[\mathfrak{B}]^-\in\Gamma'} \psi(\mathfrak{B})E_{\mathfrak{B}}|[\mathfrak{B}\mathfrak{B}^c](t^{[\mathfrak{B}]^-})$$

at $(E(\mathfrak{O}), i(\mathfrak{O}))$ gives (up to an automorphism of $W[[Z]]$) the t-expansion of the anticyclotomic measure φ_ψ^- (restricted to Γ'). Moreover, the μ-invariants $\mu(\varphi_\psi^-)$ and $\mu(\varphi_\psi)$ vanish for any choice of $\psi : Z \to \overline{\mathbb{Q}}_p^\times$ not necessarily of the form $\psi_{k,\epsilon}$.

The vanishing $\mu(\varphi_{\psi}^{-}) = 0$ implies $\mu(\varphi_{\psi}) = 0$ by Lemma 9.8. To conclude $\mu(\varphi_{\psi}^{-}) = 0$, we only need to compute $\mu(\varphi_{\psi}^{-}|_{\Gamma'})$ (the μ-invariant of φ_{ψ}^{-} restricted to Γ'), as $\mu(\varphi_{\psi}^{-}|_{\Gamma'}) = 0$ clearly implies $\mu(\varphi_{\psi}^{-}) = 0$.

Proof. We first show that the t-expansion of $\mathcal{E}$ gives (up to an automorphism of $W[[Z]]$) the t-expansion of the Katz measure. We said "up to an automorphism of $W[[Z]]$" for the following reason: In the definition of the level structure $i(\mathfrak{A})$, $x \in \mathbb{Q}_p/\mathbb{Z}_p$ is sent to $2\delta x \in M_{\mathfrak{p}}/\mathfrak{A}$. This has the following harmless effect: The t-expansion of $E(\Phi_{+}^{\circ}((x;y)\,\mathrm{diag}[2\delta,(2\delta)^{-1}]))$ actually coincides with the t-expansion of the measure. The variable change $(x;y) \mapsto (x;y)\,\mathrm{diag}[2\delta,(2\delta)^{-1}]$ corresponds to the automorphism $z \mapsto 2\delta z$ of the topological space Z (since $2\delta = \sqrt{-D}$ is prime to p), which gives rise to an automorphism of $W[[Z]]$. So we forget about the effect of this unit 2δ.

We are going to compute the κ-derivatives $d^{\kappa}\mathcal{E}$ of the Eisenstein series at $\underline{E}(\mathfrak{A})$ for applying the t-expansion principle. Let $\psi^{(\kappa)} = \psi_{k,\epsilon}^{(\kappa)}$ be a unique Hecke character of Z such that $\psi_{k,\epsilon}^{(\kappa)}(\xi) = \xi^{(1-c)\kappa}\psi_{k,\epsilon}(\xi)$ for all $\xi \equiv 1 \mod p$, $\widehat{\psi}_{k,\epsilon}^{(\kappa)}|_{\Delta} = \psi$, and if $\mathfrak{a} = (\xi)$ is a fractional $\mathbb{Q}$-ideal, $\psi_{k,\epsilon}^{(\kappa)}(\mathfrak{a}) = \xi^{(1-c)\kappa}\psi_{k,\epsilon}(\mathfrak{a}) = \psi_{k,\epsilon}(\mathfrak{a})$, as $\xi = \xi^{c}$. We write $\langle z \rangle$ for the projection of $z \in W(\mathbb{F})^{\times}$ to the p-profinite part of $W(\mathbb{F})^{\times}$. Then we have

$$\left\langle (x;y)^{\kappa(c-1)}\phi \right\rangle^{\circ} = \langle (x^{-1}y)^{\kappa}\phi \rangle^{\circ} = \langle (xy)^{\kappa} \rangle \phi^{\circ}.$$

Recall the following definition:

$$\phi_{\mathfrak{A}}(x) = \phi(x[\mathfrak{A}]^{-1}).$$

Also recall (9.2.7)

$$\int_{\Gamma'} \phi_{\alpha^{-1}\mathfrak{B}}\, dE_{\alpha^{-1}\mathfrak{B}} = (\mathbf{E}(\mathbf{1}_{\Gamma'}\phi_{\alpha^{-1}\mathfrak{B}})|[\alpha\alpha^{c}])\,(E(\mathfrak{B}), i(\mathfrak{B})) \;\text{ if } \alpha \notin \mathbb{Z}.$$

Because of these formulas, we may replace each term

$$\psi(\alpha)\psi(\beta^{-1})E(N^{-k}\epsilon^{-1}\mathbf{1}_{\Gamma'}^{\circ}|\beta)|[\alpha\alpha^{c}]|\rho(\alpha)^{-1}$$

of (9.2.12) with

$$\begin{aligned}
&\psi(\alpha)\psi(\beta^{-1})d^{\kappa}(E(N^{-k}\epsilon^{-1}\mathbf{1}_{\Gamma'}^{\circ}|\beta)|[\alpha\alpha^{c}]|\rho(\alpha)^{-1})(\underline{E}(\mathfrak{A}))\\
&\overset{(*)}{=} \psi(\alpha)\psi(\beta^{-1})(\alpha\beta^{-1})^{\kappa(c-1)}E((xy)^{\kappa}N^{-k}\epsilon^{-1}(\mathbf{1}_{\Gamma'}^{\circ}|[\alpha\beta^{-1}]))(\underline{E}((\alpha^{-1}\beta)\mathfrak{A}))\\
&= \psi(\alpha\beta^{-1})(\alpha\beta^{-1})^{\kappa(c-1)}E(((x^{-1}y)^{\kappa}x^{k}\epsilon(\mathbf{1}_{\Gamma'}\circ(\alpha\beta^{-1})))^{\circ})(\underline{E}((\alpha^{-1}\beta)\mathfrak{A}))\\
&\overset{(**)}{=} \psi^{(\kappa)}(\alpha\beta^{-1})E((\langle (x^{-1}y)^{\kappa}x^{k}\rangle\epsilon\mathbf{1}_{\Gamma'})^{\circ}|[\alpha\beta^{-1}])(\underline{E}((\alpha^{-1}\beta)\mathfrak{A}))\\
&= \psi^{(\kappa)}(\alpha\beta^{-1})E((\langle (x^{-1}y)^{\kappa}x^{k}\rangle\epsilon\mathbf{1}_{\Gamma'})^{\circ})(\underline{E}(\mathfrak{A})), \quad (9.2.13)
\end{aligned}$$

where $\phi|[\alpha](x;y) = \phi(\alpha^{-1}x;\alpha y)$ and $\phi\circ\alpha(t) = \phi(\alpha t)$ for $t \in T$. The above equality indicated by $(*)$ follows from (9.2.1) and the formulas

$d^\kappa(q^{xy}) = (xy)^\kappa q^{xy}$ in (1.3.10) and $d^\kappa(t^a) = a^\kappa t^a$ in Proposition 7.48. Indeed, each term of the q-expansion of $E(\Phi^\circ)$ is of the form $\frac{\Phi(a,b)}{|a|}q^{ab}$; so, applying d^κ, the term becomes $\frac{(ab)^\kappa\Phi(a,b)}{|a|}q^{ab}$, which is the term of the Eisenstein series of $(xy)^\kappa\Phi(x;y)$. Note also that $(x^{-1}y)^\circ = (xy)$. The equality indicated by $(**)$ follows from the fact that $\mathbf{1}_{\Gamma'} \circ (\alpha\beta^{-1})$ is the characteristic function of $(\alpha^{-1}\beta)\Gamma'$.

Let $\mathcal{F} = \left(\mathrm{Inf}_\psi\left(\widehat{\psi}|_{\Gamma'} \cdot (x^{-1}y)^\kappa \mathbf{1}_{\Gamma'}\right)|_{Z_0}\right)^\circ = \langle (xy)^\kappa \rangle \Phi^\circ$. By the computation given in the proof of Theorem 9.4 and by Theorem 1.34, the partial L-value for the character $\psi^{(\kappa)}$ and for the ideal class of $\mathfrak{A}$ is given by

$$E(\mathcal{F})(\underline{E}(\mathfrak{A})) = d^\kappa(E(\Phi^\circ))(\underline{E}(\mathfrak{A}))$$

and

$$d^\kappa E(\underline{E}(\mathfrak{A})) = E(\mathcal{F})(\underline{E}(\mathfrak{A})) \tag{9.2.14}$$

for all $\kappa \geq 0$.

Now we apply the operator $[\mathfrak{B}\mathfrak{B}^c]$ and make a variable change: $t \mapsto t^{[\mathfrak{B}]^-}$ in (9.2.14). We may assume that $[\mathfrak{B}]^- \in \Gamma'$, since $\mathfrak{B}$ without this property does not contribute to the restriction $\varphi_\psi^-|_{\Gamma'}$. By the computation given in the proof of Theorem 9.4 [or [HT1] (4.9)], the partial L-value for the character ψ_κ and for the ideal class of $\mathfrak{B}^{-1}$ is given by, for $E(\mathcal{F})$,

$$\psi^{(\kappa)}(\mathfrak{B})E(\mathcal{F})(\underline{E}(\mathfrak{B}^{-1})) = \psi^{(\kappa)}(\mathfrak{B})d^\kappa(E(\Phi^\circ)|[\mathfrak{B}\mathfrak{B}^c](t^{[\mathfrak{B}]^-}))(\underline{E}(\mathfrak{O}))$$

and

$$d^\kappa E(\underline{E}(\mathfrak{O})) = \sum_{\mathfrak{B}\in D^-, [\mathfrak{B}]^-\in\Gamma'} \psi^{(\kappa)}(\mathfrak{B})E(\mathcal{F})(\underline{E}(\mathfrak{B}^{-1})) \tag{9.2.15}$$

for all $\kappa \geq 0$. Hence, we have

$$E(\phi)|[\mathfrak{B}\mathfrak{B}^c](\underline{E}(\mathfrak{O})) = E(\phi)(\underline{E}(\mathfrak{B}^{-1})).$$

We obtain, by the effect of the differential operator d^κ,

$$d^\kappa(E_{\mathfrak{B}}|[\mathfrak{B}\mathfrak{B}^c](t^{[\mathfrak{B}]^-}))(\underline{E}(\mathfrak{O})) = \sum_{\mathfrak{a}\in D^+} \psi^{(\kappa)}(\mathfrak{a}\mathfrak{B})E(\mathcal{F})(\underline{E}((\mathfrak{a}\mathfrak{B})^{-1})). \tag{9.2.16}$$

Combined with the argument proving Theorem 9.4 [or [HT1] (4.9)], this shows that the function in the theorem, after applying d^κ and evaluating at $\underline{E}(\mathfrak{O})$, gives rise to the value (9.1.10) for the character $\widehat{\lambda} = \widehat{\psi}^{(\kappa)}$ for all κ; so, by Mahler's theorem (Theorem 1.36), the measure corresponding the t-expansion of $\mathcal{E}$ is equal to the anticyclotomic Katz measure φ_ψ^-. Since t-expansion is the Taylor expansion with respect to d, the first assertion follows.

The Igusa tower $Ig_{\Gamma_1(N)/\mathbb{F}}$ $(p \nmid N)$ is given by $\mathrm{Isom}_{Y_1(N)^{ord}/\mathbb{F}}(\mu_{p^\infty}, \mathbf{E}[p^\infty]^\circ)$ for the universal elliptic curve $\mathbf{E}$ over the ordinary locus $Y_1(N)^{ord}$. Though Theorem 3.38 and the theorems in Sect. 3.6.1 are stated for global sections

$\mathcal{O}_{Y^{(p)}}$, they are valid for global sections of $\mathcal{O}_{Ig}$, because $Ig/Y^{(p)}$ is étale (i.e., the completion of corresponding stalks is the same by Lemma 4.7); so the statement is intact. Thus, by Theorem 3.38, in view of the independence of $[\mathfrak{B}]^-$ modulo $\mathcal{T}(\mathbb{Z}_{(p)}) = \mathfrak{O}^\times_{(p)}/\mathbb{Z}^\times_{(p)}$, the sum in (9.2.16) does not vanish modulo the maximal ideal $\mathfrak{m}_W$ of W if $E_{\mathfrak{B}}|[\mathfrak{B}\mathfrak{B}^c](t)$ for one of $\mathfrak{B} \in D^-$ does not vanish modulo $\mathfrak{m}_W$. Take the identity class; we may assume that $E_{\mathfrak{O}}(t) = E_k(\Phi^\circ)$ with $\Phi^\circ(x;y) = x^{-k}\chi(x)\chi'(y)$ for finite order characters χ and χ' of p-power conductor (indeed, our choice was $\chi = \epsilon^{-1}$ and $\chi' = \mathbf{1}$). The fact that Φ° is a product of characters as above follows from (9.2.9) and (9.2.11). Thus, Φ is also a product of characters $(x;y) \mapsto x^k\chi^{-1}(x)\chi'(y)$. Then by the computation already done for such Eisenstein series in Lemma 9.2, for primes l prime to p, $a(l, E_k(\Phi^\circ)) = \chi'(l) + \chi^{-1}(l)l^{k-1}$. We leave the reader the verification of the existence of a prime l such that $a(l, E_k(\Phi^\circ)) \not\equiv 0 \mod \mathfrak{m}_W$. Thus, $\mu(\varphi^-_\psi) = 0$, as desired, for $\psi = \psi_{k,\epsilon}$.

A general character ψ can be decomposed as $\psi_0(\psi_x \circ \pi_x)\psi^-$ for characters $\psi_x : \mathbb{Z}_p^\times/\mu_{p-1}(\mathbb{Z}_p) \to \overline{\mathbb{Q}}_p^\times$ and $\psi^- : \Gamma^- \to \overline{\mathbb{Q}}_p^\times$ as $\Gamma_M = \pi_x(\Gamma_M) \times \Gamma^-$. Since $\mu^-_\psi = \psi^- * \mu_{\psi'}$ for $\psi' = \psi_0\psi_x$, by Lemma 9.8, $\mu(\mu^-_\psi) = 0$ follows from $\mu(\psi^-_{\psi'}) = 0$. It is an easy exercise to find an infinite sequence $\{(k_j, \epsilon_j)\}_j$ for integers $0 < k_j \in \mathbb{Z}$ and characters $\epsilon_j : \mathbb{Z}_p^\times/\mathbb{Z}^\times \to \mu_{p^\infty}(\overline{\mathbb{Q}}_p)$ such that $\psi' = \lim_{j\to\infty} \psi_{k_j,\epsilon_j}$ under the supremum norm $\|\cdot\|$ on $\mathcal{C}(Z, \overline{\mathbb{Q}}_p)$. Then we conclude that $\mu(\varphi^-_\psi) = \mu(\varphi^-_{\psi'}) = \lim_j \mu(\varphi^-_{\psi_{k_j,\epsilon_j}}) = 0$.

Exercise 9.16 *Prove the claim about the sequence* $\{(k_j, \epsilon_j)\}_j$.

10

Toric Subschemes in a Split Formal Torus

In this chapter, we prove a rigidity result of C.-L. Chai, which generalizes earlier lemmas (for example, Lemma 3.12). After proving the result, we gather facts from commutative/noncommutative/Lie algebra theory used in the proof.

10.1 Rigidity of Formal Tori

As we have seen, if a power series $\Phi(T) = \Phi(t)$ $(T = t - 1)$ with $\Phi(1) = 1$ satisfies $\Phi(t^z) = \Phi(t)^z$ for all z in an open subgroup of $\mathbb{Z}_p^\times$, then $\Phi(t) = t^s$ for $s \in \mathbb{Z}_p$ (Lemmas 3.12, 3.30, and 3.45). We can state this fact more geometrically as follows: By definition, we have $\mathbb{G}_m = \mathrm{Spec}(\mathbb{Z}[t, t^{-1}])$. Note that $\widehat{\mathbb{G}}_m = \mathrm{Spf}(\widehat{W[t, t^{-1}]}) = \mathrm{Spf}(W[[T]])$ for

$$\widehat{W[t, t^{-1}]} = \varprojlim_n W[t, t^{-1}]/(t^n - 1) = \varprojlim_n W[T]/((1 + T)^n - 1) = W[[T]].$$

Let $Z \subset \widehat{\mathbb{G}}_m \times \widehat{\mathbb{G}}_m$ be a proper closed, irreducible, formal subscheme. Write the right factor $\widehat{\mathbb{G}}_m = \mathrm{Spf}(\widehat{W[t', t'^{-1}]})$. Furthermore, if the projection of Z to the first factor $\widehat{\mathbb{G}}_m$ is an isomorphism, Z is defined by an equation $t' = \Phi(t)$ for $\Phi(t) \in W[[T]]$. If Z is stable under the action of $(t, t') \mapsto (t^z, t'^z)$ for all z in an open subgroup of $\mathbb{Z}_p^\times$, we have $\Phi(t^z) = t'^z = \Phi(t)^z$. Then Lemma 3.30 tells us that $\Phi(t) = t^s$; hence, Z is a formal subtorus defined by $\{(t, t^s) | t \in \widehat{\mathbb{G}}_m\}$. The cocharacter group of $\mathbb{G}_m^n$ is isomorphic to $\mathbb{Z}^n$, which we have written as $X_*(\mathbb{G}_m^n)$ [see Example 4.58 (4)]. Then, by tautology, $\mathbb{G}_m^n = \mathbb{G}_m \otimes_{\mathbb{Z}} X_*(\mathbb{G}_m^n)$. Similarly, in the formal setting (cf. Sect. 4.4.3), putting $X_*(\widehat{\mathbb{G}}_m^n) = X(\mathbb{G}_m^n) \otimes_{\mathbb{Z}} \mathbb{Z}_p = \mathbb{Z}_p^n$, we have $\widehat{\mathbb{G}}_m^n = \widehat{\mathbb{G}}_m \otimes_{\mathbb{Z}_p} X_*(\widehat{\mathbb{G}}_m^n)$. A slightly more general version of Chai's rigidity lemma can be stated as follows (see [Ch3] and [H10a] Lemma 3.7):

H. Hida, *Elliptic Curves and Arithmetic Invariants*, Springer Monographs in Mathematics, DOI 10.1007/978-1-4614-6657-4_10,

Theorem 10.1 (C.-L. Chai) *Let $\mathbb{F}$ be an algebraic closure of $\mathbb{F}_p$. If $Z \subset \widehat{\mathbb{G}}^n_{m/\mathbb{F}}$ is a reduced equidimensional closed formal subscheme of dimension r stable under the diagonal action of all z in an open subgroup of $\mathbb{Z}_p^\times$, then*

$$Z = \bigcup_L \widehat{\mathbb{G}}_m \otimes_{\mathbb{Z}_p} L \subset \widehat{\mathbb{G}}^n_{m/\mathbb{F}},$$

where L runs over (finitely many) $\mathbb{Z}_p$-direct summands of $X_*(\widehat{\mathbb{G}}^n_m)$ of rank r.

Since Z has finitely many irreducible components, which are permuted by the action of the stabilizer of Z under the diagonal action of $\mathbb{Z}_p^\times$, each component is fixed by an open subgroup of the stabilizer. Therefore, we may assume that Z is integral (i.e., reduced irreducible). In this chapter, we give a full proof of the result as Theorem 10.6, assuming irreducibility. This theorem follows from Corollary 10.7.

One might want to expect a similar result over W in place of $\mathbb{F}$. However, there is an easy counterexample. Take a diagonal torus $\Delta \subset \widehat{\mathbb{G}}_m \times \widehat{\mathbb{G}}_m$ over W; so $\Delta = \widehat{\mathbb{G}}_m \otimes_{\mathbb{Z}_p} D$ for the diagonal $D \subset X_*(\widehat{\mathbb{G}}^2_{m/W})$. Suppose that $\mu_p(W)$ is a cyclic subgroup of order p. Then, for any $1 \neq \zeta \in \mu_p^2(W) \subset \widehat{\mathbb{G}}^2_m$, $\zeta\Delta$ is stable under the diagonal action of $U = 1 + p\mathbb{Z}_p \subset \mathbb{Z}_p^\times$. However, $\zeta\Delta$ is clearly not of the form $\bigcup_L \widehat{\mathbb{G}}_m \otimes_{\mathbb{Z}_p} L$. To save the situation, we need to have an additional assumption (for example, see Lemma 4.85, where the condition for Z to contain the identity of $\widehat{\mathbb{G}}^2_m$ is imposed).

10.1.1 Power Series Ring over a Field

Here is a weak version of desingularization in [Ch1] (6.1).

Proposition 10.2 *Let k be an algebraically closed field and R a complete, local, noetherian domain over k of dimension d with residue field k. Then there exists an injective local homomorphism $\iota : R \hookrightarrow k[[y_1, \ldots, y_d]]$ of complete local k-algebras.*

Since R is a local complete noetherian domain of dimension d, R is isomorphic to $k[[x_1, \ldots, x_r]]/P$ for a closed prime ideal P of the power series ring $k[[x_1, \ldots, x_r]]$ for $r \geq d$.

Proof. We blow up $S := \mathrm{Spec}(R)$ at the unique closed point $s = \mathrm{Spec}(R/\mathfrak{m}_s) \in S$, getting a morphism of k-formal schemes $f : X \to S$. The blowup X is given by $\mathcal{P}roj_S(\mathcal{R})$ ("Proj" relative to S; see the end of Sect. 4.1.9) of the graded R-algebra $\mathcal{R} = \bigoplus_{d=0}^{\infty} \mathfrak{m}_s^d$ with $\mathfrak{m}_s^0 = R$. We have a natural projection $X \to S$, which is clearly an isomorphism over $S \setminus \{s\}$ (because $\mathcal{R}[\frac{1}{f}]$ for $0 \neq f \in \mathfrak{m}_s$ contains a unit of positive degree) and its fiber over s is a projective scheme of positive dimension (which is essentially a divisor). See [ALG] II.7.14 for the universality of blowing up. Let $\widetilde{\mathfrak{m}}_s$ be the sheaf over S associated with the R-module $\mathfrak{m}_s$; hence, $H^0(S, \widetilde{\mathfrak{m}}_s) = \mathfrak{m}_s$. Let $\widetilde{X}$ be the normalization of X. Therefore, $\widetilde{X}$ is regular in codimension 1 (see [ALG] II.6). Write $\pi : \widetilde{X} \to X \to S$ for the projection. Then $\pi^*\widetilde{\mathfrak{m}}_s$ and $f^*\widetilde{\mathfrak{m}}_s$ are invertible sheaves (cf. [ALG] II.7.14), and hence $\pi^{-1}(s) = \widetilde{X} \times_S s$ is a divisor. Let D be the reduced part $\pi^{-1}(s)^{red}$ of the exceptional divisor $\pi^{-1}(s)$. Then D is a scheme of finite type over k, and D is an excellent scheme (see Remark 4.35), and its set of regular closed points is Zariski dense in D [see (e3) in Sect. 4.1.11]; therefore, the regular locus is a Zariski-dense open subscheme D_{reg} of D such that $\mathcal{O}_{D,x}$ is regular for all $x \in D_{reg}(k)$. Since S is (excellent) normal noetherian irreducible, $\widetilde{X}$ and S are birational (cf. [ALG] 7.16); so $\widetilde{X}$ is normal noetherian irreducible. Cover $\widetilde{X}$ by affine open subschemes U° so that on $U = U^\circ \cap D$, D is defined by a single equation $x_{U^\circ} = 0$ for $x_{U^\circ} \in \Gamma(U^\circ, \mathcal{O}_{U^\circ})$. We claim $D_{reg} \subset \widetilde{X}_{reg}$. Indeed, by the definition of D_{reg}, for each closed point $x \in D_{reg}(k) \cap U^\circ$, we can find a regular sequence $\overline{x}_2, \cdots, \overline{x}_d \in \mathcal{O}_{D,x}$ vanishing at x for $d = \dim S$ such that the completed stalk $\widehat{\mathcal{O}}_{D,x} \cong k[[\overline{x}_2, \ldots, \overline{x}_d]]$. Lift $\overline{x}_j$ to $x_j \in \mathcal{O}_{U^\circ,x}$ so that (x_j mod (x_{U°)) $= \overline{x}_j$, and let $x_1 = x_{U^\circ}$. Then we have $\widehat{\mathcal{O}}_{\widetilde{X},x} \cong k[[x_1, \ldots, x_d]]$, and hence $U \subset \widetilde{X}_{reg}$. Since the open subschemes U cover D_{reg}, we have $D_{reg} \subset \widetilde{X}_{reg}$. Pick $x \in D_{reg}(k) \subset \widetilde{X}_{reg}$; then $R \hookrightarrow \widehat{\mathcal{O}}_{\widetilde{X},x} \cong k[[x_1, \ldots, x_d]]$ (as above), which is a desired embedding.

We need the following lemma in [Ch1] (6.2):

Lemma 10.3 *Let k be a field of characteristic $p > 0$ and $q = p^r$ be a p-power ($r > 0$). Let $P(x_1, \ldots, x_m) \in k[x_1, \ldots, x_m]$ be a polynomial with coefficients in k. If there exists $n_0 \geq 0$ such that $P(c_1^{q^n}, \ldots, c_m^{q^n}) = 0$ for all integers $n \geq n_0$ for a vector $c = (c_1, \ldots, c_m) \in k^m$, we have $P(c_1^{q^n}, \ldots, c_m^{q^n}) = 0$ for all $n \in \mathbb{Z}$; in particular, $P(c_1, \ldots, c_m) = 0$.*

We give a version of Chai's proof in [Ch1].

Proof. Replacing k by its algebraic closure if necessary, we may assume that k is perfect. Thus, $\sigma : x \mapsto x^{q^{-1}}$ is an automorphism of k. Write $\mathbf{x} = (x_1, \ldots, x_m)$ and $\mathbf{x}^\alpha = \prod_{j=1}^m x_j^{\alpha_j}$ for multiindices $\alpha = (\alpha_1, \ldots, \alpha_m) \in \mathbb{N}^m$. We define the total degree $\deg(f) = \max\{\alpha_1 + \alpha_2 + \cdots + \alpha_m | a_\alpha \neq 0\}$, and consider the filtration $V_0 \subset V_1 \subset \cdots \subset V_a$ of $k[\mathbf{x}]$ given by $V_a = \{f | \deg(f) \leq a\}$ ($a \in \mathbb{N}$). The dimension of V_a over k is finite.

For each polynomial $f(\mathbf{x}) = \sum_\alpha a_\alpha \mathbf{x}^\alpha \in k[\mathbf{x}]$, we write

$$\sigma^n(f) = \sum_\alpha \sigma^n(a_\alpha) x^\alpha \in k[\mathbf{x}].$$

The association $f \mapsto \sigma^n(f)$ is a ring automorphism of $k[\mathbf{x}]$ and gives rise to an σ-linear automorphism of V_a for each a.

Write I for the ideal of the ring $k[\mathbf{x}]$ generated by polynomials $\sigma^n(P(\mathbf{x}))$ for all $n \geq n_0$. Let $\mathbf{c} = (c_1, \ldots, c_m) \in k^m$. If $\sigma(I) = I$, for any $n \in \mathbb{Z}$, $P(\mathbf{c}^{q^n}) = \sigma^{n_0 - n}(P(\mathbf{c}^{q^{n_0}})) \in I$, and hence $P(\mathbf{c}^{q^n}) = 0$. Thus, we need to prove $\sigma(I) = I$. Since V_a is stable under σ, $(I \cap V_a) \cong (\sigma(I) \cap V_a)$, and so $\dim_k(I \cap V_a) = \dim_k(\sigma(I) \cap V_a) < \infty$. Since $\sigma(I)$ is generated by polynomials

$\sigma^n(P)$ for $n \geq n_0+1$, we have $\sigma(I) \subset I$. This shows that $(\sigma(I) \cap V_a) \subset (I \cap V_a)$, and by comparing dimension over k, we conclude that $(\sigma(I) \cap V_a) = (I \cap V_a)$. Since $k[\mathbf{x}] = \bigcup_{a=0}^{\infty} V_a$, we find

$$\sigma(I) = \sigma(I) \cap \left(\bigcup_{a=0}^{\infty} V_a\right) = \bigcup_a (\sigma(I) \cap V_a) = \bigcup_a (I \cap V_a) = I \cap \left(\bigcup_{a=0}^{\infty} V_a\right) = I,$$

as desired.

We keep the notation introduced in Lemma 10.3. In particular, $q = p^r$ for $r > 0$ and n_0 is a nonnegative integer. Here is a sort of Chai's implicit function theorem for power series in characteristic p-coefficients [as it gives a uniformization of the zero set of a power series $f(\mathbf{u}, \mathbf{v})$; see also [BCM] III.4.5]:

Proposition 10.4 *Let k be a field of characteristic $p > 0$. Let $f(\mathbf{u}, \mathbf{v}) \in k[[\mathbf{u}, \mathbf{v}]]$ for multivariables $\mathbf{u} = (u_1, \ldots, u_a)$ and $\mathbf{v} = (v_1, \ldots, v_b)$. Prepare new sets of variables $\mathbf{x} = (x_1, \ldots, x_m)$ and $\mathbf{y} = (y_1, \ldots, y_m)$ and an a-tuple of power series $\mathbf{g}(\mathbf{x}) = (g_1(\mathbf{x}), \ldots, g_a(\mathbf{x}))$ without constant term [i.e., $g_j(\mathbf{x}) \in (\mathbf{x}) \subset k[[\mathbf{x}]]$] and a b-tuple of power series $\mathbf{h}(\mathbf{y}) = (h_1(\mathbf{y}), \ldots, h_b(\mathbf{y}))$ without constant term, where a (resp., b) is the number of variables in $\mathbf{u}$ (resp., $\mathbf{v}$). Pick a rapidly increasing sequence $(d_n)_{n \geq n_0}$ of integers such that $\lim_{n\to\infty} \frac{q^n}{d_n} = 0$ in real numbers $\mathbb{R}$. If, for a b-tuple of power series $\Phi_n(\mathbf{v}) = (\phi_{1,n}(\mathbf{v}), \ldots, \phi_{b,n}(\mathbf{v})) \in k[[\mathbf{v}]]^b$ with $\phi_{j,n}(\mathbf{v}) \equiv v_j^{q^n} \mod (\mathbf{v})^{d_n}$ indexed by integers $n \geq n_0$, we have*

$$f(\mathbf{g}(\mathbf{x}), \Phi_n(\mathbf{h}(\mathbf{x}))) = f(g_1(\mathbf{x}), \ldots, g_a(\mathbf{x}), \phi_{1,n}(\mathbf{h}(\mathbf{x})), \ldots, \phi_{b,n}(\mathbf{h}(\mathbf{x}))) = 0$$

as an element of $k[[\mathbf{x}]]$, then $f(\mathbf{g}(\mathbf{x}), \mathbf{h}(\mathbf{y})) = 0$ in $k[[\mathbf{x}, \mathbf{y}]]$.

Proof. Here is the proof in [Ch1]. Write $h_i(\mathbf{y}) = \sum_\alpha c_{i,\alpha} \mathbf{y}^\alpha$ for multiindices $\alpha \in \mathbb{N}^m$. Note here that $c_{i,\mathbf{0}} = 0$ for $\mathbf{0} = (0, \ldots, 0) \in k^m$. Write $\mathbf{c}$ for the infinite tuple $(c_{i,\alpha})_{1 \leq i \leq b, \alpha \in \mathbb{N}^m - \{\mathbf{0}\}}$. Then we prepare an infinite tuple of variables $\mathbf{t} = (t_{i,\alpha})_{1 \leq i \leq b, \alpha \in \mathbb{N}^m - \{\mathbf{0}\}}$ with the same index set and introduce an infinite variable power series $H_i(\mathbf{t}; \mathbf{y}) = \sum_\alpha t_{i,\alpha} \mathbf{y}^\alpha \in k[\mathbf{t}][[\mathbf{y}]]$. Write $\mathbf{H}(\mathbf{t}; \mathbf{y}) = (H_1(\mathbf{t}; \mathbf{y}), \ldots, H_b(\mathbf{t}; \mathbf{y})) \in k[\mathbf{t}][[\mathbf{y}]]^b$. Thus, $\mathbf{H}(\mathbf{c}; \mathbf{y}) = \mathbf{h}(\mathbf{y})$. We look into the infinite variable power series:

$$f(\mathbf{g}(\mathbf{x}), \mathbf{H}(\mathbf{t}; \mathbf{y})) \in k[\mathbf{t}][[\mathbf{x}, \mathbf{y}]].$$

Since each coefficient of $f(\mathbf{g}(\mathbf{x}), \mathbf{H}(\mathbf{t}; \mathbf{y}))$ in $\mathbf{x}^\alpha \mathbf{y}^\beta$ is a polynomial in $\mathbf{t}$, we may write

$$f(\mathbf{g}(\mathbf{x}), \mathbf{H}(\mathbf{t}; \mathbf{y})) = \sum_{\alpha, \beta \in \mathbb{N}^m} A_{\alpha,\beta}(\mathbf{t}) \mathbf{x}^\alpha \mathbf{y}^\beta$$

with $A_{\alpha,\beta}(\mathbf{t}) \in k[\mathbf{t}]$. By our assumption, $\phi_{j,n}(\mathbf{v}) \equiv v_j^{q^n} \mod (\mathbf{v})^{d_n}$, we have

$$f(\mathbf{g}(\mathbf{x}), \Phi_n(\mathbf{h}(\mathbf{x}))) \equiv f(\mathbf{g}(\mathbf{x}), \mathbf{h}(\mathbf{x})^{q^n}) \mod (\mathbf{x})^{d_n} \tag{10.1.1}$$

for all $n \geq n_0$. We will prove the proposition by absurdity; thus, we suppose that

$$f(\mathbf{g}(\mathbf{x}), \mathbf{h}(\mathbf{y})) = \sum_{\alpha,\beta} A_{\alpha,\beta}(\mathbf{c})\mathbf{x}^\alpha \mathbf{y}^\beta \neq 0.$$

If, for any (α, β), $A_{\alpha,\beta}(\mathbf{c}^{q^n}) = 0$ for all $n \geq n_0$, by Lemma 10.3, $A_{\alpha,\beta}(\mathbf{c}) = 0$. Since there exists, by our (absurdity) assumption that $A_{\alpha,\beta}(\mathbf{c}) \neq 0$ for some (α, β), for some (α, β), there is no integer n_0 with $A_{\alpha,\beta}(\mathbf{c}^{q^n}) = 0$ for all $n \geq n_0$. In other words, for some (α, β), there are infinitely many positive integers n such that $A_{\alpha,\beta}(\mathbf{c}^{q^n}) \neq 0$. Thus, writing $|\alpha| = \sum_{j=1}^m |\alpha_m|$, the following set

$$\left\{|\beta| \middle| \text{there exists } \alpha \text{ such that } A_{\alpha,\beta}(\mathbf{c}^{q^n}) \neq 0 \text{ for infinitely many } n \in \mathbb{N}\right\}$$

is nonempty. Then we define a positive integer M_2 by the minimal integer in the above set:

$$M_2 := \operatorname{Inf}\left\{|\beta| \middle| \exists\ \alpha \text{ such that } A_{\alpha,\beta}(\mathbf{c}^{q^n}) \neq 0 \text{ for infinitely many } n \in \mathbb{N}\right\}.$$

Define another integer M_1 by the minimal nonnegative integer in the following set:

$$\left\{|\alpha| \middle| \exists\ \beta \text{ with } |\beta| = M_2 \text{ such that } A_{\alpha,\beta}(\mathbf{c}^{q^n}) \neq 0 \text{ for infinitely many } n \in \mathbb{N}\right\}.$$

Again by Lemma 10.3, if either $|\beta| < M_2$ or $|\beta| = M_2$ and $|\alpha| < M_1$, $A_{\alpha,\beta}(\mathbf{c}) = A_{\alpha,\beta}(\mathbf{c}^{q^n}) = 0$ for all $n \in \mathbb{Z}$.

We have

$$\begin{aligned} f(\mathbf{g}(\mathbf{x}), \mathbf{h}(\mathbf{x})^{q^n}) &= f(\mathbf{g}(\mathbf{x}), \mathbf{H}(\mathbf{c}^{q^n}; \mathbf{y}^{q^n}))|_{\mathbf{y}=\mathbf{x}} \\ &= \sum_{\alpha,\beta} A_{\alpha,\beta}(\mathbf{c}^{q^n})\mathbf{x}^\alpha \mathbf{y}^{q^n\beta}|_{\mathbf{x}=\mathbf{y}} = \sum_{\alpha,\beta} A_{\alpha,\beta}(\mathbf{c}^{q^n})\mathbf{x}^{\alpha+q^n\beta}. \end{aligned} \tag{10.1.2}$$

Because of $\lim_{n\to\infty} \frac{q^n}{d_n} = 0$ in real numbers $\mathbb{R}$, we can find a positive integer n_1 such that $q^{n_1} > 2M_1$ and $q^n M_2 < d_n$ for all $n \geq n_1$. Equations (10.1.1) and (10.1.2) combined, we get, for all $n \geq n_1$,

$$\begin{aligned} 0 = f(\mathbf{g}(\mathbf{x}), \Phi_n(\mathbf{h}(\mathbf{x}))) &\overset{\text{mod } (\mathbf{x})^{d_n}}{\equiv} f(\mathbf{g}(\mathbf{x}), \mathbf{h}(\mathbf{x})^{q^n}) \\ &\overset{\text{mod } (\mathbf{x})^{M_1+q^n M_2+1}}{\equiv} \sum_{|\alpha|\geq M_1, |\beta|\geq M_2} A_{\alpha,\beta}(\mathbf{c}^{q^n})\mathbf{x}^{\alpha+q^n\beta}. \end{aligned} \tag{10.1.3}$$

This implies, in the polynomial ring $k[\mathbf{x}]$,

$$\sum_{|\alpha|=M_1, |\beta|=M_2} A_{\alpha,\beta}(\mathbf{c}^{q^n})\mathbf{x}^{\alpha+q^n\beta} = 0 \quad \text{for all } n \geq n_1. \tag{10.1.4}$$

We now show that $\alpha + q^n\beta = \alpha' + q^n\beta'$ implies $(\alpha, \beta) = (\alpha', \beta')$ as long as $|\alpha| = |\alpha'| = M_1$ and $|\beta| = |\beta'| = M_2$ and $n \geq n_1$. Indeed, $\alpha = \alpha'$ and $\beta = \beta'$ because $q^n > 2M_1$. Then (10.1.4) means $A_{\alpha,\beta}(\mathbf{c}^{q^n}) = 0$ for all $n \geq n_1$, which contradicts the assumption for M_1.

10.1.2 Linearity of Subschemes of a Formal Torus

Let G and G' be a connected linear algebraic group over $\mathbb{Q}_p$ with Lie algebra $\mathfrak{g}_{/\mathbb{Q}_p}$ and $\mathfrak{g}'_{/\mathbb{Q}_p}$, respectively (see Sect. 10.2.4). As a vector space over $\mathbb{Q}_p$, $\mathfrak{g}_{/\mathbb{Q}_p}$ is the tangent space of G at the identity element 1_G of G. Thus, any morphism $f : G \to G'$ of $\mathbb{Q}_p$-schemes sending 1_G to $1_{G'}$ induces its differential $df : \mathfrak{g} \to \mathfrak{g}'$, which is a $\mathbb{Q}_p$-linear map (as it is the dual of the pullback map $f^* : \Omega_{G'/\mathbb{Q}_p} \to \Omega_{G/\mathbb{Q}_p}$). In particular, if $G = GL_V$ as in Example 4.58 (6) for a finite-dimensional $\mathbb{Q}_p$-vector space V [i.e., G associates the R-linear automorphism group $\mathrm{Aut}_R(V \otimes_{\mathbb{Q}_p} R)$ to any commutative $\mathbb{Q}_p$-algebra R], we have $\mathfrak{g} = \mathrm{End}_{\mathbb{Q}_p}(V)$ with $[x, y] = x \circ y - y \circ x$. The commutator map $(\cdot, \cdot) : G \times G \to G$ given by $(x, y) \mapsto (x, y) := xyx^{-1}y^{-1}$ sends the identity of $G \times G$ to the identity to G, inducing the bracket product $[\cdot, \cdot] = d(\cdot, \cdot) : \mathfrak{g} \times \mathfrak{g} \to \mathfrak{g}$, which makes $\mathfrak{g}$ a Lie algebra (see Sect. 10.2 for more details).

Exercise 10.5 *Check that the two bracket products defined above coincide if $G = GL_V$.*

Let $\rho : G \to \mathrm{Aut}_{\mathbb{Q}_p}(V) = GL_V$ be a morphism of group schemes over $\mathbb{Q}_p$, which is called a rational representation of G. The space V is often written as $V(\rho)$, and $V(\rho)$ could also be called a rational representation of G [identifying the G-module $V(\rho)$ and the homomorphism ρ]. Then the differential $d\rho : \mathfrak{g} \to \mathrm{End}_{\mathbb{Q}_p}(V)$ induces a homomorphism of Lie algebras (see Sect. 10.2.4 for details about the differential of a representation). As remarked already, the endomorphism algebra $\mathfrak{gl}_V := \mathrm{End}_{\mathbb{Q}_p}(V)$ is naturally a Lie algebra under the bracket product $[x, y] = x \circ y - y \circ x$. We pick a $\mathbb{Z}_p$-lattice L of V and consider the open subgroup U of $G(\mathbb{Q}_p)$ stabilizing L (that is, $U \cdot L = L$). We write $G(\mathbb{Z}_p)$ for the maximal subgroup of $G(\mathbb{Q}_p)$ with this property: $G(\mathbb{Z}_p) = \{u \in G(\mathbb{Q}_p) | u \cdot L = L\}$.

Let k be a field of characteristic $p > 0$ and consider the formal group $\widehat{\mathbb{G}}_{m/k} = \varprojlim_n \mu_{p^n/k}$. Put $T = L \otimes_{\mathbb{Z}_p} \widehat{\mathbb{G}}_m$. Thus, for any pro-Artinian k-algebra A, we have $T(A) = L \otimes_{\mathbb{Z}_p} \widehat{\mathbb{G}}_m(A)$. Through the action on L, U acts on T by $u(\ell \otimes a) = (\rho(u)\ell) \otimes a$ for $a \in \widehat{\mathbb{G}}_m(A)$ and $\ell \in L$. Here is Chai's rigidity result for U-stable formal subschemes of T:

Theorem 10.6 (Rigidity) *Let $Z \subset T$ be an integral closed formal subscheme stable under the action of U [an open subgroup in $G(\mathbb{Z}_p)$ as above]. If the representation (ρ, G) does not contain the trivial representation $\mathbf{1}_G$ as a subquotient, then there exists a $\mathbb{Z}_p$-direct summand $L_Z \subset L$ stable under $G(\mathbb{Z}_p)$ such that $Z = L_Z \otimes_{\mathbb{Z}_p} \widehat{\mathbb{G}}_m \subset L \otimes_{\mathbb{Z}_p} \widehat{\mathbb{G}}_m = T$. In particular, $V_Z = L_Z \otimes_{\mathbb{Z}_p} \mathbb{Q}_p$ is a rational subrepresentation of V.*

Here the word "integral" means reduced and irreducible. We faithfully follow the proof given in [Ch1] (6.6).

Proof. For each $\ell \in L$, we can associate a morphism $\ell : \widehat{\mathbb{G}}_m \to T$ given by $a \mapsto \ell \otimes a$. Any morphism ϕ from $\widehat{\mathbb{G}}_m$ into T can be realized in this way for $\ell \otimes 1 = \phi(1)$. Thus, L can be canonically identified with the cocharacter group $X_*(T) = \mathrm{Hom}_{\text{formal gp}}(\widehat{\mathbb{G}}_m, T)$ and $\mathrm{End}(\widehat{\mathbb{G}}_{m/k}) = \mathbb{Z}_p$ by $\mathbb{Z}_p \ni s \mapsto (t \mapsto t^s)$ if we identify $\widehat{\mathbb{G}}_m$ with $\mathrm{Spf}(\widehat{k[t,t^{-1}]})$ (for the parameter t at $1 \in \widehat{\mathbb{G}}_m$), where $t^s = \sum_{j=0}^{\infty} \binom{s}{j} T^j$ $(T = t-1)$. Then the $\mathbb{Z}_p$-dual L^* of L is identified with the character group $X^*(T)$ by $\langle \phi, \ell^* \rangle = \ell^* \circ \phi \in \mathrm{End}_k(\widehat{\mathbb{G}}_m) = \mathbb{Z}_p$ for $\phi \in X_*(T)$ and $\ell^* \in X^*(T)$. Then $T = \mathrm{Spf}(k[[L^*]])$ [cf. Example 4.58 (4) and Sect. 4.4.3], where $k[[L^*]] = \varprojlim_n k[L^*/p^n L^*]$ for the group ring $k[L^*/p^n L^*]$ of the finite group $L^*/p^n L^*$. Take an algebraic closure $\overline{k}/k$. Suppose that $\overline{Z} = Z \times_k \overline{k}$ can be written as $L_Z \otimes \widehat{\mathbb{G}}_{m/\overline{k}}$. Then $\overline{Z} = \mathrm{Spf}(\overline{k}[[L^*/L_Z^{\perp}]])$, which clearly descends to $Z = \mathrm{Spf}(k[[L/L_Z^{\perp}]])$, proving $Z = L_Z \otimes_{\mathbb{Z}_p} \widehat{\mathbb{G}}_{m/k}$. Here $L_Z^{\perp} = \{\ell^* \in L^* | \langle L_Z, \ell^* \rangle = 0\}$, which is a $\mathbb{Z}_p$-direct summand of L^*. Thus, by replacing k with $\overline{k}$, we may assume that k is an algebraically closed field.

Replacing G with its image under ρ, we may assume that $G(\mathbb{Q}_p) \subset GL_V$. Since $G(\mathbb{Z}_p)$ is a p-adic Lie group, we have its Lie algebra $\mathfrak{g}_{/\mathbb{Z}_p}$. In other words, $\mathfrak{g}_{/\mathbb{Z}_p} = \{X \in \mathfrak{g}_{/\mathbb{Q}_p} | d\rho(X) \cdot L \subset L\}$. Then we have $\exp : p\mathfrak{g}_{/\mathbb{Z}_p} \to G(\mathbb{Q}_p)$ given by $\exp(X) = \sum_{j=0}^{\infty} \frac{X^j}{j!}$ and its inverse $\log_p$ defined on an open subgroup of $G(\mathbb{Z}_p)$. Take an integer $n_0 \geq 2$ so that $\exp(p^{n_0} X) \in U$ for all $X \in \mathfrak{g}_{/\mathbb{Z}_p}$. We proceed in several steps:

Step 1: *Write $m : T \times T \to T$ for the multiplication morphism [thus, $m(a,b) = a \cdot b$]. Then we have $m \circ (\mathrm{id}_T \times \delta_X)(Z \times Z) \subset Z$ for every $X \in \mathfrak{g}_{/\mathbb{Z}_p}$, where $\delta_X \in \mathrm{End}(T)$ is induced by $d\rho(X) \in \mathrm{End}_{\mathbb{Z}_p}(L)$ [i.e, $\delta_X(\ell \otimes a) = d\rho(X)(\ell) \otimes a$].*

Proof of step 1. Writing t for the parameter on $\widehat{\mathbb{G}}_m = \mathrm{Spf}(\widehat{k[t,t^{-1}]})$ and putting $u = t - 1$, we have $\widehat{\mathbb{G}}_m = \mathrm{Spf}(k[[u]])$, and the comultiplication $m^*_{\widehat{\mathbb{G}}_m}$: $k[[u]] \to k[[u,v]]$ is given by $u \mapsto u + v + uv$ (see Exercise 4.61). Take a basis $t_1, \ldots, t_d$ of $L^* = X^*(T)$ over $\mathbb{Z}_p$ and identify $L^* \cong \mathbb{Z}_p^d$ so that $\prod_j t_j^{s_j} \leftrightarrow (s_j) \in \mathbb{Z}_p^d$. Put $u_j = t_j - 1$ and $\mathbf{u} = (u_1, \ldots, u_d)$. By Exercise 4.61, we have $T = \mathrm{Spf}(k[[u_1, \ldots, u_d]])$ with comultiplication $m^* : k[[\mathbf{u}]] \to k[[\mathbf{u}, \mathbf{v}]]$ given by $u_j \mapsto u_j + v_j + u_j v_j$, where $\mathbf{v} = (v_1, \ldots, v_d)$ is a second copy of $\mathbf{u}$, giving the coordinate of the right factor T of the multiplication $m : T \times T \to T$. We may write $Z = \mathrm{Spf}(k[[\mathbf{u}]]/P)$ for a closed prime ideal P. Then, by Proposition 10.2, we can find an injective k-algebra homomorphism $\iota : k[[\mathbf{u}]]/P \hookrightarrow k[[\mathbf{x}]]$ for a set of variables $\mathbf{x} = (x_1, \ldots, x_m)$ with $m = \dim Z$. Take a copy $\mathbf{y}$ of $\mathbf{x}$ and a copy of $\iota : k[[\mathbf{v}]]/P \hookrightarrow k[[\mathbf{y}]]$. Let $g_i(\mathbf{x}) = \iota(u_i)$ [resp., $h_i(\mathbf{y}) = \iota(v_i)$] and set $\mathbf{g}(\mathbf{x}) = (g_1(\mathbf{x}), \ldots, g_d(\mathbf{x}))$ [resp., $\mathbf{h}(\mathbf{y}) = (h_1(\mathbf{y}), \ldots, h_d(\mathbf{y}))$]. The idea of the proof is to show that for each $f_1(\mathbf{u}) \in P$, define $f(\mathbf{u}, \mathbf{v}) := (\mathrm{id} \times \delta_X)^* \circ m^*(f_1(\mathbf{u}))$ and show that $f(\mathbf{g}(\mathbf{x}), \mathbf{h}(\mathbf{y})) = 0$ in $k[[\mathbf{x}, \mathbf{y}]]$ by Proposition 10.4

[note that $\mathbf{h}(\mathbf{y}) = \mathbf{g}(\mathbf{x})|_{\mathbf{x}=\mathbf{y}}$]. If this is the case, any power series $f_1 \in P$ kills $m \circ (\mathrm{id} \times \delta_X)(Z \times Z)$, and hence $m \circ (\mathrm{id} \times \delta_X)(Z \times Z) \subset Z$.

For each $X \in \mathfrak{g}_{/\mathbb{Z}_p}$ and $n \geq n_0$, we have $\exp(p^n X) \in U$ and

$$\rho(\exp(p^n X)) = \mathrm{id}_L + d\rho(X) \sum_{i \geq 1} \frac{p^{in}}{i!} d\rho(X)^{i-1}.$$

Let $E_{p^n X} := \sum_{i \geq 1} \frac{p^{in}}{i!} d\rho(X)^{i-1}$ as an endomorphism of V. By a well-known estimate (cf. [LFE] Sect. 1.3), we have

$$\mathrm{ord}_p(p^{in}/i!) \geq in - \frac{i}{p-1} \geq 0 \tag{10.1.5}$$

(by $n \geq n_0 \geq 2$) and $\lim_{i \to \infty} \left| \frac{p^{in}}{i!} \right|_p = 0$, and hence $E_{p^n X}$ takes L into L; so we regard $E_{p^n X} \in \mathrm{End}_{\mathbb{Z}_p}(L)$. Then $\varepsilon_{p^n X} := E_{p^n X} \otimes \mathrm{id}_{\widehat{\mathbb{G}}_m}$ is an endomorphism of $T = L \otimes_{\mathbb{Z}_p} \widehat{\mathbb{G}}_m = \mathrm{Spf}(k[[\mathbf{u}]])$. Thus, $\varepsilon_{p^n X}$ induces a corresponding k-algebra endomorphism $\varepsilon^*_{p^n X} : k[[\mathbf{u}]] \to k[[\mathbf{u}]]$. Since

$$\varepsilon_{p^n X} - p^n \,\mathrm{id}_L = \sum_{i \geq 2} \frac{p^{in}}{i!} \equiv 0 \mod p^{d_n}$$

for $d_n = \lfloor 2(n - \frac{1}{p-1}) \rfloor \sim 2n$ by (10.1.5), then, in particular, $\lim_{n \to \infty} \frac{p^n}{p^{d_n}} = 0$ in $\mathbb{R}$. Thus, if we write $\phi_{i,n,X}(\mathbf{u}) := \varepsilon^*_{p^n X}(u_i)$, we have $\phi_{i,n} \equiv u_i^{p^n} \mod (\mathbf{u})^{d_n}$. Put $\Phi_n(\mathbf{u}) = \Phi_{n,X}(\mathbf{u}) = (\phi_{1,n,X}(\mathbf{u}), \ldots, \phi_{d,n,X}(\mathbf{u})) \in k[[\mathbf{u}]]$.

Since $\rho(\exp(p^n X)) \otimes \mathrm{id}_{\widehat{\mathbb{G}}_m} \in \mathrm{Aut}_{\mathbb{Z}_p}(L \otimes \widehat{\mathbb{G}}_m) = \mathrm{Aut}_{\mathbb{Z}_p}(T)$, we have

$$\begin{aligned} &\rho(\exp(p^n X)) \otimes \mathrm{id}_{\widehat{\mathbb{G}}_m} \\ &\quad = \mathrm{id}_L \otimes \mathrm{id}_{\widehat{\mathbb{G}}_m} + (d\rho(X) \otimes \mathrm{id}_{\widehat{\mathbb{G}}_m}) \circ \varepsilon_{p^n X} = m \circ (\mathrm{id}_T \times (\delta_X \circ \varepsilon_{p^n X})). \end{aligned}$$

Now pick any $f_1(\mathbf{u}) \in P \subset k[[\mathbf{u}]]$, and recall

$$f(\mathbf{u}, \mathbf{v}) = (m \circ (\mathrm{id}_T \times \delta_X))^*(f_1(\mathbf{u})) = (\mathrm{id}_T \times \delta_X)^* \circ m^*(f_1(\mathbf{u})).$$

Recall $\iota : k[[\mathbf{u}]]/P \hookrightarrow k[[\mathbf{x}]]$ and its copy $\iota : k[[\mathbf{v}]]/P \hookrightarrow k[[\mathbf{y}]]$ made out of Proposition 10.2 applied to $R = k[[\mathbf{u}]]/P$. Via $\iota \widehat{\otimes} \iota : k[[\mathbf{u}]]/P \widehat{\otimes}_k k[[\mathbf{v}]]/P \hookrightarrow k[[\mathbf{x}, \mathbf{y}]]$, we regard $f(\mathbf{u}, \mathbf{v}) = f(\mathbf{g}(\mathbf{x}), \mathbf{h}(\mathbf{y})) \in k[[\mathbf{x}, \mathbf{y}]]$. Then, by the above computation, we find

$$f(\mathbf{g}(\mathbf{x}), \Phi_n(\mathbf{h}(\mathbf{x})) = (f_1 \circ (\rho(\exp(p^n X)) \otimes \mathrm{id}_{\widehat{\mathbb{G}}_m}))(\mathbf{u}) = 0$$

for all $n \geq n_0$, because $(\rho(\exp(p^n X)) \otimes \mathrm{id}_{\widehat{\mathbb{G}}_m})(Z) \subset Z$. We have already verified all the assumptions of Proposition 10.4 for $\mathbf{h}(\mathbf{y}) = \mathbf{g}(\mathbf{x})|_{\mathbf{x}=\mathbf{y}}$ and $\Phi_n(\mathbf{u})$; so, by Proposition 10.4, we find $f(\mathbf{u}, \mathbf{v}) = f(\mathbf{g}(\mathbf{x}), \mathbf{h}(\mathbf{y})) = 0$ in $k[[\mathbf{x}, \mathbf{y}]]$ for independent variables $\mathbf{x}$ and $\mathbf{y}$. This finishes the proof of step 1.

Step 2: *Let* $\{X_{i,j}|i = 1, \ldots, r, j = 1, \ldots, n_i\}$ *be a finite family of elements in* $\mathfrak{g}_{/\mathbb{Z}_p}$, *and write* $\delta_{i,j}$ *for* $\delta_{X_{i,j}} \in \mathrm{End}(T)$. *Take the following morphism of formal groups* $s : \overbrace{T \times T \times \cdots \times T}^{r+1} \to T$ *given by*

$$(x_0, x_1, \ldots, x_r) \mapsto x_0 + \sum_{i=1}^{r} \delta_{i,1} \circ \delta_{i,2} \circ \cdots \circ \delta_{i,n_i}(x_i).$$

Then $s(Z^{r+1}) \subset Z$. *In particular,* $\sigma(Z \times Z) \subset Z$ *for*

$$\sigma(x, y) = x + \sum_{i=1}^{r} \delta_{i,1} \circ \delta_{i,2} \circ \cdots \circ \delta_{i,n_i}(y).$$

Proof of step 2. Since $T = \mathrm{Spf}(k[[\mathbf{u}]])$, $T(k) = \{1_T\}$. Thus, $Z(k) = T(k) = \{1_T\}$. By step 1, $\delta_X(Z) = m(1_T \times \delta_X(Z)) \subset Z$ for all $X \in \mathfrak{g}_{/\mathbb{Z}_p}$, and so we find $\delta_X(Z) \subset Z$. Fix i. We show $\delta_{i,1} \circ \delta_{i,2} \circ \cdots \circ \delta_{i,n_i}(Z) \subset Z$ by induction on n_i. When $n_i = 1$, taking $X = X_{i,1}$, we have $\delta_{i,1}(Z) = \delta_X(Z) \subset Z$. When $n_i > 1$, applying this $\delta_{i,n_i}(Z) = \delta_X(Z) \subset Z$ for $X = X_{i,n_i}$, we have

$$\delta_{i,1} \circ \delta_{i,2} \circ \cdots \circ \delta_{i,n_i}(Z) \subset \delta_{i,1} \circ \delta_{i,2} \circ \cdots \circ \delta_{i,n_i-1}(Z) \subset Z,$$

where the last inclusion comes from the induction hypothesis. Suppose $r = 1$. Then we have

$$\begin{aligned} s(Z \times Z) = (m \circ (\mathrm{id}_L \times \delta_{1,1}))(Z \times \delta_{i,2} \circ \cdots \circ \delta_{i,n_i}(Z)) \\ \subset (m \circ (\mathrm{id}_L \times \delta_{1,1}))(Z \times Z) \subset Z \end{aligned}$$

because $\delta_{i,2} \circ \cdots \circ \delta_{i,n_i}(Z) \subset Z$. Thus, we have shown the result when $r = 1$. Now we proceed by induction on r. Since $z_0 + \sum_{i=1}^{r-1} \delta_{i,1} \circ \delta_{i,2} \circ \cdots \circ \delta_{i,n_i}(z_i)$ with $z_0, z_1, \ldots, z_{r-1} \in Z$ is in Z, by step 1 applied to $\delta_{r,1}$ and $Z \times \delta_{r,2} \circ \cdots \circ \delta_{r,n_r}(Z) \subset Z \times Z$, we get $z_0 + \sum_{i=1}^{r} \delta_{i,1} \circ \delta_{i,2} \circ \cdots \circ \delta_{i,n_i}(z_i)$ for any $z_0, \ldots, z_{r-1}, z_r \in Z$. This finishes the proof of the first statement of step 2.

Note that σ is just the restriction of s to $Z \times \Delta(Z)$ for the diagonal map $\Delta : Z \to Z^{r-1}$. This proves step 2.

Step 3: *We have* $m(Z \times Z) \subset Z$; *so* Z *is stable under multiplication.*

Proof of step 3. By Proposition 10.14, we can find a family of elements

$$\{X_{i,j} \in \mathfrak{g}_{/\mathbb{Z}_p}|i = 1, \ldots, r, j = 1, \ldots, n_i\}$$

such that the linear operator

$$A := \sum_{i=1}^{r} d\rho(X_{i,1}) \circ d\rho(X_{i,2}) \circ \cdots \circ d\rho(X_{i,n_i})$$

is injective on L. Write $\delta = A \otimes \mathrm{id}_{\widehat{\mathbb{G}}_m}$ for the endomorphism of T induced by A. Then $\delta : T \to T$ is an isogeny (in particular, dominant) and preserves

Z by step 2; hence, $\mathrm{id}_Z \times \delta : Z \times Z \to Z \times Z$ is dominant. Again by step 2, $m(Z \times Z) \subset m \circ (\mathrm{id}_Z \times \delta)(Z \times Z) \subset Z$. This shows step 3.

Step 4: *Z is a formal subtorus of the form $L_Z \otimes \widehat{\mathbb{G}}_m$ for a $\mathbb{Z}_p$-direct summand L_Z of L stable under the action of $G(\mathbb{Z}_p)$.*

Proof of step 4. Let $T[p^n] \subset T$ be the kernel of multiplication by p^n as a closed group subscheme. Then $Z_n := T[p^n] \cap Z = T[p^n] \times_T Z$ is a finite scheme stable under multiplication. For any Artinian local k-algebra R, $Z_n(R)$ is therefore a finite semigroup. Any finite semigroup is a group. Thus, $T[p^n] \times_T Z$ is a group scheme. Since $Z = T \times_T Z = \varprojlim_n T[p^n] \times_T Z$, we know that Z is an integral group subscheme of the split formal group T, which is a formal subtorus isomorphic to $L_Z \otimes_{\mathbb{Z}_p} \widehat{\mathbb{G}}_m$ for a $\mathbb{Z}_p$-direct summand L_Z (see [LHV] Chap. 1). Since $L_Z = X_*(Z)$, the action of U on Z induces the action of U on $L_Z = X_*(Z)$, which coincides with the original action of U on L induced on the summand L_Z (because of the faithfulness of the functor $T' \mapsto X_*(T')$ defined on k-split formal tori T' (see [LHV] Chap. 1). Thus, L_Z is stable under U.

Note that $U \subset G(\mathbb{Q}_p)$ is Zariski dense and that $\rho : G_{/\mathbb{Q}_p} \to GL_{/V}$ is a morphism of algebraic groups. Then $G(\mathbb{Q}_p)$ leaves $V_Z = \mathbb{Q}_p \cdot L_Z \subset V$ stable by the Zariski density of U. Since $L_Z = V_Z \cap L$, L_Z is stable under $G(\mathbb{Z}_p)$. This finishes the proof of the theorem.

Corollary 10.7 *Let $Z \subset T$ be a reduced closed formal subscheme stable under the action of U [an open subgroup in $G(\mathbb{Z}_p)$ as above]. If the representation (ρ, G) does not contain the trivial representation $\mathbf{1}_G$ as a subquotient, then there exists a finite index set I and $\mathbb{Z}_p$-direct summands $\{L_i \subset L\}_{i \in I}$ indexed by I stable under $G(\mathbb{Z}_p)$ such that $Z = \bigcup_{i \in I} L_i \otimes_{\mathbb{Z}_p} \widehat{\mathbb{G}}_m \subset L \otimes_{\mathbb{Z}_p} \widehat{\mathbb{G}}_m = T$.*

Proof. Let I be the set of irreducible components of Z. Then U acts on the finite set I by permutation. For each irreducible component Z_i of Z, the stabilizer $U_i \subset U$ is still an open subgroup of $G(\mathbb{Z}_p)$. Applying the above theorem to Z_i, we get the result. In particular, L_i is stable under $G(\mathbb{Z}_p)$.

10.2 Representation of Lie Algebras

Here is a summary of the results we used on the representation of Lie algebras. If we replace representations of a Lie algebra $\mathfrak{g}$ by representations of a group G on a finite-dimensional vector space, the exposition is close to the one given in [MFG] Sect. 2.1. In other words, the results presented here are also valid for group representations once we replace the statement for Lie algebras with the corresponding statements for groups. As for books on Lie algebras, see [REP] for representations on a $\mathbb{C}$-vector space, [LAG] III for Lie algebras over general fields, and [BLI] for more general cases.

10.2.1 Algebras

Let R be an algebra (which can be noncommutative). The algebra R is called *simple* if there are no two-sided ideals of R except for $\{0\}$ and R. An R-module M is called *irreducible* or *simple* if $M \neq 0$ and any R-submodule $N \subsetneq M$ is trivial. Thus, M is irreducible $\Leftrightarrow M \cong R/\mathfrak{m}$ for a maximal left ideal $\mathfrak{m}$ of R.

Let M be an R-module of finite type. Then for any given proper R-submodule M_0 of M, we consider the set S of all proper R-submodules of M containing M_0. Here the word "proper" means that $M_0 \neq M$. If X is an ordered subset of S, then $M_X = \bigcup_{N \in X} N$ is an R-submodule of M. Here the word "ordered" means that if $N, N' \in X$, we can find $N'' \in X$ such that $(N \cup N') \subset N''$. If $M_X = M$, we find an element N of X such that $M = N$ because M is finitely generated over R. This contradicts our assumption that S is made of proper submodules. Thus, we have $M_X \neq M$ and $M_X \in S$. Namely, any ordered sequence in S has an upper bound in S. Then, by Zorn's lemma, S has a maximal element. This shows the existence of a maximal proper R-submodule containing a given M_0.

Let $J(M)$ (*radical* of M) be the intersection of all proper maximal R-submodules of M. Then if M is of finite type over R, $J(M) = M$ implies that $M = 0$. Let $J = J(R)$ be the intersection of all maximal left ideals of R, which is called the *radical* of R. Then $R \neq J$. If M is irreducible, then the annihilator $\operatorname{Ann}(M) = \{r \in R | rM = 0\}$ of M is a maximal left ideal of R, because $R/\operatorname{Ann}(M) \cong M$ via $r \mapsto rm$ for any $0 \neq m \in M$. Thus, $J \subset \bigcap_{M:\text{irreducible}} \operatorname{Ann}(M)$. Pick r in the intersection. Then $r(R/\mathfrak{m}) = 0$ for any maximal left ideal $\mathfrak{m}$, since $M = R/\mathfrak{m}$ for a maximal left ideal $\mathfrak{m}$ is irreducible. Thus, $r \in rR \subset \mathfrak{m}$, and hence we have

$$J = \bigcap_{M:\text{irreducible}} \operatorname{Ann}(M). \tag{10.2.1}$$

Since $rM = 0$ implies $rxM = 0$ for all $x \in R$, $Jx \subset \cap_{M:\text{irreducible}} \operatorname{Ann}(M) = J$. Thus, J is a two-sided ideal. We claim

$$\textit{For every } r \in J,\ 1 - r \in R^\times. \tag{10.2.2}$$

Proof. Since $xr \in J$ for $x \in R$ and $r \in J$, we prove $1 - xr \in R^\times$ in place of $1 - r \in R^\times$. Suppose, on the contrary, that $1 - xr$ does not even have a left inverse. Then $R(1-xr) \neq R$, and hence, there exists a maximal left ideal $\mathfrak{m} \supset R(1-xr)$. Thus, $1-xr \in \mathfrak{m}$ and $xr \in J \subset \mathfrak{m}$, which show $1 = 1-xr+xr \in \mathfrak{m}$, a contradiction. Thus, $1 - xr$ has a left inverse. In particular, we find $s \in R$ such that $s(1 - r) = 1$. Then $s = 1 + sr = 1 - (-s)r$. Applying the above argument for $x = -s$, we find $t \in R$ such that $ts = t(1 - (-s)r) = 1$. Then $t = t(s(1 - r)) = (ts)(1 - r) = 1 - r$. This shows that $(1 - r)s = ts = 1$ and $1 - r \in R^\times$. □

Suppose that $\mathfrak{a} \subset R$ is a two-sided ideal with the property that $r \in \mathfrak{a} \Rightarrow 1 - r \in R^\times$. If $\mathfrak{a} \not\subset J$, then there exists a maximal left ideal $\mathfrak{m}$ such that

$\mathfrak{a} \not\subset \mathfrak{m}$. Then $\mathfrak{a} + \mathfrak{m} = R$ and $a + m = 1$ for $a \in \mathfrak{a}$ and $m \in \mathfrak{m}$. Then $\mathfrak{m} \supset Rm = R(1-a) = R$ because $1 - a \in R^\times$. This is a contradiction. Thus, $\mathfrak{a} \subset J$. We thus have

$$\textit{If } a \in \mathfrak{a} \Rightarrow 1 - a \in R^\times \textit{ for a left (resp., right, two-sided) ideal } \mathfrak{a}, \textit{ then } \mathfrak{a} \subset J. \tag{10.2.3}$$

By (10.2.3), we conclude

$$J = \bigcap_{\mathfrak{m}:\text{maximal right ideals}} \mathfrak{m} = \bigcap_{\mathfrak{m}:\text{maximal two-sided ideals}} \mathfrak{m}. \tag{10.2.4}$$

Lemma 10.8 (Krull–Azumaya, Nakayama)

1. *Let M be an R-module of finite type. If $M = JM$, then $M = 0$.*
2. *Let A be a commutative local ring and M be an A-module. If either the A-module M is of finite type or $\mathfrak{m}_A^N M = 0$ for a sufficiently large integer N, then $M = \mathfrak{m}_A M$ implies $M = 0$.*

This follows from the two facts (i) $J(M) = M \Rightarrow M = 0$ and (ii) $JM \subset J(M)$ under the assumption of the lemma (cf. [CRT] Theorem 2.2).

Corollary 10.9 *Let A be a local ring. Let M and N be A-modules and $f : M \to N$ be an A-linear map. Suppose either that $\mathfrak{m}_A$ is nilpotent or that N is an A-module of finite type. Then if f induces a surjection $\overline{f} : M/\mathfrak{m}_A M \to N/\mathfrak{m}_A N$, then f itself is surjective.*

Proof. Consider $X = N/f(M)$. By assumption, $X/\mathfrak{m}_A X = \mathrm{Coker}(\overline{f}) = 0$. Thus, $X = \mathfrak{m}_A X$, and hence by Lemma 10.8, $X = 0$.

10.2.2 Modules over Lie Algebras

Let E be a field of characteristic 0. A *Lie algebra* $\mathfrak{g}$ over E is a vector space over E with an E-linear Lie bracket map $[\cdot,\cdot] : \mathfrak{g} \times \mathfrak{g} \to \mathfrak{g}$ satisfying $[x,y] = -[y,x]$ and $[x,[y,z]] + [z,[x,y]] + [y,[z,x]] = 0$ for all $x, y, z \in \mathfrak{g}$.

Let $\mathfrak{g}$ be a finite-dimensional Lie algebra over E with an E-linear Lie bracket $[\cdot,\cdot] : \mathfrak{g} \times \mathfrak{g} \to \mathfrak{g}$. For an E-vector space V of finite dimension n, a representation $\rho : \mathfrak{g} \to \mathrm{End}_E(V)$ is an E-linear map with $\rho([x,y]) = \rho(x)\rho(y) - \rho(y)\rho(x)$ for all $x, y \in \mathfrak{g}$. A representation $\rho : \mathfrak{g} \to \mathrm{End}_E(V)$ is called *reducible* if $V = V(\rho)$ has a proper nontrivial subspace stable under $\mathfrak{g}$. A representation is called *irreducible* if it is not reducible. An irreducible representation with coefficients in E can be reducible as a representation into $\mathrm{End}_L(V \otimes_E L)$ for a field extension L/E. A representation $\rho : \mathfrak{g} \to \mathrm{End}_E(V)$ is called *absolutely irreducible* if $\rho_{\overline{E}}$ on $V(\rho) \otimes_E \overline{E}$ is irreducible for an algebraic closure $\overline{E}/E$.

Let $R = R(\rho)$ be the E-subalgebra of the E-linear endomorphism algebra of $V = V(\rho)$ generated over E by $\rho(\sigma)$ for all $\sigma \in \mathfrak{g}$. Then $R \subset \mathrm{End}_E(V) \cong$

$M_n(E)$, and hence R is a finite-dimensional algebra over E; so it is Artinian and noetherian. Let $J = J(R)$. Then the sequence $\{J^n\}_n$ stabilizes for a sufficiently large $n = N$. From Nakayama's lemma (Lemma 10.8), we conclude that $J^N = \{0\}$. Thus, $\bigcap_{\mathfrak{m}:\text{max. two-sided}} \mathfrak{m}^N \subset J^N = 0$. For a maximal two-sided ideal $\mathfrak{m}$ and an ideal $\mathfrak{a} \not\subset \mathfrak{m}$, we see that $\mathfrak{m}+\mathfrak{a} = R$, and hence, $\mathfrak{m}^j+\mathfrak{a}^j = R$ for all $j > 0$. Applying the Chinese remainder theorem ([CRT] Theorem 1.4), we have $R = \prod_{\mathfrak{m}} R/\mathfrak{m}^N$ for sufficiently large N, where $\mathfrak{m}$ runs over all maximal two-sided ideals of R. We write $R_{\mathfrak{m}}$ for $R/\mathfrak{m}^N$. If V is irreducible, then $J = \mathfrak{m}$ for a single maximal two-sided ideal. Then $V \neq \mathfrak{m}V$, again by Lemma 10.8. Therefore, $\mathfrak{m}V = 0$. Since R acts faithfully on V, we conclude that $\mathfrak{m} = 0$. Thus, R is a simple algebra over E. We have shown that

$$V \text{ is irreducible} \Rightarrow R \text{ is simple.}$$

Thus, the study of irreducible $\mathfrak{g}$-modules is reduced to the study of R-modules for a simple E-algebra R.

10.2.3 Semisimple Algebras

To study modules over simple algebras, we start slightly more generally. Here we only assume R to be an Artinian algebra. An R-module V is called *completely reducible* if it is a direct sum of irreducible modules. The algebra R is called *semisimple* if its radical $J = J(R)$ vanishes. Since maximal left ideals of $R/J(R)$ correspond bijectively to maximal left ideals of R by the homomorphism theorem, we see that $J(R/J(R)) = 0$. This shows that the quotient $R/J(R)$ is semisimple. Now, the following three statements are equivalent:

(SS1) R is semisimple.
(SS2) The left R-module R is completely reducible.
(SS3) Every R-module V of finite length is completely reducible.

Proof. We first prove the implication: (SS1)$\Rightarrow$ (SS2): Let Ω be the set of all maximal left ideals of R. Then $\mathfrak{m} + \mathfrak{n} = R$ for two distinct elements $\mathfrak{m}, \mathfrak{n} \in \Omega$. Then, by the Chinese remainder theorem, $\bigcap_{\mathfrak{m}\in\Omega} \mathfrak{m} = J(R) = 0$ implies that $R \cong \bigoplus_{\mathfrak{m}\in\Omega} R/\mathfrak{m}$. Since R is Artinian, Ω is a finite set. By the homomorphism theorem, R-submodules of $R/\mathfrak{m}$ correspond bijectively to left ideals between $\mathfrak{m}$ and R. This shows that $R/\mathfrak{m}$ is irreducible, since $\mathfrak{m}$ is a maximal left ideal.

(SS2)$\Rightarrow$(SS3): Since $R \cong \bigoplus_{\mathfrak{m}\in\Omega} R/\mathfrak{m}$, we have minimal left ideals $I_{\mathfrak{m}}$ indexed by $\mathfrak{m} \in \Omega$ such that $R = \bigoplus_{\mathfrak{m}\in\Omega} I_{\mathfrak{m}}$ with $I_{\mathfrak{m}} \cong R/\mathfrak{m}$ as left R-modules. Then $1 = \oplus_{\mathfrak{m}\in\Omega} e_{\mathfrak{m}}$ for $e_{\mathfrak{m}} \in I_{\mathfrak{m}}$. Multiplying the left-hand and right-hand sides by $a \in I_{\mathfrak{m}}$, we get $I_{\mathfrak{m}} \ni a = a1 = \oplus_{\mathfrak{m}\in\Omega} a e_{\mathfrak{m}}$, and therefore, $ae_{\mathfrak{n}} = \delta_{\mathfrak{m},\mathfrak{n}} a$, where $\delta_{\mathfrak{m},\mathfrak{n}} = 1$ or 0 depending on whether or not $\mathfrak{m} = \mathfrak{n}$. Thus, $I_{\mathfrak{m}} = Re_{\mathfrak{m}}$. Replacing a by $e_{\mathfrak{m}}$, we get $e_{\mathfrak{m}}^2 = e_{\mathfrak{m}}$ and $e_{\mathfrak{m}}e_{\mathfrak{n}} = 0$ if $\mathfrak{m} \neq \mathfrak{n}$. Consider an R-linear map $\varphi : I_{\mathfrak{m}} \to I_{\mathfrak{m}}v$ given by $\varphi(i) = iv$ for $v \in V$. Since $\mathrm{Ker}(\varphi)$ is an R-submodule of $I_{\mathfrak{m}}$, the irreducibility of $I_{\mathfrak{m}}$ tells us that either $\mathrm{Ker}(\varphi) = 0$ or $\mathrm{Ker}(\varphi) = I_{\mathfrak{m}}$.

Thus, either $I_{\mathfrak{m}}v \cong I_{\mathfrak{m}}$ or $I_{\mathfrak{m}}v = 0$. In particular, if $I_{\mathfrak{m}}v \neq 0$, then $\sum_{\mathfrak{m}\in\Omega} e_{\mathfrak{m}}v = v \in \sum_{\mathfrak{m}\in\Omega} I_{\mathfrak{m}}v$. From this, we conclude that $V = \sum_{v\in V-\{0\}} \sum_{\mathfrak{m}\in\Omega} I_{\mathfrak{m}}v$. Then we can find $\mathfrak{m}_j$ and v_j such that $I_{\mathfrak{m}_j}v_j \neq 0$ and $V = \bigoplus_j I_{\mathfrak{m}_j}v_j \cong \bigoplus_j I_{\mathfrak{m}_j}$ as R-modules, since V is of finite length.

(SS3)$\Rightarrow$(SS1): By (SS3), R is the direct sum of irreducible R-submodules $I_{\mathfrak{m}}$. By Lemma 10.8, $I_{\mathfrak{m}} \neq J(R)I_{\mathfrak{m}}$. Since $I_{\mathfrak{m}}$ is irreducible, $J(R)I_{\mathfrak{m}} = 0$. Then $J(R) = J(R)R = \bigoplus_{\mathfrak{m}} J(R)I_{\mathfrak{m}} = 0$. □

We now decompose $\Omega = \Omega_1 \bigsqcup \cdots \bigsqcup \Omega_\lambda$ so that $\mathfrak{m}, \mathfrak{n} \in \Omega_j \iff I_{\mathfrak{m}} \cong I_{\mathfrak{n}}$ as left R-modules. Then we write $R_j = \bigoplus_{\mathfrak{m}\in\Omega_j} I_{\mathfrak{m}}$. If $\mathfrak{m} \in \Omega_j$, $I_{\mathfrak{m}}a$ for $a \in R$ is isomorphic to either$I_{\mathfrak{m}}$ or 0, and therefore, it has to be inside R_j. Thus, $R_jR \subset R_j$; hence, R_j is a two-sided ideal, and $R = \bigoplus_j R_j$ is an algebraic direct sum. Returning to the original setting and applying the above argument to $R/J(R)$, we have

(S) *An artinian algebra R has only one maximal two-sided ideal if and only if there is a unique isomorphism class of irreducible R-modules*

since the set of maximal two-sided ideals of R naturally corresponds to that of $R/J(R)$ bijectively. We now claim that the following three assertions are equivalent (Wedderburn's theorem):

(S1) R is a simple algebra;
(S2) R is a direct sum of mutually isomorphic minimal left ideals;
(S3) $R \cong M_n(D)$ for a division algebra D,

where a *division algebra* D is an algebra such that $D - \{0\} = D^\times$.

Proof. (S1) $\Rightarrow$ (S2): This follows from (S) because minimal left ideals are all irreducible.

(S2) $\Rightarrow$ (S3): Let V be the minimal left ideal of R. Then $R \cong V^n$ as a left R-module. Then $\mathrm{End}_R(R) \cong M_n(\mathrm{End}_R(V))$. Pick $\phi : V \to V \in \mathrm{End}_R(V)$. Then $\mathrm{Ker}(\phi) = 0$ or V and $\mathrm{Im}(\phi) = V$ or 0 because of the irreducibility of V. This shows that ϕ is either bijective or the zero map; hence, $D = \mathrm{End}_R(V)$ is a division algebra. On the other hand, it is easy to see that $\phi \mapsto \phi(1)$ induces $\mathrm{End}_R(R) \cong R$, because $\phi_a(x) = xa$ gives an element in $\mathrm{End}_R(R)$ with $\phi_a(1) = a$.

(S3) $\Rightarrow$ (S1): Let $\mathfrak{a}$ be a two-sided ideal of R. If $\mathfrak{a}$ has nonzero element a, multiplying a by elementary matrices from left and right, we may assume that $a = dE_{ij}$ for $0 \neq d \in D$ with the elementary matrix E_{ij} having nontrivial (i,j)-entry. Thus, $E_{ij} = d^{-1}a \in \mathfrak{a}$. Then, again multiplying E_{ij} by elementary matrices, we find $E_{ij} \in \mathfrak{a}$ for all (i,j). Thus, $\mathfrak{a} = R$. □

Let R be a simple Artinian algebra. Let $\overline{E}$ be the center of R. Then by (S3), $\overline{E}$ is a field, which is the center of D. Suppose that $\overline{E}$ is algebraically closed. Then for any $x \in D$, $\overline{E}[x] \subset D$ is a finite field extension of $\overline{E}$. Thus, $\overline{E} = \overline{E}[x]$ and $x \in \overline{E}$. This shows that $\overline{E} = D$ and

(S4) *If R is a finite-dimensional simple algebra over an algebraically closed field $\overline{E}$, then $R \cong M_n(\overline{E})$.*

Let $\rho : \mathfrak{g} \to M_n(E)$ be an absolutely irreducible representation for a field E. Let R be the E-subalgebra of $M_n(E)$ generated by $\rho(g)$ for all $g \in \mathfrak{g}$. By absolute irreducibility, $R \otimes_E \overline{E}$ remains simple for an algebraic closure $\overline{E}$ of E. Since there is no simple algebra except for matrix algebras over an algebraically closed field, $R \otimes_E \overline{E} \cong M_{n'}(\overline{E})$ for some $n' \leq n$ if ρ is absolutely irreducible. Pick $x \neq 0$ in V. We consider a map $f : R \otimes_E \overline{E} \to V \otimes_E \overline{E}$ induced by $r \mapsto rx$. Then $Ker(f)$ is a maximal left ideal of $M_{n'}(\overline{E})$, and hence $V \otimes_E \overline{E} \cong \overline{E}^{n'}$. This shows that $n = n'$, and $R \cong M_n(E)$. We record here what we have proven:

Proposition 10.10 *Suppose that $\rho : \mathfrak{g} \to M_n(E)$ for a field E is absolutely irreducible. Let R be a E-subalgebra of $M_n(E)$ generated by $\rho(g)$ for all $g \in \mathfrak{g}$. Then*

1. (Burnside) $R \cong M_n(E)$ *for* $n = \dim_E V$.
2. (Schur) *If a linear map $f : V \to V$ commutes with ρ, then f is induced by a scalar multiplication.*

The second assertion follows from the first, as a linear endomorphism of E^n commuting with all matrices in $M_n(E)$ is a scalar.

This result shows that $V(\rho) \otimes_E X$ [or $\rho : \mathfrak{g} \to M_n(X)$] remains irreducible for an arbitrary field extension X/E if ρ is absolutely irreducible.

10.2.4 Differential of Group Representations

Let $G_{/E} = \mathrm{Spec}(\mathcal{O}_G)$ be a connected affine algebraic group. Consider E-derivations $\partial, \delta : \mathcal{O}_G \to \mathcal{O}_G$. By computation, we verify that $\Delta := [\partial, \delta] = \partial \circ \delta - \delta \circ \partial$ satisfies the derivation relation $\Delta(ab) = a\Delta(b) + b\Delta(a)$. Thus $Der_E(\mathcal{O}_G) = Der_E(\mathcal{O}_G, \mathcal{O}_G)$ is a Lie algebra. The multiplication $x \mapsto gx$ induces by pull-back a ring automorphism $g^* : \mathcal{O}_G \to \mathcal{O}_G$. If $\partial \circ g^* = g^* \circ \partial$, we call ∂ a left invariant derivation. Since $[\partial, \delta]$ is left invariant if ∂ and δ are left invariant, we define the Lie algebra $\mathfrak{g}$ of G to be the Lie algebra of left invariant derivations. By evaluation at the identity $1 \in G$, we may (and do) identify $\mathfrak{g}$ with the tangent space at 1 of $\mathfrak{g}$. There is another way of defining $\mathfrak{g}$ as the kernel of the homomorphism $G(E[\epsilon]) \to G(E))$ induced by the projection $E[\epsilon] \twoheadrightarrow E$, where $E[\epsilon] = E[x]/(x^2)$ (for the polynomial ring $E[x]$) with $\epsilon = x \mod (x^2)$. Indeed, writing $\underline{1} : \mathcal{O}_G \to E$ for the co-identity, for any $P \in \mathrm{Hom}_{ALG_{/E}}(\mathcal{O}_G, E[\epsilon]) = G(E[\epsilon])$ projecting down to $1 \in G(E)$, $P(a)$ for $a \in \mathcal{O}_G$ has the form $\underline{1}(a) + \partial(a)\epsilon$. We verify $\partial(ab) = \underline{1}(a)\partial(b) + \underline{1}(b)\partial(a)$ from $P(ab) = P(a)P(b)$, and hence ∂ is in the tangent space at 1. Similarly, if P projects down to $x \in G(E)$, writing $P(a) = \underline{x}(a) + \partial_x(a)\epsilon$, ∂_x is a tangent vector at x. From this, it is clear that the tangent space at 1 extends isomorphically to the Lie algebra of left invariant derivations.

Exercise 10.11 *Write down explicitly the Lie bracket* $[x, y]$ *for two given tangent vectors* x, y *at* $1 \in G$ *without extending* x, y *to vector fields over* G.

Let G' be another connected linear algebraic group defined over E with a morphism $\rho : G \to G'$ of group schemes over E. Write $\mathfrak{g}'$ for the Lie algebra of G'. Then its *differential* $d\rho = \rho_* : \mathfrak{g} \to \mathfrak{g}'$ is a homomorphism of Lie algebras.

Exercise 10.12 *Check that* $d\rho$ *is a homomorphism of Lie algebras.*

An E-rational representation $\rho : G_{/E} \to GL(n)_{/E}$ is a homomorphism of E-group schemes. Write $\mathfrak{gl}_n(E)$ for the Lie algebra of $GL(n)_{/E}$. Then $\mathfrak{gl}_n(E)$ can be identified with the Lie algebra of $n \times n$ matrices $M_n(E)$ with entries in E whose Lie bracket is given by $[x, y] = xy - yx$ for the matrix product xy with $x, y \in M_n(E)$. Then $d\rho : \mathfrak{g} \to \mathfrak{gl}_n(E)$ is an E-linear representation of Lie algebras. Since ρ is complex analytic after extending scalars to $\mathbb{C}/E$ (after embedding E into $\mathbb{C}$), ρ is absolutely irreducible if and only if $d\rho$ is absolutely irreducible. Over $\mathbb{C}$, the exponential map $\exp_G : \mathfrak{g}_{/\mathbb{C}} = \mathfrak{g} \otimes_E \mathbb{C} \to G(\mathbb{C})$ given by the usual formula $\exp_G(X) = \sum_{n=0}^{\infty} \frac{X^n}{n!}$ converges absolutely, giving an analytic surjection of $\mathfrak{g}_{/\mathbb{C}}$ onto $G(\mathbb{C})$ whose inverse is given by $\log_G(g) = \sum_{n=1}^{\infty} (-1)^{n+1} \frac{(g-1)^n}{n}$. Then we see easily that $\exp_{GL(n)}(d\rho(X)) = \rho(\exp_G(X))$ and $\log_{GL(n)}(\rho(g)) = d\rho(\log_G(g))$ for all $X \in \mathfrak{g}_{/\mathbb{C}}$ and $g \in G(\mathbb{C})$ (see [GME] Sect. 4.3.3 for a p-adic version of this). From this, the above equivalence on absolute irreducibility of ρ and $d\rho$ is clear.

Lemma 10.13 *Let* K *be an infinite field and* $\{V_1, \ldots, V_b\}$ *be a finite set of finite-dimensional* K*-vector spaces. Let* $A_1, \cdots, A_r$ *be* K*-linear endomorphism of* $V = \bigoplus_{j=1}^{b} V_j$ *preserving* V_j $(j = 1, \ldots, b)$. *If, for each* $j = 1, \ldots, b$, *there exists* $i = 1, \ldots, r$ *with* $\det(A_i|_{V_j}) \neq 0$, *then there exists* $(\lambda_1, \ldots, \lambda_r) \in K^r$ *such that the* K*-linear endomorphism* $\sum_{i=1}^{r} \lambda_i A_i$ *of* V *is invertible on* V.

Proof. Consider the polynomial $f(t_1, \ldots, t_r) := \det(\sum_{i=1}^{r} t_i A_i) \in k[t] = K[t_1, \ldots, t_r]$. Since A_i preserves V_j for all j, we have

$$f(t) = \prod_{j=1}^{b} \det(\sum_{i=1}^{r} t_i A_i|_{V_j}).$$

If $f(t) \neq 0$ in $K[t]$, $\mathrm{Spec}(K[t, f^{-1}])$ is a rational variety defined over K birationally equivalent to a projective space $\mathbf{P}^{r-1}$, which has Zariski densely populated K-rational points $\mathbf{P}^{r-1}(K)$ (here we use the infinity of $|K|$). Therefore, $\mathrm{Spec}(K[t, f^{-1}])$ has Zariski densely populated K-rational points. Thus, we only need to show that $f(t) \neq 0$. For each j, $f_j(t) = \det(\sum_{i=1}^{r} t_i A_i|_{V_j}) \neq 0$, because we have i_0 such that $\det(A_{i_0}|_{V_j}) \neq 0$. Thus, we get $f_j \neq 0$ and its product $f \neq 0$ as desired.

Proposition 10.14 *Let* G *be a connected linear algebraic group defined over a field* E *of characteristic* 0. *Let* V *be a finite-dimensional vector space over*

E, and let $\rho : G \to GL_V$ be an E-rational representation with differential $d\rho : \mathfrak{g} \to \mathrm{End}_E(V)$. Then the following two statements are equivalent:

1. *The trivial representation $\mathbf{1}_G$ is not a subquotient of ρ.*
2. *There exist elements $\{w_{i,j} \in \mathfrak{g}\}_{i=1,\ldots,r,j=1,\ldots,n_i}$ such that*

$$\sum_{i=1}^{r} d\rho(w_{i,1}) \circ d\rho(w_{i,2}) \circ \cdots \circ d\rho(w_{i,n_i})$$

is invertible as an endomorphism of V.

Proof. Let $F : V = V_0 \supset V_1 \supset V_2 \supset \cdots V_{l-1} \supset V_l = 0$ be the Jordan–Hölder filtration under the action of G. For a linear operator T preserving the Jordan–Hölder filtration F, we write $gr(T)$ for the linear operator on $gr(V) = \bigoplus_{j=0}^{l} V_j/V_{j+1}$. Then $gr(T)$ is invertible if and only if $\det(T) \neq 0$. Thus, we may replace V by its semisimplification under the action of G. Then it is clear that if $\mathbf{1}_G$ is a subquotient of ρ, assertion (2) fails because $d\mathbf{1}_G = 0$. This shows (2)$\Rightarrow$(1).

Suppose (1). Since we only need to compute the determinant of the linear operator in (2), we may replace V by its semisimplification. Thus, V is a direct sum of irreducible G-modules V_j: $V = \bigoplus_{j=1}^{b} V_j$. By (S3), the E-subalgebra R_j of $\mathrm{End}_E(V_j)$ generated by $d\rho(X)$ for $X \in \mathfrak{g}$ is isomorphic to $M_{n_j}(D)$ for a division algebra $D_{/E}$ (except for the case where $\rho = \mathbf{1}_G$, which is excluded by our assumption). In particular, we can find $A_j \in R_j$ of the form

$$\sum_{i=1}^{r} d\rho(w_1) \circ d\rho(w_2) \circ \cdots \circ d\rho(w_{n_i})$$

with $w_\ell \in \mathfrak{g}$ such that $\det(A_j|_{V_j}) \neq 0$. Then, applying Lemma 10.13 to $(A_1, \ldots, A_b)$, we can find λ so that $\sum_j \lambda_j A_j$ is invertible on V. This linear operator $\sum_j \lambda_j A_j$ has the form specified in (2).

11

Hecke Stable Subvariety Is a Shimura Subvariety

In this last chapter, we prove that any Hecke invariant subvariety of a product of copies of mod-p Shimura curve is a Shimura subvariety, as stated in Theorems 3.43 and 3.44. From this result, we pull out a linear independence result (Corollary 11.13) more general than Theorem 3.38.

11.1 Stability under Hecke Correspondence

Recall the action of $g \in G(\mathbb{A}^{(\infty)})$ on the Shimura curve given by the pullback of the level structure $\eta \mapsto \eta \circ g$ in Chap. 7. The action of $\tau(g)$ on a closed subvariety of Sh is called a Hecke action here because it induces Hecke operators. We study a subvariety stable under many $\tau(g)$, in particular, for g belonging to a subgroup that is the image of a torus over an imaginary quadratic field.

11.1.1 Locally Linear Subvariety

We recall the definition of locally linear subvarieties for a product of copies of Shimura curves given in [Ch1] Sect. 5 for the Siegel Hilbert modular variety. Let $Sh^{\mathrm{ord}}_{/\mathbb{F}} = Sh^{(p)}[\frac{1}{H}]$ for the ordinary locus (for the Hasse invariant H). Fix a closed point $x \in Sh^{\mathrm{ord}}_{/\mathbb{F}}$ ($\mathbb{F} = \overline{\mathbb{F}}_p$) carrying a test object $(E_x, \eta^{(p)})$ (thus, E_x has complex multiplication by an order $\mathfrak{O}_x$ of an imaginary quadratic field M with $\mathfrak{O}_x \otimes_{\mathbb{Z}} \mathbb{Z}_p \cong \mathbb{Z}_p \times \mathbb{Z}_p$ as $\mathbb{Z}_p$-algebras; see Corollaries 6.39 and 6.40). Let V be the irreducible component of $Sh^{(p)}_{/\mathbb{F}}$ containing x, and put $V^{\mathrm{ord}} = V \cap Sh^{\mathrm{ord}}$. Let $m \geq 1$ be a positive integer. Suppose that $\mathcal{Z}$ is an irreducible closed subvariety of $(V^{\mathrm{ord}})^n = V^{\mathrm{ord}} \times V^{\mathrm{ord}} \times \cdots \times V^{\mathrm{ord}}$ defined over $\mathbb{F}$.

(L1) Let $z = (z_1, \ldots, z_m)$ ($z_j \in V^{\mathrm{ord}}$) be any closed point of $\mathcal{Z}$. We say that $\mathcal{Z}$ is *locally linear at* z if the formal completion of $\mathcal{Z}$ at z is a formal subtorus of the Serre–Tate formal torus $\prod_{j=1}^{n} \widehat{V^{\mathrm{ord}}_{z_j}} \cong \widehat{\mathbb{G}}_m^n$.

(L2) We say that $\mathcal{Z}$ is locally linear if it is *locally linear* at every closed point of $\mathcal{Z}$.

H. Hida, *Elliptic Curves and Arithmetic Invariants*, Springer Monographs in Mathematics, DOI 10.1007/978-1-4614-6657-4_11,

(L3) Denote by $f : Y \to \mathcal{Z}$ the normalization of $\mathcal{Z}$. We say that $\mathcal{Z}$ is *weakly locally linear* if, for every closed point y of Y, the morphism induced by f on the formal completion $\widehat{Y}_y$ of Y along y is an isomorphism of $\widehat{Y}_y$ to a formal subtorus of the Serre–Tate formal torus $\widehat{\mathbb{G}}_m^n$ [at $f(y) \in V^n$].

In [Ch1], local linearity is called Tate linearity, and the definition of locally linear subvarieties is given for a closed subvariety of the ordinary locus of the Siegel modular variety. Since $(V^{\mathrm{ord}})^n$ has a canonical closed immersion into a Siegel modular variety, this definition is equivalent to Chai's definition. Chai conjectured that a weakly locally linear subvariety is actually locally linear (see [Ch1] 5.3.1), which has been shown to hold for weakly locally linear subvarieties of the ordinary locus of Hilbert modular varieties (see [Ch1] Theorem 8.6 and [H10a] Sect. 3.5).

If $\mathcal{Z}$ is a variety with a morphism $\pi : \mathcal{Z} \to (V^{\mathrm{ord}})^m$ and if, for a closed point $z \in \mathcal{Z}$, π induces an embedding of $\widehat{\mathcal{Z}}_z$ into the formal completion of $(V^{\mathrm{ord}})^m$ at $\pi(z)$, we can still speak of locally linearity (and weak locally linearity) at z of $\mathcal{Z}$ (we make this abuse often later).

11.1.2 An Example of Hecke Stable Subvariety

We start preparing for a proof of key results on the linear independence of arithmetic modular functions/forms (Theorem 11.12 and Corollary 11.13 below, which imply Theorem 3.38). We study the skew diagonal image of a Shimura curve into its product twisted by a transcendental automorphism of the deformation space $\widehat{S}$ over W at x. We keep the notation introduced in Chap. 7. Thus, we have $\widehat{S} = \widehat{\mathbb{G}}_m$, where $\widehat{\mathbb{G}}_m$ denotes the completion of $\mathbb{G}_m$ over W along the identity 1 in the special fiber at p. The definition of the local linearity of the previous subsection applies to this case.

Reversing the setting, let M be an imaginary quadratic field with integer ring $\mathfrak{O}$ and $x \in Ig(\mathbb{F})$ be a CM point corresponding to an ordinary elliptic curve with complex multiplication by $\mathfrak{O}$ of type (M, id_x). We may assume that the point x has expression $x = [z, g] = [z, 1] \cdot g$ for $g \in G(\mathbb{Z}_p \times \mathbb{A}^{(p\infty)})$. Indeed, we can choose the CM elliptic curve E_x so that its lattice $L = g^{-1}\mathcal{L}_z$ (which is a fractional M-ideal prime to p) is given by $\mathbb{Z} + \mathbb{Z}z$ (here, placing $\mathbb{Z}z$ at the end, we already incorporated the conjugation by J_1). By our choice, $\mathcal{L}_z \otimes_{\mathbb{Z}} \mathbb{Z}_p = \mathfrak{O}_p = \mathfrak{O}_{\mathfrak{p}} \oplus \mathfrak{O}_{\mathfrak{p}^c}$; so we may choose g with $\eta_x = \eta_z \circ g$ so that $g_p \in G(\mathbb{Z}_p)$ is the matrix of change of basis from $(-1, z) \in \mathbb{Q}_p^2$ to the basis $(1_{\mathfrak{p}}, 1_{\mathfrak{p}^c}) \in \mathbb{Q}_p^2$ for the idempotents $1_{\mathfrak{p}} \in \mathfrak{O}_{\mathfrak{p}}$ and $1_{\mathfrak{p}^c} \in \mathfrak{O}_{\mathfrak{p}^c}$. The level-$p^\infty$ structure η_p^{ord} (of E_x) sends $a \in \mathbb{Z}_p$ to $a \cdot 1_{\mathfrak{p}^c}$, identifying $E_x[p^\infty]^{\mathrm{et}}$ with $M_{\mathfrak{p}}/\mathfrak{O}_{\mathfrak{p}^c}$ by $\mathcal{L}_z \otimes_{\mathbb{Z}} \mathbb{Z}_p = \mathfrak{O}_{\mathfrak{p}} \oplus \mathfrak{O}_{\mathfrak{p}^c}$.

For this choice $x = [z, g]$, we recall the representation $r_x : T_x \to G$ defined over $\mathbb{A}^{(\infty)}$ given at the beginning of Sect. 7.1.8 and the quotient torus $\mathcal{T}(\mathbb{Z}_{(p)}) = T_x(\mathbb{Z}_{(p)})/T(\mathbb{Z}_{(p)})$ defined just above (7.2.5). As studied in Lemma 7.43, the image $r_x(\mathcal{T}(\mathbb{Z}_{(p)}))$ gives the stabilizer of $x = [z, g]$ in $G(\mathbb{Z}_p \times \mathbb{A}^{(p\infty)})/Z(\mathbb{Z}_{(p)})$. We simply write ρ for r_x hereafter.

We call an open compact subgroup K in $G(\mathbb{A}^{(\infty)})$ *maximal at* p if $K = K_p \times K^{(p)}$ with $K_p = GL_2(\mathbb{Z}_p)$ and $K^{(p)} \subset G(\mathbb{A}^{(p\infty)})$. Let V_K for K maximal at p be the geometrically irreducible component containing x in the reduction $(Sh^{(p)}/K)_{/\mathbb{F}}$ modulo p of the Kottwitz model. Let $V = \varprojlim_K V_K$ for K running through open compact subgroups of $G(\mathbb{A}^{(\infty)})$ maximal at p. Strictly speaking, the point x gives rise to a projective system of points $x_K \in V_K(\mathbb{F})$ (the image of x in V_K), but we write this point as $x \in V_K(\mathbb{F})$.

The formal completion $\widehat{S}$ of V along x is isomorphic to $\widehat{\mathbb{G}}_m$, whose automorphism group is isomorphic to $\mathbb{Z}_p^\times$. Through the injective homomorphism (7.2.5): $\alpha \mapsto \alpha^{1-c}$, we regard $\mathcal{T}(\mathbb{Z}_{(p)})$ as a subgroup of $\mathbb{Z}_p^\times$, identifying $\mathfrak{O}_\mathfrak{p}$ with $\mathbb{Z}_p$ by the inclusion $\mathbb{Z} \hookrightarrow \mathfrak{O}$.

Let $\mathcal{O}_{V,x} = \varinjlim_K \mathcal{O}_{V_K,x}$ be the stalk of V at x, and let $S = \mathrm{Spec}(\mathcal{O}_{V,x})$. The local ring $\mathcal{O}_{V,x}$ is a dense subring of the affine ring $\widehat{\mathcal{O}}_S$ of $\widehat{S}$.

Take $a_1, \ldots, a_m \in \mathbb{Z}_p^\times$. By the action of a_j on $\widehat{S}$ (and hence on $\widehat{\mathcal{O}}_S$) given by $t \mapsto t^{a_j}$, we have an algebra homomorphism

$$\phi : \overbrace{\mathcal{O}_{V,x} \otimes_\mathbb{F} \cdots \otimes_\mathbb{F} \mathcal{O}_{V,x}} \to \widehat{\mathcal{O}}_S \text{ sending } f_1 \otimes \cdots \otimes f_m \text{ to } \prod_{j=1}^m a_j(f_j) \in \widehat{\mathcal{O}}_S. \tag{11.1.1}$$

If a_j's ($j = 1, \ldots, m$) are pairwise distinct modulo $\mathcal{T}(\mathbb{Z}_{(p)})$, we would like to prove that ϕ is injective. Thus, for a nonconstant modular function $f \in \mathcal{O}_{V,x}$, $\{a_1(f), \ldots, a_m(f)\}$ are linearly independent over $\mathbb{F}$. Thus, we study $\mathrm{Ker}(\phi)$ for $a_1, \ldots, a_m \in \mathbb{Z}_p^\times$.

Since $\rho(\mathcal{T}(\mathbb{Z}_{(p)}))$ fixes x (Lemma 7.43), $\mathcal{T}(\mathbb{Z}_{(p)})$ acts on $\mathcal{O}_S$ by ring automorphisms, and by Proposition 7.44, this action is compatible (and commutes) with the action of $\mathbb{Z}_p^\times$ via the embedding $\mathcal{T}(\mathbb{Z}_{(p)}) \hookrightarrow \mathbb{Z}_p^\times$. Thus, we have $\phi(\alpha(f_1) \otimes \cdots \otimes \alpha(f_m)) = \alpha^{1-c}(\phi(f_1 \otimes \cdots \otimes f_m))$ for all $\alpha \in \mathcal{T}(\mathbb{Z}_{(p)})$. In other words, the closed subscheme $\mathrm{Spec}(\mathrm{Im}(\phi)) \subset S^m = S \times \cdots \times S$ is stable under the diagonal action of $\mathcal{T}(\mathbb{Z}_{(p)})$ on S^m. Thus, we study in the following couple of propositions the (local) structure of a closed subscheme of S^m stable under the diagonal action of $\mathcal{T}(\mathbb{Z}_{(p)})$. After determining the structure of such formal subschemes, we will globalize the result to reach our desired conclusion of the injectivity of ϕ if the a_j's are independent [i.e., pairwise distinct modulo $\mathcal{T}(\mathbb{Z}_{(p)})$].

Since $\mathrm{Ker}(\phi)$ is a prime ideal of $\mathcal{O}_{V,x} \otimes \cdots \otimes \mathcal{O}_{V,x}$ stable under the diagonal action of $\mathcal{T}(\mathbb{Z}_{(p)})$, it is induced by an irreducible closed (pro-)subscheme $X \subset V^m$ passing through $x^m = (x, x, \ldots, x)$. In other words, X is the Zariski closure in V^m of $\mathrm{Spec}(\mathcal{O}_{V,x} \otimes \cdots \otimes \mathcal{O}_{V,x}/\mathfrak{b})$ for the prime ideal $\mathfrak{b} = \mathrm{Ker}(\phi)$. More precisely, for any open affine subscheme $U = \mathrm{Spec}(R) \subset V^m$ containing x^m (so $R \hookrightarrow \mathcal{O}_{V^m,x}$), we have $U \cap X = \mathrm{Spec}(R/R \cap \mathfrak{b})$. By definition, X is stable under the diagonal action of $\mathcal{T}(\mathbb{Z}_{(p)})$. A subvariety of V^m stable under the diagonal action of $\mathcal{T}(\mathbb{Z}_{(p)})$ is called a *Hecke stable* subvariety.

11.1.3 Stability under Toric Action, Local Study

We take a more general setting than the example given above, specified as follows (we use the following notation throughout).

(N0) Let $S = \mathrm{Spec}(\mathcal{O}_{V,x})_{/\mathbb{F}}$ and $S_K = \mathrm{Spec}(\mathcal{O}_{V_K,x})_{/\mathbb{F}}$ with their formal completion $\widehat{S}$ and $\widehat{S}_K$ along x isomorphic to $\widehat{\mathbb{G}}_m$.

(N1) For a prime ideal $\mathfrak{b} \subset \mathcal{O}_{V,x}^{\otimes m} := (\mathcal{O}_{V,x} \otimes \cdots \otimes \mathcal{O}_{V,x})$ (the m-fold tensor product) stable under a p-adically open subgroup $\mathbb{T}$ of $\mathcal{T}(\mathbb{Z}_{(p)})$, we write $\mathcal{X}_{/\mathbb{F}} = \mathrm{Spec}(\mathcal{O}_{V,x}^{\otimes m}/\mathfrak{b}) \subset S^m$ and let $\mathcal{Y} \twoheadrightarrow \mathcal{X}$ be the normalization.

(N2) $\widehat{X} \subset \widehat{S}^m$ is a formal completion of $\mathcal{X}$ along its closed point $x^m = (x, \ldots, x)$.

(N3) $X_{/\mathbb{F}}$ is the Zariski closure of $\mathcal{X}$ in V^m (so X is stable under $\mathbb{T}$ and $\widehat{X}$ is the formal completion of X along x^m). Let $Y \twoheadrightarrow X$ be the normalization. Write $X = \varprojlim_K X_K \subset V^m$ with irreducible closed subschemes $X_K \subset V_K^m$ (the image of X in V_K^m) and $Y_K \twoheadrightarrow X_K$ for the normalization of X_K (so $Y = \varprojlim_K Y_K$), where K runs over open compact subgroups of $G(\mathbb{A}^{(\infty)})$ with $K = K^{(p)} \times G(\mathbb{Z}_p)$.

Suppose $m = 1$. Since $\dim_{\mathbb{F}} V = \dim_{\mathbb{F}} Ig = 1$, the irreducible reduced $\mathbb{T}$-invariant closed subscheme X is either a single point $\{x\}$ or V itself, and hence, we conclude $\mathfrak{b} = 0$; in particular, $\mathrm{Ker}(\phi) = 0$, taking $\mathfrak{b}$ to be $\mathrm{Ker}(\phi)$.

We restate Proposition 4.88 for elliptic curves:

Lemma 11.1 *Let $E \to Z$ be an ordinary elliptic curve over a reduced excellent affine base scheme Z over $\overline{\mathbb{F}}_p$. For a closed point $s \in Z(\overline{\mathbb{F}}_p)$ and a formal completion $\widehat{Z}$ along s, if the p-divisible group $E[p^\infty] \times_Z \widehat{Z}$ splits into a product of its connected component $E[p^\infty]^\circ_{/\widehat{Z}}$ and étale quotient $E[p^\infty]^{et}_{/\widehat{Z}}$, then on an open subscheme $U \subset Z$ containing s, the p-divisible group $E[p^\infty]_{/U}$ canonically splits into the product $E[p^\infty]^\circ \oplus E[p^\infty]^{et}_{/U}$ for the connected component $E[p^\infty]^\circ_{/U}$ and the étale quotient $E[p^\infty]^{et}_{/U}$.*

Here the words "reduced excellent Z" means that $Z = \mathrm{Spec}(R)$ with excellent reduced ring R (see Sect. 4.1.11 about excellent rings).

Exercise 11.2 *Give a counterexample against the assertion of the above lemma if $Z = \mathrm{Spec}(\mathbb{Z}_{(p)})$.*

Our goal is to prove the injectivity of ϕ for general $m > 1$ in (11.1.1) under the assumption that the a_j's are pairwise distinct modulo $\mathcal{T}(\mathbb{Z}_{(p)})$ (as the result is trivial if $m = 1$). The injectivity is equivalent to $\mathrm{Spec}(\mathrm{Im}(\phi)) = S^m$; hence, we study the local property of $\mathrm{Spec}(\mathrm{Im}(\phi))$ to show that $\dim(\mathrm{Im}(\phi)) < \dim S^m$ implies the equality of two of the a_j modulo $\mathcal{T}(\mathbb{Z}_{(p)})$. The following result dealing with the local structure of $\mathrm{Spec}(\mathrm{Im}(\phi))$ when $m > 1$ is a key to prove the linear independence.

Proposition 11.3 *Let the notation be as in* (N0–3). *In particular, for a positive integer $m \geq 2$, let $\mathcal{X}$ be a closed integral subscheme of S^m containing $x^m = (x, x, \ldots, x)$ for the closed point $x \in S$, and let $\Pi_{\mathcal{Y}} : \mathcal{Y} \to \mathcal{X}$ be the normalization of $\mathcal{X}$. Write $S^m = S' \times S''$ for the first $(m-1)$-factor $S' = S^{m-1} \subset S^m$ and the last factor $S'' = S$. Suppose that the projection to S' induces a dominant morphism $\pi_{\mathcal{X}} : \mathcal{X} \to S'_{/\mathbb{F}}$. Suppose further that $\mathcal{X}$ is stable under the diagonal action of a subgroup $\mathbb{T}$ of $\mathcal{T}(\mathbb{Z}_{(p)}) \subset \mathrm{Aut}(\mathcal{O}_{V,x})$, whose p-adic closure is open in $\mathcal{T}(\mathbb{Z}_p)$. Then*

(1) *$\mathcal{Y}$ has finitely many points y over x^m and is locally linear at every point y over x^m; so $\widehat{\mathcal{Y}}_y = \widehat{\mathbb{G}}_m \otimes_{\mathbb{Z}_p} L$ for an $\mathbb{Z}_p$-direct summand L of $X_*(\widehat{S}^m)$. Moreover, the isomorphism class of L as a $\mathbb{Z}_p$-module is independent of y.*
(2) *$\mathcal{Y}$ is smooth over $\mathbb{F}$ and is flat over S'.*
(3) *Either $\mathcal{X} = S^m$ or $\mathcal{X}$ is finite over S' via $\pi_{\mathcal{X}}$. If $\mathcal{X}$ is finite over S', $\mathcal{Y}$ is finite flat over S'.*
(4) *If $\pi_{\mathcal{X}} \circ \Pi_{\mathcal{Y}}$ induces a surjection of the tangent space at one $y \in \mathcal{Y}$ over x^m onto that of S' at x^{m-1} and $\mathcal{X}$ is a proper subscheme of S^m, then $\pi_{\mathcal{X}} \circ \Pi_{\mathcal{Y}} : \mathcal{Y} \to S'$ is étale.*

Later in Corollaries 11.8 and 11.11, we prove that $\mathcal{X}$ is smooth; hence, $\mathcal{Y} = \mathcal{X}$.

Proof. Let $\widehat{S}$ be the formal completion of S along x. By Serre–Tate theory, we have $\widehat{S} \cong \widehat{\mathbb{G}}_m$. Since $\mathcal{X}$ is dominant over S', we have $\dim \mathcal{X} > 0$. Let K be an open compact subgroup of $G(\mathbb{A}^{(\infty)})$ maximal at p, and recall $S_K = \mathrm{Spec}(\mathcal{O}_{V_K,x})$. We assume that K is small so that S/S_K is étale. Consider the image of $\mathcal{X}_K$ of $\mathcal{X}$ in S^m_K. Then $\mathcal{X}_K = \mathrm{Spec}(\mathcal{R}_K)$ for an integral domain $\mathcal{R}_K$ and $\mathcal{X} = \mathrm{Spec}(\mathcal{R})$ for $\mathcal{R} = \varinjlim_K \mathcal{R}_K$. Since $\mathcal{R}_K$ is a localization of an integral domain of finite type over $\mathbb{F}$, $\mathcal{R}_K$ is excellent (see Sect. 4.1.11). Thus, its formal completion $\widehat{\mathcal{X}}_K \cong \widehat{\mathcal{X}}$ along x^m is reduced (by Corollary 4.37).

Since $\mathcal{X}$ is stable under the diagonal action of the subgroup $\mathbb{T}$ of $\mathcal{T}(\mathbb{Z}_{(p)}) \subset \mathrm{Aut}(\mathcal{O}_{V,x})$ and $\mathcal{X}$ is integral, by Corollary 10.7, $\widehat{\mathcal{X}}$ is a union of finitely many formal $\mathbb{Z}_p$-submodules of $\widehat{S}^m$ of the form $\widehat{\mathbb{G}}_m \otimes L$ for $\mathbb{Z}_p$-direct summands L of the cocharacter group $X_*(\widehat{S}^m) \cong \mathbb{Z}_p^m$: $\widehat{\mathcal{X}} = \bigcup_{L \in I} \widehat{\mathbb{G}}_m \otimes L$ for a finite index set I of $\mathbb{Z}_p$-direct summands L of $X_*(\widehat{S}^m)$. In particular, $\widehat{\mathcal{X}}$ is stable under the action of $\mathcal{T}(\mathbb{Z}_{(p)})$ (not just $\mathbb{T}$) diagonally embedded into $\mathrm{Aut}_{GSCH}(\widehat{S})^m$. From this, we claim that $\mathcal{X}$ and X are stable under $\mathcal{T}(\mathbb{Z}_{(p)})$. Indeed, for the $\mathfrak{m}_{x^m}$-adic completion $\widehat{\mathfrak{b}}$ of $\mathfrak{b}$, by [CRT] Theorem 8.1 (ii), $\widehat{\mathcal{X}} = \mathrm{Spf}(\widehat{\mathcal{O}_{V,x}^{\otimes m}}/\widehat{\mathfrak{b}})$ and $\widehat{\mathfrak{b}}$ is stable under $\mathcal{T}(\mathbb{Z}_{(p)})$ as $\widehat{\mathcal{X}} = \bigcup_L \widehat{\mathbb{G}}_m \otimes L$. Then $\mathfrak{b} = \widehat{\mathfrak{b}} \cap \mathcal{O}_{V,x}^{\otimes m}$ by faithful flatness of $\widehat{\mathcal{O}_{V,x}^{\otimes m}}$ over $\mathcal{O}_{V,x}^{\otimes m}$ (cf. [BCM] I.2.6); so $\mathfrak{b}$ is stable under $\mathcal{T}(\mathbb{Z}_{(p)})$, and hence, $\mathcal{X} = \mathrm{Spec}(\mathcal{O}_{V,x}^{\otimes m}/\mathfrak{b})$ is stable under $\mathcal{T}(\mathbb{Z}_{(p)})$. Then the Zariski closure X of $\mathcal{X}$ is stable under $\mathcal{T}(\mathbb{Z}_{(p)})$.

The normalization $\mathcal{Y} \to \mathcal{X}$ is given by $\varprojlim_K \mathcal{Y}_K$ for the normalization $\mathcal{Y}_K$ of $\mathcal{X}_K$. Naturally, the semigroup $\mathrm{End}_{SCH}(\mathcal{X})$ of endomorphisms of the scheme $\mathcal{X}$ acts on $\mathcal{Y}$; in particular, $\mathcal{T}(\mathbb{Z}_{(p)})$ acts on $\mathcal{Y}$. The formal completion $\widehat{\mathcal{Y}}_y$ along

y for each point $y \in \mathcal{Y}$ over x^m is isomorphic to the formal completion $\widehat{\mathcal{Y}}_{K,y}$ of $\mathcal{Y}_K$ along the image of y in $\mathcal{Y}_K$. The scheme $\mathcal{Y}_K$ is excellent, because $\mathcal{Y}_K$ is finite over $\mathcal{X}_K$ [see Remark 4.35 and [EGA] IV.7.8.3 (ii), (vii)]. Since $\widehat{\mathcal{Y}}_{K,y}$ is the normalization of $\widehat{\mathcal{X}}_K$ and $\widehat{\mathcal{Y}}_{K,y}$ is integral (by Corollary 4.37), points y_L of $\mathcal{Y}$ over x^m are indexed by the irreducible components of $\widehat{\mathcal{X}}$ and hence by $L \in I$ so that $\widehat{\mathcal{Y}}_{K,y_L} = \widehat{\mathbb{G}}_m \otimes L$. Since $\mathcal{Y} \to S'$ is dominant, for at least one $L_0 \in I$, the projection $L_0 \otimes_{\mathbb{Z}} \mathbb{Q} \to X_*(\widehat{S}') \otimes_{\mathbb{Z}} \mathbb{Q}$ is surjective [i.e., the image of L_0 is of finite index in $X_*(\widehat{S}')$].

Recall that we denote by X the Zariski closure of $\mathcal{X}$ in V^m and the normalization $\Pi : Y \to X$ of X. Again, $\mathrm{End}_{SCH}(X)$ acts on Y; in particular, $\mathcal{T}(\mathbb{Z}_{(p)})$ acts on Y. We have $Y = \varprojlim_K Y_K$ for the normalization Y_K of X_K. We look at the tangent sheaf Θ_Z for $Z = V^m, Y, X, S^m$, and $\widehat{S}^m$. Here Θ_Z is the Zariski sheaf dual $\mathrm{Hom}_{Z_{Zar}}(\Omega_{Z/\mathbb{F}}, \mathcal{O}_Z)$ as in Sect. 4.1.6. We have the tangent sheaf $\Theta_X \subset \Theta_{V^m}$, but Θ_X may not be locally free around x^m since X may have singularity at x^m. Let $y_L \in Y$ be a point above x^m with $\widehat{Y}_{y_L} = \widehat{\mathbb{G}}_m \otimes L$. Since $\widehat{Y}_{y_L} = \widehat{\mathbb{G}}_m \otimes L$, $\Theta_{\widehat{Y}_{y_L}} = L \otimes_{\mathbb{Z}} \mathcal{O}_{\widehat{Y}_{y_L}}$, and by the irreducibility of Y, $\mathrm{rank}_{\mathbb{Z}_p} L \otimes_{\mathbb{Z}} \mathbb{Z}_p = \mathrm{rank}_{\mathcal{O}_Y} \Theta_Y = \mathrm{rank}_{\mathbb{Z}_p} L_0 \otimes_{\mathbb{Z}} \mathbb{Z}_p$. Thus, $L \cong L_0$ as $\mathbb{Z}_p$-modules for all $L \in I$, and $\widehat{\mathcal{Y}}$ is equidimensional. This proves assertion (1).

If $L_0 \otimes_{\mathbb{Z}} \mathbb{Q} = X_*(\widehat{S}^m) \otimes_{\mathbb{Z}} \mathbb{Q}$, we get $\mathcal{X} = S^m$, and we are done. Hereafter we assume $\mathrm{rank}_{\mathbb{Z}_p} L_0 < \mathrm{rank}_{\mathbb{Z}_p} X_*(\widehat{S}^m)$ ($\Leftrightarrow L_0 \otimes_{\mathbb{Z}} \mathbb{Q} \neq X_*(\widehat{S}^m) \otimes_{\mathbb{Z}} \mathbb{Q}$); so $\mathrm{rank}_{\mathbb{Z}_p} L < \mathrm{rank}_{\mathbb{Z}_p} X_*(\widehat{S}^m)$ for all $L \in I$. Recall the decomposition $S^m = S' \times S''$ with the last factor $S'' = S$. Similarly, we decompose $V^m = V' \times V''$ for the product V' of the first $(m-1)$ copies of V and the last copy $V'' = V$. If $\pi : \widehat{\mathbb{G}}_m \otimes L \to \widehat{S}'$ is not dominant for $L \in I$, the image $\pi(\widehat{\mathbb{G}}_m \otimes L)$ is a closed formal subscheme of $\widehat{S}'$ distinct from $\widehat{S}'$. In particular, $\widehat{\mathcal{Y}} \to \widehat{S}'$ is not flat over $\pi(\widehat{\mathbb{G}}_m \otimes L) \subsetneq \widehat{S}'$. Define the nonflat locus $Y^{nfl} \subset Y$ as the Zariski closure of the set of closed points $y \in Y$ such that $\mathcal{O}_{Y,y}$ is not flat over $\mathcal{O}_{V',v}$ for the image $v \in V'$ of y. By Proposition 5.3, the nonflat locus $y_L \in Y^{nfl} \subset Y$ is a nonempty closed subscheme strictly smaller than Y, because $\widehat{\mathbb{G}}_m \otimes L_0 \subset \widehat{\mathcal{Y}} = \widehat{Y}_{y_{L_0}}$ is flat over $\widehat{S}'$ and flatness is an open property. Since the formal completion of Y^{nfl} at y_L contains $\widehat{\mathbb{G}}_m \otimes L$, $\dim Y^{nfl} = \dim Y$, and so Y cannot be irreducible (because Y^{nfl} is a closed subscheme of Y strictly smaller than Y), a contradiction. Thus, $\widehat{\mathbb{G}}_m \otimes L \to \widehat{S}'$ is flat for all $L \in I$, and hence $\widehat{\mathcal{Y}}$ is flat over $\widehat{S}'$. Since $\widehat{S}'/S'$ is faithfully flat, $\mathcal{Y}$ is flat over S'. In particular, for all $L \in I$, $\widehat{\mathbb{G}}_m \otimes L$ is dominant finite over $\widehat{S}'$. This proves assertion (2).

Since $\pi : \widehat{\mathcal{X}} \to \widehat{S}'$ induces a surjection $\pi_* : L \otimes_{\mathbb{Z}} \mathbb{Q} \to X_*(\widehat{S}') \otimes_{\mathbb{Z}} \mathbb{Q}$, the intersection $\widehat{S}'' \cap (\widehat{\mathbb{G}}_m \otimes L)$ has dimension equal to

$$\delta := \mathrm{rank}_{\mathbb{Z}_p} L_0 + \mathrm{rank}_{\mathbb{Z}_p} X_*(\widehat{S}) - \mathrm{rank}_{\mathbb{Z}_p} X_*(\widehat{S}^m) = \mathrm{rank}_{\mathbb{Z}_p} L_0 + 1 - m$$

for all $L \in I$. Since $\mathcal{X} \to S'$ is dominant, $\dim \mathcal{X} = \mathrm{rank}_{\mathbb{Z}_p} L_0 \geq m - 1$. Thus, $\delta = 0$ or 1. Then $\widehat{S}'' \cap (\widehat{\mathbb{G}}_m \otimes L)$ is a formal $\mathbb{Z}_p$-submodule isomorphic to $\widehat{\mathbb{G}}_m$ or the origin in $\widehat{S}'' = \widehat{S}$. If $\delta = 1$, we have $\mathrm{rank}_{\mathbb{Z}_p} L_0 = \mathrm{rank}\, X_*(\widehat{S}^m)$, and hence $\mathcal{X} = S^m$; so we are done. We hereafter assume that $\delta = 0$. Since $\pi : \widehat{\mathbb{G}}_m \otimes L \to \widehat{S}'$ is dominant and $\dim \widehat{\mathbb{G}}_m \otimes L = \dim \widehat{S}' = \mathrm{rank}_{\mathbb{Z}_p} L$, $\pi : \widehat{\mathbb{G}}_m \otimes L \to \widehat{S}'$ is finite flat for all $L \in I$. Thus, $\mathcal{X} \to S'$ is finite, and $\mathcal{Y}$ is finite flat over S' by the faithful flatness of $\widehat{\mathcal{Y}}_y$ over S'. This proves assertion (3).

To prove assertion (4), consider the differential sheaf $\Omega_{Y/S'}$. We may assume that $(\pi_{\mathcal{X}} \circ \Pi_{\mathcal{Y}})_* : \Theta_Y \to \Theta_{S'}$ is surjective at $y_0 = y_{L_0}$. Since $\widehat{Y}_{y_0}/\widehat{S}'$ is finite flat, $\widehat{Y}_{y_0}$ is étale over $\widehat{S}'$, and hence $\mathcal{Y}_{y_0}/S'$ is étale. Thus, $\Omega_{Y/S'}|_{\mathcal{Y}_{y_0}} = 0$ (Lemma 4.32), and hence $\Omega_{Y/S'}$ vanishes on a nonempty open subscheme of Y. Thus, the support $Y^{\mathrm{ram}} \subset Y$ of $\Omega_{Y/S'}$ is a closed subscheme of Y strictly smaller than Y. If $\mathcal{Y}_{y_L}/S'$ is not étale, $\mathcal{Y}_{y_L} \subset Y^{\mathrm{ram}}$. Thus, Y^{ram} is a closed subscheme of dimension equal to $\dim Y$; so it is an irreducible component of Y, and hence Y is reducible, a contradiction. Thus, $\mathcal{Y} \to S'$ is étale finite.

Remark 11.4 Let the notation and assumption be as in Proposition 11.3. We suppose that $m = 2$ and that $\mathcal{Y} \to \mathcal{X} \subsetneq S^2$ has two dominant projections onto the left and right factors S of S^2. We write the Serre–Tate coordinate (induced by the ordinary level structure on E_x) of the left factor (resp., the right factor) of $\widehat{S}^2$ as t (resp., t'). Then by Theorem 10.1, the formal completion $\widehat{\mathcal{Y}}_y$ of $\mathcal{Y}$ along a point y above $x^2 = (x, x)$ is canonically isomorphic to a formal subtorus of $\widehat{S}^2$ given by $\widehat{\mathbb{G}}_m \otimes L$ for a $\mathbb{Z}_p$-free direct summand L of $\mathbb{Z}_p^2$. Thus, if $\dim \mathcal{X} = \dim V$, $\widehat{\mathcal{Y}}_y$ is defined by the equation $t^u = t'^v$ for nonzero divisors $u, v \in \mathbb{Z}_p$ with $u\mathbb{Z}_p + v\mathbb{Z}_p = \mathbb{Z}_p$, and $L \subset \mathbb{Z}_p^2$ is given by $L = \{(x, y) \in \mathbb{Z}_p^2 | ux = vy\}$. If two projections are étale, (u, v) can be chosen to be $(1, a)$ for a unit $a = v/u \in \mathbb{Z}_p^\times$ (otherwise, one of the projections involves the pth power map, which cannot be étale).

Corollary 11.5 *Let $\mathfrak{b} = \mathrm{Ker}(\phi)$ for ϕ as in (11.1.1). Let $y \in Y$ be a point above x^m and let $\widehat{Y}_y$ be the formal completion of Y along y. Then $\widehat{Y}_y$ for at least one $y \in Y$ contains $\widehat{\Delta} = \{(t^{a_1}, \dots, t^{a_m}) | t \in \widehat{S}\}$, and for $i = 1, 2, \dots, m$, writing $\widehat{S}_i$ for the ith copy of $\widehat{S}$ in $\widehat{S}^m$, the projection $X_*(\widehat{Y}_y) \to X_*(\widehat{S}_i)$ is surjective, considering $\widehat{Y}_y \subset \widehat{S}^m$. In particular, if $m = 2$ and $\pi_{\mathcal{X}}$ is finite, $\mathcal{Y} \to S'$ is étale finite.*

Proof. We have the following commutative diagram:

$$\begin{array}{ccc} \mathcal{O}_{V,x} \otimes_{\mathbb{F}} \cdots \otimes_{\mathbb{F}} \mathcal{O}_{V,x} & \xrightarrow{\phi} & \widehat{\mathcal{O}}_S \\ {\scriptstyle\iota}\downarrow & & \Vert\downarrow \\ \widehat{\mathcal{O}}_{V,x} \otimes_{\mathbb{F}} \cdots \otimes_{\mathbb{F}} \widehat{\mathcal{O}}_{V,x} & \xrightarrow[\Phi]{} & \widehat{\mathcal{O}}_S, \end{array} \tag{11.1.2}$$

where $\widehat{\mathcal{O}}_{V,x} = \widehat{\mathcal{O}}_S$, $\Phi(f_1 \otimes \cdots \otimes f_m) = \prod_{j=1}^m a_j(f_j)$, and the map ι is the tensor product of the natural inclusion $\mathcal{O}_{V,x} \subset \widehat{\mathcal{O}}_{V,x}$. Thus, $\mathrm{Ker}(\Phi) \supset \iota(\mathfrak{b})$. Note that $\mathrm{Spf}(\widehat{\mathcal{O}}_{V,x} \otimes_{\mathbb{F}} \cdots \otimes_{\mathbb{F}} \widehat{\mathcal{O}}_{V,x}/\mathrm{Ker}(\Phi))$ is the skew diagonal image $\widehat{\Delta}$ in $\widehat{S}^m$. If we take the formal completion along x^m, the map ι brings $\widehat{\Delta}$ into $\widehat{\mathcal{X}}$ because $\mathrm{Ker}(\Phi) \supset \iota(\mathfrak{b})$. Thus, an irreducible component $\widehat{\mathbb{G}}_m \otimes L$ of $\widehat{\mathcal{X}}$ contains $\widehat{\Delta}$. In particular, if $m = 2$, $\pi_{\mathcal{X},*}(L)$ contains $\pi_{\mathcal{X},*}(X_*(\widehat{\Delta})) = X_*(\widehat{S}')$. Then the rest follows from assertion (4) of the above proposition.

11.1.4 Stability under Toric Action, Global Study

We keep the assumption and the notation in (N0–3) for $\mathfrak{b} = \mathrm{Ker}(\phi)$ (ϕ as in (11.1.1)). We have globalized $\mathcal{X}$, taking its Zariski closure X in V^m. We start with the simplest case where $m = 2$. There are two possibilities by the above result that $\dim \mathcal{X} = \dim V$ or $X = V^2$. The latter case implies that ϕ is injective as desired; so we are done. Assuming $\dim \mathcal{X} = \dim V$, we take the Zariski closure X of $\mathcal{X}$ in $V \times V$ and its normalization $\Pi : Y \to X$. We study X (resp., Y) as a global irreducible subvariety of the self-product $V \times V$ (resp., as a correspondence $V \leftarrow Y \to V$). We are going to show in Corollary 11.8 after two preparatory propositions that $X = Y$ and that the variety X is the graph of an automorphism of V given by an action of an element in $G(\mathbb{A}^{(p\infty)})$ [in other words, a_1/a_2 has to be in $\mathcal{T}(\mathbb{Z}_{(p)})$]. In this process of showing that X is a graph of an automorphism, we repeatedly use the fact that the diagonal action of $\mathcal{T}(\mathbb{Z}_{(p)})$ preserves X in $V \times V$ and extends to Y.

The subvariety X is a graph of an automorphism of V (as a correspondence in $V \times V$) if and only if the projections $\pi_j : X \to V$ ($j = 1, 2$) are isomorphisms. The only information we have is the following: (i) X is stable under the diagonal action of $\mathcal{T}(\mathbb{Z}_{(p)})$ (or a finite index subgroup thereof) and (ii) the formal completion $\mathcal{Y}$ at any point $y \in Y$ over $(x, x) \in X$ has isomorphic projections to the formal completion of V at x (that is, we know that the two projections $\Pi_j = \Pi \circ \pi_j$ of Y to V are étale infinitesimally around y). Thus, out of (i) and (ii), we need to show that π_j ($j = 1, 2$) are isomorphisms. We will do this by the following two steps:

Step 1. We show that Π_j is étale over a dense open subscheme of V (this is basically achieved by Propositions 11.6 and 11.7).

Step 2. We show that the two pullbacks $\mathbb{Y}_j := \Pi_j^* \mathbf{E}$ ($j = 1, 2$) of the universal elliptic curve $\mathbf{E}_{/V}$ by Π_j are isogenous over Y. Writing the prime-to-p-level structure of $\mathbb{Y}_j$ as $\boldsymbol{\eta}_j := \Pi_j^* \boldsymbol{\eta}$, for the isogeny $\varphi : \mathbb{Y}_1 \to \mathbb{Y}_2$, we have $\varphi \circ \boldsymbol{\eta}_1 = \boldsymbol{\eta}_2 \circ g$ for some $g \in G(\mathbb{A}^{(p\infty)})$, and we conclude $Y = X$ and X is the graph $\Delta_{1,g} \subset (V \times V)$ of $\tau(g)$ (Corollary 11.8).

After finishing off the case $m = 2$, we proceed by induction on m and show that $a_i/a_j \in \mathcal{T}(\mathbb{Z}_{(p)})$ (for some $i \neq j$) if $X \neq V^m$:

Step 3. Under a suitable assumption on $X \subset V^m$, by induction on m, we show that an irreducible subvariety $X \subset V^m$ [containing $(x, x, \dots, x)$] stable under the diagonal action of $\mathcal{T}(\mathbb{Z}_{(p)})$ (or its p-adically open subgroup) is contained in $V^{m-2} \times \Delta_{1,g}$ after permuting the components V. We get this result by applying step 2 to the projected image of X to the product $V \times V$ of the last two factors in V^m after a permutation (Corollary 11.11).

Then the linear independence of $\{f_{[\mathfrak{b}]}(t^{[\mathfrak{b}]^-})\}_{\mathfrak{b}}$ in Theorem 3.38 for elements $\{[\mathfrak{b}]\} = \{a_j|j = 1, \ldots, h\}$, which are mutually distinct modulo $\mathcal{T}(\mathbb{Z}_{(p)})$, follows easily from this (Corollary 11.13).

In step 2, the two elliptic curves $\mathbb{Y}_{j/Y}$ are isogenous if and only if $\mathrm{End}^{\mathbb{Q}}(\mathbb{Y}_{/Y}) = M_2(\mathbb{Q})$ for $\mathbb{Y} := \mathbb{Y}_1 \times_Y \mathbb{Y}_2$. Our argument is by contradiction, supposing $\mathrm{End}^{\mathbb{Q}}(\mathbb{Y}_{/Y}) = \mathbb{Q} \times \mathbb{Q}$. Since V_K [for an open subgroup $K \subset G(\widehat{\mathbb{Z}})$] is actually defined over a finite field, the generic fiber of $\mathbb{Y}_{/Y}$ is an abelian scheme over a field of finite type over the prime field $\mathbb{F}_p$; so, generically, we can use finiteness theorems of Tate–Zarhin on the endomorphism ring of an abelian scheme over a field of finite type over $\mathbb{F}_p$. Since π_j is étale over an open subset, we can study $\mathbb{Y}_1 \times_Y \mathbb{Y}_2$, specializing it to many CM points, and in such a way, we exhibit a contradiction against the assumption $\mathrm{End}(\mathbb{Y}_{/Y}) = \mathbb{Q} \times \mathbb{Q}$. This type of arguments is impossible just by studying $\mathcal{X}$, because the scheme $\mathcal{X}$ has only one closed point and the function field of $\mathcal{X}$ is not of finite type over $\mathbb{F}_p$.

The main tool in step 1 is the local information from the Serre–Tate coordinates we studied earlier in this chapter and Zariski's main theorem (or, equivalently, the Stein factorization of the projections $X \to V$). Adding cusps, we have a unique smooth compactification $\widetilde{V}$ of V. Note here a prerequisite of Zariski's main theorem is smoothness and properness (so having smooth compactification is important). By uniqueness, the isogeny Hecke action extends to $\widetilde{V}$. This fact is no longer true in the case of the Hilbert modular Shimura variety (studied in [H10a]) different from the Shimura curve. Each smooth toroidal compactification of a Hilbert modular variety for a totally real field $F \neq \mathbb{Q}$ depends on a chosen cone decomposition of $(F \otimes_{\mathbb{Q}} \mathbb{R})^{\times}$ stable under unit multiplication, and such a decomposition is not stable under the isogeny action; cf. [PAF] Sect. 4.1.4 or [HMI] Sect. 4.2.4. Thus, the generalization of our argument given in [H10a] is more involved.

Let $x \in V^{\mathrm{ord}}(\mathbb{F}) = (V \cap Sh^{\mathrm{ord}})(\mathbb{F})$. The fiber at x of $\mathbf{E}_{/V}$ is a test object $(E_x, \eta_x^{(p)})$. The elliptic curve $E_{x/\mathbb{F}}$ has complex multiplication by an imaginary quadratic field $M/\mathbb{Q}$ (by a Corollary 6.40), which gives an embedding $\rho : T_x(\mathbb{Z}_{(p)}) \hookrightarrow G(\mathbb{A}^{(p\infty)})$ satisfying $\alpha\eta_x^{(p)} = \eta_x^{(p)}\rho(\alpha)$. Since the test object $\underline{E}_x$ is given to us, the point of $Sh^{(p)}_{/\mathcal{W}}$ at which $\underline{E}_x$ is realized as a fiber of $\underline{\mathbf{E}}$ may not be in the neutral component, but it is in the image of the neutral component under the right action by $g \in G(\mathbb{A}^{(p\infty)})$. The point x is therefore of the form $[z, g]$, whose level structure η_x is of the form $\eta_z \circ g$ ($g \neq 1$; otherwise, the image of the $\mathbb{Q}$-anisotropic torus T_x under r_z cannot be diagonal at p in $G(\mathbb{A}^{(\infty)})$). Recall $\rho = r_x : T_x(\mathbb{Z}_{(p)}) \hookrightarrow G(\mathbb{A}^{(p\infty)})$ in Sect. 7.1.8 with $\mathrm{Im}(\rho)$ in $G(\mathbb{A}^{(p\infty)})/Z(\mathbb{Z}_{(p)})$, giving the stabilizer of the point $x = [z, g]$ (Lemma 7.43).

We start with the more general setting of (N0–3) with $m = 2$: Let

$$X_{/\mathbb{F}} \subset V \times V$$

be an irreducible subscheme with $(x,x') \in X^{\mathrm{ord}}(\mathbb{F})$ $[X^{\mathrm{ord}} = X \cap (V^{\mathrm{ord}} \times V^{\mathrm{ord}})]$ stable under the diagonal action of a p-adically open subgroup $\mathbb{T}$ in $\mathcal{T}(\mathbb{Z}_{(p)})$. As we mentioned already (in the sketch of step 1), for the use of Zariski's main theorem, we need smoothness and properness of our varieties; hence, we write $\widetilde{V}$ for a smooth compactification of V, $\widetilde{X}$ for the Zariski closure of X in $\widetilde{V} \times \widetilde{V}$, and $\widetilde{\Pi} : \widetilde{Y} \to \widetilde{X}$ for the normalization $\widetilde{Y}$ of $\widetilde{X}$. The action of $\mathbb{T}$ on X extends to Y. For open compact subgroups $K_1, K_2 \subset G(\mathbb{A}^{(p\infty)})$, we write V_{12} for $V_{K_1} \times V_{K_2}$, and we define X_{12} for the image of X in V_{12} and $\widetilde{X}_{12}$ for the image of $\widetilde{X}$ in $\widetilde{V}_{12} := \widetilde{V}_{K_1} \times \widetilde{V}_{K_2}$. We write $\widetilde{Y}_{12}$ (resp., Y_{12}) for the normalization of $\widetilde{X}_{12}$ (resp., X_{12}). Thus, $\widetilde{Y} = \varprojlim_{K_1 \times K_2} \widetilde{Y}_{12}$. We suppose the following [see Proposition 11.3 (3)]:

(DE) *The two projections $\Pi_1, \Pi_2 : Y \to V$ are finite at a point $y \in Y$ above a point (x,x') in $V(\mathbb{F}) \times V(\mathbb{F})$ fixed by the diagonal action of $T_x(\mathbb{Z}_{(p)})$.*

Since x and x' are fixed by $\rho(T_x(\mathbb{Z}_{(p)}))$, E_x and $E_{x'}$ have complex multiplication by the same imaginary quadratic field M. Since elliptic curves over $\mathbb{C}$ (and hence over any characteristic-0 fields) with complex multiplication by a given imaginary quadratic field are isogenous, applying this to canonical lift (with Serre–Tate coordinate $= 1$) of E_x and $E_{x'}$, they are isogenous and have complex multiplication by M. This isogeny property also follows from Honda–Tate theory [Ho] and [T2] because M is generated over $\mathbb{Q}$ by the same Frobenius endomorphism ϕ [with $(\phi) = \mathfrak{p}^m$ if E_x and $E_{x'}$ are defined over $\mathbb{F}_{p^m}$]. We also have $\dim_{\mathbb{F}} Y = \dim_{\mathbb{F}} X = \dim_{\mathbb{F}} V$ and $\dim_{\mathbb{F}} \widetilde{Y}_{12} = \dim \widetilde{X}_{12} = \dim \widetilde{V}_K$. If Π_j is not étale at (x,x'), $\Pi_{j,*}(X_*(\widehat{Y}_y)) \subset X_*(\widehat{S}) = \mathbb{Z}_p$ is an $\mathbb{Z}_p$-submodule of finite index, and we find $\alpha \in T_x(\mathbb{Z}_{(p)})$ such that $\alpha^{1-c} X_*(\widehat{S}) = \Pi_{j,*}(X_*(\widehat{Y}_y))$ in $X_*(\widehat{S}) = \mathbb{Z}_p$. Then $\rho(\alpha^{-1}) \circ \Pi_j$ is étale [by Proposition 11.3 (4)]. The action of $\rho(\alpha)$ for a p-adic nonunit α^{1-c} is not an automorphism of V but a "radiciel" endomorphism of V. Indeed, $\left(\begin{smallmatrix} 1 & 0 \\ 0 & p \end{smallmatrix}\right)$ acts on $Sh^{(p)}$ as the relative Frobenius map of degree p, and hence if $r_x(\alpha)_p K = \left(\begin{smallmatrix} 1 & 0 \\ 0 & p \end{smallmatrix}\right) K$, the action of $\rho(\alpha)$ coincides with the Frobenius map composed with $r_x(\alpha^{(p)})$ on Sh_K. Since any statement concerning the underlying topological space of our schemes is not affected by "radiciel" endomorphisms, we may assume (for such statements)

(A) Π_1 and Π_2 are étale finite at any point $y \in Y$ above (x,x'),

by modifying Π_j by $\rho(\alpha)$ for $\alpha \in M^\times$. This condition follows from Proposition 11.3 (4) and Corollary 11.5 in the case of our interest: $\mathfrak{b} = \mathrm{Ker}(\phi)$.

As already remarked, we will show that X is the graph of the action by an element $g \in G(\mathbb{A}^{(p\infty)})$. In this process, we may also assume the following condition without losing generality:

(B) $x = x'$.

By moving X under the automorphism $1 \times \rho(b^{(p)})$ of $V \times V$ for a suitable $b \in M_{\mathbb{A}}^\times$, we can bring x' to x, and hence we may assume (B).

Let $V_j = \varprojlim_{K_j} V_{K_j}$ (resp., $\widetilde{V}_j = \varprojlim_{K_j} \widetilde{V}_{K_j}$) $(j = 1, 2)$ be the jth factor in $V \times V$ (resp., in $\widetilde{V} \times \widetilde{V}$). We now study the finite locus ${}^{\mathrm{fi}}V_j = \varprojlim_{K_j} {}^{\mathrm{fi}}V_{K_j}$ in V_j of the projection $\varPi_j : Y \to V_j$ for $j = 1, 2$. By definition, the nonfinite locus ${}^{\mathrm{nfi}}\widetilde{V}_{K_j}$ is the closure of all closed points $v \in \widetilde{V}_{K_j}$ such that $\mathcal{O}_{\widetilde{Y}_{K_j},y}$ is not finite over $\mathcal{O}_{\widetilde{V}_{K_j},v}$ for at least one point $y \in \varPi_j^{-1}(v)$. We put ${}^{\mathrm{fi}}\widetilde{V}_{K_j} = \widetilde{V}_{K_j} - {}^{\mathrm{nfi}}\widetilde{V}_{K_j}$. Similarly, we define the nonflat locus ${}^{\mathrm{nfl}}\widetilde{V}_{K_j} \supset {}^{\mathrm{nfi}}\widetilde{V}_{K_j}$ by the closure of all closed points $v \in \widetilde{V}_{K_j}$ such that $\mathcal{O}_{\widetilde{Y}_{K_j},y}$ is not flat over $\mathcal{O}_{\widetilde{V}_{K_j},v}$ for at least one point $y \in \varPi_j^{-1}(v)$. Since a flat morphism is an open map, by Corollary 5.11, ${}^{\mathrm{nfl}}\widetilde{V}_{K_j}$ is a closed subvariety of $\widetilde{V}_{K_j}$ of (at least) codimension 2; so ${}^{\mathrm{nfl}}\widetilde{V}_{K_j} = \emptyset$ as $\dim \widetilde{V}_{K_j} = 1$. Thus, $\widetilde{\varPi}_j$ is everywhere finite flat.

Similarly, we define ${}^{\mathrm{et}}V_{K_j}$ by the maximal open subscheme of V_{K_j} over which $\varPi_j$ is étale (the cuspidal divisors ramify in $\widetilde{V}_j$ over $\widetilde{V}_{K_j}$; so the étale locus is in V_j). We write ${}^{\mathrm{et}}V_{K_j} = V_{K_j} \cap {}^{\mathrm{et}}\widetilde{V}_{K_j}$, and $V^{\mathrm{et}} = {}^{\mathrm{et}}V_1 \cap {}^{\mathrm{et}}V_2$ (in other words, over V^{et}, $\varPi_1$ and $\varPi_2$ are *both* étale).

Proposition 11.6 *Suppose* (DE). *Then we have*

1. *The morphisms* $\widetilde{\varPi}_j : \widetilde{Y} \to \widetilde{V}$ $(j = 1, 2)$ *in* $\widetilde{V}$ *are finite flat.*
2. *If* $\varPi_1 : \mathcal{Y} \to S$ *and* $\varPi_2 : \mathcal{Y} \to S$ *are étale,* $V^{\mathrm{et}} = {}^{\mathrm{et}}V_1 \cap {}^{\mathrm{et}}V_2$ *is an open dense subscheme of* V *containing* (x, x) *stable under* $\mathbb{T}$.

Proof. The first assertion is already proven, so we prove (2). As explained before, we may assume (A) and (B). The scheme Y_{12} is normal dominant finite over $\widetilde{V}_{K_j}$ (and is generically étale finite). Since $\widetilde{V}_{K_j}$ is smooth if K_j is sufficiently small, the ramified (non-étale) locus ${}^{\mathrm{ram}}\widetilde{V}_{K_j} = \widetilde{V}_{K_j} - {}^{\mathrm{et}}\widetilde{V}_{K_j}$ of Y_{12} over $\widetilde{V}_{K_j}$ is a divisor of V_{K_j}; hence, it is of at least codimension 1 for a given K_j. We take the Stein factorization $Y^{\mathrm{st}} \to V_{K_j}$ of $Y_{12} \to V_{K_j}$ defined in Sect. 5.1.3. Since $\widetilde{V}_{K_j} - V_{K_j}$ is the cuspidal divisor, the ramified (non-étale) locus ${}^{\mathrm{ram}}V_{K_j} \subset V_{K_j}$ of Y^{st} over V_{K_j} is a divisor of V_{K_j}; so it is of at least codimension 1 for a given K_j. To extend this result to the provariety V_j, we need to show that ${}^{\mathrm{ram}}V_{K'_j}$ is sent into ${}^{\mathrm{ram}}V_{K_j}$ under the projection map $V_{K'_j} \twoheadrightarrow V_{K_j}$. To show this, we look into the following commutative diagram (removing cuspidal divisors):

$$\begin{array}{ccccc}
Y'_{12} & \xrightarrow{\text{finite}} & X'_{12} & \xrightarrow{\hookrightarrow} & V_{K'_1} \times V_{K'_2} \\
{\scriptstyle \pi_Y}\downarrow{\scriptstyle \text{finite}} & & {\scriptstyle \pi_X}\downarrow{\scriptstyle \text{finite}} & & {\scriptstyle \pi}\downarrow{\scriptstyle \text{étale}} \\
Y_{12} & \xrightarrow[\text{finite}]{} & X_{12} & \xrightarrow{\hookrightarrow} & V_{K_1} \times V_{K_2}.
\end{array}$$

Taking fiber products, we get morphisms (from the commutativity of the above diagram): $Y'_{12} \to Y_{12} \times_{X_{12}} X'_{12}$, $X'_{12} \to X_{12} \times_{(V_{K_1} \times V_{K_2})} (V_{K'_1} \times V_{K'_2})$ and $Y_{12} \times_{(V_{K_1} \times V_{K_2})} (V_{K'_1} \times V_{K'_2}) \to X_{12} \times_{(V_{K_1} \times V_{K_2})} (V_{K'_1} \times V_{K'_2})$. The pullback

$X_{12} \times_{(V_{K_1} \times V_{K_2})} (V_{K_1'} \times V_{K_2'})$ of X_{12} to $V_{K_1'} \times V_{K_2'}$ is a closed subscheme of $V_{K_1'} \times V_{K_2'}$ étale over X_{12} containing (x, x). By definition, X'_{12} is the Zariski closure of $\mathcal{X}'$ (i.e., the image of $\mathcal{X}$ in $V_{K_1'} \times V_{K_2'}$), which is therefore given by the above closed subscheme of $V_{K_1'} \times V_{K_2'}$ over X_{12} containing (x, x). Then X'_{12} is a closed irreducible subscheme of $X_{12} \times_{(V_{K_1} \times V_{K_2})} (V_{K_1'} \times V_{K_2'})$ of equal dimension, and X'_{12} is the irreducible component of

$$X_{12} \times_{(V_{K_1} \times V_{K_2})} (V_{K_1'} \times V_{K_2'})$$

containing (x, x) and covering X_{12}, and π_X is étale finite. Thus, we have a commutative diagram:

$$\begin{array}{ccc} Y_{12} \times_{X_{12}} X'_{12} & \xrightarrow{\text{étale}} & Y_{12} \\ \downarrow & & \downarrow{\scriptstyle \text{normalization}} \\ X'_{12} & \xrightarrow[\text{étale}]{} & X_{12}. \end{array}$$

Since étale morphisms are isomorphisms at the level of completed local rings (by Lemma 4.7), they commute with the formation of normalization. Thus, Y'_{12} is an irreducible component of $Y_{12} \times_{X_{12}} X'_{12}$, and therefore π_Y is étale finite. Hence, the projection $V_{K_j'} \twoheadrightarrow V_{K_j}$ sends ${}^{\mathrm{ram}}V_{K_j'}$ into ${}^{\mathrm{ram}}V_{K_j}$, and ${}^{\mathrm{ram}}V_j = \varprojlim_{K_j} {}^{\mathrm{ram}}V_{K_j}$ is a closed pro-subscheme of at least codimension 1 (by Corollary 5.14) of the provariety V_j (whose image in V_{K_j} is contained in ${}^{\mathrm{ram}}V_{K_j}$). Hence, $V^{\mathrm{et}} = {}^{\mathrm{et}}V_1 \cap {}^{\mathrm{et}}V_2$ is open dense in V. By our assumption, $(x, x) \in V^{\mathrm{et}}$, and V^{et} is stable under $\mathbb{T}$, since étaleness is preserved by the action of $\mathbb{T}$. □

Let v be a closed point of $Y^{\mathrm{ord}}_{/\mathbb{F}} = Y \times_{V^2} (V^{\mathrm{ord}})^2$ above $(v_1, v_2) \in V^2$. We consider the formal completion $\widehat{Y}_v$ [resp., $\widehat{V}_{\Pi_j(v)}$] along v [resp., $\Pi_j(v)$]. If $\Pi_1 \times \Pi_2$ embeds $\widehat{Y}_v$ into $\widehat{V}_{v_1} \times \widehat{V}_{v_2}$ and the equation defining $\widehat{Y}_v$ in the product $\widehat{V}_{\Pi_1(v)} \times \widehat{V}_{\Pi_2(v)}$ is given by $t_1^a = t_2^b$ for the Serre–Tate coordinate t_j of $\widehat{V}_{v_j}$, we call Y *locally linear* at $\overline{v}$. Write $Y_{\mathrm{lin}} \subset Y^{\mathrm{ord}}(\mathbb{F})$ for the subset of all closed locally linear points.

Proposition 11.7 *Suppose* (DE)*, and let the notation be as in* (N0–3) *in* Sect. 11.1.3 *for* $m = 2$. *In particular, let* X *be a Zariski closure of* $\mathcal{X}$ *in* $V \times V$ *and* $Y \to X$ *be the normalization of* X. *The subset* $Y_{\mathrm{lin}} \subset Y$ *of locally linear points as defined above contains the set of all closed points of an open dense subscheme* Y^{lin} *in* Y^{ord}. *In other words, at each closed point* $y' \in Y^{\mathrm{lin}}(\mathbb{F})$*, the formal completion of* Y^{lin} *along* y' *is defined by a linear equation.*

In the following corollary, we will find that the subvariety X is a graph of the action of an element in $G(\mathbb{A}^{(p\infty)})$; hence, we conclude that $Y^{\mathrm{lin}} = X^{\mathrm{lin}} = X^{\mathrm{ord}}$. Since X^{ord} is locally linear at densely populated v in the image of Y^{lin}, by [Ch1] Proposition 5.3, X^{ord} is weakly locally linear, and so Y^{ord} is locally linear. We shall give here an argument (again suggested by Chai) sufficient to prove the weaker version as stated above.

Proof. Modifying Π_j by $\rho(\alpha)$ for $\alpha \in M^\times$ does not affect locally linearity at a closed point of Y over (x, x'); so, by changing Π_j by $\rho(\alpha_j) \circ \Pi_j$ for $\alpha \in M^\times$ if necessary, we assume (A) and (B) (thus, Π_j is étale for $j = 1, 2$).

An endomorphism $a_x \in \mathrm{End}(E_x)$ induces an endomorphism of the deformation space $\widehat{S} = \widehat{V}_x = \widehat{\mathbb{G}}_m$. Modifying a_x by the central action of $\mathbb{Z}_p$, we may assume that a_x is the identity on the connected component of $E_x[p^\infty]$ without affecting the endomorphism of the deformation space $\widehat{S}$ induced by a_x.

We fix an ordinary level-p structure η_p^{ord} on E_x. Identifying $\mathrm{End}_{GSCH}(\widehat{S}) = \mathbb{Z}_p$ by η_p^{ord}, the action of a_x on $\widehat{S}$ is then given by the action of a_x over the étale quotient $E_x[p^\infty]^{\mathrm{et}}$. By Proposition 11.3 (1), $\mathcal{Y}$ is locally linear at a point $y \in Y$ above (x, x), and hence, by Remark 11.4, we may assume that the formal completion $\widehat{Y}_y$ along y is defined by $t'^u = t^v$ $[u, v \in \mathbb{Z}_p = \mathrm{End}_{GSCH}(\widehat{S})]$ for the Serre–Tate coordinate (t, t') of $\widehat{S} \times \widehat{S}$ for $\widehat{S} = \widehat{V}_x$, where t and t' are associated with the ordinary level structure η_p^{ord}. Since the two projections $\Pi_1, \Pi_2 : Y \to V$ are étale at y [by (A)], $a = v/u$ is a unit in $\mathbb{Z}_p$, and $\widehat{Y}_y$ is defined by $t' = t^a$. We write $a_x \in \mathrm{End}(E_x) \otimes_{\mathbb{Z}} \mathbb{Z}_p = \mathrm{End}(E_x[p^\infty])$ for the endomorphism inducing $a \in \mathrm{End}(\widehat{S})$ as normalized above. Then a_x is a unit in $\mathrm{End}_{GSCH}(E_x[p^\infty])$.

We consider the universal elliptic curve $\mathbf{E}_{/V}$. We pull it back to $\Pi : Y \to X \subset V \times V$: $\mathbb{Y}_1 = \Pi_1^* \mathbf{E}$ and $\mathbb{Y}_2 = \Pi_2^* \mathbf{E}$. Identifying E_x with the fibers $\mathbb{Y}_{j,y}$ of $\mathbb{Y}_j$ $(j = 1, 2)$ at y, we regard the unit $a_x \in \mathrm{End}(E_x[p^\infty])$ as a homomorphism $a_x : \mathbb{Y}_{1,x}[p^\infty] = E_x[p^\infty] \to E_x[p^\infty] = \mathbb{Y}_{2,x}[p^\infty]$.

We now reduce the existence of the desired nonempty open subscheme $Y^{\mathrm{lin}} \subset Y^{\mathrm{ord}}$ to the existence of an étale irreducible covering $\widetilde{U}$ over an open dense subscheme $U \subset Y^{\mathrm{ord}}$ containing the given point $y \in Y^{\mathrm{ord}}$ such that a_x extends to an isomorphism $\widetilde{a} : \mathbb{Y}_1[p^\infty]_{/\widetilde{U}} \to \mathbb{Y}_2[p^\infty]_{/\widetilde{U}}$ of Barsotti–Tate groups over $\widetilde{U}$. Thus, supposing the existence of the open subscheme U and such an extension $\widetilde{a}$ over $\widetilde{U}$, we specify the open subscheme Y^{lin}. Shrinking $U \subset Y^{\mathrm{ord}}$ (keeping y inside U), we may assume that the projections $\Pi_j : U \to V$ $(j = 1, 2)$ are both étale (cf. Proposition 11.6), because the projections are étale at the specific point y.

By the étaleness of Π_j, the formal completion $\widehat{Y}_u$ along $u \in U$ with $\Pi(u) = (u_1, u_2) \in V \times V$ is isomorphic by Π_j $(j = 1, 2)$ to the universal deformation space of the Barsotti–Tate group $E_{u_j}[p^\infty]$ carrying the universal deformation $\mathbb{Y}_j[p^\infty]_{/\widehat{Y}_u} \cong \mathbf{E}_j[p^\infty]_{/\widehat{V}_{u_j}}$ via $\Pi_j : \widehat{Y}_u \cong \widehat{V}_{u_j}$.

Choose an ordinary level structures $\eta_{j,p}^{\mathrm{et}} : \mathbb{Q}_p/\mathbb{Z}_p \cong E_{u_j}[p^\infty]$. Then the canonical coordinate t_j of $\mathbf{E}[p^\infty]_{/\widehat{V}_{u_j}} \cong \mathbb{Y}[p^\infty]_{/\widehat{Y}_u}$ is given by $t_j = \lim_{n \to \infty} \text{“}p^n\text{”}(\eta_{j,p}^{\mathrm{et}}(p^{-n}))$ for Drinfeld's lift “p^n” in Theorem 6.44. Another level structure $\widetilde{a} \circ \eta_{1,p}^{\mathrm{et}}$ of $E_{u_2}[p^\infty]$ gives rise to the coordinate $t_2^{a_u}$ for a unit $a_u \in \mathbb{Z}_p^\times$ because $\mathrm{Aut}_{GSCH}(\widehat{Y}_u) = \mathbb{Z}_p^\times$. Thus, we get the relation $t_1 = t_2^{a_u}$ valid on $\widehat{Y}_u$, because $\widetilde{a}$ sends the $t_1 = \lim_n \text{“}p^n\text{”}(\eta_{1,p}^{\mathrm{et}}(p^{-n}))$ to $t_2^{a_u} = \lim_n \text{“}p^n\text{”}(\widetilde{a}(\eta_{1,p}^{\mathrm{et}}(p^{-n})))$. In other words, $\widehat{Y}_u$ is contained in the locally linear formal subscheme $\widehat{Y}'_u$ defined by $t_1 = t_2^{a_u}$. Since $\widehat{Y}_u \hookrightarrow \widehat{V}_{u_1} \times \widehat{V}_{u_2}$ is a smooth formal subscheme with two isomorphisms $\widehat{Y}_u \cong \widehat{V}_{u_j}$ induced by Π_j, we find $\widehat{Y}_u = \widehat{Y}'_u$, and hence, $\widehat{Y}_u$ is defined by $t_1 = t_2^{a_u}$. Thus, we may let $Y^{\mathrm{lin}} = U$.

Next we shall show that a_x extends to a morphism $\widehat{a} : \mathbb{Y}_1[p^\infty] \to \mathbb{Y}_2[p^\infty]$ of Barsotti–Tate groups over $\widehat{Y}_y$. We make a remark first. In the above computation, taking $u_1 = u_2 = x$ and $t = t_1 = t_2$, the action of $\rho(\alpha)$ (whose matrix expression is given by $\left(\begin{smallmatrix}\alpha & 0\\ 0 & \alpha^c\end{smallmatrix}\right)$ inside $GL_2(\mathbb{Q}_p)$ by our choice of the level-p structure) acts on the coordinate t as if we were making conjugate $\rho(\alpha)\begin{pmatrix}1 & \log_p(t)\\ 0 & 1\end{pmatrix}\rho(\alpha)^{-1} = \begin{pmatrix}1 & \log_p(t^{\alpha^{1-c}})\\ 0 & 1\end{pmatrix}$ by Proposition 7.44. By this remark, identifying E_x with the fibers $\mathbb{Y}_{j,x}$ of $\mathbb{Y}_j$ ($j = 1, 2$) at y, we regard the unit $a_x \in \mathrm{End}(E_x[p^\infty])$ as a homomorphism $a_x : \mathbb{Y}_{1,y}[p^\infty] = E_x[p^\infty] \to E_x[p^\infty] = \mathbb{Y}_{2,y}[p^\infty]$. The formal completion $\widehat{Y} = \widehat{Y}_y$ of Y along y is the maximal subscheme of $\widehat{S} \times \widehat{S}$ over which this a_x extends to a homomorphism $\widehat{a} : \mathbb{Y}_1[p^\infty]_{/\widehat{Y}} \to \mathbb{Y}_2[p^\infty]_{/\widehat{Y}}$ of $\widehat{Y}$-group schemes.

To find $\widehat{a}$, as above, we identify a with an element of $\mathbb{Z}_p$ by projecting a_x down to $\mathrm{End}_{BT}(E_x[p^\infty]^{\mathrm{et}}) = \mathbb{Z}_p$ for the maximal étale quotient $E_x[p^\infty]^{\mathrm{et}}$ of the p-divisible group. The isogeny action $\rho(\alpha) : \mathbf{E}_{/V} \to \mathbf{E}_{/V}$ for $\alpha \in \mathrm{End}(E_x) = \mathfrak{O}$ induces

$$\rho(\alpha) : \mathbb{Y}_1[p^n]_{\widehat{Y}} \xrightarrow[\Pi_1]{\sim} \mathbf{E}[p^n]_{/\widehat{V}_x} \xrightarrow{\alpha} \mathbf{E}[p^n]_{/\widehat{V}_x} \xrightarrow[\Pi_2^{-1}]{\sim} \mathbb{Y}_2[p^n]_{/\widehat{Y}}.$$

Choose α so that $\alpha \equiv a \mod p^n\mathfrak{O}_{\mathfrak{p}}$ for $\mathfrak{p} = \prod_{\mathfrak{p}\in\Sigma_p}\mathfrak{p}$ and $\alpha \equiv 1 \mod p^n\mathfrak{O}_{\mathfrak{p}^c}$, and write it as α_n. Therefore, the isogeny action of $\rho(\alpha_n)$ on $\mathbf{E}$ gives rise to $\alpha_n : \mathbb{Y}_1[p^n]_{/\widehat{Y}_n} \to \mathbb{Y}_2[p^n]_{/\widehat{Y}_n}$ well defined over an infinitesimal neighborhood $\widehat{Y}_n$ of y (isomorphic to the connected component of $E_x[p^n]$). Indeed, the embedding $\widehat{Y} \hookrightarrow \widehat{S} \times \widehat{S}$ is given by (t, t^a), $\rho(\alpha_n)$ sends $\mathbb{Y}_{1,v}[p^n]_{/\widehat{Y}_n} \xrightarrow[\Pi_1]{\sim} \mathbf{E}[p^n]_{\widehat{V}_n}$ to $\mathbb{Y}_{2,\rho(\alpha_n)_p(v)}[p^n]_{/\widehat{Y}_n} \xrightarrow[\Pi_2]{\sim} \mathbf{E}[p^n]_{/\widehat{V}_n}$, and $v^a = \rho(\alpha_n)_p(v)$ as long as $v \in \widehat{Y}_n$. Here $\widehat{V}_n$ is the infinitesimal neighborhood of x isomorphic to the connected component of $E_x[p^n]$), $\rho(\alpha_n)_p \in G(\mathbb{Q}_p)$ is the p-component of $\rho(\alpha_n)$, and α_n gives rise to $a|_{\mathbb{Y}_1[p^n]}$ after a base change to $\widehat{Y}_n$. By taking the limit $\widehat{a} = \lim_{n\to\infty} \alpha_n|_{\mathbb{Y}_2[p^n]_{/\widehat{Y}_n}}$ with respect to n, we find out that a_x gives rise to a unique homomorphism $\widehat{a} : \mathbb{Y}_1[p^\infty]_{/\widehat{Y}} \to \mathbb{Y}_2[p^\infty]_{/\widehat{Y}}$ of Barsotti–Tate groups over $\widehat{Y}$.

To show the existence of U, we follow Chai's argument in the proof of [Ch1] Proposition 8.4. We choose a sufficiently small open compact subgroup $K_1 \times K_2 \subset G(\widehat{\mathbb{Z}})^2$ maximal at p so that $\widehat{Y}_{12,y} = \widehat{Y}_y$. Since Y_{12}^{ord} is irreducible, we only need to find a nonempty open subscheme $U' \subset Y_{12}^{\mathrm{ord}}$ with the required property (then U is given by the pullback of U' to Y^{ord}). We have $\widehat{a}$ well defined over $\widehat{Y} = \widehat{Y}_{12,y}$. For any reduced local $\mathcal{O}_{Y_{12},y}$-algebra $\mathcal{R}$, a locally linear isomorphism $\alpha : \mathbb{Y}_1[p^\infty]_{/\mathcal{R}} \to \mathbb{Y}_2[p^\infty]_{/\mathcal{R}}$ inducing a_x on $E_x[p^\infty]$ (if it exists) is unique. Simply write $\widehat{\mathcal{O}}$ for $\widehat{\mathcal{O}}_{Y_{12},y} = \widehat{\mathcal{O}}_{Y,y}$ and $\mathcal{O}$ for $\mathcal{O}_{Y_{12},y}$. Since $\widehat{\mathcal{O}} \otimes_{\mathcal{O}} \widehat{\mathcal{O}}$ is reduced (by Lemma 4.87), the pullbacks of $\widehat{a}$ by the two projections $\pi_j : \widehat{Y} \times_{\mathcal{O}} \widehat{Y} \to \widehat{Y}$ coincide by the above uniqueness; that is, $\widehat{a}$ satisfies the descent datum (relative to $\widehat{\mathcal{O}}/\mathcal{O}$; see [GME] Sect. 1.11 for descent). By faithfully flat descent, $\widehat{a}$ descends to $\mathcal{O}$.

Since $\mathbb{Y}_j[p^\infty]$ ($j = 1, 2$) is an ordinary Barsotti–Tate group of $GL(2)$-type,

$$\mathrm{Hom}_{\mathrm{Spec}(\mathcal{O})}(\mathbb{Y}_1[p^n], \mathbb{Y}_2[p^\infty]) = \mathrm{Hom}_{\mathrm{Spec}(\mathcal{O})}(\mathbb{Y}_1[p^n], \mathbb{Y}_2[p^n])$$

is a finite flat group scheme over $\mathcal{O}$. The truncated $\widehat{a}_n = \widehat{a}|_{\mathbb{Y}_1[p^n]} : \mathbb{Y}_1[p^n] \to \mathbb{Y}_2[p^\infty]_{/\mathcal{O}}$ is defined over an open neighborhood U' of y. The functor $O \mapsto \mathrm{Hom}_O(\mathbb{Y}_1[p^n], \mathbb{Y}_2[p^n])$ (for open subschemes O of U') is represented by a locally free group scheme over U', since $\mathbb{Y}_j[p^\infty]$ ($j = 1, 2$) is an ordinary Barsotti–Tate group of $GL(2)$-type. For an element $a \in \mathrm{Hom}_O(\mathbb{Y}_1[p^n], \mathbb{Y}_2[p^n])$, take its graph $\Gamma_a \subset \mathbb{Y}_1[p^n] \times_O \mathbb{Y}_2[p^n]$. Then we have $\Gamma_a \cong \mathbb{Y}_1[p^n]$ over O, and so $\mathbb{Y}_1[p^n] \times_O \mathbb{Y}_2[p^n] = \Gamma_a \sqcup \Gamma'_a$; i.e., $\mathbb{Y}_1[p^n] \xrightarrow{\sim} \Gamma_a$ is a constant section of $\mathbb{Y}_1[p^n] \times_O \mathbb{Y}_2[p^n] \to \mathbb{Y}_1[p^n]$. Consider the Zariski closure $\overline{\Gamma}_a \subset \mathbb{Y}_1[p^n] \times_{Y_{12}^{\mathrm{ord}}} \mathbb{Y}_2[p^n]$ of Γ_a. Then the image $\overline{\Gamma}_a^{\mathrm{et}}$ of $\overline{\Gamma}_a$ in the maximal étale quotient $(\mathbb{Y}_1[p^n] \times_{Y_{12}^{\mathrm{ord}}} \mathbb{Y}_2[p^n])^{\mathrm{et}}$ is the Zariski closure of the maximal étale quotient of Γ_a in $(\mathbb{Y}_1[p^n] \times_{Y_{12}^{\mathrm{ord}}} \mathbb{Y}_2[p^n])^{\mathrm{et}}$ (which is finite flat over Y_{12}^{ord}). Thus, $\overline{\Gamma}_a^{\mathrm{et}}$ is finite étale over Y_{12}^{ord}. Applying Cartier duality, we find that $\overline{\Gamma}_a \cap (\mathbb{Y}_1[p^n] \times_{Y_{12}^{\mathrm{ord}}} \mathbb{Y}_2[p^n])^{\mathrm{mult}}$ is finite flat multiplicative over Y_{12}^{ord}, giving $\overline{\Gamma}_a^{mult}$. Thus, $\overline{\Gamma}_a \cong \mathbb{Y}_1[p^n]$ over Y_{12}^{ord}. In other words, a extends uniquely to $a \in \mathrm{Hom}_{Y_{12}^{\mathrm{ord}}}(\mathbb{Y}_1[p^n], \mathbb{Y}_2[p^n])$ by taking Zariski closure of its graph in $\mathbb{Y}_1[p^n] \times_{Y_{12}^{\mathrm{ord}}} \mathbb{Y}_2[p^n]$. So $U' := U_n$ is independent of n (and it can be taken to be Y_{12}^{ord}). Hence, we have found an open Zariski neighborhood $U' \subset Y_{12}^{\mathrm{ord}}$ of y over which we have an extension $\widetilde{a}$ of a_x. Then $U = U' \times_{Y_{12}} Y$ does the job.

Corollary 11.8 *Let the notation and the assumption be as in* Proposition 11.7. *Then X is everywhere smooth and $X = Y$. Moreover, if* (DE) *is satisfied, there exist nonzero $\alpha, \beta \in \mathfrak{O}_{(p)}$ such that X coincides with the skew diagonal*

$$\Delta_{\alpha,\beta} = \{(\rho(\alpha)(v), \rho(\beta)(v)) | v \in V\}.$$

If Π_1 and Π_2 are étale finite, we may assume that $(\alpha, \beta) = (1, \beta)$ with $\beta \in \mathfrak{O}_{(p)}^\times$.

This is a restatement of Theorem 3.43. Here we give a detailed proof following the outline sketched in Sect. 3.6.2.

Proof. If (DE) fails, $Y = X = V^2$ by Proposition 11.3 (3); hence, the assertion follows trivially. We may assume (DE), (A), and (B) as indicated after stating (DE). We follow Chai's argument, which is a version of the argument in [Ch1] Sect. 8 adjusted to our self-product of the modular curve. Since the two projections $\Pi_j : Y \to V_j$ are dominant, we have $\mathrm{End}(\mathbb{Y}_j) \otimes \mathbb{Q} = \mathbb{Q}$

for $\mathbb{Y}_j = \Pi_j^*\mathbf{E} = \mathbf{E} \times_{V_j,\Pi_j} Y$. Let $\mathbb{Y}_{/Y} = \mathbb{Y}_1 \times_Y \mathbb{Y}_2$. Thus, there are only two possibilities for $\mathrm{End}^{\mathbb{Q}}(\mathbb{Y}) = \mathrm{End}(\mathbb{Y}_{/Y}) \otimes \mathbb{Q}$: Either $\mathrm{End}^{\mathbb{Q}}(\mathbb{Y}) = \mathbb{Q} \times \mathbb{Q}$ or $\mathrm{End}^{\mathbb{Q}}(\mathbb{Y}) = M_2(\mathbb{Q})$. Suppose that $\mathrm{End}^{\mathbb{Q}}(\mathbb{Y}) = M_2(\mathbb{Q})$. By the semisimplicity of the category of abelian schemes (e.g., [GME] Theorem 4.1.19), we have two commuting idempotent $e_j \in \mathrm{End}^{\mathbb{Q}}(\mathbb{Y})$ such that $e_j(\mathbb{Y}) = \mathbb{Y}_j$. Since $\mathrm{End}^{\mathbb{Q}}(\mathbb{Y}) = M_2(\mathbb{Q})$, we can find an invertible element $\widetilde{\beta}$ in $GL_2(O_{(p)}) \subset M_2(\mathbb{Q})$ such that $\widetilde{\beta} \circ e_1 = e_2$; hence, $\widetilde{\beta} : \mathbb{Y}_1 \to \mathbb{Y}_2$ is an isogeny, whose specialization to the fiber of $\mathbb{Y}_j$ ($j = 1, 2$) at y gives rise to an endomorphism $\beta \in \mathrm{End}(E_x) \otimes \mathbb{Q}$. Thus, the isogeny $\widetilde{\beta}$ is induced by $\rho(\beta)$ [this point is explained more carefully after proving $\mathrm{End}^{\mathbb{Q}}(\mathbb{Y}) = M_2(\mathbb{Q})$].

We suppose $\mathrm{End}^{\mathbb{Q}}(\mathbb{Y}) = \mathbb{Q} \times \mathbb{Q}$ and try to get a contradiction [in order to prove that $\mathrm{End}^{\mathbb{Q}}(\mathbb{Y}) = M_2(\mathbb{Q})$]. We pick a sufficiently small $K_1 = K_2 = K \subset G(\mathbb{A}^{(p\infty)})$ maximal at p so that V_K is smooth. For the moment, we assume that K is open compact. The variety V_K is naturally defined over a finite extension $\mathbb{F}_q/\mathbb{F}_p$ as the solution of the moduli problem $\mathcal{E}'_{K,c}$ in (7.1.12) for the polarization ideal $\mathfrak{c}$ of E_x (the minimal choice of $\mathbb{F}_q$ is the residue field of $\mathcal{W} \cap k_K$ for Shimura's field k_K of the definition of $V_K \subset Sh_{K/\mathbb{Q}}$). The universal elliptic curve $\mathbf{E}_K$ is therefore defined over $V_{K/\mathbb{F}_q}$, and $\mathbf{E}_K$ is a variety of finite type over $\mathbb{F}_q$. Replacing q by its finite power, we may assume that $X_{12} \subset V_K \times V_{K/\mathbb{F}_q}$ is stable under the Galois action $\mathrm{Gal}(\mathbb{F}/\mathbb{F}_q)$, and hence it has a unique geometrically irreducible model $X_{12/\mathbb{F}_q} \subset V_K \times V_{K/\mathbb{F}_q}$ defined over $\mathbb{F}_q$.

Let $Y_{12/\mathbb{F}_q}$ be the normalization of $X_{12/\mathbb{F}_q}$. Then $Y_{12} = Y_{12/\mathbb{F}_q} \times_{\mathbb{F}_q} \mathbb{F}$. We write $\mathbb{Y}_{12/Y_{12/\mathbb{F}_q}}$ for the abelian scheme $\mathbf{E}_K^2 \times_{(V_K \times V_K)} Y_{12}$. Then $\mathbb{Y}_{12}$ is an abelian scheme over the variety $Y_{12/\mathbb{F}_q}$ of finite type over $\mathbb{F}_q$. Let η be the generic point of $Y_{12/\mathbb{F}_q}$, and write $\overline{\eta}$ for the geometric point over η and $\mathbb{F}_q(\overline{\eta})^{sep}$ for the separable algebraic closure $\mathbb{F}_q(\overline{\eta})^{sep}$ of $\mathbb{F}_q(\eta)$ in $\mathbb{F}_q(\overline{\eta})$. Take an odd prime $\ell \neq p$, and consider the ℓ-adic Tate module $T_\ell(\mathbb{Y}_{\overline{\eta}})$ for the generic fiber $\mathbb{Y}_{\overline{\eta}}$ of $\mathbb{Y}$. We consider the image of the Galois action $\mathfrak{G} := \mathrm{Im}(\mathrm{Gal}(\mathbb{F}_q(\overline{\eta})^{sep}/\mathbb{F}_q(\eta)))$ in $GL_{\mathbb{Z}_\ell \times \mathbb{Z}_\ell}(T_\ell(\mathbb{Y}_{\overline{\eta}}))$.

By Zarhin's result ([Z] and [DAV] Theorem V.4.7), the Zariski closure over $\mathbb{Q}$ of $\mathfrak{G}$ is a reductive subgroup of $GL_{\mathbb{Q}_\ell \times \mathbb{Q}_\ell}(T_\ell(\mathbb{Y}_{\overline{\eta}}) \otimes \mathbb{Q})$. Write $\mathcal{G}$ for the connected component of the reductive group. Then $\mathfrak{G} \cap \mathcal{G}(\mathbb{Q}_\ell)$ is an open subgroup of $\mathcal{G}(\mathbb{Q}_\ell)$. Moreover, by Zarhin's theorem, the centralizer of $\mathcal{G}$ in $GL_{\mathbb{Q}_\ell \times \mathbb{Q}_\ell}(T_\ell(\mathbb{Y}_{\overline{\eta}}) \otimes \mathbb{Q})$ is $\mathrm{End}(\mathbb{Y}) \otimes \mathbb{Q}_\ell$. Since the reductive subgroups of $GL(2)$ either are tori or contain $SL(2)$, the derived group $\mathcal{G}_1(\mathbb{Q}_\ell)$ of $\mathcal{G}(\mathbb{Q}_\ell)$ has to be $SL_2(\mathbb{Q}_\ell \times \mathbb{Q}_\ell)$. Any noncentral semisimple element in $SL_2(\mathbb{Q}_\ell)$ of infinite order generates an infinite group whose Zariski closure has a torus as its identity connected component, since $SL(2)$ has only three types of proper algebraic subgroups (i.e., tori, unipotent subgroups, and Borel subgroups).

Pick a pair (a_1, a_2) of distinct nonconjugate semisimple elements in $SL_2(\mathbb{Q}_\ell \times \mathbb{Q}_\ell)$ inside the Galois image [each belonging to $SL_2(\mathbb{Q}_\ell)$] generating an infinite group whose Zariski closure is a torus. Since Frobenius elements are dense in $\mathfrak{G}$ by (geometric) Chebotarev's density in [Se1] Theorem 7, approximating a_j by Frobenius elements, we can therefore find a set of closed points $u \in Y_{12}(\mathbb{F})$ with positive density such that the Zariski closure in $\mathcal{G}$ of the subgroup generated by the Frobenius element $Frob_u \in \mathfrak{G}$ at u with $\Pi(u) = (u_1, u_2)$ $[u_j \in V(\mathbb{F})]$ is a torus (over $\mathbb{Q}$) containing a maximal torus

$$T_u = (T_{u_1} \times T_{u_2}) \cap \mathcal{G}_1$$

(over $\mathbb{Q}$) of the derived group $\mathcal{G}_1$ of $\mathcal{G}$. In particular, the centralizer of T_u in $\mathcal{G}_1$ is itself (i.e., T_{u_1} and T_{u_2} are not $\mathbb{Q}$-rationally conjugate to each other). Then $\mathbb{Y}_u$ is isogenous to a product of two nonisogenous elliptic curves $Y_1 = E_{u_1}$ and $Y_2 = E_{u_2}$ defined over a finite field. They are nonisogenous as they correspond to distinct Frobenius conjugacy classes. The endomorphism algebra $M_j = \mathrm{End}^{\mathbb{Q}}(Y_j)$ is an imaginary quadratic field generated over $\mathbb{Q}$ by the relative Frobenius map ϕ_j induced by $Frob_u$.

The relative Frobenius map $Frob_u$ acting on $X_*(\widehat{V}_{u_1}) \cong \mathbb{Z}_p$ has only one eigenvalue $\{\phi_1^{(1-c)}\}$, which differs from the eigenvalue of $\phi_2 \in \mathrm{End}(Y_2)$ on $X_*(\widehat{V}_{u_2}) \cong \mathbb{Z}_p$. Since we have proven that over the open dense subscheme $U = Y^{\mathrm{lin}}$ of Y, the formal completion of U at $u \in U$ with $u = (u_1, u_2) \in X \subset V^2$ is canonically isomorphic to a formal subtorus $\widehat{Z} \subset \widehat{V}_{u_1} \times \widehat{V}_{u_2}$ with cocharacter group $X_*(\widehat{Z}) \cong \mathbb{Z}_p$, we may assume that our point $u = (u_1, u_2)$ as above is in the (open dense) image U_{12} of U in X_{12} (because the set of such u's has positive density). Projecting $X_*(\widehat{Z})$ down to the left and the right factors V_K, we see that the projection map $X_*(\widehat{Z}) \to X_*(\widehat{V}_{u_j})$ is actually an injection commuting with the action of $Frob_u$. Thus, $Frob_u$ has more than one eigenvalue on $X_*(\widehat{Z})$, which is a contradiction. Hence, we conclude that $\mathrm{End}^{\mathbb{Q}}(\mathbb{Y}) = M_2(\mathbb{Q})$ for any choice of small open compact subgroups K maximal at p. Passing to the limit, we may assume that $K = GL_2(\mathbb{Z}_p)$ (as we do hereafter), and we still have $\mathrm{End}^{\mathbb{Q}}(\mathbb{Y}) = M_2(\mathbb{Q})$.

As we have remarked at the beginning, $\mathrm{End}^{\mathbb{Q}}(\mathbb{Y}) = M_2(\mathbb{Q})$ implies that we have an isogeny $\widetilde{\beta} : \mathbb{Y}_1 \to \mathbb{Y}_2$ over Y. Writing $\eta_j^{(p)}$ for the prime-to-p-level structure of $\mathbb{Y}_j$, inducing the prime-to-p level structure already chosen for $E_x = \mathbb{Y}_{1,y} = \mathbb{Y}_{2,y}$ at $y \in Y$, we find that $\widetilde{\beta} \circ \eta_1^{(p)} = \eta_2^{(p)} \circ g$ for $g \in G(\mathbb{A}^{(p\infty)})$. Specializing at y, we have $g = \rho(\beta)$ for $\beta \in \mathrm{End}^{\mathbb{Q}}(E_x) = M$. Thus, $\widehat{Y}_y \subset \widehat{S} \times \widehat{S}$ is given by the equation $t' = t^{\beta^{1-c}}$ for nonzero $\beta \in \mathfrak{O}_{(p)}$ for the Serre–Tate coordinate t (resp., t') with respect to the ordinary level-p structure η_p^{ord} of $E_x = \mathbb{Y}_{1,y}$ (resp., $\mathbb{Y}_{2,y}$). By (A) (and Remark 11.4), $\beta^{1-c} \in \mathbb{Z}_p^\times$; so, we may assume that $\beta \in \mathfrak{O}_{(p)}^\times$ (and hence $\widetilde{\beta}$ is a prime-to-p isogeny).

As in the proof of Proposition 11.3 (1), we have $\widehat{X}_{(x,x)} = \bigcup_{L \in I} \widehat{\mathbb{G}}_m \otimes L$ for finitely many $\mathbb{Z}_p$-direct summands L of $X_*(\widehat{S}^2)$. As we have shown in the proof of Proposition 11.3 (1), points of Y above (x, x) are indexed by $L \in I$. Suppose that y corresponds to L. Then $\widehat{Y}_y \subset \widehat{S} \times \widehat{S}$ coincides with $\widehat{\mathbb{G}}_m \otimes L$. On the other hand, we have the skew diagonal $\Delta_\beta = \Delta_{1,\beta} = \{(z, \rho(\beta)(z)) | z \in V\} \subset V \times V$. The formal completion $\widehat{\Delta}_\beta$ along (x, x) therefore coincides with $\widehat{Y}_y$ and

$\widehat{\mathbb{G}}_m \otimes L \subset \widehat{X}_{(x,x)}$ inside $\widehat{S}^2$. Thus, $\widehat{\Delta}_\beta \subset \widehat{X}$. Again writing $\mathcal{O}_{\Delta_\beta,x^2} = \mathcal{O}_{V,x}^{\otimes 2}/\mathfrak{b}_\beta$ and $\mathcal{O}_{X,x^2} = \mathcal{O}_{V,x}^{\otimes 2}/\mathfrak{b}$, this implies $\widehat{\mathfrak{b}}_\beta \supset \widehat{\mathfrak{b}}$ for $\mathfrak{m}_{x^2}$-adic completions $\widehat{\mathfrak{b}}_\beta$ and $\widehat{\mathfrak{b}}$. Therefore, $\mathfrak{b}_\beta \supset \mathfrak{b}$ by the faithful flatness of $\widehat{\mathcal{O}_{V,x}^{\otimes 2}}$ over $\mathcal{O}_{V,x}^{\otimes 2}$; in other words, $\Delta_\beta \subset X$. By the irreducibility of X, we conclude that $X = \Delta_\beta$. Since Δ_β is smooth, $\Delta_\beta = Y$, and hence X is smooth everywhere.

If condition (A) fails, as explained after statement (DE), the morphisms $\rho(\alpha)^{-1} \circ \Pi_1$ and $\rho(\beta')^{-1} \circ \Pi_2$ for suitable nonzero $\alpha, \beta' \in \mathfrak{O}_{(p)}$ are étale, and so $(\rho(\alpha) \times \rho(\beta'))^{-1}(X) = \Delta_{1,\beta}$ by the above argument; hence, $X = \Delta_{\alpha,\beta'\beta}$. This finishes the proof.

Here are two technical lemmas, before going into the case where $m > 2$.

Lemma 11.9 *Let $N_i = A$ for a commutative ring A ($i = 1, 2, \ldots, m$). Let $N \subset N_1 \times N_2 \times \cdots \times N_m = A^m$ be an A-free submodule of A^m with $m \geq 2$. If A is a product of finitely many local rings, the projection of N to $N_i \times N_m$ is surjective for all $i = 1, 2, \ldots, m-1$, and the projection π' of N to $N' := N_1 \times N_2 \times \cdots \times N_{m-1}$ is surjective, we have $N = A^m$.*

Proof. We may assume that A is a local ring. Tensoring its residue field, by Nakayama's lemma (Lemma 10.8), we may assume that A is a field k. Suppose that $\dim N \leq m-1$. Since π' is surjective, π' is an isomorphism, and either $N \cap (N' \times z_m)$ is empty or $z \in N \cap (N' \times z_m)$ is a unique point with $\pi'(z) = z_m$. Since $N \to N_1 \times N_m$ is surjective, there exist at least two points in $N \cap (N' \times z_m)$, a contradiction. Thus, $\dim N = m$ and $N = k^m$. □

For a scheme morphism $f : Z \to Z'$, write $f(Z)$ for the Zariski closure in Z' of the image of the topological space of Z by f with reduced scheme structure. Note that $\mathcal{X}$ is affine, as it is a closed subscheme of an affine scheme $S^2 = \mathrm{Spec}(\mathcal{O}_{V,x}^{\otimes 2})$.

Lemma 11.10 *Let the assumption be as in* (N0–3). *Let $S^2 \subset S^m$ be a factor and $\pi : S^m \twoheadrightarrow S^2$ be the projection. Write $\pi(\mathcal{X}) = \mathrm{Spec}(B_0)$ for a local ring $B_0 \subset \mathcal{O}_{\mathcal{X}}$ (with maximal ideal $\mathfrak{m}_0$). Write $\widehat{\mathcal{X}} = \bigcup_{L \in I} \widehat{\mathbb{G}}_m \otimes_{\mathbb{Z}_p} L$. Then*

(1) *For $\widehat{B}_0 = \varprojlim_n B_0/\mathfrak{m}_0^n$, there exists a finite set I_1 of $\mathbb{Z}_p$-direct summands $\ell \subset X_*(\widehat{S}^2)$ such that $\mathrm{Spf}(\widehat{B}_0) = \bigcup_{\ell \in I_1} \widehat{\mathbb{G}}_m \otimes_{\mathbb{Z}_p} \ell$, and $\mathrm{rank}_{\mathbb{Z}_p} \ell = \dim B_0$.*
(2) *$\mathrm{rank}_{\mathbb{Z}_p} \pi_*(L)$ is independent of $L \in I$ and is equal to $\dim B_0 = \dim \pi(\mathcal{X})$.*

Proof. From the definition of the image $\pi(\mathcal{X})$, B_0 is given by the image of the composite $\mathcal{O}_{S^2} \xrightarrow{\pi^*} \mathcal{O}_{S^m} \twoheadrightarrow \mathcal{O}_{\mathcal{X}}$. Let $X_K \subset V_K^m$ be as in (N3) such that $\widehat{X}_K = \widehat{X}$. Then $\pi(X_K)$ is an excellent scheme since it is of finite type over a field $\mathbb{F}$ [see Sect. 4.1.11 and [EGA] IV.7.8.3 (ii)]. For the projection $\mathcal{X}_K$ of $\mathcal{X}$ to S_K^m, write $\pi(\mathcal{X}_K) = \mathrm{Spec}(B_{0,K})$. Then $B_{0,K}$ is a localization of $\pi(X_K)$ at $(x, x) \in S_K^2$ [$S_K = \mathrm{Spec}(\mathcal{O}_{V_K,x})$] and is an excellent integral local ring (Sect. 4.1.11). By [EGA] IV.7.8.3 (vii) and (x), $\widehat{B}_0 = \widehat{B}_{0,K}$ is reduced equidimensionally. Since $\pi(\mathcal{X}) \subset S^2$ is irreducible and stable under $\mathbb{T}$, $\mathrm{Spf}(\widehat{B}_0) = \bigcup_{\ell \in I_1} \widehat{\mathbb{G}}_m \otimes_{\mathbb{Z}_p} \ell$

with $\mathrm{rank}_{\mathbb{Z}_p} \ell = \dim B_0$, by Proposition 11.3 (1), for finitely many $\mathbb{Z}_p$-direct summands $\ell \subset X_*(\widehat{S}^2)$. This proves assertion (1).

We now give a proof of assertion (2). Let $\mathrm{Spec}(B)$ be the normalization of $\pi(\mathcal{X})$. Let $\pi_V : V_K^m \to V_K^2$ be the projection that induces π. Let $U = \mathrm{Spec}(\mathcal{B}_0) \subset \pi(X_K)$ be a sufficiently small affine open neighborhood of (x, x) such that its normalization $\mathcal{B}/\mathcal{B}_0$ is finite. We can find such $\mathcal{B}_0$ because B/B_0 is finite by (1). We take an affine open neighborhood $U' = \mathrm{Spec}(\mathcal{A}_0) \subset \pi_V^{-1}(U)$ of x^m such that the normalization $\mathcal{A}/\mathcal{A}_0$ is finite and $\mathcal{A}_0$ is of finite type over $\mathcal{B}_0$; hence, $\mathcal{A}$ is a noetherian domain of finite type over $\mathcal{B}$. As already explained, $\mathcal{A}$ and $\mathcal{B}$ are excellent. Let $\widetilde{U} = \mathrm{Spec}(\mathcal{B})$ and $\widetilde{U}' = \mathrm{Spec}(\mathcal{A})$. Since $\mathcal{B} \subset \mathcal{A}$ and $\mathcal{A}$ and $\mathcal{B}$ are integral domains, the morphism $\mathrm{Spec}(\mathcal{A}) \to \mathrm{Spec}(\mathcal{B})$ is generically flat. The nonflat locus $\widetilde{U}'_{nfl}$ [which is the Zariski closure of $\{P \in \widetilde{U}' | \mathcal{A}_P$ is not flat over $\mathcal{B}_{\pi'(P)}\}$] is a closed subscheme of $\widetilde{U}'$ strictly smaller than $\widetilde{U}'$. If $\mathrm{rank}_{\mathbb{Z}_p} \pi_*(L) < \mathrm{rank}_{\mathbb{Z}_p} \ell$, the formal completion $\widehat{\mathbb{G}}_m \otimes L$ of $\mathcal{A}$ along the point $y_L \in \widetilde{U}'$ corresponding to L is not flat over the formal completion $\widehat{\mathbb{G}}_m \otimes \ell$ of $\mathcal{B}$ along the image $\pi(y_L)$. Thus, $\widetilde{U}'_{nfl}$ contains a closed subscheme of maximal dimension, a contradiction against the irreducibility of $\widetilde{U}'$. Thus, $\mathrm{rank}_{\mathbb{Z}_p} \pi_*(L) = \mathrm{rank}_{\mathbb{Z}_p} \ell$, as desired.

Here is a restatement of Theorem 3.44:

Corollary 11.11 *Let the notation be as in* (N0–3) *and the assumption be as in* Proposition 11.3. *Then X is smooth everywhere, and X^{ord} is locally linear. If $\mathcal{X}$ is finite over S', X either is given by $V^{m-1} \times \{x\}$ or is identical to $V^{m-2} \times \Delta_{(\alpha,\beta)}$ for some nonzero $\alpha, \beta \in \mathfrak{O}_{(p)}$ (after permuting first $m - 1$ factors).*

We now give a detailed proof.

Proof. We use the symbols introduced in Proposition 11.3 and its proof; in particular, $S = \mathrm{Spec}(\mathcal{O}_{V,x})$ and $S^m = S' \times S'' \subset V^m$ with $S'' = S$. By Corollary 11.8 ($m = 2$), we may assume that $m > 2$. Assume that $\mathcal{X} \neq S^m$. Since $\mathcal{X}$ is irreducible, its projection to S'' is irreducible. If the projection of $\mathcal{X}$ to S'' is a proper closed subscheme in S'' (i.e., strictly smaller than S'), by $\dim S'' = 1$ (and by the fact that $\mathcal{O}_{V,x}$ is a valuation ring), a closed irreducible subscheme strictly smaller than S'' is a closed point. Thus, we find that $\mathcal{X} = S' \times \{x\}$ as $\mathcal{X} \to S'$ is dominant; hence, we are done. Thus, we may assume that the projection of $\mathcal{X}$ to S'' is dominant and that $\mathcal{X}$ is finite over S' [Proposition 11.3 (3)].

We describe a way of reducing the assertion to Corollary 11.8. Let $\Pi : Y \to X$ and $\Pi_{\mathcal{Y}} : \mathcal{Y} \to \mathcal{X}$ be the normalization. Then by Proposition 11.3 (3), $\mathcal{Y}$ is finite flat over S', and Y is locally linear at every point $y \in Y$ above $x^m \in X$ (abusing the terminology).

Pick a point $y \in Y$ above x^m, and write

$$\widehat{Y}_y = \widehat{\mathbb{G}}_m \otimes L \quad \text{and} \quad \widehat{X} = \bigcup_{L \in I} \widehat{\mathbb{G}}_m \otimes L.$$

We write $S_i = S$ be the ith component of S^m. Let $\pi_{i,m} : S^m \to S_i \times S''$ with $i < m$ be the projection. We regard $\widehat{Y}_y = \widehat{\mathbb{G}}_m \otimes L \subset \widehat{S}^m$. If $\pi_{i,m,*} : L \otimes_{\mathbb{Z}} \mathbb{Q} \to X_*(\widehat{S}_i \times \widehat{S}'') \otimes_{\mathbb{Z}} \mathbb{Q}$ is surjective for all $i < m$, by Lemma 11.9 applied to $A = \mathbb{Z}_p \otimes_{\mathbb{Z}} \mathbb{Q} = \mathbb{Q}_p$ and $N_i = X_*(\widehat{S}_i) \otimes_{\mathbb{Z}} \mathbb{Q}$, we find that $L \otimes_{\mathbb{Z}} \mathbb{Q} = X_*(\widehat{S}^m) \otimes_{\mathbb{Z}} \mathbb{Q}$, and so $\widehat{Y}_y = \widehat{S}^m$, and hence, $Y = X = V^m$, and we are done. Thus, we assume that that the projection $\pi_{i,m,*} : L \otimes_{\mathbb{Z}} \mathbb{Q} \to X_*(\widehat{S}_i \times \widehat{S}'') \otimes_{\mathbb{Z}} \mathbb{Q}$ is not surjective for an $i < m$, and

$$\mathrm{rank}_{\mathbb{Z}_p} \pi_{i,m,*}(L) < \mathrm{rank}_{\mathbb{Z}_p} X_*(\widehat{S}_i \times \widehat{S}'').$$

Recall that $f(Z)$ denotes the Zariski closure in Z' of the image of the underlying topological space of Z for a morphism $f : Z \to Z'$ of schemes. If $Z = \mathrm{Spec}(A)$ and $Z' = \mathrm{Spec}(B)$, then the topological space of $f(Z)$ is that of $\mathrm{Spec}(f^*(B))$. Write $\Pi_{i,m} := \pi_{i,m} \circ \Pi : \mathcal{Y} \to S_i \times S''$. By Lemma 11.10 (2),

$$\dim \pi_{i,m}(\mathcal{X}) = \mathrm{rank}_{\mathbb{Z}_p} \pi_{i,m,*}(L) < \mathrm{rank}_{\mathbb{Z}_p}(X_*(\widehat{S}_i \times \widehat{S}'')) = \dim(S_i \times S''). \tag{11.1.3}$$

By (11.1.3), the reduced image $\pi_{i,m}(\mathcal{X}) \subset S_i \times S''$ is an irreducible closed subscheme (strictly smaller than $S_i \times S''$) invariant under $\mathbb{T}$. Applying Corollary 11.8 to $\pi_{i,m}(\mathcal{X})$, we find that $\pi_{i,m}(\mathcal{X}) = \mathrm{Spec}(\mathcal{O}_{\Delta_{\alpha,\beta},x^2})$ and $\pi_{i,m}(X) \subset \Delta_{\alpha,\beta}$ for some nonzero $\alpha, \beta \in \mathfrak{O}_{(p)}$. Permuting indices to bring i to $m-1$, we conclude

$$X \subset \pi^{-1}_{m-1,m}(\Delta_{\alpha,\beta}) = V^{m-2} \times \Delta_{\alpha,\beta}.$$

Since $\dim X = (m-1)\dim V$ [by Proposition 11.3 (3)], we conclude by irreducibility that $X = V^{m-2} \times \Delta_{\alpha,\beta}$, as desired.

11.1.5 Linear Independence Again

Recall $\mathcal{G}_+ \subset \mathcal{G} = G(\mathbb{A}^{(\infty)})/Z(\mathbb{Q})$ in Theorem 7.18. The subgroup $\mathfrak{O}^\times_{(p)}/\mathbb{Z}^\times_{(p)} \cong \rho(\mathcal{T}(\mathbb{Z}_{(p)}))$ in $\mathcal{G}_+$ fixes $x \in Ig(\mathbb{F})$ (Lemma 7.43) and hence acts on the stalk $\mathcal{O}_{V,x}$ and the stalk $\mathcal{O}_{Ig,x}$ of x on the Igusa tower $Ig_{/V}$. The group $\mathcal{T}(\mathbb{Z}_{(p)})$ is embedded into $T(\mathbb{Z}_p) = \mathbb{Z}_p^\times$ as in (7.2.5). Then the action of $\mathcal{T}(\mathbb{Z}_{(p)})$ extends to its p-adic completion $\mathbb{Z}_p^\times = T(\mathbb{Z}_p) = \mathrm{Aut}(\widehat{S})$ for $\widehat{S} = \mathrm{Spf}(\widehat{\mathcal{O}}_{V,x}) = \mathrm{Spf}(\widehat{\mathcal{O}}_{Ig,x})$. Each $a \in \mathbb{Z}_p^\times$ acts on the formal completion $\widehat{\mathcal{O}}_{Ig,x}$ as automorphisms sending the canonical coordinate t to t^a for $a \in \mathcal{T}(\mathbb{Z}_p) = \mathbb{Z}_p^\times$. Each diagonal element $g = \mathrm{diag}[a,d] \in T^\Delta(\mathbb{Z}_p)$ for the diagonal torus $T^\Delta \subset G$ also acts on Ig by the change of level structure $\eta_p^{\mathrm{ord}} \mapsto \eta_p^{\mathrm{ord}} \circ g$. The image of $T^\Delta(\mathbb{Z}_p) \cap \mathcal{G}$ in $\mathcal{G}_+$ has trivial intersection with $\mathcal{T}(\mathbb{Z}_{(p)})$ inside $\mathcal{G}_+$, because $T_x(\mathbb{Z}_{(p)})$ is embedded diagonally in $G(\mathbb{A}^{(\infty)})$ by $\rho = r_x$, while each element of $T^\Delta(\mathbb{Z}_p)$ has only nontrivial component at p. Thus, the two actions of $a \in \mathcal{T}(\mathbb{Z}_p) = \mathbb{Z}_p^\times$ fixing x and that of $g \in T^\Delta(\mathbb{Z}_p) \cap \mathcal{G}$ moving x are compatible. In the following theorem, $a \in \mathcal{T}(\mathbb{Z}_p) = \mathbb{Z}_p^\times$ acts on $\widehat{\mathcal{O}}_{Ig,x}$ via $t \mapsto t^a$ for the canonical coordinate t. We are now ready to prove the following generalization of Theorem 3.38:

Theorem 11.12 *Let $\mathbb{F}$ be the residue field of $\mathcal{W}$ (so it is an algebraic closure of $\mathbb{F}_p$). Let $x \in Ig(\mathbb{F})$ be a closed point [which is fixed by the action of $\mathcal{T}(\mathbb{Z}_{(p)})$ embedded in $\mathcal{G}_+$ by ρ]. Let $a_1, \dots, a_m \in \mathcal{T}(\mathbb{Z}_p)$, and assume that $a_i a_j^{-1} \notin \mathcal{T}(\mathbb{Z}_{(p)})$ for all $i \neq j$. Then $a_j(\mathcal{O}_{Ig,x/\mathbb{F}})$ $(j = 1, 2, \dots, m)$ are linearly disjoint over $\mathbb{F}$ in $\widehat{\mathcal{O}}_{Ig,x/\mathbb{F}}$, where $\mathcal{O}_{Ig,x}$ is the stalk at x of the Igusa tower over V.*

Let $\mathfrak{b} = \mathrm{Ker}(\phi)$ be the kernel of the homomorphism ϕ in (11.1.1). Let $\widetilde{\mathfrak{b}}$ be the unique prime ideal of $\mathcal{O}_{Ig,x} \otimes \cdots \otimes \mathcal{O}_{Ig,x}$ over $\mathfrak{b}$. Then $\widetilde{\mathfrak{b}}$ is the kernel of the map $\widetilde{\phi}$ defined in the exactly the same way as ϕ replacing $\mathcal{O}_{V,x}$ by $\mathcal{O}_{Ig,x}$. The assertion of the theorem is equivalent to $\widetilde{\mathfrak{b}} = 0$. Since $\mathfrak{b} = 0 \Leftrightarrow \widetilde{\mathfrak{b}} = 0$ by the fact that $Ig_{/V}$ is étale over V and hence $\widehat{Ig_x} = \widehat{V}_x$ (Lemma 4.7), it is sufficient to prove $\mathfrak{b} = 0$.

Proof. We first suppose that $m = 2$. We use the symbols we introduced in the proof of the above two propositions. In particular, $V = \varprojlim_K V_K$ for K maximal at p. For simplicity, we write $\widehat{\mathcal{O}} = \mathcal{O}_{\widehat{S}/\mathbb{F}}$, $S = \mathrm{Spec}(\mathcal{O})$, and $\widehat{S} = \widehat{\mathbb{G}}_m \otimes_{\mathbb{Z}} O$. Suppose $\mathfrak{b} \neq 0$. Apply Proposition 11.3 (3) to the two projections $S \times S \to S$ and $\mathcal{X} = \mathrm{Spec}(\mathcal{R})$, which has two dominant projections onto $S = \mathrm{Spec}(\mathcal{O})$. Then the formal completion $\widehat{\mathcal{X}}$ of $\mathcal{X}$ along $x^2 = (x, x)$ is a formal torus defined by $t'^{u_2} = t^{u_1}$ for $u_1, u_2 \in \mathbb{Z}_p \cap \mathbb{Q}_p^\times$ (Corollary 11.8). By our definition of $\mathfrak{b}$, we have $u_2/u_1 = a_2/a_1 \in \mathbb{Z}_p^\times$; hence, we may choose u_1 and u_2 so that $u_1 = 1$ and $u_2 \in \mathbb{Z}_p^\times$. Let X be the schematic closure of $\mathcal{X}$ in $V \times V$. Then by Corollary 11.8, we find that $X = \Delta_{1,\alpha}$ and, hence, $u_2 = \alpha^{1-c}$ for $\alpha \in \mathfrak{O}_{(p)}^\times$. Since $\alpha^{1-c} \in \mathcal{T}(\mathbb{Z}_{(p)})$, this contradicts $a_1/a_2 \notin \mathcal{T}(\mathbb{Z}_{(p)})$. Thus, $X = V^2$, and $a_1(\mathcal{O})$ and $a_2(\mathcal{O})$ are linearly disjoint over $\mathbb{F}$.

We now deal with the case where $m > 2$. Recall

$$\phi : \overbrace{\mathcal{O} \otimes_{\mathbb{F}} \mathcal{O} \otimes_{\mathbb{F}} \cdots \otimes_{\mathbb{F}} \mathcal{O}}^{m} \to \widehat{\mathcal{O}}$$

in (11.1.1), which is the $\mathbb{F}$-algebra homomorphism given by $f_1 \otimes f_2 \otimes \cdots \otimes f_m \mapsto \prod_{j=1}^m a_j(f_j) \in \widehat{\mathcal{O}} \otimes_W \mathbb{F}$. Let $\mathcal{R} = \mathrm{Im}(\phi) \subset \widehat{\mathcal{O}} \otimes_W \mathbb{F}$. We consider $\mathcal{X} = \mathrm{Spec}(\mathcal{R})$ a closed subscheme of S^m for $S = \mathrm{Spec}(\mathcal{O})$. Take the schematic closure X of $\mathcal{X}$ in V^m. By the induction hypothesis on m, $\mathcal{X}$ surjects onto S' and S''. By Proposition 11.3 (3), $\mathcal{X}$ is either finite over S' or $\mathcal{X} = S^m$In the latter case, we are done; thus, we assume that $\mathcal{X}$ is finite over S'. Then by Corollary 11.11, there exists an index $0 < i < m$ so that $\pi_{i,m}(X) = \Delta_{\alpha,\beta}$ inside $V^2 = V_i \times V''$ for the ith factor $V_i = V$ in V^m. We have $\widehat{\mathcal{X}} \supset \widehat{\Delta}$ in Lemma 11.5, and we conclude that $\mathbb{Z}_p^\times = \mathcal{T}(\mathbb{Z}_p) \ni a_i/a_m = (\alpha/\beta)^{1-c} \in \mathcal{T}(\mathbb{Q})$, which is a contradiction [because $\mathcal{T}(\mathbb{Z}_{(p)}) = \mathcal{T}(\mathbb{Z}_p) \cap \mathcal{T}(\mathbb{Q})$]. Thus, $\mathcal{X} = S^m$ (and hence, $\mathfrak{b} = 0$) and $X = V^m$.

By the construction of the Igusa tower and p-adic modular forms in Sect. 7.2.4, if we look into the Igusa tower Ig over $Y_1(N)$ for a level N prime to p, $\mathcal{O}_{Ig/\mathbb{F}}$ is exactly the space of p-adic modular forms with coefficients in

$\mathbb{F}$. Thus, $\mathcal{O}_{Ig/\mathbb{F}}$ contains $G_k(\mathbb{F}) = \bigcup_r G_k(\Gamma_1(Np^r);\mathbb{F})$ for any fixed $k \geq 1$. Let $\mathbf{1}$ be the constant function in $\mathcal{O}_{Ig/\mathbb{F}}$ whose q-expansion is just 1 as the Hasse invariant. Thus, the above theorem applies to modular forms mod p:

Corollary 11.13 *Fix a weight $k \geq 1$. Let $a_0, \dots, a_n \in T_x(\mathbb{Z}_p)$ and suppose that $a_i a_j^{-1} \notin T_x(\mathbb{Q})$ for all $i \neq j$. Let $I \subset \{0, 1, 2, \dots, n\}$ be a subset of indices. Then if $\{\mathbf{1}, f_{ij} \in G_k(\mathbb{F})\}_j$ are linearly independent over $\mathbb{F}$ for each $i \in I$, then $\{a_i(f_{ij})\}_{i \in I, j}$ in $\mathcal{O}_{\widehat{S}/\mathbb{F}}$ are linearly independent over $\mathbb{F}$.*

References

Books

[AAF] G. Shimura, *Arithmeticity in the Theory of Automorphic Forms*. Mathematical Surveys and Monographs, vol. 82 (American Mathematical Society, Providence, RI, 2000)

[AAG] S.S. Gelbart, *Automorphic Forms on Adele Groups*. Annals of Mathematics Studies, vol. 83 (Princeton University Press, Princeton, NJ, 1975)

[ABV] D. Mumford, *Abelian Varieties*. TIFR Studies in Mathematics (Oxford University Press, New York, 1994)

[ACM] G. Shimura, *Abelian Varieties with Complex Multiplication and Modular Functions* (Princeton University Press, Princeton, NJ, 1998)

[ACS] K.-W. Lan, *Arithmetic Compactifications of PEL-Type Shimura Varieties*. London Mathematical Society Monographs, vol. 36 (Princeton University Press, Princeton, NJ, 2013)

[ALG] R. Hartshorne, *Algebraic Geometry*. Graduate Texts in Mathematics, vol. 52 (Springer-Verlag, New York, 1977)

[ALR] J.-P. Serre, *Abelian l-Adic Representations and Elliptic Curves* (A K Peters, Wellesley, MA, 1998)

[AME] N.M. Katz, B. Mazur, *Arithmetic Moduli of Elliptic Curves*. Annals of Mathematics Studies, vol. 108 (Princeton University Press, Princeton, NJ, 1985)

[AQF] G. Shimura, *Arithmetic of Quadratic Forms*. Springer Monograph in Mathematics (Springer, New York, 2010)

[ARG] G. Cornell, J.H. Silverman (eds.), *Arithmetic Geometry* (Springer-Verlag, New York, 1986)

[BCM] N. Bourbaki, *Algèbre Commutative* (Hermann, Paris, 1961–1998)

[BLI] N. Bourbaki, *Groupes et Algèbres de Lie* (Hermann, Paris, 1972–1985)

[BNT] A. Weil, *Basic Number Theory* (Springer-Verlag, New York, 1974)

[CBT] W. Messing, *The Crystals Associated to Barsotti–Tate Groups; with Applications to Abelian Schemes*. Lecture Notes in Mathematics, vol. 264 (Springer-Verlag, New York, 1972)

[CFN] J. Neukirch, *Class Field Theory* (Springer-Verlag, Berlin, 1986)

[CMA] H. Matsumura, *Commutative Algebra* (Benjamin, New York, 1970)

[CPS] G. Shimura, *Collected Papers*, vols. I, II, III, IV (Springer, New York, 2002)

H. Hida, *Elliptic Curves and Arithmetic Invariants*, Springer Monographs in Mathematics, DOI 10.1007/978-1-4614-6657-4,

[CRT] H. Matsumura, *Commutative Ring Theory.* Cambridge Studies in Advanced Mathematics, vol. 8 (Cambridge University Press, New York, 1986)

[CSM] C.-L. Chai, *Compactification of Siegel Moduli Schemes.* LMS Lecture Note Series, vol. 107 (Cambridge University Press, New York, 1985)

[DAV] G. Faltings, C.-L. Chai, *Degeneration of Abelian Varieties* (Springer-Verlag, New York, 1990)

[ECH] J.S. Milne, *Étale Cohomology* (Princeton University Press, Princeton, NJ, 1980)

[EDM] G. Shimura, *Elementary Dirichlet Series and Modular Forms.* Springer Monographs in Mathematics (Springer, New York, 2007)

[EEK] A. Weil, *Elliptic Functions According to Eisenstein and Kronecker* (Springer-Verlag, Heidelberg, 1976)

[EGA] A. Grothendieck, J. Dieudonné, *Eléments de Géométrie Algébrique.* Publications IHES, vol. 4 (1960), vol. 8 (1961), vol. 11 (1961), vol. 17 (1963), vol. 20 (1964), vol. 24 (1965), vol. 28 (1966), vol. 32 (1967)

[FAN] D. Ramakrishnan, R.J. Valenza, *Fourier Analysis on Number Fields.* Graduate Texts in Mathematics, vol. 186 (Springer-Verlag, New York, 1999)

[FGA] A. Grothendieck, *Fondements de la Géométrie Algébrique.* Séminaire Bourbaki exp. no. 149 (1956/57), 182 (1958/59), 190 (1959/60), 195 (1959/60), 212 (1960/61), 221 (1960/61), 232 (1961/62) (Benjamin, New York, 1966)

[GIT] D. Mumford, *Geometric Invariant Theory.* Ergebnisse, vol. 34 (Springer-Verlag, New York, 1965)

[GME] H. Hida, *Geometric Modular Forms and Elliptic Curves*, 2nd edn. (World Scientific, Singapore, 2011)

[HMI] H. Hida, *Hilbert Modular Forms and Iwasawa Theory* (Oxford University Press, New York, 2006)

[IAT] G. Shimura, *Introduction to the Arithmetic Theory of Automorphic Functions* (Princeton University Press, Princeton, NJ, 1971)

[ICF] L.C. Washington, *Introduction to Cyclotomic Fields.* Graduate Texts in Mathematics, vol. 83 (Springer-Verlag, New York, 1982)

[IEC] E. de Shalit, *Iwasawa Theory of Elliptic Curves with Complex Multiplication. p-Adic L Functions.* Perspectives in Mathematics, vol. 3 (Academic Press, Boston, 1987)

[LAG] J.E. Humphreys, *Linear Algebraic Groups.* Graduate Texts in Mathematics, vol. 21 (Springer-Verlag, New York, 1987)

[LAP] H. Yoshida, *Absolute CM Period.* Mathematical Surveys and Monographs, vol. 106 (American Mathematical Society, Providence, RI, 2003)

[LFE] H. Hida, *Elementary Theory of L-Functions and Eisenstein Series.* London Mathematical Society Student Texts, vol. 26 (Cambridge University Press, Cambridge, 1993)

[LHV] E.Z. Goren, *Lectures on Hilbert Modular Varieties and Modular Forms* (American Mathematical Society, Providence, RI, 2002)

[LRF] J.-P. Serre, *Linear Representations of Finite Groups.* Graduate Texts in Mathematics, vol. 42 (Springer-Verlag, New York, 1977)

[MFG] H. Hida, *Modular Forms and Galois Cohomology.* Cambridge Studies in Advanced Mathematics, vol. 69 (Cambridge University Press, Cambridge, 2000)

[MFM] T. Miyake, *Modular Forms* (Springer-Verlag, New York, 1989)

[MTV] U. Jannsen, S. Kleiman, J.-P. Serre, *Motives.* Proceedings of Symposia in Pure Mathematics, vol. 55, part 1 and 2 (American Mathematical Society, Providence, RI, 1994)

[NAA] S. Bosch, U. Güntzer, R. Remmert, *Non-Archimedean Analysis. A Systematic Approach to Rigid Analytic Geometry.* Grundlehren der Mathematischen Wissenschaften, vol. 261 (Springer-Verlag, Berlin, 1984)

[NMD] S. Bosch, W. Lütkebohmert, M. Raynaud, *Néron Models* (Springer-Verlag, New York, 1990)

[PAF] H. Hida, *p-Adic Automorphic Forms on Shimura Varieties.* Springer Monographs in Mathematics (Springer, New York, 2004)

[RAG] J.C. Jantzen, *Representations of Algebraic Groups* (Academic Press, Orlando, FL, 1987)

[REC] J. Silverman, J. Tate, *Rational Points on Elliptic Curves.* Undergraduate Texts in Mathematics (Springer-Verlag, New York, 1992)

[REP] W. Fulton, J. Harris, *Representation Theory.* Graduate Texts in Mathematics, vol. 129 (Springer-Verlag, New York, 1991)

[SGA] A. Grothendieck, *Revetements Étale et Groupe Fondamental.* Séminaire de geometrie algébrique. Lecture Notes in Mathematics, vol. 224 (Springer-Verlag, Berlin, 1971)

[SGL] H. Hida, On the search of genuine p-adic modular L-functions for $GL(n)$. Mém. Soc. Math. Fr. (N.S.) No. 67 (1996)

[TCF] M. Nagata, *Theory of Commutative Fields* (American Mathematical Society, Providence, RI, 1993)

Articles

[A] S. Ahlgren, On the irreducibility of Hecke polynomials. Math. Comput. **77**, 1725–1731 (2008)

[BDGP] K. Barré-Sirieix, G. Diaz, F. Gramain, G. Philibert, Une preuve de la conjecture de Mahler–Manin. Invent. Math. **124**, 1–9 (1996)

[BGV] B. Balasubramanyam, E. Ghate, V. Vatsal, On local Galois representations attached to ordinary Hilbert modular forms (a preprint posted in `http://www.math.tifr.res.in/~eghate/math.html`) To appear in Manuscripta Math.

[Br1] M. Brakočević, Anticyclotomic p-adic L-function of central critical Rankin–Selberg L-value. Int. Math. Res. Not. **2011**(21), 4967–5018 (2011)

[Br2] M. Brakočević, Non-vanishing modulo p of central critical Rankin–Selberg L-values with anticyclotomic twists, preprint (posted in arXiv:1010.6066)

[BrG] A. Brown, E. Ghate, Endomorphism algebras of motives attached to elliptic modular forms. Ann. Inst. Fourier (Grenoble) **53**, 1615–1676 (2003)

[Bu] A. Burungale, On the μ-invariant of cyclotomic derivative of Katz p-adic L-function, preprint, 2012

[BuH] A. Burungale, M.-L. Hsieh, The vanishing of the μ-invariant of p-adic Hecke L-functions for CM fields. Int. Math. Res. Not. (2012): rns019. doi:10.1093/imrn/rns019

[C] H. Carayol, Sur les représentations ℓ-adiques associées aux formes modulaires de Hilbert. Ann. Sci. Éc. Norm. Sup. 4th series **19**, 409–468 (1986)

[Cd] B. Conrad, Several approaches to non-Archimedean geometry, in *p-Adic Geometry*. University Lecture Series, vol. 45 (American Mathematical Society, Providence, RI. 2008), pp. 9–63

[Ch1] C.-L. Chai, Families of ordinary abelian varieties: canonical coordinates, p-adic monodromy, Tate-linear subvarieties and Hecke orbits, preprint 2003 (posted at: `www.math.upenn.edu/~chai`)

[Ch2] C.-L. Chai, Monodromy of Hecke-invariant subvarieties. Pure Appl. Math. Q. **1**(Special Issue: In memory of Armand Borel), 291–303 (2005)

[Ch3] C.-L. Chai, A rigidity result for p-divisible formal groups. Asian J. Math. **12**, 193–202 (2008)

[Ch4] C.-L. Chai, Methods for p-adic monodromy. J. Inst. Math. Jussieu **7**, 247–268 (2008)

[Cn1] G. Chenevier, On the infinite fern of Galois representations of unitary type. Ann. Sci. Éc. Norm. Sup. (4) **44**, 963–1019 (2011)

[Cn2] G. Chenevier, The infinite fern and families of quaternionic modular forms. Lecture Notes at the Galois Trimester, 2010 (posted at `http://www.math.polytechnique.fr/~chenevier/coursihp.html`)

[CnC] G. Chenevier, L. Clozel, Corps de nombres peu ramifiés et formes automorphes autoduales. J. Am. Math. Soc. **22**, 467–519 (2009)

[CV1] C. Cornut, V. Vatsal, CM points and quaternion algebras. Doc. Math. **10**, 263–309 (2005)

[CV2] C. Cornut, V. Vatsal, Nontriviality of Rankin–Selberg L-functions and CM points, in *L-Functions and Galois Representations*. London Mathematical Society Lecture Note Series, vol. 320 (Cambridge University Press, Cambridge, 2007), pp. 121–186

[Cz] P. Colmez, Invariants L et dérivées de valeurs propres de Frobenius. Astérisque **331**, 13–28 (2010)

[CzS] P. Colmez, L. Schneps, p-Adic interpolation of special values of Hecke L-functions. Compos. Math. **82**, 143–187 (1992)

[D1] P. Deligne, Formes modulaires et représentations l-adiques, Sém. Bourbaki, Exp. 335, 1969

[D2] P. Deligne, Variété abeliennes ordinaires sur un corps fini. Invent. Math. **8**, 238–243 (1969)

[D3] P. Deligne, Travaux de Shimura, Sem. Bourbaki, Exp. 389. Lect. Notes Math. **244**, 123–165 (1971)

[D4] P. Deligne, La conjecture de Weil. I. Publ. IHES **43**, 273–307 (1974)

[D5] P. Deligne, Variétés de Shimura: Interprétation modulaire, et techniques de construction de modéles canoniques. Proc. Symp. Pure Math. **33.2**, 247–290 (1979)

[D6] P. Deligne, Valeurs des fonctions L et périodes d'intégrales. Proc. Symp. Pure Math. **33.2**, 313–346 (1979)

[DRa] P. Deligne, M. Rapoport, Les schémas de modules de courbes elliptiques. Lect. Notes Math. **349**, 143–316 (1973)

[DRi] P. Deligne, K.A. Ribet, Values of abelian L-functions at negative integers over totally real fields. Invent. Math. **59**, 227–286 (1980)

[DS] P. Deligne, J.-P. Serre, Formes modulaires de poids 1. Ann. Sci. Éc. Norm. Sup. 4th series **7**, 507–530 (1974)

[DiFG] F. Diamond, M. Flach, L. Guo, The Tamagawa number conjecture of adjoint motives of modular forms. Ann. Sci. Éc. Norm. Sup. (4) **37**, 663–727 (2004)

[DoHI] K. Doi, H. Hida, H. Ishii, Discriminant of Hecke fields and twisted adjoint L-values for $GL(2)$. Invent. Math. **134**, 547–577 (1998)

[Em] M. Emerton, p-Adic families of modular forms (after Hida, Coleman, and Mazur). Astérisque **339**, 31–61 (2011)

[FG] B. Ferrero, R. Greenberg, On the behavior of p-adic L-functions at $s = 0$. Invent. Math. **50**, 91–102 (1978)

[Fi1] T. Finis, Divisibility of anticyclotomic L-functions and theta functions with complex multiplication. Ann. Math. **163**, 767–807 (2006)

[Fi2] T. Finis, The μ-invariant of anticyclotomic L-functions of imaginary quadratic fields. J. Reine Angew. Math. **596**, 131–152 (2006)

[Fo] J.-M. Fontaine, Il n'y a pas de variété abélienne sur **Z**. Invent. Math. **81**, 515–538 (1985)

[FW] B. Ferrero, L. Washington, The Iwasawa invariant μ_p vanishes for abelian number fields. Ann. Math. **109**, 377–395 (1979)

[GeJ] S. Gelbart, H. Jacquet, A relation between automorphic representations of GL(2) and GL(3). Ann. Sci. Éc. Norm. Sup. (4) **11**, 471–542 (1978)

[GhV] E. Ghate, V. Vatsal, On the local behaviour of ordinary $\mathbb{I}$-adic representations. Ann. Inst. Fourier (Grenoble) **54**, 2143–2162 (2004)

[Gi1] R. Gillard, Fonctions L p-adiques des corps quadratiques imaginaires et de leurs extensions abéliennes. J. Reine Angew. Math. **358**, 76–91 (1985)

[Gi2] R. Gillard, Remarques sur l'invariant mu d'Iwasawa dans le cas CM, Sém. Théorie Nombres, Bordeaux **3**, 13–26 (1991)

[Gr] R. Greenberg, Trivial zeros of p-adic L-functions. Contemp. Math. **165**, 149–174 (1994)

[GS] R. Greenberg, G. Stevens, p-Adic L-functions and p-adic periods of modular forms. Invent. Math. **111**, 407–447 (1993)

[Gs] B.H. Gross, On the factorization of p-adic L-series. Invent. Math. **57**, 83–95 (1980)

[GsK] B.H. Gross, N. Koblitz, Gauss sums and the p-adic Γ-function. Ann. Math. **109**, 569–581 (1979)

[GZ] B. Gross, D. Zagier, Heegner points and derivatives of L-series. Invent. Math. **84**, 225–320 (1986)

[H81a] H. Hida, Congruences of cusp forms and special values of their zeta functions. Invent. Math. **63**, 225–261 (1981)

[H81b] H. Hida, On congruence divisors of cusp forms as factors of the special values of their zeta functions. Invent. Math. **64**, 221–262 (1981)

[H82] H. Hida, Kummer's criterion for the special values of Hecke L-functions of imaginary quadratic fields and congruences among cusp forms. Invent. Math. **66**, 415–459 (1982)

[H85] H. Hida, A p-adic measure attached to the zeta functions associated with two elliptic modular forms. I. Invent. Math. **79**, 159–195 (1985)

[H86a] H. Hida, Iwasawa modules attached to congruences of cusp forms. Ann. Sci. Éc. Norm. Sup. 4th series **19**, 231–273 (1986)

[H86b] H. Hida, Galois representations into $GL_2(\mathbb{Z}_p[[X]])$ attached to ordinary cusp forms. Invent. Math. **85**, 545–613 (1986)

[H88a] H. Hida, Modules of congruence of Hecke algebras and L-functions associated with cusp forms. Am. J. Math. **110**, 323–382 (1988)

[H88b] H. Hida, On p-adic Hecke algebras for GL_2 over totally real fields. Ann. Math. **128**, 295–384 (1988)

[H94] H. Hida, On the critical values of L-functions of $GL(2)$ and $GL(2) \times GL(2)$. Duke Math. J. **74**, 431–529 (1994)

[H98] H. Hida, Automorphic induction and Leopoldt type conjectures for $GL(n)$. Asian J. Math. **2**, 667–710 (1998)

[H04a] H. Hida, Greenberg's $\mathcal{L}$-invariants of adjoint square Galois representations. Int. Math. Res. Not. **59**, 3177–3189 (2004)

[H04b] H. Hida, Non-vanishing modulo p of Hecke L-values, in *Geometric Aspects of Dwork Theory*, ed. by A. Adolphson, F. Baldassarri, P. Berthelot, N. Katz, F. Loeser (Walter de Gruyter, Berlin, 2004), pp. 731–780 (a preprint version posted at `www.math.ucla.edu/~hida`)

[H06a] H. Hida, Anticyclotomic main conjectures. Doc. Math. Extra Volume Coates, 465–532 (2006)

[H06b] H. Hida, Automorphism groups of Shimura varieties of PEL type. Doc. Math. **11**, 25–56 (2006)

[H07a] H. Hida, Non-vanishing modulo p of Hecke L-values and application, in *L-Functions and Galois Representations*. London Mathematical Society Lecture Note Series, vol. 320 (Cambridge University Press, Cambridge, 2007), pp. 207–269

[H07b] H. Hida, On a generalization of the conjecture of Mazur–Tate–Teitelbaum. Int. Math. Res. Not. **2007**, 49 p. (2007), article ID rnm102. doi:10.1093/imrn/rnm102

[H09a] H. Hida, Irreducibility of the Igusa tower. Acta Math. Sin. Engl. Ser. **25**, 1–20 (2009)

[H09b] H. Hida, Quadratic exercises in Iwasawa theory. Int. Math. Res. Not. **2009**, 912–952 (2009). doi:10.1093/imrn/rnn151

[H10a] H. Hida, The Iwasawa μ-invariant of p-adic Hecke L-functions. Ann. Math. **172**, 41–137 (2010)

[H10b] H. Hida, Central critical values of modular Hecke L-functions. Kyoto J. Math. **50**, 777–826 (2010)

[H11a] H. Hida, Hecke fields of analytic families of modular forms. J. Am. Math. Soc. **24**, 51–80 (2011)

[H11b] H. Hida, Vanishing of the μ-invariant of p-adic Hecke L-functions. Compos. Math. **147**, 1151–1178 (2011)

[H11c] H. Hida, Constancy of adjoint $\mathcal{L}$-invariant. J. Number Theor. **131**, 1331–1346 (2011)

[H12a] H. Hida, A finiteness property of abelian varieties with potentially ordinary good reduction. J. Am. Math. Soc. **25**, 813–826 (2012)

[H12b] H. Hida, Image of Λ-adic Galois representations modulo p. Invent. Math. (2012). doi:10.1007/s00222-012-0439-7

[H13a] H. Hida, Local indecomposability of Tate modules of non CM abelian varieties with real multiplication. J. Am. Math. Soc. **26**, 853–877 (2013) (a preprint version posted in `www.math.ucla.edu/~hida`)

[H13b] H. Hida, Big Galois representations and p-adic L-functions, preprint, 2012, 51 p. (a preprint version posted in `www.math.ucla.edu/~hida`)

[H13c] H. Hida, Hecke fields of Hilbert modular analytic families, to appear in the memorial volume of Piatetskii-Shapiro from Contemporary Math. preprint, 2013, 31 p. (a preprint version posted in `www.math.ucla.edu/~hida`)

[HM] H. Hida, Y. Maeda, Non-abelian base-change for totally real fields. Special Issue of Pac. J. Math. in memory of Olga Taussky Todd, 189–217 (1997)

[HT1] H. Hida, J. Tilouine, Anticyclotomic Katz p-adic L-functions and congruence modules. Ann. Sci. Éc. Norm. Sup. 4th series **26**, 189–259 (1993)

[HT2] H. Hida, J. Tilouine, On the anticyclotomic main conjecture for CM fields. Invent. Math. **117**, 89–147 (1994)

[Ho] T. Honda, Isogeny classes of abelian varieties over finite fields. J. Math. Soc. Jpn. **20**, 83–95 (1968)

[Hs1] M.-L. Hsieh, On the non-vanishing of Hecke L-values modulo p. American Journal of Mathematics, **134**, 1503–1539 (2012)

[Hs2] M.-L. Hsieh, On the μ-invariant of anticyclotomic p-adic L-functions for CM fields. J. Reine Angew. Math. (in press). doi:10.1515/crelle-2012-0056, `http://www.math.ntu.edu.tw/~mlhsieh/research.htm`)

[Hs3] M.-L. Hsieh, Eisenstein congruence on unitary groups and Iwasawa main conjecture for CM fields (preprint posted at `http://www.math.ntu.edu.tw/~mlhsieh/research.htm`)

[Hz1] A. Hurwitz, Ueber die Entwicklungscoefficienten der lemniscatischen Functionen. Göttingen Nachr., 273–276 (1897)

[Hz2] A. Hurwitz, Ueber die Entwickelungscoefficienten der lemniscatischen Functionen. Math. Ann. **51**, 196–226 (1899)

[I] J. Igusa, Kroneckerian model of fields of elliptic modular functions. Am. J. Math. **81**, 561–577 (1959)

[K1] N.M. Katz, Serre–Tate local moduli, in *Surfaces Algébriques*. Lecture Notes in Mathematics, vol. 868 (Springer-Verlag, Berlin, 1978), pp. 138–202

[K2] N.M. Katz, p-Adic L-functions for CM fields. Invent. Math. **49**, 199–297 (1978)

[Kh1] C. Khare, Serre's modularity conjecture: The level one case. Duke Math. J. **134**, 557–589 (2006)

[Kh2] C. Khare, Serre's conjecture and its consequences. Jpn. J. Math. **5**, 103–125 (2010)

[KhW] C. Khare, J.-P. Wintenberger, Serre's modularity conjecture. I, II. I: Invent. Math. **178**, 485–504 (2009); II: Invent. Math. **178**, 505–586 (2009)

[Ki1] M. Kisin, Overconvergent modular forms and the Fontaine–Mazur conjecture. Invent. Math. **153**, 373–454 (2003)

[Ki2] M. Kisin, Geometric deformations of modular Galois representations. Invent. Math. **157**, 275–328 (2004)

[KiW] L.J.P. Kilford, G. Wiese, On the failure of the Gorenstein property for Hecke algebras of prime weight. Exp. Math. **17**, 37–52 (2008)

[Ko] R. Kottwitz, Points on Shimura varieties over finite fields. J. Am. Math. Soc. **5**, 373–444 (1992)

[L] S. Lang, Sur les séries L d'une variété algébrique. Bull. Soc. Math. France **84**, 385–407 (1956)

[Lx] J.H. Loxton, On two problems of R. M. Robinson about sum of roots of unity. Acta Arith. **26**, 159–174 (1974)

[Mh] K. Mahler, An interpolation series for continuous functions of a p-adic variable. J. Reine Angew. Math. **199**, 23–34 (1958) (Correction: J. Reine Angew. Math. **208**, 70–72 (1961))

[Mz1] B. Mazur, Courbes elliptiques et symboles modulaires. Séminaire Bourbaki, 24ème année (1971/1972), Exp. No. 414. Lecture Notes in Mathematics, vol. 317 (Springer-Verlag, Berlin, 1973), pp. 277–294

[Mz2] B. Mazur, Deforming Galois representations, in *Galois Group Over* $\mathbb{Q}$. MSRI Publications, vol. 16 (Springer-Verlag, New York, 1989), pp. 385–437

[MzT] B. Mazur, J. Tilouine, Représentations galoisiennes, différentielles de Kähler et "conjectures principales." Publ. IHES **71**, 65–103 (1990)

[MzTT] B. Mazur, J. Tate, J. Teitelbaum, On p-adic analogues of the conjectures of Birch and Swinnerton-Dyer. Invent. Math. **84**, 1–48 (1986)

[MzW] B. Mazur, A. Wiles, Class fields of abelian extensions of **Q**. Invent. Math. **76**, 179–330 (1984)

[Mo] B. Moonen, Serre–Tate theory for moduli spaces of PEL type. Ann. Sci. Éc. Norm. Sup. (4) **37**, 223–269 (2004)

[O] A. Ogus, Hodge cycles and crystalline cohomology, in *Hodge Cycles, Motives, and Shimura Varieties*, Chapter VI. Lecture Notes in Mathematics, vol. 900 (Springer-Verlag, Berlin, 1982), pp. 357–414

[Ri1] K.A. Ribet, P-adic interpolation via Hilbert modular forms. Proc. Symp. Pure Math. **29**, 581–592 (1975)

[Ri2] K.A. Ribet, On l-adic representations attached to modular forms. Invent. Math. **28**, 245–275 (1975)

[Ri3] K.A. Ribet, On l-adic representations attached to modular forms. II. Glasgow Math. J. **27**, 185–194 (1985)

[Ri4] K.A. Ribet, Abelian varieties over $\mathbb{Q}$ and modular forms, in *Algebra and Topology 1992 (Taejŏn)* (Korea Advanced Institute of Science and Technology, Taejŏn, 1992), pp. 53–79

[Ru1] K. Rubin, The "main conjectures" of Iwasawa theory for imaginary quadratic fields. Invent. Math. **103**, 25–68 (1991)

[Ru2] K. Rubin, More "main conjectures" for imaginary quadratic fields, in *Elliptic Curves and Related Topics*. CRM Proceedings and Lecture Notes, vol. 4 (American Mathematical Society, Providence, RI, 1994), pp. 23–28

[Sch] A.J. Scholl, Motives for modular forms. Invent. Math. **100**, 419–430 (1990)

[Scn] L. Schneps, On the μ-invariant of p-adic L-functions attached to elliptic curves with complex multiplication. J. Number Theor. **25**, 20–33 (1987)

[Se1] J.-P. Serre, Zeta and L functions, in *Arithmetical Algebraic Geometry.* Proceedings of Conference Purdue University, 1963 (Harper & Row, New York, 1965), pp. 82–92 (Œuvres II 249–259, No. 64)

[Se2] J.-P. Serre, Formes modulaires et fonctions zêta p-adiques. Lect. Notes Math. **350**, 191–268 (1973) (Œuvres III 95–172, No. 97)

[ST] J.-P. Serre, J. Tate, Good reduction of abelian varieties. Ann. Math. **88**, 452–517 (1968) (Serre's Œubres II 472–497, No. 79)

[Sh1] G. Shimura, Correspondances modulaires et les fonctions ζ de courbes algébriques. J. Math. Soc. Jpn. **10**, 1–28 (1958) ([58a] in [CPS] I).

[Sh2] G. Shimura, On analytic families of polarized abelian varieties and automorphic functions. Ann. Math. **78**, 149–192 (1963) ([63b] in [CPS] I)

[Sh3] G. Shimura, Moduli and fibre system of abelian varieties. Ann. Math. **83**, 294–338 (1966) ([66b] in [CPS] I)

[Sh4] G. Shimura, On canonical models of arithmetic quotients of bounded symmetric domains. Ann. Math. **91**, 144–222 (1970); II, **92**, 528–549 (1970) ([70a–b] in [CPS] II)

[Sh5] G. Shimura, On the holomorphy of certain Dirichlet series. Proc. Lond. Math. Soc. (3) **31**, 79–98 (1975) ([75a] in [CPS] II)

[Sh6] G. Shimura, On some arithmetic properties of modular forms of one and several variables. Ann. Math. **102**, 491–515 (1975) ([75c] in [CPS] II)

[Sh7] G. Shimura, The special values of the zeta functions associated with cusp forms. Commun. Pure Appl. Math. **29**, 783–804 (1976) ([76b] in [CPS] II)

[Sh8] G. Shimura, On the periods of modular forms. Math. Ann. **229**, 211–221 (1977) ([77d] in [CPS] II)

[Sh9] G. Shimura, The special values of the zeta functions associated with Hilbert modular forms. Duke Math. J **45**, 637–679 (1978) (a new version with some mathematical addenda and typographical correction incorporated: [78c] in [CPS] III)

[Sn1] W. Sinnott, On the μ-invariant of the Γ-transform of a rational function. Invent. Math. **75**, 273–282 (1984)

[Sn2] W. Sinnott, On a theorem of L. Washington. Astérisque **147–148**, 209–224 (1987)

[SU] C. Skinner, E. Urban, The Iwasawa main conjecture for GL_2, to appear in Inventiones Math. (posted at `http://www.math.columbia.edu/~urban/EURP.html`)

[SW] C. Skinner, A. Wiles, Residually reducible representations and modular forms. Publ. IHES **89**, 5–126 (2000)

[T1] J. Tate, p-Divisible groups, in *Proceedings of a Conference on Local Fields*, Driebergen, 1966 (Springer-Verlag, Berlin, 1967), pp. 158–183

[T2] J. Tate, Class d'isogénies des variétés abéliennes sur un corps fini (d'après Honda). Séminaires Bourbaki 318, Novembre 1966.

[T3] J. Tate, A review of non-Archimedean elliptic functions, in *Elliptic Curves, Modular Forms, & Fermat's Last Theorem*. Series in Number Theory I (International Press, Boston, 1995), pp. 162–184

[Ta] R. Taylor, On the meromorphic continuation of degree two L-functions. Doc. Math., Extra Volume: John Coates' Sixtieth Birthday, 729–779 (2006)

[Ti] J. Tilouine, Sur la conjecture principale anticyclotomique. Duke Math. J. **59**, 629–673 (1989)

[V1] V. Vatsal, Uniform distribution of Heegner points. Invent. Math. **148**, 1–46 (2002)

[V2] V. Vatsal, Special values of anticyclotomic L-functions. Duke Math. J. **116**, 219–261 (2003)

[V3] V. Vatsal, Special values of L-functions modulo p, in *International Congress of Mathematicians*, vol. II (European Mathematical Society, Zürich, 2006), pp. 501–514

[Wa] J.-L. Waldspurger, Sur les valeurs de certaines fonctions L-automorphes en leur centre de symétrie. Compos. Math. **54**, 173–242 (1985)

[Ws] L. Washington, The non-p–part of the class number in a cyclotomic $\mathbb{Z}_p$-extension. Invent. Math. **49**, 87–97 (1978)

[We1] A. Weil, Numbers of solutions of equations in finite fields. Bull. AMS **55**, 497–508 (1949) (Œuvres I, [1949])

[We2] A. Weil, Exercices dyadiques. Invent. Math. **27**, 1–22 (1974) (Œuvres III, [1974e])

[Y] R. Yager, p-Adic measures on Galois groups. Invent. Math. **76**, 331–343 (1984)

[Z] J.G. Zarhin, Endomorphisms of abelian varieties over fields of finite characteristics. Math. USSR Izvestija **9**(2), 255–260 (1975)

[Zh] B. Zhao, Local indecomposability of Hilbert modular Galois representations, preprint, 2012 (posted on web: arXiv:1204.4007v1 [math.NT])

Symbol Index

H. Hida, *Elliptic Curves and Arithmetic Invariants*, Springer Monographs in Mathematics, DOI 10.1007/978-1-4614-6657-4,

Statement Index

H. Hida, *Elliptic Curves and Arithmetic Invariants*, Springer Monographs in Mathematics, DOI 10.1007/978-1-4614-6657-4,

Subject Index

H. Hida, *Elliptic Curves and Arithmetic Invariants*, Springer Monographs in Mathematics, DOI 10.1007/978-1-4614-6657-4,

MIX
Papier aus verantwortungsvollen Quellen
Paper from responsible sources
FSC® C105338

If you have any concerns about our products,
you can contact us on
ProductSafety@springernature.com

In case Publisher is established outside the EU,
the EU authorized representative is:
Springer Nature Customer Service Center GmbH
Europaplatz 3, 69115 Heidelberg, Germany

Printed by Libri Plureos GmbH
in Hamburg, Germany